Exploration Seismology

Exploration Seismology

S E C O N D   E D I T I O N

R. E. SHERIFF
Professor, Geosciences Department,
University of Houston, Houston, Texas

L. P. GELDART
Former Coordinator,
Canadian International Development Agency Program for Brazil

CAMBRIDGE
UNIVERSITY PRESS

CAMBRIDGE UNIVERSITY PRESS
Cambridge, New York, Melbourne, Madrid, Cape Town, Singapore, São Paulo

Cambridge University Press
The Edinburgh Building, Cambridge CB2 2RU, UK

Published in the United States of America by Cambridge University Press, New York

www.cambridge.org
Information on this title: www.cambridge.org/9780521462822

First published 1982
Second edition 1995
Re-issued in this digitally printed monochrome version 2006

A catalogue record for this publication is available from the British Library

Library of Congress Cataloguing in Publication data

Sheriff, Robert E.
 Exploration seismology / R. E. Sheriff, L. P. Geldart. —
2nd ed.
  p. cm.
 Includes bibliographical references.
 ISBN 0-521-46282-7. — ISBN 0-521-46826-4 (pbk.)
 1. Seismic prospecting. I. Geldart, L. P. II. Title.
 TN269.S52415 1994   94-4153
 622'.1592—dc20   CIP

ISBN-13 978-0-521-46826-8 paperback
ISBN-10 0-521-46826-4 paperback

Contents

Preface

Many improvements have occurred in the seismic method since the publication of the first edition in 1982. Concepts that were of academic concern then have since become practical tools and some of the new concepts we have added may become tomorrow's tools.

We want this book to be a reference work as well as a textbook and guide for practicing geophysicists. These three objectives are not always compatible and readers will often skip over portions that do not fit their current needs, hopefully later referring back to skipped portions. For those readers who do skip about, we have cross-referenced sections, equations, and figures. For those who are rusty with their mathematics, we express concepts in words as well as by equations.

Our special interest is in the interpretation of geophysical data, but an interpreter needs to have a thorough understanding of geophysical principles in order to determine the validity of his data and the possibility that features he sees are artifacts of acquisition or processing. To this end, we have tried to emphasize seismic fundamentals.

We give a systematic derivation of relationships from first principles, except for a few cases where the derivations are excessively lengthy or involve higher mathematics, in which instances we refer the reader to other sources. As the preface to our first edition states, "A reader willing to take the mathematics on faith should be able to jump over the equations and still see the implications of the mathematical conclusions, which we have endeavored to explain in words rather than merely letting the equations speak for themselves. We have been encouraged by Sir Harold Jeffreys' preface to *The Earth* (1924):

> If the geologist cannot follow a part of the book, I hope he will omit it and go on. . . . Mathematically trained readers, with few exceptions, will do the same.

Our treatment of seismic theory has been expanded. We have expanded partitioning at interfaces, the heart of most seismic applications, and made it a separate chapter. Anisotropy, AVO, Stoneley waves, and tube waves, merely mentioned in our first edition, are now discussed in more detail. 3-D methods and 3-D interpretation now occupies an entire chapter. Techniques somewhat out of the mainstream, such as

VSP, *S*-wave methods, and channel waves are now expanded into a full chapter on specialized techniques, and we have added new topics such as tomography and geostatistics. One chapter is devoted to nonpetroleum applications, including not only coal and engineering seismic work, but also the growing areas of groundwater, environmental, and reservoir geophysics. Our final chapter on mathematical background is again intended to refresh a reader's forgotten mathematical concepts.

Terms defined are indicated by *italics*. Precision of terminology is often one of the clearest indications of a person's degree of understanding and we have made special effort to define and use the specialized vocabulary of seismology precisely.

This book is structured to be as consistent as possible. Mathematical conventions, definitions, and the symbols used throughout the book are listed immediately following this preface. We have tried to conform to accepted practices and nomenclature, but we have had to use some symbols for several purposes.

Each chapter begins with an overview to orient the reader as to the various topics to be discussed and their relationship to each other. All chapters except the first end with problems; these endeavor to elucidate aspects not treated in the text and develop proofs and additional relationships. Each problem has been designed to illustrate a specific point and we have included hints where we anticipate students may have difficulty in knowing how to proceed. Enlarging the book's figures to use with the problems involving them should not prove difficult with the ready availability of enlarging copiers.

We acknowledge the assistance of many people in the preparation of this book and express our thanks to them. In addition to those cited in our first edition (Harry Mayne, Dan Skelton, Bill Laing, O. Leenhardt, Bruce Frizelle, Howard Taylor, Thomas Thompson, and Willis Reed), we particularly thank Barbara Barnes, Leslie Denham, Brian Evans, Tony Lauhoff, Dereck Palmer, and Margaret Sheriff for their help with this edition. We also express appreciation to the University of Houston Geosciences Department and the Curtin University of Technology (where one of us (RES) worked on this book while the Haydn Williams Fellow).

R. E. Sheriff
L. P. Geldart

Mathematical conventions and symbols

General rules and definitions

General functions

g_t Function of the discrete variable $t = n\Delta$, $n = 0, \pm 1, \pm 2, \ldots$.

$g(t)$ Function of a continuous variable t

$g(t) * h(t)$ Convolution of $g(t)$ with $h(t)$

$\hat{g}(\zeta), \hat{G}(\omega)$ Functions involving the cepstrum transform: $\log G(\omega) = \hat{G}(\omega) \leftrightarrow \hat{g}(\zeta)$

$G(\nu)$ Transform of $g(t)$ to a function of frequency ν; the arguments ν or ω indicate Fourier transform, s Laplace transform, z z-transform

$\phi_{gh}(\tau)$ Correlation of $g(t)$ with $h(t)$ as function of the time shift τ; a cross-correlation if $g \neq h$, $\phi_{gg}(\tau)$ = autocorrelation

Special functions

$\text{box}_a[t]$ Boxcar of unit height and width a, centered at $t = 0$

$\text{comb}[t]$ Series of equally spaced unit impulses

$\text{sgn}[t]$ Sign of $t = -1$ for $t < 0$, $+1$ for $t > 0$

$\text{sinc}[t]$ $(1/t) \sin t$

$\text{step}[t]$ Unit step function, $\text{step}(t) = 0$ for $t < 0$, $+1$ for $t > 0$

$\delta[t], \delta_t$ Unit impulse at $t = 0$

Mathematical conventions

$\approx$ Approximately equal to

$\leftrightarrow$ Denotes corresponding functions in different domains, the arguments ν, ω, s, z indicating the type of transform; thus, $g(t) \leftrightarrow G(\nu)$ and $g(t) \leftrightarrow G(\omega)$ indicate Fourier transforms, $g(t) \leftrightarrow G(s)$ a Laplace transform, $g_t \leftrightarrow G(z)$ a z-transform; $g(x, y) \leftrightarrow G(\ell, \theta)$ a Radon transform; lowercase letters indicate the time domain, uppercase the frequency domain.

$[a, \overset{\downarrow}{b}, c, d]$ Denotes a time sequence consisting of the elements a, b, c, d, with the superscribed arrow indicating the value associated with $t = 0$ (b in this instance); values not otherwise specified are zero.

$\mathbf{A}$ Vector quantity, magnitude is $|\mathbf{A}|$

$\mathbf{A} \cdot \mathbf{B}, \mathbf{A} \times \mathbf{B}$ Scalar and vector products of $\mathbf{A}$ and $\mathbf{B}$

$\mathcal{A}$ Matrix with elements a_{ij}

$\mathcal{A}^{\mathrm{T}}$ Transpose of matrix $\mathcal{A}$

$\arg(\omega)$ Argument of ω

$\mathfrak{B}$ Principal value of

∇ Del, the vector operator, $\mathbf{i}(\partial/\partial x) + \mathbf{j}(\partial/\partial y) + \mathbf{k}(\partial/\partial z)$

∇^2 Laplacian operator, $\partial^2/\partial x^2 + \partial^2/\partial y^2 + \partial^2/\partial z^2$

$\nabla \phi$ Gradient of ϕ (grad ϕ)

$\nabla \cdot \mathbf{A}$ Divergence of $\mathbf{A}$ (div $\mathbf{A}$)

$\nabla \times \mathbf{A}$ Curl of $\mathbf{A}$ (curl $\mathbf{A}$)

$\det(a)$ Determinant with elements a_{ij}

$\blacktriangle\{\ \}$ Difference operator

$D\{\ \}$ Derivative operator

$D^{-1}\{\ \}$ Integration operator

$E\{\ \}$ Delay operator

$\exp(x)$ e^x

$g(0+), g(0-)$ Value of g when approaching 0 from right, left

$\overline{G(z)}, G(z^{-1})$ Complex conjugate of $G(z)$

$P(e_j)$ Probability of e_j

$\text{Re}\{g(t)\}, \text{Im}\{g(t)\}$ Real, imaginary parts of $g(t)$

$|w|, |\mathbf{W}|$ Absolute value or modulus of w, $\mathbf{W}$

$\dot{x}, \dot{y}, \dot{z}$ Time derivatives of x, y, z

$\displaystyle\prod_{i=0}^{n} a_i$ Product $a_0 a_1 a_2 a_3 \cdots a_n$

$\displaystyle\sum_{i=0}^{n} g_i$ Sum $g_0 + g_1 + \cdots + g_n$

$\displaystyle\sum_{k} g_k$ Sum of g_k over appropriate values of k

Latin symbols

a Element spacing, rate of increase of velocity with depth, area, width of a boxcar

a, b, c Constants

a_i, b_i Angles of incidence

a_n, b_n, c_n Fourier-series coefficients

A, B Constants

A, B, C Amplitudes of waves or of displacements

$A(t)$ Amplitude of envelope, amplitude attributed to reflectivity

$A(\nu), B(\nu)$ Amplitude spectrum

$\mathcal{A}, \mathcal{B}$ Amplitudes of displacement potential functions, matrices

$c_{\mathcal{V}}, c_{\mathcal{P}}$ Specific heat at constant volume, pressure

C Incompressibility $= 1/k$, volume fraction of clay

D Displacement potential function for diffraction, distance

D_s, D_w	Depth of source, weathering
e_i	Error in ith output
e_t	Impulse response of sequence of reflectors
E	Sum of errors squared, Young's modulus, energy, energy density, energy ratio
E_s, E_d	Elevation of surface, datum
E_R, E_T	Fraction of energy reflected, transmitted
$f(t)$, f_t	Filter in time domain, function of t
$f_H(t)$, f_t^H	High-pass filter
$f_L(t)$, f_t^L	Low-pass filter
$f_u(t)$	Response of filter to unit step in time domain
F	Array response, constant, magnitude of force, depth-conversion factor
$F(v)$, $F(\omega)$, $F(z)$	Filter in frequency domain
$F_u(s)$	Transform of response to unit step
$\mathbf{F}$	Force, force per unit mass, vector wave function
g	Acceleration of gravity
$g(t)$, g_t	Seismic trace in time domain, input trace, function of t
$g_\perp(t)$	Quadrature trace
$\hat{g}(\zeta)$	Cepstrum of $\hat{G}(\omega)$ or $\hat{G}(z)$
$G(v)$, $G(\omega)$, $G(z)$	Seismic trace in frequency domain
h	Damping factor, distance to reflector or refractor, depth, thickness
$h(t)$, h_t	Output in time domain
$\hbar(t)$, $\hbar_t$	Desired output, complex trace
H	Magnetic field strength, depth
$H(v)$, $H(\omega)$, $H(z)$	Output in frequency domain
$\mathcal{H}(v)$, $\mathcal{H}(\omega)$, $\mathcal{H}(z)$	Desired output in frequency domain
i	Current, index
i_0	Angle of raypath at source point
i_t	Inverse filter in time domain
$\mathbf{i}$, $\mathbf{j}$, $\mathbf{k}$	Unit vectors in x-, y-, z-directions
I	Intensity
$I(v)$	Transform of inverse filter
j	$(-1)^{1/2}$
j	Index
k	Constant, bulk modulus, index
K	Force on geophone coil per unit current, effective elastic modulus
ℓ	Distance in Radon transform
ℓ, m, n	Direction cosines relative to x-, y-, z-axes
ℓ_1, ℓ_2, ℓ_4	Criteria in optimum filtering
L	Kinetic energy per unit volume, length, prediction lag, self inductance
L_k, M_k	Delay at location k due to structure, normal-moveout error
m	Mass
m, n	Constants, exponential decay parameters, integers
M	Effective elastic modulus, width of fringe zone
n	Integer, number of layers

n_t	Impulse response of near-surface zone
N	Noise
p, p'	Raypath parameter
P	Power
p_t	Impulse response of waveshape-modifying factors in the earth
$\mathscr{P}$	Pressure
Q	Quality factor, probability
r	Distance, integer, radius, radial coordinate
r, s	Receiver, source coordinates
r, R	Radius, resistance
r_t	Additive noise
R, R_i	Reflection coefficient of ith interface, displacement potential function for reflection
R'	Resistivity
R_k, S_k	Delay due to geophone, source at location k
$R(v)$, $R(\omega)$	Real part of Fourier transform = cosine transform
s	Distance, Laplace transform parameter = $\sigma + j\omega$
s_t	Impulse response of source
S	Entropy density, semblance, signal, spring constant
S_w	Water saturation
$\mathscr{S}$	Area, surface
t	Time, traveltime
t_0	Traveltime for geophone at the source
t_p, t_{iu}, t_{id}	Refraction intercepts
t_{ij}	Time shift between traces i and j
t_{uh}	Uphole time
Δt	Time interval
Δt_0, Δt_s, Δt_g	Near-surface corrections
Δt_c	Differential weathering correction
$\Delta t_d/\Delta x$	Dip moveout
Δt_{NMO}	Normal moveout (NMO)
T	Period, transmission coefficient
ΔT	Specific transit time = $1/V$
u, v, w	Displacements in x-, y-, z-directions
$\dot{u}$, $\ddot{u}$	Time derivatives of u, $\partial u/\partial t$, $\partial^2 u/\partial t^2$
U	Group velocity
v	Amplitude of geophone velocity
v_t	Vibroseis™ input to ground
V	Interval velocity, phase velocity, velocity of a wave
$\overline{V}$	Equivalent average velocity
V_a	Apparent velocity
V_{rms}	rms velocity
V_s	Stacking velocity
V_u, V_d	Apparent velocity in updip, downdip directions
V_H, V_W	Velocity below, in the low-velocity layer (LVL)
V_L, V_R, V_S, V_t	Velocity of Love, Rayleigh, Stoneley, tube waves
$\mathscr{V}$	Volume
w	Weighting factors
w_t	Embedded (equivalent) wavelet, impulse

	response of water layer, downgoing waveform
W	S-wave acoustic impedance
x	Displacement, distance, offset
x', x_c	Critical distance, crossover distance for a refraction
$X(\nu)$, $X(\omega)$	Imaginary part of Fourier transform = $-$(sine transform)
z	Depth, z-transform parameter
Z	P-wave acoustic impedance

Greek symbols

α	P-wave velocity, angle, angle of approach
α_n	Fourier-series coefficients
β	S-wave velocity, angle
γ	Phase or phase difference, specific heat ratio, potential function, source density
γ_n	Phase shift
$\gamma(t)$	Instantaneous phase
$\gamma(\nu)$, $\gamma(\omega)$	Phase spectrum
Γ	Geophone transduction constant, measure of simplicity
δ	Angle of S-wave, delay time, logarithmic decrement
δ_s, δ_g	Delay time associated with shotpoint, geophone
$\delta(t)$	Unit impulse at $t = 0$
Δ	Dilatation, increment, sampling interval
ε	Eccentricity, phase shift upon reflection, strain
ε_{xx}, ε_{xy}	Normal, shearing strains
$\boldsymbol{\varepsilon}$	Error matrix
ζ	Argument of wave function, quefrency
$\boldsymbol{\zeta}$	Vector displacement = $u\mathbf{i} + v\mathbf{j} + w\mathbf{k}$
η	Absorption coefficient
$\boldsymbol{\eta}$	Outward-drawn unit normal
θ	Angle, angle of P-wave, argument of complex quantity, polar coordinate, spherical coordinate (colatitude)
θ_c	Critical angle
θ_x	Angle of rotation about x-axis
$\boldsymbol{\Theta}$	Vector rotation = $\theta_x\mathbf{i} + \theta_y\mathbf{j} + \theta_z\mathbf{k}$
κ	Angular wavenumber = 2π(wavenumber) $= 2\pi/\lambda$
κ_a	2π(apparent wavenumber)
κ_N	2π(Nyquist wavenumber)
λ	Lamé constant, length ratio, constant, wavelength, weighting factor

λ_a	Apparent wavelength
λ_N	Nyquist wavelength
$\lambda_\parallel$, $\lambda_\perp$	Lamé constants for transversely isotropic medium
μ	Mass ratio, mass/unit length, rigidity (shear) modulus (a Lamé constant)
$\mu_\parallel$, $\mu_\perp$, μ^*	Rigidity moduli for transversely isotropic medium
ν	Frequency = $\omega/2\pi = 1/T$
ν_0	Natural frequency, fundamental frequency
$\nu_i(t)$	Instantaneous frequency
ν_N	Nyquist frequency
$\boldsymbol{\nu}$	Vector potential
ξ	Dip, distance from origin to moving point on a curve
Ξ	Strike
ρ	Density, radius (of curvature)
σ	Convergence factor, Poisson's ratio, standard deviation, strength per unit length
σ_{xy}	Stress in x-direction on surface perpendicular to y-axis
τ	Damping factor, tension
Υ	Scalar wave function, source density
ϕ	Angle, loss angle, magnetic flux, porosity, P-wave displacement potential function, spherical coordinate (longitude)
$\phi_{gh}(\tau)$	Correlation of $g(t)$ with $h(t)$ as function of time shift τ
Φ	Transform of Φ, transform of P-wave displacement potential function
$\Phi_{gh}(\omega)$	Cross-energy spectrum (transform of $\phi_{gh}(\tau)$)
χ, $\boldsymbol{\chi}$	S-wave displacement potential function
ψ	Wavefunction, disturbance
$\psi(x, z, t)$	Wavefunction in x, z, and t dimensions
$\psi^*(x, z, t^*)$	Wavefunction in moving coordinate system
$\Psi(\kappa_x, \kappa_z, \omega)$	transform of wavefunction
$\Psi^*_{xz}(\kappa_x, \kappa_z, t)$	Transform of wavefunction with respect to x, z
ω	Angular frequency = $2\pi\nu$
ω_0	Natural frequency
ω_N	Nyquist frequency
$\boldsymbol{\Omega}$	Vector potential function for rotation

1
Introduction

Overview

Exploration seismology deals with the use of artificially generated elastic waves to locate mineral deposits (including hydrocarbons, ores, water, geothermal reservoirs, etc.), archaeological sites, and to obtain geological information for engineering. Exploration seismology provides data that, when used in conjunction with other geophysical, borehole, and geological data and with concepts of physics and geology, can provide information about the structure and distribution of rock types. Usually, seismic exploration is part of a commercial venture and, hence, economics is an ever-present concern. Seismic methods alone cannot determine many of the features that make for a profitable venture and, even when supplemented by other data, a unique interpretation is rarely evident. Seismic exploration usually stops long before unambiguous answers are obtained and before all has been learned that might possibly be learned, because in someone's judgment further information is better obtained in some other way, such as by drilling a well. Seismic methods are in continual economic competition with other methods.

Almost all oil companies rely on seismic interpretation for selecting the sites for exploratory oil wells. Despite the indirectness of the method – most seismic work results in the mapping of geological structure rather than finding petroleum directly – the likelihood of a successful venture is improved more than enough to pay for the seismic work. The enormous detail produced by 3-D techniques has opened up a huge reservoir engineering potential. Likewise, seismic methods are important in groundwater searches and in civil engineering, especially to measure the depth to bedrock in connection with the construction of large buildings, dams, highways, and harbor surveys, and to determine whether blasting will be required in road cuts, if potential hazards such as limestone caves or forgotten mine workings underlie building sites, if tunnels or mine drifts are likely to encounter water-filled zones, or if faults are present that might be hazards to a nuclear power plant. On the other hand, seismic techniques have found little application in direct exploration for minerals because they do not produce good definition where interfaces between different rock types are highly irregular. However, they are useful in locating features such as buried channels in which heavy minerals may be accumulated.

Exploration seismology is an offspring of earthquake seismology. When an earthquake occurs, the earth is fractured and the rocks on opposite sides of the fracture move relative to one another. Such a rupture generates seismic waves that travel outward from the fracture surface. These waves are recorded at various sites using seismographs. Seismologists use the data to deduce information about the nature of the rocks through which the earthquake waves traveled.

Exploration seismic methods involve basically the same type of measurements as earthquake seismology. However, the energy sources are controlled and movable and the distances between the source and the recording points are relatively small. Much seismic work consists of *continuous coverage*, where the response of successive portions of earth is sampled along lines of profile. Explosives and other energy sources are used to generate the seismic waves and arrays of seismometers or geophones are used to detect the resulting motion of the earth. The data are usually recorded in digital form on magnetic tape so that computer processing can be used to enhance the signals with respect to the noise, extract the significant information, and display the data in such a form that a geological interpretation can be carried out readily.

The basic technique of seismic exploration consists of generating seismic waves and measuring the time required for the waves to travel from the source to a series of geophones, usually disposed along a straight line directed toward the source. From a knowledge of traveltimes to the various geophones and the velocity of the waves, one attempts to reconstruct the paths of the seismic waves. Structural information is derived principally from paths that fall into two main categories: *head-wave* or *refracted* paths in which the principal portion of the path is along the interface between two rock layers and hence is approximately horizontal, and *reflected* paths in which the wave travels downward initially and at some point is reflected back to the surface, the overall path being essentially vertical. For both types of path, the traveltimes depend upon the physical properties of the rocks and the attitudes of the beds. The objective of seismic exploration is to deduce information about the rocks, especially about the attitudes of the beds, from the observed arrival times and (to a limited extent) from variations in amplitude, frequency, and waveform.

A brief outline of the seismic reflection and refraction methods is given first (§1.1); this explanation ignores complications and variations, which are the subjects of future chapters.

Exploration seismology is a fairly young activity, having begun only about 1923. The history of seismic exploration is summarized in §1.2. The seismic method is by far the most important geophysical technique in terms of capital expenditure (§1.3) and number of geophysicists involved. The predominance of the seismic method over other geophysical methods is due to various factors, the most important of which are the high accuracy, high resolution, and great penetration of which the method is capable. Seismic literature is discussed in §1.4.

1.1 Outline of seismic methods

1.1.1 Seismic reflection method

Seismic techniques have changed considerably within recent years and many variations exist. The technique described in what follows provides a background to the understanding of subsequent discussions; the reasons for various steps and various modifications of techniques will be described in subsequent chapters.

Assume a land crew using an explosive charge as the energy source. The first step after determining proper locations is the drilling of a vertical hole in the earth at the *source point,* the hole diameter being perhaps 10 or 12 cm and the depth usually between 6 and 30 m. A *charge* of 1 to 25 kg of explosive is armed with an electric blasting *cap* and then placed near the bottom of the hole. Two wires extend from the cap to the surface, where they are connected to a *blaster,* which is used to send an electrical current through the wires to the cap, which then explodes, initiating the explosion of the dynamite (the *shot*).

Two *cables* 2 to 4 km long are laid out in a straight line extending each way from the hole about to be fired. The cables contain many pairs of electrical conductors, each pair terminating in an electrical connector at both ends of the cable. In addition, each pair of wires is connected to one of several outlets spaced at intervals of 25 to 100 m along the cable. Several *geophones* (*seismometers*) are connected to each of these outlets so that each pair of wires in the cable carries the output energy of a *group* of geophones back to the recording instruments. Because of the small spacing between the geophones in the group attached to one pair of wires, the whole group is approximately equivalent to a single fictitious geophone located at the center of the group. Usually, 48 or more geophone groups are located at equal intervals along the cable. When the dynamite charge is exploded, each geophone group generates a signal that depends upon the motion of the ground in the vicinity of the group. The net result is the generation of signals furnishing information about the ground motion at a number of regularly spaced points (the *group centers*) along a straight line passing through the source.

The electrical signals from the geophone groups go to an equal number of amplifiers. These amplifiers increase the overall signal strength and partially eliminate (*filter out*) parts of the input deemed to be undesirable. The outputs from the amplifiers along with accurate timing signals are recorded on magnetic tape and on paper records. Thus, the recorded data consist of several *traces,* each trace showing how the motion of one geophone group varies with time after the source activation.

The data are usually processed to attenuate noise vis-à-vis reflected energy based on characteristics that distinguish them from each other, and the data are displayed in a form suitable for interpretation.

Events, that is, arrivals of energy that vary systematically from trace to trace and that are believed to represent reflected energy, are identified on the records. The *arrival times* (the interval between the source instant and the arrival of the energy at a geophone group, also known as the *traveltime*) of these events are measured for various geophone groups. The location and attitude of the interface that gave rise to each reflection event are then calculated from the arrival times. Seismic velocity enters into the calculation of the location and attitude of the interfaces. The results are combined into cross-sections and contour maps that represent the structure of the geological interfaces responsible for the events. Patterns in the seismic data are sometimes interpreted in terms of stratigraphic features or as indicators of hydrocarbons. However, the presence or absence of hydrocarbons or other minerals is usually inferred from the structural information.

We have introduced a number of terms used in a specialized sense in seismic work (indicated by *italics*), for example, sourcepoint, group, trace, events, and arrival time. Exploration seismology abounds in such technical terms. We shall henceforth use italics to indicate that we are defining a term; we shall follow the definitions given in the *Encyclopedic Dictionary of Exploration Geophysics* (Sheriff, 1991) for seismic terms and the *Glossary of Geology* (Bates and Jackson, 1987) for geologic terms.

1.1.2 Seismic refraction method

The principal difference between reflection and refraction methods is that for refraction, the distance between source and geophones is large relative to the depths of the interfaces being mapped, whereas it is small or comparable to the depths for reflection. Consequently, the travel paths in refraction work are predominantly horizontal, whereas for reflection work, they are predominantly vertical. Head waves or refractions (see §3.5) enter and leave a high-velocity bed at the critical angle and only a bed with velocity significantly higher than any bed above it can be mapped. Consequently, the applications of refraction methods are more restricted than those of reflection. (It should be noted that refraction is used in two different senses in seismology, to refer to the bending of raypaths due to changes in velocity and in the present sense of involving head waves. The classical mapping of high-velocity masses such as salt domes is also classed as a refraction method, although refraction at the critical angle is not necessarily involved; see §11.1.2.)

Refraction exploration generally involves greater distances than reflection work, so stronger sources are required. Because distributed in-line geophones would attenuate the head waves that have an appreciable horizontal component of motion, geophones are either bunched together or distributed perpendicular to the source–geophone line. Otherwise, however, the same equipment can often be used.

1.2 History of seismic exploration

1.2.1 Historical sources

This account is based mainly on articles by Barton (1929); Heiland (1929a, 1929b); Mintrop (1931); Shaw, Bruckshaw, and Newing (1931); Rosaire and Lester (1932); DeGolyer (1935); Rosaire (1935); Leet (1938); Weatherby (1940); Vajk (1949); Schriever (1952); Born (1960); McGee and Palmer (1967); Elkins (1970); Laing and Searcy (1975); Owen (1975); Petty (1976); Sweet (1978); Green (1979); Bates, Gaskell, and Rice (1982), and Karcher (1987), supplemented by conversations with individuals who were personally involved in early geophysical work. Recent historical articles that offer interesting glimpses into seismic history include Barrington (1982); Clark (1982, 1983, 1984a, 1984b, 1985, 1990a, 1990b); Mayne (1982); Proubasta (1982, 1983a, 1983b, 1983c, 1984, 1985a, 1985b, 1986a, 1986b, 1991); Robertson (1986); Robinson (1985); Sheriff (1985, 1988); Wilcox (1990); Keppner (1991), and Proffitt (1991).

1.2.2 Preliminary events

Geophysical exploration for oil began with the torsion balance, which was developed by Baron Roland von Eötvös about 1888 (Vajk, 1949). Although gravity surveys with the torsion balance were made in Europe on a limited scale, beginning about 1900, to map geologic structures, the first extensive surveys for petroleum objectives were in the United States and Mexico in the 1920s. In December 1922, a survey of the known Spindletop salt dome in Texas gave a gravity anomaly, but subsequent surveys were disappointing until 1924 when the Nash Dome was discovered. This resulted in the first geophysical oil discovery in January 1926. Through 1929 sixteen salt domes found by torsion balance surveys subsequently resulted in hydrocarbon discoveries (Sweet, 1978).

The theory of seismic waves might be dated from Robert Hooke's law enunciated in 1678, but most of the theory of elasticity was not developed until the 1800s. Baron Cauchy's memoir on wave propagation won the Grand Prix of the French Institute in 1818

and S. D. Poisson showed theoretically the separate existence of *P*- and *S*-waves around 1828. C. G. Knott (1899) presented a paper on the propagation of seismic waves and their reflection and refraction, and Emil Wiechert and Karl Zoeppritz (1907) published their work on seismic waves. Lord Rayleigh (1885), A. E. H. Love in 1911 (see Love, 1927), and R. Stoneley (1924) developed the theories of the surface waves that bear their names.

Robert Mallet (1848, 1851) began experimental seismology by measuring the speed of seismic waves using black powder as the energy source and a disturbance of the surface of a bowl of mercury as the detector. Mallet obtained very low velocities; probably low sensitivity allowed him to see only the later cycles of Rayleigh waves, then unknown. H. L. Abbot (1878) measured *P*-wave velocities using essentially the same type of detectors but a very large explosion. John Milne (1885) and T. Gray used a falling weight as a source (as well as explosives) in a series of seismic-wave studies using two seismographs in line, probably the first seismic spread. Otto Hecker (1900) used nine mechanical horizontal seismographs in line to record both *P*- and *S*-waves.

The possibility of employing the seismograph to define subsurface conditions was first put forward by Milne in 1898 (Shaw, Bruckshaw, and Newing, 1931).

As an earthquake wave travels from strata to strata, if we study its reflection and changing velocity in transit, we may often be led to the discovery of certain rocky structures buried deep beneath our view, about which without the help of such waves it would be hopeless ever to attain any knowledge.... Earthquakes are gigantic experiments which tell the elastic moduli of rocks as they exist in nature, and when properly interpreted may lead to the proper comprehension of many ill-understood phenomena.

L. P. Garret in 1905 suggested the use of seismic refraction to find salt domes but suitable instruments had not yet been developed (DeGolyer, 1935).

1.2.3 Early applications to petroleum exploration

After the sinking of the Titanic by an iceberg in 1912, Reginald A. Fessenden worked on inventions for iceberg detection. Among the methods was the use of acoustic waves in water, and an outcome of this was the first (U.S.) patent (fig. 1.1) on the application of seismic waves to exploration, applied for in 1914 and issued in 1917, entitled "Method and apparatus for locating ore bodies." Fessenden's patent said:

The invention described herein relates to methods and apparatus whereby, being given or having ascertained two or more of the following quantities, i.e., time, distance, intensity and medium, one or more of the remaining quantities may be determined.

He proposed using sources and detectors in water-filled holes, and locating ore bodies by both the use of reflections from them and by variations they introduce in traveltime measurements between holes. His

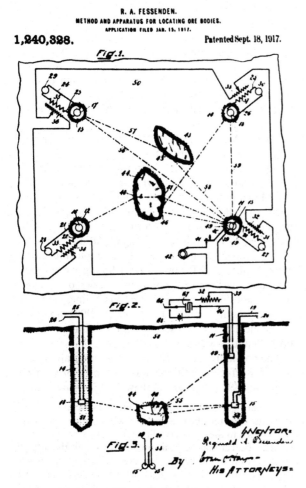

R. A. FESSENDEN.
METHOD AND APPARATUS FOR LOCATING ORE BODIES.
APPLICATION FILED JAN. 15, 1917.

1,240,328.   Patented Sept. 18, 1917.

Fig. 1.1   First page of Fessenden's patent.

patent was subsequently challenged (unsuccessfully) by (among others) Mintrop (1931) because Fessenden used "acoustic" waves rather than "seismic" waves and because his use of boreholes for sources and detectors did not accord with subsequent practice.

Ludger Mintrop in Germany in 1914 devised a seismograph with which he could make observations of explosion-generated waves with sufficient accuracy to make exploration feasible.

The Germans and Allies both experimented during World War I with the use of three or more mechanical seismographs to locate enemy artillery, but airwaves generally proved more satisfactory than seismic waves for this purpose. Among those involved in these experiments were Mintrop and the Americans R. A. Fessenden, E. A. Eckhardt, W. P. Haseman, J. C. Karcher, and Burton McCollum. These six were predominant in the development of commercial application of seismic waves after the war. McCollum attributed the idea of applying seismic methods to petroleum exploration to Haseman (unpublished "Recollections re McCollum" by R. L. Palmer). Mintrop's work was clearly independent and Fessenden

was apparently only brought into application efforts about 1925.

Mintrop in 1919 applied for a German patent on "Method for the determination of rock structures," which was issued in 1926. Mintrop's patent said:

Where the problem is to obtain . . . the approximate composition of the strata, the divining rod has been used as is well known. However, . . . it has not yet been possible to ascertain a connection of unique meaning between the indication of the divining rod and the geologic particularities of the subsoil. . . . According to my invention . . . the connection of mechanical waves with the characteristic properties of the strata is much more immediate . . . mechanical waves are artificially generated . . . by detonating a certain amount of explosives, their elastic propagation through the various formations is recorded by a seismometer located at a suitable distance . . . from the records of the latter the velocities of the various waves and the depth to which they penetrated can be determined, which allows conclusions as to the succession, thickness, density as well as the direction of the strike and dip of rock formations.

John William Evans and Willis B. Whitney in 1920 applied for a British patent on "Improvements in and relating to means for investigating the interior of the Earth's crust" which was issued in 1922. Their patent said:

The present invention . . . is characterized in that the sound waves . . . are received simultaneously or approximately so at a plurality (at least two . . .) of receiving stations . . . for the following reasons: Even in the simplest case when it is known that the stratum to be examined is horizontal there are two unknown quantities namely (1) the average velocity of the reflected wave . . . and (2) the depth of the reflecting stratum and therefore two equations . . . and two observations are consequently necessary.

Despite their rather complete grasp of reflection seismology, this patent does not figure prominently in subsequent developments, which concentrated on refraction.

Udden (1920) wrote in the *Bulletin of the American Association of Petroleum Geologists* (AAPG) (and illustrated with fig. 1.2):

it ought to be possible, with present refinements in physical apparatus and their use, to construct an instrument that would record the reflections of earth waves started at the surface, as they encounter such a well-marked plane of difference in hardness and elasticity as that separating the Bend and Ellenberger formations (in North-central Texas). . . . A seismic wave might be started by an explosion at the surface of the Earth, and a record of the emerged reflection of this wave . . . might be registered on an instrument placed at some distance from the point of explosion. . . . It ought to be possible to notice the point at which the first reflection from the Ellenberger appears. . . . With a map of the surface of the Ellenberger, it seems to me that millions of dollars worth of drilling could be eliminated.

In 1920, the Geological Engineering Company was founded by Haseman, Karcher, Eckhardt, and McCollum to apply seismic exploration to finding petroleum. Karcher had recorded a seismic reflection

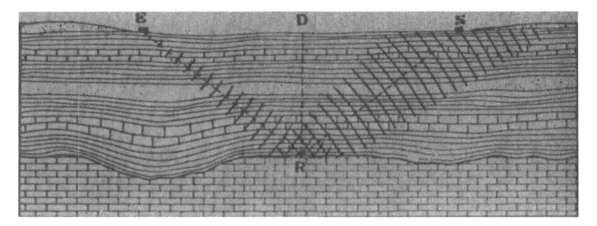

Fig. 1.2 Reflection expected from the contact between the Bend formation and the underlying Ellenburger limestone. (From Udden, 1920.)

from waves generated by artillery at the Indian Head test range in Maryland in 1917 and in a quarry (fig. 1.16a) in Washington, D.C., in 1919 (Karcher, 1987). They converted an oscillograph into a three-trace recorder and constructed electrodynamic geophones from radiotelephone receivers. In June 1921, Karcher, Haseman, I. Perrine, and W. C. Kite at Belle Isle (Oklahoma City) obtained a clear reflection from the contact between the Sylvan shale and the Viola limestone (fig. 1.3). About five months of reflection and refraction experimentation were carried out. One experiment involved dropping dynamite from an airplane in an attempt to obtain more nearly plane waves (Karcher had tried using aerial fireworks as a source in 1919). The company ran out of funds when a surplus of oil forced the price down to 15¢/barrel. The principals returned to their former jobs, except for McCollum. He agreed to settle with the company's creditors in return for the company's patents and equipment.

During 1920–1, Mintrop (Keppner, 1991) shot refraction lines across two known salt domes in northern Germany and discovered another, the Meissendorf dome, although it had no commercial significance. In 1921, he founded Seismos to do geophysical exploration and subsequently wrote a number of pamphlets promoting refraction exploration. In 1922, Seismos tried seismic methods in Sweden for mining objectives and in Holland for coal mapping.

Everett Lee DeGolyer wrote on October 3, 1922, to J. B. Body in London (the following three extracts are from DeGolyer's papers in the library of Southern Methodist University, Dallas):

You will remember that during the past summer Dr. Barton, of the Amerada Petroleum Corporation's staff, spent some time in Europe receiving instruction in the use of Eötvös torsion balances, and while there made several visits to Germany to investigate other physical methods of approach to geologic problems.

One of the methods which interested him very much, and which he seemed to think had considerable possibility, was the Seismic method. . . . I should like to suggest that it be called to Dr Erb's [Shell's] attention with the recommendation that he consider its availability for use in the Mecatepec–Papantla District [Mexico]. . . .

Body wrote to DeGolyer on December 14, 1922:

You will have seen my cable No. 88. . . . "Negotiating with Seismos from Hannover for using Mexico their method measuring with seismograph transmission waves caused by explosions thereby determining depth positions subterranean Tamasopo also outline salt domes. Method gave satisfactory results central Europe and are assured can be used Mexican conditions. Our intention is send out party. . . ."

Seismos party 1 began work in the Golden Lane area of Mexico for Mexican Eagle (Shell) in 1923. The contract for this work provided:

Seismos bind themselves to organize an expedition in order to carry out the . . . investigations. . . . This expedition shall consist of 2 seismologists and 1 mechanic, all of them experts with thoroughly up-to-date technical knowledge and possessing such zeal and sense of duty as is requisite for the success of their work. [They were] to be equipped with two complete seismic field-stations with the necessary instruments . . . to carry out in Mexico during 25 days effectual observations . . . in a geologically known territory. . . . Upon arrival the expedition shall confer with the local manager . . . who shall decide where and when their operations shall be carried out and to the solutions of which geological problems same shall be applied, on the understanding that as far as purely scientific questions are concerned . . . they shall use their own discretion. . . .

Compensation was to be U.S.$600 for the two seismologists together and $150 for the mechanic during the time the expedition was in Mexico and $500 for the instruments. If Mexican Eagle so elected, the contract could be replaced "by a new contract concluded for an indefinite period and for observations in Mexican regions geologically unknown." In this case, monthly compensation was to be increased to $800 and $250 for the men and $1000 for the instruments. The geological problem involved finding high-velocity lime-

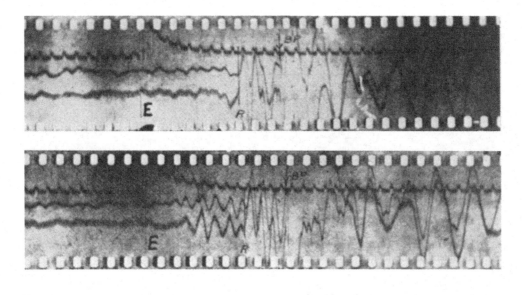

(a)

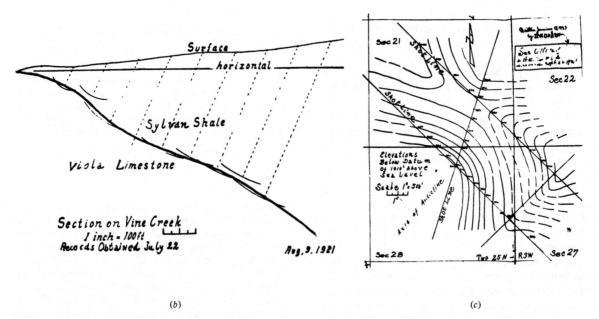

(b)   (c)

Fig. 1.3 First application of the reflection seismograph to exploration. (From Schriever, 1952.) (a) Two reflection records made in September 1921: E marks the explosion time, R the reflection from the Viola limestone, and BP the airwave (blastphone). (b) First depth section, at Vines Branch, Oklahoma, August 9, 1921. (c) First seismic structure map, near Ponca City, Oklahoma, September 1921.

stone reefs under a shale cover, a situation for which refraction appeared to be ideal.

Seismos party 2 began work in Oklahoma and Texas for Marland Oil Company (a predecessor of Conoco), also in 1923. Seismos party 1 moved to Texas to work for Gulf Oil in 1924 and in June discovered the Orchard Dome southwest of Houston, which is usually considered to be the first seismic (refraction) hydrocarbon discovery, a claim disputed by McCollum (see what follows). Early refraction records are shown in figs. 1.4a and 1.4b.

The Seismos crews used a mechanical seismograph (fig. 1.5) consisting of a mass suspended by a horizontal leaf spring with a natural frequency of about 10 Hz. The only amplification was mechanical and optical, and recording was done by directing a beam of light onto a mirror connected to the mass by a hair so that the mirror rotated when the mass moved, and then onto a strip of photographic paper moved by a hand crank turned by the observer. Shotpoint-to-seismograph distance was surveyed and a blastphone (fig. 1.6) was used to record the airwave to find the

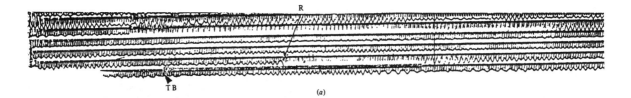

(a)

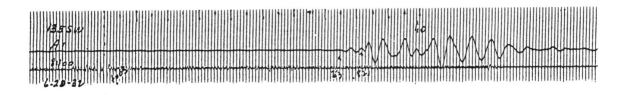

(b)

Fig. 1.4 Early refraction records. (Courtesy of Conoco.) (a) Record obtained with mechanical seismograph, 1924 or 1925; the recording was helical about a drum so that traces at the right end continue again at the left end. TB = shot instant obtained by radio and R = refraction arrival. (b) Refraction record, Texas, June 1932.

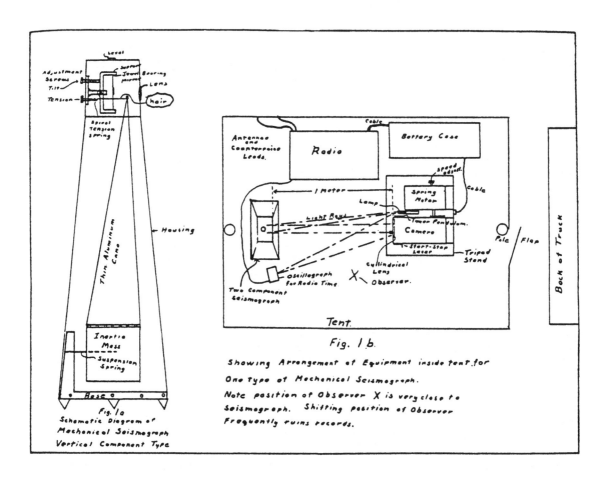

Fig. 1.5 Mintrop's mechanical seismograph. Movement of the case with respect to the inertial mass tilted the aluminum cone, pulling on the hair and rotating the mirror. (From Malamphy, 1929.)

Fig. 1.6 Blastphone used to detect the airwave for determining shot–detector distance. The diaphragm *d* is a pie tin and the transducer a carbon granule microphone. (Photographed at the Museum of the Geophysical Society of Houston.)

shot instant. (Subsequently, radio was used to determine the shot instant and the airwave arrival to find the shot-to-detector distance.) The overall sensitivity and precision were low and profiles were only $3\frac{1}{2}$ miles long, which gave limited penetration so that Seismos crews missed a number of domes at moderate depths. L. P. Garrett of Gulf Oil, for whom Seismos was working, developed the fan-shooting method (§11.1.2) about 1925, which increased the effectiveness in locating salt domes. By 1929, the refraction method had found fifty salt domes that resulted in hydrocarbon discoveries (Sweet, 1978). During the same period, "geology and accident" discovered one dome (Barton 1929: 616).

Following the failure of the Geological Engineering Company in 1922, McCollum obtained the backing of Atlantic Refining and formed McCollum Geological Exploration to carry out refraction work. New equipment was built, and in 1924 both reflection and refraction was used in the Tampico area of Mexico. The first well drilled on a seismic location, La Gatero No. 4, was dry, although the seismic prediction was correct. In May 1924, the Zacamixtle 199 well in the Golden Lane area succeeded in finding oil, to dispute the claim of the Orchard Dome in Texas as the first seismic discovery. However, the Mexican well was noncommercial at the time because of its remote location (Owen, 1975). In 1928 the Atlantic–McCollum joint venture was dissolved; McCollum and Atlantic divided the four sets of instruments and McCollum formed McCollum Exploration Company.

The Marland Oil Company had supported two months of the 1921 Geological Engineering Company reflection experimentation (which was unsuccessful) and had brought Seismos party 2 to Texas in 1923. The Seismos party failed to find any salt domes for Marland. In 1925, Marland hired Haseman, Eckhardt, Eugene McDermott and others to develop a more sensitive electrical seismograph. The Marland field party began exploration in 1926, replacing the Seismos party. The equipment worked well, but Marland never recorded a salt dome discovery.

1.2.4 The Geophysical Research Corporation

DeGolyer was at first disappointed with the refraction method, but success by Seismos crews working for Gulf changed his mind and he began to search for personnel to develop seismic methods. He learned of Karcher's 1921 experiments and in May 1925, Amerada, Rycade (an Amerada subsidiary) and Karcher formed the Geophysical Research Corporation (GRC). They acquired Fessenden's patent and his services as consultant.

GRC built an electrical seismograph that was much more sensitive than the Seismos mechanical seismograph. The detector was a variable-reluctance type and the amplifier was resistance-coupled using a vacuum tube. The oscillograph used two galvanometers and the recording film was hand-cranked. Timing lines were obtained by shining a light through slits attached to the prongs of a 50-Hz tuning fork. The time-break (shot instant) was transmitted by interrupting a CW (continuous-wave) transmitter.

GRC fielded seven field parties in 1926 and refraction exploration greatly expanded. A GRC crew under E. E. Rosaire was forbidden to shoot profiles more than 3½ miles in length, which had become the standard distance since it had been successful for Mintrop, but the observers "got lost" and discovered the Port Barre salt dome (Sweet, 1978). Thereafter, the standard distance became 6 miles. Refraction at the time was used as a reconnaissance method and was usually followed by detailing with a torsion balance (and later gravimeter) survey.

GRC party 6, an experimental crew, tried reflection work in Kansas in 1926. They soon moved to Texas and obtained usable reflection records from the caprock of the Nash salt dome. Other GRC parties also experimented at recording reflections. In 1927, party 6 moved to the Seminole Basin of Oklahoma, an area ideally suited to reflection work, where they soon found a structure that became the first discovery by the reflection method, the Maud Field (in 1928). This success was quickly followed by others, and by 1930, the reflection method began to take over from the refraction method. Early reflection records are shown in fig. 1.16.

1.2.5 Other activities in the 1920s

Humble Oil Company, at the instigation of Wallace E. Pratt, established a geophysical department in 1924 under Dr. N. H. Ricker, and the following year fielded two refraction crews using mechanical seismographs designed by O. H. Truman (Carlton, 1946). These crews began using a telephone line to carry the time-break, but before the end of 1925, they used radio for both communications and transmission of the time-break.

Frank Rieber in 1924 obtained funding (from General Petroleum, Standard Oil of California, Associated, and Shell) for a refraction survey in the San Joaquin valley of California. This survey was unsuccessful in obtaining deep information. Rieber carried out other surveys in California in 1927–8, but his company failed in 1930. In 1932, he began work on reflection instruments, and in subsequent years introduced a number of instrumental innovations.

In 1925, the Petty Geophysical Engineering Company was formed by Dabney E. Petty and Olive Scott Petty (and other family members). They felt they could easily improve on Mintrop's mechanical seismograph, and in 1926 fielded a crew equipped with condenser-type geophones (fig. 1.7) and vacuum-tube amplifiers. They used string galvanometers (fig. 1.8)

Fig. 1.7 The Petty prototype geophone. The "steady mass" m is on a long beam hinged (h) at the left and supported by a strong spring s. At the right end, the beam is attached to one plate of an air-gap condenser c. Movement of the case with respect to the steady mass changes the separation of the condenser plates, the change in capacity being proportional to the displacement. The dimensions are $48 \times 32 \times 15$ cm. (Photographed at the Museum of the Geophysical Society of Houston.)

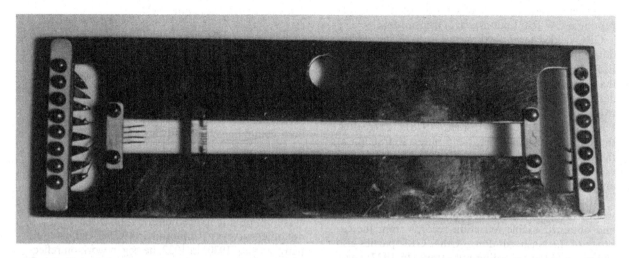

Fig. 1.8 String-galvanometer harp. Currents pass through fine wires (some are broken on this harp) held taut by small springs, causing the wires to be deflected in a magnetic field. Shadows of the wires were focused onto photographic paper. A string-galvanometer record is shown in fig. 1.16e. (Photographed at the Museum of the Geophysical Society of Houston.)

and a camera with photographic paper pulled along by a spring motor; shadows cast by the moving strings on the paper were recorded. The Pettys did appreciable experimentation to find a quicker, easier way to locate salt domes. They discovered that a salt forerunner (fig. 1.9), which could be distinguished by its amplitude, could be used to tell if a salt dome had been encountered even without knowing the shot-to-detector distance. They also found that the Rayleigh-wave pattern changed when a salt dome intervened and used this fact when they could not get a readable *P*-wave. The increased sensitivity of their equipment and their interpretational ingenuity allowed them to survey with smaller shots than others used.

In 1927, the first well velocity measurements were made. A geophone lowered 5000 ft (1500 m) into a Gulf well in Kansas recorded the traveltime from a shot at the surface. Also in 1927, C. A. Heiland established the first course in exploration geophysics at the Colorado School of Mines.

McCollum successfully mapped the Barbers Hill Dome by reflection in 1928 using 100 detectors at each station spaced to attenuate horizontal waves, but the use of multiple detectors was too cumbersome and so was abandoned; it was revived about the mid-1930s with four to six detectors per group.

In the late 1920s, seismic exploration began to move abroad: to Persia (Iran) and Venezuela in 1927, to Australia in 1929, to the Netherlands East Indies in 1930. Donald C. Barton (1929), who subsequently became first president of the Society of Exploration Geophysicists (SEG), described the early methods:

For work with the mirage [refraction] method, a troop rather commonly consists of one firing unit, two, three or four receiving units, a squad of hole diggers, a chief of party, a "landman," a calculator, and in some cases a crew of surveyors, and in some a hole-filling crew. . . . The firing unit . . . is equipped with the necessary apparatus to fire the charge, . . . with meteorological apparatus and with a sending and receiving wireless set, which is used to communicate with the receiving units and to send out the instant of the explosion. . . . The receiving unit . . . is equipped with a seismograph, . . . a wireless sending and receiving set, and meteorological apparatus. . . . To set up a station, a 3 inch hole is dug to a depth of 3 ft, the geophone cable is reeled out and the geophone is dropped down the hole. . . . Each receiving unit . . . signals that tentatively it is ready to receive. . . . When all . . . have signaled their o.k. for firing, the firing master sets his wireless sending out a continuous wave note, . . . waits a standard short interval and then fires the charge. The wireless key is held down by a circuit which goes around the dynamite. The explosion instantaneously breaks the circuit, causes the release of the wireless key and the instantaneous [*sic*] cutting off of the wireless note. . . . The average charges used . . . range from 40 to 250 lbs. For the same shots, Seismos Gesellschaft would use two to three times as large a charge. . . .

In the reflection type of shooting, the charge is very much smaller. . . . The practice . . . [is to place the] main charge 17 to 25 ft down a 6 inch hole and the auxiliary charge at the surface. The latter is used to produce an air wave. The holes are dug with hand augers. . . . The distance between the firing point and the seismograph . . . ranges from 1.2 to 1.8 times the depth of the formation which it is desired to map. . . .

A scout from a rival company not uncommonly is set to watch the troop and to report their activity to his company and especially to report anything to indicate that possibly they may have picked up a salt dome. He often gets to be on good terms with the troop, but at critical moments they go to all sorts of strategy to outwit him.

1.2.6 Early geophysical case history

The exploration of Cameron Meadows field, a salt dome in Louisiana, documented by McGuckin (1945), provides interesting comparisons of early exploration methods. Attention was first called to this area because of "sulphur gas" seeps ("boils") in the marsh. The area has "no recognizable physiographic

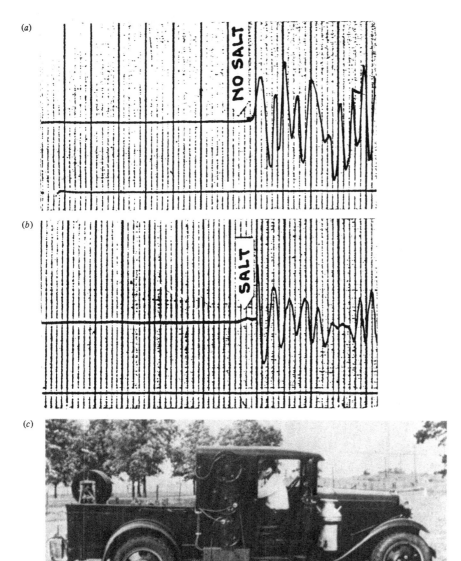

Fig. 1.9 Refraction recording, 1930. (a) Normal (no salt) record; (b) record showing salt forerunner; (c) recording truck in use about this time. (Parts (a) and (b) are from Petty, 1976; part (c) is courtesy of Haliburton Geophysical Services.)

expression . . . [except that] Old North Bayou . . . may possibly be deflected slightly," according to the *Louisiana Geological Bulletin* No. 6 in 1935.

A series of geophysical surveys were carried out from 1926 to 1943. They had different objectives and covered different areas. Hence, the maps shown in fig. 1.10 were to different scales but they have been enlarged or reduced to facilitate comparisons.

The Seismos Company used its one-channel mechanical seismograph to shoot refraction over this area in 1926 as part of a larger survey. Seismos generally built tiny islets of sandbags to support its detector above the water. The report said: "The surface indications around the shell hill of tent location 145, its sud-

den uplift out of a very deep marsh, and a gas seep south of it, lead to the conclusion that it might be caused by the presence of harder layers. However, the investigations . . . show clearly that no harder layers occur down to considerable depths. . . . there may be a possibility to find a structure in sections 21 and 28 of this region." The main part of the Cameron Meadows field was subsequently discovered in section 21. Seismos interpreted (correctly) SW-dip southwest of the recording station (fig. 1.10a). Source–detector distances of less than 3.5 miles did not provide very deep penetration.

A reconnaissance torsion balance survey in 1927 (fig. 1.10b) indicated a large closed gravity minimum

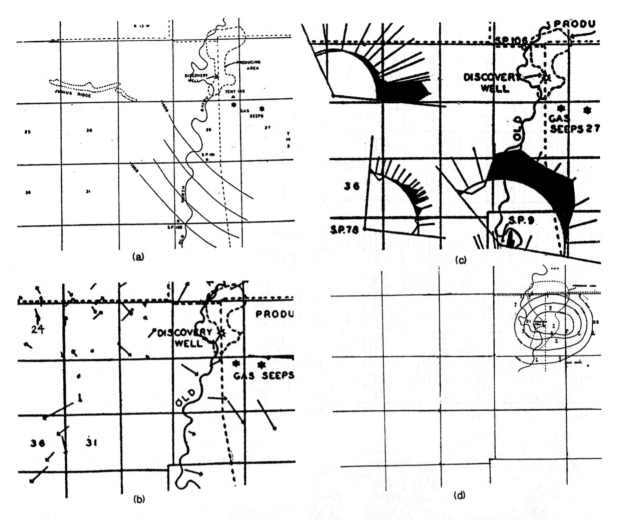

Fig. 1.10 A sequence of maps (to approximately the same scale) of the Cameron Meadows dome based on the early geophysical surveys. (From McGuckin, 1945: 1–16.) (a) Seismos mechanical refraction map, 1926. (b) Torsion-balance reconnaissance map, 1927, showing gravity gradients. (c) Geophysical Research Corp. (GRC) refraction fan map, 1928; the filled-in areas indicate leads. (d) GRC correlation refraction map, 1929. (e) Geophysical Service, Inc., dip reflection map, 1933. (f) Map of the top of salt from drilling, showing wells that encountered salt. (g) Top-salt map from McCollum refraction shooting into a well, 1942. (h) Top-salt profile from McCollum refraction survey into a well and from well data; the deepest map of the 1942 Petty continuous profiling survey is also indicated. (i) Petty Geophysical Engineering Co. continuous profiling reflection map to define faulting on the south flank, 1942. (j) Robert H. Ray, Inc., gravimeter survey, 1943, showing also residual gravity (dashed contours).

centered in section 24 at the left edge of the fig. 1.10 maps. An unsuccessful well was drilled on this minimum; the Cameron Meadows area lay 3 miles to the east on the eastern flank of the minimum.

A refraction fan shot in 1928 from SP-9 (fig. 1.10c) indicated a salt lead (in the direction of the blacked-in area), but a second fan from SP-78 was interpreted as failing to confirm the lead, and it was concluded that no salt dome was present. However, additional fans shot to average distances of 8 miles in 1929 (one in the upper left corner of fig. 1.10c) suggested a salt dome in section 21. A follow-up correlation reflection survey lasting 17 days in July 1929 (fig. 1.10d) mapped the closure on which the discovery well, completed in 1931, was drilled. A dip-method survey (fig. 1.10e)

was run in 1933 to serve as a guide for further drilling. A map of the top salt based on 15 wells is shown in fig. 1.10f. The jump correlations in the correlation and dip-method surveys precluded the mapping of faults.

Because the refraction fans had suggested an extension to the southeast where the seeps were located, a series of radial lines were shot in 1942 into a geophone located near the salt in an abandoned well on the northeast flank of the dome (the method is described in §11.1.3). The resulting map of the salt dome, shown in fig. 1.10g, can be compared with fig. 1.10f. The cross-section of fig. 1.10h shows the general agreement between seismic and well results.

Continuous profiling was carried out in 1942 to map the southeastern nosing that had been indicated

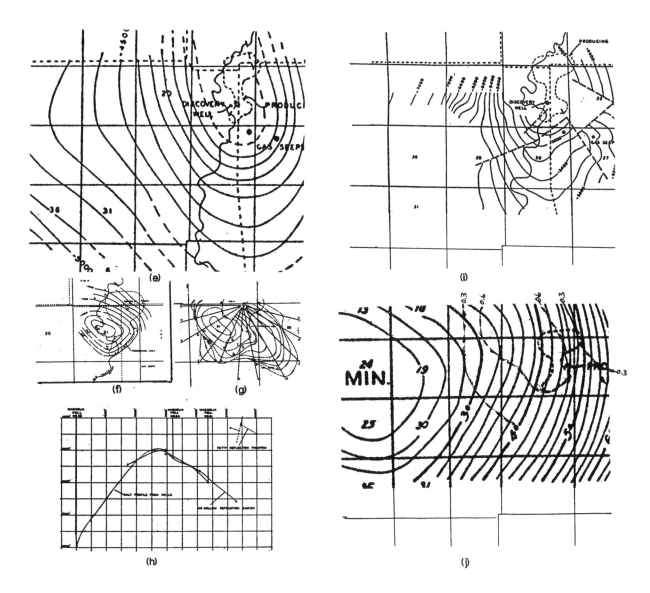

by shooting into the well. This survey defined faulting and the deepest map, a phantom several thousand feet above the salt, is shown in fig. 1.10i. No reflections were recorded in the mainly shale section below this phantom, although the shale section also contains some prospective sands.

In 1943, a gravimeter survey (fig. 1.10j) conducted over the area confirmed the earlier torsion balance results. It showed the producing area at the SE-edge of a residual gravity anomaly.

1.2.7 Development of the geophysical contracting industry

Burton McCollum in 1922 applied for a patent on "Method and apparatus for determining the contour of subterranean strata," which was issued in 1928 along with two other patents on variations of seismic methods. McCollum sold two of his patents to the Texas Company in 1928 and seven more between 1929

and 1935. These and other patents were transferred to the Texas Development Company, which tried to collect royalties from others but was mainly unsuccessful. In 1934, they sued Sun Oil Company for patent infringement. Almost the entire petroleum industry joined with Sun in the defense; the matter was settled out of court in 1937. The settlement involved companies forming a Seismic Immunities Group and granting each other royalty-free licenses of their patents and of patents for which they might file within a year of withdrawal from the group. Initially, 64 patents were involved, including 2 of Mintrop's, 10 of McCollum's, 2 of Harvey C. Hayes', 8 of Fessenden's, and 2 of Karcher's. Several payment schemes could be elected; one involved a lifetime payment of $10,000 per party, a party being defined as either (a) a single recording unit with no more than 12 traces or (b) up to four recording units where shot-to-detector distance exceeded 2 miles (to cover respectively reflection and refraction work). The Mayne CDP patent (see

§1.2.9) was one of the last important patents involved before the group disbanded entirely.

In 1929, a new Amerada president decided that GRC would no longer do reflection work for other companies (Karcher, 1987). Although Petty and McCollum offered independent alternatives to accomplish geophysical work and some oil companies, such as Humble and Gulf, operated their own crews, the oil companies in general encouraged the formation of more new geophysical enterprises. Thus, the early 1930s saw the advent of many geophysical contractors, including most that dominate today. Most of these were formed by people who left GRC since it dominated the industry until then; some of these are shown in fig. 1.11. In addition, a few companies (such as Rogers and General) were formed by people who left other companies such as Petty, and still other companies (such as Heiland) were formed without clear connections to preceding industry.

Most of the major geophysical contractors through the 1980s can be traced directly back to spinoffs from GRC. Haliburton Geophysical Services was the successor to Geophysical Service Inc. (GSI), which was formed in 1930 (as well as successor to other companies, including Petty). Schlumberger is successor to Seismograph Service Corp. (SSC) formed in 1931 (and to Seismos and Prakla). Teledyne is successor to Independent Exploration formed in 1932. Western Geophysical Company, formed in 1933, is still independent; it has now taken over Haliburton Geophysical Services. Grant Geophysics is successor to United Geophysical, formed in 1935. Compagnie Génerale de Géophysique (CGG) is successor to SGRM, which began refraction work in France in 1930.

The field work of Seismos declined to zero in 1931, but Seismos revived refraction work in Germany in 1934 and began reflection work about the same time. Prakla was founded in 1936 and subsequently merged with Seismos to form Prakla–Seismos. The rapid growth of exploration geophysics was almost entirely due to private enterprise with intense rivalry and competition and extreme secrecy between the individual companies involved. From the early 1930s to the early 1990s, no single company dominated geophysical exploration. However, by 1994 only two large contracting companies, Schlumberger and Western, continued (see fig. 1.24).

1.2.8 Evolution of reflection equipment and methods

The first GRC reflection work in 1926 employed the same two-galvanometer arrangement used in refraction work, but a third galvanometer was added soon. A four-channel system was built in 1928 and before long six-channel instruments were in use; the standard was 6 to 8 in 1937 and by 1940 most crews were 10 to 12 channel. The number of channels has continued to increase (fig. 1.12). For many years after World War II, 24 channels were standard, then in the late 1960s,

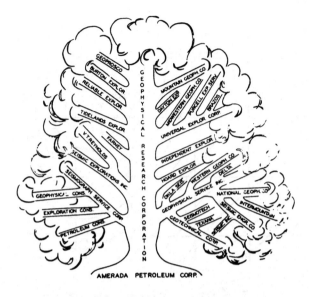

Fig. 1.11 The "Amerada Tree," a diagram drawn in 1950 to show geophysical contracting companies formed by people who left GRC. (W. J. Zwart and K. M. Lawrence located this historic document.)

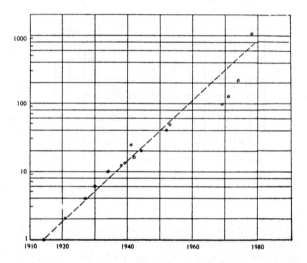

Fig. 1.12 Growth of multichannel recording. The starting point in 1914 is Mintrop's first portable seismograph. The vertical scale gives the first known instance of use of numbers of channels. The use of group recorders in the 1980s effectively removed an upper limit on the number of channels except for operational practicality. In one instance, 4000 channels reputedly were used.

48 channels became common, and today (1994) most crews use 120 to 240 channels and some use appreciably more. Digitization at the geophone (§7.5.2) has now effectively removed limits on the number of channels that can be recorded.

The mechanical seismograph was soon replaced with electrical geophones and vacuum-tube amplifiers. The early electrical geophones were mostly of three types: capacitance, variable-reluctance, and moving-coil electrodynamic; oil damping was generally used. Early electrical geophones (fig. 1.13) had to have high sensitivity because of the high noise level of available vacuum tubes. For the variable reluctance and moving-coil types, this meant large magnets because of the low permeability of the magnetic materials then available. As better magnetic materials and lower-noise vacuum tubes became available, the electromagnetic geophone increased in sensitivity and decreased in weight (from some 15 kg to a few hundred grams), electromagnetic damping replaced oil damping, and the electromagnetic type eventually became dominant (for land work). As a result of these improvements, multiple geophones per channel became practical; this usage was introduced in 1933 and was common practice by 1937.

The gain of early instruments was constant and repeated shots were usually required so that reflections at several arrival times could be mapped. Sometimes the gain was manually changed during the recording by the operator turning a switch. About 1932, automatic gain control was developed, first by changing the grid bias with time after the time-break, later by a feedback circuit. Amplifiers increased in gain, sophistication (initial suppression, automatic gain, mixing, etc.), and reliability. A 10-channel recorder from 1931, the first to change frequency response with time, is shown in fig. 1.14. Some timing wheels are shown in fig. 1.15 and reflection records are shown in fig. 1.16. About 1950, recording instruments became sufficiently reliable that the observer could "do a day's work rather than instrument repair and adjustment." The Southwestern Industrial Electronics (SIE) Company's P-11 recording system was a major advance in

reliability. A chronology of some instrument developments is given in table 1.1.

The need for weathering corrections (§8.8.2) was recognized very early and shallow refraction shots were often made for this purpose. The first mapping was done by correlating reflections on widely separated profiles (fig. 1.17). Barton (1929) wrote:

a depth determination is made by each shot and to map the dip, folding or faulting of the surface, . . . it is necessary only to scatter "shots" over the area to be mapped and draw structure-contours or profiles from the results. Practically the application of the method is somewhat uncertain. . . . The impossibility of recognizing the reflecting bed is a serious disadvantage. . . .

The correlation method did not work well in the Gulf Coast because the area lacks distinctive reflections.

In 1929, T. I. Harkins

noticed that abnormal stepouts [dip moveout – see §4.1.2] were rather characteristic of the (Darrow dome) area and that these abnormal stepouts reversed. He correctly attributed this phenomenon to dipping beds. [Rosaire and Adler, 1934]

Soon dip shooting was carried out along continuous lines of traverse.

Although from the earliest days crews carried out surveys in water-covered areas, the methods were basically those for land crews, improvised for use in water. Petty (1976) describes a survey in Chacahoula Swamp, Louisiana, in 1926 (fig. 1.18), and in 1927, GRC fielded two crews for work in water-covered areas (Rosaire and Lester, 1932). Sidney Kaufman (personal communication), in following up an on-shore lead in 1938, took his Shell shallow-water crew seaward 4 miles into 65 feet of water. Surveyors on-

Fig. 1.13 Early geophones in the Museum of the Geophysical Society of Houston. The geophones weigh (left to right, back row) 6.1, 8.7, 7.9, 6.7, (front row) 8.8, and 0.8 kg. At the lower right is a modern phone (30 g) for comparison.

Fig. 1.14   Petty 10-channel recorder from 1931. Vacuum tubes V amplified the current, which then passed through a "harp" H (of the type shown in fig. 1.8); light from a source L passed by the harp and was focused onto photographic paper in the takeup magazine M. A timing wheel T (see fig. 1.15) driven by a clock interrupted the light to give the timing lines. (Photographed at the Museum of the Geophysical Society of Houston.)

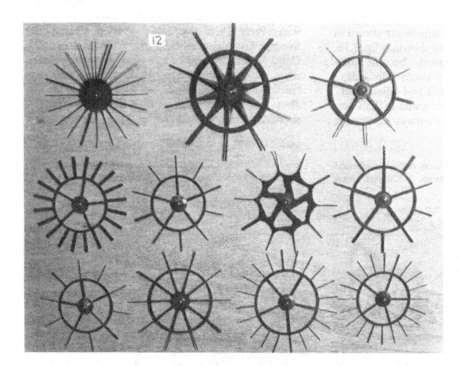

Fig. 1.15   Early timing wheels ("paddle wheels"). The wheels were rotated by a clock motor so that they cut the light beam to produce timing line shadows. Observers often had individual-ized wheels that characterized their records. (Photographed at the Museum of the Geophysical Society of Houston.)

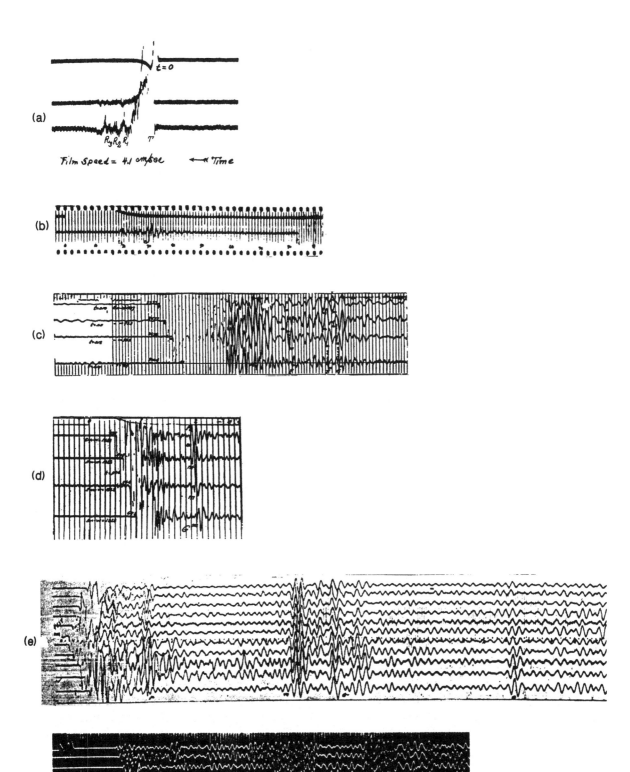

Fig. 1.16 Early reflection records. (a) First record made in a rock quarry near Washington, D.C., April 12, 1919, by Karcher. Time increases from right to left; the upper trace is the time-break; the lower traces are the geophone response at different gains. Three reflections are marked. (From Schriever, 1952.) (b) Record showing reflection from caprock of the Nash Dome, Texas, 1926; the charge is 1 pound; the offset is 950 ft. (From Weatherby, 1940.) (c) 1929 three-channel Oklahoma record showing Viola reflection. (From Weatherby, 1945.) (d) 1929 record from Cimarron anhydrite in Kansas. (e) 1934 California record with AGC. (From Weatherby, 1940.) (f) 1938 string-galvanometer record from Mississippi.

Table 1.1 *Chronology of seismic instrumentation and methods*

1914	Mintrop's mechanical seismograph		1952	Analog magnetic recording*
1917	Fessenden patent on seismic method		1953	Vibroseis recording*
1921	Seismic reflection work by Geological Engineering Co.			Weight-dropping
			1954	Continuous velocity logging
1923	Refraction exploration by Seismos in Mexico and Texas		1955	Moveable magnetic heads
			1956	Central data processing
1925	Fan-shooting method		1961–2	Analog deconvolution and velocity filtering
	Electrical refraction seismograph		1963	Digital data recording*
	Radio used for communications and/or time-break		1965	Air-gun seismic source
			1967	Depth controllers on marine streamer
1926	Reflection correlation method		1968	Binary gain
1927	First well velocity survey		1969	Velocity analysis
1929	Reflection dip shooting			Transit satellite positioning
1931	Reversed refraction profiling		1971	Instaneous floating-point amplifier
	Use of uphole phone		1972	Surface-consistent statics
	Truck-mounted drill			Bright spot as hydrocarbon indicator
1932	Automatic gain control		1974	Digitization in the field
	Interchangeable filters		1975	Seismic stratigraphy
1933	Use of multiple geophones per group		1976	Three-dimensional surveying
1936	Rieber sonograph; first reproducible recording			Image-ray migration (depth migration)
			1984	Amplitude variation with offset
1939	Use of closed loops to check misties			Determining porosity from amplitude
1942	Record sections			DMO (dip-moveout) processing
	Mixing		1985	Interpretation workstations
1944	Large-scale marine surveying		1986	Towing multiple streamers
	Use of large patterns		1988	S-wave exploration
1947	Marine shooting with Shoran			Autopicking of 3-D volumes
1950	Common-midpoint method*		1989	Dip and azimuth displays
1951	Medium-range radionavigation		1990	Acoustic positioning of streamers
				GPS satellite positioning

*The acceptance of these methods is shown in fig. 1.21. Dates are approximate; secrecy and competition often involved development and use of the same feature by several companies without public disclosure.

shore directed locations and this imposed the 4-mile limit. The survey was conducted from three 35-foot fishing boats. The instruments were eight-channel using one land geophone per channel bolted to an 18-inch steel plate to keep it upright on the sea floor.

Extensive marine operations did not appear until 1944 when Superior and Mobil began refraction fan-shooting for salt domes offshore Louisiana (Jack Lester, personal communication). A survey to map the offshore extension of Los Angeles basin fields was also carried out about this time (C. C. Bates, personal communication). A surveyor onshore gave instructions to keep lines straight while wire paying out through a counter gave the distance; buoys were set to indicate locations. As work progressed farther offshore, the chaining continued on a compass bearing. Sighting on shot plumes (water sent up into the air by the shot explosion) both visually and with radar was also used. Surveying was the principal operational problem and often constituted the major cost. Shoran came into use about 1946, followed by Raydist about 1951. The early refraction and reflection work used geophones planted on the bottom. About 1946, reflection work began using a 12-channel bottom drag

cable with gimble-mounted geophones. The floating streamer was first used in 1949–50. Both the radionavigation methods and the floating streamer were based on World War II developments. Proffitt (1991) summarizes the history of marine work.

1.2.9 Reproducible recording, the common-midpoint method, and nonexplosive sources

Frank Rieber (1936) proposed the "Sonograph" method of recording seismic data (fig. 1.19) so that it could be "played back." His oscillograph recorded in variable density on film. On playback, variations in the intensity of a light beam that passed through the film were detected by a photocell. Rieber used the Sonograph to determine the variation of reflection amplitude with apparent dip.

Despite Rieber's pioneering work, reproducible recording did not become practical until the introduction of magnetic-tape recording. Commercial recording and playback equipment became available about 1952. The principal advantage of magnetic-tape recording was thought to be the ability to replay with different filters. About 1955, moveable heads allowed static and normal-moveout corrections (§6.1) to be applied. The growth of analog magnetic-tape recording is shown in fig. 1.21.

A very important postwar development was the use of record sections for interpretation. Individual seismic records had been laid out adjacent to each other in the interpretation process for a long time (fig. 1.20), but the large size of individual records and variations in paper speed and developing quality made it difficult to obtain a synoptic view. Normal moveout (§4.1.1), irregularities in recording or spreads, and the wiggle-trace display mode added to the difficulties. Gulf Oil and Carter (now part of Exxon) and perhaps Shell apparently led in developing variable-density or variable-area displays with uniform horizontal scale and display amplitude. Carter bought Rieber's equipment for this use (among others) about 1946.

Common-midpoint (CMP, originally called CRP or common-reflection point, later CDP or common-depth point; see §8.3.3) recording was invented by Harry Mayne (Petty Geophysical) in 1950 as a way of attenuating noise that could not be handled by the use of arrays. Magnetic-tape recording made CMP practical, and CMP recording began about 1956, but it did

not become used extensively until the early 1960s (fig. 1.21) when its ability to attenuate multiples (§6.3.2) and other kinds of noise led to rapid adoption. Today, its use is nearly universal.

Magnetic-tape recording also permitted the addition of traces and thus the use of weaker sources because records from several weak sources could be added together to get the effect of a stronger source. McCollum introduced the use of a dropped weight, the Thumper, as a seismic source about 1953. Weight-dropping expanded seismic work in areas of difficult shothole drilling, such as West Texas, and in desert areas where water for drilling is scarce.

A variety of surface sources for use on land were also developed besides weight-dropping. The most ingenious of these, the Vibroseis™ method (see §7.3.1; a list of trademarks and the companies that hold them is given in app. B) was developed by John M. Crawford, William Doty, and Milford Lee and first used in 1953. Surface sources are now used for about half the land work and Vibroseis is the predominant surface source. Several alternatives to the use of dynamite as a source in the marine environment were developed about 1965. They were generally cheaper and more

Fig. 1.17 Portion of a dip map that resulted from dip shooting, January 1935. Dips were expressed in feet per mile and the arrow lengths indicated the spacing for 50-ft contours. (Courtesy of Conoco.)

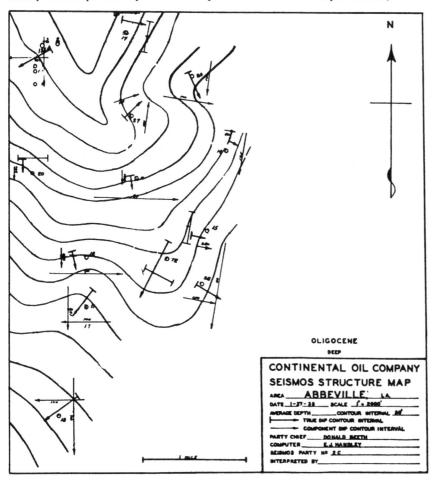

(a)

(b)

Fig. 1.18   Early refraction work. (From Petty, 1976.) (a) D. E. Petty washes a refraction record in Chacahoula Swamp; the geophone is on the cypress stump in the background. (b) Petty (in boat) with his crew in Chacahoula Swamp.

efficient and adaptable to CMP recording. Consequently, they rapidly replaced dynamite as a marine source (fig. 1.21). In addition, they were environmentally acceptable because they did not injure marine life.

Although some sophisticated playback processing was done with magnetic-tape recording and some dig-ital processing was done on analog data, the full potential of data processing was not achieved until digital recording was introduced in the 1960s. Digital recording not only resulted in higher fidelity, but also in the large-scale application of the digital computer in the processing and interpretation of seismic data. The "digital revolution" was probably the most far-

Fig. 1.19 Rieber's sonograph, 1936. Field data were recorded on film in variable-density mode. In playback, the total light through a slit was summed to give a single output trace. By changing the slit angle, data with various angles of approach (also called apparent dip; see §4.1.2) could be emphasized, each slit angle giving an additional trace. Thus, the sonogram record displayed amplitude in the angle-of-approach versus arrival-time domain. (Two views from advertisements in *Geophysics*, vols. 1 and 2.)

reaching development in seismic exploration since the pioneering days. For example, obtaining useful data in the North Sea is almost impossible without deconvolution (§9.5).

1.2.10 Recent history

Since the development of the common-midpoint, Vibroseis, and digital-processing methods, a succession of incremental improvements have expanded many fold the amount of geologic information extractable from seismic data. In consequence, the seismic method is being applied in ways not previously contemplated.

The prevailing concept prior to the 1970s was that noise was so much stronger than the seismic signal that only structural information could be extracted from seismic data on a practicable basis; consequently, the emphasis was almost exclusively on noise attenuation in order to obtain more accurate traveltimes. Recognition in the early 1970s that a hydrocarbon accumulation sometimes changed the reflectivity enough to allow its "direct" detection and mapping led to more accurate recording and interpretation of amplitude information. In the mid-1970s, recognition of depositional patterns in seismic data further changed attitudes and gave rise to the belief that much of what had previously been regarded as noise was actually geologic signal. The development of three-dimensional (3-D) methods in the late 1970s began to resolve the interpretation ambiguities inherent in in-terpolating between seismic lines and also provided significant noise reduction. 3-D reveals so much structural and stratigraphic detail that its use in reservoir engineering is growing very rapidly (see fig. 1.21).

It had long been recognized that the combined use of S- and P-waves would permit extracting more useful geologic information than from P-waves alone. The development of S-wave sources resulted in appreciable experimentation in the early 1980s, the primary objectives being hydrocarbon and lithology identification. Measurement of amplitude variation with offset (AVO), which gives much the same information as the combined use of S- and P-waves but at lower cost, resulted in some curbing of S-wave studies. Today, S-waves (and three-component recording) are being studied mainly to measure fracture orientation and intensity.

The 1917 Fessenden patent contemplated seismic measurements in boreholes, and in the late 1920s, a borehole geophone began to be used to define a salt dome flank (as an extension of the refraction method) and to measure seismic velocity. Gal'perin (1974) described Soviet development of vertical seismic profiling (VSP) during the 1960–70 period. The use of borehole measurements for purposes beyond simply velocity measurement expanded in the early 1980s. Tomographic borehole-to-borehole and borehole-to-surface methods have been undergoing rapid development since the mid-1980s but are still experimental.

Three-dimensional surveying requires more channels for its efficient execution and involves more logis-

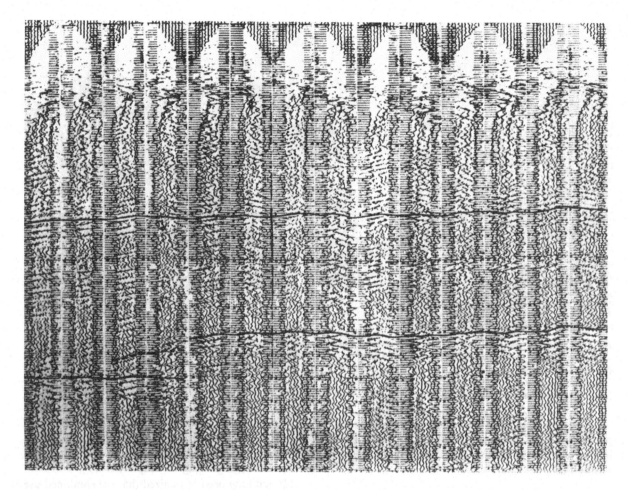

Fig. 1.20 Early record section made by splicing individual records together. The records are made with a 10-trace camera and are exceptionally uniform for the period. The lower marked horizon is clearly cut by a large fault. (Courtesy of Conoco.)

tical problems in the field. These have led to increased use of digitization at the geophone and telemetry methods for recording information.

Marine seismic data acquisition has been rendered more efficient by towing more than one source and more than one streamer offset by paravanes so that several lines of data can be acquired on a single pass of a ship. The use of improved radiopositioning systems has resulted in continual refinement in the precision of locating acquisition points at sea. Marine seismic exploration has made extensive use of Transit satellite positioning and now extensively uses the Global Positioning System (GPS), both developed by the U.S. Navy. The GPS is also being used for land surveying. Ships involved in 3-D work now have very elaborate positioning systems for determining, in real time within a few meters, the locations of all of the survey elements (source and each hydrophone group in each streamer for every energy release) in the local coordinate system. This information is needed in order to bin and correct data properly and also to advise if the survey is obtaining the requisite uniformity of coverage while still on the prospect.

The power of computers increased enormously in the late 1980s; at the same time, costs decreased and the speed increased so much that it became practical to provide interpreters with their own computers (*workstations*) to help with interpretation. Workstations solve many of the data-handling problems that historically have consumed so much of an interpreter's time, allowing more cross-checking and displaying of data in ways that make features of interest more visible, and generally allowing for a more complete and accurate interpretation.

The interpreter can also carry out simple data processing at the workstation. As computers become even more powerful and rapid, it is expected that interpreters will be able to tailor processing more to the characteristics and problems in their particular data. This may somewhat close the gap between processors and interpreters that developed with digital processing about 1960 (fig. 1.22).

Ships acquiring marine 3-D data carry an enor-

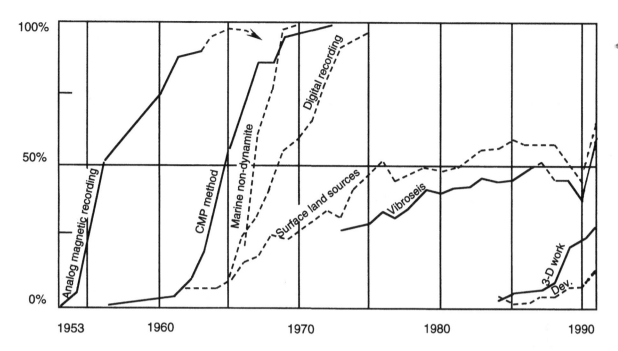

Fig. 1.21   Percentage of seismic activity involving various techniques. (Data from SEG annual *Geophysical Activity Reports*, pre-1981 data are for U.S. activity, post-1980 for worldwide activity, 3-D data from Dutt, 1992, adjusted according to judgment expressed in Goodfellow, 1991.)

mous amount of computer power. Marine acquisition is highly automated and shipboard computers are capable of elaborate processing. Workstations permit a preliminary interpretation to be made quickly.

The major users of seismic data, especially 3-D but also VSP and other seismic methods, have been changing from the exploration to the production departments of oil companies. The precision and detail of 3-D seismic data make it an increasingly cost-effective tool in field development and production. It is becoming increasingly recognized that most reservoirs are quite heterogeneous and seismic data provide about the only way of determining how they change horizontally. This is especially important in the marine environment where drilling and investment costs are so large, but it is also true on land.

The use of seismic methods in coal studies has grown rapidly since longwall mining became the major extraction method for deep coal deposits (§14.2). Some of the most rapidly growing areas of seismic applications are in engineering (§14.1), groundwater, and waste disposal (§14.3).

1.3 Geophysical activity

1.3.1 The future of exploration seismology

The petroleum industry has long dominated applied geophysics. A graph of the number of seismic crews searching for oil and gas in the United States (fig. 1.23) presents a discouraging picture. However, much of this decline in the number of crews is offset by in-

creases in the productivity (and cost) of a field crew, especially of a marine crew.

Much uncertainty exists today as to the future of geophysical activity. The majority view is that new technology will continue to create opportunities in the geophysical industry and the level of petroleum activity will not fall much farther; however, a return to the high activity of the 1980s is thought to be unlikely. Petroleum-related opportunities for geophysicists should continue at about the current level and opportunities in nonpetroleum areas should continue to grow.

Although seismic crew activity has been long regarded as a leading indicator of the health of the petroleum (and energy) sector, it may be losing validity as an economic predictor because of changes in technology, attitudes, and world politics. The petroleum industry is beginning to shift drilling funds into 3-D seismology. Environmental concerns are changing the demand part of the hydrocarbon supply/demand equation. For years, the United States dominated the petroleum industry, but in the 1970s, much of the economic control shifted to the OPEC countries. In the 1980s, rising North Sea production changed the supply/demand balance in Europe, and recently the revelation of enormous reserves and decayed infrastructure in the former Soviet Union has introduced new uncertainty. Future changes may be dominated by unpredictable crises, just as in the past applied geophysics was very much affected by events such as the cold war and conflicts in the Middle East.

Petroleum and geophysical service companies are

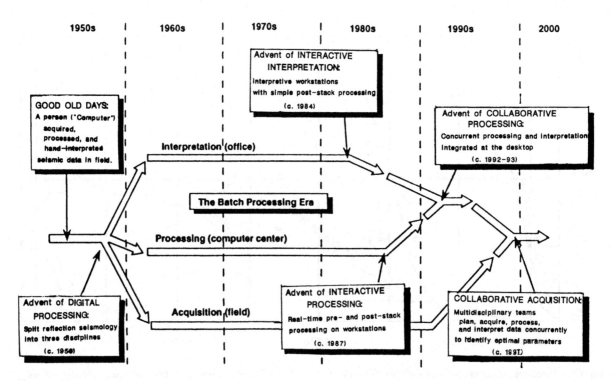

Fig. 1.22   Concept showing the specialization that followed the advent of digital processing and the projected unifying resulting from the increased power of workstations. (From Landmark Graphics Corp., 1992.)

reorganizing (see fig. 1.24) and repositioning to accommodate expected changes. Major petroleum companies are shifting their interests outside the United States and Canada, where they perceive greater opportunities. This has left behind a wealth of opportunities, but small companies have been slow to take advantage of these because of restricted funding.

An often-promulgated concept is that the Earth's petroleum resources have topped out and an exponential decline is inevitable. Past predictions of the timing of this decline have invariably been wrong because generally they have not allowed for technologic improvements. Lindseth (1990) found that the amount of oil discovered per well has been nearly constant (fig. 1.25) rather than declining sharply with time; the periodic arrival of new technology presumably arrests the declines. Thus, new technology (which almost everyone predicts) should continue to sustain the geophysical industry.

1.3.2 History of seismic activity

Geophysical activity is tabulated each year by the Society of Exploration Geophysicists (SEG). Admittedly, not all activity is reported (the most important omission has been activity in the [former] Soviet Union and China), some is reported on different bases (some reports include only acquisition and some include processing and/or interpretation), accounting practices differ (reported university work often does

not include equipment costs or costs of student labor), and totals lump both small-scale and large-scale operations. Nevertheless, the reports are regarded as relatively accurate and permit judgments with regard to trends. The latest annual report (as of early 1994) is for the year 1991 (Riley, 1993). In addition, monthly reports of seismic crew activity are published in *The Leading Edge*. The following is based on these reports and Dutt (1992).

The number of seismic crews is shown in fig. 1.23b along with the number of wildcat wells drilled. The number of seismic crews was long regarded as a leading indicator of the petroleum-industry economy, but major changes in the early 1990s have changed this situation (see also fig. 1.27). The sharp upturns in prices and activity in the 1970s and early 1980s resulting from the Middle East oil crisis did not produce comparable increases in hydrocarbon discoveries (see fig. 1.25).

The mean wellhead costs of oil and natural gas adjusted for inflation are shown in fig. 1.23a. Economics along with technology have been the governing factors in seismic activity. A surplus of oil about 1937 produced a decline in activity that lasted until the United States became involved in World War II. A doubling of petroleum prices between 1945 and 1948 resulted in an increase in seismic activity. However, major finds of oil in the Middle East after World War II resulted in another world surplus of oil and prices dropped. From 1948 to 1973, the price of petroleum

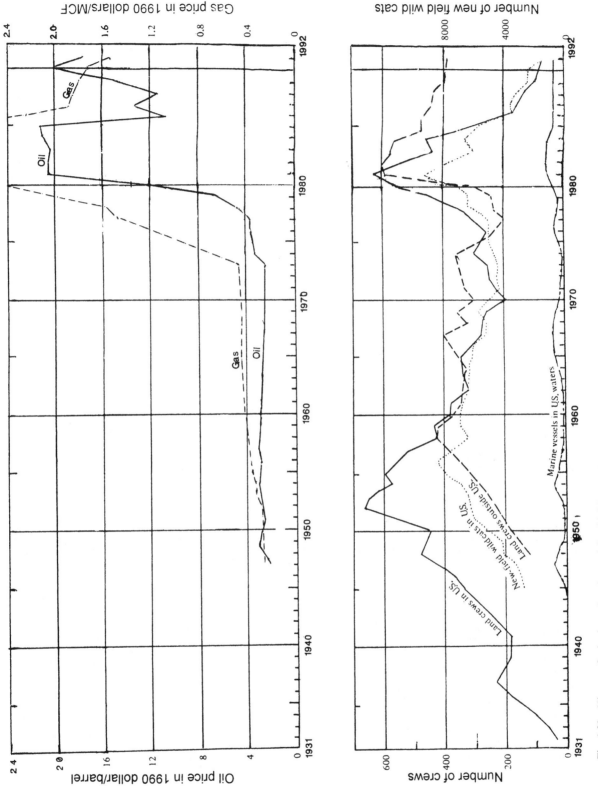

Fig. 1.23 History of seismic exploration activity. (a) Mean U.S. wellhead price of crude oil and natural gas adjusted for inflation (data from the American Petroleum Institute). (b) Mean number of seismic crews (data from SEG *Geophysical Activity Reports*) and new field wildcat wells drilled (data from AAPG activity reports); activity in some areas, especially the former Soviet Union and China, is not included.

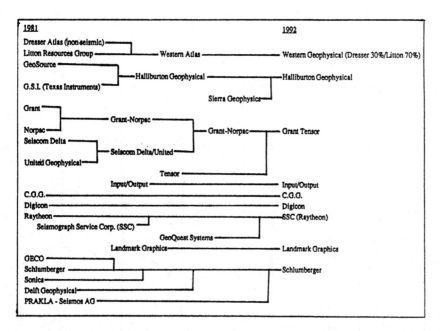

Fig. 1.24   Changes in the geophysical contracting industry. In late 1992, SSC (Raytheon) was taken over by Schlumberger, and in 1993, Haliburton was taken over by Western. Only major con-tracting companies are shown. In a sense, this updates the Am-erada Tree shown in fig. 1.11. (From Dutt, 1992.)

remained almost constant and activity generally de-clined for most of this period, the decline being slowed and occasionally temporarily reversed by new devel-opments. Natural gas reserves peaked about 1970 and, for a while thereafter, exploration had gas rather than oil as an objective. In 1973–4, oil prices increased sharply as a result of the limiting of supplies by the OPEC (Organization of Petroleum Exporting Coun-tries) cartel. The price increases as well as policy changes caused by uncertainties about dependence on foreign supplies stimulated seismic exploration. Con-cerns about secure oil supplies diminished as large re-serves were found in the North Sea and other "safe" areas and geophysical activity went into the decline (especially in the United States) that continues today.

Activity outside the United States increased rather steadily until 1958 and then leveled off. During the period 1958–74 the geography of activity changed sev-eral times in response to political and economic fac-tors and discoveries in new areas. Activity in Latin America declined sharply after 1959 because of dis-couraging results and political changes in several countries. The discovery of significant hydrocarbon reserves in North Africa, the beginning of North Sea exploration, nationalization threats in Indonesia, the opening of tropical African waters to exploration, and repeated political disruptions in North Africa and the Middle East were probably the most significant of the events. International activity since 1981 (fig. 1.26) did not decline nearly as steeply as in the United States, and since 1987, activity outside North America has been increasing steadily.

The history of seismic activity is also illustrated in

fig. 1.27. The number of field crews (especially on land) has been declining since the early 1980s but the volume of new data acquired since 1988 has been in-creasing because of the shift from 2-D to 3-D data ac-quisition.

1.3.3 Data for 1991

Expenditures for geophysical data acquisition and processing in 1991 were U.S.$ 2250 million (Riley, 1993). While the activity data are not strictly compa-rable with previous periods, they indicate that expen-ditures were up 3% from 1990 and 54% from 1987, when expenditures plunged after the Middle East oil crisis; however, they are only 53% of the peak expen-ditures of U.S.$ 4168 million in 1982. These expendi-tures are shown by objectives in table 1.2 and by areas in table 1.3 and fig. 1.26. Activity in the United States continued to decline by 25% of the 1990 figure, now standing at 16% of worldwide activity, compared with double this percentage in 1987.

Seismic work constituted almost 97% of the re-ported geophysical expenditure and 99% of this work had petroleum objectives. Although table 1.2 shows only 13.6% of expenditures had petroleum develop-ment (as opposed to petroleum exploration) objec-tives, Riley (1993) gives several reasons why he be-lieves this figure is considerably understated. Goodfellow (1991: 62) in the report for 1990 stated that "we believe that a considerably larger proportion . . . was actually spent on development . . . [probably] 24 percent," compared with a reported 6.2%. Several recent spectacular successes in defining reservoirs

Table 1.2 *Worldwide 1991 expenditures by objectives (U.S.$ × 10³)*

Object	Land	Transition	Marine	Airborne	Borehole	Percent
Petroleum						96.7
Exploration	1 189 500.	6 100.	817 900.	6 600.	1 860.	83.1
Development	252 600.	27 900.	49 300.		900.	13.6
Minerals	16 100.			11 900.	690.	1.2
Environmental	3 100.			140.	180.	0.1
Engineering	34 400.			15.		1.4
Geothermal	400		900.			0.1
Groundwater	2 200.				140.	0.1
Oceanography	10.		900.			< 0.1
Research	9 000.	300.	500.	80.	320.	0.4
Total	1 507 300.	34 300.	869 500.	18 700.	4 000.	
Percent	62.0	1.4	35.7	0.8	0.2	

Note: Because of differences in the manner of reporting, there are minor inconsistencies in some of the numbers in this and the following tables.

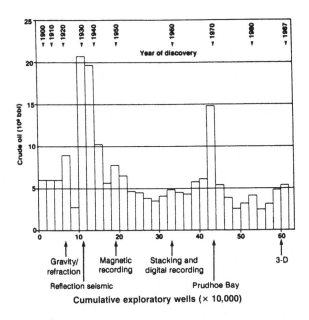

Fig. 1.25 Oil-finding efficiency indicated by new oil found per well drilled. (After Lindseth, 1990.)

ber of crews (fig. 1.27a). Increased exploration activity in the late 1970s brought about by threatened supplies from the Mideast produced a substantial increase in unit costs (fig. 1.28), but unit costs have been declining since 1980, the decrease being especially impressive when adjusted for inflation.

Nonpetroleum seismic statistics are given in table 1.5. These figures are probably much more incomplete than the petroleum statistics because many small companies are involved. Environmental and engineering work generally involve detailed surveying of small areas, whereas oceanographic work is apt to involve cursory surveys of large tracts of ocean, and crews may range from 2 to 3 people to as many as 50 or so. Nevertheless, the statistics do give some idea of the scope of geophysical work and work in these areas probably will increase significantly in the years ahead.

Regarding the seismic sources used for acquiring

have been attributed to this type of work (Sheriff, 1992). Oil companies are now resurveying older finds with newer seismic technology in both the marine and land environments, resulting in increasing their production and proven reserves substantially (Abriel et al., 1991).

Table 1.4 lists petroleum seismic activity statistics. Although many different kinds of work are averaged together in these statistics, the figures are probably representative of the costs and productivity of present-day full-size geophysical crews. Marine work produces much more data than land work and the ratio of the volume of new marine data to new land data (fig. 1.27b) is quite different from the ratio of the num-

Table 1.3 *Worldwide 1991 expenditures by areas (U.S.$ × 10³)*

Area	Costs	Percent of total	Percent change 1990–1
United States	371 200.	16.4	−36
Canada	133 800.	5.9	−5
Mexico	44 600.	2.0	+700
South America	287 600.	12.7	+47
Europe	566 700.	25.1	+58
Africa	236 800.	10.5	+9
Middle East	312 300.	13.8	+119
Far East	202 000.	8.9	−24
Australia–New Zealand	86 300.	3.8	−27
International	18 900.	0.8	−12.
Total	2 250 300.		

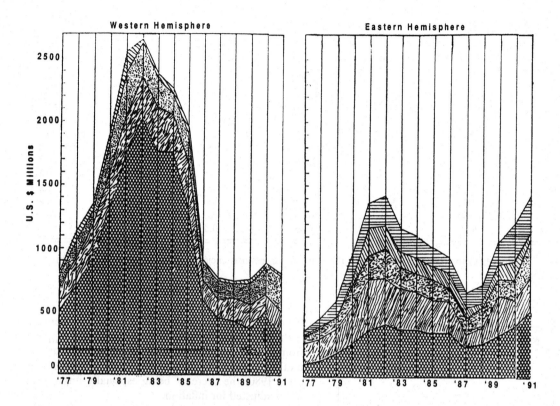

Fig. 1.26 Expenditures on geophysical petroleum exploration and development by areas. (After Riley, 1993.) (a) Western Hemisphere; the areas from the bottom upward represent the United States, Canada, South America, and Mexico. (b) Eastern Hemisphere; the areas from the bottom upward represent Europe, Africa, the Middle East, Australia, and the Far East.

Table 1.4 *Petroleum 1991 seismic activity statistics*

	Land	Transition zone	Marine
Acquisition costs (U.S.$ $\times 10^3$)	1 442 100.	34 000.	867 200.
Line miles	242 000	24 700	1 367 000
Line kilometers	389 000	39 700	2 199 000
Crew-months	2 780	61	947
Average miles/ month	87	405	1 440
Average km/ month	140	651	2 320
Average cost/ month (U.S.$)	519 000	557 000	916 000
Average cost/ mile (U.S.$)	4 600	1 400	630
Average cost/ km (U.S.$)	2 860	870	390
Cost of 3-D/ mile^2 (U.S.$)	36 000		27 300
Cost of 3-D/ km^2 (U.S.$)	13 900		10 500

the land miles, explosives constituted 34.6%, vibrators 59.2%, land air guns 0.6%, and weight drop 0.1%. Air guns were the only marine source reported for petro-

leum work. The percentage of land work that was 3-D cannot be calculated because some numbers are reported in line miles and some in square miles surveyed. Although some marine work was reported in square miles surveyed, 72.5% of the line miles reported were 3-D and so the total percentage exceeds this value.

1.4 The literature of exploration seismology

A. S. Eve and D. A. Keys (1928) wrote in their preface to *Applied Geophysics,* "we know of no book in English which deals with the theoretical and practical sides of all of the many schemes of exploration now available." This was only 4 years after the first discovery of hydrocarbons based on seismic refraction. Eve and Keys noted that "in 1928 there were thirty or more groups or 'troops' at work . . . , each consisting of three to five trained men, with an equal number of helpers." Extreme secrecy was common at this time and their book gives only a brief sketch of methods. As late as the early 1950s, some "black-box" elements remained, that is, details as to how they worked were not disclosed.

Literature on earthquake seismology preceded that dealing with prospecting applications. H. Jeffreys' classic *The Earth* appeared in 1924 (3rd ed. in 1952). L. D. Leet's *Practical Seismology and Seismic Pros-*

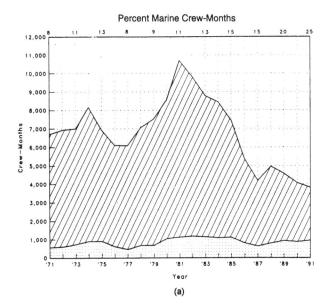

Percent Marine Crew-Months

(a)

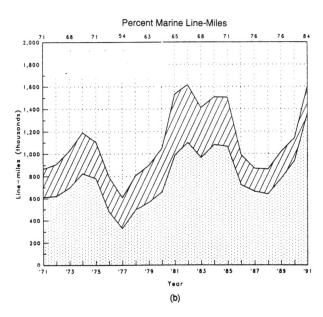

Percent Marine Line-Miles

(b)

Fig. 1.27 Worldwide land and marine acquisition. The stippled area indicates land, the diagonal slashed marine. (From Riley, 1993.) (a) Number of field crew-months; (b) volume of data acquired.

pecting (1938) combined earthquake and exploration seismology.

Although geophysical literature is published in several languages, the seismologist who reads English is especially fortunate in that almost all important references are in this language. Most of the important papers and books that have appeared in other languages have either English equivalents or English translations. Furthermore, most of the important technical papers are contained in two journals, *Geophysics,* published by the Society of Exploration Geophysicists (SEG), and *Geophysical Prospecting,* published by the

European Association of Exploration Geophysicists (EAEG).

The Society of Economic Geophysicists was founded in Houston in 1930; the name was changed that same year to the Society of Petroleum Geophysicists and in 1937 to the Society of Exploration Geophysicists. It continues to be the largest professional geophysical society today. The society began publication of *Geophysics* in 1936. Prior to this, papers were published in issues of the *AAPG Bulletin* and *Physics;* many of the most important papers prior to 1936 were republished in *Early Geophysical Papers* in 1947. The European Association of Exploration Geophysicists was founded in 1951 and began publishing *Geophysical Prospecting* in 1953.

The unrefereed magazines published by these two societies, *The Leading Edge* and the *First Break,* provide survey articles, interpretation case histories, and information about newer topics. The Canadian and Australian Societies of Exploration Geophysics also publish journals that are more along the lines of *The Leading Edge* and the *First Break* than of *Geophysics* and *Geophysical Prospecting.* Other journals that often contain important articles are published in Europe, India, and elsewhere. The *Bulletin of the American Association of Petroleum Geologists* often contains important papers on interpretive applications of geophysics. The geophysical literature of basic seismology also often contains papers of interest to exploration seismologists. The most important non-English journals are Russian and Chinese.

A *Cumulative Index of Geophysics* is published every few years (most recently as a supplement to the March 1990 issue of *Geophysics*); it lists the papers in the publications of most of the foregoing societies except for those of the American Association of Petroleum Geologists. The cumulative index is also available on a computer disk, which also lists the expanded abstracts of papers given at the annual meetings. This computer disk can be searched for key words. The most important papers from *Geophysics* are reprinted in the 25th and 50th anniversary volumes (Classic papers of the past 25 years, 1985) and important exploration seismic papers from various journals are reprinted in three volumes of the *Treatise of Petroleum Geology Reprint Series* (Beaumont and Foster, 1989). A series of 14 (as of 1994) reprint volumes dealing with various subjects and a number of other geophysics books are published by the Society of Exploration Geophysicists.

A multitude of books on various aspects of seismic exploration are available today. In the first edition, we were able to list most of the important books on aspects of seismic exploration, but today there are so many that it is not feasible to do this. Many are referenced in subsequent chapters. Seismic technology today embraces so much signal processing and computer technology as well as geology that a reading list would include many works that are not specifically geophysical. Particular mention should be made of

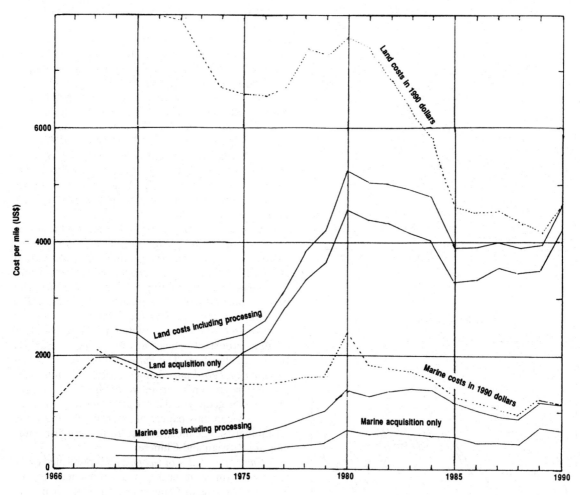

Fig. 1.28 Seismic costs per mile (1980 and pre-1980, Western Hemisphere; post-1980, worldwide). The dotted curves show costs adjusted for inflation. (Data from SEG *Geophysical Activity Reports.*)

Table 1.5 *Nonpetroleum 1991 seismic activity*

Survey type	Type work	Expenditures (U.S.$ × 10³)	Cost/mile	Cost/km
Minerals	*P*-wave reflection	5 290.	3 150.	1 960.
	S-wave reflection	1 050.	6 900.	4 300.
	Refraction	170.	2 900.	1 800.
Environmental	*P*-wave reflection	820.	7 300.	4 600.
	S-wave reflection	60.		
Engineering	*P*-wave reflection	31.	1 500.	940.
	S-wave reflection	0.2	1 260.	780.
	Refraction	12.	6 100.	3 800.
Groundwater	Reflection	410.	4 300	2 600.
	Refraction	110.	1 600.	1 000.
Geothermal	Passive	0.3		
Oceanography	Reflection	920.	90.	60.
Research	*P*-wave reflection	4 910.	2 900.	1 800.
	S-wave reflection	150.	50 000.	31 000.
	Refraction	15.	15 000.	9 300.

the safety and environmental guides published by the International Association of Geophysical Contractors (IAGC).

References

Abbot, H. L. 1878. On the velocity of transmission of earth waves. *Amer. J. Sci. Arts,* Ser. 3, **15**: 178–84.

Abriel, W. L., P. S. Neale, J. S. Tissue, and R. M. Wright. 1991. Modern technology in an old area. Bay Marchand revisited. *The Leading Edge,* **10(6)**: 21–35.

Barrington, T. 1982. Cecil Green. *The Leading Edge,* **1(1)**: 16–23.

Barton, D. C. 1929. The seismic method of mapping geologic structure. In *Geophysical Prospecting,* pp. 572–624. New York: American Institute of Mining and Metallurgical Engineers.

Bates, C. C., T. F. Gaskell, and R. B. Rice. 1982. *Geophysics in the Affairs of Man.* Oxford: Pergamon Press.

Bates, R. L., and J. A. Jackson. 1987, *Glossary of Geology,* 3d ed. Falls Church, VA: American Geological Institute.

Beaumont, E. A., and N. H. Foster. 1989. *Geophysics I: Seismic Methods; Geophysics II: Tools for Seismic Interpretation; Geophysics III: Geologic Interpretation of Seismic Data,* Treatise of Petroleum Geology, Reprint Series Nos. 12, 13, and 14. Tulsa: American Association of Petroleum Geologists.

Born, W. T. 1960. A review of geophysical instrumentation. *Geophysics,* **25**: 77–91.

Carlton, D. P. 1946. *The History of the Geophysics Department.* Houston: Humble Oil and Refining Co.

Clark, R. D. 1982. Gerald Westby. *The Leading Edge,* **1(1)**: 28–35.

Clark, R. D. 1983. Sidney Kaufman. *The Leading Edge,* **2(7)**: 22–7.

Clark, R. D. 1984a. T. I. Harkins. *The Leading Edge,* **3(4)**: 14–18.

Clark, R. D. 1984b. C. Hewitt Dix. *The Leading Edge,* **3(8)**: 14–17.

Clark, R. D. 1985. Enders Robinson. *The Leading Edge,* **4(2)**: 16–20.

Clark, R. D. 1990a. Theodor Krey. *The Leading Edge,* **9(4)**: 13–17.

Clark, R. D. 1990b. Kenneth E. Burg. *The Leading Edge,* **9(10)**: 13–16.

Classic papers of the past 25 years. 1985. *Geophysics,* **50**: 1797–2280.

DeGolyer, E. 1935. Notes on the early history of applied geophysics in the petroleum industry. *Trans. Soc. Pet. Geophys.,* **6**: 1–10. (Reprinted in *Early Geophysical Papers of the Society of Exploration Geophysicists,* pp. 245–54. Tulsa: Society of Exploration Geophysicists, 1947.)

Dutt, G. B. 1992. *Seismic overview.* New Orleans: Howard, Weil, Labouisse, Friedrichs Inc.

Elkins, T. A. 1970. *A Brief History of Gulf's Geophysical Prospecting.* Pittsburgh: Gulf Research and Development Co.

Eve, A. S., and D. A. Keys. 1928. *Applied Geophysics.* Cambridge: Cambridge University Press.

Gal'perin, E. I. 1974. *Vertical Seismic Profiling.* Tulsa: Society of Exploration Geophysicists.

Goodfellow, K. 1991. Geophysical activity in 1990. *The Leading Edge,* **10(11)**: 45–72.

Green, C. H. 1979. John Clarence Karcher, 1894–1978, father of the reflection seismograph. *Geophysics,* **44**: 1018–21.

Hecker, O. 1990. Ergebnisse de Messung von Bodenbewegungen bei einer Sprengung. *Gerland's Beiträge zur Geophysik,* **4**: 98–104.

Heiland, C. A. 1929a. Modern instruments and methods of seismic prospecting. In *Geophysical Prospecting,* pp. 625–53. New York: American Institute of Mining and Metallurgical Engineers.

Heiland, C. A. 1929b. Geophysical methods of prospecting – Principles and recent successes. *Quart. Col. Sch. Mines,* **24(1)**.

Jeffreys, H. 1952. *The Earth,* 3d ed. Cambridge: Cambridge University Press.

Karcher, J. C. 1987. The reflection seismograph: Its invention and use in the discovery of oil and gas fields. *The Leading Edge,* **6(11)**: 10–20.

Keppner, G. 1991. Ludger Mintrop. *The Leading Edge,* **10(9)**: 21–8.

Knott, C. G. 1899. Reflexion and refraction of elastic waves, with seismological applications. *Phil. Mag.,* **48**: 64–97.

Laing, W. E., and F. Searcy. 1975. *Geophysics – The First Fifty Years.* Houston: Conoco.

Landmark Graphics Corp. 1992. *The Coming Reunion of Seismic Interpretation and Processing.* Houston: Landmark Graphics Corp.

Leet, L. D. 1938. *Practical Seismology and Seismic Prospecting.* New York: Appleton-Century.

Lindseth, R. O. 1990. The new wave in exploration geophysics. *The Leading Edge,* **9(12)**: 9–15.

Love, A. E. H. 1927. *Some Problems of Geodynamics.* Cambridge: Cambridge University Press.

Malamphy, M. C. 1929. Factors in design of portable field seismographs. *Oil Weekly,* March 22, 1929.

Mallet, R. 1848. On the dynamics of earthquakes; being an attempt to reduce their observed phenomena to the known laws of wave motion in solids and fluids. *Trans. Roy. Irish Acad.,* **21**: 50–106.

Mallet, R. 1851. Second report on the facts of earthquake phenomena. *BAAS,* **21**: 272–320.

Mayne, W. H. 1982. The evolution of geophysical technology. *The Leading Edge,* **1(1)**: 75–80.

McGee, J. E., and R. L. Palmer. 1967. Early refraction practices. In *Seismic Refraction Prospecting,* A. W. Musgrave, Ed. Tulsa: Society of Exploration Geophysicists.

McGuckin, G. M. 1945. History of the geophysical exploration of the Cameron Meadows Dome, Cameron Parish, Louisiana. *Geophysics,* **10**: 1–16.

Milne, J. 1895. Seismic experiments. *Trans. Seis. Soc. Jpn.,* **8**: 1–82.

Mintrop, L. 1931. *On the History of the Seismic Method for the Investigation of Underground Formations and Mineral Deposits.* Hanover, Germany: Seismos.

Owen, E. W. 1975. *Trek of the Oil Finders: A History of Exploration for Petroleum,* AAPG Memoir 6. Tulsa: American Association of Petroleum Geologists.

Petty, O. S. 1976. *Seismic Reflections.* Houston: Geosource.

Proffitt, J. M. 1991. A history of innovation in marine seismic data acquisition. *The Leading Edge,* **10(3)**: 24–30.

Proubasta, D. 1982. O. S. Petty. *The Leading Edge,* **1(7)**: 16–24.

Proubasta, D. 1983a. John Hollister. *The Leading Edge,* **2(7)**: 14–19.

Proubasta, D. 1983b. Henry Salvatori. *The Leading Edge,* **2(8)**: 14–22.

Proubasta, D. 1983c. John Crawford. *The Leading Edge,* **2(12):** 16–26.

Proubasta, D. 1984. Remembrance of geophysical things past. *The Leading Edge,* **3(10):** 32–8.

Proubasta, D. 1985a. Sven Treitel. *The Leading Edge,* **4(2):** 24–8.

Proubasta, D. 1985b. Harry Mayne. *The Leading Edge,* **4(7):** 18–24.

Proubasta, D. 1986a. Erik Jonsson. *The Leading Edge,* **5(6):** 14–23.

Proubasta, D. 1986b. Enders Robinson and the shot heard round the geophysical world. *The Leading Edge,* **5(9):** 14–17.

Proubasta, D. 1991. Maurice Ewing. *The Leading Edge,* **10(3):** 15–20.

Rayleigh, Lord. 1885. On waves propagated along the plane surface of an elastic solid. *Proc. London Math. Soc.,* **17:** 4–11.

Rieber, F. 1936. A new reflection system with controlled directional sensitivity. *Geophysics,* **1:** 97–106.

Riley, D. C. 1993. Special report: Geophysical activity in 1991. *The Leading Edge,* **12:** 1094–1117.

Robertson, H. 1986. Everette Lee DeGolyer. *The Leading Edge,* **5(11):** 14–21.

Robinson, E. A. 1985. A historical account of computer research in seismic data processing, 1949–1954. *The Leading Edge,* **4(2):** 40–5.

Rosaire, E. E. 1935. On the strategy and tactics of exploration for petroleum. *J. Soc. Pet. Geophys.,* **6:** 11–26. (Reprinted in *Early Geophysical Papers of the Society of Exploration Geophysicists,* pp. 255–70. Tulsa: Society of Exploration Geophysicists, 1947.)

Rosaire, E. E., and J. L. Adler. 1934. Applications and limitations of dip shooting. *Bull. AAPG,* **18:** 19–32.

Rosaire, E. E., and O. C. Lester, Jr. 1932. Seismological discovery and partial detail of Vermillion Bay salt dome. *Bull. AAPG,* **16:** 51–9. (Reprinted in *Early Geophysical Papers of the Society of Exploration Geophysicists,* pp. 381–9. Tulsa: Society of Exploration Geophysicists, 1947.)

Schriever, W. 1952. Reflection seismograph prospecting – How it started. *Geophysics,* **17:** 936–42.

Shaw, H., J. M. Bruckshaw, and S. T. Newing. 1931. *Applied Geophysics.* London: His Majesty's Stationery Office.

Sheriff, R. E. 1985. History of geophysical technology through advertisements in Geophysics. *Geophysics,* **50:** 2299–2408.

Sheriff, R. E. 1988. Processing and interpretation of seismic reflection data: An historical précis. *The Leading Edge,* **7(1):** 40–2.

Sheriff, R. E. 1991. *Encyclopedic Dictionary of Exploration Geophysics,* 3d ed. Tulsa: Society of Exploration Geophysicists.

Sheriff, R. E., ed. 1992. *Reservoir Geophysics.* Tulsa: Society of Exploration Geophysicists.

Stoneley, R. 1924. Elastic waves at the surface of separation of two solids. *Proc. Roy. Soc. (London),* **A-106:** 416–28.

Sweet, G. E. 1978. *History of Geophysical Prospecting.* Sudbury, England: Spearman.

Udden, J. A. 1920. Suggestions of a new method of making underground observations. *Bull. AAPG,* **4:** 83–5. (Reprinted in *Geophysics,* **16:** 715–16.)

Vajk, R. 1949. Baron Roland Eötvös. *Geophysics,* **14:** 6–9.

Weatherby, B. B. 1940. History and development of seismic prospecting. *Geophysics,* **5:** 215–30.

Weatherby, B. B. 1945. Early seismic discoveries in Oklahoma. *Geophysics,* **10:** 345–67.

Wiechert, E., and K. Zoeppritz. 1907. Uber Erdbebenwellen. *Nachrichten von der Königlichen Gesellschaft der Wissenschaften zur Göttingen,* pp. 415–549. Berlin.

Wilcox, S. W. 1990. Reminiscence of the second decade of seismic prospecting. *The Leading Edge,* **9(8):** 42–5.

2
Theory of seismic waves

Overview

The seismic method utilizes the propagation of waves through the earth. To introduce the basic concepts of wave motion, we first discuss waves on a stretched string (§2.1.1) and introduce definitions of phase, frequency, wavelength, and other terms dealing with periodicity. Because wave propagation depends upon the elastic properties of the rocks, we next discuss some of the basic concepts of elasticity. (For more thorough treatments, see Saada, 1974, or Landau and Lifshitz, 1986.)

The size and shape of a solid body can be changed by applying forces to the external surface of the body. These external forces are opposed by internal forces, which resist the changes in size and shape. As a result, the body tends to return to its original condition when the external forces are removed. Similarly, a fluid resists changes in size (volume) but not changes in shape. This property of resisting changes in size or shape and of returning to the undeformed condition when the external forces are removed is called *elasticity*. A perfectly elastic body is one that recovers completely after being deformed. Many substances including rocks can be considered perfectly elastic without appreciable error provided the deformations are small, as they are in seismic surveys.

The theory of elasticity relates the forces that are applied to the external surface of a body to the resulting changes in size and shape. The relations between the applied forces and the deformations are most conveniently expressed in terms of the concepts of stress and strain. Strain, a change in shape or dimensions, is generally proportional to the stress (force per unit area) that produces it, as stated in Hooke's law. The constant of proportionality is called an elastic constant, or modulus, and moduli for different types of stress and strain are interrelated.

Section 2.2 concerns seismic-wave motion. Newton's second law of motion, that an unbalanced force on a mass produces an acceleration, is used to derive two forms of the wave equation. The wave equation is expressed in vector as well as the more conventional scalar notation. Methods of including a source of disturbance and Kirchhoff's theorem are also given in this section.

Plane- and spherical-wave solutions to the wave equation are given next. Waves are disturbances that travel through the medium. The concepts of wave-

fronts and raypaths are introduced, as is the more general Huygens' principle approach.

The two forms of the wave equation that had been derived earlier are related to two types of disturbances that can travel through the body of solids (§2.4). These involve changes in volume (*P*-waves) and rotations (*S*-waves). Discussion of potential functions, from which particle displacements and velocities can be derived, follows. At interfaces, both stresses and particle displacements must be continuous; these boundary conditions are discussed in §2.4.4.

Surface waves are examined next. Rayleigh waves are important because of the ground-roll noise that they produce on seismic records. Love, Stoneley, and tube waves are encountered occasionally.

Most seismic theory assumes that media are isotropic, that is, their properties are the same regardless of the direction of measurement. Anisotropy (§2.6) of several types has been observed; however, anisotropic effects are usually small. The most important exceptions requiring study are those of transverse isotropy because of layering and fracturing.

Section 2.7 examines what happens to seismic body waves as they travel in the earth. Intensity decreases because of geometrical spreading (divergence) and absorption (and partitioning at interfaces; see chap. 3). Divergence is the most important factor affecting the change of intensity for the first few kilometers, but eventually absorption becomes dominant. Absorption increases approximately linearly with frequency and hence changes the waveshape with distance. Various expressions for absorption are interrelated. Dispersion and the concepts of group and phase velocity are discussed, although dispersion is not an important factor in seismic exploration.

Reflection and refraction are discussed in §2.7.5. Diffraction (§2.8), the scattering of waves at discontinuities, involves somewhat complex mathematics. However, the construction of diffraction wavefronts using Huygens' principle is fairly straightforward and nonmathematical.

2.1 Theory of elasticity

2.1.1 Waves on a stretched string

As an introduction to seismic waves in three dimensions, we consider the one-dimensional wave in a stretched string because many basic concepts of wave motion can be more simply illustrated in this way. Parts of the following discussion will be treated later in a broader context.

We assume an ideal case where the mass of the string per unit length, μ, is negligibly small in comparison with the tension, τ, in the string, that the string when at rest is along the x-axis, and that the displacements, ψ, which are parallel to the y-axis, are small in comparison with the length of the string so that angles α_1 and α_2 are also small (fig. 2.1a). Because these angles are not equal, the tension produces a net force in the y-direction (the net force in the x-direction is negligible) on an element of the string, Δx, equal to $\tau(\sin \alpha_2 - \sin \alpha_1) \approx \tau(\tan \alpha_2 - \tan \alpha_1) \approx \tau(\partial\psi/\partial x|_{x_2} - \partial\psi/\partial x|_{x_1}) \approx \tau \Delta(\partial\psi/\partial x)$. Newton's second law of motion states that this force equals the product of the mass $\mu \Delta x$ and the acceleration $\partial^2\psi/\partial t^2$. Dividing both sides by $\tau_0 \Delta x$ and taking the limit as $\Delta x \to 0$ gives the one-dimensional wave equation:

$$\partial^2\psi/\partial x^2 = (\mu/\tau)\partial^2\psi/\partial t^2 = (1/V^2)\partial^2\psi/\partial t^2, \quad (2.1)$$

where $V = (\tau/\mu)^{1/2}$ (compare with eq. (2.45)). Equation (2.1) shows that V has dimensions of distance/time, or velocity. The wave equation relates variation in space (the left side) with variation in time (the right side).

The general solution of eq. (2.1) (also called d'Alembert's solution; see §2.2.5) is

$$\psi(x, t) = \psi_1 (x - Vt) + \psi_2 (x + Vt), \quad (2.2)$$

where ψ_1 and ψ_2 are arbitrary functions, ψ_1 is a disturbance moving in the positive x-direction with increase of time, ψ_2 a disturbance moving in the negative x-direction, and V is the velocity of propagation along the string (see the following).

Fourier analysis (§9.1.2) shows that any waveform (within reason) can be represented by a superposition of harmonic (sinusoidal) waves, so we do not lose generality by confining our attention to harmonic waves. Thus, we consider a harmonic solution of eq. (2.1) in the form

$$\psi = A \cos [(2\pi/\lambda)(x - Vt)]. \quad (2.3)$$

The waveform is harmonic with ψ varying between $+A$ and $-A$; A is the *amplitude*. If we look at the wave passing a fixed point in space (fig. 2.1b), *period* T is the time between successive repetitions of the waveform; *frequency* $\nu = 1/T$ is the number of waves per unit time. If we look at the waveform at some moment of time (fig. 2.1c) the distance between successive repetitions of the waveform is the *wavelength* λ and $1/\lambda$ is the *wavenumber* or number of waves per unit distance. Multiplying $1/T$ and $1/\lambda$ by 2π, we get the *angular frequency* $\omega = 2\pi/T = 2\pi\nu$ and the *angular wavenumber* $\kappa = 2\pi/\lambda$. Because ν is the number of waves passing a fixed point per unit time and each wave has length λ, velocity V must be given by the equation

$$V = \nu\lambda. \quad (2.4)$$

The argument of the cosine in eq. (2.3), namely,

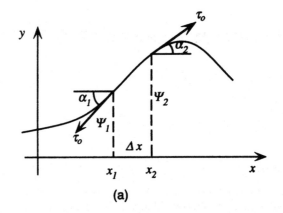

(a)

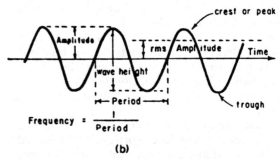

(b)

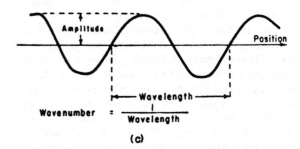

(c)

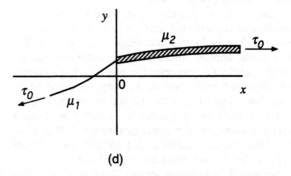

(d)

Fig. 2.1 Waves on a stretched string. (a) Portion of the string showing the relation between displacement and tension; (b) representation of the wave in time; (c) representation of the wave in space; (d) the effect of change in mass/unit length.

$(2\pi/\lambda) (x - Vt) = \kappa(x - Vt) = (\kappa x - \omega t)$, is called the *phase*. In eq. (2.3), the phase is zero at the origin; at times we add a fixed phase angle γ_0 so that the phase becomes $\kappa x - \omega t + \gamma_0$.

Returning to the stretched string, if the mass/unit length changes abruptly from μ_1 to μ_2 at some point, say, $x = 0$ (fig. 2.1d), certain boundary conditions (§2.4.4) must hold, namely, both the displacement and the y-component of the tension in the string must be continuous, that is, neither changes in value as we go through the junction. These conditions can be expressed by the equations

$$\begin{aligned} \psi_{\text{left}} &= \psi_{\text{right}}, \\ \tau(\partial\psi/\partial x)_{\text{left}} &= \tau(\partial\psi/\partial x)_{\text{right}}. \end{aligned} \qquad (2.5)$$

We take the incident wave as $A_i \cos(\kappa_1 x - \omega t)$ coming from the left and the wave passing on to the right (the transmitted wave) as $A_t \cos(\kappa_2 x - \omega t)$; however, we cannot satisfy eqs. (2.5) with these two waves only and we must postulate a reflected wave going to the left, $A_r \cos(\kappa_1 x + \omega t)$. Substituting into eqs. (2.5), we find that the boundary conditions will be satisfied provided

$$\left.\begin{aligned} A_i + A_r &= A_t, \\ \kappa_1 A_i - \kappa_1 A_r &= \kappa_2 A_t. \end{aligned}\right\} \qquad (2.6a)$$

Equations (2.6a) can be solved for A_r and A_t:

$$\left.\begin{aligned} R &= A_r/A_i = (\kappa_2 - \kappa_1)/(\kappa_2 + \kappa_1), \\ T &= A_t/A_i = 2\kappa_1/(\kappa_2 + \kappa_1), \end{aligned}\right\} \qquad (2.6b)$$

where R and T are called the *reflection coefficient* (or *reflectivity*) and the *transmission coefficient*, respectively (see also §3.2).

If the string is fixed at $x = 0$, the effect is the same as if $\mu_2 = \infty$; then $T = 0$, so no wave is transmitted, and $R = +1$, which means that the reflected wave is the same as the incident one except that the direction of travel is reversed. The two waves interfere (§2.3.2) at the fixed end to produce perfect cancellation, hence, zero movement (*node*). If both ends are fixed, perfect cancellation must occur at both ends, so these are nodes.

When a string fixed at both ends is vibrating at its lowest frequency, called the fundamental (ν_0), the displacement has its maximum amplitude at the midpoint (*antinode*). The wave pattern is fixed, so the wave is said to be *stationary,* or *standing.* If the string length is L, $L = \lambda/2$ and $\nu_0 = V/\lambda = V/2L$. The string can vibrate in a number of patterns called *modes* or *eigenstates,* the frequencies being harmonics (multiples) of the fundamental, that is, $\nu_i = n\nu_0$, $n = 1, 2, 3, \ldots$. In each case, the ends of the string are nodes and $L = n\lambda/2 = (2n)\lambda/4$.

If the left end of the string is fixed and the right end free, we set $\kappa_2 = 0$ and get $R = -1$. The end of the string is an antinode, $L = \lambda/4$, $\nu_0 = V/4L$, and the harmonics are $\nu = (2n + 1)\nu_0$ and $L = (2n + 1)\lambda/4$. The two cases of a string fixed at one end only and fixed at both ends are analogous to organ pipes closed at one end only and closed at both ends (Logan, 1987; see also §13.3).

2.1.2 Stress

Stress is defined as force per unit area. Thus, when a force is applied to a body, the stress is the ratio of the force to the area on which the force is applied. If the force varies from point to point, the stress also varies, and its value at any point is found by taking an infinitesimally small element of area centered at the point and dividing the total force acting on this area by the magnitude of the area. If the force is perpendicular to the area, the stress is said to be a *normal stress* (or *pressure*). In this book, positive values correspond to tensile stresses (the opposite convention of signs is sometimes used). When the force is tangential to the element of area, the stress is a *shearing stress.* When the force is neither parallel nor perpendicular to the element of area, it can be resolved into components parallel and perpendicular to the element; hence, any stress can be resolved into component normal and shearing stresses.

If we consider a small element of volume inside a stressed body, the stresses acting upon each of the six faces of the element can be resolved into components, as shown in fig. 2.2 for the two faces perpendicular to the x-axis. Subscripts denote the x-, y-, and z-axes, respectively, and σ_{yx} denotes a stress parallel to the y-axis acting upon a surface perpendicular to the x-axis. When the two subscripts are the same (as with σ_{xx}), the stress is a normal stress; when the subscripts are different (as with σ_{yx}), the stress is a shearing stress.

When the medium is in static equilibrium, the stresses must be balanced. This means that the three stresses, σ_{xx}, σ_{yx}, and σ_{zx}, acting on face $OABC$ must be equal and opposite to the corresponding stresses shown on opposite face $DEFG$, with similar relations for the remaining four faces. In addition, a pair of shearing stresses, such as σ_{yx}, constitute a couple tending to rotate the element about the z-axis, the magnitude of the couple being

$$\text{force} \times \text{lever arm} = (\sigma_{yx}\, dy\, dz)\, dx.$$

If we consider the stresses on the other four faces, we find that this couple is opposed solely by the couple due to the pair of stresses σ_{xy} with magnitude $(\sigma_{xy}\, dx\, dz)\, dy$. Because the element is in equilibrium,

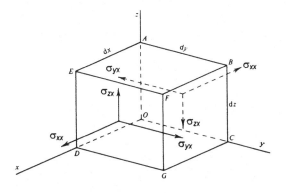

Fig. 2.2 Components of stress on faces perpendicular to the x-axis.

the total moment must be zero; hence $\sigma_{xy} = \sigma_{yx}$. In general, we must have

$$\sigma_{ij} = \sigma_{ji}. \qquad (2.7)$$

2.1.3 Strain

When an elastic body is subjected to stresses, changes in shape and dimensions occur. These changes, which are called *strains,* can be resolved into certain fundamental types.

Consider rectangle *PQRS* in the *xy*-plane (see fig. 2.3). When the stresses are applied, let *P* move to *P'*, *PP'* having components *u* and *v*. If the other vertices *Q*, *R*, and *S* have the same displacement as *P*, the rectangle is merely displaced as a whole by the amounts *u* and *v;* in this case, there is no change in size or shape, and no strain exists. However, if *u* and *v* are different for the different vertices, the rectangle will undergo changes in size and shape, and strains will exist.

Let us assume that $u = u(x, y)$ and $v = v(x, y)$. Then the coordinates of the vertices of *PQRS* and *P'Q'R'S'* are as follows:

$P(x, y)$: $P'(x + u, y + v)$;
$Q(x + dx, y)$:

$$Q'\left(x + dx + u + \frac{\partial u}{\partial x} dx, \; y + v + \frac{\partial v}{\partial x} dx\right);$$

$S(x, y + dy)$:

$$S'\left(x + u + \frac{\partial u}{\partial y} dy, \; y + dy + v + \frac{\partial v}{\partial y} dy\right);$$

$R(x + dx, y + dy)$:

$$R'\left(x + dx + u + \frac{\partial u}{\partial x} dx + \frac{\partial u}{\partial y} dy,\right.$$
$$\left. y + dy + v + \frac{\partial v}{\partial x} dx + \frac{\partial v}{\partial y} dy\right).$$

In general, the changes in *u* and *v* are much smaller than the quantities d*x* and d*y;* accordingly, we shall assume that the terms $(\partial u/\partial x)$, $(\partial u/\partial y)$, and so on are small enough that powers and products can be neglected. With this assumption, we see the following:

1. *PQ* increases in length by the amount $(\partial u/\partial x)$ d*x* and *PS* by the amount $(\partial v/\partial y)$ d*y;* hence $\partial u/\partial x$ and $\partial v/\partial y$ are the fractional increases in length in the direction of the axes.
2. The infinitesimal angles δ_1 and δ_2 are equal to $\partial v/\partial x$ and $\partial u/\partial y$, respectively.
3. The right angle at *P* decreases by the amount $\delta_1 + \delta_2 = \partial v/\partial x + \partial u/\partial y$.
4. The rectangle as a whole has been rotated counterclockwise through the angle $(\delta_1 - \delta_2)/2 = (\partial v/\partial x - \partial u/\partial y)/2$.

Strain is defined as the relative change (that is, the fractional change) in a dimension or shape of a body. The quantities $\partial u/\partial x$ and $\partial v/\partial y$ are the relative increases in length in the directions of the *x*- and *y*-axes, and are referred to as *normal strains*. The quantity $\partial v/\partial x + \partial u/\partial y$ is the amount by which a right angle in

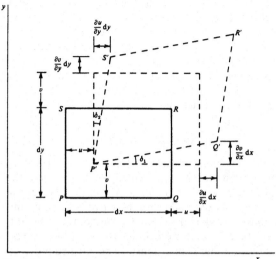

Fig. 2.3   Analysis of two-dimensional strain.

the *xy*-plane is reduced when the stresses are applied, hence, is a measure of the change in shape of the medium; it is known as a *shearing strain* and will be denoted by the symbol ε_{xy}. The quantity $(\partial v/\partial x - \partial u/\partial y)/2$, which represents a rotation of the body about the *z*-axis, does not involve change in size or shape and hence is not a strain; we shall denote it by the symbol θ_z.

Extending this analysis to three dimensions, we write (u, v, w) as the components of displacement of a point $P(x, y, z)$. The elementary strains are thus

$$\text{Normal strains} \quad \left. \begin{aligned} \varepsilon_{xx} &= \frac{\partial u}{\partial x}, \\ \varepsilon_{yy} &= \frac{\partial v}{\partial y}, \\ \varepsilon_{zz} &= \frac{\partial w}{\partial z}; \end{aligned} \right\} \qquad (2.8)$$

$$\text{Shearing strains} \quad \left. \begin{aligned} \varepsilon_{xy} = \varepsilon_{yx} &= \frac{\partial v}{\partial x} + \frac{\partial u}{\partial y}, \\ \varepsilon_{yz} = \varepsilon_{zy} &= \frac{\partial w}{\partial y} + \frac{\partial v}{\partial z}, \\ \varepsilon_{zx} = \varepsilon_{xz} &= \frac{\partial u}{\partial z} + \frac{\partial w}{\partial x}. \end{aligned} \right\} \qquad (2.9)$$

In addition to these strains, the body is subjected to simple rotation about the three axes given by

$$\left. \begin{aligned} \theta_x &= \frac{\partial w/\partial y - \partial v/\partial z}{2}, \\ \theta_y &= \frac{\partial u/\partial z - \partial w/\partial x}{2}, \\ \theta_z &= \frac{\partial v/\partial x - \partial u/\partial y}{2}. \end{aligned} \right\} \qquad (2.10)$$

Equations (2.10) can be written in vectorial form (see §15.1.2(a) and 15.1.2(c)):

$$\mathbf{\Theta} = \theta_x \mathbf{i} + \theta_y \mathbf{j} + \theta_z \mathbf{k} = \frac{\nabla \times \boldsymbol{\zeta}}{2}, \qquad (2.11)$$

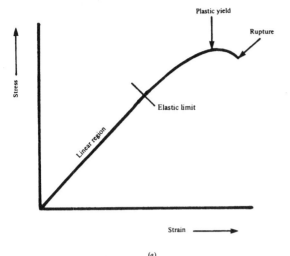

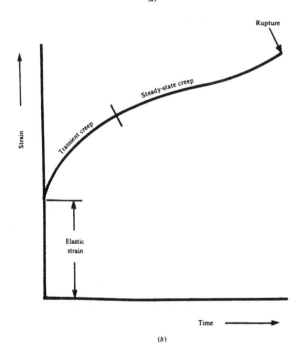

Fig. 2.4 Stress–strain–time relationships. (a) Stress versus strain; (b) strain versus time.

change in volume per unit volume Δ is

$$\Delta = \varepsilon_{xx} + \varepsilon_{yy} + \varepsilon_{zz} = \frac{\partial u}{\partial x} + \frac{\partial v}{\partial y} + \frac{\partial w}{\partial z} = \nabla \cdot \boldsymbol{\zeta}.$$

(2.12)

2.1.4 Hooke's law

In order to calculate the strains when the stresses are known, we must know the relationship between stress and strain. When the strains are small, this relation is given by *Hooke's law*, which states that a given strain is directly proportional to the stress producing it. The strains involved in seismic waves are usually less than 10^{-8} except very near the source, so that Hooke's law holds. When several stresses exist, each produces strains independently of the others; hence, the total strain is the sum of the strains produced by the individual stresses. This means that each strain is a linear function of all of the stresses and vice versa. This linearity has important implications that will be utilized later: It allows us to represent curved wavefronts as a superposition of plane waves, for example, in p–τ transforms (§9.1.5 and 9.11.1), to express a reflected wavetrain as a superposition of individual reflections (the convolutional model), and to justify many aspects of seismic data processing.

In general, Hooke's law leads to complicated relations. Stress and strain can both be regarded as second-order (3×3) matrices so that the Hooke's law proportionality relating them is a fourth-order tensor. Stress and strain can also be looked on as (1×6) matrices (as in eq. (2.15)) and the Hooke's law proportionality as a 6×6 matrix whose elements are elastic constants (Landau and Lifshitz, 1986: 32–51). Symmetry considerations immediately reduce the number of independent constants to 21. However, when the medium is *isotropic*, that is, when properties do not depend upon direction, it can be expressed in the following relatively simple form (Love, 1944: 102):

$$\sigma_{ii} = \lambda\Delta + 2\mu\varepsilon_{ii} \qquad (i = x,\, y,\, z), \qquad (2.13)$$

$$\sigma_{ij} = 2\mu\varepsilon_{ij} \qquad (i,\, j = x,\, y,\, z;\; i \neq j). \qquad (2.14)$$

These equations are often expressed as a matrix equation, $\boldsymbol{\sigma} = \mathbf{C\varepsilon}$:

$$
\begin{vmatrix} \sigma_{xx} \\ \sigma_{yy} \\ \sigma_{zz} \\ \sigma_{xy} \\ \sigma_{yz} \\ \sigma_{zx} \end{vmatrix}
=
\begin{vmatrix}
\lambda+2\mu & \lambda & \lambda & 0 & 0 & 0 \\
\lambda & \lambda+2\mu & \lambda & 0 & 0 & 0 \\
\lambda & \lambda & \lambda+2\mu & 0 & 0 & 0 \\
0 & 0 & 0 & \mu & 0 & 0 \\
0 & 0 & 0 & 0 & \mu & 0 \\
0 & 0 & 0 & 0 & 0 & \mu
\end{vmatrix}
\begin{vmatrix} \varepsilon_{xx} \\ \varepsilon_{yy} \\ \varepsilon_{zz} \\ \varepsilon_{xy} \\ \varepsilon_{yz} \\ \varepsilon_{zx} \end{vmatrix}
$$

(2.15)

The equation is sometimes written $\boldsymbol{\varepsilon} = \mathbf{S\sigma}$, where $\mathbf{S} = \mathbf{C}^{-1}$. Components of C (or S) are sometimes called *stiffness* (or *compliance*) components.

The quantities λ and μ are known as *Lamé's constants*. If we write $\varepsilon_{ij} = \sigma_{ij}/\mu$, it is evident that ε_{ij} is

where $\boldsymbol{\zeta} = u\mathbf{i} + v\mathbf{j} + w\mathbf{k}$ is the vector displacement of point $P(x, y)$, and $\mathbf{i}, \mathbf{j}, \mathbf{k}$ are unit vectors in the x-, y-, z- directions, respectively.

The changes in dimensions given by the normal strains result in volume changes when a body is stressed. The change in volume per unit volume is called the *dilatation* and represented by Δ. If we start with a rectangular parallelepiped with edges dx, dy, and dz in the unstrained medium, in the strained medium the dimensions are d$x(1 + \varepsilon_{xx})$, d$y(1 + \varepsilon_{yy})$, and d$z(1 + \varepsilon_{zz})$, respectively; hence the increase in volume is approximately $(\varepsilon_{xx} + \varepsilon_{yy} + \varepsilon_{zz})\,\mathrm{d}x\,\mathrm{d}y\,\mathrm{d}z$. Because the original volume was $(\mathrm{d}x\,\mathrm{d}y\,\mathrm{d}z)$, we see that the smaller the larger μ is. Hence, μ is a measure of the resistance to shearing strain and is often referred to as the *modulus of rigidity, incompressibility,* or *shear modulus.*

Although Hooke's law has wide application, it does not hold for large stresses. When the stress is increased beyond an *elastic limit* (fig. 2.4a), Hooke's law no longer holds and strains increase more rapidly. Strains resulting from stresses that exceed this limit do not entirely disappear when the stresses are removed. With further stress, a plastic yield point may be reached at which plastic flow begins and the plastic yielding may result in decreasing the strain. Some materials do not pass through a plastic flow phase but rupture first. Rocks usually rupture at strains $\sim 10^{-3}$–10^{-4}.

Some materials also have a time-dependent behavior to stress (fig. 2.4b). When subjected to a steady stress, such materials creep until eventually they rupture. Creep strain does not disappear if the stress is removed.

2.1.5 Elastic constants

Although Lamé's constants are convenient when we are using eqs. (2.13) and (2.14) other elastic constants are also used. The most common are *Young's modulus* (E), *Poisson's ratio* (σ), and the *bulk modulus* (k) (the symbol σ is more or less standard for Poisson's ratio – the subscripts should prevent any confusion with the stress σ_{ij}). To define the first two, we consider a medium in which all stresses are zero except σ_{xx}. Assuming σ_{xx} is positive (that is, a tensile stress), dimensions parallel to σ_{xx} will increase and dimensions normal to σ_{xx} will decrease; this means that ε_{xx} is positive (elongation in the x-direction) whereas ε_{yy} and ε_{zz} are negative. Also, we can show (see problem 2.1a) that $\varepsilon_{yy} = \varepsilon_{zz}$. We now define E and σ by the relations

$$E = \sigma_{xx}/\varepsilon_{xx}, \qquad (2.16)$$

$$\sigma = -\varepsilon_{yy}/\varepsilon_{xx} = -\varepsilon_{zz}/\varepsilon_{xx}, \qquad (2.17)$$

with the minus sign inserted to make σ positive.

To define the bulk modulus k, we consider a medium acted upon only by a pressure $\mathcal{P}$; this is equivalent to the stresses

$$\sigma_{xx} = \sigma_{yy} = \sigma_{zz} = -\mathcal{P}, \qquad \sigma_{xy} = \sigma_{yz} = \sigma_{zx} = 0.$$

Pressure $\mathcal{P}$ causes a decrease in the volume $\Delta\mathcal{V}$ and a *dilatation* $\Delta = \Delta\mathcal{V}/\mathcal{V}$; k is defined as the ratio of the pressure to the dilatation that it causes, that is,

$$k = -\mathcal{P}/\Delta, \qquad (2.18)$$

with the minus sign inserted to make k positive. Sometimes the *compressibility*, $1/k$, is used as an elastic constant rather than the bulk modulus.

By substituting the preceding values in Hooke's law, we can obtain the following relations between E, σ, and k and Lamé's constants, λ and μ (see problems 2.1b and 2.1c):

$$E = \frac{\mu(3\lambda + 2\mu)}{\lambda + \mu}, \qquad (2.19)$$

$$\sigma = \frac{\lambda}{2(\lambda + \mu)}, \qquad (2.20)$$

$$k = \tfrac{1}{3}(3\lambda + 2\mu). \qquad (2.21)$$

In nonviscous fluids, the shear modulus $\mu = 0$, and hence $k = \lambda$. Because we have not previously given a specific name to λ, we may call it the *fluid incompressibility*. By eliminating different pairs of constants among the three equations, many different relations can be derived expressing one of the five constants in terms of two others (see problem 2.2).

The elastic constants are defined in such a way that they are positive numbers. As a consequence of this, σ must have values between 0 and 0.5 (this follows from eq. (2.20), because both λ and μ are positive and hence $\lambda/(\lambda + \mu)$ is less than unity). Values range from 0.05 for very hard, rigid rocks to about 0.45 for soft, poorly consolidated materials. Liquids have no resistance to shear and hence for them $\mu = 0$ and $\sigma = 0.5$. For most rocks, E, k, and μ lie in the range from 20 to 120 GPa (2×10^{10} to 12×10^{10} N/m²), E generally being the largest and μ the smallest of the three. Tables of elastic constants of rocks have been given by Birch (1966). (See also problem 2.4.)

Most of the preceding theory assumes an isotropic medium. In fact, rocks are usually in layers with different elastic properties, these properties often varying with direction. Nevertheless, in discussing wave propagation, we generally ignore such differences and treat sedimentary rocks as isotropic media; when one does so, the results are useful and to do otherwise leads to extremely complex and cumbersome mathematical equations, except for the case of *transversely isotropic media*, that is, media in which the properties are the same in one plane but different along the normal to the plane. Some rocks, especially shales, are transversely isotropic, and more importantly, a series of parallel beds, each of which is isotropic, but where the properties vary from bed to bed, behaves as though it is transversely isotropic (Postma, 1955; Uhrig and van Melle, 1955). Anisotropy is discussed in §2.6.

2.1.6 Strain energy

When an elastic medium undergoes deformation, work is done and an equivalent amount of potential energy is stored in the medium; this energy is intimately related to elastic wave propagation.

If the stress σ_{xx} results in a displacement ε_{xx}, we assume that the stress is increased uniformly from zero to σ_{xx}, and hence the average stress is $\sigma_{xx}/2$. Thus,

$$E = \text{work done per unit volume}$$
$$= \text{energy per unit volume}$$
$$= \sigma_{xx}\varepsilon_{xx}/2.$$

Summing the effects of all the independent stresses and using eqs. (2.13) and (2.14) gives (Love, 1944: 100)

$$
\begin{aligned}
E &= \tfrac{1}{2}\sum_i\sum_j \sigma_{ij}\varepsilon_{ij} \\
&= \tfrac{1}{2}(\sigma_{xx}\varepsilon_{xx} + \sigma_{yy}\varepsilon_{yy} + \sigma_{zz}\varepsilon_{zz} + \sigma_{xy}\varepsilon_{xy} \\
&\quad + \sigma_{yz}\varepsilon_{yz} + \sigma_{zx}\varepsilon_{zx}) \\
&= \tfrac{1}{2}\left[\sum_i (\lambda\,\Delta + 2\mu\varepsilon_{ii})\varepsilon_{ii} + \mu\sum_i\sum_j \varepsilon_{ij}^2\right], \quad (i \neq j) \\
&= \tfrac{1}{2}\lambda\,\Delta^2 + \mu(\varepsilon_{xx}^2 + \varepsilon_{yy}^2 + \varepsilon_{zz}^2) \\
&\quad + \tfrac{1}{2}\mu(\varepsilon_{xy}^2 + \varepsilon_{yz}^2 + \varepsilon_{zx}^2).
\end{aligned} \tag{2.22}
$$

Note that eq. (2.22) gives

$$
\begin{aligned}
\partial E/\partial\varepsilon_{xx} &= \lambda\,\Delta + 2\mu\varepsilon_{xx} = \sigma_{xx}, \\
\partial E/\partial\varepsilon_{xy} &= \mu\varepsilon_{xy} = \sigma_{xy},
\end{aligned}
$$

hence,

$$
\partial E/\partial\varepsilon_{ij} = \sigma_{ij} \qquad (i, j = x, y, z). \tag{2.23}
$$

2.2 Wave equations

2.2.1 Scalar wave equation

Up to this point, we have been discussing a medium in static equilibrium. We shall now remove this restriction and consider what happens when the stresses are not in equilibrium. In fig. 2.2, we now assume that the stresses on the rear face of the element of volume are as shown in the diagram but that the stresses on the front face are, respectively,

$$
\sigma_{xx} + \frac{\partial\sigma_{xx}}{\partial x}\,\mathrm{d}x, \qquad \sigma_{yx} + \frac{\partial\sigma_{yx}}{\partial x}\,\mathrm{d}x, \qquad \sigma_{zx} + \frac{\partial\sigma_{zx}}{\partial x}\,\mathrm{d}x.
$$

Because these stresses are opposite to those acting on the rear face, the net (unbalanced) stresses are

$$
\frac{\partial\sigma_{xx}}{\partial x}\,\mathrm{d}x, \qquad \frac{\partial\sigma_{yx}}{\partial x}\,\mathrm{d}x, \qquad \frac{\partial\sigma_{zx}}{\partial x}\,\mathrm{d}x.
$$

These stresses act on a face having an area $(\mathrm{d}y\,\mathrm{d}z)$ and affect the volume $(\mathrm{d}x\,\mathrm{d}y\,\mathrm{d}z)$; hence, we get for the net forces per unit volume in the directions of the x-, y-, and z-axes the respective values

$$
\frac{\partial\sigma_{xx}}{\partial x}, \qquad \frac{\partial\sigma_{yx}}{\partial x}, \qquad \frac{\partial\sigma_{zx}}{\partial x}.
$$

Similar expressions hold for the other faces; hence, we find for the total force in the direction of the x-axis the expression

$$
\frac{\partial\sigma_{xx}}{\partial x} + \frac{\partial\sigma_{xy}}{\partial y} + \frac{\partial\sigma_{xz}}{\partial z}.
$$

Newton's second law of motion states that the unbalanced force equals the mass times the acceleration; thus, we obtain the equation of motion along the x-axis:

$$
\begin{aligned}
\rho\frac{\partial^2 u}{\partial t^2} &= \text{unbalanced force in the } x\text{-direction on} \\
&\quad\ \text{a unit volume} \\
&= \frac{\partial\sigma_{xx}}{\partial x} + \frac{\partial\sigma_{xy}}{\partial y} + \frac{\partial\sigma_{xz}}{\partial z},
\end{aligned} \tag{2.24}
$$

where ρ is the density (assumed to be constant). Similar equations can be written for the motion along the y- and z-axes.

Equation (2.24) relates the displacements to the stresses. We can obtain an equation involving only displacements by using Hooke's law to replace the stresses with strains and then expressing the strains in terms of the displacements, using eqs. (2.8), (2.9), (2.12), (2.13), and (2.14). Thus,

$$
\begin{aligned}
\rho\frac{\partial^2 u}{\partial t^2} &= \frac{\partial\sigma_{xx}}{\partial x} + \frac{\partial\sigma_{xy}}{\partial y} + \frac{\partial\sigma_{xz}}{\partial z} \\
&= \lambda\frac{\partial\Delta}{\partial x} + 2\mu\frac{\partial\varepsilon_{xx}}{\partial x} + \mu\frac{\partial\varepsilon_{xy}}{\partial y} + \mu\frac{\partial\varepsilon_{xz}}{\partial z} \\
&= \lambda\frac{\partial\Delta}{\partial x} + \mu\left[2\frac{\partial^2 u}{\partial x^2} + \left(\frac{\partial^2 v}{\partial x\partial y} + \frac{\partial^2 u}{\partial y^2}\right)\right. \\
&\quad \left. + \left(\frac{\partial^2 w}{\partial x\partial z} + \frac{\partial^2 u}{\partial z^2}\right)\right] \\
&= \lambda\frac{\partial\Delta}{\partial x} + \mu\nabla^2 u + \mu\frac{\partial}{\partial x}\left(\frac{\partial u}{\partial x} + \frac{\partial v}{\partial y} + \frac{\partial w}{\partial z}\right) \\
&= (\lambda + \mu)\frac{\partial\Delta}{\partial x} + \mu\nabla^2 u,
\end{aligned} \tag{2.25}
$$

where $\nabla^2 u$ is the *Laplacian* of $u = \partial^2 u/\partial x^2 + \partial^2 u/\partial y^2 + \partial^2 u/\partial z^2$ (see eq. (15.14)). By analogy, we can write the equations for v and w:

$$
\rho\frac{\partial^2 v}{\partial t^2} = (\lambda + \mu)\frac{\partial\Delta}{\partial y} + \mu\,\nabla^2 v, \tag{2.26}
$$

$$
\rho\frac{\partial^2 w}{\partial t^2} = (\lambda + \mu)\frac{\partial\Delta}{\partial z} + \mu\,\nabla^2 w. \tag{2.27}
$$

To obtain the wave equation, we differentiate these three equations with respect to x, y, and z, respectively, and add the results together. This gives

$$
\begin{aligned}
\rho\frac{\partial^2}{\partial t^2}\left(\frac{\partial u}{\partial x} + \frac{\partial v}{\partial y} + \frac{\partial w}{\partial z}\right) &= (\lambda + \mu)\left(\frac{\partial^2\Delta}{\partial x^2} + \frac{\partial^2\Delta}{\partial y^2} + \frac{\partial^2\Delta}{\partial z^2}\right) \\
&\quad + \mu\,\nabla^2\left(\frac{\partial u}{\partial x} + \frac{\partial v}{\partial y} + \frac{\partial w}{\partial z}\right),
\end{aligned}
$$

that is,

$$
\rho\frac{\partial^2\Delta}{\partial t^2} = (\lambda + 2\mu)\,\nabla^2\Delta
$$

or

$$\frac{1}{\alpha^2}\frac{\partial^2\Delta}{\partial t^2} = \nabla^2\Delta, \left.\begin{array}{c} \\ \\ \\ \\ \end{array}\right\} \qquad (2.28)$$

where

$$\alpha^2 = (\lambda + 2\mu)/\rho.$$

By subtracting the derivative of eq. (2.26) with respect to z from the derivative of eq. (2.27) with respect to y, we get

$$\rho\frac{\partial^2}{\partial t^2}\left(\frac{\partial w}{\partial y} - \frac{\partial v}{\partial z}\right) = \mu\,\nabla^2\left(\frac{\partial w}{\partial y} - \frac{\partial v}{\partial z}\right),$$

that is,

$$\frac{1}{\beta^2}\frac{\partial^2\theta_x}{\partial t^2} = \nabla^2\theta_x, \left.\begin{array}{c} \\ \\ \\ \\ \end{array}\right\} \qquad (2.29)$$

where

$$\beta^2 = \mu/\rho.$$

By subtracting appropriate derivatives, we obtain similar results for θ_y and θ_z. Equations (2.28) and (2.29) are different examples of the *wave equation*, which we can write in the general form

$$\frac{1}{V^2}\frac{\partial^2\psi}{\partial t^2} = \nabla^2\psi, \qquad (2.30)$$

where V is a constant.

2.2.2 Vector wave equation

The wave equation can also be obtained using vector methods. Equations (2.25), (2.26), and (2.27) are equivalent to the *vector wave equation*:

$$\rho\frac{\partial^2\zeta}{\partial t^2} = (\lambda + \mu)\,\nabla\Delta + \mu\,\nabla^2\zeta. \qquad (2.31)$$

If we take the divergence of eq. (2.31) and use eqs. (2.12) and (15.14) we get eq. (2.28). Taking the curl of eq. (2.31) and using eq. (2.11) and problem 15.7 gives the vector wave equation for S-waves (see §2.4.1),

$$\frac{1}{\beta^2}\frac{\partial^2\Theta}{\partial t^2} = \nabla^2\Theta, \qquad (2.32)$$

which is equivalent to the three scalar equations,

$$\frac{1}{\beta^2}\frac{\partial^2\theta_i}{\partial t^2} = \nabla^2\theta_i \qquad (i = x, y, z). \qquad (2.33)$$

2.2.3 Wave equation including source term

The foregoing discussion of the wave equation has made no mention of the sources of the waves, and in fact, the equations discussed are only valid in a source-free region. Sources can be taken into account

in two ways in general: (a) include in the wave equation terms that represent the forces generating the waves or (b) surround the point of observation P by a closed surface $\mathcal{S}$ and regard the effect at P as being given by a volume integral throughout the interior of $\mathcal{S}$ to take into account sources inside $\mathcal{S}$ plus a surface integral over $\mathcal{S}$ to give the effect of sources outside $\mathcal{S}$ (see §2.2.4). To apply the first method, we note that eqs. (2.25), (2.26), and (2.27) are equivalent to Newton's second law, and these three equations are combined in eq. (2.31). Therefore, a source can be taken into account by adding to the right-hand side of eq. (2.31) the term $\rho\mathbf{F}$, where $\mathbf{F}$ is the external nonelastic force per unit mass (often called *body force*) that gives rise to the wave motion. Thus, eq. (2.31) becomes

$$\rho\frac{\partial^2\zeta}{\partial t^2} = (\lambda + \mu)\,\nabla\Delta + \mu\,\nabla^2\zeta + \rho\mathbf{F}. \qquad (2.34)$$

Taking the divergence and curl of eq. (2.34) and using eq. (15.14) and problem 15.7 gives

$$\frac{\partial^2\Delta}{\partial t^2} = \alpha^2\,\nabla^2\Delta + \nabla\cdot\mathbf{F}, \qquad (2.35)$$

$$\frac{\partial^2\Theta}{\partial t^2} = \beta^2\,\nabla^2\Theta + \nabla\times\mathbf{F}/2. \qquad (2.36)$$

These equations are difficult to solve as they stand. The solution is greatly simplified by using the Helmholtz separation method, which involves expressing both ζ and $\mathbf{F}$ in terms of new scalar and vector functions. Thus, we write

$$\zeta = \nabla\phi + \nabla\times\chi, \quad \nabla\cdot\chi = 0, \qquad (2.37)$$
$$\mathbf{F} = \nabla Y + \nabla\times\Omega, \quad \nabla\cdot\Omega = 0. \qquad (2.38)$$

Then, using problem 15.7, we obtain

$$\left.\begin{array}{l} \Delta = \nabla\cdot\zeta = \nabla^2\phi, \\ 2\Theta = \nabla\times\zeta = -\nabla^2\chi, \\ \nabla\cdot\mathbf{F} = \nabla^2 Y, \\ \nabla\times\mathbf{F} = -\nabla^2\Omega. \end{array}\right\} \qquad (2.39)$$

Substituting in eqs. (2.35) and (2.36), we get

$$\nabla^2\left(\alpha^2\nabla^2\phi + Y - \frac{\partial^2\phi}{\partial t^2}\right) = 0,$$

$$\nabla^2\left(\beta^2\nabla^2\chi + \Omega - \frac{\partial^2\chi}{\partial t^2}\right) = 0.$$

Whenever ϕ, χ, Y, or Ω contain powers of x, y, and z higher than the first, these equations can only be satisfied for all values of x, y, and z if the expressions inside the parentheses are identically zero at all points. Because a linear function of x, y, and z corresponds to a uniform translation and/or rotation of the medium, we can ignore this possibility and write (Savarensky, 1975: 199)

$$\frac{\partial^2\phi}{\partial t^2} = \alpha^2\nabla^2\phi + Y, \qquad (2.40)$$

$$\frac{\partial^2 \chi}{\partial t^2} = \beta^2 \nabla^2 \chi + \Omega. \qquad (2.41)$$

2.2.4 Kirchhoff's theorem

Method (b) referred to in §2.2.3 is in fact an extension of method (a). It uses the superposition concept (which follows from the linearity expressed in Hooke's law). We regard the wave motion at a point P as the superposition of the waves from all sources R within some volume $\mathcal{V}$ surrounding P plus the waves radiated by points Q on the surface $\mathcal{S}$ surrounding the volume (which takes into account any disturbances from sources outside the volume). We adjust the times for these sources so that their effects all arrive at P at the same instant t_0. We take $Y(x, y, z, t_R)$ in eq. (2.40) as the source density (body force/unit volume) inside $\mathcal{S}$ and specify $\phi(x, y, z, t_Q)$ for each point on the surface $\mathcal{S}$, t_R, and t_Q being the *retarded times* $(t_0 - r/V)$, where V is the velocity, and r is the distance between P and the sources R or Q, that is, r/V is the time for the wave to travel from R or Q to P. Thus, we specify the wave motion at different points at different times such that the waves from all points arrive at P at the same instant t_0. The result, known as *Kirchhoff's theorem* (or formula) (Ewing, Jardetzky, and Press, 1957: 16), is

$$4\pi\phi_P(x, y, z, t_0) = \iiint_{\mathcal{V}} \left(\frac{Y}{r}\right) d\mathcal{V}$$
$$+ \iint_{\mathcal{S}} \left\{ \left(\frac{1}{Vr}\right)\left(\frac{\partial r}{\partial \eta}\right)\left[\frac{\partial \phi}{\partial t}\right] - [\phi]\frac{\partial(1/r)}{\partial \eta} \right.$$
$$\left. + \left(\frac{1}{r}\right)\left[\frac{\partial \phi}{\partial \eta}\right] \right\} d\mathcal{S}, \qquad (2.42)$$

where η is the outward-drawn unit normal, and the square brackets denote functions evaluated at point Q at time $t_Q = t_0 - r/V$; $[\phi]$ is often referred to as a *retarded potential*. If we assume that each source emits spherical waves (§2.2.6) of the form $(1/r)e^{-j\omega(r/V - t)}$ (see eqs. (2.55) and (2.56)), eq. (2.42) becomes (Savarensky, 1975: 234)

$$4\pi\phi_P(x, y, z, t_0) = \left(\frac{1}{V^2}\right)\iiint_{\mathcal{V}} \left(\frac{Y}{r}\right) d\mathcal{V}$$
$$+ \iint_{\mathcal{S}} \left\{ \xi\left[\frac{\partial \phi}{\partial \eta}\right] - [\phi]\frac{\partial \xi}{\partial \eta} \right\} d\mathcal{S}, \qquad (2.43)$$

where in the integrand,

$$[\phi] = (1/r)e^{j\omega(t_0 - r/V)} = \xi e^{j\omega t_0}, \qquad \xi = (1/r)e^{-j\omega r/V}, \qquad (2.44)$$

and ω is the angular frequency (see §2.1.1).

Because we started from eq. (2.40), eqs. (2.42) to (2.43) are valid for P-waves (see §2.4.1). However, we could just as well have started from eq. (2.41) and ob-

tained identical results for S-waves. Thus, the previous equations refer to either P- or S-waves.

2.2.5 Plane-wave solutions

Let us consider first the case where ψ is a function only of x and t, so that eq. (2.30) reduces to

$$\frac{1}{V^2}\frac{\partial^2 \psi}{\partial t^2} = \frac{\partial^2 \psi}{\partial x^2}. \qquad (2.45)$$

Any function of $(x - Vt)$,

$$\psi = f(x - Vt), \qquad (2.46)$$

is a solution of eq. (2.45) (see problem 2.5a) provided that ψ and its first two derivatives are finite and continuous. This solution (known as d'Alembert's solution) furnishes an infinite number of particular solutions (for example, $e^{k(x - Vt)}$, $\sin (x - Vt)$, $(x - Vt)^3$, where we must exclude points at which these functions and their first three derivatives cease to exist or are discontinuous). The answer to a specific problem consists of selecting the appropriate combination of solutions that also satisfies the boundary conditions for the problem.

A *body wave* is defined as a "disturbance" that travels through the medium and carries energy (Logan, 1987: 230). In our notation, the disturbance ψ is a volume change when $\psi = \Delta$ and a rotation when $\psi = \theta_i$. Obviously, the disturbance in eq. (2.46) is traveling along the x-axis. We shall now show that it travels with a speed equal to the quantity V.

In fig. 2.5a the certain part of the wave has reached point P_0 at time t_0. If the coordinate of P_0 is x_0, then the value of ψ at P_0 is $\psi_0 = f(x_0 - Vt_0)$. If this same portion of the wave reaches P_1 at time $t_0 + \Delta t$, then we have for the value of ψ at P_1

$$\psi_1 = f[x_0 + \Delta x - V(t_0 + \Delta t)].$$

But, because this is the same portion of the wave that was at P_0 at time t_0, we must have $\psi_0 = \psi_1$, that is,

$$x_0 - Vt_0 = x_0 + \Delta x - V(t_0 + \Delta t).$$

Thus, the quantity V is equal to $\Delta x/\Delta t$ and is therefore the speed with which the disturbance travels. The reciprocal of velocity, $1/V$, is called *slowness*.

A function of $(x + Vt)$, for example, $\psi = g(x + Vt)$, is also a solution of eq. (2.45). It denotes a wave traveling in the negative x-direction. The general solution of eq. (2.45),

$$\psi = f(x - Vt) + g(x + Vt), \qquad (2.47)$$

represents two waves traveling along the x-axis in opposite directions with velocity V.

The quantity $x \pm Vt$ (or a constant times these expressions; see §2.1.1) is the phase. The surfaces on which the wave motion is the same, that is, the surfaces on which the phase has the same value, are known as *wavefronts*. In the case we are considering, ψ is independent of y and z, and so the disturbance is

the same everywhere on a plane perpendicular to the
x-axis; the wavefront is therefore plane and the wave
is a *plane wave*. Note that the wave is traveling in the
direction normal to the wavefront; this holds for all
waves in isotropic media. A line denoting the direction
of travel of the wave energy is called a *raypath*.

Plane waves are easier to visualize and to treat
mathematically than more complicated waves. Moreover, curved wavefronts can be approximated as
closely as desired by a superposition of plane waves.

It is convenient at times to have an expression for a
plane wave traveling along a straight line inclined at
an angle to each of the axes. Assume that the wave is
traveling along the x'-axis, which has direction cosines (ℓ, m, n) relative to the x-, y-, and z-axes (fig.
2.6). Then, at a point P on the x'-axis at a distance x'
from the origin, we have

$$x' = \ell x + my + nz,$$

where the coordinates of P are (x, y, z). Then,

$$\psi = f(\ell v + my + nz - Vt) \\ + g(\ell x + my + nz + Vt). \quad (2.48)$$

2.2.6 Spherical-wave solutions

In addition to plane waves, we shall have occasion to
use another important type of wave, the *spherical
wave*, where the wavefronts are a series of concentric
spherical surfaces. We express eq. (2.30) in spherical
coordinates (r, θ, ϕ), where θ is the colatitude, and ϕ
the longitude (see problem 2.6b).

$$\frac{1}{V^2}\frac{\partial^2\psi}{\partial t^2} = \frac{1}{r^2}\left[\frac{\partial}{\partial r}\left(r^2\frac{\partial\psi}{\partial r}\right) + \frac{1}{\sin\theta}\frac{\partial}{\partial\theta}\left(\sin\theta\frac{\partial\psi}{\partial\theta}\right) \\ + \frac{1}{\sin^2\theta}\frac{\partial^2\psi}{\partial\phi^2}\right]. \quad (2.49)$$

We consider only the special case when the wave motion is independent of θ and ϕ, hence is a function
only of r and t. Then we get the simplified equation

$$\frac{1}{V^2}\frac{\partial^2\psi}{\partial t^2} = \frac{1}{r^2}\frac{\partial}{\partial r}\left(r^2\frac{\partial\psi}{\partial r}\right). \quad (2.50)$$

A solution of the foregoing equation is

$$\psi = (1/r)f(r - Vt) \quad (2.51)$$

(see eq. (2.46)). Obviously,

$$\psi = (1/r)g(r + Vt)$$

is also a solution and the general solution of eq. (2.50)
(see problem 2.5c) is

$$\psi = (1/r)f(r - Vt) + (1/r)g(r + Vt), \quad (2.52)$$

in which the first term represents a wave expanding
outward from a central point and the second term a
wave collapsing toward the central point.

When r and t are fixed, $(r - Vt)$ is constant and
hence ψ is constant. Thus, at the instant t, the wave

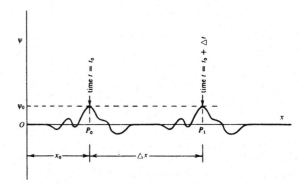

Fig. 2.5   Illustrating the velocity of a wave.

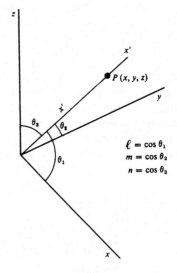

Fig. 2.6   Wave direction not along an axis.

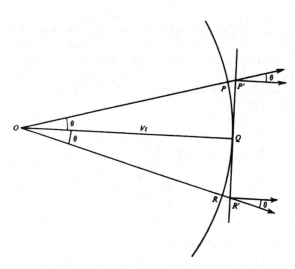

Fig. 2.7   Relation between spherical and plane waves.

has the same value at all points on the spherical surface of radius r. The spherical surfaces are therefore
wavefronts and the radii are rays. Obviously, the rays

are normal to the wavefronts as in the case of plane waves. (This is not always the case in anisotropic media.)

As the wave progress outward from the center, the radius increases by the amount V during each unit of time. Eventually, the radius becomes very large and the portion of the wavefront near any particular point will be approximately plane. If we consider fig. 2.7, we see that the error that we introduce when we replace the spherical wavefront PQR by the plane wavefront $P'QR'$ is due to the divergence between the true direction of propagation given by the direction of the radius and the assumed direction normal to the plane. By taking OQ very large or PR very small (or both), we can make the error as small as desired. Because plane waves are easy to visualize and also the simplest to handle mathematically, we generally assume that conditions are such that the plane-wave assumption is valid.

2.3 General aspects of waves

2.3.1 Harmonic waves

In §2.2.5 and 2.2.6, we discussed the geometrical aspects of waves, that is, how they depend on the space coordinates. We now consider the time dependence of wave motion.

The simplest time variation that a wave can have is *harmonic* (sinusoidal), equivalent to simple harmonic motion. In general, waves are more complex than this, but the methods of Fourier analysis (§ 15.2) allow us to represent almost any complex wave as a superposition of harmonic waves. Harmonic waves, because of their simplicity, can be regarded as the time equivalent of plane waves in space.

Adding $\pi/2$ to the phase in eq. (2.3) changes cosine to sine, so harmonic waves can be written in either form. Some of the commonest forms are the following:

$$\begin{aligned}
\psi &= A \cos [(2\pi/\lambda)(x - Vt)] = A \cos \kappa(x - Vt) \\
&= A \cos (\kappa x - \omega t) \\
&= A \cos 2\pi(x/\lambda - \nu t) = A \cos 2\pi(x/\lambda - t/T) \\
&= A \cos \omega(x/V - t),
\end{aligned}$$

$$\left. \begin{aligned} \end{aligned} \right\} \tag{2.53}$$

$$\begin{aligned}
\psi &= A \cos \kappa(\ell x + my + nz - Vt) \\
&= A \sin [\kappa(\ell x + my + nz - Vt) + \pi/2],
\end{aligned} \right\} \tag{2.54}$$

$$\psi = (A/r) \cos \kappa(r - Vt) + (B/r) \cos \kappa(r + Vt). \tag{2.55}$$

Equation (2.53) represents a plane wave traveling in the $+x$-direction, eq. (2.54) a plane wave moving along a straight line with direction cosines (ℓ, m, n), and eq. (2.55) a spherical wave expanding from and collapsing toward the origin.

Equation (15.45) enables us to combine the cosine and sine expressions for a harmonic wave; thus, if we write

$$\psi = A e^{j\omega[(\ell x + my + nz)/V - t]} = A e^{j\omega(r/V - t)}, \tag{2.56}$$

we can get either the cosine or sine form by taking the real or imaginary part of ψ.

The quantities (ℓ, m, n) in eq. (2.54) represent the direction cosines of the ray. In problem 15.9a, we show that $\ell^2 + m^2 + n^2 = 1$. Although ordinarily each of the cosines has a maximum value of unity, satisfying the wave equation requires only that the sum of the squares be unity. If we admit pure imaginary numbers, some of the "direction cosines" can be greater than unity. Let us take in fig. 2.6, $\theta_1 = j\theta$, $\theta_2 = \frac{1}{2}\pi$, $\theta_3 = \frac{1}{2}\pi - j\theta$, θ being real and positive; then

$$\ell = \cos j\theta = \cosh \theta, \qquad m = 0,$$
$$n = \cos (\tfrac{1}{2}\pi - j\theta) = \sin j\theta = j \sinh \theta,$$
$$\ell^2 + m^2 + n^2 = \cosh^2 \theta - \sinh^2 \theta = 1,$$
$$\psi = A e^{-(\omega z/V) \sinh \theta} e^{j\omega[(x/V) \cosh \theta - t]}. \tag{2.57}$$

This represents a plane wave traveling parallel to the x-axis with velocity $V/\cosh \theta < V$ and amplitude $A e^{-(\omega z/V)\sinh \theta}$. If we had taken $\theta_1 = -j\theta$, this would give a wave traveling in the x-direction with amplitude decreasing upward in the negative z-direction. Because the amplitude decreases exponentially with z, these waves are called *evanescent waves*. We shall refer again to these waves in §2.7.5.

In exploration seismology, the range of frequencies recorded with appreciable energy is generally from about 2 to 120 Hz, and the dominant frequencies lie in a narrower range from 15 to 50 Hz for reflection work and from 5 to 20 Hz for refraction work. Because velocities generally range from 1.6 to 6.5 km/s, dominant wavelengths range from about 30 to 400 m for reflection work and from 80 to 1300 m for refraction.

2.3.2 Wave interference

If two waves are superimposed, they interfere with each other; the interference is *constructive* if they tend to add and *destructive* if they tend to cancel. When the two waves are harmonic and have the same frequencies and wavelengths (hence the same velocities), their amplitudes sometimes add together and sometimes cancel (at least partially); thus, they form a new wave of the same frequency and wavelength with different amplitude and phase-shifted. When several harmonic waves with different amplitudes, frequencies, and/or wavelengths are added together, the results are usually very complex; constructive interference occurs when the phases are nearly the same (e.g., §2.7.4 and 13.3), otherwise destructive interference results in at least some attenuation. If the waves are not harmonic, they can be resolved by Fourier analysis (§9.1 and 15.2) into harmonic components that can then be added to determine the nature of the interference.

If we add two harmonic waves of equal amplitudes (A) and velocities but slightly different frequencies (see problem 2.7), the sum is $B \cos (\kappa_0 x - \omega_0 t)$, where $B = 2A \cos (\Delta\kappa x - \Delta\omega t)$, κ_0 and ω_0 being average values, and $\Delta\kappa$ and $\Delta\omega$ half of the differences between the values for the two waves. We regard B as the variable amplitude of the resultant wave; at a fixed point, B varies between $\pm 2A$ at the rate of $\Delta\omega/2\pi$ times per second, that is, slowly in comparison with the wave frequency ω_0. This phenomenon is called *beating*.

2.3.3 Huygens' principle

The solutions of the wave equation given by eqs. (2.47) and (2.52) are restricted to plane and spherical waves. On the other hand, the Kirchhoff formula is valid for any type of body wave. As expressed in eq. (2.42) (assuming no sources inside $\mathscr{S}$), it states that the effect at a point P is the sum of effects that took place earlier at all points on a surface $\mathscr{S}$ enclosing P, allowance being made for the time for these effects to travel from $\mathscr{S}$ to P. Thus, each point on $\mathscr{S}$ behaves as though it were a new wave source.

To obtain Huygens' principle, we take $\mathscr{S}$ coincident with that portion of the wavefront that we wish to take into account in finding the effect at P, and then complete the closed surface by passing it through space where the effect has not yet arrived so that ϕ is zero over this part.

Huygens' principle is important in understanding wave travel and is often useful in drawing successive positions of wavefronts. *Huygens' principle* states that every point on a wavefront can be regarded as a new source of waves. The physical rationale behind this is that each particle located on a wavefront has moved from its equilibrium position in approximately the same manner, that the elastic forces on neighboring particles are thereby changed, and that the resultant of the changes in force because of the motion of all the points on the wavefront thus begins to produce the motion that forms the next wavefront. In this way, Huygens' principle helps explain how information about seismic disturbances is communicated in the earth. Specifically, given the location of a wavefront at a certain instant, future positions of the wavefront can be found by considering each point on the first wavefront as a new wave source. In fig. 2.8, AB is the wavefront at the time t_0 and we wish to find the wavefront at a later time $t_0 + \Delta t$. During the interval Δt, the wave will advance a distance $V \Delta t$, V being the velocity (which may vary from point to point). We select points on the wavefront, P_1, P_2, P_3, and so on, from which we draw arcs of radii $V \Delta t$. Provided we select enough points, the envelope of the arcs ($A'B'$) will define as accurately as we wish the position of the

wavefront at time $t_0 + \Delta t$. Except on the envelope, the elemental waves interfere destructively with each other so that their effects cancel. When AB is plane, and V constant, we need draw only two arcs and the straight-line tangent to the two arcs defines the new wavefront. Note that Huygens' principle gives only phase information; it does not give amplitudes.

2.4 Body waves

2.4.1 P-waves and S-waves

Up to this point, our discussion of wave motion has been based upon eq. (2.30). The quantity ψ has not been defined; we have merely inferred that it is some disturbance that is propagated from one point to another with speed V. However, in a homogeneous isotropic medium, eqs. (2.28) and (2.29) must be satisfied. We can identity the functions Δ and θ_i with ψ and conclude that two types of waves can be propagated in a homogeneous isotropic medium, one corresponding to changes in the dilatation Δ and the other to one or more components of the rotation given in eq. (2.11).

The first type is variously known as a *dilatational, longitudinal, irrotational, compressional,* or *P-wave,* the latter name being given because this type is usually the first (primary) event on an earthquake recording. The second type is referred to as the *shear, transverse, rotational,* or *S-wave* (because it is usually the second major event observed on earthquake records). The P-wave has the velocity α in eq. (2.28) and the S-wave the velocity β in eq. (2.29), that is,

$$\alpha = \left(\frac{\lambda + 2\mu}{\rho}\right)^{1/2} = \left(\frac{M}{\rho}\right)^{1/2}, \qquad (2.58)$$

$$\beta = \left(\frac{\mu}{\rho}\right)^{1/2}, \qquad (2.59)$$

where M is the *P-wave modulus*. Because the elastic constants are always positive, α is always greater than β. Using eq. (2.20), we see that

$$\frac{\beta}{\alpha} = \left(\frac{\mu}{\lambda + 2\mu}\right)^{1/2} = \left(\frac{0.5 - \sigma}{1 - \sigma}\right)^{1/2} \qquad (2.60)$$

(see fig. 2.9). As σ decreases from 0.5 to 0, β/α increases from 0 to its maximum value, $1/\sqrt{2}$; thus, the velocity of the S-wave ranges from 0 to 70% of the velocity of the P-wave.

For fluids, μ is zero and hence β is also zero; therefore S-waves do not propagate in fluids. Using eq. (2.21), we see that for a fluid, $\lambda = k$; hence,

$$\alpha = (k/\rho)^{1/2}. \qquad (2.61)$$

The seismic velocity in actual rocks depends on many factors, including porosity, lithology, cementation, depth, age, pressure regime, interstitial fluids, etc., which are discussed in chap. 5. The velocity of water-saturated sedimentary rocks is generally in the 1.5 to 6.5 km/s range, increasing with loss of porosity,

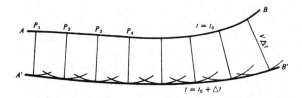

Fig. 2.8   Using Huygens' principle to locate new wavefronts.

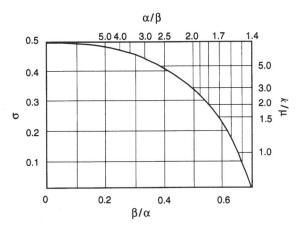

Fig. 2.9   β/α as a function of Poisson's ratio σ and k/μ.

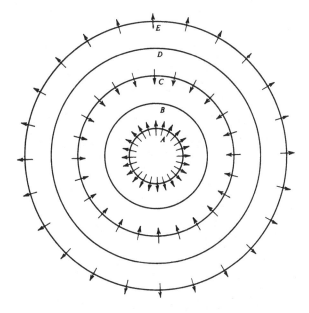

Fig. 2.10   Displacements for a spherical *P*-wave.

cementation, depth, and age. (Velocity versus depth relations for three situations are shown in fig. 11.28.) The velocity of *P*-waves in water is approximately 1.5 km/s. *P*-wave velocity is lowered, often markedly, when a gas replaces water as the interstitial fluid. This is especially important in the near-surface, generally above the water table, where a low-velocity layer (LVL, also called the weathered layer) typically has a velocity in the 0.4 to 0.8 km/s range, occasionally as low as 150 m/s, sometimes as high as 1.2 km/s.

Let us investigate the nature of the motion of the medium corresponding to the two types of wave motion. Consider a spherical *P*-wave of the type given by eq. (2.51). Figure 2.10 shows wavefronts drawn at quarter-wavelength intervals, *t* being chosen so that $\kappa V t$ is a multiple of $\pi/2$. The arrows represent the direction of motion of the medium at the wavefront. The medium is undergoing maximum compression at *B* (that is, the dilatation Δ is a minimum) and mini-

mum compression (maximum Δ) at the wavefront *D;* particle velocity is zero at each of these points.

We can visualize the plane-wave situation by imagining that the radius in fig. 2.10 has become very large so that the wavefronts are practically plane surfaces. The displacements will be everywhere perpendicular to these planes so that there will no longer be convergence or divergence of the particles of the medium as they move back and forth parallel to the direction of propagation of the wave. Such a displacement is longitudinal, which explains why *P*-waves are sometimes called longitudinal waves. *P*-waves are the dominant waves involved in seismic exploration. A plane *P*-wave is illustrated in fig. 2.11a.

To determine the motion of a medium during the passage of an *S*-wave, we return to eq. (2.29) and consider the case where a rotation θ_z, which is a function of *x* and *t* only, is being propagated along the *x*-axis. We have

$$\frac{1}{\beta^2}\frac{\partial^2\theta_z}{\partial t^2} = \frac{\partial^2\theta_z}{\partial x^2}.$$

Because

$$2\theta_z = \frac{\partial v}{\partial x} - \frac{\partial u}{\partial y} = \frac{\partial v}{\partial x}$$

from eq. (2.10), we see that the wave motion consists

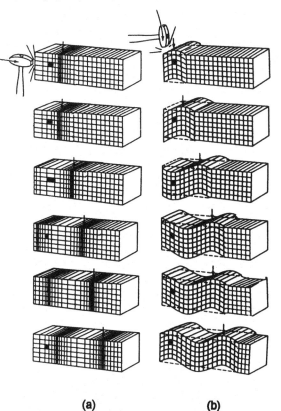

(a)   **(b)**

Fig. 2.11   Motion during passage of plane body waves. (After *Earth,* 2d ed., by F. Press and B. Siever, p. 424. Copyright 1974 by W. H. Freeman and Company; reprinted with permission.) (a) *P*-wave; (b) *S*-wave.

solely of a displacement v of the medium in the y-direction, v being a function of both x and t. Because v is independent of y and z, the motion is the same everywhere in a plane perpendicular to the x-axis; thus, the case we are discussing is that of a plane S-wave traveling along the x-axis (fig. 2.11b).

We can visualize the foregoing relations using fig. 2.12. When the wave arrives at P, it causes the medium in the vicinity of P to rotate about the axis $Z'Z''$ (parallel to the z-axis) through an angle ε. Because we are dealing with infinitesimal strains, ε must be infinitesimal and we can ignore the curvature of the displacements and consider that points such as P' and P'' are displaced parallel to the y-axis to the points Q' and Q''. Thus, as the wave travels along the x-axis, the medium is displaced transversely to the direction of propagation, hence the name transverse wave. Moreover, because the rotation varies from point to point at any given instant, the medium is subjected to varying shearing stresses as the wave moves along; this accounts for the name shear wave.

Because we might have chosen to illustrate θ_y in fig. 2.12 instead of θ_z, it is clear that shear waves have two degrees of freedom, unlike P-waves, which have only one – along the radial direction. In practice, S-wave motion is usually resolved into components parallel and perpendicular to the surface of the ground, these are known respectively as SH- and SV-waves. When the wave is traveling neither horizontally nor vertically, the motion is resolved into a horizontal [SH] component and a component in the vertical plane through the direction of propagation. Henceforth, S-wave will mean SV-wave unless otherwise noted.

Because the two degrees of freedom of S-waves are independent, we can have an S-wave that involves motion in only one plane, for example, SH or SV motion; such a wave is said to be *plane-polarized*. We can also have a wave in which the SH and SV motion have the same frequency and a fixed phase difference; such a wave is *elliptically polarized*. Polarization of S-waves is a factor in their exploration use (see §13.1). Note that we cannot have a spherically symmetrical S-wave (analogous to the P-wave illustrated in fig. 2.10). S-wave amplitude must vary with direction.

In the case of a medium that is not homogeneous and isotropic, it may not be possible to resolve wave motion into separate P- and S-waves. However, inhomogeneities and anisotropy in the earth are small enough that the assumption of separate P- and S-waves is valid for practical purposes.

2.4.2 Displacement and velocity potentials

Solutions of the wave equations such as those in eqs. (2.48) and (2.52) furnish expressions for Δ and θ. However, often we need to know the displacements u, v, w, or the velocities $\dot{u} = \mathrm{d}u/\mathrm{d}t$, $\dot{v}$, $\dot{w}$, and reference to eqs. (2.8) to (2.12) shows that these are not easily found given only values of Δ and θ_i. This difficulty is often resolved by using potential functions $\phi(x, y, z, t)$

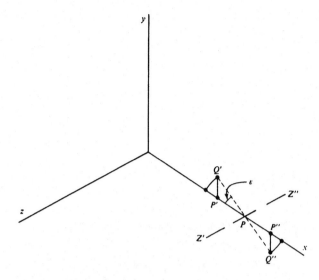

Fig. 2.12   Rotation of medium during passage of an S-wave.

and $\chi(x, y, z, t)$, which are solutions of the P- and S-wave equations, respectively, and which are so chosen that u, v, w (or $\dot{u}$, $\dot{v}$, $\dot{w}$) can be found by differentiation.

A simple example of such functions is the following:

$$\chi = 0, \qquad \nabla\phi = \zeta = (u\mathbf{i} + v\mathbf{j} + w\mathbf{k}),$$

so that

$$\left. u = \frac{\partial\phi}{\partial x}, \qquad v = \frac{\partial\phi}{\partial y}, \qquad w = \frac{\partial\phi}{\partial z} . \right\} \quad (2.62)$$

This procedure is valid only if it corresponds with Δ being a solution of the P-wave equation. Because ζ is a solution and $\Delta = \nabla \cdot \zeta = \nabla^2\phi$, Δ is also a solution (because derivatives of a solution are also solutions). Setting $\chi = 0$ is equivalent to saying that S-waves do not exist and this choice of potential functions is suitable for discussing wave motion in fluids.

For wave motion in three-dimensional solids, ϕ and χ can be defined so that

$$\zeta = \nabla\left(\phi + \frac{\partial\chi}{\partial z}\right) - \nabla^2\chi\mathbf{k}. \quad (2.63)$$

This ensures that Δ and Θ are solutions of the P- and S-wave equations, respectively (see problem 2.9a).

For two-dimensional wave motion in the xz-plane, ϕ and χ can be defined by

$$\left. \zeta = \nabla\phi + \nabla \times \chi, \qquad \chi = -\chi\mathbf{j}, \right.$$
$$\left. u = \frac{\partial\phi}{\partial x} + \frac{\partial\chi}{\partial z}, \qquad w = \frac{\partial\phi}{\partial z} - \frac{\partial\chi}{\partial x}. \right\} \quad (2.64)$$

It is easy to show that eqs. (2.12) and (2.11) can be expressed as

$$\left. \Delta = \nabla \cdot \zeta = \nabla^2\phi, \right.$$
$$\left. 2\Theta = \nabla \times \zeta = \nabla^2\chi\mathbf{j}, \right\} \quad (2.65)$$

so that Δ and Θ are again solutions of the P- and S-wave equations (see problem 2.9b).

Because the wave equations are still valid if both sides are differentiated with respect to time t, it follows that velocity potentials will be obtained in each of the preceding cases if u, v, w, and ζ are replaced with $\dot{u}$, $\dot{v}$, $\dot{w}$, and $\dot{\zeta}$.

2.4.3 Wave equation in fluid media

In fluids, only P-waves are propagated and we are generally interested in pressure variations rather than displacements or velocities, as in solid media. Equation (2.62) can be expressed in terms of pressure $\mathcal{P}$. We redefine ϕ in the form

$$\nabla\phi = \dot{u}\mathbf{i} + \dot{v}\mathbf{j} + \dot{w}\mathbf{k}, \qquad \dot{u} = \frac{\partial u}{\partial t}, \qquad \text{etc. (2.66)}$$

In eq. (2.24), we set

$$\sigma_{xy} = \sigma_{yz} = \sigma_{zx} = 0, \qquad \sigma_{xx} = \sigma_{yy} = \sigma_{zz} = -\mathcal{P};$$

hence, using eq. (2.24), we get

$$\rho\frac{\partial^2 u}{\partial t^2} = -\frac{\partial\mathcal{P}}{\partial x} = \text{acceleration along the } x\text{-axis, (2.67)}$$

and similarly for the y- and z-axes. Adding the three components of acceleration gives

$$\rho\,\nabla\frac{\partial\phi}{\partial t} = -\nabla\mathcal{P}.$$

Ignoring the additive constant due to hydrostatic pressure (because we are interested only in pressure variations),

$$\mathcal{P} = -\rho\,\frac{\partial\phi}{\partial t} = \mathrm{j}\omega\rho\phi, \qquad (2.68)$$

if we consider only harmonic waves of the form

$$\phi = A\mathrm{e}^{\mathrm{j}(\kappa r - \omega t)} = A\mathrm{e}^{\mathrm{j}\omega(r/\alpha - t)}$$

(see eq. (2.56)). Thus, both ϕ and $\mathcal{P}$ satisfy the P-wave equation as in eq. (2.28), the velocity reducing to $\alpha = (k/\rho)^{1/2}$ in fluids (see eq. (2.61)).

In the case of a gas, k depends upon the way the gas is compressed, isothermally or adiabatically (that is, with no transfer of heat during the wave passage). For sound waves in air, the compression is essentially adiabatic so that the pressure and volume obey the law,

$$\mathcal{P}\mathcal{V}^\gamma = \text{constant}, \qquad \gamma = c_{\mathcal{P}}/c_v \approx 1.4 \text{ for air, (2.69)}$$

where $c_{\mathcal{P}}$ and c_v are the specific heats at constant pressure and volume, respectively (Hsieh, 1975: 54–5; Lapedes, 1978). Equation (2.18) can be written

$$k = -\frac{\Delta\mathcal{P}}{\Delta\mathcal{V}/\mathcal{V}} = -\frac{\mathcal{V}\mathrm{d}\mathcal{P}}{\mathrm{d}\mathcal{V}},$$

where $\Delta\mathcal{P}$ is the pressure change created by the wave.

By using the adiabatic law, logarithmic differentiation of eq. (2.69) gives $k = \gamma\mathcal{P}$, and hence

$$\alpha = (\gamma\mathcal{P}/\rho)^{1/2}. \qquad (2.70)$$

2.4.4 Boundary conditions

When a wave arrives at a surface separating two media having different elastic properties, it gives rise to reflected and refracted waves as described in §3.1.1. The relationships between the various waves can be found from the relations between the stresses and displacements on the two sides of the interface. At the boundary between two media, the stresses and displacements must be continuous.

Two neighboring points R and S, which lie on opposite sides of the boundary as shown in fig. 2.13 will in general have different values of normal stress. This difference results in a net force that accelerates the layer between them. However, if we choose points closer and closer together, the stress values must approach each other and in the limit when the two points coincide on the boundary, the two stresses must be equal. If this were not so, the infinitesimally thin layer at the boundary would be acted upon by a finite force and hence have an acceleration that would approach infinity as the two points approach each other. Because the same reasoning applies to a tangential stress, we see that the normal and tangential components of stress must be continuous (cannot change abruptly) at the boundary.

The normal and tangential components of displacement must also be continuous. If the normal displacement were not continuous, one medium would either separate from the other, leaving a vacuum in between, or else would penetrate into the other so that the two media would occupy the same space. If the tangential displacement were not continuous, the two media would move differently on opposite sides of the boundary and one would slide over the other. Such relative motion is assumed to be impossible and so displacement must be continuous.

When one or both of the solid media are replaced by a fluid or a vacuum, the boundary conditions are reduced in number (see problem 2.10).

2.4.5 Waves from a spherical source

The potential function $\phi = (1/r)f(t - r/V)$ is a solution to the wave equation when there is spherical symmetry (see eq. (2.51)); hence, the radial displacement $u(r, t)$ is

$$u(r,\,t) = \frac{\partial\phi}{\partial r} = -\left(\frac{1}{r^2}\right)f\left(t - \frac{r}{V}\right)$$
$$+ \left(\frac{1}{r}\right)\frac{\partial}{\partial r}\left[f\left(t - \frac{r}{V}\right)\right] \qquad (2.71)$$

(using eq. (2.62) with the x-axis in the radial direction). For harmonic waves, the two terms have equal importance at a distance $r = \lambda/2\pi$, but the first term decays rapidly in importance at greater distances. The

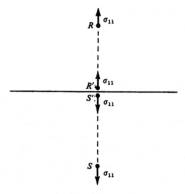

Fig. 2.13   Continuity of normal stress.

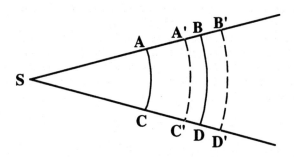

Fig. 2.14   Radial displacement involves shape distortion. When the radial displacement decreases with distance from source S, sector $ABDC$ becomes thinner and approaches a rectangular plate.

second term is the *far-field effect*, whereas the *near-field effect* depends on both terms. This distinction is important when calculating a far-field waveshape from near-field recordings. Note that radial motion involves shape distortion (fig. 2.14) and therefore shear strain.

Equation (2.71) can be used to derive the wave motion created by symmetrical displacement of the medium outward from a point source. When the wave is created by very high pressures, as in an explosion of dynamite, the wave equation is not valid near the source because the medium does not obey Hooke's law there; this difficulty is usually resolved by surrounding the source by a spherical surface of radius r_0 such that the wave equation is valid for $r \geq r_0$, then specifying the displacement or pressure on this surface due to the source.

Let us consider the case where the displacement $u(r, t)$ is to be found, given the displacement $u_0(t)$ of the surface $r = r_0$. We let $\zeta = t - (r - r_0)/V$ and write

$$\phi(r, t) = (1/r)f(\zeta), \qquad \zeta \geq 0, r \geq r_0, \atop = 0, \qquad\qquad \zeta < 0; \quad (2.72)$$

then

$$u(r, t) = \frac{\partial \phi}{\partial r} = -\left[\frac{1}{r^2}f(\zeta) + \frac{1}{rV}\frac{df(\zeta)}{d\zeta}\right]. \quad (2.73)$$

At $r = r_0$, $\zeta = t$ and $u(r, t) = u_0(t)$, where $u_0(t)$ depends on the specific source:

$$u_0(t) = -\left[\frac{1}{r_0^2}f(t) + \frac{1}{r_0 V}\frac{df(t)}{dt}\right]. \quad (2.74)$$

Using these values and multiplying both sides of eq. (2.74) by the integrating factor e^{Vt/r_0}, we get

$$\frac{d}{dt}[e^{Vt/r_0}f(t)] = e^{Vt/r_0}\left[\frac{df(t)}{dt} + \frac{V}{r_0}f(t)\right]$$
$$= -r_0 V u_0(t)e^{Vt/r_0},$$
$$f(t) = -r_0 V e^{-Vt/r_0}\int_0^t u_0(t)e^{Vt/r_0}\,dt. \quad (2.75)$$

Note that the lower limit of the integral means that $t = 0$ is the instant at which the wave first reaches the surface r_0, $u_0(t)$ being zero before this.

To carry the calculation further, we must know $u_0(t)$. Let us approximate an explosion by the expression

$$u_0(t) = ke^{-at}, \qquad t \geq 0, a > 0, \atop = 0 \qquad\qquad t < 0. \quad (2.76)$$

Then

$$f(t) = -r_0 V e^{-Vt/r_0}\int_0^t ke^{(V/r_0 - a)t}\,dt$$
$$= \frac{r_0 Vk}{V/r_0 - a}(e^{-Vt/r_0} - e^{-at}).$$

We replace t in this expression by $\zeta = t - (r - r_0)/V$ and eq. (2.73) becomes

$$u(r, t) = \frac{\partial \phi}{\partial r} = \frac{r_0 k}{r(V/r_0 - a)}\left[\frac{V}{r_0}e^{-V\zeta/r_0} - ae^{-a\zeta}\right.$$
$$\left. - \frac{V}{r}e^{-V\zeta/r_0} + \frac{V}{r}e^{-a\zeta}\right] \quad (2.77)$$
$$\approx \frac{r_0 k}{r(V/r_0 - a)}\left(\frac{V}{r_0}e^{-V\zeta/r_0} - ae^{-a\zeta}\right), \qquad r \gg r_0, \quad (2.78)$$

the latter equation giving the far-field solution.

The fact that eqs. (2.77) and (2.78) are valid only for $\zeta > 0$ (see eq. (2.72)) merely means that $u(r, t)$ is zero until $t = (r - r_0)/V$, that is, until the disturbance reaches the point. At this instant, $\zeta = 0$ and $u(r, t) = k(r_0/r)$; hence the initial displacement is the same as that of the surface r_0 except that it is reduced by the factor r_0/r, that is, $u(r, t)$ falls off inversely as the distance (see §2.7.1 and eq. (2.109). Moreover, $u = 0$ at $t = \infty$ and also when (see eq. (2.77))

$$V(1/r_0 - 1/r)e^{-V\zeta/r_0} + (V/r - a)e^{-a\zeta} = 0,$$

that is, when

$$t = \frac{r - r_0}{V} + \frac{1}{V/r_0 - a}\ln\frac{V(r - r_0)}{r_0 r(a - V/r)}.$$

Provided $V/r_0 > a > V/r$, this equation has a real positive root and $u(r, t)$ will vanish, that is, the displacement must change sign. Because V/r_0 is large in practice and V/r rapidly becomes small, in general the unidirectional pulse in eq. (2.76) gives rise to an oscillatory wave.

By using different expressions for $u_0(t)$ in eq. (2.75) or by specifying $\mathcal{P}_0(t)$, the pressure at the cavity, we can investigate the wave motion for various spherically symmetrical sources (see Blake, 1952; Savarensky, 1975: 243–55). By finding the limit as a in eq. (2.76) goes to zero (see problem 2.12), we get the result for a unit step, step (t); then the results for other inputs can be found using convolution techniques (see §15.4.1).

2.5 Surface waves

2.5.1 General

The wave equations for P- and S-waves in terms of the potential functions of eq. (2.64) are

$$\nabla^2\phi = (1/\alpha^2)\, \partial^2\phi/\partial t^2, \qquad (P\text{-wave}) \qquad (2.79)$$

$$\nabla^2\chi_V = (1/\beta^2)\, \partial^2\chi_V/\partial t^2, \qquad (SV\text{-wave}) \qquad (2.80)$$

$$\nabla^2\chi_H = (1/\beta^2)\, \partial^2\chi_H/\partial t^2, \qquad (SH\text{-wave}) \qquad (2.81)$$

where the S-wave potential has been replaced with the functions χ_V and χ_H corresponding to SV- and SH-components. If we consider plane waves traveling in the direction of the x-axis in an infinite homogeneous medium, solutions of these equations are of the form $e^{j\kappa(x - Vt)}$, $V = \alpha$ or β. However, other solutions are possible when the infinite medium is divided into different media. When the xy-plane separates two media, solutions of the form $e^{\pm\kappa z}e^{j\kappa(x - Vt)}$ exist under certain conditions. These solutions correspond to plane waves traveling parallel to the x-axis with velocity V and amplitude decreasing exponentially with distance from the xy-plane (in a semiinfinite medium; see §2.5.2 to 2.5.4). Such waves are called *surface waves* because they are "tied" to the surface and diminish as they get farther from the surface.

2.5.2 Rayleigh waves

The most important surface wave in exploration seismology is the *Rayleigh wave*, which is propagated along a free surface of a solid. Although a "free" surface means contact with a vacuum, the elastic constants and density of air are so low in comparison with values for rocks that the surface of the earth is approximately a free surface. *Ground roll* is the term commonly used for Rayleigh waves.

We take the free surface as the xy-plane with the z-axis positive downward. The boundary conditions (§2.4.4) require that $\sigma_{zz} = 0 = \sigma_{xz}$ at $z = 0$ (see problem 2.10), that is, two conditions must be satisfied, and so we require two parameters that can be adjusted. Therefore, we assume that both P- and SV-components exist (SH-motion is parallel to the xy-plane and so is not involved in the boundary conditions) and adjust their amplitudes to satisfy the boundary conditions.

Appropriate potentials are

$$\phi = Ae^{-m\kappa z}e^{j\kappa(x - V_Rt)}, \qquad \chi_V = Be^{-n\kappa z}e^{j\kappa(x - V_Rt)}, \qquad (2.82)$$

where m and n must be real positive constants so that the wave decreases in amplitude away from the surface; V_R is, of course, the velocity of the Rayleigh wave. Substituting ϕ and χ_V in eqs. (2.79) and (2.80) gives

$$m^2 = (1 - V_R^2/\alpha^2), \qquad n^2 = (1 - V_R^2/\beta^2). \qquad (2.83)$$

Because m and n are real, $V_R < \beta < \alpha$, so that the velocity of the Rayleigh wave is less than that of the S-wave.

We next apply the boundary conditions. Using the results of problem 2.11, we get for $z = 0$

$$\left.\begin{array}{l} \sigma_{zz} = \lambda\,\nabla^2\phi + 2\mu\left(\dfrac{\partial^2\phi}{\partial z^2} - \dfrac{\partial^2\chi_V}{\partial x\,\partial z}\right) = 0, \\[2mm] \sigma_{xz} = \mu\left(2\dfrac{\partial^2\phi}{\partial x\,\partial z} + \dfrac{\partial^2\chi_V}{\partial z^2} - \dfrac{\partial^2\chi_V}{\partial x^2}\right) = 0. \end{array}\right\} (2.84)$$

Substituting eq. (2.82) into the foregoing and setting $z = 0$ gives

$$[(\lambda + 2\mu)m^2 - \lambda]A + 2jn\mu B = 0$$

and

$$-2jmA + (n^2 + 1)B = 0.$$

We can use eqs. (2.58), (2.59), and (2.83) to write the first result in the form

$$(2\beta^2 - V_R^2)A + 2jn\beta^2B = 0.$$

Eliminating the ratio B/A from the two equations gives

$$(2 - V_R^2/\beta^2)(n^2 + 1) = 4mn;$$

hence,

$$V_R^6 - 8\beta^2V_R^4 + (24 - 16\beta^2/\alpha^2)\beta^4V_R^2 + 16(\beta^2/\alpha^2 - 1)\beta^6 = 0. \qquad (2.85)$$

Because the left side of eq. (2.85) is negative for $V_R = 0$ and positive for $V_R = +\beta$, a real root must exist between these two values, this root giving the Rayleigh wave velocity V_R. However, we cannot find this root without knowing β/α.

For many rocks, $\sigma \approx 1/4$, that is, $(\beta/\alpha)^2 \approx 1/3$ from eq. (2.60). If we use this value, the three roots of eq. (2.85) are $V_R^2 = 4\beta^2, 2(1 \pm 1/\sqrt{3})\beta^2$. Because V_R/β must be less than unity, the only permissible solution is

$$V_R^2 = 2(1 - 1/\sqrt{3})\beta^2, \qquad \text{or} \qquad V_R = 0.919\beta.$$

We now find that $V_R/\alpha = 0.531$, $m = 0.848$, $n = 0.393$, and $B/A = +1.468j$; hence,

$$\phi = Ae^{-0.848\kappa z}\,e^{j\kappa(x - V_R t)},$$
$$\chi_V = 1.468jAe^{-0.393\kappa z}\,e^{j\kappa(x - V_R t)}.$$

Using eq. (2.64), we get for the displacements at the surface

$$u = 0.423j\kappa Ae^{j\kappa(x - V_R t)}, \qquad w = 0.620\kappa Ae^{j\kappa(x - V_R t)}.$$

Taking the real part of the solution (which corresponds to a displacement at the source of $\cos \omega t$ (see eq. (15.45)), we obtain finally

$$u = -0.423\kappa A \sin \kappa(x - V_R t),$$
$$w = 0.620\kappa A \cos \kappa(x - V_R t). \qquad (2.86)$$

At a given point on the surface, a particle describes an ellipse in the vertical xz-plane, as shown in fig. 2.15a, the horizontal axis being about two-thirds the vertical axis. The angle θ is given by

$$\tan \theta = -w/u = 1.465 \cot \kappa(x - V_R t). \quad (2.87)$$

As t increases, $\cot \kappa(x - V_R t)$ and θ increase, that is, P moves around the ellipse in a counterclockwise (retrograde) direction for a wave moving from left to right.

Rayleigh-wave velocity as a function of Poisson's ratio is shown in fig. 2.16. Because V_R as given by eq. (2.85) is independent of frequency, Rayleigh waves on the surface of a homogeneous medium do not exhibit dispersion (see §2.7.4). Field observations (figs. 2.15b and 2.15d) agree roughly with the type of motion shown in fig. 2.15a, differences being attributed to the Earth being layered and anisotropic rather than an ideal, homogeneous, isotropic medium. Measurements also show that Rayleigh waves are dispersive (Dobrin, 1951). Rayleigh waves are low-velocity, low-frequency waves with a spectrum that is not sharply peaked, and hence involve a broad range of wavelengths. Because $m\kappa$ and $n\kappa$ determine the penetration (penetration showing the exponential falloff predicted by eq. (2.82) is illustrated in fig. 2.15c), there is a large variation of penetration for different frequency components, most of the energy being confined to a zone one or two wavelengths thick. Because the elastic constants vary considerably near the surface, especially at the base of the LVL (see §5.3.2), the velocity varies with wavelength, the waves are dispersive, and the shape of the wavetrain changes with distance.

2.5.3 Stoneley waves

If an infinite medium is divided by the xy-plane into two different semiinfinite media, four boundary conditions must be satisfied, so we need four parameters to adjust. We take the potential functions in the form:

Medium (1):

$$\phi_1 = A_1 e^{m_1 \kappa z} e^{j\kappa(x - V_S t)}, \qquad \chi_1 = B_1 e^{n_1 \kappa z} e^{j\kappa(x - V_S t)};$$

Medium (2):

$$\phi_2 = A_2 e^{m_2 \kappa z} e^{j\kappa(x - V_S t)}, \qquad \chi_2 = B_2 e^{n_2 \kappa z} e^{j\kappa(x - V_S t)},$$

where m_i, n_i are real, positive constants, and V_S is the velocity. Substituting ϕ_i, χ_i in eqs. (2.79) and (2.80), we find that

$$m_i^2 = 1 - (V_S/\alpha_i)^2, \qquad n_i^2 = 1 - (V_S/\beta_i)^2 \qquad (i = 1, 2) \qquad (2.88)$$

Because m_i and n_i are real, V_S must be less than the smaller of β_1 and β_2.

The boundary conditions required that $w_1 = w_2$, $u_1 = u_2$, $\sigma_{zz}|_1 = \sigma_{zz}|_2$, $\sigma_{xz}|_1 = \sigma_{xz}|_2$ at $z = 0$. The results of problem 2.11 show that these conditions lead to equations involving first and second derivatives of the potentials, z being set equal to zero after the differentiation; consequently, all terms will have the factors $e^{j\kappa(x - V_S t)}$ and either κ or κ^2, and these factors will cancel out, so we can ignore them; at the same time, the exponential term in z will become unity. Moreover, differentiation with respect to x and z is equivalent to multiplying the potentials by $j\kappa$ and $\pm m_i\kappa$ or $\pm n_i\kappa$, respectively; by eliminating κ, differentiation becomes equivalent to multiplying by j and $\pm m_i$, $\pm n_i$. The four boundary conditions now give

$$m_1 A_1 - jB_1 = -m_2 A_2 - jB_2,$$
$$jA_1 + n_1 B_1 = jA_2 - n_2 B_2,$$
$$\lambda_1(-1 + m_1^2)A_1 + 2\mu_1(m_1^2 A_1 - jn_1 B_1)$$
$$= \lambda_2(-1 + m_2^2)A_2 + 2\mu_2(m_2^2 A_2 + jn_2 B_2),$$
$$\mu_1[2jm_1 A_1 + (n_1^2 + 1)B_1]$$
$$= \mu_2[-2jm_2 A_2 + (n_2^2 + 1)B_2].$$

If we transfer all terms to the left-hand side, we have a set of four homogeneous equations (§15.1.1); these have a nontrivial solution only if the determinant of the coefficients vanishes. Setting the determinant equal to zero and using eqs. (2.58), (2.59), and (2.88), we get the following equation for V_S:

$$V_S^4[(\rho_2 - \rho_1)^2 - (\rho_1 m_2 + \rho_2 m_1)(\rho_1 n_2 + \rho_2 n_1)]$$
$$+ 4V_S^2(\mu_1 - \mu_2)[\rho_2(1 - m_1 n_1) - \rho_1(1 - m_2 n_2)]$$
$$+ 4(\mu_1 - \mu_2)^2(1 + m_1 n_1)(1 + m_2 n_2) = 0. \qquad (2.89)$$

Equation (2.89) was first given by Stoneley (1924). Scholte (1947) studied the properties of the equation and found that a solution always exists when one of the media is a fluid but, when both media are solids, a solution exists only when $\beta_1 \approx \beta_2$ and the ratios ρ_1/ρ_2 and μ_1/μ_2 fall within the narrow limits shown in fig. 2.17. These waves are a type of generalized Rayleigh waves and are usually called *Stoneley waves*.

Stoneley waves are often present in borehole seismic surveys (§2.5.5). Stoneley-wave frequencies are far below those used in sonic logging (§5.4.3), but they fall within the frequency range of VSPs (§13.4) and Stoneley waves constitute an important source of coherent noise in VSP surveys.

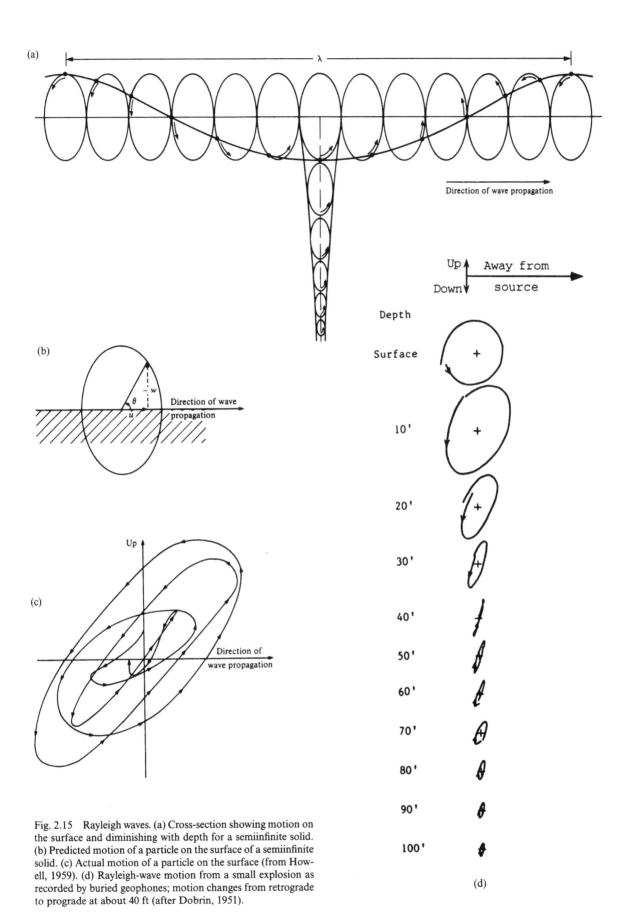

(a)

λ

Direction of wave propagation

Up ↑   Away from
Down ↓   source

Depth

Surface

10'

20'

30'

40'

50'

60'

70'

80'

90'

100'

(b)

θ
w
u
Direction of wave
propagation

(c)

Up

Direction of
wave propagation

Fig. 2.15 Rayleigh waves. (a) Cross-section showing motion on the surface and diminishing with depth for a semiinfinite solid. (b) Predicted motion of a particle on the surface of a semiinfinite solid. (c) Actual motion of a particle on the surface (from Howell, 1959). (d) Rayleigh-wave motion from a small explosion as recorded by buried geophones; motion changes from retrograde to prograde at about 40 ft (after Dobrin, 1951).

(d)

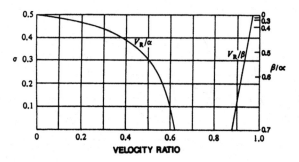

Fig. 2.16   Rayleigh-wave velocity, V_R, as a function of Poisson's ratio, σ.

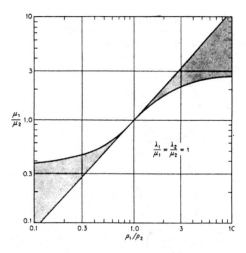

Fig. 2.17   Conditions for the existence of a Stoneley wave; solutions exist within the shaded area. (After Scholte, 1947.)

2.5.4 Love waves

Love waves (Love, 1911) are surface waves consisting of *SH*-motion parallel to an interface. They exist only when a semiinfinite medium is overlain by an upper layer of finite thickness terminating at a free surface. We take the lower interface as the *xy*-plane and the free surface as the parallel plane $z = -h$; the *SH*-motion is in the *y*-direction. The only component of displacement is v, so $\boldsymbol{\zeta} = v\mathbf{j}$ in eq. (2.11), hence, v satisfies the wave equation and we can dispense with the potential χ_H of eq. (2.81) and use v instead.

The boundary conditions (see problem 2.10) require that $\sigma_{yz} = 0$ at $z = -h$ and that σ_{yz} and v be continuous at $z = 0$; thus, we need three parameters to adjust and so we write the following:

Medium (1):

$$v_1 = (Ae^{\eta_1\kappa z} + Be^{-\eta_1\kappa z})e^{j\kappa(x - V_L t)}, \qquad -h \leq z \leq 0.$$

Medium (2):

$$v_2 = Ce^{-\eta_2\kappa z}e^{j\kappa(x - V_L t)}, \qquad 0 \leq z \leq +\infty.$$

Substituting v_1 and v_2 in the wave equations gives

$$\left.\begin{array}{l} \eta_1^2 = 1 - (V_L/\beta_1)^2, \\ \eta_2^2 = 1 - (V_L/\beta_2)^2. \end{array}\right\} \qquad (2.90)$$

We must have η_2 real and positive so that $e^{-\eta_2\kappa z} \to 0$ as $z \to +\infty$; therefore, $V_L < \beta_2$. However, η_1 is unrestricted because z is finite in the upper layer.

Applying the boundary conditions, we have at $z = -h$, $\sigma_{yz} = 0 = \mu_1\varepsilon_{yz}|_1$, so $\partial v_1/\partial z = 0$ from eq. (2.9), that is,

$$Ae^{-\eta_1\kappa h} - Be^{\eta_1\kappa h} = 0.$$

At $z = 0$, $\sigma_{yz}|_1 = \sigma_{yz}|_2$, $v_1 = v_2$, so

$$\mu_1\eta_1(A - B) = -\mu_2\eta_2 C,$$

and

$$A + B = C.$$

By setting $a = e^{-2\eta_1\kappa h}$, $b = \mu_2\eta_2/\mu_1\eta_1$, these equations become

$$\left.\begin{array}{l} aA - B = 0, \\ A - B + bC = 0, \\ A + B - C = 0. \end{array}\right\} \qquad (2.91)$$

For these homogeneous equations to have a nontrivial solution, the determinant of the coefficients must vanish (see §15.1.1), that is,

$$\begin{vmatrix} a & -1 & 0 \\ 1 & -1 & b \\ 1 & +1 & -1 \end{vmatrix} = 0,$$

or

$$\begin{aligned} -b &= (1 - a)/(1 + a) = (1 - e^{-2\eta_1\kappa h})/(1 + e^{-2\eta_1\kappa h}) \\ &= (e^{\eta_1\kappa h} - e^{-\eta_1\kappa h})/(e^{\eta_1\kappa h} + e^{-\eta_1\kappa h}) \\ &= \tanh \eta_1\kappa h = -(\mu_2\eta_2/\mu_1\eta_1). \end{aligned}$$

But $\tanh x$ is positive for all real values of x, also η_2 is real and positive; therefore, η_1 must be imaginary, that is, $\eta_1 = j\zeta$, where ζ is real. Because $\tanh jx = j \tan x$, we now get

$$\mu_2\eta_2 = \mu_1\zeta \tan \zeta\kappa h. \qquad (2.92)$$

From eq. (2.90), we have

$$V_L^2/\beta_1^2 = 1 - \eta_1^2 = 1 + \zeta^2,$$

so that $V_L > \beta_1$. Thus,

$$\beta_2 > V_L > \beta_1,$$

and the *S*-wave velocity must be higher in the deeper layer than in the surface layer, V_L then being in between the two velocities.

Because $\kappa = 2\pi/\lambda = \omega/V_L$, as the frequency increases from zero, $\tan \kappa\zeta h$ increases and approaches infinity; thus for eq. (2.92) to hold, as the frequency increases, ζ must approach zero and V_L must approach β_1. Conversely, as κ approaches zero, ζ approaches its maximum value and V_L approaches β_2. Hence, at high frequencies, the Love-wave velocity approaches the velocity of *S*-waves in the surface layer, and as the frequency approaches zero, the Love-wave velocity approaches the *S*-wave velocity in the lower layer (Dobrin, 1951).

The expression for v_1 can be written

$$v_1 = (Ae^{j\kappa\zeta z} + Be^{-j\kappa\zeta z})e^{j\kappa(x - V_L t)}$$
$$= A(e^{j\kappa\zeta z} + ae^{-j\kappa\zeta z})e^{j\kappa(x - V_L t)}$$

on using eq. (2.91). Therefore,

$$v_1 = A(e^{j\kappa\zeta z} + e^{-2j\kappa\zeta h}e^{-j\kappa\zeta z})e^{j\kappa(x - V_L t)}$$
$$= A[e^{j\kappa\zeta(z+h)} + e^{-j\kappa\zeta(z+h)}]e^{j\kappa(x - \zeta h - V_L t)}$$
$$= 2A[\cos \kappa\zeta(z + h)]e^{j\kappa(x - \zeta h - V_L t)} \qquad (2.93)$$

on taking the real part of the amplitude. We see that v_1 vanishes on planes where

$$\kappa\zeta(z + h) = (r + 1/2)\pi \qquad (r \text{ integral}) \quad (2.94)$$

(recall that h is positive and z is negative in the upper layer); these planes are called *nodal planes* (see §2.1.1). Nodal planes are characteristic of normal-mode propagation (§13.3) and indeed Love waves can be explained in terms of normal-mode propagation (Grant and West, 1965: 81–5).

2.5.5 Tube waves

Waves traveling in a fluid-filled borehole or on the walls of a borehole in the direction of the axis (*tube waves*) are of considerable interest in velocity surveys in wells (§5.4.2), in vertical seismic profiling (§13.4), and in sonic logging (§5.4.3). Because they have mainly only 1 degree of freedom (along the axis), their amplitude decreases slowly with distance. Sometimes several modes of tube waves are present and often the mechanisms of their generation and the nature of their motion are not clear. Tube waves have the potential of furnishing information about the elastic properties and permeability of the surrounding formations.

Most tube-wave energy travels axially, but radial motion is also involved in some modes. A pressure geophone or one hanging freely in the borehole will sense the maximum tube-wave effects in the borehole fluid, whereas a geophone clamped to the borehole wall will sense much smaller motion.

The classical tube wave is merely a *P*-wave propagating in the fluid, the borehole wall expanding and contracting as the pressure wave passes. We assume a homogeneous fluid in a cylindrical borehole penetrating a homogeneous isotropic medium (fig. 2.18). Using $\mathcal{P}$ for the pressure and w for the displacement, Newton's second law, net force = mass × acceleration, applied to a volume element of the fluid, $\mathcal{V} = \pi r^2 \Delta z$, is

$$\left(\frac{\partial \mathcal{P}}{\partial z} \Delta z\right) \pi r^2 = -(\rho \pi r^2 \Delta z) \frac{\partial^2 w}{\partial t^2},$$

or

$$\frac{\partial \mathcal{P}}{\partial z} = -\rho \frac{\partial^2 w}{\partial t^2}. \qquad (2.95)$$

From eq. (2.18),

$$\mathcal{P} = -k\Delta = -k\Delta\mathcal{V}/\mathcal{V}.$$

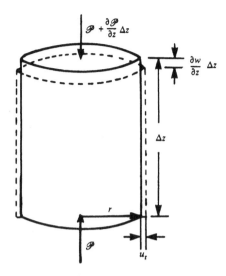

Fig. 2.18   Changes involved in passage of a tube wave.

The change in volume $\Delta\mathcal{V}$ is due to expansion both along the axis and radially, that is,

$$\Delta\mathcal{V} = \pi r^2 \frac{\partial w}{\partial z} \Delta z + (2\pi r u_r)\Delta z,$$

where u_r is the change in the radius of the hole. Thus, we get

$$\mathcal{P} = -k\left(\frac{\partial w}{\partial z} + \frac{2u_r}{r}\right). \qquad (2.96)$$

Lamb (1960: §157) derived the following relation between u_r and $\mathcal{P}$ for an annulus of inner and outer radii r and R, where E, σ, and μ are respectively Young's modulus, Poisson's ratio, and the shear modulus for the annulus material:

$$\frac{u_r}{r} = \frac{\mathcal{P}}{E} \frac{(1 + \sigma)(R^2 + r^2) - 2\sigma r^2}{R^2 - r^2}.$$

If we let $R \to \infty$, we obtain for a cylindrical hole in an infinite medium

$$u_r/r = \mathcal{P}(1 + \sigma)/E = \mathcal{P}/2\mu$$

(using problem 2.2). Substitution in eq. (2.96) gives

$$\mathcal{P}\left(\frac{1}{k} + \frac{1}{\mu}\right) = -\frac{\partial w}{\partial z},$$

and substitution of this result in eq. (2.95) gives the wave equation:

$$\frac{\partial^2 w}{\partial z^2} = \left(\frac{1}{V_T^2}\right)\frac{\partial^2 w}{\partial t^2}, \qquad V_T^2 = \frac{1}{\rho}\left(\frac{1}{k} + \frac{1}{\mu}\right)^{-1}. \quad (2.97)$$

White (1965: 153–6; 1983: 139–91) discusses tube waves in greater detail.

Cheng and Toksoz (1981) discuss two other tube-wave modes. One is a Stoneley wave (§2.5.3) propagating along the borehole wall and dying away exponentially in the formation surrounding the borehole; this is the dominant tube-wave mode in VSP work. The other tube-wave mode is *pseudo-Rayleigh waves,* guided waves (§13.3) confined largely to the fluid, also dying away exponentially in the surrounding formation. Both waves are dispersive (§2.7.4).

Cheng and Toksoz calculated dispersion curves for both modes (fig. 2.19a). The Stoneley wave is slightly dispersive with both group and phase velocities close to $0.9\alpha_f$, where α_f is the *P*-wave velocity in the borehole fluid (see eq. (2.61)). Pseudo-Rayleigh waves cannot exist below a minimum frequency (where their velocity equals the *S*-wave velocity of the surrounding rock, β_r) and their group velocity passes through a minimum, which results in an Airy phase (see §13.3 and fig. 13.19). Several modes may exist (see eq. (13.1)). Pseudo-Rayleigh waves are not a factor in ordinary seismic work (fig. 2.19a shows a 10-kHz low-frequency cutoff), but they are involved in sonic logging. At higher frequencies, the velocities of both Stoneley and pseudo-Rayleigh waves approach the *S*-wave velocity in the medium surrounding the borehole.

Cheng and Toksoz calculated "synthetic microseismograms" for various circumstances, one of which is shown in fig. 2.19b; fig. 2.19c shows an observed waveform.

Hardage (1985) discusses the role of tube waves in VSP surveys. Figure 2.20a shows prograde elliptical motion in an axial plane. The radial motion is zero at the center of the hole and maximum at the borehole wall, where it is continuous (fig. 2.20b), but it decays rapidly in the surrounding formation. The axial component of motion is relatively constant in the fluid but is discontinuous at the borehole wall where its amplitude decreases by a factor as large as several hundred. This explains why geophones should be clamped to the borehole wall.

Tube waves are reflected at impedance changes, just as other acoustic waves are (§3.2). When the borehole cross-sectional area changes from a_1 to a_2, the reflection (R) and transmission (T) coefficients are (Hardage, 1985: 86–7)

$$R = \frac{a_2 - a_1}{a_2 + a_1}, \qquad T = \frac{2a_1}{a_2 + a_1} \qquad (2.98)$$

(compare with eqs. (3.14) and (3.15)). At the top of the borehole fluid and the bottom of the hole, $R = -1$ and $+1$, respectively. Tube waves are also reflected at a geophone sonde and where casing changes. Figure 2.21 shows several reflected tube waves.

Tube waves can be generated by almost anything that disturbs the borehole fluid. The most common source is a Rayleigh wave passing over the top of the borehole; thus, tube waves are uncommon in marine VSP surveys and, in land surveys, lowering the borehole fluid level often lessens tube-wave generation. Tube waves initially have the same spectrum as the generating source and their spectrum changes slowly because there is little absorption in the borehole fluid.

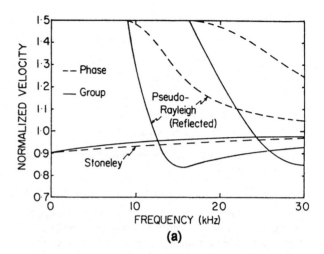

(a)

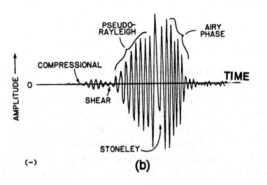

(b)

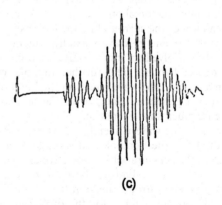

(c)

Fig. 2.19 Stoneley and pseudo-Rayleigh waves in a fluid-filled borehole. (From Cheng and Toksoz, 1981: 1045.) (a) Dispersion curves for the Stoneley wave and the first two pseudo-Rayleigh modes are shown for $\beta_r/\alpha_f = 1.5$, where β_r is the *S*-wave velocity in the surrounding rock, and α_f is the *P*-wave velocity in the borehole fluid (velocities normalized by dividing by the fluid velocity). (b) Calculated signature for a broadband source showing the pseudo-Rayleigh wave, Stoneley wave, and Airy phase; $\alpha_r = 5.94$ km/s, $\beta_r = 3.05$ km/s, $\rho_r = 2.30$ g/cm³, $\alpha_f = 1.83$ km/s, $\rho_f = 1.20$ g/cm³, hole radius = 10.2 cm (after Hardage, 1985: 75). (c) Observed tube-wave signature.

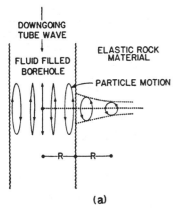

(a)

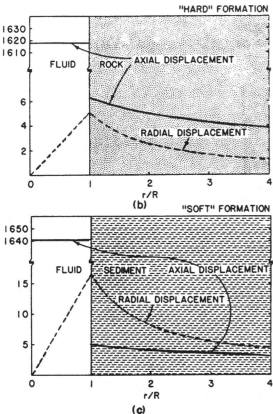

(b)

(c)

Fig. 2.20 Wave motion for a tube wave. (From Hardage, 1985: 78, 79.) (a) Prograde elliptical motion of fluid particles during passage of a tube wave (ellipticity is greater than shown here). (b) Axial and radial displacements for hard formation, $\nu = 82$ Hz; and (c) for soft formation, $\nu = 74$ Hz.

Consequently, tube waves often have appreciable energy in the signal range even after considerable travel.

2.6 Anisotropic media

2.6.1 Types of anisotropy

Anisotropy is a general term denoting variation of a physical property depending on the direction in which it is measured. Seismic anisotropy is evidenced by a

variation of seismic velocity with the direction in which it is measured or with wave polarization (§13.1.6). The general elasticity matrix relating stress σ_{ij} to strain ε_{kl} (the generalized form of the 6×6 matrix in eq. (2.15)) can contain at most 21 independent constants because of symmetry considerations, but Winterstein (1990: 1084–5) says that only 18 of these can be truly independent. The number of independent constants depends on the symmetry of the system (Love, 1944: 99).

A number of different types of symmetry (*symmetry systems*) can exist. Classically, eight systems are defined (Love, loc. cit.; Landau and Lifshitz, 1986; Saada, 1974), but some writers define subsystems as well; for example, Winterstein (1990: 1083–5) lists 11 systems plus subsystems in discussing cracks. Anisotropy types are associated with the symmetry systems. At seismic wavelengths, however, the only anisotropy types reported are transverse isotropy (hexagonal symmetry), orthorhombic anisotropy, and monoclinic anisotropy.

Transverse isotropy involves elastic properties that are the same in any direction perpendicular to an axis but are different parallel to this axis. Two important types of transverse isotropy are observed: that with a nearly vertical symmetry axis (*thin-layer anisotropy*) and that with a nearly horizontal axis (*azimuthal anisotropy*) (Bush and Crampin, 1987). Transverse isotropy is the most important type of anisotropy encountered; it is discussed further in §2.6.2.

Orthorhombic anisotropy is equivalent to a superposition of thin-layer anisotropy and azimuthal anisotropy. It arises because a vertical fracture system has

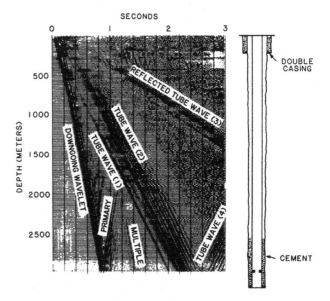

Fig. 2.21 VSP record showing several tube waves. Tube wave (1) is generated at the base of the surface casing; (2) is generated at the surface by a Rayleigh wave; (3) is a reverberation of wave (2) between the well sonde and the surface; and (4) is a reflection from the bottom of the borehole. (From Hardage, 1985: 88.)

been superimposed on a horizontally layered system. VSP data from the Paris Basin have been interpreted using an orthorhombic model (Bush and Crampin, 1987; MacBeth, 1990). Layering anisotropy is usually much stronger than fracture anisotropy so that the overall effect may be difficult to distinguish from thin-layer anisotropy. *Monoclinic anisotropy* can be produced by superimposing tilted fractures on a layered medium (Schoenberg and Muir, 1989). Examples of monoclinic anisotropy have been observed in the field (Crampin, McGonigle, and Bamford, 1980; Winterstein and Meadows, 1990).

The stress–strain relationships require 5 independent elastic moduli for transverse isotropy, 9 for orthorhombic anisotropy, and 13 for monoclinic anisotropy, compared with only 2 for the isotropic case.

2.6.2 Transverse isotropy

Taking the z-axis as the axis of symmetry, Love (1944: 160–1) showed that for transverse isotropy, Hooke's law reduces to the following:

$$\left. \begin{array}{l} \sigma_{xx} = (\lambda_\parallel + 2\mu_\parallel)\varepsilon_{xx} + \lambda_\parallel\varepsilon_{yy} + \lambda_\perp\varepsilon_{zz}, \\[4pt] \sigma_{yy} = \lambda_\parallel\varepsilon_{xx} + (\lambda_\parallel + 2\mu_\parallel)\varepsilon_{yy} + \lambda_\perp\varepsilon_{zz}, \\[4pt] \sigma_{zz} = \lambda_\perp\varepsilon_{xx} + \lambda_\perp\varepsilon_{yy} + (\lambda_\perp + 2\mu_\perp)\varepsilon_{zz}, \end{array} \right\} \quad (2.99)$$

$$\left. \begin{array}{l} \sigma_{xy} = \mu_\parallel\varepsilon_{xy}, \\[4pt] \sigma_{yz} = \mu^*\varepsilon_{yz}, \\[4pt] \sigma_{zx} = \mu^*\varepsilon_{zx}, \end{array} \right\} \quad (2.100)$$

where the five independent constants are $\lambda_\parallel$ and $\mu_\parallel$, $\lambda_\perp$ and $\mu_\perp$, and μ^*.

Layering and parallel fracturing tend to produce transverse isotropy. A sequence of isotropic layers (such as sedimentary bedding) produces thin-layer anisotropy for wavelengths appreciably larger than the layer thicknesses ($\lambda > 8d$, where d is layer thickness; see Ebrom et al., 1990). The symmetry axis is perpendicular to the bedding with the velocities of P- and S-waves that involve motion parallel to the bedding larger than those involving motion perpendicular to the bedding. The velocity parallel to the bedding is greater because the higher-velocity members carry the first energy, whereas for wave motion perpendicular to the bedding, each member contributes in proportion to the time taken to traverse it.

Nonhorizontal fracturing and microcracks produce azimuthal anisotropy with a symmetry axis perpendicular to the fracturing (fractures often are somewhat parallel and vertical). The velocity of waves that involve motion parallel to the fracturing (S_1) is larger than that of waves with motion perpendicular to the fracturing (S_2). If the motion is neither parallel nor perpendicular to the fracturing, an S-wave splits into two waves with orthogonal polarizations (fig. 2.22): one ($S_\parallel$) traveling at the S_1 velocity, the other ($S_\perp$) at the S_2 velocity; this is called *shear-wave splitting* or *bi-*

refringence (Crampin, 1981). For horizontal thin-layer anisotropy, the two waves are the qSP-waves (that is, quasi-S-waves having displacement parallel to the symmetry axis) and SR-waves (displacement in radial directions). For azimuthal asymmetry, they are sometimes called qSV- and SH-waves.

In anisotropic media, pure S- and P-waves may exist only in certain directions. In transversely isotropic media, SV- and P-modes of propagation are coupled (see §2.6.3). Wavefronts are not in general orthogonal to the directions of wave propagation. *Phase velocity* is velocity perpendicular to a surface of constant phase (a wavefront), and *group velocity,* the velocity with which the energy travels (§2.7.4), is in a different direction (see fig. 2.23). The surfaces for SV-wavefronts may have cusps.

Anisotropy is often described by the fractional difference between the maximum and minimum velocities for a given wave surface, i.e., $(V_{max} - V_{min})/V_{max}$, sometimes by the ratio of maximum and minimum velocities, V_{max}/V_{min}.

Uhrig and van Melle (1955) give a table showing anisotropy values of 1.2 to 1.4 for rocks at the surface and 1.1 to 1.2 for sediments at depths of 2.1 to 2.4 km in west and central Texas. Stoep (1966) found average values between 1.00 and 1.03 for Texas Gulf Coast sediments. Ségonzac and Laherrére (1959) obtained values from 1.00 for sandstones to 1.08 to 1.12 for limestones and 1.15 to 1.20 for anhydrites from the northern Sahara.

2.6.3 Wave equation for transversely isotropic media

When media are not isotropic, the mathematics become more complex the more anisotropic the medium. However, the case of a transversely isotropic medium can be treated without great difficulty. We consider waves in the xz-plane, where the symmetry axis is along the z-axis. Derivatives with respect to y are zero, but S-waves may involve motion in the y-direction. We substitute eqs. (2.99) and (2.100) into eq. (2.24), and using eqs. (2.8) and (2.7), we get the wave equations for transversely isotropic media:

$$\rho\frac{\partial^2 u}{\partial t^2} = \frac{\partial\sigma_{xx}}{\partial x} + \frac{\partial\sigma_{xz}}{\partial z}$$

$$= \frac{\partial}{\partial x}\left[(\lambda_\parallel + 2\mu_\parallel)\frac{\partial u}{\partial x} + \lambda_\perp\frac{\partial w}{\partial z}\right] + \frac{\partial}{\partial z}\left[\mu^*\left(\frac{\partial u}{\partial z} + \frac{\partial w}{\partial x}\right)\right]$$

$$= (\lambda_\parallel + 2\mu_\parallel)\frac{\partial^2 u}{\partial x^2} + \mu^*\frac{\partial^2 u}{\partial z^2} + (\lambda_\perp + \mu^*)\frac{\partial^2 w}{\partial z\,\partial x},$$

$$(2.101)$$

$$\rho\frac{\partial^2 v}{\partial t^2} = \mu_\parallel\frac{\partial^2 v}{\partial x^2} + \mu^*\frac{\partial^2 v}{\partial z^2}, \quad (2.102)$$

$$\rho\frac{\partial^2 w}{\partial t^2} = (\lambda_\perp + \mu^*)\frac{\partial^2 u}{\partial x\,\partial z} + \mu^*\frac{\partial^2 w}{\partial x^2} +$$

$$(\lambda_\perp + 2\mu_\perp)\frac{\partial^2 w}{\partial z^2}. \quad (2.103)$$

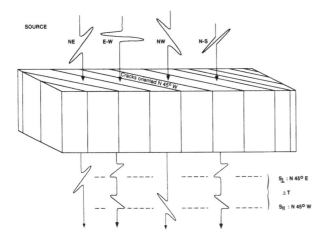

Fig. 2.22 *S*-wave propagation in a fractured medium with cracks oriented N45°W. For an *S*-wave traveling parallel to fracturing, the velocity (S_2) is slower for a component involving motion perpendicular to the fracturing than for one involving motion parallel to the fracturing (velocity S_1).

Note the dependence in eqs. (2.101) and (2.103) on derivatives of both u and w; *P*- and *SV*-waves are said to be *coupled*. The *SH*-wave governed by eq. (2.102) is, however, independent of the other two.

We simplify the problem by assuming a plane wave traveling in the *xz*-plane in the direction of increasing x and decreasing z, the angle between the raypath and the x-axis being θ. We now use the potential functions of eq. (2.64) in the form

$$\phi = Ae^{j\omega\zeta}, \qquad \chi = Be^{j\omega\zeta},$$

where

$$\zeta = (\ell x - nz)/V - t, \qquad \ell = \cos\theta, \qquad n = \sin\theta.$$

Then

$$u = \frac{\partial\phi}{\partial x} + \frac{\partial\chi}{\partial z} = \left(\frac{j\omega}{V}\right)(\ell A - nB)e^{j\omega\zeta},$$

$$w = \frac{\partial\phi}{\partial z} - \frac{\partial\chi}{\partial x} = -\left(\frac{j\omega}{V}\right)(nA + \ell B)e^{j\omega\zeta}.$$

When we substitute these into eqs. (2.101) and (2.103), the following factors appear in every term and hence can be ignored: $j\omega/V$, $(j\omega)^2$, and $e^{j\omega\zeta}$. Equations (2.101) and (2.103) become

$$[\rho V^2\ell - (\lambda_\| + 2\mu_\|)\ell^3 - (\lambda_\perp + 2\mu^*)\ell n^2]A$$
$$- [\rho V^2 n - (\lambda_\| + 2\mu_\| - \lambda_\perp - \mu^*)\ell^2 n - \mu^* n^3]B = 0,$$

$$[\rho V^2 n - (\lambda_\perp + 2\mu_\perp)n^3 - (\lambda_\perp + 2\mu^*)\ell^2 n]A$$
$$+ [\rho V^2\ell + (\mu^* - 2\mu_\perp)\ell n^2 - \mu^*\ell^3]B = 0.$$

Writing

$$\alpha_\|^2 = (\lambda_\perp + 2\mu_\|)/\rho, \qquad \alpha_\perp^2 = (\lambda_\perp + 2\mu_\perp)/\rho,$$

$$\alpha^{*2} = (\lambda_\perp + 2\mu^*)/\rho, \qquad \beta^{*2} = \mu^*/\rho,$$

these become

$$(V^2 - \alpha_\|^2\ell^2 - \alpha^{*2}n^2)\ell A$$
$$- [V^2 - (\alpha_\|^2 - \alpha^{*2} + \beta^{*2})\ell^2 - \beta^{*2}n^2]nB = 0,$$

$$(V^2 - \alpha_\perp^2 n^2 - \alpha^{*2}\ell^2)nA$$
$$+ [V^2 - (\alpha_\perp^2 - \alpha^{*2} + \beta^{*2})n^2 - \beta^{*2}\ell^2]\ell B = 0.$$

Eliminating A and B gives the following quadratic equation in V^2:

$$\frac{[V^2 - (\alpha_\|^2 - \alpha^{*2} + \beta^{*2})\ell^2 - \beta^{*2}n^2]n}{(V^2 - \alpha_\|^2\ell^2 - \alpha^{*2}n^2)\ell}$$
$$= \frac{-[V^2 - (\alpha_\perp^2 - \alpha^{*2} + \beta^{*2})n^2 - \beta^{*2}\ell^2]\ell}{(V^2 - \alpha_\perp^2 n^2 - \alpha^{*2}\ell^2)n},$$

or

$$[V^2 - (\alpha_\|^2 - \alpha^{*2} + \beta^{*2})\ell^2 - \beta^{*2}n^2]$$
$$\times (V^2 - \alpha_\perp^2 n^2 - \alpha^{*2}\ell^2)n^2$$
$$+ [V^2 - (\alpha_\perp^2 - \alpha^{*2} + \beta^{*2})n^2 - \beta^{*2}\ell^2]$$
$$\times(V^2 - \alpha_\|^2\ell^2 - \alpha^{*2}n^2)\ell^2 = 0. \qquad (2.104)$$

The solution has been given by Stoneley (1949), Grant and West (1965: 42), and White (1965: 46). The roots are always real and positive and approach α and β of eqs. (2.58) and (2.59) as the anisotropy approaches zero. When the wave is traveling vertically, $\ell = 0$, $n = 1$, and $V = \alpha_\perp$ or β^* for vertically traveling *P*- or *SH*-waves. When $\ell = 1$, $n = 0$, $V = \alpha_\|$ or β^*, corresponding to horizontally traveling *P*- or *SH*-waves. However, when the wave is traveling at an angle to the vertical, the roots are complicated functions of the elastic constants and the motion is not separated into distinct *P*- and *S*-waves.

2.7 Effects of the medium on wave propagation

2.7.1 Energy density and geometrical spreading

Probably the single most important feature of any wave is the energy associated with the motion of the medium as the wave passes through it. Usually, we are not concerned with the total energy of a wave but rather with the energy in the vicinity of the point where we observe it; the *energy density* is the energy per unit volume.

Consider a spherical harmonic *P*-wave for which the radial displacement for a fixed value of r is given by

$$u = A\cos(\omega t + \gamma),$$

where γ is a phase angle. The displacement u ranges from $-A$ to $+A$. Because displacement varies with time, each element of the medium has a velocity, $\dot{u} =$

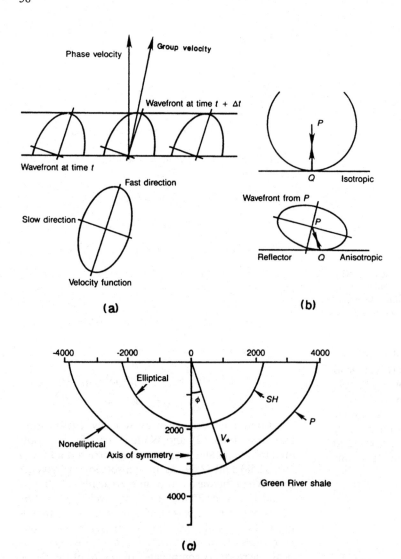

(a)

(b)

(c)

Fig. 2.23 Wavefronts in anisotropic media. (a) Application of Huygens' principle to an anisotropic medium illustrates direction and magnitude differences between phase and group velocities. (b) Fermat's principle applied to a reflection for a coincident source and receiver shows that a reflection may not occur at a right angle to the reflector. (c) SH-wavefronts (surfaces of constant phase for a point source) in transversely isotropic media are elliptical; however, P- and SV-wavefronts are not elliptical except in special instances. $V_H > V_V$ with vertical axis of symmetry. V_ϕ is group velocity as a function of the angle with the symmetry axis.

$\partial u/\partial t$, and an associated kinetic energy. The kinetic energy δE_k contained within each element of volume $\delta\mathcal{V}$ is

$$\delta E_k = \tfrac{1}{2}(\rho\delta\mathcal{V})\dot{u}^2,$$

The kinetic energy per unit volume is

$$\frac{\delta E_k}{\partial\mathcal{V}} = \tfrac{1}{2}\rho\dot{u}^2 = \tfrac{1}{2}\rho\omega^2 A^2 \sin^2(\omega t + \gamma).$$

This expression varies from zero to a maximum of $\tfrac{1}{2}\rho\omega^2 A^2$.

The wave also involves potential energy resulting from the elastic strains created during the passage of the wave. As the medium oscillates back and forth, the energy is converted back and forth from kinetic to potential forms, the total energy remaining fixed. When a particle is at zero displacement, the potential energy is zero and the kinetic energy is a maximum, and when the particle is at its extreme displacement, the energy is all potential. Because the total energy equals the maximum value of the kinetic energy, the energy density E for a harmonic wave is

$$E = \tfrac{1}{2}\rho\omega^2 A^2 = 2\pi^2\rho\nu^2 A^2. \qquad (2.105)$$

Thus, we see that the energy density is proportional to the first power of the density of the medium and to the second power of the frequency and amplitude of the wave. (See Braddick, 1965, for a different derivation of eq. (2.105).)

We are also interested in the rate of flow of energy and we define the *intensity* as the quantity of energy that flows through a unit area normal to the direction of wave propagation in unit time. Take a cylinder of infinitesimal cross-section, $\delta \mathscr{S}$, whose axis is parallel to the direction of propagation and whose length is equal to the distance traveled in the time, δt. The total energy inside the cylinder at any instant t is $EV \delta t \, \delta \mathscr{S}$; at time $t + \delta t$ all of this energy has left the cylinder through one of the ends. Dividing by the area of the end of the cylinder, $\delta \mathscr{S}$, and by the time interval, δt, we get I, the amount of energy passing through unit area in unit time:

$$I = EV. \qquad (2.106)$$

For a harmonic wave, this becomes

$$I = \tfrac{1}{2}\rho V \omega^2 A^2 = 2\pi^2 \rho V \nu^2 A^2. \qquad (2.107)$$

In fig. 2.24, we show a spherical wavefront diverging from a center O. By drawing sufficient radii, we can define two portions of wavefronts, $\mathscr{S}_1$ and $\mathscr{S}_2$, of radii r_1 and r_2, such that the energy that flows outward through the spherical cap $\mathscr{S}_1$ in 1 second must be equal to that passing outward through the spherical cap $\mathscr{S}_2$ in 1 second (because the energy is moving only in the radial direction). The flow of energy per second is the product of the intensity and the area; hence,

$$I_1 \mathscr{S}_1 = I_2 \mathscr{S}_2.$$

Because the areas $\mathscr{S}_1$ and $\mathscr{S}_2$ are proportional to the square of their radii, we get

$$I_2/I_1 = \mathscr{S}_1/\mathscr{S}_2 = (r_1/r_2)^2.$$

Moreover, it follows from eq. (2.106) that E is proportional to I and hence

$$I_2/I_1 = E_2/E_1 = (r_1/r_2)^2. \qquad (2.108)$$

Thus, geometrical spreading causes the intensity and the energy density of spherical waves to decrease inversely as the square of the distance from the source (Newman, 1973). This is called *spherical divergence*.

For a plane wave, the rays do not diverge and hence the intensity of a plane wave is constant. Figure 2.24 could represent a cross-section of a cylindrical wave, that is, a wave generated by a very long linear source, arcs $\mathscr{S}_1$ and $\mathscr{S}_2$ being cylindrical wavefronts. Because the arcs are proportional to the radii, *cylindrical divergence* causes the intensity to vary inversely as the radius. Thus, we can write

$$I_2/I_1 = E_2/E_1 = (r_1/r_2)^m, \qquad (2.109)$$

where $m = 0, 1,$ or 2 according as the wave is plane, cylindrical, or spherical.

Ratios of intensity, energy, or power are usually expressed in decibels, the value in dB being $10 \log_{10}$ of the intensity, energy, or power ratio. Because these vary as the square of the amplitude, dB is also given as $20 \log_{10}$ of the amplitude ratio. The natural log of the amplitude ratio (in nepers) is also used (see problem 2.17).

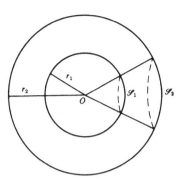

Fig. 2.24 Dependence of intensity upon distance.

The foregoing assumes constant velocity, whereas velocity usually increases with depth, producing more rapid spreading. A factor of $V_s^2 t$ is often used (§9.8), where V_s is the stacking velocity (§5.4.1). The term "spherical divergence" is still used in this situation even though wavefronts may not be spherical.

2.7.2 Absorption

(a) General. In the preceding section, we considered variations of the energy distribution as a function of geometry. Implicit in the discussion was the assumption that none of the wave energy was transformed into other forms. In reality, as the wave motion passes through the medium, the elastic energy associated with the wave motion is gradually absorbed by the medium, reappearing ultimately in the form of heat. This process is called *absorption* and is responsible for the eventual complete disappearance of the wave motion (see also §6.5). Toksoz and Johnston (1981) summarize much of the literature regarding absorption.

The measurement of absorption is very difficult, mainly because it is not easy to isolate absorption from other effects making up attenuation (see §6.5.2). Moreover, absorption varies with frequency, so that it is not clear how laboratory measurements apply to seismic wave travel in the earth.

(b) Expressions for absorption. The decrease of amplitude due to absorption appears to be exponential with distance for elastic waves in rocks. Thus, we can write for the decrease in amplitude because of absorption

$$A = A_0 e^{-\eta x}, \qquad (2.110)$$

where A and A_0 are values of the amplitudes of a plane wavefront at two points a distance x apart, and η is the *absorption coefficient*.

Other measures of absorption are based on the decrease in amplitude with time; to relate these to η, we assume a cyclic waveform:

$$A = A_0 e^{-ht} \cos 2\pi \nu t, \qquad (2.111)$$

and make measurements at a fixed location; h is called

the *damping factor*. The *logarithmic decrement (log dec)* δ is defined by

$$\delta = \ln\left(\frac{\text{amplitude}}{\text{amplitude 1 cycle later}}\right). \quad (2.112)$$

It can be expressed in terms of the damping factor as

$$\delta = hT = h/\nu = 2\pi h/\omega, \quad (2.113)$$

where T is the period; δ is measured in nepers. *Quality factor Q* can be defined as

$$\begin{aligned}Q &= 2\pi/(\text{fraction of energy lost per cycle})\\ &= 2\pi(E/\Delta E), \quad (2.114)\end{aligned}$$

where ΔE is energy loss. Because energy is proportional to amplitude squared, $E = E_0 e^{-2ht}$ and $\Delta E/E_0 = 2h\,\Delta t$. Setting $\Delta t = T$, we get $\Delta E/E_0 = 2hT = 2\delta$ and

$$Q = \pi/hT = \pi/\delta. \quad (2.115)$$

If n is the number of oscillations for the amplitude to decrease by the factor e, then $e^{hnT} = e$, $n = 1/hT$, and

$$Q = \pi n. \quad (2.116)$$

Still another manner of expressing Q is $Q = \cot\phi$, where ϕ is the *loss angle*.

During one period, a wave travels one wavelength so that if the loss of energy is due to absorption only, the attenuation factor is $hT = \eta\lambda$ (from eqs. (2.110) and (2.111)), and we can interrelate η, δ, and Q:

$$Q = \pi/\eta\lambda = \omega/2\eta V = \pi/\delta. \quad (2.117)$$

Absorption in the form given by eq. (2.110) appears naturally in solutions of the type given in eq. (2.56) if we permit the elastic constants to be complex numbers. Real elastic constant values correspond to media without absorption and complex values imply exponential absorption. Complex values of λ and μ result in *complex velocity* values (see eqs. (2.58) and (2.59)). If the $1/V$ in eq. (2.56) is replaced with $1/V + j\eta/\omega$, then

$$\psi = Ae^{j\omega[r(1/V + j\eta/\omega) - t]} = Ae^{-\eta r}e^{j\omega[r/V - t]},$$

which agrees with eq. (2.110).

2.7.3 Relative importance of absorption and spreading

To compare the loss by absorption with the loss of intensity by geometrical spreading (see eq. (2.108)), we have calculated the losses in going various distances from a point 200 m from the source assuming $\eta = 0.15$ dB/λ. The results shown in table 2.1 were calculated using the following relations:

Absorption:
$$\begin{aligned}\text{intensity loss in dB} &= 10\log_{10}(I_0/I)\\ &= 20\log_{10}(A_0/A)\\ &= 0.3(x/\lambda) = 0.3(x_s - 200)/\lambda\\ &= 0.3\nu(x_s - 200)/2000,\end{aligned}$$

Spreading:
$$\begin{aligned}\text{intensity loss in dB} &= 10\log_{10}(I_0/I)\\ &= 20\log_{10}(x_s/200),\end{aligned}$$

where x_s is the distance to the source. The table shows that losses by spreading are more important than losses by absorption for low frequencies and short distances from the source. As the frequency and distance increase, absorption losses increase and eventually become dominant.

The increased absorption at higher frequencies results in change of waveshape with distance. Peg-leg multiples (§6.3.2b) and possibly other mechanisms also produce waveshape changes. Figure 2.25 shows the energy decreasing with distance and with frequency; the frequency-dependent attenuation is greater than expected from absorption alone.

2.7.4 Dispersion; group velocity

Velocities V, α, and β, which appear in §2.2 and subsequent sections, are phase velocities because they are the distances traveled per unit time by a point of constant phase (for example, a trough) of a simple wave such as those in eqs. (2.53) to (2.55). This is not necessarily the velocity with which a pulse travels, called the *group velocity*. For the wavetrain shown in fig. 2.26a, we can determine the group velocity U by drawing the envelope of the pulse (the double curve ABC, $AB'C'$) and measuring the distance that the envelope travels in unit time. The phase velocity V is given approximately by the rate of advance of a distinct "phase break," as indicated in figs. 2.26a and 2.26b, but to find V accurately, we should decompose the pulse into its frequency components by Fourier analysis (see eq. (15.113)) and measure the speed of each component $A(\omega)e^{j\gamma(\omega)}$.

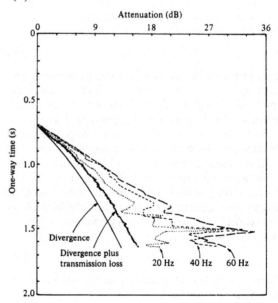

Fig. 2.25 Loss of amplitude as a function of one-way traveltime, based on measurements with a geophone clamped in borehole with the source at the surface. The curves labeled "divergence" and "divergence plus transmission loss" are calculated from sonic-log data allowing for loss of energy in transmission through reflecting interfaces. The 20-, 40-, 60-Hz curves show attenuation at those frequencies. (Courtesy of SSC.)

Table 2.1 *Energy losses by absorption and spreading*
(η = 0.15 dB/wavelength and V = 2.0 km/s)

	Frequency (ν)	Distance from shotpoint (x_s)			
		1200 m	2200 m	4200 m	8200 m
Absorption	1 Hz	0.075 dB	0.15 dB	0.3 dB	0.6 dB
	3	0.22	0.45	0.9	1.8
	10	0.75	1.5	3	6
	30	2.2	4.5	9	18
	100	7.5	15	30	60
Spreading	All	16	21	26	32

If phase velocity V is the same for all frequencies in the pulse, the pulse shape does not change and $U = V$. However, if the velocity varies with frequency, the different components travel with different speeds, the pulse shape changes, and $U \neq V$, that is, the medium is *dispersive*. Consider the two harmonic waves shown in fig. 2.26c that travel at slightly different velocities, the higher-frequency one being faster than the lower-frequency one. The "pulse" that results from their interference changes waveshape and travels at a velocity different from either of them (greater than either in this instance). If we write $A(\omega)e^{j(\kappa x - \omega t)}$ for a plane-wave component traveling along the x-axis, κ being $\kappa(\omega)$, in general, $A(\omega)$ varies slowly whereas the phase ($\kappa x - \omega t$) varies rapidly. When adjacent frequency components are added together, the net result is usually approximately zero because of destructive interference. However, when the phase ($\kappa x - \omega t$) varies slowly, constructive interference occurs in the vicinity of a point (x, t). The condition for this is

$$\frac{\mathrm{d}}{\mathrm{d}\omega}[\kappa(\omega)x - \omega t] = 0 = x\frac{\mathrm{d}\kappa(\omega)}{\mathrm{d}(\omega)} - t,$$

Point (x, t) will move with velocity U, where $U = \mathrm{d}x/\mathrm{d}t$ (see §2.2.5). Hence, differentiation of the above equation gives

$$U = \frac{\mathrm{d}x}{\mathrm{d}t} = \left[\frac{\mathrm{d}\kappa(\omega)}{\mathrm{d}\omega}\right]^{-1} = \left\{\frac{\mathrm{d}}{\mathrm{d}\omega}\left[\frac{\omega}{V(\omega)}\right]\right\}^{-1}$$
$$= \left[\frac{1}{V(\omega)} - \frac{\omega}{V^2(\omega)}\frac{\mathrm{d}V(\omega)}{\mathrm{d}\omega}\right]^{-1}.$$

The derivative is small, so

$$U \approx V + \omega\frac{\mathrm{d}V}{\mathrm{d}\omega} = V + \nu\frac{\mathrm{d}V}{\mathrm{d}\nu} = V - \lambda\frac{\mathrm{d}V}{\mathrm{d}\lambda},$$
(2.118)

where $V, \omega, \nu, \lambda, \mathrm{d}V/\mathrm{d}\omega, \mathrm{d}V/\mathrm{d}\nu$, and $\mathrm{d}V/\mathrm{d}\lambda$ are average values for the range of frequencies making up the principal part of the pulse. (See problem 2.7 for a more elementary derivation of eq. (2.118).)

When V decreases with frequency, we have *normal dispersion*, and V is larger than U, where the envelope travels slower than the individual cycles; this is the usual case with ground roll. When V increases with

frequency, we have *inverse dispersion* and the opposite is true (as in fig. 2.26a).

Dispersion of body waves is a consequence of most theories proposed to account for absorption. Aki and Richards (1980: 170–2) show that the assumption of constant Q, which most data indicate is the situation in solid earth materials, and no dispersion are inconsistent because they lead to noncausality. Their argument follows. Starting with a plane-wave impulse at $x = 0 = t$ and using eqs. (15.127) and (15.136), we have

$$\psi(x, t) = \delta(t - x/V) \leftrightarrow e^{-j\omega x/V}.$$

Adding attenuation corresponding to constant Q (see eqs. (2.110) and (2117)), the right side becomes $e^{(-j\omega x/V - |\omega|x/2VQ)}$; we must use the absolute value of ω to avoid having the amplitude increase with increasing x when ω is negative). Using eq. (15.109) to revert to the time domain, we have

$$\psi(x, t) = \frac{1}{2\pi}\left[\int_{-\infty}^{0} e^{-\omega(jx/V - x/2VQ)}e^{j\omega t}\,\mathrm{d}\omega\right.$$
$$\left. + \int_{0}^{+\infty} e^{-\omega(jx/V + x/2VQ)}e^{j\omega t}\,\mathrm{d}\omega\right]$$
$$= \frac{1}{2\pi}\left[\int_{-\infty}^{0} e^{\omega(A+jB)}\,\mathrm{d}\omega + \int_{0}^{+\infty} e^{\omega(-A+jB)}\,\mathrm{d}\omega\right],$$

where $A = x/2VQ$, and $B = (t - x/V)$. Integration gives

$$\psi(x,t) = (1/2\pi)2A/(A^2 + B^2)$$
$$= \frac{x/2\pi VQ}{(x/2VQ)^2 + (t - x/V)^2}.$$
(2.119)

This function is shown in fig. 2.27; it has a maximum value at $t = x/V$ and is roughly symmetrical. Observations show instead a sharper rise near x/V followed by a slower decay. The function of eq. (2.119) has a finite value for $t < 0$, $x \neq 0$, and is therefore noncausal, showing that our assumptions are inconsistent. We conclude, therefore, that absorption necessarily requires that V varies with ω, that is, dispersion must exist.

Dispersion is important for several reasons, perhaps the most important being that the energy of a pulse travels with the velocity U (except where there is appreciable absorption; see Brillouin, 1960: 98–100). Dispersion of seismic body waves has not been definitively observed over the wide range of frequencies from hertz to megahertz. Most rocks simply exhibit little variation of velocity with frequency in the seismic frequency range. Ward and Hewitt (1977) found the same velocity at 35 Hz as at 55 Hz in a monofrequency well survey to about 800 m. Futterman (1962) shows that the dispersion expected for seismic body waves is small for usual situations. Dispersion is, however, important in connection with surface waves (see §2.5) and channel waves (§13.3) as well as other phenomena.

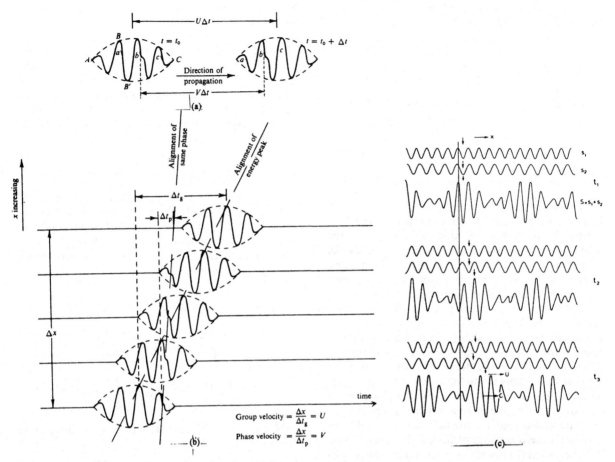

Fig. 2.26   Comparison of group and phase velocities. (a) Definition of group velocity U and phase velocity V; (b) arrival of a dispersive wave at different geophones; (c) two sine waves of slightly different frequency and velocity traveling from left to right form beats; the envelope travels with the group velocity $U = \Delta\omega/\Delta\kappa$ and points of constant phase within the beats with the phase velocities $V_1 = \omega_1/\kappa_1$ and $V_2 = \omega_2/\kappa_2$ (from Gerkens, 1989: 37).

2.7.5 Reflection and refraction; Snell's law

Whenever a wave encounters an abrupt change in the elastic properties, as when it arrives at a surface separating two beds, part of the energy is *reflected* and remains in the same medium as the original energy; the balance of the energy is *refracted* into the other medium with an abrupt change in the direction of propagation occurring at the interface. Reflection and refraction are fundamental in exploration seismology and we shall discuss these in some detail.

We can derive the familiar laws of reflection and refraction using Huygens' principle. Consider a plane wavefront AB incident on a plane interface as in fig. 2.28 (if the wavefront is curved, we merely take A and B sufficiently close together that AB is a plane to the required degree of accuracy; however, see also §2.8.1); AB occupies the position $A'B'$ when A arrives at the surface; at this instant, the energy at B' still must travel the distance $B'R$ before arriving at the interface. If $B'R = V_1 \Delta t$, then Δt is the time interval between the arrival of the energy at A' and at R. By Huygens' principle, during time Δt, the energy that reached A' will have traveled either upward a distance $V_1 \Delta t$ or

downward a distance $V_2 \Delta t$. By drawing arcs with center A' and lengths equal to $V_1 \Delta t$ and $V_2 \Delta t$, and then drawing the tangents from R to these arcs, we locate the new wavefronts, RS and RT in the upper and lower media. The angle at S is a right angle and $A'S = V_1 \Delta t = B'R$; therefore, the triangles $A'B'R$ and $A'SR$ are equal, with the result that the *angle of incidence* θ_1 is equal to the *angle of reflection* θ_1'; this is the *law of reflection*. For the refracted wave, the angle at T is a right angle and we have

$$V_2 \Delta t = A'R \sin \theta_2$$

and

$$V_1 \Delta t = A'R \sin \theta_1,$$

hence,

$$\frac{\sin \theta_1}{V_1} = \frac{\sin \theta_2}{V_2} = p. \qquad (2.120)$$

Angle θ_2 is called the *angle of refraction* and eq. (2.120) is the *law of refraction*, also known as *Snell's law*. The angles are usually measured between the raypaths and a normal to the interface, but these angles are the

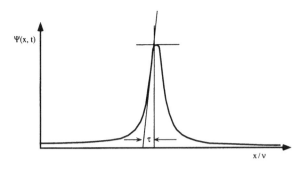

Fig. 2.27   Illustrating eq. (2.119).

same as those between the interface and the wavefronts in isotropic media. The laws of reflection and refraction can be combined in the single statement: at an interface, the quantity $p = (\sin \theta_i)/V_i$ has the same value for the incident, reflected, and refracted waves. The quantity p is called the *raypath parameter*. It will be shown in §3.1.1 that Snell's law also holds for wave conversion from P- to S-waves (and vice versa) upon reflection or refraction. The generalized form of Snell's law (eq. (3.1)) will be understood in future references to Snell's law.

When the medium consists of a number of parallel beds, Snell's law requires that the quantity p have the same value everywhere for all reflected and refracted rays resulting from a given initial ray.

The foregoing derivation assumed a planar surface and therefore specular reflection. If the surface includes bumps of height d, reflected waves from them will be ahead of those from the rest of the surface by $2d$. These can be neglected where $2d/\lambda < \frac{1}{4}$ (the "Rayleigh" criterion), i.e., when $d < \lambda/8$. Most interfaces satisfy this criterion for ordinary seismic waves. For oblique reflection, the criteria are less stringent and reflection can be regarded as specular from relatively rough surfaces.

When V_2 is less than V_1, θ_2 is less than θ_1. However, when V_2 is greater than V_1, θ_2 reaches 90° when $\theta_1 = \sin^{-1}(V_1/V_2)$. For this value of θ_1, the refracted ray is traveling along the interface. The angle of incidence for which $\theta_2 = 90°$ is the *critical angle*, θ_c; obviously,

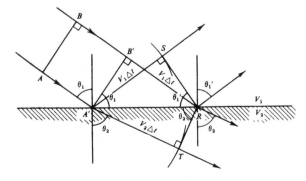

Fig. 2.28   Reflection and refraction of a plane wave.

$$\sin \theta_c = V_1/V_2. \qquad (2.121)$$

For angles of incidence greater than θ_c, it is impossible to satisfy Snell's law (using real angles) because $\sin \theta_2$ cannot exceed unity and *total reflection* occurs. This does not mean that 100% of the energy is reflected as P-waves, however, because converted S-waves (see §3.1.1) and evanescent waves (see §2.3.1) are generated.

Noting the method used to derive eq. (2.57) in §2.3.1, we write Snell's law for the case $\theta_1 > \theta_c$ (see fig. 2.29) in the form

$$\sin \theta_2 = (V_2/V_1) \sin \theta_1 = \sin (\tfrac{1}{2}\pi - j\theta)$$
$$= \cos j\theta = \cosh \theta = \ell,$$
$$n = \cos \theta_2 = \sin j\theta = j \sinh \theta;$$

hence, eq. (2.56) becomes

$$\psi = A e^{-(\omega z/V)\sinh \theta} e^{j\omega[(x/V)\cosh \theta - t]}. \qquad (2.122)$$

If we take θ negative in fig. 2.29, the only change is in the sign of the first exponential on the right-hand side. Thus, just as in the case of eq. (2.57), evanescent waves can exist on both sides of the interface and their amplitudes decrease as we go away from the interface. The rate of attenuation is proportional to $\sinh \theta$, which has its maximum value at the grazing angle, $\theta_1 = \frac{1}{2}\pi$. The introduction of imaginary angles to satisfy[2] Snell's law for angles exceeding the critical angle means that the reflection coefficient (§3.2) will be complex and phase shifts will occur (see problem 3.6b) that will be complicated functions of the angle of incidence.

Snell's law is very useful in determining raypaths and arrival times and in deriving reflector position from observed arrival times, but it does not give information about the amplitudes of the reflected and transmitted waves. This subject is taken up in chap. 3.

2.8 Diffraction

2.8.1 Basic formulas

In discussing reflection and refraction, we stated that when an interface is curved we merely have to select a portion sufficiently small that it can be considered a plane. However, such a simplification is not always possible, for example, when the radius of curvature of an interface is less than a few wavelengths or the reflector is terminated by a fault, pinchout, unconformity, etc. In such cases, the simple laws of reflection and refraction are no longer adequate because the energy is *diffracted* rather than reflected or refracted. Because seismic wavelengths are often 100 m or more, many geologic features give rise to diffractions.

The mathematical treatment of diffraction is complex and we shall give only a brief summary of a simplified treatment due to Trorey (1970). We shall assume a coincident source and receiver (see Trorey, 1977, for the noncoincident case) and constant velocity. Using ϕ in place of ψ, we take the Laplace transform of the wave equation, eq. (2.30), obtaining

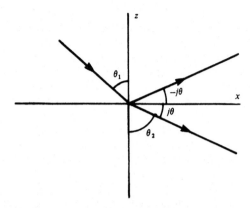

Fig. 2.29　Imaginary angles of reflection and refraction.

$$\nabla^2\phi = (1/V^2)\,\partial^2\phi/\partial t^2 \leftrightarrow \nabla^2\Phi = (s/V)^2\Phi,$$

where $\Phi(x, y, z, s)$ is the the Laplace transform of $\phi(x, y, z, t)$ (see §15.3), and the double-headed arrow indicates equivalence in different domains. Note that we are assuming that ϕ and $\partial\phi/\partial t$ are zero at $t = 0$ for all x, y, z.

The solution of this equation for a point source at the origin is

$$\Phi = (c/r)e^{-sr/V}, \qquad (2.123)$$

where r is the distance from the source to the point of observation, and V is the wave velocity. (This can be verified by direct substitution, noting that $r^2 = x^2 + y^2 + z^2$, $\partial r/\partial x = x/r$, etc.) In general, c should include the Laplace transform of the input waveform at the source, but in effect we have taken the transform to be unity, that is, the source is $c\,\delta(t)$ (eq. (15.180)). The results for other types of sources can be found by time-domain convolution (see eqs. (15.201) and (15.202)).

In a source-free region, the P-wave potential function ϕ is given by eq. (2.42) with $Y = 0$; hence, we can get another expression for ϕ by taking the Laplace transform of eq. (2.42), the result being

$$4\pi\Phi = \iint_{\mathscr{S}} e^{-sr/V}\left\{[\Phi]\left[\frac{s}{rV}\frac{\partial r}{\partial\eta} - \frac{\partial(1/r)}{\partial\eta}\right] + \frac{1}{r}\left[\frac{\partial\Phi}{\partial\eta}\right]\right\}d\mathscr{S} \tag{2.124}$$

The factor $e^{-sr/V}$ arises because ϕ in the integrand of eq. (2.42) is evaluated at time $t = t_0 - r/V$, whereas Φ is the transform of $\phi(x, y, z, t)$ (see eq. (15.187)).

2.8.2 Diffraction effect of part of a plane reflector

We shall now calculate the diffraction effect of an area $\mathscr{S}$ that is part of a plane reflector $z = h$ (see fig. 2.30a), both source and detector being at the origin. We enclose the origin with a hemisphere of infinite radius with center $(0, 0, h)$, the base of which is the plane $z = h$. In order to apply eq. (2.124), we replace the source with its image at $(0, 0, 2h)$, thus making the hemisphere a source-free region. We ignore absorption and

assume a constant reflection coefficient over $\mathscr{S}$, so that c in eq. (2.123) is constant. Clearly, $1/r = 0 = \Phi$ over the hemisphere, hence the contribution to the integral over the hemisphere is zero. We can also set $\Phi = 0$ over the portions of the plane $z = h$ except for the portion $\mathscr{S}$ whose effect we wish to evaluate.

We now substitute eq. (2.123) in eq. (2.124), noting that r in eq. (2.123) is now r_0 in fig. 2.30a because the source is now at the image point O'; hence

$$\frac{\partial r}{\partial\eta} = \frac{\partial r}{\partial z} = \frac{z}{r} = \frac{h}{r};$$

$$\frac{\partial r_0}{\partial\eta} = \frac{\partial r_0}{\partial z} = \frac{-h}{r_0}; \qquad \frac{\partial(1/r)}{\partial\eta} = -\frac{h}{r^3};$$

$$\frac{\partial\Phi}{\partial\eta} = \frac{\partial\Phi}{\partial r_0}\frac{\partial r_0}{\partial\eta} = -\frac{c}{r_0}e^{-sr_0/V}\left(\frac{1}{r_0} + \frac{s}{V}\right)\left(\frac{-h}{r_0}\right)$$

$$= \frac{ch}{r^2}e^{-sr/V}\left(\frac{1}{r} + \frac{s}{V}\right),$$

where we have set $r_0 = r$ after differentiating. The result is

$$2\pi\Phi = ch\iint_{\mathscr{S}} e^{-2sr/V}\left(\frac{1}{r^4} + \frac{s}{Vr^3}\right)d\mathscr{S}. \tag{2.125}$$

This surface integral can be transformed into a contour integral as follows. In fig. 2.30b, the element of area is $\rho\,d\rho\,d\theta$ in polar coordinates; because $r^2 = \rho^2 + h^2$, $\rho\,d\rho = r\,dr$, and, therefore,

$$2\pi\Phi = ch\int_{\theta}\int_{r} e^{-2sr/V}\left(\frac{1}{r^3} + \frac{s}{Vr^2}\right)dr\,d\theta. \tag{2.126}$$

If we integrate the first term by parts with respect to r, we obtain

$$\int_{r_1}^{r_2}\frac{e^{-2sr/V}}{r^3}\,dr = \frac{e^{-2sr/V}}{-2r^2}\Big|_{r_1}^{r_2} - \int_{r_1}^{r_2}\frac{se^{-2sr/V}}{Vr^2}\,dr.$$

Substituting in eq. (2.126), we get

$$4\pi\Phi = ch\oint\left[(1/r_1^2)e^{-2sr_1/V} - (1/r_2^2)e^{-2sr_2/V}\right]d\theta. \tag{2.127}$$

If the z-axis does not cut $\mathscr{S}$, we can get $r = \xi$ for points on the boundary, giving

$$\Phi = -(ch/4\pi)\oint(1/\xi^2)e^{-2s\xi/V}d\theta, \tag{2.128}$$

where A traverses the boundary of $\mathscr{S}$ in the counterclockwise direction.

If the z-axis cuts $\mathscr{S}$ (see fig. 2.30c), $r_1 = h =$ constant in eq. (2.127), so that

$$\Phi = (c/2h)e^{-2sh/V} - (ch/4\pi)\oint(1/\xi^2)e^{-2s\xi/V}d\theta, \tag{2.129}$$

where A again traverses the boundary in the counterclockwise direction. If $\mathscr{S}$ includes the entire xy-plane,

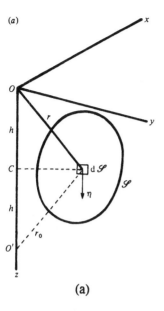

(a)

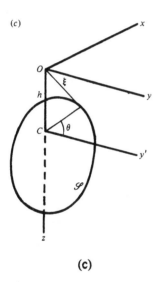

(c)

Fig. 2.30 Diffraction effect of a plane area $\mathscr{S}$. (After Trorey, 1970.) (a) Calculation using a surface integral; (b) calculation using a line integral where the origin is not over the area; (c) calculation when the origin is over the area; (d) evaluating the line integral when ξ is multivalued.

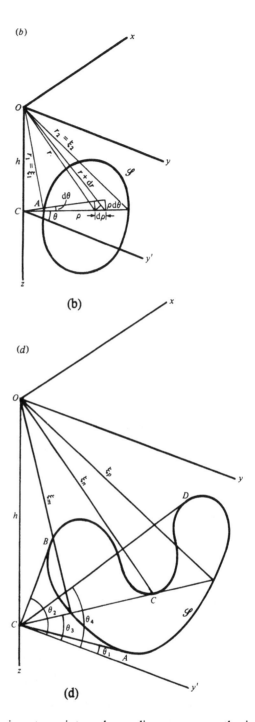

(b)

(d)

i.e., we have a continuous plane reflector, $\xi = \infty$ and the integral vanishes. The first term in this equation thus represents the simple reflection from this plane and the integral represents the diffracted wave. Comparison of eqs. (2.128) and (2.129) shows that the diffracted wave is given by the same expression in both cases.

An important point to note is that both the reflection and diffraction terms in eqs. (2.128) and (2.129) are derived from the integral in eq. (2.125), where the integration is carried over the entire surface. When we use rays and think of reflection and diffraction as oc-

curring at a point or along a line, we are greatly simplifying the actual phenomena. In fact, both reflections and diffractions are the resultants of energy that return from all parts of the surface. From this point of view a reflection is merely a special type of diffraction, a point of view that has interesting practical applications (see §9.12.2).

2.8.3 Time-domain solution for diffraction

We now obtain the time-domain solution of eqs. (2.128) and (2.129). The inverse transform of the reflection term on the right-hand side of eq. (2.129) gives the impulse $(c/2h)\,\delta(t - 2h/V)$, that is, a repetition of the input at the source after a delay of $2h/V$, which is the two-way traveltime from the source to the plane surface, with the amplitude falling off inversely as the distance. Thus, the reflection has the same waveshape as the source. The diffraction terms can be found as follows: We write $t = 2\xi/V$, which is the two-way traveltime from the source to the variable point on the boundary. Equation (2.128) now becomes

$$\Phi = \frac{ch}{\pi V^2}\oint \frac{e^{-st}}{t^2}\,d\theta = \frac{ch}{\pi V^2}\oint \frac{e^{-st}}{t^2}\frac{d\theta}{dt}\,dt. \quad (2.130)$$

We must pay careful attention to the limits of integration because ξ, and hence t, is in general a multivalued function of θ; for example, when $\theta = \theta_3$ in fig. 2.30d, ξ can have any of the values ξ_m, ξ_n, or ξ_p. To avoid difficulties, the integral is calculated as the point of integration goes from A to B (θ from θ_1 to θ_2), then from B to C, C to D, and finally D to A, the proper values of ξ (and t) being used along each segment of the path. Along a given portion of the path, say, between $t = t_1$ and $t = t_2$ ($t_2 > t_1$), we have

$$\left. \Phi = \frac{ch}{\pi V^2}\int_{t_1}^{t_2}\frac{e^{-st}}{t^2}\left(\frac{d\theta}{dt}\right)dt = \int_0^{+\infty}\phi(t)e^{-st}\,dt, \right\}$$

Comparing the right-hand integral with eq. (15-178a), we conclude that $\quad\quad\quad (2.131)$

$$\left.\begin{aligned} \phi(t) &= 0 & (t < t_1),\\ &= (ch/\pi V^2 t^2)(d\theta/dt) & (t_1 < t < t_2),\\ &= 0 & (t > t_2). \end{aligned}\right\}$$

The derivative, $d\theta/dt$, is finite except when ξ is constant, such as where $\mathscr{S}$ is bounded by an arc of a circle with its center at the origin for which case $dt = 0$. In this special case, eq. (2.130) gives

$$\Phi = (ch/\pi V^2 t_0^2)e^{-st_0}(\theta_i - \theta_j),$$

where θ_i and θ_j fix the end points of the circular arc, and t_0 is the two-way traveltime to the arc. The inverse transform is

$$\phi = (ch/\pi V^2 t_0^2)(\theta_i - \theta_j)\delta(t - t_0).$$

When $d\theta/dt$ is finite, we can get the time-domain solution for the diffracted wave by dividing the boundary of the area $\mathscr{S}$ so that the two-way traveltime t between P and the boundary is a single-valued function of θ in each part, calculating ϕ for each part from eq. (2.131), then summing the various ϕ's to get ϕ for the diffracted wave.

2.8.4 Diffraction effect of a half-plane

A simple illustration of the method is the calculation of the diffraction effect in the important case of a horizontal half-plane at a depth h, the edge being parallel to the x-axis and a distance y_0 from it. By referring to fig. 2.31, the edge of the half-plane is BD, where in

fact B and D are at infinity, so that θ increases from $-\tfrac{1}{2}\pi$ to $+\tfrac{1}{2}\pi$ as the point of integration A traverses the boundary in a clockwise direction. The result (see problem 2.19) is

$$\left.\begin{aligned} \phi &= \frac{2(ch/\pi V^2)(1/t^2)(t_y t)}{(t^2 + t_y^2 - t_r^2)(t^2 - t_r^2)^{1/2}}\\ &= \frac{(4chy_0/\pi V^3 t)}{(t^2 + t_y^2 - t_r^2)(t^2 - t_r^2)^{1/2}} & (t > t_r),\\ &= 0 & (t < t_r), \end{aligned}\right\}\quad (2.132)$$

where $t = 2\xi/V$, $t_r = 2r/V$, and $t_y = 2y_0/V$.

This value of $\phi(t)$ gives the diffracted wave recorded at point $P(0, 0, 0)$ as the result of an impulse, $c\,\delta(t)$, applied at the same point. If the input is $cg(t)$ instead of $c\,\delta(t)$, Φ will have the factor $G(s)$ and the response becomes $\phi(t) * g(t)$ (see eq. (15.195)).

Equation (2.132) gives the diffraction effect whether P is off the plane, as in fig. 2.31, or over the plane; because y_0 changes sign as P passes over the edge, the diffracted wave undergoes a 180° phase shift as P passes over the edge. Moreover, if we write D for the value of ϕ for the diffracted wave observed when P is infinitesimally close to the edge and to the left of it, the total effect observed when P is the same distance to the right of the edge will be $R - D$, R being the value of the reflection term in eq. (2.129). Because $\phi(t)$ is continuous,

$$R - D = D \quad\text{or}\quad D = \tfrac{1}{2}R. \quad (2.133)$$

Thus, the maximum amplitude of the diffraction from a half-plane is half the amplitude of the reflected wave (as observed far from the edge). Figure 2.32 shows what is expected from a half-plane based on eq. (2.132). As the edge of the reflector is approached, the diffraction gains in amplitude whereas $(R - D)$ decreases in amplitude until at the edge $D = \tfrac{1}{2}R$ and the

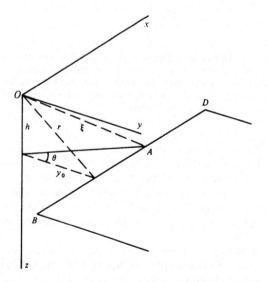

Fig. 2.31   Calculation of the diffraction from a half-plane.

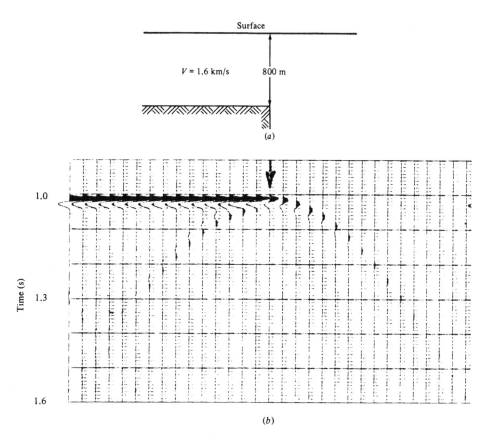

Fig. 2.32 Seismic response of a half-plane. (After Trorey, 1970.) (a) Model and (b) computed seismic record for coincident sources and geophones. The arrowhead indicates the location of the edge of the half-plane.

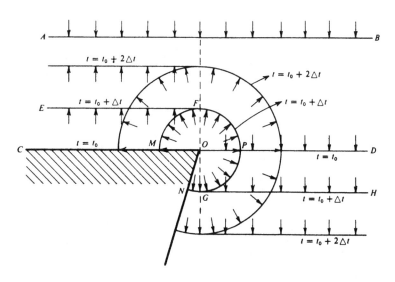

Fig. 2.33 Diffracted wavefronts for a faulted bed.

sum is $\frac{1}{2}R$. The phase reversal of the diffraction before the edge is reached (the *backward branch* of the diffraction) from that beyond the edge (the *forward branch*) is evident in fig. 2.32.

2.8.5 Using Huygens' principle to construct diffracted wavefronts

The surface integral in eq. (2.125) shows that the diffraction effect at a point is the sum of effects arising from the entire diffracting surface. This suggests the use of Huygens' principle to construct diffracted wavefronts, and this is the case for points more than a few wavelengths away from the diffracting source. Figure 2.33 illustrates this construction for a faulted reflector. We assume a plane wavefront *AB* incident normally on the faulted bed *CO*, the position of the wavefront when it reaches the surface of the bed at $t = t_0$ being *COD*. At $t = t_0 + \Delta t$, the portion to the right of *O* has advanced to position *GH*, whereas the portion to the left of *O* has been reflected and has reached position *EF*. We might have constructed wavefronts *EF* and *GH* by selecting a large number of centers in *CO* and *OD* and drawing arcs of length $V\Delta t$; *EF* and *GH* would then be determined by the envelopes of these arcs. However, for the portion *EF*, there would be no centers to the right of *O* to define the envelope, whereas for portion *GH* there would be no centers to the left of *O* to define the envelope. Thus, *O* marks the transition point between centers that give rise to the upward-traveling wavefront *EF* and centers that give rise to the downward-traveling wavefront *GH*; arc *FPG* with center *O* is the diffracted wavefront originating at *O* and connecting the two wavefronts, *EF* and *GH*. The diffracted wavefront also extends into the geometrical shadow area *GN* and into region *FM*.

The characteristics of diffractions in various situations are discussed further in §6.3.1.

Problems

2.1 (a) If σ_{xx} is the only nonzero normal stress, that is, if $\sigma_{yy} = \sigma_{zz} = 0$, use Hooke's law, eq. (2.13), to show that strain $\varepsilon_{yy} = \varepsilon_{zz}$, and verify eq. (2.20) for Poisson's ratio.
(b) By adding the three equations in part (a) for σ_{xx}, σ_{yy}, and σ_{zz}, derive eq. (2.19) for Young's modulus.
(c) Pressure $\mathcal{P}$ is equivalent to stresses $\sigma_{xx} = \sigma_{yy} = \sigma_{zz} = -\mathcal{P}$. Substituting these relationships into eq. (2.13), derive eq. (2.21) for the bulk modulus.
2.2 The entries in table 2.2 express the quantities at the heads of the columns in terms of the pairs of elastic constants or velocities at the left ends of the rows. The first three entries in the ninth row are eqs. (2.19) to (2.21). Starting with these and eqs. (2.58) and (2.59), derive the other relations in the table.

2.3 (a) Firing an air gun (§7.4.3) in water creates a pressure transient a small distance away with peak pressure of 5 atmospheres (5×10^5 Pa). If the compressibility of water is 4.5×10^{-10}/Pa, what is the peak energy density?
(b) If the same wave is generated in rock with $\lambda = \mu = 3 \times 10^{10}$ Pa, what is the peak energy density? Assume a symmetrical *P*-wave with $\varepsilon_{xx} = \varepsilon_{yy} = \varepsilon_{zz}$, $\varepsilon_{ij} = 0$ for $i \neq j$.
2.4 To illustrate the interrelationship and magnitude of the elastic constants, complete Table 2.3. Note that these values apply to specific specimens; the values for rocks range considerably, especially as porosity and pressure change.
2.5 (a) Verify that $\psi = f(x - Vt)$ and $\psi = g(x + Vt)$ in eqs. (2.46) and (2.47) are solutions of the one-dimensional wave equation, eq. (2.45). (*Hint:* Let $\zeta = x - Vt$ and show that

$$\frac{\partial \psi}{\partial x} = \frac{df}{d\zeta}\frac{\partial \zeta}{\partial x} = \frac{df}{d\zeta} \equiv f'$$

etc.)
(b) Verify that $\psi = f(\ell x + my + nz - Vt) + g(\ell x + my + nz + Vt)$ in eq. (2.48) is a solution of the plane-wave equation, eq. (2.30).
(c) Using the same technique, show that $\psi = (1/r)f(r - Vt) + (1/r)g(r + Vt)$ in eq. (2.52) satisfies the wave equation in spherical coordinates, eq. (2.50).
2.6 (a) Show that the wave equation, eq. (2.30), can be written in cylindrical coordinates ($x = r \cos \theta$, $y = r \sin \theta$, and $z = z$; see fig. 2.34a) as

$$\frac{\partial^2 \psi}{\partial r^2} + \frac{1}{r}\frac{\partial \psi}{\partial r} + \frac{1}{r^2}\frac{\partial^2 \psi}{\partial \theta^2} + \frac{\partial^2 \psi}{\partial z^2} = \frac{1}{V^2}\frac{\partial^2 \psi}{\partial t^2}.$$

(b) Verify that eq. (2.49) is the wave equation in spherical coordinates by substituting the following coordinate transformation (see fig. 2.34b) into eq. (2.30):

$$x = r \sin \theta \cos \phi,$$
$$y = r \sin \theta \sin \phi,$$
$$z = r \cos \theta.$$

(For an easier solution, see problem 15.8.)
2.7 A pulse composed of two different frequency components, $\omega_0 \pm \Delta\omega$, can be represented by factors involving the sum and differences of the frequencies. If they have equal amplitudes, we can write for the two components

$$A \cos (\kappa_1 x - \omega_1 t), \qquad A \cos (\kappa_2 x - \omega_2 t),$$

where $\omega_1 = \omega_0 + \Delta\omega$, $\omega_2 = \omega_0 - \Delta\omega$, $\kappa_0 = 2\pi/\lambda_0 = \omega_0/V$, $\kappa_1 \approx \kappa_0 + \Delta\kappa \approx (\omega_0 + \Delta\omega)/V$, and $\kappa_2 \approx \kappa_0 - \Delta\kappa \approx (\omega_0 - \Delta\omega)/V$.
(a) Show that the pulse is given approximately by the expression $\mathcal{B} \cos (\kappa_0 x - \omega_0 t)$, where $\mathcal{B} = 2A \cos \Delta\kappa\{x - (\Delta\omega/\Delta\kappa)t\}$.
(b) Why do we regard $\mathcal{B}$ as the amplitude? Show that the envelope of the pulse is the graph of $\mathcal{B}$ plus its reflection in the *x*-axis.

Table 2.2 *Relations between elastic constants and velocities (isotropic media)*

	Young's modulus, E	Poisson's ratio, σ	Bulk modulus, k	Shear modulus, μ	Lamé constant, λ	P-wave velocity, α	S-wave velocity, β	Velocity ratio, β/α
(E, σ)			$\dfrac{E}{3(1-2\sigma)}$	$\dfrac{E}{2(1+\sigma)}$	$\dfrac{E\sigma}{(1+\sigma)(1-2\sigma)}$	$\left[\dfrac{E(1-\sigma)}{(1+\sigma)(1-2\sigma)\rho}\right]^{1/2}$	$\left[\dfrac{E}{2(1+\sigma)\rho}\right]^{1/2}$	$\left[\dfrac{1-2\sigma}{2(1-\sigma)}\right]^{1/2}$
(E, k)		$\dfrac{3k-E}{6k}$		$\dfrac{3kE}{9k-E}$	$3k\left(\dfrac{3k-E}{9k-E}\right)$	$\left[\dfrac{3k(3k+E)}{\rho(9k-E)}\right]^{1/2}$	$\left[\dfrac{3kE}{(9k-E)\rho}\right]^{1/2}$	$\left(\dfrac{E}{3k+E}\right)^{1/2}$
(E, μ)		$\dfrac{E-2\mu}{2\mu}$	$\dfrac{\mu E}{3(3\mu-E)}$		$\mu\left(\dfrac{E-2\mu}{3\mu-E}\right)$	$\left[\dfrac{\mu(4\mu-E)}{(3\mu-E)\rho}\right]^{1/2}$	$\left(\dfrac{\mu}{\rho}\right)^{1/2}$	$\left(\dfrac{3\mu-E}{4\mu-E}\right)^{1/2}$
(σ, k)	$3k(1-2\sigma)$			$\dfrac{3k}{2}\left(\dfrac{1-2\sigma}{1+\sigma}\right)$	$3k\left(\dfrac{\sigma}{1+\sigma}\right)$	$\left[\dfrac{3k(1-\sigma)}{\rho(1+\sigma)}\right]^{1/2}$	$\left[\dfrac{3k(1-2\sigma)}{2\rho(1+\sigma)}\right]^{1/2}$	$\left[\dfrac{1-2\sigma}{2(1-\sigma)}\right]^{1/2}$
(σ, μ)	$2\mu(1+\sigma)$		$\dfrac{2\mu(1+\sigma)}{3(1-2\sigma)}$		$\mu\left(\dfrac{2\sigma}{1-2\sigma}\right)$	$\left[\left(\dfrac{2\mu}{\rho}\right)\left(\dfrac{1-\sigma}{1-2\sigma}\right)\right]^{1/2}$	$\left(\dfrac{\mu}{\rho}\right)^{1/2}$	$\left[\dfrac{1-2\sigma}{2(1-\sigma)}\right]^{1/2}$
(σ, λ)	$\lambda\dfrac{(1+\sigma)(1-2\sigma)}{\sigma}$		$\lambda\left(\dfrac{1+\sigma}{3\sigma}\right)$	$\lambda\left(\dfrac{1-2\sigma}{2\sigma}\right)$		$\left[\left(\dfrac{\lambda}{\rho\sigma}\right)(1-\sigma)\right]^{1/2}$	$\left[\dfrac{\lambda}{\rho}\left(\dfrac{1-2\sigma}{2\sigma}\right)\right]^{1/2}$	$\left[\dfrac{1-2\sigma}{2(1-\sigma)}\right]^{1/2}$
(k, μ)	$\dfrac{9k\mu}{3k+\mu}$	$\dfrac{3k-2\mu}{2(3k+\mu)}$			$k-2\mu/3$	$\left(\dfrac{k+4\mu/3}{\rho}\right)^{1/2}$	$\left(\dfrac{\mu}{\rho}\right)^{1/2}$	$\left(\dfrac{\mu}{k+4\mu/3}\right)^{1/2}$
(k, λ)	$9k\left(\dfrac{k-\lambda}{3k-\lambda}\right)$	$\dfrac{\lambda}{3k-\lambda}$		$\dfrac{3}{2}(k-\lambda)$		$\left(\dfrac{3k-2\lambda}{\rho}\right)^{1/2}$	$\left[\dfrac{3(k-\lambda)}{2\rho}\right]^{1/2}$	$\left[\dfrac{1}{2}\left(\dfrac{k-\lambda}{k-2\lambda/3}\right)\right]^{1/2}$
(μ, λ)	$\mu\left(\dfrac{3\lambda+2\mu}{\lambda+\mu}\right)$	$\dfrac{\lambda}{2(\lambda+\mu)}$	$\lambda+\dfrac{2}{3}\mu$			$\left(\dfrac{\lambda+2\mu}{\rho}\right)^{1/2}$	$\left(\dfrac{\mu}{\rho}\right)^{1/2}$	$\left(\dfrac{\mu}{\lambda+2\mu}\right)^{1/2}$
(α, β)	$\rho\beta^2\left(\dfrac{3\alpha^2-4\beta^2}{\alpha^2-\beta^2}\right)$	$\dfrac{\alpha^2-2\beta^2}{2(\alpha^2-\beta^2)}$	$\rho\left(\alpha^2-\dfrac{4}{3}\beta^2\right)$	$\rho\beta^2$	$\rho(\alpha^2-2\beta^2)$			

(a)

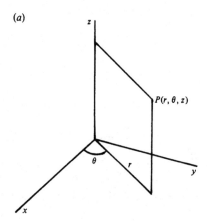

$P(r, \theta, z)$

(b)

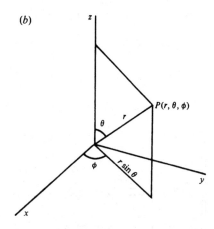

$P(r, \theta, \phi)$

Fig. 2.34   Coordinate systems. (a) Cylindrical coordinates; (b) spherical coordinates.

(c) Show that the envelope moves with the group velocity U, where

$$U = \frac{\Delta\omega}{\Delta\kappa} \approx \frac{d\omega}{d\kappa} \approx V - \lambda\frac{dV}{d\lambda} \approx V + \omega\frac{dV}{d\omega}$$

(see fig. 2.26).

2.8 The magnitudes of period T, frequency ν, wavelength λ, and angular wavenumber κ are important in practical situations. Calculate T, λ, and κ for the situations shown in table 2.4.

2.9 (a) Show that the potential function $\boldsymbol{\zeta} = \boldsymbol{\nabla}(\phi + \partial\chi/\partial z) - \nabla^2\chi\mathbf{k}$ of eq. (2.63) requires that Δ and θ_z be solutions of the P- and S-wave equations, eqs. (2.28) and (2.29), respectively. (*Hint:* Recall that ϕ and χ are solutions of the wave equation and calculate $\boldsymbol{\nabla} \cdot \boldsymbol{\zeta}$ and $\boldsymbol{\nabla} \times \boldsymbol{\zeta}$ using eqs. (15.12) and (15.13).)

(b) Show that the dilatation and rotation (eqs. (2.65)) can be derived from $\boldsymbol{\zeta}$ in eq. (2.64) using eq. (15.14) and the results of problem 15.7.

2.10 Justify on physical grounds the following boundary conditions for different combinations of media in contact at an interface:

(a) Solid–fluid: normal stress and displacement are continuous, tangential stress in the solid vanishes at the interface.

(b) Solid–vacuum: normal and tangential stresses in the solid vanish at the interface.

(c) Fluid–fluid: normal stresses and displacements are continuous.

(d) Fluid–vacuum: normal stress in the fluid vanishes at the interface.

2.11 Using eqs. (2.8), (2.9), (2.12), (2.13), (2.14), and (2.64), show that the boundary conditions at the xy-plane separating two semiinfinite solids require, for a wave in the xz-plane, the continuity of the following potential functions:

$$(\phi_z - \chi_x), \qquad (\phi_x + \chi_z),$$
$$\lambda\nabla^2\phi + 2\mu(\phi_{zz} - \chi_{xz}),$$
$$\mu(2\phi_{xz} + \chi_{zz} - \chi_{xx}),$$

where the subscripts denote partial differentiation (these are respectively the normal and tangential displacements, normal and tangential stresses).

2.12 A source of seismic waves produces a step displacement on a spherical cavity of radius r_0 enclosing the source of the form

$$\text{step}_0\,(t) = 0 \ (t < 0),$$
$$= k \ (t \geq 0).$$

Show from eq. (2.77) that the displacement is given by

$$u = \frac{r_0^2 k}{r}\left[\left(\frac{1}{r_0} - \frac{1}{r}\right)e^{-Vt/r_0} + \frac{1}{r}\right].$$

Is the motion oscillatory? What is the final (permanent) displacement at distance r?

2.13 Show that for harmonic waves of the form $\phi = (A/r)\cos\omega(r/V - t)$, the two terms in eq. (2.71), which decay at different rates, are of equal importance at distance $r = \lambda/2\pi$.

2.14 Equations (2.86) and (2.87) for a Rayleigh wave are valid at the surface $z = 0$ for $\sigma = 1/4$.

(a) Show that at depths $z \neq 0$, the expressions for the x- and z-displacements u and w are

$$u = \kappa A(-e^{-0.848\kappa z} + 0.577e^{-0.393\kappa z})\sin\kappa(x - V_R t),$$
$$w = \kappa A(-0.848e^{-0.848\kappa z} + 1.468e^{-0.393\kappa z})$$
$$\times \cos\kappa\,(x - V_R t).$$

(b) What are the values of u, w, and θ when $z = 1/2\kappa$? When $z = 1/\kappa$?

(c) Is the motion retrograde for all values of z? (*Hint:* Note that for the motion to change direction, the amplitude of either u or w must pass through zero.)

(d) What are the values of V_R, the Rayleigh-wave velocity, when $\sigma = 0.4$ and when $\sigma = 0.2$? What are the corresponding values of the constants in part (a)?

2.15 Assume three geophones so oriented that one records only the vertical component of a seismic wave, another records only the horizontal component in the direction of the source, and the third only the hori-

Table 2.3 *Example of magnitudes of elastic constants and velocities*

	Water	Stiff mud	Sandstone	Limestone	Granite
Young's modulus, E ($\times 10^9$ Pa)			16	54	50
Bulk modulus, k ($\times 10^9$ Pa)	2.1				
Rigidity modulus, μ ($\times 10^9$ Pa)					
Lamé's λ constant ($\times 10^9$ Pa)					
Poisson's ratio, σ	0.5	0.43	0.34	0.25	0.20
Density, ρ (g/cm³)	1.0	1.5	1.9	2.5	2.7
P-wave velocity, α (km/s)	1.5	1.6			
S-wave velocity, β (km/s)					

Table 2.4 *Magnitudes of T, λ, and κ*

	α (km/s)	For $\nu = 15$ Hz			For $\nu = 60$ Hz		
		T (s)	λ (m)	κ (m⁻¹)	T (s)	λ (m)	κ (m⁻¹)
Weathering (min.)	0.1						
Weathering (avg.)	0.5						
Water	1.5						
Poorly consolidated sands–shales at 0.75 km	2.0						
Tertiary clastics at 3.00 km	3.3						
Porous limestone	4.3						
Dense limestone	5.5						
Salt	4.6						
Anhydrite	6.1						

zontal component at right angles to this. Assume a simple waveshape and draw the response of the three geophones for the following cases:
(a) A *P*-wave traveling directly from the source to the geophones.
(b) A *P*-wave reflected from a deep horizon.
(c) An *S*-wave generated by reflection of a *P*-wave at an interface.
(d) A Rayleigh wave generated by the source.
(e) A Love wave.
Compare the relative magnitudes of the components for short and long offsets.
2.16 (a) A tube wave has a velocity of 1.05 km/s. The fluid in the borehole has a bulk modulus of 2.15×10^9 Pa and density 1.20 g/cm³. The wall rock has $\sigma = 0.25$ and $\rho = 2.5$ g/cm³. Calculate μ and α for the wall rock. (b) Repeat for $V_T = 1.20$ km/s and 1.30 km/s. What do you conclude about the accuracy of this method for determining μ?
2.17 The natural logarithm of the ratio of amplitudes is measured in nepers. Show that 1 neper = 8.686 dB.
2.18 A refraction seismic wavelet assumed to be essentially harmonic with frequency 40 Hz is found to have amplitudes of 5.00 and 4.57 mm on traces 2.50 and 3.00 km from the source. Assuming a velocity of 3.20 km/s, constant subsurface conditions, and ideal recording conditions, what is the ratio of the amplitudes on a given trace of the first and fourth cycles?

What percentage of the energy is lost over three cycles? What is the value of h?
2.19 Using eq. (2.131), the general equation for the diffraction from a plane surface, verify eq. (2.132) for the diffraction effect of a half-plane.

References

Aki, K., and P. G. Richards. 1980. *Quantitative Seismology: Theory and Methods.* San Francisco: W. H. Freeman.

Birch, F. 1966. Compressibility; elastic constants. In *Handbook of Physical Constants,* S. P. Clark, ed., GSA Memoir 97.

Blake, F. C. 1952. Spherical wave propagation in solid media. *J. Acoust. Soc. Amer.,* **24:** 211–15.

Braddick, H. J. J. 1965. *Vibrations, Waves, and Diffractions.* New York: McGraw-Hill.

Brillouin, L. 1960. *Wave Propagation and Group Velocity.* New York: Academic Press.

Bush, I., and S. Crampin. 1987. Observations of EDA and PTL anisotropy in shear-wave VSP. *Expanded Abstracts, 57th Annual International Meeting of the Society of Exploration Geophysicists,* pp. 646–59. Tulsa: Society of Exploration Geophysicists.

Cheng, C. H., and M. N. Toksoz. 1981. Elastic wave propagation in fluid-filled borehole and synthetic acoustic logs. *Geophysics,* **46:** 1042–53.

Crampin, S. 1981. A review of wave propagation in anisotropic and cracked elastic media. *Wave Motion,* **3:** 343–91.

Crampin, S., R. McGonigle, and D. Bamford. 1980. Estimating crack parameters from observations of P-wave velocity anisotropy. *Geophysics,* **45:** 345–60.

Dobrin, M. B. 1951. Dispersion in seismic waves. *Geophysics,* **16:** 63–80.

Ebrom, D. A., R. H. Tatham, K. K. Sekharan, J. A. McDonald, and G. H. F. Gardner. 1990. Dispersion and anisotropy in laminated versus fractured media: An experimental comparison. *Expanded Abstracts, 60th Annual International Meeting of the Society of Exploration Geophysicists,* pp. 1416–19. Tulsa: Society of Exploration Geophysicists.

Ewing, W. M., W. S. Jardetzky, and F. Press. 1957. *Elastic Waves in Layered Media.* New York: McGraw-Hill.

Futterman, W. I. 1962. Dispersive body waves. *J. Geophys. Res.,* **67:** 5279–91.

Gerkens, J. C. d'Arnaud. 1989. *Foundation of Exploration Geophysics.* Amsterdam: Elsevier.

Grant, F. S., and G. F. West. 1965. *Interpretation Theory in Applied Geophysics.* New York: McGraw-Hill.

Hardage, B. A. 1985. *Vertical Seismic Profiling, Part A: Principles,* 2d ed. London: Geophysical Press.

Howell, B. 1959. *Introduction to Geophysics.* New York: McGraw-Hill.

Hsieh, J. S. 1975. *Principles of Thermodynamics.* New York: McGraw-Hill.

Lamb, H. 1960. *Statics.* New York: Cambridge University Press.

Landau, L. D., and E. M. Lifshitz. 1986. *Theory of Elasticity,* 3d ed. Oxford: Pergamon Press.

Lapedes, D. N., ed. 1978. *McGraw-Hill Dictionary of Physics and Mathematics.* New York: McGraw-Hill.

Logan, J. D. 1987. *Applied Mathematics.* New York: John Wiley.

Love, A. E. H. 1911. *Some Problems of Geodynamics.* New York: Dover.

Love, A. E. H. 1944. *A Treatise on the Mathematical Theory of Elasticity.* New York: Dover.

MacBeth, C. 1990. Inversion of shear-wave polarizations for anisotropy using three-component offset VSPs. *Expanded Abstracts, 60th Annual International Meeting of the Society of Exploration Geophysicists,* pp. 1404–6. Tulsa: Society of Exploration Geophysicists.

Newman, P. 1973. Divergence effects in a layered earth. *Geophysics,* **38:** 481–8.

Postma, G. W. 1955. Wave propagation in a stratified medium. *Geophysics,* **20:** 780–806.

Press, F., and R. Siever. 1978. *Earth,* 2d ed. San Francisco: W. H. Freeman.

Saada, A. S. 1974. *Elasticity: Theory and Applications.* Oxford: Pergamon Press.

Savarensky, A. 1975. *Seismic Waves.* Moskow: MIR.

Schoenberg, M., and F. Muir. 1989. A calculus for finely layered media. *Geophysics,* **54:** 581–9.

Scholte, J. C. 1947. The range of existence of Rayleigh and Stoneley waves. *Royal Astron. Soc. Monthly Notices Geophys. Supp.,* Ser. A, **106:** 416–28.

Ségonzac, P. D., and J. Laherrére. 1959. Application of the continuous velocity log to anisotropy measurements in Northern Sahara: Results and consequences. *Geophys. Prosp.,* **7:** 202–17.

Stoep, P. M. 1966. Velocity anisotropy measurements in wells. *Geophysics,* **31:** 900–16.

Stoneley, R. 1924. Elastic waves at the surface of separation of two solids. *Proc. Roy. Soc. (London),* A-**106:** 416–28.

Stoneley, R. 1949. The seismological implications of aeolotropy in continental structures. *Monthly Notices, Roy. Astron. Soc. Geophys. Supp.,* **5:** 343–53.

Toksoz, M. N., and D. H. Johnston. 1981. *Seismic Wave Attenuation,* SEG Geophysical Reprint Series 2. Tulsa: Society of Exploration Geophysicists.

Trorey, A. W. 1970. A simple theory for seismic diffractions. *Geophysics,* **35:** 762–84.

Trorey, A. W. 1977. Diffractions for arbitrary source–receiver locations. *Geophysics,* **42:** 1177–82.

Uhrig, L. F., and F. A. van Melle. 1955. Velocity anisotropy in stratified media. *Geophysics,* **20:** 774–9.

Ward, R. W., and M. R. Hewitt. 1977. Monofrequency borehole traveltime survey. *Geophysics,* **42:** 1137–45.

White, J. E. 1965. *Seismic Waves – Radiation, Transmission, and Attenuation.* New York: McGraw-Hill.

White, J. E. 1983. *Underground Sound – Application of Seismic Waves.* Amsterdam: Elsevier.

Winterstein, D. F. 1990. Velocity anisotropy terminology for geophysicists. *Geophysics,* **55:** 1070–88.

Winterstein, D. F., and M. A. Meadows. 1990. Shear-wave polarizations and subsurface stress directions at Lost Hills Field. *Expanded Abstracts, 60th Annual International Meeting of the Society of Exploration Geophysicists,* pp. 1431–4. Tulsa: Society of Exploration Geophysicists.

3

Partitioning at an interface

Overview

The partitioning of energy at interfaces is the central phenomenon of seismic exploration. Boundary conditions permit calculating how wave energy is divided among reflected and transmitted waves. The most easily understood approach, in terms of displacements, yields Zoeppritz' equations, and the calculation in terms of potentials yields Knott's equations. Both approaches are given here, even though they accomplish the same purpose, because of the importance of partitioning.

For the simple but important case of normal incidence (§3.2), which is commonly assumed in most reflection work, Zoeppritz' (or Knott's) equations reduce to the familiar equation for the normal reflection coefficient. This equation states that the reflection amplitude, compared with the incident amplitude, varies directly as the change in acoustic impedance (the product of velocity and density). Examples of reflection coefficient magnitudes are given.

Reflection at nonnormal incidence (§3.3) leads to wave conversion and amplitude changes, especially near the critical angle. The Zoeppritz equations, while exact, do not give a feeling for how amplitudes depend on the various factors involved. Several approximations have been made in an effort to achieve an equation form that gives more insight into the changes expected for various situations. In particular, the variation of amplitudes with angle (§3.4), more commonly expressed as amplitude variation with offset (AVO), has become important in the last few years as an indicator of hydrocarbon gas.

Refraction methods are usually based on a simple head-wave concept (§3.5) (although the waves involved are, in fact, much more complex); the simple concept yields correct traveltimes and provides a basis for the interpretation of head-wave traveltimes.

3.1 Application of boundary conditions

3.1.1 General

The boundary conditions described in §2.4.4 lead to rather complex relations for reflection and refraction at an interface. The nature of the two media fixes the densities and elastic constants and thus the velocities. The angles of reflection and refraction are fixed in terms of the velocities, as will be shown. The only variables remaining to satisfy the boundary conditions are the amplitudes of the waves generated. When both media are solids, there are four equations resulting from the boundary conditions, so that we must have four variables. A P-wave (or an S-wave) incident on an interface separating two solids must in general generate reflected and refracted S-waves as well as reflected and refracted P-waves. Thus, for an incident P-wave, as shown in fig. 3.1, we have reflected and refracted P-waves at angles θ_1 and θ_2 and reflected and refracted S-waves at angles δ_1 and δ_2. The waves whose modes change at an interface (the reflected and refracted S-waves in the foregoing example) are called *converted waves*.

Note that S-waves have 2 degrees of freedom, and motion perpendicular to the plane containing the incident wave and the normal to the interface is not involved in conversion from P- to S-waves nor vice versa. Where the interface is horizontal, this is equivalent to saying that incident P-waves can generate reflected and refracted P- and SV-waves but not SH-waves, that incident SV-waves can generate P- and SV-waves, but that incident SH-waves generate only reflected and refracted SH-waves.

Snell's law can be extended to cover converted waves. In fig. 2.28, we change θ_1' to δ_1, the angle of reflection of the converted S-wave. Writing α_1 and β_1 for the P- and S-wave velocities, we have

$$\sin \theta_1 = B'R/A'R = \alpha_1 \, \Delta t/A'R,$$
$$\sin \delta_1 = A'S/A'R = \beta_1 \, \Delta t/A'R,$$

hence,

$$\frac{\sin \delta_1}{\beta_1} = \frac{\sin \theta_1}{\alpha_1}.$$

Following the same procedure for the refracted S-wave, we arrive at the general form of Snell's law:

$$\frac{\sin \theta_1}{\alpha_1} = \frac{\sin \delta_1}{\beta_1} = \frac{\sin \theta_2}{\alpha_2} = \frac{\sin \delta_2}{\beta_2} = p. \qquad (3.1)$$

Note that p is the component of the slowness of each ray in fig. 3.1 parallel to the interface. We conclude, therefore, that wave conversion, reflection, and transmission do not change the component of the slowness parallel to the interface.

3.1.2. Zoeppritz' equations

These equations (Zoeppritz, 1919) determine the amplitudes of the reflected and refracted waves at a plane

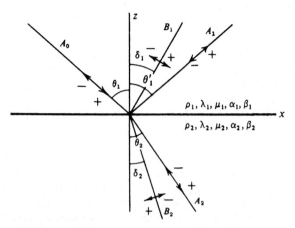

Fig. 3.1 Waves generated at a solid–solid interface by an incident P-wave. Displacement polarities assumed in §3.1.2 are indicated.

interface for an incident P-wave. We write A_0, A_1, A_2, B_1, and B_2 for the displacement amplitudes of the incident, reflected, and refracted P-waves, and for the reflected and refracted S-waves (see fig. 3.1), the plus signs indicating the positive directions of displacements. We consider waves traveling in the xz-plane, the interface being the xy-plane. We can write for the incident wave (compare with eq. (2.56))

$$\psi_0 = A_0 e^{j\omega(\ell x + nz)/\alpha_1}$$
$$= A_0 e^{j\omega(x \sin\theta_1 - z \cos\theta_1)/\alpha_1} = A_0 e^{j\omega p(x - z \cot\theta_1)},$$

where we have omitted the factor $e^{-j\omega t}$ because we shall not be differentiating with respect to time and the factor will cancel out in the boundary conditions. We can now write for the various waves in fig. 3.1,

$$\left.\begin{array}{lll} \psi_0 = A_0 e^{j\omega\zeta_0}, & \psi_1 = A_1 e^{j\omega\zeta_1}, & \psi_2 = A_2 e^{j\omega\zeta_2}, \\ & \psi_1' = B_1 e^{j\omega\zeta_1'}, & \psi_2' = B_2 e^{j\omega\zeta_2'}, \end{array}\right\} (3.2a)$$

where

$$\left.\begin{array}{l} \zeta_0 = p(x - z \cot\theta_1), \\ \zeta_1 = p(x + z \cot\theta_1), \quad \zeta_2 = p(x - z \cot\theta_2), \\ \zeta_1' = p(x + z \cot\delta_1), \quad \zeta_2' = p(x - z \cot\delta_2) \end{array}\right\} (3.2b)$$

(the minus signs mean that the wave propagation is in the direction of the negative z-axis).

Next we resolve the displacements ψ_i, ψ_i' into components along the x- and z-axes. This gives

$$\begin{array}{ll} u_1 = & A_0 \sin\theta_1 \, e^{j\omega\zeta_0} + A_1 \sin\theta_1 \, e^{j\omega\zeta_1} + B_1 \cos\delta_1 \, e^{j\omega\zeta_1'}, \\ u_2 = & A_2 \sin\theta_2 \, e^{j\omega\zeta_2} - B_2 \cos\delta_2 \, e^{j\omega\zeta_2'}, \\ w_1 = & -A_0 \cos\theta_1 \, e^{j\omega\zeta_0} + A_1 \cos\theta_1 \, e^{j\omega\zeta_1} - B_1 \sin\delta_1 \, e^{j\omega\zeta_1'}, \\ w_2 = & - A_2 \cos\theta_2 \, e^{j\omega\zeta_2} - B_2 \sin\delta_2 \, e^{j\omega\zeta_2'}. \end{array}$$

The boundary conditions require that when $z = 0$,

$$w_1 = w_2, \qquad u_1 = u_2, \qquad \sigma_{zz}|_1 = \sigma_{zz}|_2, \qquad \sigma_{xz}|_1 = \sigma_{xz}|_2.$$

At $z = 0$, all of the exponential factors reduce to $e^{j\omega px}$ so that the exponential factors can be omitted when formulating the boundary conditions. The first two boundary conditions give

$$-A_0 \cos\theta_1 + A_1 \cos\theta_1 - B_1 \sin\delta_1$$
$$= -A_2 \cos\theta_2 - B_2 \sin\delta_2, \qquad (3.3)$$
$$A_0 \sin\theta_1 + A_1 \sin\theta_1 + B_1 \cos\delta_1$$
$$= A_2 \sin\theta_2 - B_2 \cos\delta_2. \qquad (3.4)$$

By using eqs. (2.8), (2.9), (2.13), and (2.14), the last two boundary conditions become

$$\lambda_1(\partial u_1/\partial x + \partial w_1/\partial z) + 2\mu_1(\partial w_1/\partial z)$$
$$= \lambda_2(\partial u_2/\partial x + \partial w_2/\partial z) + 2\mu_2(\partial w_2/\partial z),$$
$$\mu_1(\partial u_1/\partial z + \partial w_1/\partial x)$$
$$= \mu_2(\partial u_2/\partial z + \partial w_2/\partial x).$$

Because

$$\partial\psi_i/\partial x = j\omega p\psi_i, \qquad \partial\psi_i'/\partial x = j\omega p\psi_i',$$
$$\partial\psi_i/\partial z = j\omega p(\pm\cot\theta_i)\psi_i, \quad \partial\psi_i'/\partial z = j\omega p(\pm\cot\delta_i)\psi_i',$$

we can drop the factor $j\omega p$ and consider that $\partial/\partial x = 1$ and $\partial/\partial z = \pm\cot\theta_i$ or $\pm\cot\delta_i$. We now get for the normal stress condition

$$\lambda_1(A_0 \sin\theta_1 + A_1 \sin\theta_1 + B_1 \cos\delta_1)$$
$$+ (\lambda_1 + 2\mu_1)(A_0 \cos\theta_1 \cot\theta_1$$
$$+ A_1 \cos\theta_1 \cot\theta_1 - B_1 \cos\delta_1)$$
$$= \lambda_2(A_2 \sin\theta_2 - B_2 \cos\delta_2)$$
$$+ (\lambda_2 + 2\mu_2)(A_2 \cos\theta_2 \cot\theta_2 + B_2 \cos\delta_2);$$

then

$$[\lambda_1 \sin\theta_1 + (\lambda_1 + 2\mu_1)\cos\theta_1 \cot\theta_1](A_0 + A_1)$$
$$- (2\mu_1 \cos\delta_1)B_1$$
$$= [\lambda_2 \sin\theta_2 + (\lambda_2 + 2\mu_2)\cos\theta_2 \cot\theta_2]A_2$$
$$+ (2\mu_2 \cos\delta_2)B_2,$$

or

$$\left[\frac{(\lambda_1 + 2\mu_1) - 2\mu_1 \sin^2\theta_1}{\sin\theta_1}\right](A_0 + A_1) - (2\mu_1 \cos\delta_1)B_1$$
$$= \left[\frac{(\lambda_2 + 2\mu_2) - 2\mu_2 \sin^2\theta_2}{\sin\theta_2}\right]A_2 + (2\mu_2 \cos\delta_2)B_2.$$

By using eqs. (2.58), (2.59), and (3.1),

$$\left(\frac{\rho_1\alpha_1^2 - 2\rho_1\beta_1^2 \sin^2\theta_1}{p\alpha_1}\right)(A_0 + A_1) - (2\rho_1 \beta_1^2 \cos\delta_1)B_1$$
$$= \left(\frac{\rho_2\alpha_2^2 - 2\rho_2\beta_2^2 \sin^2\theta_2}{p\alpha_2}\right)A_2 + (2\rho_2 \beta_2^2 \cos\delta_2)B_2.$$

Hence,

$$\frac{\rho_1\alpha_1^2(1 - 2\sin^2\delta_1)}{p\alpha_1}(A_0 + A_1) - \frac{2\rho_1\beta_1(\sin\delta_1 \cos\delta_1)}{p}B_1$$
$$= \frac{\rho_2\alpha_2^2(1 - 2\sin^2\delta_2)}{p\alpha_2}A_2 + \frac{2\rho_2\beta_2(\sin\delta_2 \cos\delta_2)}{p}B_2.$$

Finally,

$$(A_0 + A_1)Z_1 \cos 2\delta_1 - B_1 W_1 \sin 2\delta_1$$
$$= A_2 Z_2 \cos 2\delta_2 + B_2 W_2 \sin 2\delta_2, \qquad (3.5)$$

where $Z_i = \rho_i \alpha_i$ and $W_i = \rho_i \beta_i$; Z_i and W_i are called *acoustic impedances*.

The last boundary condition gives

$$\mu_1\{[(-A_0 + A_1)\cos\theta_1 + B_1\cos\delta_1\cot\delta_1]$$
$$+ [(-A_0 + A_1)\cos\theta_1 - B_1\sin\delta_1]\}$$
$$= \mu_2\{(-A_2\cos\theta_2 + B_2\cos\delta_2\cot\delta_2)$$
$$+ (-A_2\cos\theta_2 - B_2\sin\delta_2)\},$$

that is,

$$\mu_1\{(-A_0 + A_1)2\cos\theta_1 + B_1(\cos\delta_1\cot\delta_1 - \sin\delta_1)\}$$
$$= \mu_2\{-2A_2\cos\theta_2 + B_2(\cos\delta_2\cot\delta_2 - \sin\delta_2)\}.$$

This becomes

$$(\rho_1\beta_1^2/p\beta_1)\,[(-A_0 + A_1)2\cos\theta_1\sin\delta_1 + B_1\cos2\delta_1]$$
$$= (\rho_2\beta_2^2/p\beta_2)(-2A_2\cos\theta_2\sin\delta_2 + B_2\cos2\delta_2),$$

or

$$(\beta_1/\alpha_1)W_1(-A_0 + A_1)\sin2\theta_1 + W_1 B_1\cos2\delta_1$$
$$= -(\beta_2/\alpha_2)W_2 A_2\sin2\theta_2 + W_2 B_2\cos2\delta_2. \quad (3.6)$$

Collecting together the results of applying the boundary conditions, we have the four Zoepprtiz equations:

$$(-A_0 + A_1)\cos\theta_1 - B_1\sin\delta_1$$
$$= -A_2\cos\theta_2 - B_2\sin\delta_2, \quad (3.3)$$

$$(A_0 + A_1)\sin\theta_1 + B_1\cos\delta_1$$
$$= A_2\sin\theta_2 - B_2\cos\delta_2, \quad (3.4)$$

$$(A_0 + A_1)Z_1\cos2\delta_1 - B_1 W_1\sin2\delta_1$$
$$= A_2 Z_2\cos2\delta_2 + B_2 W_2\sin2\delta_2, \quad (3.5)$$

$$(-A_0 + A_1)(\beta_1/\alpha_1)W_1\sin2\theta_1 + B_1 W_1\cos2\delta_1$$
$$= -A_2(\beta_2/\alpha_2)W_2\sin2\theta_2 + B_2 W_2\cos2\delta_2. \quad (3.6)$$

Thus, for a *P*-wave of given amplitude A_0 incident at the angle θ_1 on a plane interface separating two media with given values of ρ, μ, α, and β, Snell's law determines the angles θ_i and δ_i, whereas Zoeppritz' equations fix the reflected and refracted amplitudes A_i and B_i. Similar equations can be derived for an incident *S*-wave. For a fluid medium, $B_i = 0$ because only *P*-waves are propagated.

3.1.3. Knott's equations

Knott (1899) was the first to derive equations giving the amplitudes of reflected and refracted waves generated at a plane interface. He used the displacement potential functions ϕ and χ of eq. (2.64) in the form:
Medium (1):

$$\phi_1 = \mathscr{A}_0 e^{j\omega\zeta_0} + \mathscr{A}_1 e^{j\omega\zeta_1},$$
$$\chi_1 = \mathscr{B}_1 e^{j\omega\zeta_1'},$$

Medium (2):

$$\phi_2 = \mathscr{A}_2 e^{j\omega\zeta_2},$$
$$\chi_2 = \mathscr{B}_2 e^{j\omega\zeta_2'},$$

where ζ_0, ζ_1, ζ_2, ζ_1', and ζ_2' have the same meaning as in eq. (3.2b). As in §3.1.2, the time factor $e^{-j\omega t}$ is omitted, derivatives with respect to x and z are taken as

equivalent to multiplication by 1 and by $\pm \cot\theta_i$ and $\pm \cot\delta_i$, respectively, and the common factor $e^{j\omega px}$ is omitted after the differentiation. Problem 2.11 gives the expressions for the normal and tangential displacements and stresses in terms of derivatives of ϕ and χ.

The first boundary condition requires the continuity of normal displacements at the interface, that is,

$$\left(\frac{\partial\phi}{\partial z} - \frac{\partial\chi}{\partial x}\right)_1 = \left(\frac{\partial\phi}{\partial z} - \frac{\partial\chi}{\partial x}\right)_2, \qquad z = 0,$$

hence,

$$(-\mathscr{A}_0 + \mathscr{A}_1)\cot\theta_1 - \mathscr{B}_1 = -\mathscr{A}_2\cot\theta_2 - \mathscr{B}_2.$$

The next condition is that the shear displacements be equal:

$$\left(\frac{\partial\phi}{\partial x} + \frac{\partial\chi}{\partial z}\right)_1 = \left(\frac{\partial\phi}{\partial x} + \frac{\partial\chi}{\partial z}\right)_2, \qquad z = 0,$$

or

$$(\mathscr{A}_0 + \mathscr{A}_1) + \mathscr{B}_1\cot\delta_1 = \mathscr{A}_2 - \mathscr{B}_2\cot\delta_2.$$

The continuity of normal stress requires that

$$\lambda\nabla^2\phi + 2\mu\left(\frac{\partial^2\phi}{\partial z^2} - \frac{\partial^2\chi}{\partial x\,\partial z}\right)$$

be continuous; thus,

$$\lambda_1(\mathscr{A}_0 + \mathscr{A}_1)(1 + \cot^2\theta_1)$$
$$+ 2\mu_1[(\mathscr{A}_0 + \mathscr{A}_1)\cot^2\theta_1 - \mathscr{B}_1\cot\delta_1]$$
$$= \lambda_2\mathscr{A}_2(1 + \cot^2\theta_2) + 2\mu_2(\mathscr{A}_2\cot^2\theta_2 + \mathscr{B}_2\cot\delta_2).$$

By using eqs. (2.58), (2.59), and (3.1), this becomes

$$\mu_1(\cot^2\delta_2 - 1)(\mathscr{A}_0 + \mathscr{A}_1) - 2\mu_1\mathscr{B}_1\cot\delta_1$$
$$= \mu_2(\cot^2\delta_2 - 1)\mathscr{A}_2 + 2\mu_2\mathscr{B}_2\cot\delta_2.$$

Continuity of the tangential stress means that

$$\mu\left(2\frac{\partial^2\phi}{\partial x\,\partial z} + \frac{\partial^2\chi}{\partial z^2} - \frac{\partial^2\chi}{\partial x^2}\right)$$

is continuous; thus,

$$\mu_1\,[2(-\mathscr{A}_0 + \mathscr{A}_1)\cot\theta_1 + \mathscr{B}_1(\cot^2\delta_1 - 1)]$$
$$= \mu_2\,[-2\mathscr{A}_2\cot\theta_2 + \mathscr{B}_2(\cot^2\delta_2 - 1)].$$

If we substitute $a_i = \cot\theta_i$, $b_i = \cot\delta_i$, and $c_i = b_i^2 - 1$, the preceding equations become

$$-a_1\mathscr{A}_0 + a_1\mathscr{A}_1 - \mathscr{B}_1 = -a_2\mathscr{A}_2 - \mathscr{B}_2, \quad (3.7)$$

$$\mathscr{A}_0 + \mathscr{A}_1 + b_1\mathscr{B}_1 = \mathscr{A}_2 - b_2\mathscr{B}_2, \quad (3.8)$$

$$\mu_1 c_1\mathscr{A}_0 + \mu_1 c_1\mathscr{A}_1 - 2\mu_1 b_1\mathscr{B}_1$$
$$= \mu_2 c_2\mathscr{A}_2 + 2\mu_2 b_2\mathscr{B}_2, \quad (3.9)$$

$$-2\mu_1 a_1\mathscr{A}_0 + 2\mu_1 a_1\mathscr{A}_1 + \mu_1 c\mathscr{B}_1$$
$$= -2\mu_2 a_2\mathscr{A}_2 + \mu_2 c_2\mathscr{B}_2. \quad (3.10)$$

(Note that $\mathscr{A}_i$ and $\mathscr{B}_i$ in these equations are the amplitudes of the potential functions ϕ and χ, not of the displacements as in eqs. (3.3) to (3.6)). These equations can also be obtained by subsituting equations of the form of eq. (3.13) into eqs. (3.3) to (3.6).

3.1.4. Distribution of energy

Knott's equations have a very interesting property. If we multiply corresponding sides of the first and third equations and also corresponding sides of the second and fourth equations, and then add these products, the result is

$$(\rho_1 \cot \theta_1)\mathscr{A}_1^2 + (\rho_1 \cot \delta_1)\mathscr{B}_1^2 + (\rho_2 \cot \theta_2)\mathscr{A}_2^2$$
$$+ (\rho_2 \cot \delta_2)\mathscr{B}_2^2 = (\rho_1 \cot \theta_1)\mathscr{A}_0^2 \quad (3.11)$$

Because the first and third Knott equations relate to the normal displacement and stress on the two sides of the interface whereas the second and fourth relate to the tangential displacement and stress, the products have the dimensions of energy per unit area. From this we make the correct surmise that eq. (3.11) gives the distribution of energy among the various reflected and refracted waves. To demonstrate this, we repeat the derivation of eq. (2.105) in terms of the potential function ϕ (χ does not enter here because it concerns S-waves). For the incident P-wave, we have for the kinetic energy per unit volume E_K

$$\frac{\delta E_K}{\delta \mathscr{V}} = \frac{1}{2}\rho_1 \left[\left(\frac{\partial u}{\partial t}\right)^2 + \left(\frac{\partial w}{\partial t}\right)^2 \right]$$
$$= \frac{1}{2}\rho_1 \left[\left(\frac{\partial^2 \phi}{\partial x \partial t}\right)^2 + \left(\frac{\partial^2 \phi}{\partial z \partial t}\right)^2 \right].$$

Using eqs. (3.2) and noting that we must take $\partial/\partial x = j\omega p$, $\partial/\partial z = -(j\omega \cos \theta_1)/\alpha_1$, and $\partial/\partial t = -j\omega$, we have for $z = 0$,

$$\frac{\delta E_K}{\delta \mathscr{V}} = \frac{1}{2}\rho_1 \left\{ (+\omega^2 p\mathscr{A}_0)^2 \right.$$
$$\left. + \left[\frac{+\omega^2(\cos \theta_1)\mathscr{A}_0}{\alpha_1} \right]^2 \right\} e^{j\omega[(x \sin \theta_1)/\alpha_1 - t]}.$$

Taking the maximum of the real part, we find the expression for the energy density E, that is,

$$E = \frac{1}{2}\rho_1 \omega^4 (\mathscr{A}_0/\alpha_1)^2. \quad (3.12)$$

Comparison with eq. (2.105) shows that

$$\mathscr{A}_0 = (\alpha_1/\omega)A_0, \quad (3.13)$$

where A_0 is the amplitude of the displacement in the direction of propagation.

The energy brought up to a unit area of the interface per unit time by the incident P-wave will be the energy in a cylinder of length α_1, of unit cross-section (measured parallel to the interface), and inclined at an angle θ_1 to the normal to the surface, that is,

(Volume of cylinder) $\times E = (\alpha_1 \cos \theta_1)(\rho_1 \omega^4 \mathscr{A}_1^2/2\alpha_1^2)$
$$= \frac{1}{2}p\rho_1 \omega^4 \mathscr{A}_0^2 (\cot \theta_1).$$

Because similar expressions must hold for the energy carried away by the other waves, we see that we have only to multiply each term in eq. (3.11) by $\frac{1}{2}\rho\omega^4$ to obtain the distribution of energy among the various waves.

3.2 Partitioning at normal incidence

Zoepprtiz' equations reduce to a very simple form for normal incidence. The curves change slowly for small angels of incidence (say, up to 15°), so the results for normal incidence have wide application. For a P-wave at normal incidence, there are no tangential stresses and displacements; hence, $B_1 = B_2 = 0$ and eqs. (3.3) to (3.6) reduce to

$$A_1 + A_2 = A_0,$$
$$Z_1 A_1 - Z_2 A_2 = -Z_1 A_0.$$

The solution of these equations is

$$R = \frac{A_1}{A_0} = \frac{\alpha_2\rho_2 - \alpha_1\rho_1}{\alpha_2\rho_2 + \alpha_1\rho_1} = \frac{Z_2 - Z_1}{Z_2 + Z_1} \approx \frac{\Delta Z}{2Z}$$
$$\left. \approx \frac{1}{2}\Delta(\ln Z) \approx \frac{1}{2}(\Delta\alpha/\alpha + \Delta\rho/p), \right\} \quad (3.14)$$

$$T = \frac{A_2}{A_0} = \frac{2\alpha_1\rho_1}{\alpha_2\rho_2 + \alpha_1\rho_1} = \frac{2Z_1}{Z_2 + Z_1}. \quad (3.15)$$

Equations (3.14) and (3.15) give the reflection coefficient R and the transmission coefficient T. Equation (3.14) shows that the amplitudes of a sequence of isolated reflections are a record of changes in the log of acoustic impendances, the viewpoint taken in seismic log manufacture (§5.4.5). The fractions of energy reflected and transmitted are given by E_R and E_T, respectively (which are also sometimes called reflection and transmission energy coefficients):

$$E_R = \frac{\alpha_1\rho_1\omega^2 A_1^2}{\alpha_1\rho_1\omega^2 A_0^2} = \left(\frac{Z_2 - Z_1}{Z_2 + Z_1}\right)^2 = R^2, \quad (3.16)$$

$$E_T = \frac{\alpha_2\rho_2\omega^2 A_2^2}{\alpha_1\rho_1\omega^2 A_0^2} = \frac{4Z_1 Z_2}{(Z_2 + Z_1)^2} = \frac{Z_2}{Z_1} T^2. \quad (3.17)$$

Obviously, $E_R + E_T = 1$. Note that eqs. (3.16) and (3.17) are unchanged if Z_1 and Z_2 are interchanged; hence, the energy partition does not depend upon which medium contains the incident wave. When $Z_2/Z_1 = 1$, $R = E_R = 0$, and all the energy is transmitted; note that this does not require that $\rho_1 = \rho_2$ and $\alpha_1 = \alpha_2$. As the impedance contrast approaches zero or infinity, T approaches zero and R approaches unity; thus, the farther the impedance contrast is from unity, the stronger the reflected energy.

Table 3.1 shows how the reflected energy varies for impedance contrasts such as may be expected within the earth. Because both density and velocity contrasts are small for most of the interfaces encountered, only

Table 3.1 *Energy reflected at interface between two media*

Interface	First medium		Second medium		Z_1/Z_2	R	E_R
	Velocity	Density	Velocity	Density			
Sandstone on limestone	2.0	2.4	3.0	2.4	0.67	0.2	0.040
Limestone on sandstone	3.0	2.4	2.0	2.4	1.5	−0.2	0.040
Shallow interface	2.1	2.4	2.3	2.4	0.93	0.045	0.0021
Deep interface	4.3	2.4	4.5	2.4	0.97	0.022	0.0005
"Soft" ocean bottom	1.5	1.0	1.5	2.0	0.50	0.33	0.11
"Hard" ocean botom	1.5	1.0	3.0	2.5	0.20	0.67	0.44
Surface of ocean (from below)	1.5	1.0	0.36	0.0012	3800	−0.9994	0.9988
Base of weathering	0.5	1.5	2.0	2.0	0.19	0.68	0.47
Shale over water sand	2.4	2.3	2.5	2.3	0.96	0.02	0.0004
Shale over gas sand	2.4	2.3	2.2	1.8	1.39	−0.16	0.027
Gas sand over water sand	2.2	1.8	2.5	2.3	0.69	0.18	0.034

All velocities in km/s, densities in g/cm³; the minus signs indicate 180° phase reversal.

a small portion of the energy is reflected at any one interface; this is illustrated by the first four lines in table 3.1. The "sandstone-on-limestone" interface is about as large a contrast as is apt to be encountered, whereas the "shallow interface" and "deep interface" figures are much more typical of most interfaces in the earth; hence, usually appreciably less than 1% of the energy is reflected at any interface. The major exceptions involve the bottom and surface of the ocean and the surface and base of the weathering (see §5.3.2). A much larger proportion of the energy can be reflected from these, and hence they are especially important in the generation of multiple reflections (§6.3.2) and other phenomena with which we shall deal later.

Note that although the energy fractions E_R and E_T do not depend on which side of an interface the wave is incident, this is not true of the reflected amplitude $\mathscr{A}_1$ because interchanging Z_1 and Z_2 in eq. (3.14) changes the sign of the ratio $\mathscr{A}_1/\mathscr{A}_0$. A negative value of $\mathscr{A}_1$ means that the reflected wave is 180° out-of-phase with the incident wave; thus, for an incident wave $\mathscr{A}_0 \cos \omega t$ the reflected wave is $\mathscr{A}_1 \cos (\omega t + \pi)$. In table 3.1, phase reversal occurs for the situations where Z_1 exceeds Z_2

3.3 Partitioning at nonnormal incidence

Turning now to the general case where the angle of incidence is not necessarily 0°, fig. 3.2 shows energy partition as functions of the angle of incidence for certain values of parameters. Many curves would be required to show the variations of energy partitioning as a function of incident angle because of the many parameters that can be varied: incident P-, SH-, or SV-wave, P-wave velocity ratio, density ratio, and S-wave

velocities in each medium (or the equivalent of defining Poisson's ratio for each medium).

Figure 3.2a shows the partitioning of energy as a function of the angle of incidence when a P-wave is incident in the high-velocity medium for a P-wave velocity ratio $\alpha_2/\alpha_1 = 0.5$, a density ratio $\rho_2/\rho_1 = 0.8$, $\sigma_1 = 0.30$, and $\sigma_2 = 0.25$. For small incident angles, all of the energy is in the reflected or transmitted P-waves, E_{RP} and E_{TP}, respectively, and hence there are essentially no S-waves. As the incident angle increases, some of the energy goes into reflected and transmitted S-waves, E_{RS} and E_{TS}, respectively, mostly at the expense of the reflected P-wave. Note that at intermediate angles of incidence, the reflected S-wave carries more energy than the reflected P-wave. Such converted waves (waves resulting from the conversion of P-waves to S-waves or vice versa at an interface) are sometimes recorded at long offsets where they are evidenced by alignments that disappear as one tries to follow them to shorter offsets (see $(R)_{PS}$ in fig. 6.28b). As grazing incidence is approached, the energy of the reflected P-wave increases until at grazing incidence all of the energy is in the reflected P-wave.

The opposite situation is shown in fig. 3.2.b, where $\alpha_2/\alpha_1 = 2.0$, $\rho_2/\rho_1 = 0.5$, $\sigma_1 = 0.30$, and $\sigma_2 = 0.25$. Because $Z_1 = Z_2$, the P-wave reflection coefficient is essentially zero for small incident angles. As the incident angle increases, S-wave energy increases. As the critical angle for P-waves is approached, the transmitted P-wave energy falls rapidly to zero and no transmitted P-wave exists for larger incident angles. Also, as the critical angle for P-waves is approached, both reflected P-wave and reflected S-wave become very strong; such a buildup in reflection strength near the critical angle is called *wide-angle reflection*. Sometimes it is possible to make use of this phenomenon to map reflectors using long offsets where they cannot be fol-

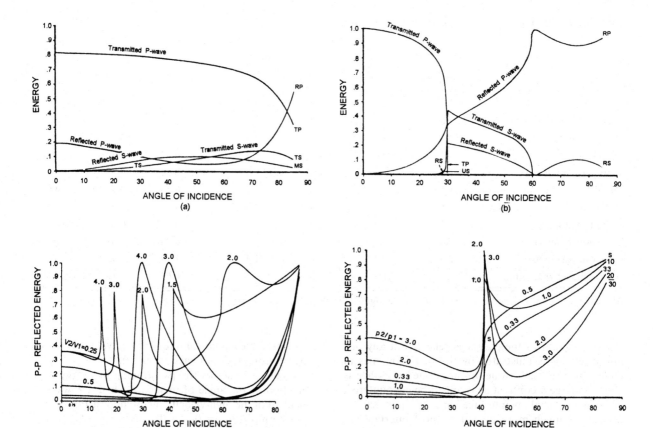

Fig. 3.2 Partitioning of energy between transmitted and reflected waves as a function of angle of incidence for the case of an incident P-wave. (From Tooley, Spencer, and Sagoci, 1965, except (c) from Denham and Palmeira, 1984.) (a) Case where the velocity in the incident medium is larger; $\alpha_2/\alpha_1 = 0.5$, $\rho_2/\rho_1 = 0.8$, $\sigma_1 = 0.3$, and $\sigma_2 = 0.25$. (b) Case where the velocity in the incident medium is smaller; $\alpha_2/\alpha_1 = 2.0$, $\rho_2/\rho_1 = 0.5$, $\sigma_1 = 0.3$, and $\sigma_2 = 0.25$. (c) Fraction of energy reflected as a P-wave for various P-wave velocity ratios; $\rho_2/\rho_1 = 1.0$ and $\sigma_1 = \sigma_2 = 0.25$. (d) Fraction of energy reflected as a P-wave for various density ratios and $\alpha_2/\alpha_1 = 1.5$ and $\sigma_1 = \sigma_2 = 0.25$.

lowed at short offsets (Meissner, 1967). As the critical angle for S-waves is approached, the transmitted S-wave falls to zero.

If we had not had a density contrast but otherwise the values had been as indicated in fig. 3.2b, there would have been a reflected P-wave at small incident angles whose fractional energy would have decreased slightly as the incident angle increased.

Figure 3.2c shows the P-wave reflection coefficient for various P-wave velocity ratios when $\rho_1 = \rho_2$ and $\sigma_1 = \sigma_2 = 0.25$. The reflected energy is zero for a velocity ratio of 1 (no impedance contrast) and increases both as the ratio becomes larger than 1 and as it becomes smaller than 1. The two peaks for $\alpha_2/\alpha_1 > 1$ occur at the critical angles for P- and S-waves, respectively. Figure 3.2d shows the energy of the reflected P-wave for various density contrasts when $\alpha_2/\alpha_1 = 1.5$ and $\sigma_1 = \sigma_2 = 0.25$.

Koefoed (1962) gives 100 tables of the longitudinal and transverse reflection and transmission coefficients and the phase shifts for angles greater than the critical angle (see §2.7.5) for incident longitudinal waves.

3.4 Variation of amplitude with angle (AVA)

The four Zoeppritz equations, eqs. (3.3) to (3.6), contain four unknowns, A_1, B_1, A_2, and B_2. Dividing through by A_0, we can solve for the four reflection and transmission coefficients $R_P = A_1/A_0$, $R_S = B_1/A_0$, $T_P = A_2/A_0$, and $T_S = B_2/A_0$, using either Cramer's rule (eq. (15.3b)) or matrices (eq. (15.22)). Aki and Richards (1980: chap. 5) derived expressions for these for P- and S-waves incident on each of the five combinations of solid, fluid, and vacuum half-spaces. The expressions are, however, quite complex.

In the following, we assume that $\Delta\rho/\rho$, $\Delta\alpha/\alpha$, and $\Delta\beta/\beta$ are all small, ρ, α, and β being averages (note that $\Delta\sigma/\sigma$ need not be small; see problem 3.10). These conditions are satisfied for almost all sedimentary situations where $R_0 < 0.2$ (R_0 is the reflection coefficient at normal incidence); in this case, the changes in the raypath direction are small. To show this we let $\theta_2 = \theta_1 + \Delta\theta$, $\theta = (\theta_1 + \theta_2)/2 \approx \theta_1$; then

$$\sin\theta_2 = \sin(\theta_1 + \Delta\theta) = \sin\theta_1\cos\Delta\theta$$
$$+ \cos\theta_1\sin\Delta\theta \approx \sin\theta_1 + \Delta\theta\cos\theta_1.$$

By using Snell's law:

$$\sin \theta_2/\sin \theta_1 \approx 1 + \Delta\theta \cot \theta \approx \alpha_2/\alpha_1$$
$$\approx (\alpha_1 + \Delta\alpha)/\alpha_1 \approx 1 + \Delta\alpha/\alpha,$$

so $\Delta\theta = (\Delta\alpha/\alpha)\tan \theta$.

Aki and Richards (1980: eq. (5.44)) expand the terms in the exact expressions for the reflection and transmission coefficients for a P-wave incident on a solid–solid interface, assuming that the squares and products of differentials are sufficiently small that they can be dropped. For the reflected and transmitted P-waves, they get

$$R_P \approx \frac{1}{2}\left[1 - 4\left(\frac{\beta^2}{\alpha^2}\sin^2\theta\right)\frac{\Delta\rho}{\rho} + \frac{1}{2}\sec^2\theta\left(\frac{\Delta\alpha}{\alpha}\right)\right.$$
$$\left. - 4\left(\frac{\beta^2}{\alpha^2}\right)\sin^2\theta\left(\frac{\Delta\beta}{\beta}\right)\right], \qquad (3.18)$$

$$T_P \approx 1 - \frac{1}{2}\left(\frac{\Delta\rho}{\rho}\right) + \left(\frac{1}{2}\sec^2\theta - 1\right)\left(\frac{\Delta\alpha}{\alpha}\right). \qquad (3.19)$$

These equations are valid when (a) $\Delta\alpha/\alpha$, $\Delta\beta/\beta$, and $\Delta\rho/\rho$ are small (hence, $\Delta\theta$ and $\Delta\delta$ are also small), (b) $\theta < 80°$ if $\alpha_2 < \alpha_1$, or (c) $\theta < 10°$ if $\alpha_2 > \alpha_1$. Hilterman (private communication) believes condition (c) can be relaxed to $\theta < \theta_c - 10°$ if $\alpha_2 > \alpha_1$, where θ_c is the critical angle. The form of the equations allows one to see the separate effects of changes in density and P- or S-wave velocities, which are difficult to see from the exact expressions. Equation (3.18) is frequently used to find the amplitude variation with offset (AVO) for reflected P-waves.

Shuey (1985) decided that Poisson's ratio was the elastic constant most directly related to the variation of R_P with θ, and therefore used eq. (2.60) to replace β with σ. Taking the log of eq. (2.60) and differentiating, we have

$$\frac{\Delta\beta}{\beta} = \frac{\Delta\alpha}{\alpha} + \frac{1}{2}\Delta\sigma\left(\frac{1}{1-\sigma} - \frac{2}{1-2\sigma}\right).$$

Substituting into eq. (3.18) and dividing by R_0, we have

$$R_P/R_0 \approx 1 + P\sin^2\theta + Q(\tan^2\theta - \sin^2\theta),$$

where $R_0 \approx (\Delta\alpha/\alpha + \Delta\rho/\rho)/2$ from eq. (3.14),

$$P = \left[Q - \frac{2(1+Q)(1-2\sigma)}{1-\sigma}\right] + \frac{\Delta\sigma}{R_0(1-\sigma)^2},$$

$$Q = \frac{\Delta\alpha/\alpha}{\dfrac{\Delta\alpha}{\alpha} + \dfrac{\Delta\rho}{\rho}} = \frac{1}{1 + \dfrac{\Delta\rho/\rho}{\Delta\alpha/\alpha}}. \qquad \left.\right\} \ (3.20)$$

The last term in eq. (3.20) is almost always positive and thus increases R_P at very large angles of incidence ($> 30°$), but it is small for the angles usually involved in reflection seismology. The middle term in eq. (3.20) governs R_P at intermediate angles; this term can have either sign and it is mainly the factor $\Delta\sigma/R_0$ that determines whether R_P decreases or increases with θ. R_P can change sign with increasing θ if the middle term in eq. (3.20) has polarity opposite to that of R_0. In some young clastic sections, $\Delta\rho/\rho \approx -\Delta\alpha/\alpha$, which causes calculation difficulties. For $\theta < 30°$, eq. (3.20) can be further simplified (Shuey, 1985: 612) to

$$R_P/R_0 \approx 1 + P\theta^2. \qquad (3.21)$$

Hilterman (private communication) rewrites eq. (3.20) in the form

$$R_P \approx R_0\left[1 - 4\left(\frac{\beta}{\alpha}\right)^2\sin^2\theta\right] + \frac{\Delta\sigma}{(1-\sigma)^2}\sin^2\theta$$
$$+ R_0\frac{\Delta\alpha}{2\alpha}\left[\tan^2\theta - 4\left(\frac{\beta}{\alpha}\right)^2\sin^2\theta\right], \qquad (3.22a)$$

where most of the dependence on σ is in the middle term. The relative contributions of the three terms in eq. (3.22a) are shown in fig. 3.3 for one situation. Hilterman further approximates eq. (3.22a) as

$$R_P \approx R_0 \cos^2\theta + 2.25 \Delta\sigma \sin^2\theta. \qquad (3.22b)$$

Ostrander (1984) applied these results to practical cases (fig. 3.4). The reflection coefficient becomes more negative with increasing incident angles (fig. 3.4b) and vice versa (fig. 3.4c). For practical reflection cases, three possible results exist:

1. For little change in Poisson's ratio (fig. 3.4a), the amplitude decreases with increasing incident angle regardless of the polarity of the reflection coefficient.
2. For (a) a positive reflection coefficient and an increase in Poisson's ratio (which is apt to be true for a gas/water contact or the base of a gas sand embedded in shale), or (b) a negative reflection coefficient and decrease in Poisson's ratio (which is apt to be true for the top of a gas sand embedded in shale), the amplitude increases with incident angle.
3. For (a) a positive reflection coefficient and a decrease in Poisson's ratio, or (b) a negative reflection coefficient and an increase in Poisson's ratio, the amplitude decreases with incident angle at first and then the waveform reverses polarity and the amplitude increases with opposite polarity. This is apt to be true for high-impedance reservoirs.

Where a porous sandstone encased in shale is water-saturated, Poisson's ratio is apt to be slightly smaller for sand than for shale, but when the pore space is filled with gas, Poisson's ratio for sandstone is apt to be much smaller than for shale. Consequently, the variation of amplitude with angle of incidence (AVA) is often regarded as a hydrocarbon indicator

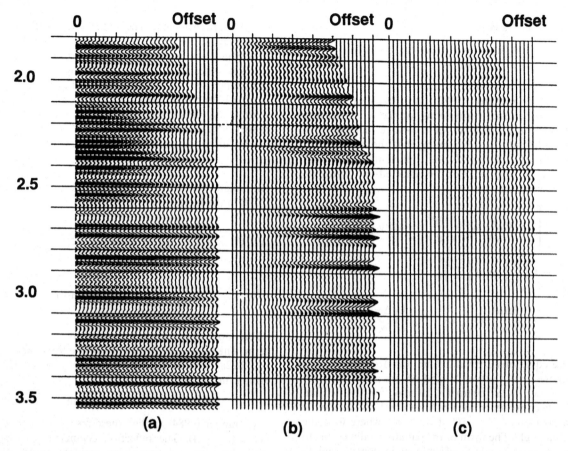

Fig. 3.3 Contribution of the three terms in eq. (3.22a) as a function of offset, for data from a well offshore Vermillion Parish, Louisiana. (a), (b), and (c) refer to the first, second, and third terms, respectively. (Courtesy of Geophysical Development Co.)

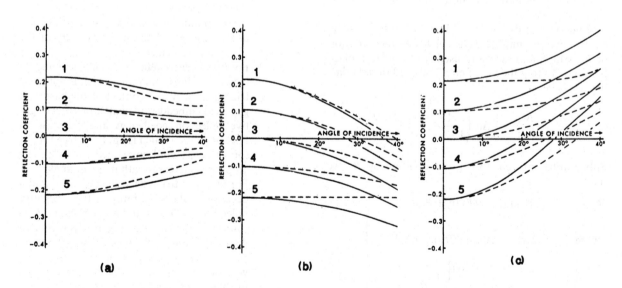

Fig. 3.4 Variation of a *P*-wave reflection coefficient with angle of incidence. For curves 1, $\alpha_2/\alpha_1 = \rho_2/\rho_1 = 1.25$; for 2, 1.11; for 3, 1.0; for 4, 0.9; and for 5, 0.8. (From Ostrander, 1984.) (a) No change in Poisson's ratio at the interface (solid curves, $\sigma_1 = \sigma_2 = 0.3$; dashed, $\sigma_1 = \sigma_2 = 0.2$). (b) Decreasing Poisson's ratio (solid, $\sigma_1 = 0.4$, $\sigma_2 = 0.1$; dashed, $\sigma_1 = 0.3$, $\sigma_2 = 0.1$). (c) Increasing Poisson's ratio (solid, $\sigma_1 = 0.1$, $\sigma_2 = 0.4$; dashed, $\sigma_1 = 0.1$, $\sigma_2 = 0.2$).

(§10.8). A common situation in young clastic sediments (the "bright-spot" case, §10.8) is for a sand to have an acoustic impedance nearly the same as surrounding shale when liquid-filled but a much lower acoustic impedance when gas-filled. The result is a strong negative reflection marking the top of a gas sand and a strong positive reflection at the base (or at the fluid contact between the gas- and liquid-filled portions); both reflections then increase in amplitude with incident angle, as shown in fig. 3.5. However, changes in parameter values can alter these responses. Poisson's ratio can change where no reservoir is pres-

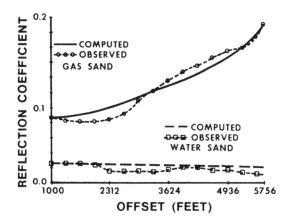

Fig. 3.5 Variation of amplitude with offset for a gas sand and a water sand. (From Yu, 1985.)

ent and also where only a little gas is present, so that an AVO anomaly does not necessary indicate a commercial reservoir. The measurement of AVO is also fraught with measurement and processing difficulties (see Castagna and Backus, 1993; Allen and Peddy, 1993).

3.5 Head waves

In refraction seismology, we make use of waves that have been refracted at the critical angle (figs. 3.6a and 3.6b); these waves are called *head waves, conical waves,* or merely "refractions." The name "conical waves" arises because the wavefronts obtained when we rotate fig. 3.6b about the vertical axis OL are conical surfaces generated by wavefronts such as RQ. Head waves were first postulated by Mohorovičić in 1909 to explain earthquake observations.

In fig. 3.6a, we see a P-wave incident on the refracting horizon at the critical angle θ_c. After refraction, it travels along the interface in the lower medium. This produces an oscillatory motion parallel to and immediately below the interface (as shown by the double-headed arrow just below the interface). Because relative motion between the two media is not possible, the upper medium is forced to move in phase with the lower medium. The disturbance in the upper medium travels along the interface with the same velocity V_2 as the refracted wave just below the interface.

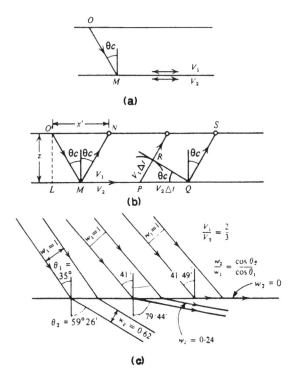

Fig. 3.6 Head waves. (a) Motion at the interface; (b) wavefront emerging from refractor at the critical angle; (c) changes in beam width upon refraction.

Let us assume that these disturbances represented by the arrows reach point P in fig. 3.6b at time t. According to Huygens' principle, P then becomes a center from which a wave spreads out into the upper medium. After a further time interval Δt, this wave has a radius of $V_1 \Delta t$ while the wave moving along the refractor has reached Q, PQ being equal to $V_2 \Delta t$. Drawing the tangent from Q to the arc of radius $V_1 \Delta t$, we obtain the wavefront RQ. Hence, the passage of the refracted wave along the interface in the lower medium generates a plane wave traveling upward in the upper medium at the angle θ, where

$$\sin \theta = V_1 \Delta t / V_2 \Delta t = V_1/V_2$$

Thus, we see that $\theta = \theta_c$ so that the two inclined portions of the path are symmetrically disposed with respect to the normal to the refractor.

In §3.1.4, we showed that the energy carried by a wave was proportional to the cross-section of the beam; however, the beam width of the wave from M to Q in fig. 3.6b is zero, so head waves should have zero energy density and therefore should not exist. However, head waves do exist and frequently are very strong. The apparent conflict between theory and observation is due to the assumption that the source is at infinity so that the incident wave is plane. For a source at a finite distance from the interface, the incident wave is spherical and the situation is changed dramatically.

The propagation of spherical waves in layered me-

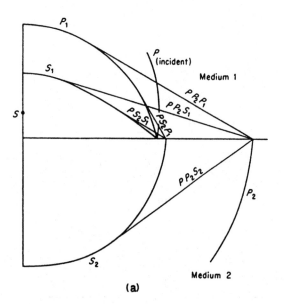

(a)

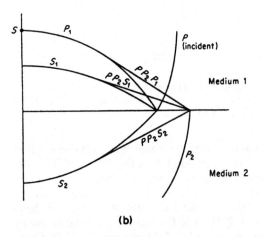

(b)

Fig. 3.7 Head waves. (After Cagniard, 1962.) (a) At an interface where $\alpha_1 < \beta_2$; (b) at an interface where $\alpha_1 > \beta_2 > \beta_1$.

dia has been discussed by many writers. Sommerfeld (1909) dealt with the propagation of electromagnetic waves generated by a source at an interface and with a source above the interface (Sommerfeld, 1949: 237–46). Joos and Teltow (1939) showed that Sommerfeld's results applied to elastic waves. Jeffreys (1926) was the first to show clearly that the wave equation predicted the existence of head waves. Ewing, Jardetzky, and Press (1957: §3.3) used Sommerfeld's results to develop the theory of head waves.

The most complete account of head waves is that of Cagniard (1962). He assumed a source giving a steady-state displacement of $(1/r)e^{j(\kappa r - \omega t)}$ and used Laplace transform theory (§15.3) to obtain solutions of the wave equation in terms of complex integrals; convolution (§9.2) then gives solutions for other types of inputs. The mathematics involved is very complex and evaluation of the integrals is difficult; nevertheless, we regard the problem of the existence of head waves as resolved. Grant and West (1965: §6.3) and Dix (1954)

summarize the mathematics involved and Bortfeld (1962a, 1962b) gives solutions for special cases.

The numbers and types of head waves predicted by Cagniard's results depend upon the relative values of the velocities. Figure 3.7 shows the head waves for an incident P-wave for two cases: (a) $\beta_2 > \alpha_1$ and (b) $\alpha_1 > \beta_2 > \beta_1$. In the first case, five head waves exist, the maximum possible number; four are in the upper medium, one in the lower. In the second case, there are two head waves in the upper medium and one in the lower.

Problems

3.1 Derive the following results:
(a) The displacements of a free surface for an incident P-wave of amplitude A_0 are

$$u/A_0 = [2/(m + n)](m \sin \theta + \cos \delta)e^{j\omega(px - t)},$$
$$w/A_0 = [-2/(m + n)](n \cos \theta + \sin \delta) e^{j\omega(px - t)},$$

where $m = (\beta/\alpha)\tan 2\delta$ and $n = (\alpha/\beta)\cos 2\delta/\sin 2\theta$. (*Hint*: The displacements of a free surface are not restricted, so eqs. (3.3) and (3.4) have no meaning. Set $A_2 = 0 = B_2$ in eqs. (3.5) and (3.6) and express u and w in terms of A_1/A_0 and B_1/A_0.)
(b) For normal incidence on a free surface,

$$u/A_0 = 0, \qquad w/A_0 = -2 \qquad (z = 0).$$

(c) At the free surface of a solid, where $\theta = 45°$, $\alpha = 3$ km/s, $\beta/\alpha = 1/\sqrt{2}$; then

$$u/A_0 = 1.793, \qquad w/A_0 = -1.035.$$

(d) At the surface of the ocean,

$$u/A_0 = 0, \qquad w/A_0 = -2 \cos \theta.$$

3.2 For an SH-wave (an incident SH-component perpendicular to the paper in fig. 3.1), write the boundary conditions and find the amplitudes of all reflected and refracted waves. The absence of P-waves is important in S-wave studies.
3.3 (a) Derive Knott's equations and Zoepprtiz' equations for a P-wave incident on a liquid–solid interface when the incident wave is (i) in the liquid and (ii) in the solid.
(b) Calculate the amplitude of the reflected and transmitted P- and S-waves where an incident P-wave strikes the interface from a water layer ($\alpha = 1.5$ km/s, $\beta = 0$, $\rho = 1.0$ g/cm³) at 20° when the sea floor is (i) "soft" ($\alpha = 2.0$ km/s, $\beta = 1.0$ km/s, $\rho = 2.0$ g/cm³); and (ii) "hard" ($\alpha = 4.0$ km/s, $\beta = 2.5$ km/s, $\rho = 2.5$ g/cm³).
(c) Repeat part (b) for an angle of incidence of 30°.
3.4 Derive the Zoeppritz equations for an incident SV-wave and (b) an incident SH-wave.
3.5 Show that the maximum amplitude of an incident wave and its reflection at the surface of the ocean occurs at the depth $\lambda/(4 \cos \theta)$, where θ is the angle of incidence, by expressing pressure $\mathcal{P}$ in the form used in eq. (3.2) and applying appropriate boundary conditions.

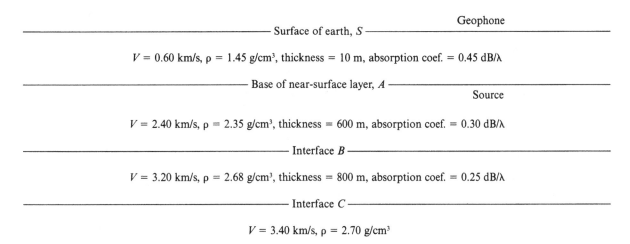

Fig. 3.8 A layered model.

3.6 (a) Using eq. (2.56) to represent a plane wave incident on a plane interface, show that a complex coefficient of reflection, $R = a + jb$, $a^2 + b^2 < 1$, R being defined by eq. (3.14), corresponds to a reduction in amplitude by the factor $(a^2 + b^2)^{1/2}$ and an advance in phase by $\tan^{-1}(b/a)$.

(b) Show that an imaginary angle of refraction, θ_2 (see §2.7.5) in eqs. (3.3) to (3.6) leads to a complex value of R, and hence to phase shifts.

3.7 Calculate the reflection and transmission coefficients, R and T of eqs. (3.14) and (3.15), for a sandstone–shale interface for the following:

(a) $V_{ss} = 2.43$, $V_{sh} = 2.02$ km/s, $\rho_{ss} = 2.08$, and $\rho_{sh} = 2.23$ g/cm³;

(b) $V_{ss} = 3.35$, $V_{sh} = 3.14$ km/s, $\rho_{ss} = 2.21$, and $\rho_{sh} = 2.52$ g/cm³

(c) What are the corresponding values in nepers and in decibels?

3.8 Assume horizontal layering, as shown in fig. 3.8, and a source just below interface A.

(a) Calculate (ignoring absorption and divergence) the relative amplitudes and energy densities for the primary reflections from B and C and the multiples (see §6.3.2) BSA, BAB, and BSB (where the letters denote the interfaces involved). Compare traveltimes, amplitudes, and energy densities of these five events.

(b) Recalculate for 15- and 75-Hz waves allowing for absorption.

(c) Recalculate amplitudes for the 15-Hz wave allowing also for divergence. Normalize values by letting the divergence effects of reflection B be unity.

(d) Summarize your conclusions regarding (i) the relative importance of multiples versus primaries and (ii) the relative importance of different attenuation mechanisms.

3.9 Show that when angles in the Zoeppritz equations, eqs. (3.3) to (3.6), are small (so that the squares and products are negligible), eqs. (3.14) and (3.15) are still valid and

$$\frac{B_1}{A_0} = \frac{2W_2q + 4Z_1r}{(W_1 + W_2)(Z_1 + Z_2)},$$

$$\frac{B_2}{A_0} = \frac{2W_1q - 4Z_1r}{(W_1 + W_2)(Z_1 + Z_2)},$$

$$q = Z_1\theta_2 - Z_2\theta_1,$$

$$r = W_1\delta_1 - W_2\delta_2.$$

3.10 In §3.4, we stated that $\Delta\sigma/\sigma$ is not necessarily small when $\Delta\alpha/\alpha$, $\Delta\beta/\beta$, and $\Delta\rho/\rho$ are all small; verify this statement. (*Hint:* Use eq. (2.60).)

3.11 How would you recalibrate the scale to change a plot showing amplitude variation with offset (AVO) into a plot of amplitude variation with angle (AVA)? What will be the effect if velocity increases with depth?

References

Aki, K., and P. G. Richards. 1980. *Quantitative Seismology: Theory and Methods*, Vol. 1. San Francisco: W. H. Freeman.

Allen, J. L., and C. P. Peddy. 1993. *Amplitude Variation with Offsetes: Gulf Coast Studies*. Tulsa: Society of Exploration Geophysicists.

Bortfeld, R. 1962a. Exact solution of the reflection and refraction of arbitrary spherical compressional waves at liquid–liquid interfaces and at solid–solid interfaces with equal shear velocities and equal densities. *Geophys. Prosp.*, **10**: 35–67.

Bortfeld, R. 1962b. Reflection and refraction of spherical compressional waves at arbitrary plane interfaces. *Geophys. Prosp.*, **10**: 517–38.

Cagniard, L. 1962. *Reflection and Refraction of Progressive Seismic Waves*, E. A. Flynn and C. H. Dix, trans. New York: McGraw-Hill.

Costagna, J. P., and M. M. Backus. 1993. *Offset-Dependent Reflectivity – Theory and Practice of AVO Analysis*. Tulsa: Society of Exploration Geophysicists.

Denham, L. R., and R. A. R. Palmeira. 1984. Discussion on reflection and transmission of plane compressional waves. *Geophysics*, **49**: 2195.

Dix, C. H. 1954. The method of Cagniard in seismic pulse problems. *Geophysics,* **19:** 722–38.

Ewing, W. M., W. S. Jardetzky, and F. Press. 1957. *Elastic Waves in Layered Media.* New York: McGraw-Hill.

Grant, F. S., and G. F. West. 1965. *Interpretation Theory in Applied Geophysics.* New York: McGraw-Hill.

Jeffreys, H. 1926. On compressional waves in two superposed layers. *Proc. Camb. Phil. Soc.,* **22:** 472–81.

Joos, G., and J. Teltow. 1939. Zur Deutung der Knallwellenausbreitung an der Trennschicht zweier Medien. *Physik. Z.,* **40:** 289–93.

Knott, C. G. 1899. Reflexion and refraction of elastic waves with seismological applications: *Phil Mag.,* **48:** 64–97.

Koefoed, O. 1962. Reflection and transmission coefficients for plane longitudinal incident waves. *Geophys. Prosp.,* **10:** 304–51.

Meissner, R. 1967. Exploring deep interfaces by seismic wide-angle measurements. *Geophys. Prosp.,* **15:** 598–617.

Ostrander, W. J. 1984. Plane-wave reflection coefficients for gas sands at nonnormal angles of incidence. *Geophysics,* **49:** 1637–48.

Shuey, R. T. 1985. A simplification of the Zoeppritz equations. *Geophysics,* **50:** 609–14.

Sommerfeld, A. 1909. Uber die Ausbreitung der Wellen in der drahtlosen Telegraphie. *Ann. Phys.,* **28:** 665–736.

Sommerfeld, A. 1949. *Partial Differential Equations in Physics.* New York: Academic Press.

Tooley, R. D., T. W., Spencer, and H. F. Sagoci. 1965. Reflection and transmission of plane compressional waves. *Geophysics,* **30:** 552–70.

Yu, G. 1985. Offset-amplitude variation and controlled-amplitude processing. *Geophysics,* **50:** 2697–708.

Zoeppritz, K. 1919. Uber reflexion und durchgang seismischer Wellen durch Unstetigkerlsfläschen. *Uber Erdbebenwellen VII B, Nachrichten der Königlichen Gesellschaft der Wissenschaften zu Göttingen, Math. Phys.,* **K1:** 57–84.

4

Geometry of seismic waves

Overview

This chapter uses a geometrical-optics approach to derive the basic relationships between traveltime and the locations of reflecting/refracting interfaces; most structural interpretation relies on such an approach.

The accurate interpretation of reflection data requires a knowledge of the velocity at all points along the reflection paths. However, even if we had such a detailed knowledge of the velocity, the calculations would be tedious; often we assume a simple distribution of velocity that is close enough to give useable results. The simplest assumption, which is made in §4.1, is that the velocity is constant between the surface and the reflecting bed. Although this assumption is rarely even approximately true, it leads to simple formulas that give answers that are within the required accuracy in many instances.

The basic problem in reflection seismic surveying is to determine the position of a bed that gives rise to a reflection on a seismic record. In general, this is a problem in three dimensions. However, the dip is often very gentle and the direction of profiling is frequently nearly along either the direction of dip or the direction of strike. In such cases, a two-dimensional solution is generally used. The arrival time-versus-offset relation for a plane reflector and constant velocity is hyperbolic. The distance to the reflector can be found from the reflection arrival time at the source point if the velocity is known. The variation of arrival time as a geophone is moved away from the source, called normal moveout, provides the most important criterion for identifying reflections and a method of determining velocity. The dip is found from differences in arrival times of a reflection at different locations after correction for normal moveout; dip moveout is related to dip and also to the angle of approach of wavefronts at the surface and to apparent velocity. Reflector dip and strike can be found from the components of dip moveout at the intersection of seismic lines. Reflecting points move updip as source–receiver offset increases, so that the traces in a common-midpoint gather do not have common reflecting points.

Section 4.2 deals with reflection raypaths where velocity changes vertically; this results in changes in raypath direction. One solution in some situations is to use equivalent average velocity. For parallel velocity layers, the slope of the traveltime curve gives (in the

limit as $x \to 0$) the root-mean-square (rms) velocity. Vertical velocity is often expressed as a function of arrival time or depth. Where velocity is linear with depth, wavefronts are spherical and raypaths are arcs of circles, facts that can be used in graphical plotting of depth sections.

Section 4.3 concerns the geometry of head-wave paths as used in refraction exploration. In most cases, we assume a series of beds, each having a constant velocity, the velocity increasing as we go to deeper beds, and then we derive formulas relating traveltime, offset, depth, dip, and velocities. The cases considered include a single horizontal refractor, several horizontal refractors, and a single dipping refractor. Velocities can be found from slopes of the traveltime-versus-offset curves, depths from the intercepts of projections to the source point, and dip from the differences in depth at two source locations. Where velocity increases with depth, diving waves eventually return to the surface even where reflection is not involved. Refraction paths in the case of a linear increase in overburden velocity are also considered.

4.1 Reflection paths for constant velocity

4.1.1 Horizontal reflector, normal moveout

The simplest two-dimensional problem is that of zero dip illustrated in the lower part of fig. 4.1. The reflecting bed, AB, is at a depth h below source S. Energy leaving S along the direction SC will be reflected in such a direction that the angle of reflection equals the angle of incidence.

Although the reflected ray CR can be determined by laying off an angle equal to α at C, it is easier to use *image point I*, which is located on the same normal to the reflector as S and as far below the bed as S is above. If we join I to C and prolong the straight line to R, CR is the reflected ray (because CD is parallel to SI, making all the angles marked α equal).

Denoting the average velocity by V, traveltime t for the reflected wave is $(SC + CR)/V$. However, $SC = CI$, so that IR is equal in length to the actual path, SCR. Therefore, $t = IR/V$, and in terms of x, the source-to-geophone distance (*offset*), we can write

$$V^2 t^2 = x^2 + 4h^2, \qquad (4.1)$$

85

or

$$V^2t^2/4h^2 - x^2/4h^2 = 1. \qquad (4.2)$$

Thus, the traveltime curve is a hyperbola, as shown in the upper part of fig. 4.1.

The geophone at R will also record the *direct wave,* which travels along the path SR. Because SR is always less than $SC + CR$, the direct wave arrives first. The traveltime is $t_D = x/V$, and the traveltime curves are the straight lines OM and ON passing through the origin with slopes of $\pm 1/V$.

When distance x becomes very large, the difference between SR and $SC + CR$ becomes small, and the reflection traveltime approaches the direct-wave traveltime asymptotically.

The location of the reflecting bed is determined by measuring t_0, the traveltime for a geophone at the sourcepoint. Setting $x = 0$ in eq. (4.1), we see that

$$h = \tfrac{1}{2}Vt_0. \qquad (4.3)$$

Equation (4.1) can be written

$$t^2 = x^2/V^2 + 4h^2/V^2 = x^2/V^2 + t_0^2. \qquad (4.4)$$

If we plot t^2 against x^2 (instead of t versus x, as in fig. 4.1), we obtain a straight line of slope $1/V^2$ and intercept t_0^2. This forms the basis of a well-known scheme for determining V, the "$X^2 - T^2$ method"; this will be described in §5.4.4a.

We can solve eq. (4.1) for t, the traveltime measured on the seismic record. Generally $2h$ is appreciably larger than x, so that we can use a binomial expansion (§15.1.4c) as follows:

$$\begin{aligned} t &= (2h/V)[1 + (x/2h)^2]^{1/2} = t_0[1 + (x/Vt_0)^2]^{1/2} \\ &= t_0[1 + \tfrac{1}{2}(x/Vt_0)^2 - \tfrac{1}{8}(x/Vt_0)^4 + \cdots]. \qquad (4.5) \end{aligned}$$

If t_1, t_2, x_1, and x_2 are two traveltimes and offsets, we have to the first approximation

$$\Delta t = t_2 - t_1 \approx (x_2^2 - x_1^2)/2V^2t_0. \qquad (4.6)$$

In the special case where one geophone is at the sourcepoint, Δt is known as the *normal moveout* (NMO), which we shall denote by Δt_{NMO}.

$$\Delta t_{NMO} \approx x^2/2V^2t_0 \approx x^2/4Vh. \qquad (4.7)$$

At times, we retain another term in the expansion (see also problem 4.1c):

$$\begin{aligned} \Delta t_{NMO}^* &\approx x^2/2V^2t_0 - x^4/8V^4t_0^3 \\ &= (x^2/2V^2t_0)[1 - (x/4h)^2]. \qquad (4.8) \end{aligned}$$

From eq. (4.7), we note that the normal moveout increases as the square of the offset x, inversely as the square of the velocity, and inversely as the first power of the traveltime (or depth – see eq. (4.3)). Thus, reflection curvature increases rapidly as we go to more distant geophones; at the same time, the curvature becomes progressively less with increasing record time.

The concept of normal moveout is extremely important. It is the principal criterion by which we de-

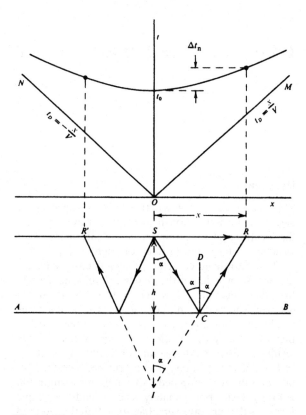

Fig. 4.1   Traveltime curve for a horizontal reflector.

cide whether an event observed on a seismic record is a reflection or not. If the normal moveout differs from the value given by eq. (4.7) by more than the allowable experimental error, we are not justified in treating the event as a reflection. One of the most important quantities in seismic interpretation is the change in arrival time caused by dip; to find this quantity, we must eliminate normal moveout. Normal moveout must also be eliminated before "stacking" (adding together) common-midpoint records (see §8.3.3). Finally, eq. (4.7) can be used to find V by measuring x, t_0, and Δt_{NMO}; this forms the basis of the T–ΔT method of finding velocity (see §5.4.4b) and also of velocity analysis (§9.7). Brown (1969) discusses refinements to handle dip and long offset.

4.1.2 Dipping reflector; dip moveout

When the bed is dipping in the direction of the profile, we have the situation shown in fig. 4.2, ξ being the dip, and h the distance normal to the bed. To draw the raypath for the reflection arriving at geophone R, we join image point I to R by a straight line, cutting the bed at C. The path is then SCR, and t is equal to $(SC + CR)/V$; because $SC + CR = IR$, application of the cosine law to triangle SIR gives

$$\begin{aligned} V^2t^2 &= IR^2 \\ &= x^2 + 4h^2 - 4hx \cos\left(\tfrac{1}{2}\pi + \xi\right) \\ &= x^2 + 4h^2 + 4hx \sin\xi. \qquad (4.9) \end{aligned}$$

On completing the squares, we obtain

$$\frac{V^2 t^2}{(2h \cos \xi)^2} - \frac{(x + 2h \sin \xi)^2}{(2h \cos \xi)^2} = 1.$$

Thus, as before, the traveltime curve is a hyperbola, but the axis of symmetry is now the line $x = -2h \times \sin \xi$ instead of the t-axis. This means that t has different values for geophones symmetrically placed on opposite sides of the sourcepoint, unlike the case for zero dip.

Setting x equal to 0 in eq. (4.9) gives the same value for h as in eq. (4.3); note, however, that h is not measured vertically as it was in the earlier result. We call points C, C', C'' in fig. 4.2, where the angles of incidence and reflection are equal, *reflecting points*. (These are sometime called "depth points," but this term is also used for the point on the surface midway between source and receiver; we call the latter a *midpoint*, and to avoid confusion we shall avoid the term "depth point.") The updip displacement of reflecting points compared to midpoints for dipping reflectors is important in migrating data (§9.10.2) and in the common-midpoint method (§8.3.3).

To obtain the dip, ξ, we solve for t in eq. (4.9) by assuming that $2h$ is greater than x and expanding as in the derivation of eq. (4.5). Then

$$t = \frac{2h}{V} \left(1 + \frac{x^2 + 4hx \sin \xi}{4h^2} \right)^{1/2}$$
$$\approx t_0 \left(1 + \frac{x^2 + 4hx \sin \xi}{8h^2} \right), \qquad (4.10)$$

using only the first term of the expansion. The simplest method of finding ξ is from the difference in traveltimes for two geophones equally distant from and on opposite sides of the source. Letting x in fig. 4.2 have the values $+\Delta x$ for the downdip geophone and $-\Delta x$ for the updip geophone and denoting the equivalent traveltimes by t_1 and t_2, we get

$$t_1 \approx t_0 \left[1 + \frac{(\Delta x)^2 + 4h \, \Delta x \sin \xi}{8h^2} \right],$$

$$t_2 \approx t_0 \left[1 + \frac{(\Delta x)^2 - 4h \, \Delta x \sin \xi}{8h^2} \right],$$

$$\Delta t_d = t_1 - t_2 \approx t_0 \left(\frac{\Delta x \sin \xi}{h} \right) \approx \frac{2\Delta x}{V} \sin \xi.$$

Dip ξ is given by

$$\sin \xi \approx \frac{1}{2} V \left(\frac{\Delta t_d}{\Delta x} \right). \qquad (4.11)$$

The quantity $\Delta t_d / \Delta x$ is called the *dip moveout*. (Note that dimensionally, dip moveout is time/distance, whereas normal moveout is time. Note also that

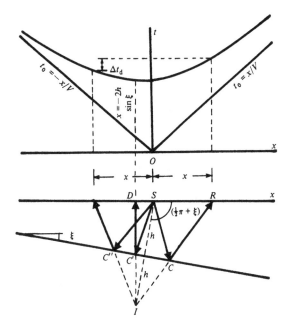

Fig. 4.2   Traveltime curve for a dipping reflector

DMO or dip-moveout processing [§9.10.2] involves different concepts.) For small angles, ξ is approximately equal to $\sin \xi$, so that the dip is directly proportional to Δt_d under these circumstances. To obtain the dip as accurately as possible, we use as large a value of Δx as the data quality permits; for symmetrical spreads (§8.3.1), we measure dip moveout between the geophone groups at the opposite ends of the spread, Δx then being half the spread length.

Dip moveout can also be measured by the time difference between t_0 at different sourcepoints. As shown in fig. 4.3, $\Delta t_d = t_{01} - t_{02}$ and

$$\sin \xi = \frac{1}{2} V \left(\frac{t_{01} - t_{02}}{\Delta x} \right), \qquad (4.12)$$

where Δx is the distance between sourcepoints. When we measure dip on a record section (§8.8.3), Δx is the distance between any two convenient points.

It should be noted that normal moveout was eliminated in the derivation of eq. (4.11). The terms in $(\Delta x)^2$ that disappeared in the subtraction represent the normal moveout.

Figure 4.4 illustrates diagrammatically the relation between normal moveout and dip moveout. Diagram (A) represents a reflection from a dipping bed; the alignment is curved and unsymmetrical about the sourcepoint. Diagram (B) shows what would have been observed if the bed had been horizontal; the alignment is curved symmetrically about the source position owing to the normal moveout. The latter ranges from 0 to 13 ms (1 millisecond = 10^{-3} s = 1 ms, the unit of time commonly used in seismic work) at an offset of 400 m. Diagram (C) was obtained by subtracting the normal moveouts shown in (B) from the arrival times in (A). The resulting alignment shows

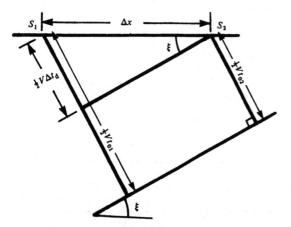

Fig. 4.3   Geometry involved in dip moveout measured between sourcepoints or on record sections.

the effect of dip alone; it is straight and has a time difference between the outside curves of 10 ms, that is, $\Delta t_d = 10$ ms when $\Delta x = 400$ m. Thus, we find that the dip is $2500(10 \times 10^{-3}/800) = 0.031$ rad $= .1.8°$.

The method of normal-moveout removal illustrated in fig. 4.4 was used to demonstrate the difference between normal moveout and dip moveout. If we require only the dip moveout, Δt_d, we merely subtract the traveltimes for the two outside geophones in (A).

Frequently, we do not have a symmetrical spread and we find the dip moveout by removing the effect of normal moveout. As an example, refer to fig. 4.4, curve (D), which shows a reflection observed on a spread extending from $x = -133$ m to $x = +400$ m. Let $t_0 = 1.225$ s, $t_1 = 1.223$ s, $t_2 = 1.242$ s, and $V = 2800$ m/s. From eq. (4.7), we get for Δt_{NMO} at offsets of 133 and 400 m, respectively, the values 1 ms and 8 ms (rounded off to the nearest millisecond because this is usually the precision of measurement on seismic records). Subtracting these values, we obtain for the corrected arrival times $t_1 = 1.222$ and $t_2 = 1.234$; hence, the dip moveout is $12/(533/2)$ ms/m. The corresponding dip is $\xi = 2800(12 \times 10^{-3}/533) = 0.063$ rad $= 3.6°$.

An alternative to the preceding method is to use the arrival times at $x = -133$ m and $x = +133$ m, thus obtaining a symmetrical spread and eliminating the need for calculating normal moveout. However, doing this would decrease the effective spread length from 533 m to 266 m and thereby reduce the accuracy of the ratio $(\Delta t_d/\Delta x)$.

The *apparent velocity* V_a of a wavefront is the ratio of the distance (Δx) between two points on a surface (usually, the surface of the ground) to the difference in arrival times (Δt) for the same event at the two points. It is given by

$$V_a = \Delta x/\Delta t = V_0/\sin \alpha, \qquad (4.13a)$$

where α is the *angle of approach* (fig. 4.5); α is sometimes called *apparent dip*. We can divide this equation by the frequency to give

$$\lambda_a = \lambda/\sin \alpha = 2\pi/\kappa_a, \qquad (4.13b)$$

where λ_a is the *apparent wavelength*, and $\kappa_a/2\pi$ the *apparent wavenumber*. Equation (4.13a) is somewhat similar to eqs. (4.11) and (4.12), but it has a different significance, because it gives the direction of travel of a plane wave as it reaches the spread, V being the velocity between C and the surface. In eqs. (4.11) and (4.12), V is the average velocity (§4.2.2) down to the reflector, and ξ is the angle of dip. Because $\sin \alpha$ can be very small, the apparent velocity V_a (and λ_a) can be very large, and for energy approaching vertically, $V_a = \infty$.

4.1.3 Cross-dip

When the profile is at an appreciable angle to the direction of dip, the determination of the latter becomes a three-dimensional problem and we use the methods of solid analytical geometry. In fig. 4.6, we take the xy-plane as horizontal with the z-axis extending vertically downward. Line OP of length h is perpendicular to a dipping plane bed that outcrops (that is, intersects the xy-plane) along line MN if extended sufficiently.

We write θ_1, θ_2, θ_3 for the angles between OP and the x-, y-, and z-axes, and ℓ, m, n for the direction cosines of OP. The angle Ξ between MN and the x-axis is the direction of strike of the bed while $\theta_3 = \xi$, the angle of dip.

The path of a reflected wave arriving at geophone R on the x-axis can be found using image point I. The line joining I to R cuts the reflector at Q; hence, OQR is the path. Because $OQ = QI$, line IR is equal to Vt, t being the traveltime for the geophone at R. The coordinates of I and R are respectively $(2h\ell, 2hm, 2hn)$ and $(x, 0, 0)$; hence, we have

$$
\begin{aligned}
V^2 t^2 &= (IR)^2 \\
&= (x - 2h\ell)^2 + (0 - 2hm)^2 + (0 - 2hn)^2 \\
&= x^2 + 4h^2(\ell^2 + m^2 + n^2) - 4h\ell x \\
&= x^2 + 4h^2 - 4h\ell x,
\end{aligned}
$$

because $\ell^2 + m^2 + n^2 = 1$ (problem 15.9a).

When $x = 0$, we obtain the same relation between h and t_0 as in eq. (4.3). Proceeding as in the derivation of eq. (4.10), we get for the approximate value of t,

$$t \approx t_0 \left(1 + \frac{x^2 - 4h\ell x}{8h^2}\right).$$

By subtracting the arrival times at two geophones located on the x-axis at $x = \pm\Delta x$, we find

$$
\begin{aligned}
\Delta t_x &\approx t_0(\ell\,\Delta x/h) \\
&\approx 2\ell\,\Delta x/V.
\end{aligned}
$$

$$\ell = \cos\theta_1 \approx \frac{1}{2}V\left(\frac{\Delta t_x}{\Delta x}\right). \qquad (4.14)$$

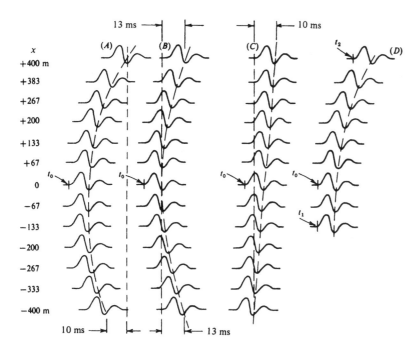

Fig. 4.4   Relation between normal moveout and dip moveout. For curves (A), (B), and (C), $t_0 = 1.000$ s and $\overline{V} = 2500$ m/s. For

curve (D), $t_0 = 1.225$ s, $t_1 = 1.223$ s, $t_2 = 1.242$ s, and $\overline{V} = 2800$ m/s. $\overline{V}$ is the average velocity.

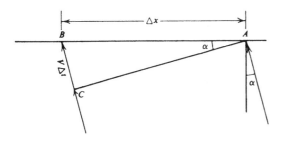

Fig. 4.5   Finding the angle of approach of a wave.

If we also have a spread along the y-axis (cross-spread), we get

$$m = \cos \theta_2 \approx \frac{1}{2} V \left(\frac{\Delta t_y}{\Delta y} \right),\qquad (4.15)$$

where Δt_y is the time difference ("cross-dip") between geophones a distance $2\,\Delta y$ apart and symmetrical about the source. Because

$$n = \cos \xi = [1 - (\ell^2 + m^2)]^{1/2}$$
$$\sin \xi = (1 - n^2)^{1/2} = (\ell^2 + m^2)^{1/2}$$
$$= \frac{1}{2} V \left[\left(\frac{\Delta t_x}{\Delta x} \right)^2 + \left(\frac{\Delta t_y}{\Delta y} \right)^2 \right]^{1/2}. \qquad (4.16)$$

The components of dip moveout, $\Delta t_x/\Delta x$ and $\Delta t_y/\Delta y$, are also called apparent dips.

To find the strike Ξ, we start from the equation of a plane (that is, the reflector) that has a perpendicular from the origin of length h and direction cosines (ℓ, m, n), namely (see problem 15.9b),

$$\ell x + my + nz = h.$$

Setting $z = 0$ gives the equation of the line of intersection of the reflector and the surface; this strike line has the equation

$$\ell x + my = h.$$

The intercepts of this line on the x- and y-axes are h/ℓ and h/m. Referring to fig. 4.7, we find that

$$\tan \Xi = \frac{h/m}{h/\ell} = \frac{\ell}{m}$$
$$= \frac{(\Delta t_x/\Delta x)}{(\Delta t_y/\Delta y)}. \qquad (4.17)$$

Consider the case where the profile lines are not perpendicular, for example, where they are in the $\mathbf{r}_1$ and $\mathbf{r}_2$ directions of fig. 4.8a and the dip is in the $\mathbf{r}_0$ direction. We express the dip moveout as the vector $(dt/dx)\mathbf{r}_0 = \mathbf{AO}$; the component of dip moveout on the line in the $\mathbf{r}_2$ direction is thus $(dt/dx)\mathbf{r}_0 \cdot \mathbf{r}_2 = (dt/dx)\cos \beta = OB$ (see problem 4.2a). The converse problem of finding the total dip moveout from measurements of the components of dip moveout OB and OC can be done graphically, as shown in fig. 4.8b (see also problem 4.2b), or mathematically as follows. We take one profile along the x-axis and the other along the y'-axis at an angle α to the x-axis. By taking the length of a symmetrical spread along the y'-axis as $2\Delta y'$, the coordinates of the ends of the spread (relative to the x-, y-axes) are $\pm \Delta y' \cos \alpha$, $\pm \Delta y' \sin \alpha$. Then

$$V^2 t_{\pm}^2 = (2h\ell \pm \Delta y' \cos \alpha)^2$$
$$+ (2hm \pm \Delta y' \sin \alpha)^2 + (2hn)^2$$
$$= (\Delta y')^2 + 4h^2 \pm 4h\Delta y'(\ell \cos \alpha + m \sin \alpha).$$

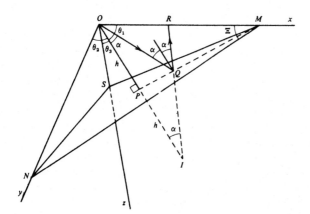

Fig. 4.6   Three-dimensional view of a reflection path for a dipping bed.

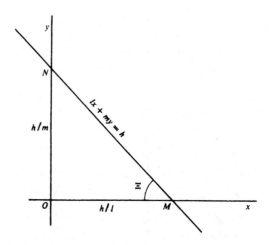

Fig. 4.7   Determination of strike.

The dip moveout along this line, $\Delta t'/\Delta y'$, is

$$\Delta t'/\Delta y' = 2(\ell \cos \alpha + m \sin \alpha)/V. \quad (4.18)$$

Because α is known, ℓ can be found from $\Delta t/\Delta x$ and m from eq. (4.18).

4.1.4 Reflection points for offset receivers

When the source and receiver are coincident and the velocity is constant, the locus of a reflecting point R for constant traveltime as the dip ξ varies is a circle (fig. 4.9a). However, when the source and receiver are offset by x, the locus is an ellipse (fig. 4.9b) with the source and geophone at the foci; this follows from the definition of an ellipse because $SR + RG = Vt$ is a constant. The equation for the traveltime t is given by eq. (4.9), namely,

$$(Vt)^2 = 4h^2 + 4s^2 + 8hs \sin \xi.$$

Expressing this in terms of the depth at the midpoint M, $h' = h + s \sin \xi$:

$$(Vt)^2 = 4(h')^2 + 4s^2 \cos^2\xi \quad (4.19)$$

(compare with problem 4.3); Levin (1971) writes this equation:

$$t^2 = 4(h')^2/V^2 + 4s^2/V_{NMO}^2, \quad (4.20)$$

where $V_{NMO} = V/\cos \xi$; thus, $V_{NMO} > V$.

In fig. 4.9b, the reflection point R has moved updip by $PR = \Delta L$. To determine ΔL, we find the coordinates (x_0, z_0) and (x_1, z_1) of points P and R. Because MP is parallel to SI, x_0 is $s - h'\ell$ and z_0 is $h'n$, ℓ and n being direction cosines of SI, and $2s$ the source–geophone distance. If (x, z) is a point on the line joining I and G, we must have

$$\frac{2s - x}{z} = \frac{2h\ell + 2s}{2hn}, \qquad \frac{2s - x}{s + h\ell} = \frac{z}{hn} = k,$$

where k is a parameter that fixes the location of the point (x, z) along IG. To get k, we use the fact that IG

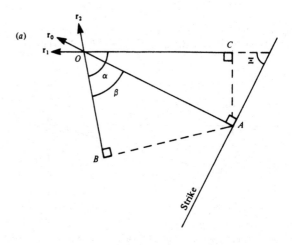

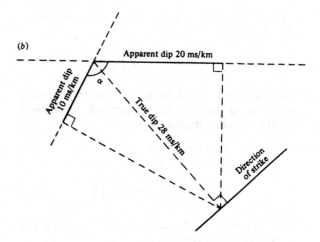

Fig. 4.8   Determining dip and strike from nonperpendicular observations. (a) Relation between the point of observation O and the reflecting point A (A is always updip from O). (b) Example of a graphical solution.

cuts the reflector at R, so (x, z) must satisfy the equation of the reflecting plane (see problem 15.9b):

$$-\ell x + nz = h.$$

Substituting the previous values for (x, z), we get

$$-\ell[2s - k(h\ell + s)] + n(khn) = h,$$

that is, $k = (h + 2\ell s)/(h + \ell s) = (h + 2\ell s)/h'$. Therefore, the coordinates of R are (see problem 4.11).

$$\begin{aligned} x_1 &= 2s - k(h\ell + s) = 2s - (h\ell + s)(h + 2\ell s)/h' \\ &= x_0 - \ell n^2 s^2/h', \end{aligned} \tag{4.21a}$$

and

$$z_1 = k(hn) = (h + 2\ell s)(hn)/h' = z_0 - \ell^2 ns^2/h'. \tag{4.21b}$$

Finally,

$$\begin{aligned} (\Delta L)^2 &= (x_0 - x_1)^2 + (z_0 - z_1)^2 \\ &= (s^2/h')^2(\ell^2 n^4 + \ell^4 n^2) \\ &= (s^2/h')^2(\ell^2 n^2), \end{aligned}$$

and

$$\Delta L = (s^2/h')\sin \xi \cos \xi = (s^2/2h')\sin 2\xi. \tag{4.21c}$$

If we wish to stack data elements that have the common reflection point R, we have to stack updip from the midpoint by the distance Δx, where

$$\Delta x = \Delta L/\cos \xi = (s^2/h')\sin \xi. \tag{4.22a}$$

The updip offset Δx changes the zero-offset time by

$$\Delta t = 2 \Delta x \sin \xi/V = 2(s^2/h'V)\sin^2\xi. \tag{4.22b}$$

The DMO (dip moveout) correction (§9.10.2) accommodates this updip movement of the reflecting point as offset increases.

4.2 Vertical velocity gradient and raypath curvature

4.2.1 Effect of velocity variation

The assumption of constant velocity is not valid in general, the velocity usually changing as we go from one point to another. In petroleum exploration, we are usually dealing with more or less flat-lying bedding and the changes in seismic velocity as we move horizontally are for the most part small, being the result of slow changes in density and elastic properties of the beds. These horizontal variations are generally much less rapid than the variations in the vertical direction where we are going from bed to bed with consequent lithological changes and increasing pressure with increasing depth. Because the horizontal changes are gradual, they can often be taken into account by dividing the survey area into smaller areas within each of which the horizontal variations can be ignored and the same vertical velocity distribution used. Such areas are often large enough to include several structures of the size of interest in oil exploration so that

changes from one velocity function to another do not necessarily impose a serious burden upon the interpreter.

4.2.2 Equivalent average velocity

Vertical variations in velocity can be taken into account in various ways. One of the simplest is to use a modification of the constant-velocity model. We assume that the actual section existing between the surface and a certain reflecting horizon can be replaced with an equivalent single layer of constant velocity $\overline{V}$ equal to the average velocity between the surface and the reflecting horizon; $\overline{V}$ is the *equivalent average velocity*. This velocity is usually given as a function of depth (or of t_0, which is nearly the same except when the dip is large). Thus, the section is assigned a different constant velocity for each of the reflectors below it. Despite this inconsistency, the method is useful and is extensively applied. The variation of the average velocity with t_0 is found using one of the methods described in §5.4. For the observed values of the arrival time t_0, we select the average velocity $\overline{V}$ corresponding to this reflector; using the values of t_0, the dip moveout, $\Delta t_d/\Delta x$, and $\overline{V}$, we calculate the depth h and the dip ξ using eqs. (4.3) and (4.11).

4.2.3 Velocity layering

When the velocity is constant, eq. (4.1) shows that a graph of t^2 versus x^2 is a straight line with slope $1/V^2$. If the velocity varies in the vertical direction, raypaths will bend as required by Snell's law (eq. (3.1)). A commonly used method to take into account vertical velocity variations is to replace the actual velocity distribution with a number of horizontal layers of different velocities, the velocity being constant within each layer. We can approximate any vertical velocity changes as closely as desired by using enough layers. A graphical method using a wavefront chart can be used to find the depth and dip of a reflecting interface; the preparation and use of these charts will be discussed in §8.8.3.

In effect, we replace actual raypaths with a series of line segments that are straight within each layer but undergo abrupt changes in direction at the boundaries between layers. Larger portions of travelpaths are spent in the higher-velocity layers as the source-geophone distance increases. The result is that a graph of t^2 versus x^2 is slightly curved, as shown in fig. 4.10b.

Dix (1955) showed that eq. (4.4) can still be used except that the slope of the x^2–t^2 curve at $x = 0$ yields the inverse of the rms velocity squared, $1/V_{rms}^2$. We approximate the x^2–t^2 curve by the straight line

$$t^2 = x^2/V_{rms}^2 + t_0^2;$$

hence,

$$dt/dx = x/V_{rms}^2 t. \tag{4.23}$$

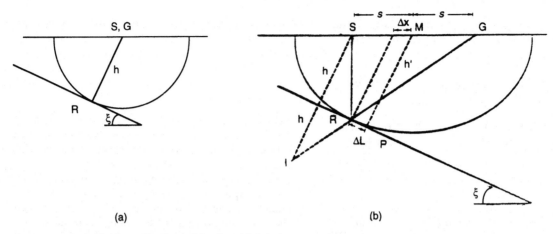

(a)   (b)

Fig. 4.9 Loci of reflection points for various dips. (a) Coincident source and geophone; (b) geophone offset from source.

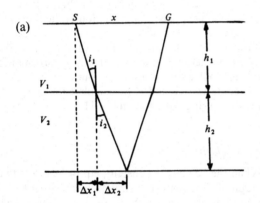

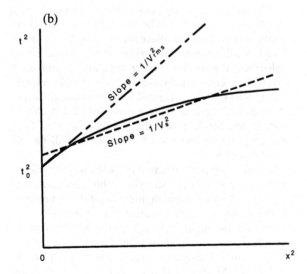

Fig. 4.10 Derivation of the formula for x^2–t^2 velocity in two-layer medium. (a) Reflection path. (b) x^2–t^2 curve (the curvature is exaggerated). The reciprocal of the slope of the dashed line (tangent to the curve at $x = 0$) gives the rms velocity. The best-fit straight line for some portion of the curve is what is often measured; the slope of this line (shown dashed) is the reciprocal of the square of the "stacking velocity" V_s (§5.4.4a); it depends on the portion being fit.

The angle of approach, i_1, is given by

$$\sin i_1 = V_1 \frac{dt}{dx} = \frac{V_1 x}{V_{rms}^2 t} \qquad (4.24)$$

using eq. (4.23). Also, writing Δt_i for the vertical traveltime through the ith bed, and, because x is small, we have

$$\begin{aligned}\tfrac{1}{2}x &= \Delta x_1 + \Delta x_2 = h_1 \tan i_1 + h_2 \tan i_2 \\ &\approx V_1 \Delta t_1 \sin i_1 + V_2 \Delta t_2 \sin i_2 \\ &\approx (V_1^2 \Delta t_1 + V_2^2 \Delta t_2)\sin i_1/V_1 \\ &\approx (V_1^2 \Delta t_1 + V_2^2 \Delta t_2)(x/V_{rms}^2 t)\end{aligned}$$

from eq. (4.24) (note that x cancels here because we have assumed it to be small). Because $t \approx 2(\Delta t_1 + \Delta t_2)$, we get

$$V_{rms}^2 = \frac{\sum\limits_{i=1}^{2} V_i^2 \Delta t_i}{\sum\limits_{i=1}^{2} \Delta t_i}.$$

This equation can be generalized for n horizontal beds (Dix, 1955), giving

$$t^2 = \frac{x^2}{V_{rms}^2} + t_0^2, \qquad (4.25)$$

$$V_{rms}^2 = \frac{\sum\limits_{i=1}^{n} V_i^2 \Delta t_i}{\sum\limits_{i=1}^{n} \Delta t_i}. \qquad (4.26)$$

Shah and Levin (1973) give higher-order approximations necessary to get more accuracy for large values of x.

4.2.4 Effect of variable velocity on raypath direction

Changes in the direction of rays at interfaces are determined by Snell's law (eq. (3.1)). For planar parallel layering (fig. 4.11), the angle of emergence from a layer equals the angle of entry into the next layer and the raypath parameter $p = (\sin i)/V = (\sin i_0)/V_0 = \Delta t/\Delta x$ (see eq. (4.13a)) specifies ray direction, that is, p is constant along any ray and is fixed by the direction in which the ray left the source. Note that $1/V$ is the *slowness* and p is the component of slowness parallel to the interface, hence, the component of slowness parallel to the interface is constant for each ray.

In earthquake studies, it is often assumed that the earth is divided into concentric spherical shells (layers) of constant velocity, as in fig. 4.12. In this case, the angle of entry into a layer is not equal to the angle of exit from that layer, that is, $i_2 \neq i_2^*$. However, because $OP = r_2 \sin i_2 = r_3 \sin i_2^*$, using Snell's law shows that $(r_2 \sin i_2)/V_2 = (r_3 \sin i_3)/V_3$. Thus, in this case, direction can be specified by a raypath parameter p':

$$p' = (r_n \sin i_n)/V_n. \qquad (4.27)$$

At times, the assumption is made that the velocity varies in a systematic continuous manner and therefore can be represented by a velocity function. The actual velocity usually varies extremely rapidly over short intervals, as shown by sonic logs (see §5.4.3); however, if we integrate these changes over distances of a wavelength or so (30–100 m), we obtain a function that is generally smooth except for discontinuities at marked lithological changes. If the velocity discontinuities are small, we are often able to represent the velocity distribution with sufficient accuracy by a smooth velocity function. The path of a wave traveling in such a medium is then determined by two integral equations.

To derive the equations, we assume that the medium is divided into a large number of thin beds in each of which the velocity is constant; on letting the number of beds go to infinity, the thickness of each bed becomes infinitesimal and the velocity distribution becomes a continuous function of depth. Referring to fig. 4.11, we have for the nth bed

$$\frac{\sin i_n}{V_n} = \frac{\sin i_0}{V_0} = p,$$
$$V_n = V_n(z),$$
$$\Delta x_n = \Delta z_n \tan i_n,$$
$$\Delta t_n = \frac{\Delta z_n}{V_n \cos i_n}.$$

In the limit when n becomes infinite, we get

$$\frac{\sin i}{V} = \frac{\sin i_0}{V_0} = p, \qquad V = V(z),$$
$$\frac{dx}{dz} = \tan i, \qquad \frac{dt}{dz} = \frac{1}{V \cos i}, \qquad (4.28)$$
$$x = \int_0^z \tan i \, dz, \qquad t = \int_0^z \frac{dz}{V \cos i};$$

hence,

$$x = \int_0^z \frac{pV \, dz}{[1 - (pV^2)]^{1/2}}, \qquad (4.29)$$

$$t = \int_0^z \frac{dz}{V[1 - (pV)^2]^{1/2}}. \qquad (4.30)$$

Because V is a function of z, eqs. (4.29) and (4.30) furnish two integral equations relating x and t to the depth z. These equations can be solved by numerical methods when we have a table of values of V at various depths.

4.2.5 Linear increase of velocity with depth

Sometimes we can express V as a continuous function of z and integrate eqs. (4.29) and (4.30). One case of considerable importance is that of a linear increase of velocity with depth, namely,

$$V = V_0 + az,$$

where V_0 is the velocity at the horizontal datum plane, V is the velocity at a depth z below the datum plane, and a is a constant whose value is generally between 0.3/s and 1.3/s.

If we introduce a new variable $u = pV = \sin i$, then $du = p \, dV = pa \, dz$, and we can solve for x and t as follows (p is the raypath parameter):

$$x = \frac{1}{pa} \int_{u_0}^u \frac{u \, du}{(1 - u^2)^{1/2}} = \frac{1}{pa} (1 - u^2)^{1/2} \Big|_u^{u_0} = \frac{1}{pa} \cos i \Big|_i^{i_0}$$
$$= \frac{1}{pa} (\cos i_0 - \cos i), \qquad (4.31)$$

$$t = \frac{1}{a} \int_{u_0}^u \frac{du}{u(1 - u^2)^{1/2}} = \frac{1}{a} \ln \left[\frac{u}{1 + (1 - u^2)^{1/2}} \right] \Big|_{u_0}^u$$
$$= \frac{1}{a} \ln \left[\frac{\sin i}{\sin i_0} \left(\frac{1 + \cos i_0}{1 + \cos i} \right) \right] = \frac{1}{a} \ln \left(\frac{\tan \frac{1}{2} i}{\tan \frac{1}{2} i_0} \right); \qquad (4.32)$$

hence,

$$i = 2 \tan^{-1}(e^{at} \tan \tfrac{1}{2} i_0), \qquad (4.33)$$

$$z = (V - V_0)/a = (\sin i - \sin i_0)/pa. \qquad (4.34)$$

The parametric equations (4.31) and (4.34) give the coordinates x and z, the parameter i being related to the one-way traveltime t by eq. (4.32) or (4.33).

The raypath given by eqs. (4.31) and (4.34) is a circle; this can be shown by calculating the radius of curvature ρ, which turns out to be a constant:

$$\rho = [1 + (x')^2]^{3/2}/x'',$$

where

$$x' = \frac{dx}{dz} = \tan i, \text{ using eqs. (4.31) and (4.34)},$$
$$x'' = \frac{d^2x}{dz^2} = \frac{d}{di}(\tan i)\frac{di}{dz} = \sec^2 i \frac{di}{dz}$$
$$= pa \sec^2 i, \text{ using eq. (4.34)}.$$

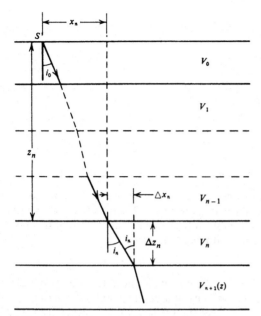

Fig. 4.11   Raypath where velocity varies with depth.

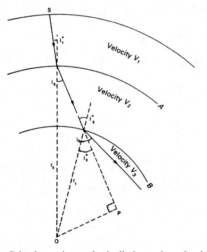

Fig. 4.12   Seismic ray in a spherically layered earth with construction to show the geometric significance of the ray parameter.

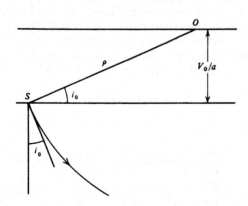

Fig. 4.13   Circular ray leaving the source at the angle i_0.

Hence,

$$\rho = \frac{(1 + \tan^2 i)^{3/2}}{pa \sec^3 i} = \frac{1}{pa} = \left(\frac{V_0}{a}\right)\frac{1}{\sin i_0} = \text{constant}.$$

Figure 4.13 shows a ray leaving the source at the angle i_0. The center, O, of the circular ray lies above the surface a distance $\rho \sin i_0$, that is, V_0/a. Because this is independent of i_0, the centers of all rays lie on the same horizontal line. This line is located where the velocity would be zero if the velocity function were extrapolated up into the air (because $z = -V_0/a$ at this elevation).

To determine the shape of the wavefront, we make use of fig. 4.14. The raypaths SA and SB are circular arcs with centers O_1 and O_2, respectively. If we continue the arcs upwards to meet the vertical through S at point S', line O_1O_2 bisects $S'S$ at right angles. Next, we select any point C on the downward extension of $S'S$ and draw the tangents to the two arcs, CA and CB. From plane geometry, we know that the square of the length of a tangent to a circle from an external point (for example, CA^2) is equal to the product of the two segments of any chord drawn from the same point ($CS \cdot CS'$ in fig. 4.14). Using both circles, we see that

$$CS \cdot CS' = CA^2 = CB^2,$$

hence, $CA = CB$. Thus, a circle with center C and radius $R = CA$ cuts the two raypaths at right angles. Because SA and SB can be any raypaths and a wavefront is a surface that meets all rays at right angles, the circle with center C must be the wavefront that

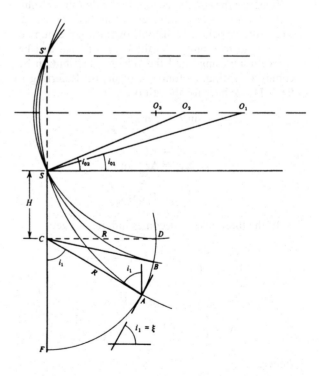

Fig. 4.14   Construction of wavefronts and raypaths for linear increase of velocity.

passes through A and B. Even though arc SA is longer than SB, the greater path length is exactly compensated for by the higher velocity at the greater depth of raypath SA.

We can draw the wavefront for any value of t if we can obtain the values of H and R in fig. 4.14. Thus, the quantities H and R are equal to the values of z and x for a ray that has $i = \frac{1}{2}\pi$ at time t, that is, SD in the diagram. Substitution of $i = \frac{1}{2}\pi$ in eqs. (4.31), (4.33), and (4.34) yields

$$\tan \tfrac{1}{2}i_0 = e^{-at}, \quad \sin i_0 = \operatorname{sech} at, \quad \cos i_0 = \tanh at,$$

$$\left. \begin{aligned}
H &= (1/pa)(1 - \sin i_0) \\
  &= (V_0/a)[(1/\sin i_0) - 1] \\
  &= (V_0/a)(\cosh at - 1), \\
R &= (1/pa)\cos i_0 = (V_0/a)\cot i_0 \\
  &= (V_0/a)\sinh at.
\end{aligned} \right\} \quad (4.35)$$

Equation (4.35) shows that the center of the wavefront moves downward and the radius becomes larger as time increases.

Field measurements yield values of the arrival time at the source t_0 and angle of approach $\Delta t/\Delta x$. Because the ray that returns to the sourcepoint must have encountered a reflecting horizon normal to the raypath and retraced its path back to the point of origin, the dip is equal to angle i_1 at time $t = \frac{1}{2}t_0$. Thus, to locate the segment of reflecting horizon corresponding to a set of values of t_0 and $\Delta t/\Delta x$, we make the following calculations:

$$(a) \quad t = \tfrac{1}{2}t_0,$$

$$(b) \quad i_0 = \sin^{-1}\!\left(V_0 \frac{\Delta t}{\Delta x}\right),$$

$$(c) \quad i_1 = 2 \tan^{-1}(e^{at} \tan \tfrac{1}{2}i_0),$$

$$(d) \quad H = (V_0/a)(\cosh at - 1),$$

$$(e) \quad R = (V_0/a)\sinh at.$$

With these values, we find C, lay off the radius R at the angle i_1, and draw the reflecting segment perpendicular to the radius, as shown at the point A in fig. 4.14. This method is easily adapted to a simple plotting machine (Daly, 1948) or to wavefront charts (Agocs, 1950).

Refraction studies involving linear increase of overburden velocity are discussed in §4.3.6.

4.3 Refraction paths

4.3.1 General

Refraction seismology involves the study of head waves (§3.5) using primarily first arrivals, the equivalent of first breaks in reflection seismology (see, however, second arrivals, §11.2). For a head wave to be generated, the velocity below an interface must be higher than that above it; accordingly, we shall assume in the following sections that the velocity increases downward monotonically. However, this is not always the case, and problems sometimes result from a *hidden* (*blind*) *zone*, a layer whose velocity is lower than that

of the overlying bed so that it never carries a head wave (see §11.2).

4.3.2 Single horizontal refractor

For the case of a single horizontal refracting horizon, we can readily derive a formula expressing the arrival time in terms of the offset, the depth, and the velocities. In fig. 4.15, the lower part shows a horizontal plane refractor separating two beds of velocities V_1 and V_2, where $V_2 > V_1$. For a geophone at R, the path of the refracted wave is $OMPR$, θ_c being the critical angle. The traveltime t can be written

$$\begin{aligned}
t &= \frac{OM}{V_1} + \frac{MP}{V_2} + \frac{PR}{V_1} = \frac{MP}{V_2} + 2\frac{OM}{V_1} \\
  &= \frac{x - 2h \tan \theta_c}{V_2} + \frac{2h}{V_1 \cos \theta_c} \\
  &= \frac{x}{V_2} + \frac{2h}{V_1 \cos \theta_c}\left(1 - \frac{V_1}{V_2}\sin \theta_c\right) \\
  &= \frac{x}{V_2} + \frac{2h \cos \theta_c}{V_1}, \quad (4.36)
\end{aligned}$$

where we have used the relation $\sin \theta_c = V_1/V_2$ in the last step. This equation can also be written

$$t = (x/V_2) + t_1, \quad (4.37)$$

where

$$t_1 = (2h \cos \theta_c)/V_1, \quad (4.38)$$

or

$$h = \tfrac{1}{2}V_1 t_1/\cos \theta_c.$$

Obviously, the head wave will not be observed at offsets less than the *critical distance*, OQ in fig. 4.15, writing x' for the critical distance,

$$\begin{aligned}
x' &= OQ = 2h \tan \theta_c = 2h \tan [\sin^{-1}(V_1/V_2)] \\
   &= 2h [(V_2/V_1)^2 - 1]^{1/2}. \quad (4.39)
\end{aligned}$$

The relation between x'/h and V_2/V_1 is shown in fig. 4.16. As the ratio V_2/V_1 increases, x' decreases. When V_2/V_1 equals 1.4, x' is equal to $2h$. As a rule of thumb, offsets should be greater than twice the depth to the refractor to observe refractions without undue interference from shallower head waves.

Equations (4.36) and (4.37) represent a straight line of slope $1/V_2$ and *intercept time* t_1. This is illustrated in fig. 4.15, where OMQ, $OMP'R'$, $OMPR$, and $OMP''R''$ are a series of refraction paths and DWS the corresponding time–distance curve. Note that this straight-line equation does not have physical meaning for offsets less than x' because the refracted wave does not exist for such values of x; nevertheless, we can project the line back to the time axis to find t_1.

The problem to be solved usually is to find the depth h and the two velocities V_1 and V_2. The slope of the direct-wave time–distance curve is the reciprocal of V_1 and the same measurement for the refraction event gives V_2. We can then calculate the critical angle

θ_c from the relation $\theta_c = \sin^{-1}(V_1/V_2)$, and use the intercept time, t_1, to calculate h from eq. (4.38).

In fig. 4.15, the time–distance curves for the reflection from the interface AP'' and for the direct path are represented by the hyperbola CDE and the straight line OF, respectively. Because the path OMQ can be regarded either as a reflection or as the beginning of the refracted wave, the reflection and refraction time–distance curves must coincide at $x = x'$, that is, at point D. Moreover, differentiating eq. (4.1) to obtain the slope of the reflection time–distance curve at $x = x'$, we find

$$\left[\frac{dt}{dx}\right]_{x=x'} = \left[\frac{x}{V_1^2 t}\right]_{x=x'} = \frac{1}{V_1}\left(\frac{OQ}{OM + MQ}\right)$$

$$= \frac{1}{V_1}\left(\frac{\tfrac{1}{2}OQ}{OM}\right) = \frac{1}{V_1}\sin\theta_c = \frac{1}{V_2}.$$

We see, therefore, that the reflection and refraction curves have the same slope at D, and, consequently, the refraction curve is tangent to the reflection curve at $x = x'$.

Comparing reflected and refracted waves from the same horizon arriving at the same geophone, we note that the refraction arrival time is always less than the reflection arrival time (except at D). The intercept time t_1 for the refraction is less than the arrival time t_0 for the reflection at the sourcepoint because

$$t_1 = (2h/V_1)\cos\theta_c, \qquad t_0 = 2h/V_1;$$

hence, $t_1 < t_0$.

Starting at the point Q, we see that the direct wave arrives ahead of the reflected and refracted waves because its path is the shortest of the three. However,

part of the refraction path is traversed at velocity V_2, so that as x increases, eventually the refraction wave will overtake the direct wave. In fig. 4.15, these two traveltimes are equal at the point W. If the offset corresponding to W is x_c, we have

$$\frac{x_c}{V_1} = \frac{x_c}{V_2} + \frac{2h}{V_1}\cos\theta_c,$$

$$\therefore h = \frac{x_c}{2}\left(1 - \frac{V_1}{V_2}\right)\Big/\cos\theta_c$$

$$= \frac{x_c}{2}\left(\frac{V_2 - V_1}{V_2}\right)\frac{V_2}{(V_2^2 + V_1^2)^{1/2}}$$

$$= \frac{x_c}{2}\left(\frac{V_2 - V_1}{V_2 + V_1}\right)^{1/2}. \qquad (4.40)$$

This relation is sometimes used to find h from measurements of the velocities and the *crossover distance* x_c. However, usually we can determine t_1 more accurately than x_c and hence eq. (4.38) provides a better method of determining h. The relation between x_c/h and V_2/V_1 is shown in fig. 4.16.

4.3.3 Several horizontal refractors

Where all layers are horizontal, eq. (4.36) an be generalized to cover the case of more than one refracting horizon. Consider the situation in fig. 4.17, where we have three layers of velocities, V_1, V_2, and V_3. Whenever $V_2 > V_1$, we have the refraction path $OMPR$ and corresponding time–distance curve WS, just as we had in fig. 4.15. If $V_3 > V_2 > V_1$, travel by a refraction path in V_3 will eventually overtake the refraction in V_2. The refraction paths such as $OM'M''P''P'R'$ are fixed by Snell's law:

$$\frac{\sin\theta_1}{V_1} = \frac{\sin\theta_{c2}}{V_2} = \frac{1}{V_3},$$

where θ_{c2} is the critical angle for the lower horizon and θ_1 is less than the critical angle for the upper horizon. The expression for the traveltime curve ST is obtained as before:

$$t = \frac{OM' + R'P'}{V_1} + \frac{M'M'' + P'P''}{V_2} + \frac{M''P''}{V_3}$$

$$= \frac{2h_1}{V_1\cos\theta_1} + \frac{2h_2}{V_2\cos\theta_{c2}}$$

$$\quad + \frac{x - 2h_1\tan\theta_1 - 2h_2\tan\theta_{c2}}{V_3}$$

$$= \frac{x}{V_3} + \frac{2h_2}{V_2\cos\theta_{c2}}\left(1 - \frac{V_2}{V_3}\sin\theta_{c2}\right)$$

$$\quad + \frac{2h_1}{V_1\cos\theta_1}\left(1 - \frac{V_1}{V_3}\sin\theta_1\right)$$

$$= \frac{x}{V_3} + \frac{2h_2}{V_2}\cos\theta_{c2} + \frac{2h_1}{V_1}\cos\theta_1 = \frac{x}{V_3} + t_2.$$

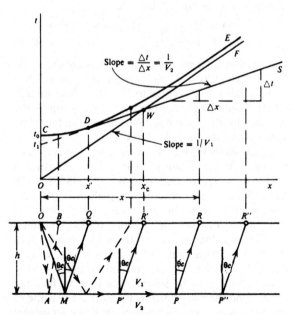

Fig. 4.15 Relation between reflection and refraction raypaths and traveltime curves.

(4.41)

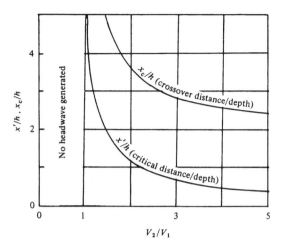

Fig. 4.16 Relation between critical distance x', crossover distance x_c, and velocity contrast.

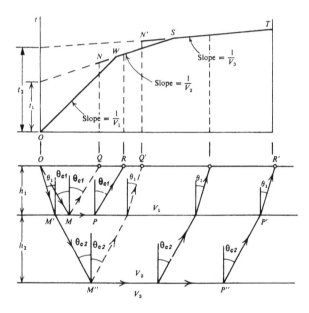

Fig. 4.17 Raypaths and traveltime curves for two horizontal refractors.

Thus, the time–distance curve for this refraction is also a straight line whose slope is the reciprocal of the velocity just below the refracting horizon and whose intercept is the sum of terms of the form $2h_i \cos \theta_1/V_i$, each layer above the refracting horizon contributing one term. We can generalize for n layers:

$$t = \frac{x}{V_n} + \sum_{i=1}^{n-1} \frac{2h_i}{V_i} \cos \theta_i , \qquad (4.42)$$

where $\theta_i = \sin^{-1}(V_i/V_n)$. This equation can be used to find the velocities and thicknesses of each of a series of horizontal refracting layers, each of constant velocity higher than any of the layers above it, provided each layer contributes enough of the time–distance

curve to permit it to be analyzed correctly. We can find all of the velocities (hence, the angles θ_1 also) by measuring the slopes of the various sections of the time–distance curve and then get the thicknesses of the layers from the intercepts

$$h_n = \frac{V_n}{2 \cos \theta_{cn}} \left(t_n - \sum_{i=1}^{n-1} \frac{2h_1 \cos \theta_i}{V_i} \right). \qquad (4.43)$$

4.3.4 Effect of refractor dip

The simple situations on which eqs. (4.36) to (4.43) are based are frequently not valid. One of the most serious defects is the neglect of dip because dip changes the refraction time–distance curve drastically. The lower part of fig. 4.18 shows a vertical dip section through a refracting horizon. Let t be the traveltime for the refraction path $OMPO'$. Then, we have

$$\begin{aligned} t &= \frac{OM + O'P}{V_1} + \frac{MP}{V_2} \\ &= \frac{h_d + h_u}{V_1 \cos \theta_c} + \frac{OQ - (h_d + h_u)\tan \theta_c}{V_2} \\ &= \frac{x \cos \xi}{V_2} + \frac{h_d + h_u}{V_1} \cos \theta_c. \end{aligned} \qquad (4.44)$$

If we place the source at O and a detector at O', we are "shooting downdip." In this case, it is convenient to have t in terms of the distance from the source to the refractor h_d; hence, we eliminate h_u using the relation

$$h_u = h_d + x \sin \xi.$$

Writing t_d for the downdip traveltime, we obtain

$$\begin{aligned} t_d &= (x/V_2)\cos \xi + (x/V_1)\cos \theta_c \sin \xi + (2h_d/V_1)\cos \theta_c \\ &= (x/V_1)\sin (\theta_c + \xi) + (2h_d/V_1)\cos \theta_c \\ &= (x/V_1)\sin (\theta_c + \xi) + t_{1d}, \end{aligned}$$
where
$$t_{1d} = (2h_d/V_1)\cos \theta_c. \qquad \left.\begin{aligned}\\ \\ \\ \\ \end{aligned}\right\} \qquad (4.45)$$

The result for shooting in the updip direction is similarly obtained by eliminating h_d:

$$t_u = (x/V_1)\sin (\theta_c - \xi) + t_{1u},$$
where
$$t_{1u} = (2h_u/V_1)\cos \theta_c. \qquad \left.\begin{aligned}\\ \\ \end{aligned}\right\} \qquad (4.46)$$

Note that the downdip traveltime from O to O' is equal to the updip traveltime from O' to O; this source point to source point traveltime is called the *reciprocal time* and is denoted by t_r. The concept that traveltime along a path is the same regardless of the direction of travel is an example of the *principle of reciprocity*.

These equations can be expressed in the same form as eq. (4.37):

$$t_d = (x/V_d) + t_{1d}, \qquad (4.47)$$

$$t_u = (x/V_u) + t_{1u}, \qquad (4.48)$$

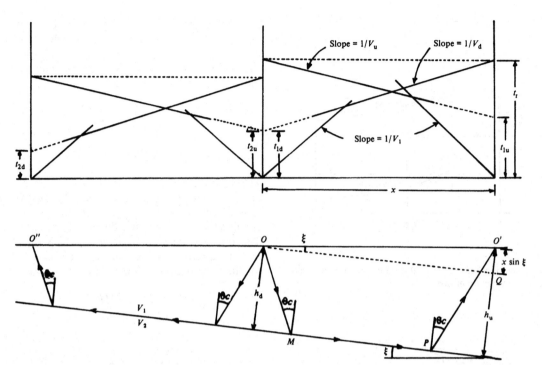

Fig. 4.18 Raypaths and traveltime curves for a dipping refractor.

where

$$V_d = V_1/\sin(\theta_c + \xi), \qquad V_u = V_1/\sin(\theta_c - \xi). \tag{4.49}$$

V_d and V_u are apparent velocities and are given by the reciprocals of the slopes of the time–distance curves.

For reversed profiles, such as shown in fig. 4.18, eq. (4.49) can be solved for the dip ξ and the critical angle θ_c (and hence for the refractor velocity V_2):

$$\left.\begin{array}{l} \theta_c = \frac{1}{2}[\sin^{-1}(V_1/V_d) + \sin^{-1}(V_1/V_u)], \\[4pt] \xi = \frac{1}{2}[\sin^{-1}(V_1/V_d) - \sin^{-1}(V_1/V_u)]. \end{array}\right\} \tag{4.50}$$

The distances to the refractor, h_d and h_u, can then be found from the intercepts using eqs. (4.45) and (4.46).

Equation (4.49) can be simplified where ξ is small enough that we can approximate by letting $\cos \xi \approx 1$ and $\sin \xi = \xi$. With this simplification eq. (4.49) becomes

$$V_1/V_d = \sin(\theta_c + \xi) \approx \sin\theta_c + \xi\cos\theta_c,$$

$$V_1/V_u = \sin(\theta_c - \xi) \approx \sin\theta_c - \xi\cos\theta_c;$$

hence,

$$\sin\theta_c = (V_1/V_2) \approx \frac{1}{2}V_1\{(1/V_d) + (1/V_u)\},$$

so that

$$1/V_2 \approx \frac{1}{2}[(1/V_d) + (1/V_u)]. \tag{4.51}$$

An even simpler approximate formula for V_2 (although slightly less accurate) can be obtained by applying the binomial theorem (§15.1.4c) to eq. (4.49)

and assuming that ξ is small enough that higher powers of ξ are negligible:

$$\begin{aligned} V_d &= (V_1/\sin\theta_c)(\cos\xi + \cot\theta_c \sin\xi)^{-1} \\ &\approx V_2(1 - \xi\cot\theta_c), \\ V_u &\approx V_2(1 + \xi\cot\theta_c); \end{aligned}$$

hence,

$$V_2 \approx \frac{1}{2}(V_d + V_u). \tag{4.52}$$

4.3.5 Diving waves

It is obvious that raypaths will eventually return to the surface whenever the velocity increases with depth. The waves traveling by such raypaths are called *diving waves*. Symmetry shows that for horizontal velocity layering, the angle of emergence is i_0 (fig. 4.19a); at the deepest point on the raypath (h_m), $i = 90°$ and $p = 1/V_m$, that is, p is the slowness at the deepest point on the raypath for a diving ray. We can rewrite eqs. (4.29) and (4.30) for this situation as

$$x = 2\int_0^{h_m} \frac{pV(z)\,\mathrm{d}z}{\{1-[pV(z)]^2\}^{1/2}}, \tag{4.53a}$$

$$t = 2\int_0^{h_m} \frac{\mathrm{d}z}{V(z)\{1-[pV(z)]^2\}^{1/2}}; \tag{4.53b}$$

the doubling factor of 2 arising because of the raypath from h_m back to 0. If x, t measurements are available for diving ways from a common source and if the velocity has increased with depth monotonically, then eqs. (4.53) can be solved numerically for $V(z)$.

For a linear increase of velocity, eqs. (4.53) become (see problem 4.20a)

$$x = (2V_0/a)\cot i_0 = (2V_0/a)\sinh(at/2), \quad (4.54a)$$

$$t = (2/a) \ln[\cot(i_0/2)], \quad (4.54b)$$

and the maximum depth of penetration is

$$h_m = (V_0/a)[\cosh(at/2) - 1]. \quad (4.55)$$

For the case of concentric spherical layering (fig. 4.19b), eqs. (4.53) become (see problem 4.20b)

$$\Delta = 2\int_{R_e-h_m}^{R_e} \frac{[p'V(r)/r]\,dr}{r\{1 - [p'V(r)/r]^2\}^{1/2}}, \quad (4.56a)$$

$$t = 2\int_{R_e-h_m}^{R_e} \frac{dr}{V(r)\{1 - [p'V(r)/r]^2\}^{1/2}}, \quad (4.56b)$$

where Δ is the angle subtended at the center of the Earth by raypath SD, and R_e is the radius of the Earth. Richter (1958: app. VI) gives a numerical solution for $V(r)$ in eqs. (4.56). Using ΔDEF in fig. 4.19b, we can write eq. (4.27) as

$$p' = R_e(\sin i_0)/V_0 = V_0 \,\delta t_0/R_e\,\delta\Delta = \delta t_0/\delta\Delta_{DE}. \quad (4.57)$$

At the deepest point, $p' = r_m/V_m$, so

$$p' = r_m/V_m = (R_e/V_0)\sin i_0 = \delta t/\delta\Delta_{DE}. \quad (4.58)$$

If we plot a curve of traveltime t versus angular distance Δ for an earthquake event at various stations, the slope, $\Delta t/\delta\Delta$, gives r_m/V_m.

With diving waves, two velocity situations require special attention. A velocity gradient in a layer that is substantially higher than the gradient above the bed, as in fig. 4.20a, causes a very sharp increase in raypath curvature and a folding back of the time–distance curve. Such a triplication of branches of the time–distance curve (fig. 4.20c) is usually difficult to see because the later branches become lost in the later cycles of the earlier arrivals. Unless all branches are correctly recognized, errors will be made in solving the time–distance observations for $V(z)$.

The other situation leading to errors in determining $V(z)$ is that of a velocity inversion (fig. 4.20b). The inversion may produce a gap in the time–distance curve, as in fig. 4.20c, and this gap may not be recognized because diffraction tends to fill it in. Although fig. 4.20 shows situations for planar layers, similar (but more complex) situations occur with concentric spherical layers in earthquake studies.

4.3.6 Linear increase in velocity above a refractor

The case of a high-velocity layer overlain by a layer in which the velocity increases linearly with depth (fig. 4.21) is of considerable practical importance. The relation between t and x for a horizontal refractor can be found as follows:

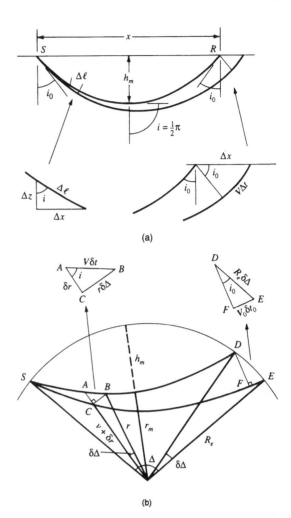

Fig. 4.19 Raypaths for increase in velocity with depth. (a) Planar velocity layering; (b) concentric velocity layering.

$$t = t_{SN} + t_{NP} + t_{PR}$$
$$= 2t_{SN} + (x - 2MN)/V_m.$$

Noting that $(\sin i_0)/V_0 = (\sin\theta_c)/V_c = 1/V_m$, we find from eq. (4.32)

$$t_{SN} = \frac{1}{a}\ln\left(\frac{\tan\frac{1}{2}\theta_c}{\tan\frac{1}{2}i_0}\right)$$

$$= \frac{1}{a}\ln\left[\left(\frac{V_c}{V_m + (V_m^2 - V_c^2)^{1/2}}\right)\left(\frac{V_m + (V_m^2 - V_0^2)^{1/2}}{V_0}\right)\right]$$

$$= \frac{1}{a}\left[\cosh^{-1}\left(\frac{V_m}{V_0}\right) - \cosh^{-1}\left(\frac{V_m}{V_c}\right)\right],$$

where use has been made of the identity $\cosh^{-1}x = \ln[x + (x^2 - 1)^{1/2}]$. From eq. (4.31), we get

$$MN = (1/pa)(\cos i_0 - \cos\theta_c)$$
$$= (1/pa)\{[1 - (V_0/V_m)^2]^{1/2} - [1 - (V_c/V_m)^2]^{1/2}\}.$$

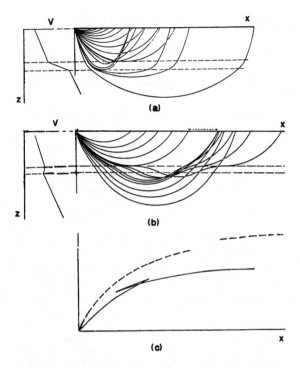

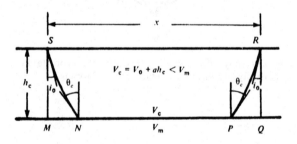

Fig. 4.20 Velocity situations making it difficult to determine time–distance curves. (a) Region with large velocity gradient. (b) Region with velocity inversion. (c) Time–distance curves for situation (a) (solid line) and (b) (dashed line); diffractions will complicate these curves.

tion at $t_0 = 2.358$ s, given that the velocity $\overline{V} = 2.90$ km/s.

(b) Typical errors in x, $\overline{V}$, t_0 might be 0.6 m, 0.2 km/s, and 5 ms. Calculate the corresponding errors in Δt_{NMO} approximately. What do you conclude about the accuracy of Δt_{NMO} calculations?

(c) Show that the more accurate NMO equation, eq. (4.8), can be written

$$\Delta t^*_{NMO} \approx \Delta t_{NMO}(1 - \Delta t_{NMO}/2t_0),$$

in terms of the first-order value of Δt_{NMO} given by eq. (4.7). Taking into account the errors in x, $\overline{V}$, t_0, when is this equation useful?

4.2 (a) Show that the quantity dt/dx can be considered as a vector or component of a vector according as dt corresponds to the total dip or component of dip.

(b) Using fig. 4.22, verify that the construction of fig. 4.8b gives the same results as eq. (4.18). (*Hint:* Express ℓ, m, and OC in terms of OA.)

4.3 Show that the equation for a dipping reflection, eq. (4.9), becomes

$$(Vt)^2 = (2x \cos \xi)^2 + 4h_c^2$$

(Gardner, 1947), where h is replaced by h_c, the slant depth at the midpoint between the source and receiver (see fig. 4.23).

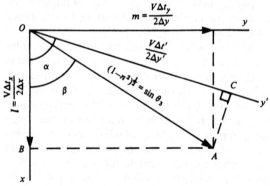

Fig. 4.22 Combining dip components.

Fig. 4.21 Refraction with a linear increase in velocity in the upper layer.

Substituting in the first expression for t gives

$$t = (x/V_m) + t_0, \qquad (4.59)$$

where t_0 = intercept time:

$$t_0 = (2/a)\{[\cosh^{-1}(V_m/V_0) - \cosh^{-1}(V_m/V_c)] - [1 - (V_0/V_m)^2]^{1/2} + [1 - (V_c/V_m)^2]^{1/2}\}. \quad (4.60)$$

The slope of the head-wave traveltime curve gives V_m. A curve is plotted of t_0 against h_c (or V_c) for given values of V_0 and a, and h_c and V_c are read from this curve for particular measurements of t_0.

Problems

4.1 (a) Calculate the normal moveout Δt_{NMO} for geophones 600 and 1200 m from the source for a reflec-

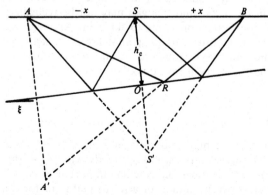

Fig. 4.23 Derivation of X^2–T^2 relation for a dipping bed.

4.4 (a) Using the dip-moveout equation, eq. (4.11), and the results of problem 4.3, verify the following result (due to Favre according to Dix, 1955):

$$\tan \xi \approx t/t_{AB}^2 - t_0^2)^{1/2},$$

where ξ = dip, $t = t_{SA} - t_{SB}$, t_{AB} = traveltime between source A and receiver B, t_0 = traveltime at source S (see fig. 4.23).
(b) Using eq. (4.9), show that

$$\sin \xi = V^2(t_{SA}^2 - t_{SB}^2)/8h_c x.$$

(c) Under what condition is the result in part (b) the same as eq. (4.11) and also consistent with part (a)?

4.5 The expression for dip in terms of dip moveout, eq. (4.11), involves the approximation of dropping higher-order terms in the quadratic expansion used to get eq. (4.10). What is the effect on eq. (4.11) if an additional term is carried in this expansion? What is the percentage change in dip?

4.6 In fig. 10.5b, the reflection time at the top is 1.0 s and the depth 1500 m, the reflection time at the bottom is 1.4 s, the interval velocity between events 1 and 2 is 3300 m/s, and the trace spacing is 100 m. Calculate the depth and dip of the three reflectors.

4.7 A well encounters a horizon at a depth of 3 km with a dip of 7°. Sources are located 200 m updip from the well with a geophone at depths of 1.0 to 2.6 km at intervals of 400 m. Plot the raypaths and traveltime curves for the primary reflection from the 3-km horizon and its first multiple at the surface. Assume V = 3.0 km/s. (*Hint:* See fig. 6.33.)

4.8 The numbered ticks at the top of fig. 8.5 are 1 km apart.
(a) Select two fairly steeply dipping reflections, assume velocities (fairly high in this area), and determine the approximate dips.
(b) Figure 8.5 is a migrated section; by what horizontal distances are the reflecting points for these events displaced, that is, how far did they migrate? (*Hint:* See §8.8.3 and fig. 8.30.)

4.9 (a) Sources B and C are respectively 600 m north and 500 m east of source A. Traveltimes at A, B, and C for a certain reflection are t_0 = 1.750, 1.825, and 1.796 s. What are the dip and strike of the horizon, V being 3.25 km/s?
(b) What are the changes in dip and strike if line AC has the bearing N80°E?

4.10 (a) Two intersecting seismic spreads have bearings N10°E and N140°E. If the first spread shows an event at t_0 = 1.760 s with dip moveout of 56 ms/km and the same event on the second spread has a dip moveout of 32 ms/km, find the true dip, depth, and strike, assuming that (i) both dips are down to the south and west, and (ii) dip on the first spread is down to the south and the other is down to the southeast. The average velocity is 3 km/s.
(b) Calculate the position of the reflecting point (migrated position) for each spread in (i) as if the cross information had not been available and each had been assumed to indicate total moveout; compare with the

results of part (a). Would the errors be more serious or less serious if the calculations were made for the usual situation where the velocity increases with depth?

4.11 Verify the derivation of the expression for k in §4.1.4 and of eq. (4.21a).

4.12 Given the velocity–depth data shown in fig. 5.19, what problems would you expect using simple functional-fit relations in the different areas?

4.13 (a) Calculate $\bar{V}$ and V_{rms} down to each of the interfaces in table 4.1. Why do they differ (give a geometrical explanation)?
(b) Plot $\bar{V}$ and V_{rms} versus depth and versus traveltime and determine the best-fit straight lines for the four cases. What are the main problems in approximating data with functional fits?

Table 4.1 *Layered model*

Depth (km)	Velocity (km/s)
0–1.00	2.00
1.00–2.50	3.00
2.50–2.80	6.00
2.80–4.80	4.00

4.14 (a) Assuming flat bedding, calculate depths corresponding to t_0 = 1.0, 2.0, 2.1, and 3.1 s using the velocity functions for $\bar{V}$ and V_{rms} determined in problems 4.13a and 4.13b. What errors are introduced relative to the depths given in table 4.1?
(b) Using the velocity data in table 4.1, trace a nonvertical ray through the various layers and find the arrival times and reflecting points of flat reflectors at each of the interfaces.

4.15 (a) Repeat the calculations of problem 4.14a assuming dip moveout of 104 ms/km and find the dip in each case.
(b) Trace rays assuming the velocity is constant at the values of $\bar{V}$ and V_{rms} calculated in problem 4.13. Find the arrival times and reflecting points of reflectors at each of the interfaces.

4.16 Figure 7.45 shows part of a seismic record where the geophone group spacing is 50 m, the offset to the near groups being 50 m and that of the far groups 600 m.
(a) What is the velocity of the first-breaks?
(b) Assuming that the source is below the base of the LVL and that the LVL velocity is 500 m/s, how thick is the LVL?
(c) Arrival times at the sourcepoint for two reflections are given as 0.475 and 0.778 s; what are the average velocities to these reflectors?
(d) For these reflections, the arrival-time differences between the far traces in opposite directions from the source point are given as +0.005 for both reflections. What are the dips of these reflectors?
(e) What is the dominant frequency of these reflections (approximately)?

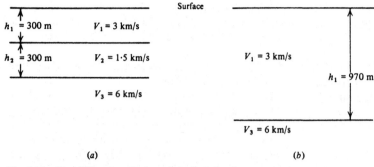

Fig. 4.24   Two different geologic sections that give the same refraction time–distance curves.

4.17 (a) Given the velocity function $V = 1.60 + 0.60z$ km/s (z in km), find the depth, dip, and offset of the point of reflection when $t_0 = 4.420$ s and $\Delta t/\Delta x = 0.155$ s/km. What interpretation would you give of the result?

(b) If the ray continued without reflection, when and where would it emerge? What moveout would be observed at the recording spread? Calculate the maximum depth of penetration.

4.18 (a) Show that the two geological sections illustrated in fig. 4.24 produce the same time–distance curves.

(b) What would be the apparent depth to the lower interfaces in figs. 4.24a and 4.24b if $V_3 = 3.15$ km/s instead of 6 km/s?

4.19 Figure 4.25 shows a refraction profile recorded as a ship firing an air gun moved away from a sonobuoy. Identify the direct wave through the water and use its traveltimes to give the source-to-sonobuoy distances (assume 1.5 km/s as the velocity in water).

(a) Identify distinctive head-wave arrivals, determine their velocities, intercept times, and depths of the refractors assuming flat bedding and no velocity inversions.

(b) What is the water depth? Identify multiples and explain their probable travel paths. (The data in the upper right corner result from paging (§8.6.3) and actually belong below the bottom of the record.)

4.20 (a) Verify eqs. (4.54) and (4.55). (*Hint:* Use eqs. (4.31) to (4.34).)

(b) Derive eqs. (4.56). (*Hint:* In fig. 4.19b, ΔABC gives $(V\delta t)^2 = (\delta r)^2 + (r\,\delta\Delta)^2$; using eq. (4.27), show that $p' = (r/V)^2\,(\delta\Delta/\delta t)$; eliminating first δt, then $\delta\Delta$, and integrating gives eqs. (4.56a) and (4.56b).

4.21 If the velocity function in problem 4.17 applies above a horizontal refractor at a depth of 2.40 km, where the refractor velocity is 4.25 km/s, plot the traveltime–distance curve.

4.22 Given that situations (a) through (h) in fig. 4.26 involve the same two rock types, draw the appropriate time–distance curves. Diagram (c) shows two cases for dip in opposite directions. In figs. (i) and (j), the velocity in the lower medium varies laterally according to the density of the shading.

4.23 Barton (1929) discusses shooting into a geo-

phone placed in a borehole (fig. 4.27) as a means of determining where the bottom of the borehole is located.

(a) Given that A, B, D, and E are equidistant from W in the cardinal directions and assuming straight-line travel paths at the velocity V and that the traveltimes from D and E are equal, derive expressions for CC' and CW in fig. 4.27a in terms of the traveltimes from A and B, $t_{AC'}$ and $t_{BC'}$.

(b) What are the values of $t_{AC'}$ and $t_{BC'}$ for $V = 2.500$ km/s, $AW = BW = CC' = 1000$ m, $CW = 200$ m?

(c) How sensitive is the method, that is, what are $\Delta(CC')/\Delta t_{AC'}$, and $\Delta(CW)/\Delta t_{AC'}$? For the specific situation in part (b), how much change is there in WC and CC' per millisecond error in $t_{AC'}$?

(d) Modify the assumptions in part (b) by taking the velocity as 1.5 km/s for the first 500 m and 3.5 km/s for the lower 500 m. What are the actual traveltimes now and how would these be interpreted assuming the straight-path assumption in part (a)?

4.24 Sources A and B are located at the ends of a 225-m spread of 16 geophones. Using the data in table 4.2, find the velocities, dip, and depth to the refractor.

Table 4.2 *Refraction profile*

x_A (m)	t_A (ms)	t_B (ms)	x_B (m)
0	0	98	225
15	10	92	210
30	21	87	195
45	30	81	180
60	41	75	165
75	50	71	150
90	59	65	135
105	65	60	120
120	70	52	105
135	73	46	90
150	78	43	75
165	81	37	60
180	85	31	45
195	89	21	30
210	94	10	15
225	98	0	0

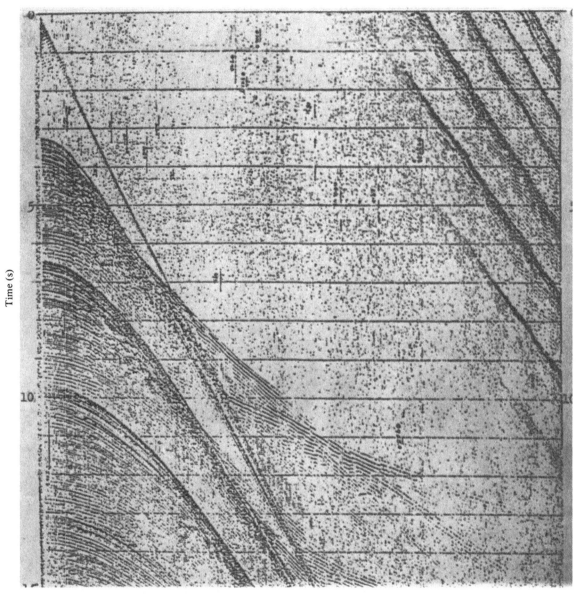

Fig. 4.25 Sonobuoy refraction profile in Baffin Bay. Source
was a 1000-in.³ air gun. (Courtesy of Fairfield Industries.)

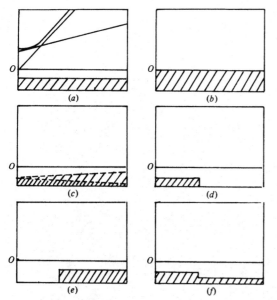

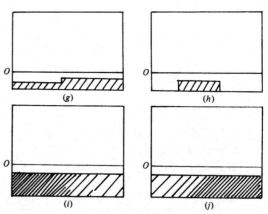

Fig. 4.26 Time–distance curves for various two-layer configurations. This figure is adapted from Barton (1929) in the first publication in English on the seismic method. The part above O of each diagram provides space for a curve of arrival time versus distance for the model shown in cross-section below O. In each case, the velocity in the cross-hatched portion is higher than that above. Part (a) has been completed to show what is expected. In (c), two alternatives are given so two sets of curves should be drawn. In (i) and (j), refractor velocity varies horizontally and is proportional to the shading density.

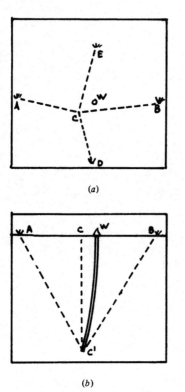

Fig. 4.27 Mapping a crooked borehole by measuring traveltimes to a geophone at C' in the borehole. (From Barton, 1929.) (a) Plan view; (b) vertical section AWB.

References

Agocs, W. B. 1950. Comparison charts for linear increase of velocity with depth. *Geophysics,* **15:** 226–36.

Barton, D. C. 1929. The seismic method of mapping geologic structure. In *Geophysical Prospecting,* pp. 572–624. New York: American Institute of Mining and Metallurgical Engineers.

Brown, R. J. S. 1969. Normal-moveout and velocity relations for flat and dipping beds and for long offsets. *Geophysics,* **34:** 180–95.

Daly, J. W. 1948. An instrument for plotting reflection data on the assumption of a linear increase of velocity. *Geophysics,* **13:** 153–7.

Dix, C. H. 1955. Seismic velocities from surface measurements. *Geophysics,* **20:** 68–86.

Gardner, L. W. 1947. Vertical velocities from reflection shooting. *Geophysics,* **12:** 221–8.

Levin, F. K. 1971. Apparent velocity from dipping interface reflections. *Geophysics,* **36:** 510–16.

Richter, C. F. 1958. *Elementary Seismology.* San Francisco: W. H. Freeman.

Shah, P. M., and F. K. Levin. 1973. Gross properties of time–distance curves. *Geophysics,* **38:** 643–56.

5
Seismic velocity

Overview

Knowledge of velocity values is essential in determining the depth, dip, and horizontal location of reflectors and refractors, in determining whether certain things like head waves and velocity distortions occur, and in ascertaining the nature of rocks and their interstitial fluids from velocity measurements.

We develop a heuristic appreciation of the factors that affect seismic velocity from a conceptual model of a sedimentary rock. F. Gassmann, M. A. Biot, and J. Geertsma developed a model for a fluid-filled porous rock, and G. H. P. Gardner, L. W. Gardner, and A. R. Gregory hypothesized that microcracks in nonporous rocks lower velocity. Fracturing also generally lowers velocity.

Lithology is the most obvious factor we would expect to control velocity. However, velocity ranges are so broad and there is so much overlap that velocity alone does not provide a good basis for distinguishing lithology. Sand velocities, for example, can be smaller or larger than shale velocities, and the same is true for densities; both velocity and density play important roles in seismic reflectivity.

Porosity appears to be the most important single factor in determining a rock's velocity, and the dependence of porosity on depth of burial and pressure relationships makes velocity sensitive to these factors also. Velocity is generally lowered when gas or oil replaces water as the interstitial fluid, sometimes by so much that amplitude anomalies result from hydrocarbon accumulations.

The near-surface layer of the earth usually differs markedly from the remainder of the earth in velocity and some other properties. This makes the near-surface low-velocity layer (LVL) especially important; our determinations of depths, attitudes, and continuity of deeper events are affected as reflections pass through this layer. In arctic areas, a zone of permanently frozen earth, permafrost, distorts deeper events because of an exceptionally high velocity. Fluid pressure that exceeds that of a column of fluid extending to the surface ("normal" pressure) lowers seismic velocity; this is used to predict abnormal pressures. Gas hydrates that form in the sediments just below the sea floor in deep water also produce velocity changes.

Velocity terminology is often misused and causes much confusion. Section 5.4.1 attempts to clarify the precise meaning of average, root-mean-square, stacking, interval, Dix, phase, group, apparent, and other velocity terms.

Seismic velocity is measured in boreholes by sonic logs (§5.4.3) (and by vertical seismic profiling discussed in §13.4). Velocity is also measured by surface seismic data because of the dependence of normal moveout on velocity. The reflection-coefficient equation can be used to obtain velocity information from amplitudes, a form of inversion.

5.1 Model of a sedimentary rock

5.1.1 A pack of uniform spheres

Seismic velocity as given by eqs. (2.58) and (2.59) relates to a homogeneous medium, but sedimentary rocks are far from homogeneous. These equations can be written, for solid media,

$$\alpha^2 = (\lambda + 2\mu)/\rho, \qquad \beta^2 = \mu/\rho,$$

and for fluid media,

$$\alpha^2 = \lambda/\rho, \qquad \beta^2 = 0;$$

hence, in general,

$$V = (K/\rho)^{1/2}, \tag{5.1}$$

where K is the effective elastic parameter. Thus, the dependence of V upon the elastic constants and density appears to be straightforward. In fact, the situation is much more complicated because K and ρ are interrelated, both depending to a greater or lesser degree upon the material and structure of the rock, the lithology, porosity, interstitial fluids, pressure, depth, cementation, degree of compaction, and so on. The most notable inhomogeneity of sedimentary rocks is that they are porous, containing fluid-filled spaces within them. *Porosity* is simply the pore volume per unit volume. Wang and Nur (1992b) discuss theories relating seismic velocity to the composition of rocks.

The simplest rock model consists of identical spheres arranged in a cubic pattern (fig. 5.1a) with the matrix subjected to a compressive pressure $\mathcal{P}$. If the radius of the spheres is R, the force F pressing two adjacent spheres together is the total force acting on a layer of $n \times n$ spheres (that is, $(2Rn)^2\mathcal{P}$) divided by the number of spheres (n^2), or $F = 4R^2\mathcal{P}$. This force causes a point of contact to become a circle of contact of radius r and the centers to move closer together a distance s (see figs. 5.1b and 5.1c), r and s being related to R, F, and the elastic constants E, σ of the spheres by Hertz' equations (see Timoshenko and Goodier, 1951: 372–7):

$$\left.\begin{array}{l} r = [3(1 - \sigma^2)RF/4E]^{1/3}, \\[2mm] s = [9(1 - \sigma^2)^2F^2/2RE^2]^{1/3}. \end{array}\right\} \qquad (5.2)$$

When a P-wave passes, $\mathscr{P}$ changes by $\Delta\mathscr{P}$, resulting in changes $\Delta F = 4R^2 \Delta\mathscr{P}$ and $\Delta s = -2R\varepsilon$, where ε is the strain in the direction of F (see fig. 5.1d). Thus, the effective elastic modulus K is given by

$$K = -\frac{\Delta\mathscr{P}}{\varepsilon} = \frac{1}{2R}\frac{\Delta F}{\Delta s} = \left[\frac{3E^2\mathscr{P}}{8(1-\sigma^2)^2}\right]^{1/3}$$

on differentiating eq. (5.2). The average density is the weight of a sphere divided by the volume of the circumscribed cube, that is, $\bar{\rho} = \frac{4}{3}(\pi R^3\rho)/(2R)^3 = \frac{1}{6}\pi\rho$, ρ being the density of the material of the spheres. Thus, we get for the P-wave velocity, V_{cubic},

$$V_{\text{cubic}} = (K/\bar{\rho})^{1/2} = [81E^2\mathscr{P}/(1 - \sigma^2)^2\pi^3\rho^3]^{1/6}. \qquad (5.3)$$

Gassmann (1951) calculated the velocity for a hexagonal packing of identical spheres (fig. 5.2) under a pressure produced by the weight of a thickness z of overlying spheres; he obtained for a vertical ray

$$V_{\text{hex}} = [128E^2gz/(1 - \sigma^2)^2\pi^2\rho^2]^{1/6}, \qquad (5.4)$$

where g is the acceleration of gravity. Because $\mathscr{P}$ is nearly proportional to z, eqs. (5.3) and (5.4) give the same variation of velocity with depth. Faust (1953) found an empirical formula for velocity in terms of depth of burial z and formation resistivity R', that is consistent with eqs. (5.3) and (5.4):

$$V_P = 900(zR')^{1/6}, \qquad (5.5)$$

V_P being in m/s, z in m, and R' in Ω.m. However, the deviations of individual measurements were very large, indicating the presence of other factors that have not been taken into account.

Random packs of well-sorted particles have porosities in the range of 45–50%, but under pressure, the particles deform at the contacts, and as a result the density increases and the porosity decreases (Sheriff, 1977; the elastic constants also change – see §5.2.5).

5.1.2 Expectations based on the model

What velocity relationships might we expect based on the foregoing model of a rock? Clearly, porosity will be an important factor in velocity because it should affect both the effective elasticity and the density, and indeed it is often said that porosity is the most important factor in determining the velocity of a sedimentary rock. The contact area between spheres is not proportional to the pressure forcing them together, so we may expect that the pressure dependence of velocity will not be linear but will diminish with increasing

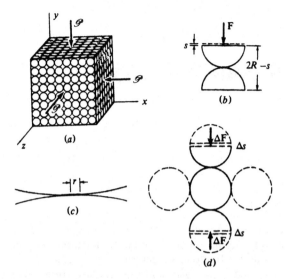

Fig. 5.1 Effects of compression on a cubic packing of spheres. (After White, 1965.) (a) Cubic packing; (b) force causes centers to move closer together; (c) force causes point contact to become circular area of contact; (d) effect of change in force.

pressure (or with depth of burial). Fluid filling the interstices in a rock may be expected to resist the effects of the overburden, that is, the overburden weight tends to squeeze out the porosity whereas the interstitial fluid tends to preserve the porosity. Thus, the effective pressure on a rock will be the difference between the overburden pressure and the fluid pressure, the *differential pressure*. If the pore space is connected to the surface, the fluid pressure should be that of a column of porefluid extending to the surface whereas the overburden pressure is the weight of the overlying rocks. Where this is true, the pressure is said to be *normal*. However, if the pore fluid cannot escape to allow the grain-to-grain contacts to adjust to normal pressure, then some of the overburden weight will be supported by the interstitial fluid and we will have an *overpressured* situation. An overpressured rock will "feel" the same differential pressure as it would at some shallower depth, where it would have a lower velocity, and, hence, we expect overpressuring to lower the velocity.

We would not expect the deformation of a rock under high pressures to be elastic. Hence, if a rock from which the porosity has been squeezed out by depth of burial should be uplifted, we would not expect porosity to return, except for a small amount because of some remaining elasticity. The porosity of a rock might be expected to depend on the maximum stresses it has endured since formation, that is, porosity may depend on the maximum depth of burial rather than on the present depth.

Gas as a formation fluid is much more compressible than a liquid and hence gas in the pore space should lower the velocity much more than oil or water. In fact, gas is so compressible that the presence of just a

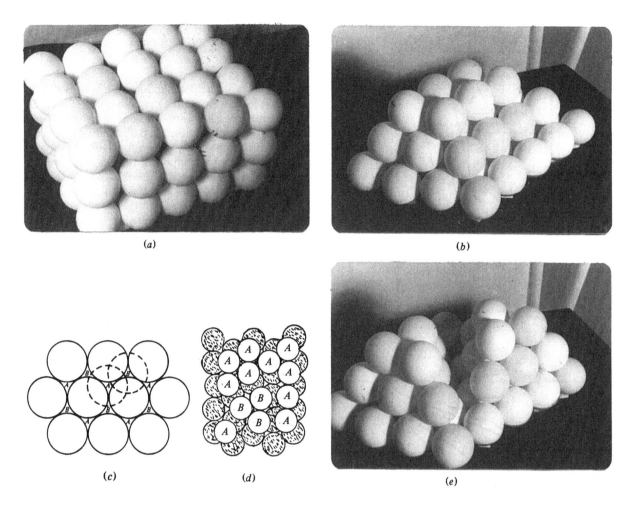

Fig. 5.2   Close packing of uniform spheres. (a) Cubic packing (as in fig. 5.1a), an arrangement that is not gravitationally stable. (b) Hexagonal packing, gravitationally stable and the densest packing possible. (c) First layer of a hexagonal stack, showing two classes of sites (*A* and *B*), adjacent sites of which cannot both be occupied at the same time (for example, the two dashed locations). (d) Second layer of spheres showing how occupying some *A* and some *B* sites leaves extra space in between. (e) Hexagonal stack with left side occupying *A* sites and right portion *B* sites; the consequence is increased porosity. The random choice of *A* and *B* sites leads to a completely random pack after a few layers.

small amount of gas should lower the effective elasticity nearly as much as a large amount, and hence we expect the effect of gas on velocity to be very nonlinear. Gas in the pore space would affect density as well as effective elasticity; if we gradually introduce gas into the pore space, the first small amount of gas should have a large effect on the numerator of eq. (5.1), but additional gas will have much less effect, whereas the effect on the density term in the denominator will be linear with the amount of gas. Thus, as the amount of gas is gradually increased, we expect at first a sharp decrease in velocity and thereafter a gradual steady increase in velocity. The near-surface weathering layer being generally above the water table, we expect it to have exceptionally low velocity. Because *S*-waves do not travel through fluids, the nature of the pore fluid should have little effect on *S*-waves compared to that on *P*-waves; however, it will still have a minor influence because of its effect on the density. By changing *P*-wave velocity much more than *S*-wave velocity, the presence of gas will change the effective value of Poisson's ratio and hence change amplitude-versus-offset relationships.

Cementation and pressure-induced recrystalization would be expected to decrease porosity. Very few of the things that might happen to rocks increase porosity (see fig. 5.3a). Hence, generally, we may expect porosity to decrease (and velocity to increase) with increase in depth of burial (fig. 5.3b), cementation, age, as sorting becomes poorer, and so on.

The major failure in expectations is that an increase in density usually does not lower velocity, as might be expected from eq. (5.1). Phenomona that change the density usually change the effective elasticity more, so

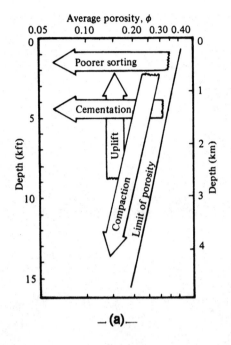

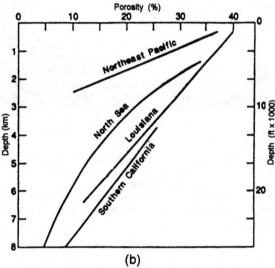

Fig. 5.3 Factors affecting porosity. (a) Porosity in a clastic rock decreases with depth of burial (compaction), cementation, and poorer sorting, but is essentially unchanged by uplift. The "limit-of-porosity" line refers to normally pressured situations and ignores possible secondary porosity. (After Zieglar and Spotts, 1978.) (b) Porosity–depth curves. (From Atkins and McBride, 1992; reprinted with permission.)

that the explicit density term in the denominator gives the wrong implication.

5.1.3 Gassmann, Biot, Geertsma equations

To obtain a useful formula for the velocity of a fluid-filled porous rock, the effects of porosity and the pore fluid must be taken into account. Gassmann (1951) derived expressions for the effective bulk modulus as-

suming that relative motion between the fluid and rock is negligible. We shall follow the account given by White (1983: 57–63).

The rock is assumed to be a porous skeleton or framework with the pore fluid moving in unison with the rock so that there are no viscous energy losses. To distinguish various components of the system, we use the following notation for the bulk moduli: k_f, k_s, k^*, and k refer respectively to the fluid filling the pore space, the material comprising the skeleton, average values for the skeleton plus empty spaces, and the fluid-filled skeleton. We use ϕ for the porosity and $C = 1/k$ for the incompressibility. We assume that the saturated rock is isotropic and that the fluid has no effect on the shear modulus, so $\mu = \mu^*$. The average density is simply the volume-weighted average:

$$\rho = \phi\rho_f + (1 - \phi)\rho_s. \tag{5.6}$$

We consider a cube of the saturated rock and apply an incremental pressure $\Delta\mathscr{P}$. We assume that the pores are interconnected so that the fluid pressure is that applied to the pore openings on the cube faces (however, no fluid enters or leaves the cube because there is no fluid motion relative to the rock). We write for the total pressure

$$\Delta\mathscr{P} = \Delta\mathscr{P}^* + \Delta\mathscr{P}_f. \tag{5.7}$$

From the definition of dilatation, we can write eq. (2.18) in the form

$$-\Delta\mathscr{V}/\mathscr{V} = C\,\Delta\mathscr{P}.$$

Thus, the pressure $\Delta\mathscr{P}_f$ changes the fluid volume by $-\Delta\mathscr{V}_f/\mathscr{V} = \phi C_f\,\Delta\mathscr{P}_f$. But $\Delta\mathscr{P}_f$ also compresses the material of the skeleton, so $-\Delta\mathscr{V}_{sf}/\mathscr{V} = (1 - \phi)C_s\,\Delta\mathscr{P}_f$. Finally, $\Delta\mathscr{P}^*$ compresses the skeleton so that $-\Delta\mathscr{V}_{ss}/\mathscr{V} = C_s\,\Delta\mathscr{P}^*$. Adding these three effects, we find for the total volume change

$$(-\Delta\mathscr{V}/\mathscr{V}) = [\phi C_f + (1 - \phi)C_s]\,\Delta\mathscr{P}_f + C_s\,\Delta\mathscr{P}^*. \tag{5.8}$$

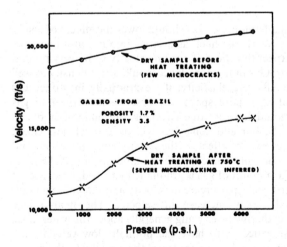

Fig. 5.4 Effect of microcracks on velocity of gabbro. (From Gardner, Gardner, and Gregory, 1974.)

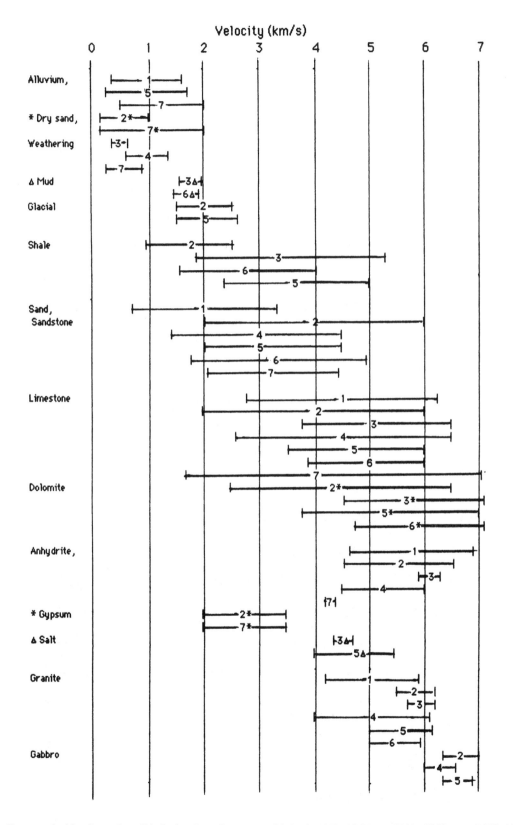

Fig. 5.5 *P*-wave velocities for various lithologies. Data from
(1) Grant and West (1965); (2) Kearey and Brooks (1984); (3)
Lindseth (1979); (4) Mares (1984); (5) Sharma (1976); (6) Sheriff
and Geldart (1983); and (7) Waters (1987).

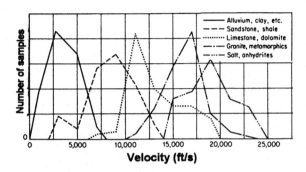

Fig. 5.6 Histogram of velocity values tabulated in Birch (1942) for different lithologies. (From Grant and West, 1965.)

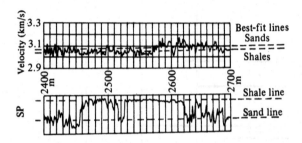

Fig. 5.7 Portion of *SP*- and velocity logs for a well in the U.S. Gulf Coast. The *SP*-values distinguish sands from shales. (After Sheriff, 1978.)

We obtain another equation for $-\Delta \mathcal{V}/\mathcal{V}$ by considering that $\Delta \mathcal{P}^*$ produces a relative volume change in the skeleton plus pores equal to $C^* \Delta \mathcal{P}^*$ and $\Delta \mathcal{P}_f$ results in a relative change in the skeleton material $C_s \Delta \mathcal{P}_f$ (the volume change of the fluid is taken care of in the term $C^* \Delta \mathcal{P}^*$). Adding, we find that

$$(-\Delta \mathcal{V}/\mathcal{V}) = C_s \Delta \mathcal{P}_f + C^* \Delta \mathcal{P}^*. \quad (5.9)$$

Equations (5.8) and (5.9) are now solved for $\Delta \mathcal{P}_f$ and $\Delta \mathcal{P}^*$ with the result

$$\Delta \mathcal{P}_f = (-\Delta \mathcal{V}/\mathcal{V})(C_s - C^*)/[\phi C^*(C_s - C_f)$$
$$+ C_s(C_s - C^*)],$$
$$\Delta \mathcal{P}^* = (-\Delta \mathcal{V}/\mathcal{V})\phi(C_s - C_f)/[\phi C^*(C_s - C_f)$$
$$+ C_s(C_s - C^*)].$$

Adding and using eq. (5.7), we have for the effective bulk modulus

$$k = \frac{1}{C} = \frac{\Delta \mathcal{P}}{-\Delta \mathcal{V}/\mathcal{V}} = \frac{\phi(C_s - C_f) + (C_s - C^*)}{\phi C^*(C_s - C_f) + C_s(C_s - C^*)},$$
$$= \frac{\phi(1/k_s - 1/k_f) + (1/k_s - 1/k^*)}{\phi(1/k^*)(1/k_s - 1/k_f) + (1/k_s)(1/k_s - 1/k^*)}. \quad (5.10)$$

Multiplying numerator and denominator by k^* gives

$$k = [k^*\phi(1/k_s - 1/k_f) + (k^*/k_s - 1)]/[\phi(1/k_s - 1/k_f) + (1/k_s)(k^*/k_s - 1)].$$

Adding and subtracting $(k^*/k_s)(k^*/k_s - 1)$ in the numerator, we get

$$k = k^* + \frac{(1-k^*/k_s)^2}{\phi(1/k_f - 1/k_s) + (1/k_s)(1 - k^*/k_s)}. \quad (5.11)$$

Thus, k equals k^* for the skeleton plus a term that depends in part upon the fluid filling the pores. Because $M = k + 4\mu/3$ (see eq. (2.58) and table 2.2), we can add $4\mu/3$ to both sides and get

$$M = M^* + \frac{(1 - k^*/k_s)^2}{\phi(1/k_f - 1/k_s) + (1/k_s)(1 - k^*/k_s)} \quad (5.12)$$

Because $\alpha^2 = M/\rho$, the P-wave velocity depends upon the fluid bulk modulus and the porosity as well as the rock properties. On the other hand, the fluid influences β only through the density (see eq. (5.6)).

One might expect the coupling between the rock skeleton and the fluid to be greater at low frequencies; Gassman's equation is therefore called the low-frequency solution. Biot (1956) assumed that the fluid could flow through the pore spaces to give a "high-frequency solution"; this introduced the additional factors of fluid viscosity and matrix permeability. Biot also defined the low-frequency range of the applicability of the Gassmann equation as

$$\nu < 0.1(\varepsilon\phi/2\pi\kappa\rho_f), \quad (5.13)$$

where ε is fluid viscosity, and κ is matrix permeability.

Geertsma and Smit (1961) derived an equation

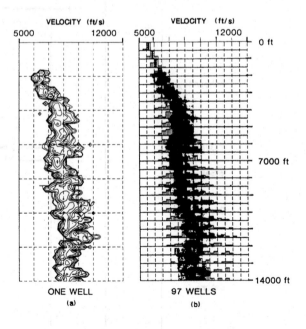

Fig. 5.8 Shale velocities in the Ship Shoal region, offshore Louisiana. (From Hilterman, 1990.) (a) A velocity analysis in one well; (b) histograms showing velocities in 97 wells throughout the area.

from the Biot equations where wavelength is greater than pore size:

$$\alpha = \left\{ \left[\left(\frac{1}{C} + \frac{4\mu}{3} \right) + \frac{(1 - C_s/C)^2}{(1 - \phi - C_s/C)C_s + \phi C_f} \right] \frac{1}{\bar{\bar{\rho}}} \right\}^{1/2}.$$

(5.14)

This equation gives values similar to eq. (5.12). These equations fit experimental data reasonably well considering how many variables are usually not known precisely.

5.1.4 Model of a nonporous rock

The foregoing sections basically explain observed velocity variations as attributable mainly to changes in the porosity and the fluid filling the porosity. However, nonporous rocks also show variation of velocity with pressure and other parameters.

Gardner, Gardner, and Gregory (1974) hypothesized that nonporous rocks have minute voids ("microcracks") that result in lowering the velocity. Generally, rocks are composed of many minerals that have different temperature coefficients of expansion, so that a temperature change will create stresses and open up microcracks. To test their hypothesis, they determined the velocity–pressure response for a gabbro with only 1.7% porosity and then heated the gabbro to 750°C and cooled it, after which they repeated the velocity–pressure measurement (fig. 5.4). The lowering of the velocity–pressure curve is presumed to be due to the creation of new microcracks. After being subjected to pressure, the sample returned to a higher velocity when the pressure was lowered; presumably, the pressure healed some of the microcracks. Probably, repetition of the pressure cycle would heal more microcracks and elevate the velocity–pressure curve still more, approaching more closely the preheating curve. One might also expect the heat-treatment-induced stresses to gradually dissipate with time so that the curve would climb gradually.

The inclusion of fluid in microcracks greatly increases the P-wave velocity, but leaves the S-wave velocity nearly unchanged (Nur and Simmons, 1969).

5.2 Experimental data on velocity

5.2.1 General

Velocity can be determined from measurements (a) in situ (see §5.4) or (b) on samples in a laboratory. Press (1966) lists measurements of both types. Care has to be taken that measurements on samples are not distorted by changes in the sample conditions; many early measurements gave misleading values because they were made on desiccated or otherwise altered samples. Gregory (1977) discusses laboratory measurements and gives a number of references from outside the usual geophysical literature. Reports of velocity measurements in the literature are numerous, and

in the following sections, we cite only those believed to be representative and that give insight into the interrelationship of factors.

The usual way to determine the effects of various factors is to observe what happens when we let them vary one at a time; we then assume that when more than one factor changes, the effect will be the same as if the effects changed sequentially. However, the factors are not independent; thus, for example, changes in external (overburden) pressure (or depth of burial) are apt to change the interstitial-fluid pressure, the porosity, and the density. Also ordinary descriptions of rocks often ignore the facts that they have various structures and are heterogeneous in composition. Thus, interpretation of experimental data regarding the parameters governing rock velocity becomes difficult and the data in the literature involve appreciable scatter.

Despite the central role that velocity plays in interpretation and the fact that it is often the principal source of uncertainty, much of the literature (Press, 1966; Robie et al., 1966; Christensen, 1989; Nur and Wang, 1989) ignores the factors affecting velocity, and others give such broad ranges that the data are not very useful.

5.2.2 Effect of lithology

Lithology is probably the most obvious factor affecting velocity and some of the data from the literature are summarized in fig. 5.5. The most impressive aspects of this figure are the ranges of values (some measurements extend beyond the ranges shown) and the tremendous overlap of the values for differing lithologies. These suggest that velocity is not a good criterion for determining lithology except in a general sense. High velocity for sedimentary rocks generally indicates carbonates and low-velocity sands or shales, whereas intermediate velocity can indicate either. The broad ranges for each of the lithologies illustrate that many other variables are involved, especially porosity and age. The Grant and West (1965) histograms (fig. 5.6) of the data from Birch (1942) also show broad ranges and overlap, as do the data tabulated by Press (1966) and Christensen (1989).

Velocity measurements are sometimes used to discriminate between sandstone and shale in areas of clastic deposition. Sands and shales in boreholes are usually identified on the basis of self-potential (SP) or gamma-ray logs; fig. 5.7 shows part of SP and sonic logs in a well that was part of a large study involving many wells in the U.S. Gulf Coast region. Regression analysis found a difference between the best-fit velocity lines for the sand data and the shale data, but the scatter of individual values exceeds the differences between the best-fit lines. Statistical predictions, such as of the overall sand/shale ratio, sometimes are satisfactory when based on reasonably good local data,

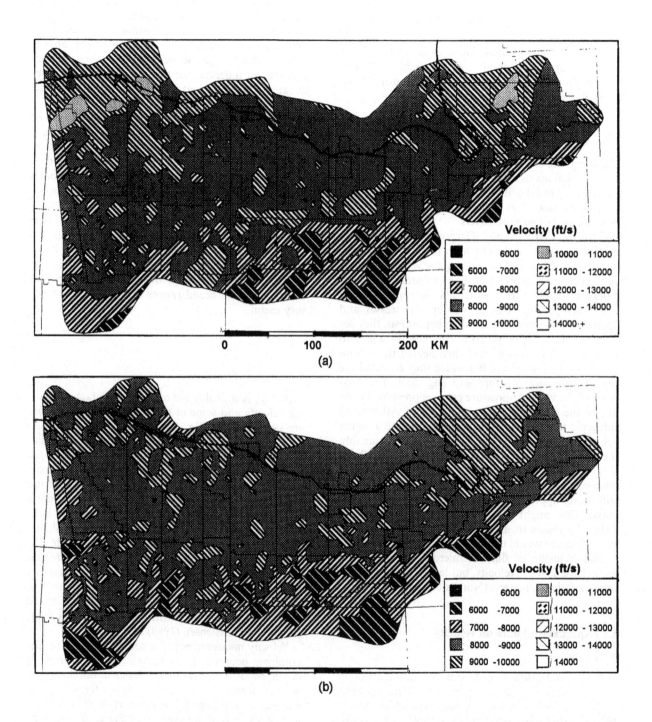

Fig. 5.9 Maps of average sand and shale velocities and densities for the depth interval from 7000 to 8000 ft for offshore Louisiana. The coastline shows near the top of these maps and the block markings show the Louisiana offshore lease systems. (Courtesy of the Geophysical Development Corp.) (a) Sand velocity; (b) shale velocity; (c) sand density; (d) shale density.

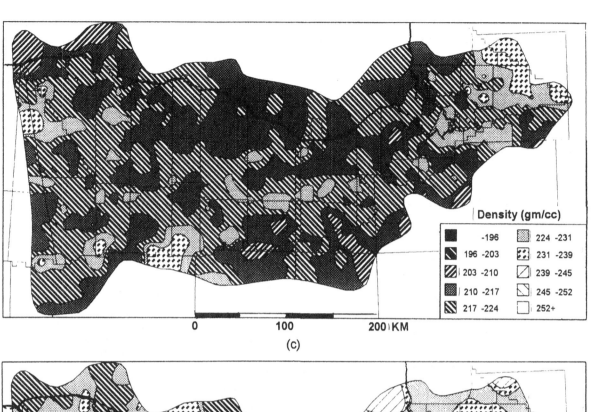

(c)

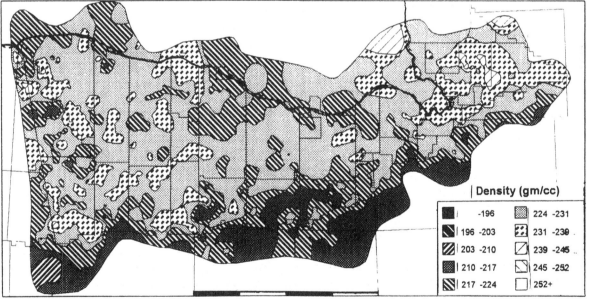

(d)

whereas predictions for specific samples are little better than guesses.

Hilterman (1990) found much variation in shale properties, presumably because of variations in grain size and cementation. The velocities of shales seen in one well (fig. 5.8a) almost tracked histograms of average shale velocities from 97 wells within the area (fig. 5.8b), both showing fairly broad ranges of values. Shale members are more continuous than sands and the strongest and most continuous reflectors are often caused by shale–shale rather than shale–sand contrasts.

In studies of U.S. Gulf Coast wells, Hilterman found that curves of velocities and densities against depth for sands versus shales vary considerably from area to area; this variability is illustrated in the maps of fig. 5.9. Because of the variability, Hilterman prefers to base synthetic seismogram studies on edited data from nearby wells rather than to use generalized values.

Figure 5.10 shows the dependence of reflection coefficients on the density and velocity differences between sands and shales according to his studies. In very young sediments, sand–shale acoustic impedance contrasts are caused mainly by density, rather than velocity, differences, but in older and more deeply buried sediments, velocity differences dominate. In the Plio-Pleistocene, a shale-to-sand reflection is generally negative, but for the Lower Miocene, it becomes positive at greater depths. Where the density and velocity contributions have opposite polarity and roughly equal magnitude, reflections are very weak (as in overpressured Upper Miocene section, fig. 5.10b).

Hilterman also found (fig. 5.11) that Poisson's ratio σ decreases with increase of velocity for both sands and shales, clean sands having appreciably smaller σ values than shales. This implies that water-filled sands may show an increase in amplitude with offset. The σ contrast between sands and shales becomes smaller as the clay content of sands increases.

Sonic logs (§5.4.3) and density logs (Telford, Geldart, and Sheriff, 1990: §11.7.2 and 11.8.3) often result in poor synthetic seismogram matches to observed data in the Gulf Coast area. Hilterman believes that sonic logs indicate sand velocities that are too high because they measure an invaded-zone velocity that exceeds the velocity of uninvaded sand. The editing of sonic-log data for synthetic seismogram manufacture (§6.2.1) attempts to correct for this. The use of deep induction logs (which depend on porosity, like the sonic log; see Telford et al., loc. cit.: 652–4) to give the acoustic impedance for synthetic seismogram manufacture often results in better matches to actual seismic records.

Sandstones often contain appreciable clay filling the pore spaces, and clay content is the next most important factor (after porosity) in determining velocities. Han, Nur, and Morgan (1986) say that the reduction of P-wave velocity when pores are clay-filled is about 30% of that when fluid-filled, and the factor for

S-wave velocity is about 40%.

A graph of S-wave velocity for different lithologies shows spreads comparable to those for P-waves except that the data are much sparser. By cross-plotting P-wave slowness against S-wave slowness, Pickett (1963) found that the domains of different lithologies separated (fig. 5.12a) but some authors quote values well outside the indicated ranges (Hamilton, 1971). The ratio of P- to S-wave velocities (β/α) is thus to some extent indicative of lithology, as illustrated also in fig. 5.12b, and S-wave surveying has been employed to determine lithology. Hamilton summarizes this usage of β/α data; he notes that there is general agreement that $\beta/\alpha < 0.5$ for unconsolidated sands, but he notes that consolidated rocks do not always have $\beta/\alpha > 0.5$. The data for shales are still very sparse; Hamilton (loc. cit.) quotes values ranging from 0.08 to 0.36, but some authors believe that the range for shales overlaps those of other lithologies to such an extent that lithology identification by β/α measurement is no longer as promising as once thought. The P- to S-wave velocity ratio in sandstones generally decreases as both porosity and clay content increase (Han et al., 1986).

5.2.3 Effect of density

The density of a rock is simply a volume-weighted average of the densities of the rock constituents. The densities of the minerals that constitute most sedimentary rocks (table 5.1) encompass a relatively narrow range of about ±7% (halite excepted). The major reason why rocks vary in density ρ is because they vary in porosity (see eq. (5.6)). Histograms of density (fig. 5.13) resemble those of seismic velocity (fig. 5.6). The densities of igneous and metamorphic rocks are generally higher than those of sedimentary rocks because they have low porosity.

Seismic velocity appears to be proportional to mean atomic weight (Birch, 1961), determined by dividing the molecular weight by the number of atoms. This is shown in fig. 5.14. Most of the relatively abundant minerals have mean atomic weights around 20 (table 5.1). Metallic ores generally have higher mean atomic weights, for example, 30.4 for ilmenite, 31.9 for hematite, and 33.1 for magnetite.

Gardner et al. (1974) graphed velocity against density (fig. 5.15) and found that the major sedimentary lithologies defined a relatively narrow swath across the graph. The principal exceptions are the evaporites (anhydrite, gypsum, salt) and carbonaceous rocks (coal, peat, lignite). They determined an empirical equation relating velocity and density, often called *Gardner's rule:*

$$\rho = aV^{1/4}, \qquad (5.15)$$

where density ρ is in g/cm^3, $a = 0.31$ when velocity V is in m/s and $a = 0.23$ when V is in ft/s. This equation is often used to obtain density values in synthetic seismogram construction or in inversion.

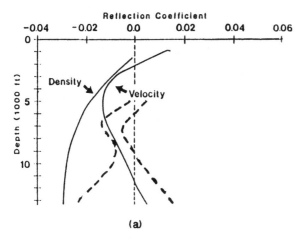

(a)

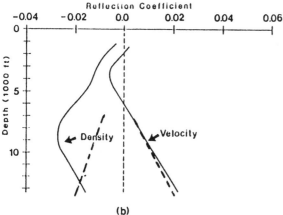

(b)

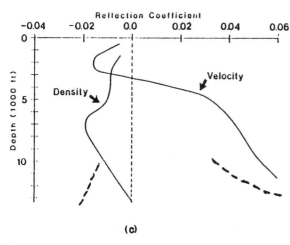

(c)

Fig. 5.10 Sand–shale reflection coefficients at normal incidence attributed to differences in density and velocity values of sand and shale, Gulf of Mexico Tertiary. The solid curves are for normal pressures and the dashed ones for overpressured conditions. (Courtesy of the Geophysical Development Corp.) (a) Pliocene and Pleistocene; (b) Upper Miocene; (c) Lower Miocene.

5.2.4 Effect of porosity

As previously stated, porosity is often the most important factor in determining a rock's velocity. An equation analogous to eq. (5.6) is often used:

$$\Delta t = \phi \, \Delta t_f + (1 - \phi) \, \Delta t_m, \qquad (5.16a)$$

where Δt is the specific transit time (slowness), Δt_f and Δt_m the specific transit times of the pore fluid and rock matrix, respectively. In terms of velocity V, this equation is

$$\frac{1}{V} \equiv \phi \frac{1}{V_f} + (1 - \phi) \frac{1}{V_m}. \qquad (5.16b)$$

Equations (5.16), the *time-average equations,* were developed by Wyllie, Gregory, and Gardner (1958) (see fig. 5.16). However, unlike eq. (5.6), which is rigorous, eqs. (5.16) are statistical and empirical. They make no allowance for the structure of a rock matrix, the connectivity of pore spaces, cementation, or past history, all of which might be expected to affect velocity. Equations (5.16) are used extensively in well-log interpretation, often with values (table 5.2) for Δt_f and Δt_m (or V_f and V_m) that are empirically determined to give the best fit over a range of interest rather than the actual slowness (or velocity) values, and the fit may be poor outside the intended range, for example, for poorly consolidated high-porosity sediments.

It should be noted that the interstitial water in shales is mostly bound water rather than free water in pore spaces; nevertheless, the volume fraction occupied by this water is usually treated as porosity.

Equations (5.16) are sometimes generalized by adding terms for the volume fractions occupied by other constituents. For example, Han et al. (1986) found that adding terms for clay content reduced the scatter from 6.6 to 2.8% for P-wave velocity and from 10.3 to 5.1% for S-wave velocity. However, they also found that they could fit velocity measurements better than those of slowness; their equations are

$$\left.\begin{aligned} \alpha &= 5.59 - 6.93\phi - 2.18C(\pm 2.1\%) \text{ km/s} \\ &= 18.3 - 2.27\phi - 7.2C \text{ kft/s}, \end{aligned}\right\} \quad (5.17a)$$

$$\left.\begin{aligned} \beta &= 3.52 - 4.91\phi - 1.89C(\pm 4.3\%) \text{ km/s} \\ &= 11.5 - 16.1\phi - 6.2C \text{ kft/s}, \end{aligned}\right\} \quad (5.17b)$$

Table 5.1 *Density of representative sedimentary rock minerals (after Robie et al., 1966)*

Mineral	Formula	Density (g/cm³)	Mean atomic weight
Calcite	$CaCO_3$	2.71	20.0
Dolomite	$CaMg(CO_3)_2$	2.87	18.4
Anhydrite	$CaSO_4$	2.96	22.7
Halite	$NaCl$	2.16	29.2
Quartz (α)	SiO_2	2.68	20.0
Albite	$NaAlSi_3O_8$	2.62	20.2
Orthoclase	$KAlSi_3O_8$	2.55	21.4
Kaolinite	$Al_2Si_2O_5(OH)_4$	2.60	15.2
Muscovite	$KAl_2(AlSi_3O_{10})(OH)_2$	2.83	19.0

Many natural minerals vary in composition and hence in density. Kaolinite and muscovite are included as representative of clay minerals.

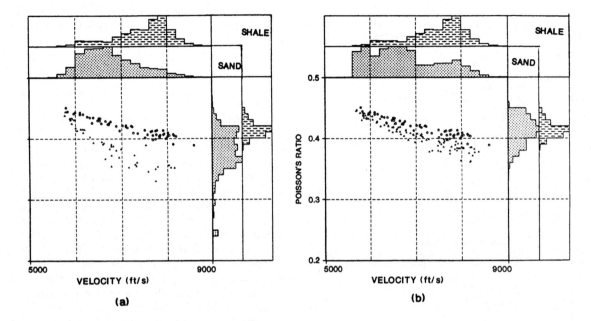

Fig. 5.11   *P*-wave velocity versus Poisson's ratio for Gulf Coast sands and shales. Triangles indicate sand values, circles shale values. (Courtesy of the Geophysical Development Corp.) (a) Shales and clean sands; (b) shales and dirty (shaly) sands.

Table 5.2 *Matrix velocities commonly used in sonic-log interpretation*

	V_m		Δt_m	
	km/s	kft/s	μs/m	μs/ft
Unconsolidated sand	<5.2	<17.0	>193	>58.8
Sandstone	5.5–9.0	18.0–19.5	182 or 167	55.5 or 51.0
Shale	1.8–4.9	6.0–16.0	205–550	62.5–167
Limestone	6.4–7.0	21.0–23.0	156–143	47.6–43.5
Dolomite	7.0	23.0	139	43.5
Anhydrite	6.1	20.0	164	50.0
Salt	4.6	15.0	218–220	66.7–67.0
Gypsum	5.5	18.0	182	55.6
Granite	6.1	20.0	164	50.0
Casing	57.4	17.5	187	57.0

where porosity ϕ and clay content C are volume fractions.

5.2.5 Effects of depth of burial and pressure

Porosity generally decreases with increasing depth of burial (or overburden pressure) and hence velocity increases with depth. The elastic constants also depend on the pressure because of the structure of sedimentary rocks, which are not homogeneous as elasticity theory assumes.

The rocks of the Louisiana Gulf Coast are generally relatively undisturbed clastic rocks whose conditions are similar to the rock model described in §5.1.1. Gregory (1977) gives velocity versus depth data for Gulf Coast sands and shales under normal pressure conditions (fig. 5.17). The use of a 1/4 exponent gives a better fit than the 1/6 exponent of eq. (5.5).

The pore spaces in rocks are filled with a fluid under a pressure, which is usually different from that resulting from the weight of the overlying rocks; the effective pressure on the granular matrix is the difference between the overburden and fluid pressures. Normal fluid pressure is that of a column of fluid extending to the surface. Where formation fluids are overpressured, the differential pressure becomes that appropriate to a shallower depth and the velocity tends to be that of the shallower depth (§5.3.4). Laboratory measurements (fig. 5.18) show that velocity is essentially constant when the overburden and fluid pressures are changed, provided the differential pressure remains constant. Abnormal fluid pressure con-

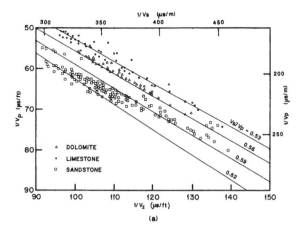

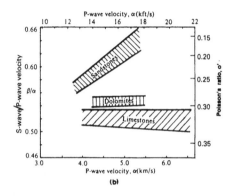

Fig. 5.12 Relation between S- and P-wave velocities (V_s and V_p) for various lithologies. (a) Cross-plot of laboratory measurements (after Pickett, 1963). (b) Use of S- and P-wave velocity ratio (β/α) as an indicator of lithology (from Sheriff, 1989; 388).

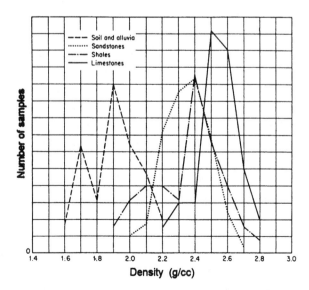

Fig. 5.13 Histogram of density values tabulated in Birch (1942) for different lithologies. (From Grant and West, 1965.)

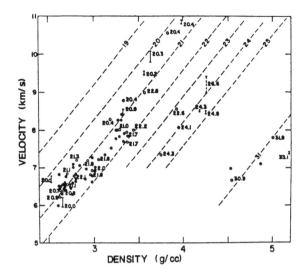

Fig. 5.14 Velocity at 10 kilobars versus density for silicates and oxides. The numbers refer to mean atomic weights. (From Birch, 1961.)

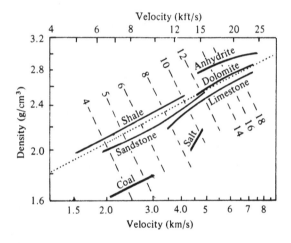

Fig. 5.15 P-wave velocity–density relationship for different lithologies (the scale is log–log). The dotted line shows eq. (5.15) and the dashed lines show constant acoustic impedance (kg/s. m² × 10⁶). After Gardner et al. 1974; and Meckel and Nath, 1977.)

stitutes a severe hazard in drilling wells and one use of seismic velocity measurements is in predicting such zones (see §5.3.4).

The variation of velocity with depth, often referred to as the velocity function (§4.2.4), is frequently a reasonably systematic increase as we go to greater depths. Velocity versus depth relationships for several areas are shown in fig. 5.19.

Gardner et al. (1974: 775–6) state: "With increasing depth the velocity increases partly because the pressure increases and partly because cementation occurs at the grain-to-grain contacts. Cementation is the more important factor." Their graph for sands is

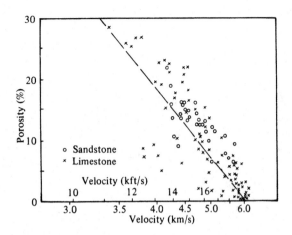

Fig. 5.16 Velocity–porosity relationship. The horizontal scale is linear in transit time ($1/V$). The dashed line is the time-average eq. (5.16b) for V_m = 5.94 km/s (19.5 kft/s) and V_f = 1.62 km/s (5.32 kft/s). (After Wyllie et al., 1958.)

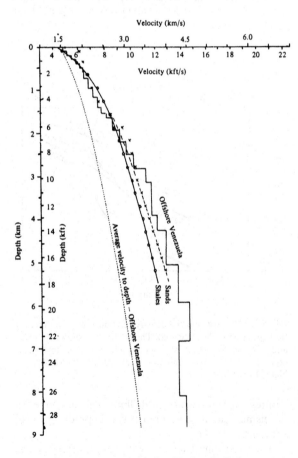

Fig. 5.17 Velocity–depth relationship for Gulf Coast sands and shales. Best-fit quadratic curves are also shown. The step graph shows data for offshore Venezuela, where conditions are similar. The dotted curve shows average velocity versus depth for the Venezuelan data. (Data from Gregory, 1977.)

shown in fig. 5.20; the curves for Ottawa sand and glass beads subjected to pressure indicate what they believe would happen if sands were buried without consolidation or cementation. The increase in velocity

with depth for in-situ sediments is more rapid until they become consolidated; below about 6000 ft they roughly follow the time-average equation. Some authors interpret data as showing a simpler curve (e.g., Faust, 1951), that is, they regard consolidation as a more continuous process.

An example of the variation with depth of α, β, and β/α in a predominantly sand–shale section is shown in fig. 5.21. Gardner and Harris (1968) consider values of $\beta/\alpha < 0.5$ as indicating water-saturated unconsolidated sand.

5.2.6 Effects of age, frequency, and temperature

An early form (Faust, 1951) of eq. (5.5) included the age of the rock as a factor in determining velocity. In fig. 5.22, each data point is the average of many values. Older rocks generally have higher velocities than younger rocks, but most geophysicists agree that age is merely a measure of the net effect of many geologic processes, that is, older rocks have had longer time to be subjected to cementation, tectonic stresses, and so on, which decrease porosity. The history of rocks varies so much in time and space that the time factor must be only approximate. Time-dependent strain may play some part, but how large a part is not known.

Experimental data generally support the conclusion that dispersion (variation of velocity with frequency) is small over the range from hertz to megahertz. We expect velocity to change with frequency because of absorption (§2.7.2), the manner of change depending on the absorption mechanism. Nur and Wang (1989:

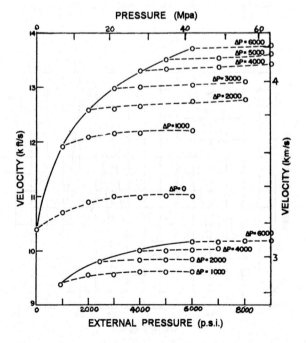

Fig. 5.18 P-wave velocity for two sandstone cores versus confining pressure where the differential pressure is held constant at the indicated values (in psi). (After Hicks and Berry, 1956.)

318–19) discuss dispersion in rocks. There is some dispersion in liquid-saturated rocks, but not in dry rocks, from which Spencer (1981) concludes that dispersion is due to fluid movement along pore surfaces. Han (1987) observed some dispersion, which decreases with increasing porosity and increases with clay content; dispersion also decreases with pressure. Wang (1988) found that dispersion increases with fluid viscosity, but it is still small in rocks saturated with light oil. Murphy (1985) found that α increased by 15% between 2 and 200 kHz.

Velocity appears to vary slightly with temperature (fig. 5.23), decreasing by 5 to 6% for an increase of 100°C. However, velocity in heavy crude oil and tar varies considerably with temperature and the same applies to rocks saturated with them (fig. 5.24) (Wang and Nur, 1988). This forms the basis for the monitoring of enhanced oil-recovery programs based on thermal stimulation, that is, steam and fire floods.

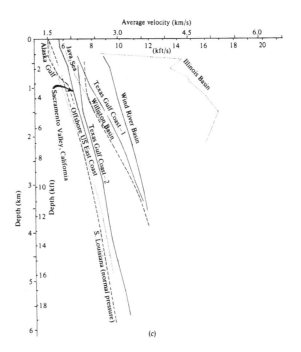

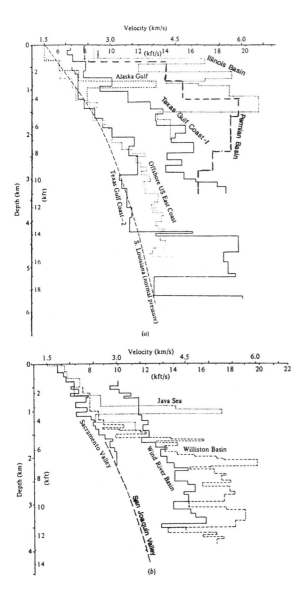

Fig. 5.19 Velocity–depth relationships for selected wells. (a) Data from Gulf of Alaska Cost-B2 well, offshore U.S. East Coast, wells in Tyler (#1) and Dewitt (#2) counties in Texas Gulf Coast, Illinois Basin, and Permian Basin. (b) Data from Sacramento Valley, Yolo Co., Calif.; Central Valley, Calif. (from Stulken, 1941); Wind River Basin, Fremont Co., Wyo.; Williston Basin, Divide Co., N.D.; and the Java Sea. (c) Average velocity to various depths for the data in parts (a) and (b).

The velocity in water-saturated rocks increases markedly as temperature is lowered through the freezing point (fig. 5.25). As the temperature drops, the liquid in the larger pores freezes first, the salinity of the liquid controlling the freezing curve. At a slightly lower temperature, the liquid in the smaller pores freezes.

5.2.7 Effect of interstitial fluid

Porous rocks are almost always saturated with fluids, generally salt water, the pores in oil and gas reservoirs being filled with varying amounts of water, oil, and gas. The replacement of water by oil or gas changes the bulk density and the elastic constants, and hence also the P-wave velocity and the reflection coefficient. These changes are sometimes sufficient to indicate the presence of gas or oil. Horizontal variations in reflection amplitude, velocity, frequency, and other factors are sometimes important indicators of oil and gas accumulations (see §10.8). The low velocities when gas fills the pore space at least partially explain the low velocities observed in the weathered (LVL) layer (§5.3.2) and why its lower boundary is so often the water table.

The nature of the interstitial fluid does not change the shear modulus appreciably and hence S-wave ve-

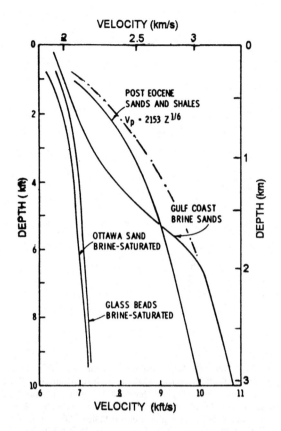

Fig. 5.20 Velocity–depth for an unconsolidated sand pack and for actual sands. The dash–dot curve shows curve shapes indicated by some authors.

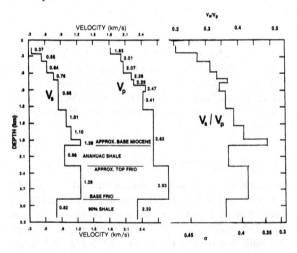

Fig. 5.21 Velocity–depth for a well in Chambers Co., Texas. Note the decrease in the ratio of P- to S-wave velocities in shales. (After Lash, 1980; and Tatham and McCormack, 1991.)

every amplitude anomaly was associated with a commercial gas or oil field. Domenico (1974), applying the Geertsma equation (5.14), showed that only a small amount of gas in the pore space produced a large decrease of velocity (fig. 5.27a) and a large change in reflectivity (fig. 5.27b). The Geertsma formula allows for the fluid compressibility as well as the density and the elastic moduli of the matrix material. Domenico (1976, 1977) partially verified the theoretical results with laboratory experiments.

5.2.8 Summary of factors affecting velocity

Tatham and McCormack (1991) summarized the effects of various parameters in fig. 5.28, including two parameters that we did not discuss: pore shape and anisotropy. If the pores are approximated by ellipsoids, the aspect ratio is the smallest axis divided by the largest. The aspect ratio of circular pores is unity, whereas small aspect ratios denote very elongate pores. Velocity decreases when pores are elongate. In transversely isotropic media, the velocity depends on the direction of particle motion with respect to the axis of symmetry (§2.6.2).

5.3 Application of velocity concepts

5.3.1 Introduction

An understanding of the factors affecting velocity helps us forsee the kinds of velocity variations to expect in an area and hence the velocity distortions to expect in seismic data (§10.5). In the U.S. Gulf Coast area, a younger section is encountered at a given depth as one goes seaward because of the seaward regional dip. Lithologic changes are relatively small and the maximum pressures to which the rocks have been subjected are the existing pressures, which depend mainly on depth, not age. Hence, the velocity (fig. 5.9) does not vary greatly from area to area, and lateral velocity changes are relatively small, although still large enough to modify structures significantly. In contrast, areas subject to recent structural deformation and uplift, such as many mountainous areas, exhibit rapid lateral variations of velocity. In such areas, many of the rocks have been subjected to greater stresses by burial to greater depths than at present, and the result is rapid lateral changes in velocity that profoundly affect seismic interpretation.

Empirical data suggest that the maximum depth to which a rock has been buried is a measure of the irreversible effect on porosity and is therefore an important parameter in determining porosity. In short, porosity is often determined principally by the existing differential pressure and the maximum depth of burial.

The irreversible change in porosity (and consequently in velocity) with depth of burial has been used to determine the maximum depth at which a section formerly lay. If the velocity–depth relationship for a

locity changes only slightly (mainly because the density changes). The ratio of P- to S-wave velocity (α/β) has been used as a method of distinguishing the fluid filling the pore space (fig. 5.26 and §13.1.1).

Some early successes in locating hydrocarbons by increased reflection amplitude led to expectations that

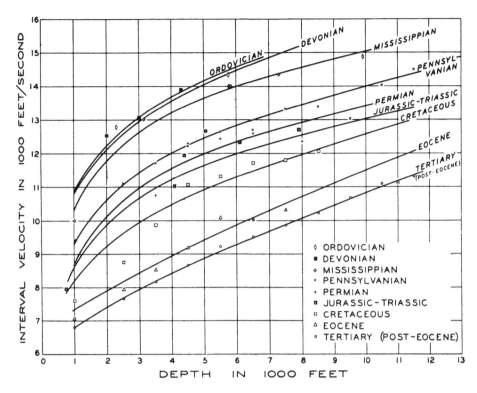

Fig. 5.22 Velocity versus age and depth of burial. (From Faust, 1951.)

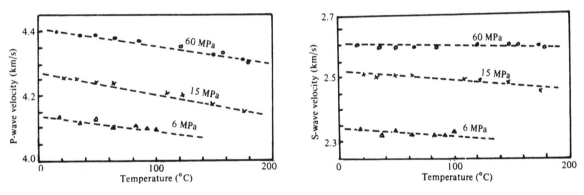

Fig. 5.23 Velocity in brine-saturated Berea sandstone as a function of temperature and pressure. (After Timur, 1977.)

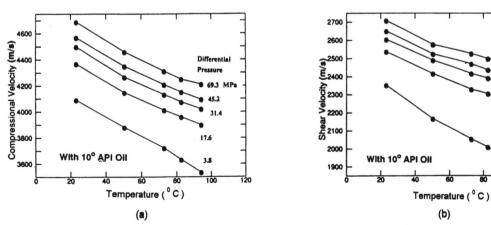

(a)

(b)

Fig. 5.24 Velocity in Berea sandstone saturated with 10° API oil as a function of temperature. (From Wang and Nur, 1992a.)

(a) P-wave velocity and (b) S-wave velocity.

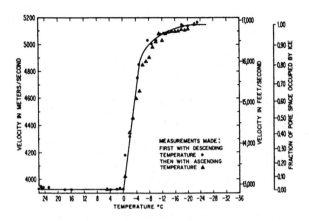

Fig. 5.25 *P*-wave velocity in Berea sandstone as the temperature passes through the freezing point. (From Timur, 1968.)

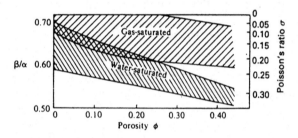

Fig. 5.26 Relation of *S*- and *P*-wave velocities and porosity for gas- and water-saturated rock. (From Sheriff, 1989; with data from Gregory, 1976.)

given lithology can be established in an area not subjected to uplift, the maximum depth of burial may be approximated from the observed velocity–depth relationship and hence the amount of the uplift can be inferred. In fig. 5.29, the shale and limestone regression lines (curves *A* and *C*) represent measurements on "pure" shales and limestones that are believed to be at their maximum depth of burial. Curve *B*, which is obtained from these curves by interpolation, is the predicted curve based on the relative amounts of shale and limestone actually present and assuming the rocks to be at their maximum depth of burial. The displacement in depth required to fit this curve to the actual observations is presumed to indicate the amount of uplift that has occurred. This technique can sometimes be used to determine if rocks have ever been buried deeply enough to acquire the high temperatures required for hydrocarbon generation (see §10.1.1).

5.3.2 The weathered or low-velocity layer

Seismic velocities that are lower than the velocity in water usually imply that gas (air or methane resulting from the decomposition of vegetation) fills at least some of the pore space (Watkins, Walters, and Godson, 1972). Such low velocities are usually seen only

on land near the surface in a zone called the *weathered layer* or the *low-velocity layer,* often abbreviated *LVL.* This layer, which is usually 4 to 50 m thick, is characterized by seismic velocities that are not only low (usually between 250 and 1000 m/s), but at times highly variable. Frequently, the base of the LVL coincides roughly with the water table, indicating that the low-velocity layer corresponds to the aerated zone above the water-saturated zone, but this is not always the case. In areas of seasonal fluctuation of the water table, leaching and redeposition of minerals may produce the effect of double-weathering layers. Double-weathering effects sometimes result from a perched water table or changes at the base of glacial drift that is at a different depth than the water table. In sandy desert areas where there may be no definite water table, the LVL may grade continuously into sediments with normal velocity. In subarctic areas, muskeg swamp is mushy with low velocity in summer and frozen with high velocity in winter (see also §5.3.3). In other areas, the nature of the low-velocity layer and the problems associated with it change considerably with season. Obviously, the term "weathering" as used by geophysicists differs from the geologist's "weathering," which denotes the disintegration of rocks under the influence of the elements.

The importance of the low-velocity layer is fivefold: (1) the absorption of seismic energy is high in this zone; (2) the low velocity and the rapid changes in velocity have a disproportionately large effect on traveltimes; (3) because of the low velocity, wavelengths are short and hence much smaller features produce significant scattering and other noise; (4) the marked velocity change at the base of the LVL sharply bends seismic rays so that their travel through the LVL is nearly vertical regardless of their direction of travel beneath the LVL; and (5) the very high-impedance contrast at the base of the LVL makes it an excellent reflector, important in multiple reflections and in mode conversion. Because of the first factor, records where the sources are within this layer often are of poor quality; shots in boreholes are often placed 5 to 10 m below the LVL. Methods of investigating the low-velocity layer are discussed in §8.5 and methods of correcting for its effects in §8.8.2.

In some areas where there is significant compaction with depth within the low-velocity layer, the increase in the velocity *V* with depth *z* approximates

$$V = az^{1/n}, \qquad (5.18)$$

where *a* and *n* are empirically derived constants. Blondeau and Swartz developed the *Blondeau method* for determining the vertical traveltime to a datum when the velocity obeys eq. (5.18) (Duska, 1963; Musgrave and Bratton, 1967). If the first-break time–distance curve is nearly a straight line when plotted on log–log paper, the method is applicable for making weathering corrections (§8.8.2). The line's slope gives *n*. The calculation procedure is discussed in problem

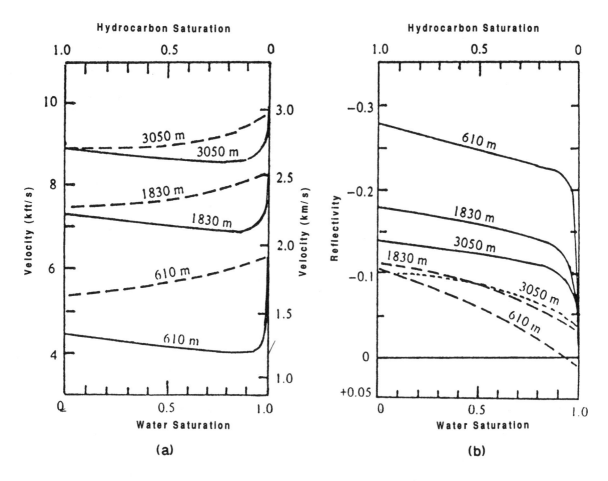

Fig. 5.27 Effect of gas/water or oil/water saturation on velocity. Solid curves are for gas, dashed for oil. (After Domenico, 1974.) (a) *P*-wave velocity versus saturation, and (b) reflection coefficient for gas/oil sands overlain by shale.

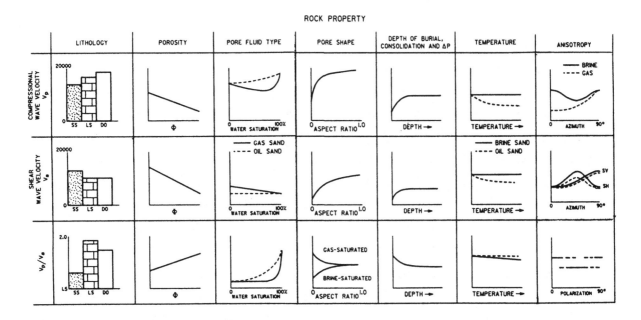

Fig. 5.28 Summary of effects of different rock properties on *P*- and *S*-wave velocities and their ratios. (After Tatham and McCormack, 1991.)

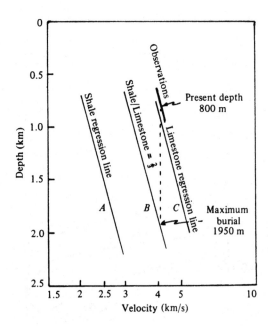

Fig. 5.29 Finding maximum depth of burial from velocity. (From Jankowsky, 1970.)

8.21. This method has been applied in glacial-drift areas.

5.3.3 Permafrost

The temperature of near-surface rocks is usually about the mean annual temperature for the location, and in arctic and some subarctic areas this temperature is below the freezing point. Seismic velocity generally increases markedly when the pore fluid in a rock freezes (fig. 5.25). In muskeg areas where the near-surface material is essentially swamp when not frozen and rich in undecayed vegetation, the velocity may increase from 1.8 km/s or lower to 3.0 to 3.8 km/s upon freezing. Timur (1968) reports Berea sandstone velocity changing from 3.9 to 5.2 km/s, Spergen limestone from 4.4 to 5.7 km/s, and black shale from 3.6 to 3.9 km/s upon freezing. The amount by which the velocity changed was roughly proportional to the porosity.

The portion of the section that is frozen year-round is called *permafrost*. There is usually a layer above it that thaws in the summer and the general increase of temperature with depth imposes a lower limit. Permafrost thickness varies from tens of centimeters to a kilometer. Where it is very thick, the velocity near its base may decrease with depth gradually until velocities are normal for the rock type. Where the permafrost is relatively thin, the decrease in velocity at its base may be fairly abrupt.

A body of water on the surface usually does not freeze deeper than a few meters and the water insulates the sediments lying below it from the cold so that permafrost is often absent under water bodies. The lateral change from normal velocities under lakes and rivers to high permafrost velocities on adjacent land areas can be very abrupt and can produce the appearance of major fictitious structures deeper in the section. Permafrost is found on the continental shelf of the Arctic Ocean where the present sea floor was exposed during a period of lowered sea level (Hofer and Varga, 1972). The thermal conductivity of earth materials is so low that thermal equilibrium has not yet been reestablished.

Whereas refraction at the base of the low-velocity layer tends to make raypaths traverse the layer more nearly vertically, thereby simplifying corrections for the layer, refraction at a permafrost boundary makes raypaths more oblique and increases the travel within the permafrost, complicating analysis/correction procedures. The effect is greater with long-offset traces, which generally have a larger horizontal component of travel. Also, the base of permafrost is often gradational and not necessarily horizontal. The result is that our ability to correct for permafrost effects is often poor. To complicate the problem, we usually cannot determine accurately permafrost thickness and velocity variations.

Another phenomenon associated with permafrost is *frost breaks* (ice breaks) that result from cracking of the ice outward from the source (fig. 5.30). These sudden energy releases occur abruptly at various times after the source activation and involve appreciable energy release, so that their effect is that of repeated erratic extra sources that may obscure reflections from the primary shot. Frost breaks are less likely to occur as the source energy decreases, so one may have to use smaller sources than otherwise desirable and increase the amount of stacking to compensate (Rackets, 1971).

5.3.4 Abnormal-pressure detection

"Normal" pressure for rocks is the situation where the pressure of the fluid in the rock's pore space is that of a hydrostatic head equal to the depth of burial. If the density of the fluid is ρ_f, normal fluid pressure $\mathscr{P}_f = \rho_f z$, where z is the depth. Drillers often speak of the pressure gradient, $d\mathscr{P}_f/dz = \rho_f$, which is about 10 kPa/m or 0.45 psi/ft for $\rho_f = 1.04$ g/cm^3 (gradients between 0.48 and 0.43 psi/ft are usually regarded as "normal"). The pressure gradient due to the rock overburden is about $d\mathscr{P}_m/dz = 22.5$ kPa/m or 1.0 psi/ft (for $\rho_m = 2.3$ g/cm^3). The effective stress on a rock (as discussed in §5.2.5) is due to the differential pressure gradient $\Delta(d\mathscr{P}/dz) = d\mathscr{P}_m/dz - d\mathscr{P}_f/dz = 12.5$ kPa/m or 0.55 psi/ft.

Abnormal or overpressure situations (subnormal pressures are also occasionally encountered) result from a sealing of formations as they are buried so that the formation liquid cannot escape to allow the formation to compact under the increased overburden pressure (Plumley, 1980). In effect, part of the weight of the overburden is transferred from the rock matrix

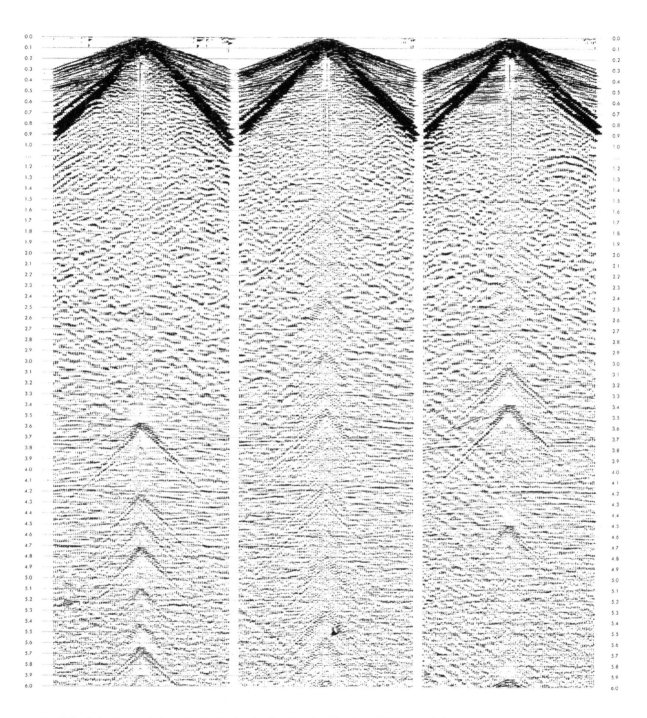

Fig. 5.30   Seismic records from the Arctic showing frost breaks. (Courtesy of Petrocanada.)

to the fluid in the pore spaces. Consequently, the rock "feels" that it is under a differential pressure appropriate to some shallower depth and the velocity of the rock is that of the shallower depth (Dutta, 1987).

The deeper portions of many depositional sequences involve fine-grained sediments where the permeability was not sufficient to allow the interstitial water to escape during compaction and abnormal pressures are common. This is especially true in young Tertiary basins where deposition has been fairly rapid, such as the U.S. Gulf Coast, the Niger and Mackenzie deltas, and along the continental slopes in many areas. Abnormal-pressure formations may behave as viscous fluids lacking shear strength and become involved in diapiric flow (see fig. 10.23) or become weak detachment zones for faulting (Gretener, 1979).

Sonic and induction well logs (Telford, Geldart, and Sheriff, 1990: §11.2.6) in a borehole encountering high pressure (fig. 5.31) often show a fairly abrupt onset of high pressure; it is sometimes more gradual. The high pressure in this well shifts the trendline toward higher transit times (lower velocity); the magnitude of this shift is a measure of the overpressure (Reynolds, 1970, 1973).

Reflections within or below an abnormally pressured section may permit calculating interval velocities by velocity analysis (§9.7.3) (Keyser et al., 1991); curves for two situations showing overpressuring are shown in fig. 5.32. Analyses for overpressure are usually plotted as transit time ($\Delta t = 1/V$) versus depth to make them more comparable to sonic logs. Analysis often includes predicting the mud weight that will be required when drilling (fig. 5.32b). Mud weight of about 9.2 pounds/gallon produces normal pressure in a wellbore.

The tendency in picking velocity analyses (§9.7) is to honor only velocity data that show a monotonic increase of stacking velocity with depth, so that conventional interpretations are apt to ignore velocity inversions, which often indicate abnormal-pressure zones. Thus, data are apt to be stacked with too high a velocity, causing deterioration of reflections within such zones, and often the zones appear rather dead. Multiples (§6.3.2), of course, are also usually evidenced by low stacking velocities and hence may make the interpretation of abnormal-pressure zones difficult.

The prediction of abnormal pressure is of considerable importance in drilling plans to minimize the possibilities of blowouts and other drilling problems such as "gas cut," "heaving" shale, and "bridging," indicated in fig. 5.31. Abnormal-pressure zones are also of importance in predicting reserves because gas reservoirs within them can contain exceptionally large amounts of gas for the reservoir volume.

When searching for abnormal pressures, velocity surveys (§5.4.2) are usually run with smaller increments than when the objective is primarily to determine the stacking velocity. Aud (1976) contends that velocity-scan increments should be 50 ft/s and time

increments 10 ms; he also argues that data should be at least 12-fold with offset sufficient to give at least 100 ms of normal moveout. He often picks events spaced only 100 ms apart. The technique of averaging data for a number of adjacent midpoints generally leads to less noise in measurements, but also to poorer identification of velocity values with specific events. Because of the uncertainties in the values determined from any single velocity analysis, weighted averaging of the results of several adjacent analyses improves reliability.

5.3.5 Gas-hydrate effects

Reflections that cut across the bedding are sometimes seen on deep-water seismic data a short distance below the sea floor, as in fig. 5.33. These are often attributed to gas hydrates, icelike crystalline lattices of water molecules in which gas molecules are trapped physically. These are stable under the temperature and pressure conditions found just below the sea floor in deep water. They apparently can form where the gas concentration exceeds that necessary to saturate the interstitial water. The velocity in methane-hydrate sediments is about 2.0 to 2.2 km/s (Tucholke, Bryan, and Ewing, 1977). A "base of gas-hydrate reflection" usually is roughly conformable to the sea floor and the depth of the reflection beneath the sea floor corresponds roughly to the limit of stability of methane hydrate (fig. 5.34). The reflection is thus interpreted as marking the interface between hydrate and gas trapped by the overlying hydrate. The gas trapped in this way may someday be an energy resource.

5.4 Measurement of velocity

5.4.1 Velocity terminology

Many adjectives are used preceding the word "velocity," the speed with which a wave travels. Velocity in heterogeneous media depends on the proportions and distributions of the media and the direction of a wave (and amplitude for very large disturbances). Terms such as *average velocity* $\overline{V}$ (the distance traveled divided by the time required to traverse the path) depend on the raypath. Likewise, the *root-mean square (rms) velocity* V_{rms} refers to a specific raypath; rms velocities are typically a few percent larger than corresponding average velocities.

$$\overline{V} = \frac{\int_0^t V(t)\, dt}{\int_0^t dt}, \qquad V_{rms}^2 = \frac{\int_0^t V^2(t)dt}{\int_0^t dt}. \quad (5.19)$$

Although average and rms velocity have meaning only with respect to a particular path, a vertical path and horizontally layered media having velocities V_i, thicknesses Δz_i, and traveltimes through the layers Δt_i are

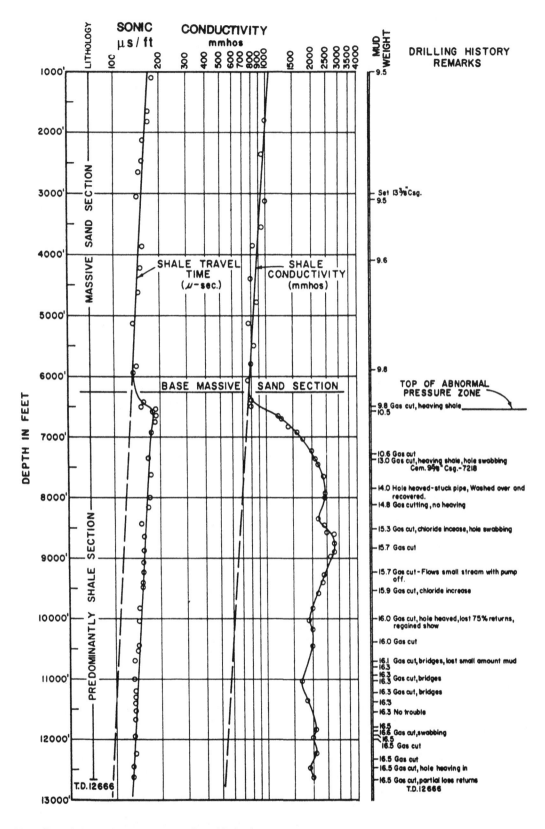

Fig. 5.31   Effect of abnormal pressure on sonic and induction
(conductivity) logs of an offshore U.S. Gulf Coast well. (From
MacGregor, 1965; reprinted with permission.)

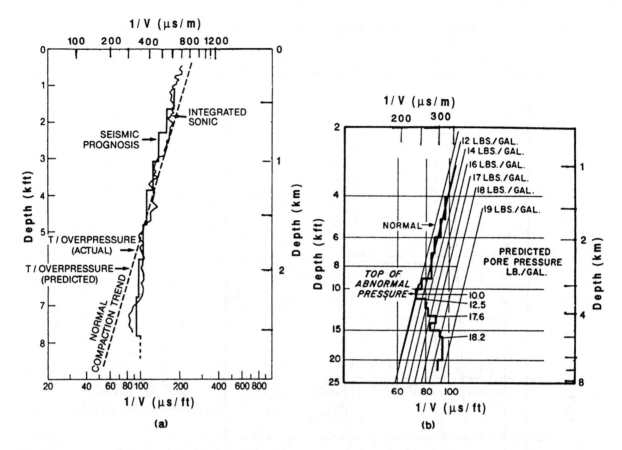

Fig. 5.32   Lowering of seismic velocity as indicator of overpressure. (a) After Reynolds, 1973; (b) after Pennebaker, 1968.

often implied, that is, depth divided by the traveltime vertically to that depth (straight raypaths). Then, by replacing the integrals with sums, we obtain

$$\overline{V} = \frac{\sum\limits_{f} V_f\,\Delta t_f}{\sum\limits_{f} \Delta t_f} = \frac{\sum\limits_{f} \Delta z_f}{\sum\limits_{f} \Delta t_f}, \quad V_{\text{rms}}^2 = \frac{\sum\limits_{f} V_f^2\,\Delta t_f}{\sum\limits_{f} \Delta t_f}; \quad (5.20)$$

the former is eq. (5.24), the latter eq. (4.26). *Stacking velocity* V_s (also called *NMO velocity*) is the velocity value determined by velocity analysis (see eq. (5.23) and §9.7); these measurements are usually made by finding the best-fit hyperbola to data that are not perfectly hyperbolic. For isotropic horizontal layers, $V_s = V_{\text{rms}}$ in the limit as the source–receiver distance approaches zero (see fig. 4.10).

Interval velocity V_i is the average velocity over some interval of the travel path. The interval may be the distance over which a sonic log measures or the interval between parallel horizontal reflections. The latter is also called the *Dix velocity* given by eq. (5.25), which can be written as

$$V_i^2 = \frac{V_n^2 t_n - V_{n-1}^2 t_{n-1}}{t_n - t_{n-1}}, \quad (5.21)$$

where V_n is the rms velocity and t_n is the zero-offset

arrival time corresponding to the nth reflection.

Because wavefronts are surfaces of constant phase, perpendicular distances between them yield *phase velocity*. *Apparent velocity* refers to the apparent speed of a given phase in a particular direction, usually the spread direction; thus, apparent velocity is greater than instantaneous velocity by the secant of the angle between the direction of wave travel and the direction in which the apparent velocity is measured. *Group velocity* was discussed in §2.6.2 and 2.7.4.

5.4.2 Conventional well surveys

The most direct methods of determining velocity require the use of a deep borehole. Three types of well surveys are used: the "conventional" method of "shooting a well," a variation called vertical seismic profiling (VSP), described in §13.4, and sonic logging, described in the next section.

Shooting a well consists of suspending a geophone or hydrophone in the well by means of a cable and recording the time required for energy to travel from a source near the wellhead down to the geophone (see fig. 5.35). Air guns in the mud pit or in the water for marine wells are often used as energy sources. The geophone is specially constructed to withstand im-

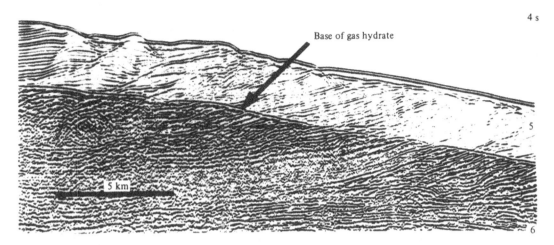

4 s

Base of gas hydrate

5 km

5

6

Fig. 5.33   Seismic line on the Blake Outer Ridge, offshore U.S. southeast coast, showing base of gas-hydrate reflection. Water depth is from 3000 to 3600 m. (From Shipley et al., 1979.)

mersion under the high temperatures and pressures encountered in deep wells. A mechanical arm presses the geophone against the borehole wall to assure coupling. The cable has a threefold role: it supports the geophone, it serves to measure the depth of the geophone, and it carries electrical conductors that bring the geophone output to the surface, where it is recorded. Sources are located at points near the wellhead. The geophone is moved between activations of the source so that the results are a set of traveltimes from the surface down to various depths. The geophone depths are chosen to include the most important geological markers, such as tops of formations and unconformities, and also intermediate locations so that the interval between successive measurements is small enough to give reasonable accuracy (often 200 m apart).

Results of a typical well survey are shown in fig. 5.36. The vertical traveltime, t, to the depth, z, is obtained by multiplying the observed time by the factor $z/(z^2 + x^2)^{1/2}$ to correct for the slant distance. The average velocity between the surface and the depth z is then given by the ratio z/t. Figure 5.36 shows the average velocity $\bar{V}$ and the vertical traveltime t plotted as functions of z. If we subtract the depths and times for two sources, we find the *interval velocity* V_i, the average velocity in the interval $z_m - z_n$, by means of the formula

$$V_i = \frac{z_m - z_n}{t_m - t_n}. \qquad (5.22)$$

Time measurements often have an accuracy of only 2 ms and consequently interval velocity values are not as accurate as traveltime–depth data.

Shooting a well gives the average velocity with good accuracy. It is, however, expensive because the cost includes not only the one-half to one day's time of the seismic crew, but also the cost of standby time for the well (which often far exceeds the seismic cost). Poten-

tial damage to the well is another factor that discourages shooting wells; while the survey is being run, the well must stand without a drill stem in the hole and hence is vulnerable to cave-in, blowout, or other serious damage. A further disadvantage in new exploration areas is that seismic surveys are often completed before the first well is drilled.

5.4.3 Velocity (sonic) logging

The continuous-velocity survey makes use of one or two pulse generators and two or four detectors, all located in a single unit called a *sonde*, which is lowered into the well. Figure 5.37a shows the borehole-compensated sonic-logging sonde developed by Schlumberger. It consists of two sources of seismic pulses, S_1 and S_2, and four detectors, R_1 to R_4, the

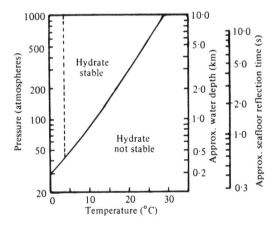

Fig. 5.34   Limit of stability of methane hydrate in water containing 3.5% NaCl. The horizontal distance between the dotted vertical line (indicating sea-floor temperature of 3°C) and the stability-limit line is roughly proportional to the thickness of the gas-hydrate zone. (After Tucholke et al., 1977.)

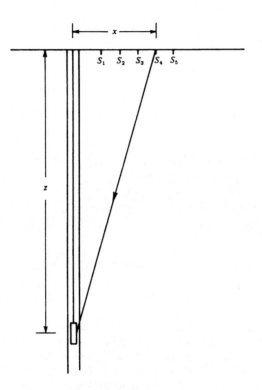

Fig. 5.35 Shooting a well for velocitity.

"span" distances from R_1 to R_3 and from R_2 to R_4 often being 61 cm (2 ft). The spacing of the "borehole-compensated" sonde is 1.22 m (4 ft), on the "long-spaced" sonde 2.44 m (8 ft). With the longer spacing, there is greater likelihood that the velocity measured will be that of unaltered formation. The velocity is found by measuring the traveltime difference for a pulse traveling from S_1 to R_2 and R_4, similarly for a pulse going from S_2 to R_3 and R_1, then taking the average of the differences. The sonde is run in boreholes filled with drilling mud, which has a seismic velocity of roughly 1500 m/s; however, the first energy arrivals are the P-waves that have traveled in the rock surrounding the borehole. Errors arising from variations in borehole size or mud-cake thickness near the transmitters are effectively eliminated by measuring the difference in arrival time between two receivers; errors resulting from such variations near the receivers are reduced by averaging the results from the two pairs of receivers. The sonic log (fig. 5.37b) shows as a function of depth the transit time divided by the span (expressed in μs/ft), the result being the reciprocal of the P-wave velocity in the formation.

The traveltime interval between sonic-log receivers is measured by a device that automatically registers the arrival of the signal at each of the two receivers and measures the time interval between the two. Because the signal at the receiver is not a sharp pulse but instead is a wavetrain, the detector is actuated by the first peak (or trough) that exceeds a certain threshold value. At times, the detector is not actuated by the same peak (or trough) at the two receivers, and hence

the increment of traveltime will be in error. This effect, called *cycle skip*, usually can be detected and allowed for because the error is exactly equal to the known interval between successive cycles in the pulse (Kokesh and Blizard, 1959).

The accuracy of sonic-log values is often rather poor, as evidenced by frequent disagreements between regular and long-spaced logs. This fact should be realized by geophysicists, who are inclined to believe sonic and distrust seismic data when faced with disagreements. Sonic logs often suffer from inadequate penetration, hole caving, alterations with time after drilling the borehole, and other factors. The borehole-compensated sonic log generally does a good job of compensating for sonde tilt and minor hole irregularities. Errors because of cycle skip can usually be recognized and compensated. Measurements may be affected by invasion of the borehole fluid into formations so that the velocity measured is not that of the unaltered formation. Long-spaced tools help ensure penetration beyond the invaded zone; they use a different scheme for compensation for sonde tilt and hole-size irregularities. Dispersion, because of differences between the P-wave velocity at the frequencies of sonic logs (ordinarily 20 kHz) and seismic frequencies, usually is quite small, probably less than 1%.

The instantaneous velocity fluctuates rapidly in many formations, as seen in fig. 5.37b. While the velocity distribution, if considered in detail, is an extremely irregular function, the wavelengths used in seismic exploration are so long (generally greater than 30 m) that the rapid fluctuations are not significant in determining the path of waves.

The sonic log is automatically integrated to give total traveltime, which is then shown on the log as a function of depth by means of ticks at intervals of 1 ms. There is a tendency for small systematic errors to

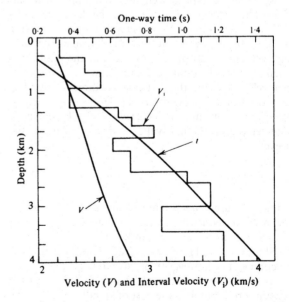

Fig. 5.36 Plot of a well-velocity survey.

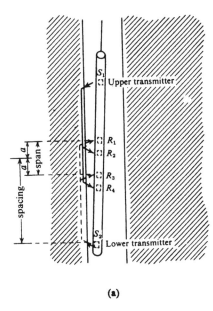

(a)

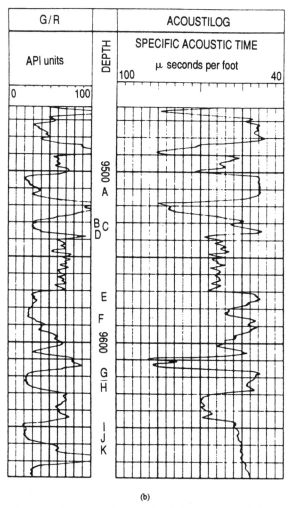

(b)

Fig. 5.37 Sonic logging. (a) Borehole-compensated logging sonde (courtesy of Schlumberger). (b) BHC Acoustilog survey in a dolomite–shale sequence. The left curve is gamma-ray response (courtesy of Atlas Wireline Services Division of Western Atlas).

accumulate in the integrated result. *Check shots* (an abbreviated "conventional" well survey (§5.4.2)) can be made at the base and top of the sonic log (and occasionally in between) so that the effect of the cumulative error can be reduced by distributing the difference in a linear manner. The sonde may include a well seismometer of the type used in shooting a well to facilitate taking check shots.

Sonic logs are used for porosity determination because porosity appears to be the dominant factor in seismic velocity. Although sonic logs are of great value to the geophysicist, they are usually not run with the geophysical uses in mind and hence often do not produce all the information that the geophysicist wants. For example, check shots are not necessary for porosity determination and, therefore, are often omitted; the log usually does not cover the entire hole depth and sonic-log data are rarely available for the shallow part of the borehole. Thus, the sonic-log data are usually incomplete so that using such logs for velocity control involves assumptions where the data are missing. Of course, head-wave travel is not achieved where the formation has lower velocity than the borehole mud, which is apt to be the case in the upper part of the borehole. Depending on the use to which sonic-log data are to be put, editing and correcting may be needed. Editing involves detailed comparisons of different logs to locate improbable values and replacing them with values believed to be more reasonable. Editing is often essential if good synthetic seismograms (§6.2.1) are to be made from log data. Missing shallow data are especially a problem in preparing synthetic seismograms from sonic-log data.

An array-sonic tool (fig. 5.38) employing downhole digitizing is now coming into use. Compared with a conventional tool employing two transmitters and two receivers, it contains eight additional receivers spaced 6 in. (15 cm) apart that record full waveforms (fig. 5.39a). The high-frequency P-wave traveling in the formation arrives first, usually with very low amplitude, to be followed by a higher-frequency, stronger S-wave and then a lower-frequency, very high-amplitude Stoneley wave. The Stoneley wave travels along the borehole wall and sometimes provides indications of fractures and permeability changes. Arrivals are picked by computer to yield plots of the P-, S-, and Stoneley-wave traveltimes (fig. 5.39b). Sometimes a "Poisson's ratio log" (fig. 5.40) is also computed. This information is proving useful in studies to evaluate and optimize production from reservoirs.

Several other variations of acoustic borehole logs are used sometimes though they have little application in seismic work. The amplitude of S-waves is sometimes displayed on a "fracture-finder log," high amplitude indicating the absence of fractures. The P-wave amplitude when the sonde is inside the casing is displayed in the "cement-bond log"; the amplitude is high when the casing is hanging freely because the energy is trapped in the steel casing, and low when bonded to the formation by cement. The full wave-

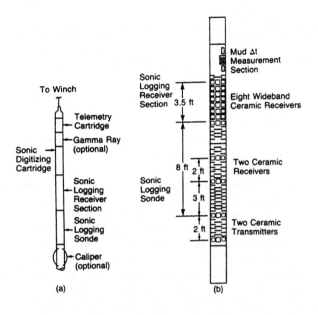

Fig. 5.38 Schematic of an array sonic-logging sonde. (From Charnock, 1990.)

form is sometimes displayed, and this log is also used to measure the quality of bonding the casing to the formation.

5.4.4 Velocity from traveltime-offset measurements

(a) X^2–T^2 method. The arrival time of reflected energy depends not only on the reflection depth and velocity above the reflector, but also on offset distance. Several methods (including the velocity-analysis methods that will be described in §9.7) utilize this dependence on offset as a means of measuring the velocity. Two classical methods, X^2–T^2 and T–ΔT, are central to surface velocity-measurement methods, even though both methods have fallen into disuse.

The X^2–T^2 method is based on eq. (4.4). We write

$$t^2 = x^2/V_s^2 + t_0^2 \qquad (5.23)$$

When we plot t^2 as a function of x^2 (fig. 5.41), if the velocity is constant, we get a straight line whose slope is $1/V_s^2$ and whose intercept is t_0^2, from which we can determine the corresponding depth. If velocity changes are not extreme, the x^2–t^2 curve can be approximated by a straight line. This is equivalent to fitting a portion of the curve of fig. 4.10b by a straight line. The slope of this line gives the *stacking velocity,* V_s, the velocity assumed in CMP stacking (§8.3.3).

When we have horizontal velocity layering and horizontal reflectors, V_s is nearly the same as the rms velocity V_{rms} in eq. (4.26). Under different circumstances,

$$V_s = V \text{ for constant velocity and horizontal reflector,}$$

$V_s = V/\cos\xi$ for constant velocity and reflector dip ξ,

$V_s \approx V_{rms}$ for horizontal velocity layering and reflectors,

$V_s \approx V_{rms}/\cos\xi$ for dipping but parallel velocity layering and reflectors.

In the general case, there is no simple relationship between V_s and V_{rms}. The stacking velocity V_s is used in correcting CMP data before stacking (hence its name) even where its relation to the velocity distribution is very complicated or not known. We sometimes use the equivalent average velocity $\overline{V}$ (§4.2.2) for depth determination (see eq. (5.20)):

$$\overline{V} = \frac{\sum_{i=1}^{n} V_i \, \Delta t_i}{\sum_{i=1}^{n} \Delta t_i}; \qquad (5.24)$$

This equation also assumes horizontal velocity layering (and vertical raypath).

When the regular seismic profile does not have a sufficiently large range of x-values to enable us to find the velocity with the accuracy required, special long-offset profiles can be acquired (Dix, 1955), but, with the longer offsets used in CMP work, they are rarely required. An X^2–T^2 survey can give velocities accurate within a few percent.

Once velocities have been determined to two successive parallel reflectors using eq. (4.26), the interval velocity can be found from the Dix equation. Writing V_L for the rms velocity to the nth reflector and V_U for the rms velocity to the reflector above it, eq. (4.26) gives

$$\sum_{1}^{n} V_i^2 \, \Delta t_i = \sum_{1}^{n-1} V_i^2 \, \Delta t_i + V_n^2 \, \Delta t_n = V_L^2 \sum_{1}^{n} \Delta t_i,$$
$$\sum_{1}^{n-1} V_i^2 \, \Delta t_i = V_U^2 \sum_{1}^{n-1} \Delta t_i.$$

Subtracting and dividing both sides by Δt_n then gives the Dix velocity (see eq. (5.21)):

$$V_n^2 = (V_L^2 \sum_{1}^{n} \Delta t_i - V_U^2 \sum_{1}^{n-1} \Delta t_i)/\Delta t_n. \qquad (5.25)$$

Note that this equation implies that the travel paths to the $(n-1)$th and nth reflectors are essentially identical except for the additional travel between the two reflectors.

When the two reflectors are not parallel or when the offset is large, this condition is not satisfied and the Dix equation may give meaningless results.

(b) T–ΔT method. The T–ΔT method is based upon eq. (4.7), which can be written in the form

$$V = \frac{x}{(2t_0 \, \Delta t_n)^{1/2}}. \qquad (5.26)$$

With symmetrical spreads Δt_n can be calculated from the arrival times of a reflection event at the source (t_0) and at the outside geophone groups, t_1 and t_k. Dip

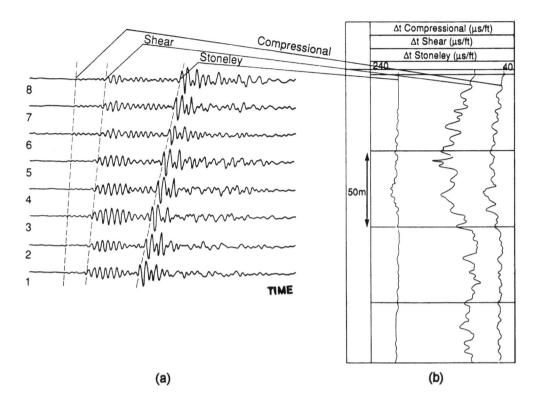

(a)   **(b)**

Fig. 5.39 Array sonic logging. (From Charnock, 1990.) (a) Waveform recorded by the upper eight receivers in fig. 5.38 showing the P-, S-, and Stoneley-wave arrivals. (b) Log of the velocities of the three waves.

moveout is eliminated by averaging the moveouts on the opposite sides of the source point:

$$\Delta t_n = \tfrac{1}{2}[(t_1 - t_0) + (t_k - t_0)] \qquad (5.27)$$
$$= \tfrac{1}{2}(t_1 + t_k) - t_0.$$

The values of Δt_n given by this equation are subject to large errors, mainly because of uncertainties in the near-surface corrections. To get useful results, large numbers of measurements must be averaged in the hope that weathering variations and other uncertainties will be sufficiently reduced (Swan and Becker, 1952).

(c) Best-fit approaches. Most velocity determination is done in data processing, which is discussed in §9.7. These methods are based on either (1) finding the hyperbola that best fits coherent events assumed to be primary reflections within some given space and time window, or (2) finding which stacking velocity produces the "best" stacked section. Such measurements are generally sufficiently accurate for stacking but not always for the lithologic conclusions sometimes drawn from them.

(d) Measurements on diving waves. Where the velocity increases monotonically with depth and velocity layering is parallel to the surface, $x–t$ data can be used to determine the velocity as a function of depth for diving waves (§4.3.5) by solution of eq. (4.53), (4.54),

or (4.56). However, the complete $x–t$ curve is often not available so that this method is rarely used except to determine the velocity in the Earth's mantle and core where other methods cannot be used.

5.4.5. Measurements based on reflection amplitude

In concept at least, with amplitude preserved in recording and processing, acoustic impedance changes are proportional to seismic amplitudes. The reflection coefficient equation at normal incidence is given by eq. (3.14):

$$R_\perp = \frac{Z_{i+1} - Z_i}{Z_{i+1} + Z_i} \approx \tfrac{1}{2}\,\Delta(\ln Z) \approx \tfrac{1}{2}kA, \qquad (5.28)$$

where Z_i and Z_{i+1} are acoustic impedances on opposite sides of the interface giving rise to a reflection, $\Delta(\ln Z)$ is the change in the logarithm of the acoustic impedance, and A is the amplitude of the reflection; k is called the *scaler,* a constant depending on system gain. Applying this relationship to real data, we assume that the amplitude is not affected by noise or the shape of the seismic wavelet. Although we cannot achieve these conditions, we can sometimes get close enough to get useful acoustic impedance data.

Using eq. (5.28) to give reflection coefficients from impedance data is an example of a direct problem, whereas obtaining impedance information from the amplitude measurements is an inverse problem and an example of *inversion*. Trace inversion is preceded by

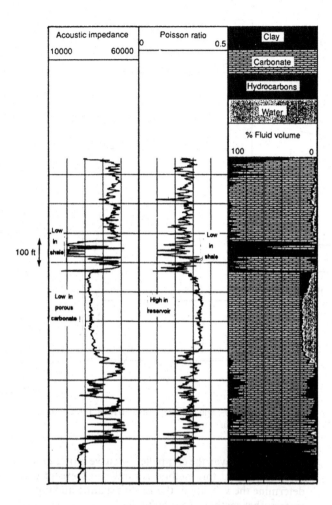

Fig. 5.40   Acoustic impedance and Poisson's ratio logs. Shown to the right is an interpretation of the lithology and fluid content for a Middle East carbonate reservoir. (From Charnock, 1990.)

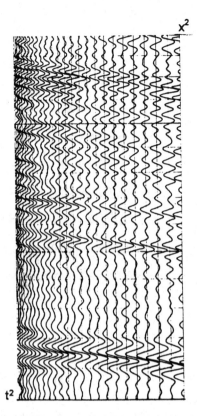

Fig. 5.41   Seismic gather plotted on x^2–t^2 scale. (From Waters, 1987: 214; reprinted by permission of John Wiley & Sons, Inc.)

$$\int_0^t \Delta[\ln Z(t)] = \int_0^t d[\ln Z(t)] = \ln Z(t) - \ln Z(0)$$

$$= \int_0^t kA(t)\, dt, \qquad (5.30)$$

$$Z(t) = Z(0) \exp\left[k\int_0^t A(t)\, dt\right].$$

Neither the scaler k nor the constant of integration $Z(0)$ can be obtained from the seismic data alone. Because we have data only within a limited passband, we usually let $Z(0)$ represent the missing "low-frequency components" in addition to the starting value; it is usually made time-dependent rather than constant. The contributions of low frequencies are evident in fig. 5.42. Velocity determined from normal moveout (§9.7) can sometimes give part of the missing information (perhaps values as low as 7 Hz if we can calculate for intervals as small as 150 ms); however, moveout-derived interval velocity is often unreliable, especially when calculated for small time intervals. Better results are achieved by using logs in a nearby well where the geology is expected to be similar. Missing high frequencies simply limit the amount of detail derivable.

A seismic log is often expressed as a velocity–time series, that is, as a *synthetic sonic log*. This requires either knowledge of the density distribution or an assumption about the relation between the velocity and the density. Commonly, we assume either that the density is constant, in which case it drops out, or that

processing to remove as much of the noise as possible and the effects of the embedded wavelet. If we also have knowledge of density, we can solve for velocity rather than acoustic impedance.

Equation (5.28) can be solved for the acoustic impedance below an interface in terms of that above the interface:

$$Z_{i+1} \approx \frac{1 + \frac{1}{2}kA_i}{1 - \frac{1}{2}kA_i} Z_i. \qquad (5.29)$$

If the density and velocity, ρ_1 and V_1, respectively, are known for the shallowest layer, then $Z_1 = \rho_1 V_1$ is known and Z_2 can be found from amplitude A_1 of the first reflection, Z_3 can be found from A_2, and so on. The resulting recursively derived value of the acoustic impedance as a function of arrival time or depth is called a *synthetic acoustic-impedance log* or *seismic log*.

A more common way for solving for Z as a function of time $Z(t)$ or of depth $Z(z)$ is to integrate eq. (5.28):

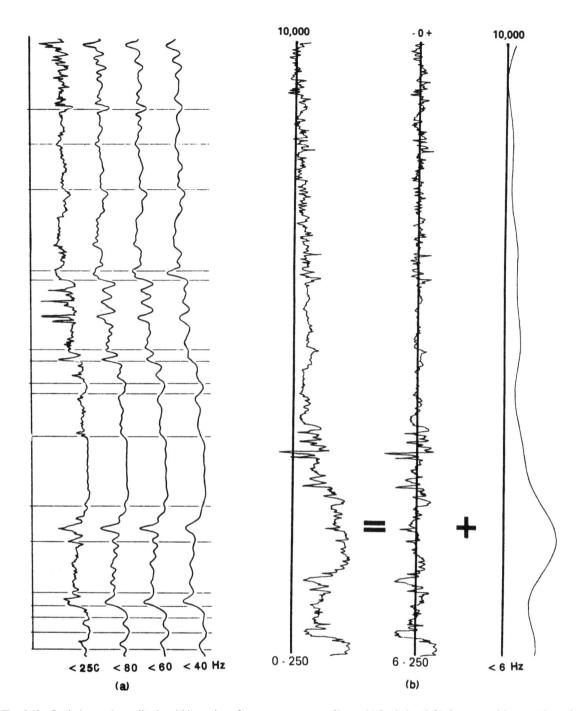

10,000   **· 0 +**   **10,000**

< 25C   < 80   < 60   < 40 Hz   **0 · 250**   **6 · 250**   **< 6 Hz**

(a)   **(b)**

Fig. 5.42 Sonic log and amplitude within various frequency bands. (From Lindseth, 1979.) (a) Sonic log with various high-cut filters. (b) Sonic log (left) decomposed into portions above and below 6 Hz.

Gardner's rule, eq. (5.15), holds. In the latter case, we have

$$Z = \rho(V)V = aV^{5/4},$$

$$V(t) = V(0) \exp \left[(4k/5) \int_0^t A(t) \, dt \right], \qquad (5.31)$$

which is the same form as eq. (5.30) except that a new constant, $4k/5$, replaces k. Because not all lithologies follow the same velocity–density relation (see fig.

5.15), we should not expect quantitative agreement. In particular, evaporites (anhydrite, gypsum, salt) and carbonaceous sediments (coal, lignite) depart considerably from the velocity–density relation for other sedimentary lithologies.

Synthetic acoustic-impedance (or sonic) logs emphasize the acoustic impedance (or velocity) of beds rather than the contrasts at the interfaces, making it easier to relate to rock properties. Using seismic logs, we often see features that we would otherwise miss even though the synthetic logs merely represent a re-arrangement of the seismic information rather than new information.

The principal limitation with synthetic logs is the assumption of a linear relation between reflectivity and amplitude, which in effect assumes a noise-free seismic record (showing the effect of primary reflections only) plus recording and processing that have preserved amplitudes faithfully. The successful manufacture of seismic logs requires excellent reflection records. The additional limitations involving the determining of $V(0)$ and k are minimized where the application is to interpolate velocity information between wells or to extrapolate velocity information in the immediate vicinity of well control. Synthetic sonic logs constitute a powerful tool for locating stratigraphic changes, porosity changes and hydrocarbon accumulation under these circumstances (Lindseth, 1979).

A black-and-white synthetic-log display is shown in fig. 5.43. There is one synthetic-log trace for each seismic trace input. The horizontal scale for each trace is linear in transit time (the reciprocal of velocity); the transit-time scale is shown for one trace (dark curve at the left). This trace shows the usual increase of velocity with depth. An actual sonic log has been superimposed at the right and another synthetic-log trace (dark curves) for comparison. Synthetic sonic-log traces do not show the detail of sonic logs because the seismic trace is deficient in high frequencies, synthetic logs often cannot follow sharp velocity changes faithfully, and they often drift because the low-frequency "constant" is incomplete. They also often show periodicity because the embedded wavelet has not been completely deconvolved.

It is difficult to read transit-time values from synthetic sonic-log displays such as shown in fig. 5.43 and velocity or transit times are usually color-encoded and superimposed (plate 1).

5.4.6 Other sources of velocity information

In addition to determining velocity from measurements in boreholes, traveltime offsets, or amplitudes as discussed in the preceding sections, velocity-dependent processing potentially provides another source of velocity information. One may vary the velocity in processing in order to maximize the coherence (§6.1) or consistency of the resulting picture. One example would be to vary the velocity distribution *(migration velocity)* so that the most coherent migrated structure (§9.12) results. Mills et al. (1993) built a velocity model where the velocity changed at migrated reflections layer by layer. After determining the velocity in shallower layers, they varied the velocity in the next layer in a trial-and-error manner to determine which would match most closely the moveouts observed on selected gathers.

Velocity information is potentially given by other processes such as general inversion methods (§13.9).

However, the techniques to accomplish this on a practical basis have not yet been worked out. Some velocity information can also be obtained from types of measurements that do not depend on reflection travelpaths such as head-wave velocity or surface-wave dispersion.

5.5 Uses of velocity data

Velocity information is used in many processing and interpretation situations, and in most of these, the accuracy of the velocity data is not as good as is needed. Table 5.3 lists some of the uses of different "kinds" of velocity; this list is based on Al-Chalabi (1979).

Problems

5.1 What physical fact determines the "limit-of-porosity" line in fig. 5.3a? What is implied for measurements that fall to the right of this line?

5.2 Figures 5.12a and 5.12b are based on different experimental data. Show the compatibility or incompatibility of these figures.

5.3 (a) Assume that sandstone is composed only of grains of quartz, limestone only of grains of calcite, and shale of equal quantities of kaolinite and muscovite. For sandstone, limestone, and shale saturated with salt water ($\rho = 1.03$ g/cm³), what porosities are implied by the densities shown in fig. 5.13?

(b) What velocities would be expected for these values according to Gardner's rule (eq. (5.15))? Where do these values plot on fig. 5.5?

(c) From the graph of fig. 5.3b, what densities would you expect at 7500 ft and how do these compare with figs. 5.9c and 5.9d?

5.4 Assume that the velocity in calcite is 6.86 km/s and in quartz 5.85 km/s. What velocities should be expected for 10, 20, and 30% porosity in (a) limestone composed only of calcite; (b) sandstone composed only of quartz? Where do these values plot on figs. 5.5 and 5.16?

5.5 (a) Why do the velocity–depth curves for the various areas shown in figs. 5.17 and 5.19 depart from each other? Incorporate your knowledge of the geology of the various areas in your answer.

(b) Plot the shale and limestone values from fig. 5.29 for depths of 1000 and 2000 m on fig. 5.17. How do they compare?

5.6 (a) Assume a subsiding area without uplift activity. A shale is normally pressured until it reaches a depth of burial of 1400 m, at which point it becomes cut off from fluid communication, that is, interstitial fluid can no longer escape. If it is found at a depth of 2000 m, what velocity and what fluid pressure would you expect? If at a depth of 3000 m?

(b) Assume a shale buried to 3000 m and then uplifted to 2000 m, being normally pressured all the time. What velocity and fluid pressure would you expect?

(c) Assume the shale in part (a) is buried to 3000 m and then uplifted to 2000 m, without fluid communi-

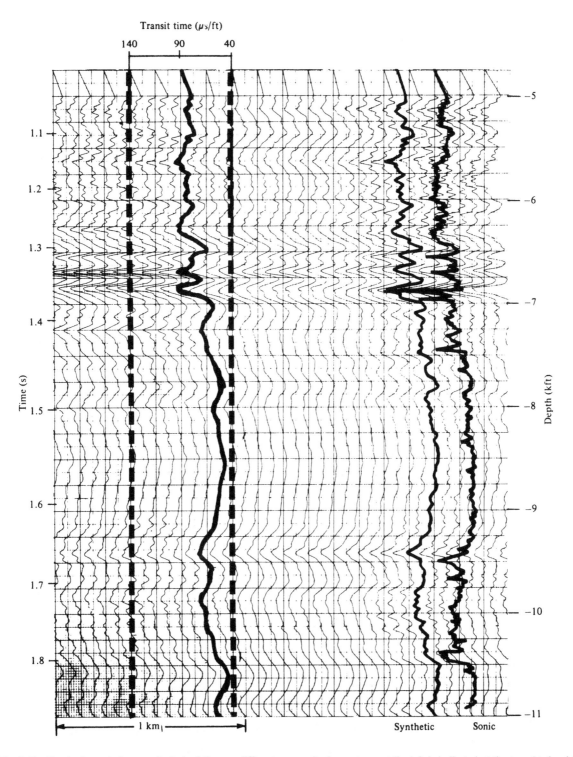

Fig. 5.43 Synthetic sonic logs, each derived from a different seismic trace. The vertical scale is linear with depth and the horizontal scale with specific transit time, the reference values for each trace moving according to the trace spacing. The scale for the heavy trace at the left is indicated at the top. At the right, one heavy trace shows a synthetic log along with an actual sonic log for comparison. (Courtesy of Technica.)

Table 5.3 *Uses of velocity data*

Velocity	Main uses	Precision requirements
Stacking velocity	Stacking of seismic sections	Modest to low
	Preliminary migration processing	Modest to low
	rms velocity estimation	Dependent on situation
rms velocity	Estimation of migration velocity	Generally modest
	Interval-velocity estimation	Dependent on situation
	Average-velocity estimation	Dependent on situation
Interval velocity	Gross lithologic and stratigraphic studies	High to modest
	General interpretation purposes	Modest to low
	Age estimation	High to modest
	Detection of abnormal pressure	High to modest
	Ray tracing	Dependent on situation
	Migration processing	Generally modest
	Average-velocity estimation	Dependent on situation
Average velocity	Depth conversion	Generally modest
	General interpretation purposes	Modest to low

Precision requirements:
   high = 0.1 to 1.0%
   modest = 1 to 5%
   low > 5%

After Al-Chalabi (1979).

cation being established. What velocity and fluid pressure would you expect? What if uplifted to 1000m?

5.7 By comparing figs. 5.19 and 5.36, what can you deduce about the nature of the rocks in the well for fig. 5.36?

5.8 What shale velocities are consistent with the oil–sand data shown in fig. 5.27b? (Determine velocities for two values of water saturation for each of the three depths, neglecting differences between sand and shale densities.)

5.9 (a) Figure 5.17 shows velocity versus depth for normally pressured shales. How do the velocities shown in fig. 5.31 above and below the "top abnormal pressure" compare with the curve? What depth would correspond to normal pressure for the overpressured shale? What porosity would you expect for the overpressured shale?

(b) Plot the velocities for 100% water saturation from fig. 5.27 on fig. 5.17. How do they compare?

5.10 Assume that raypaths have an angle of approach of 10°, 20°, and 30° in the subweathering with a velocity of 2400 m/s.

(a) For a weathered layer 10 m thick with a velocity of 500 m/s, how do traveltimes through the weathering compare with that for a vertically traveling ray? What is the horizontal component of the raypath in the weathering?

(b) For permafrost 100-m thick with a velocity of 3600 m/s, answer the questions in part (a).

5.11 In the early days of refraction exploration for salt domes, sketches were drawn indicating that the angle of approach to the surface should have a large horizontal component, but measurements with three-component seismographs showed very little horizontal component and controversy arose therefore over whether the travelpaths could be as drawn. Explain the apparent discrepancy based on your concept of the actual earth.

5.12 (a) Assume six horizontal layers each 300 m thick and having a constant velocity within each layer (fig. 5.44a), the successive layers having velocities of 1.5, 1.8, 2.1, 2.4, 2.7, and 3.0 km/s. Ray-trace through the model to determine offset distances and arrival times for rays that make an angle of incidence with the base of the 3.0 km/s layer (at A) of 0, 10, 20, and 30°. Calculate stacking velocity from each pair of values (six calculations) and compare with the average and rms velocities.

(b) Repeat assuming the layers dip 20°, as shown in fig. 5.44b for reflecting point B.

(c) By trial and error, shift the reflecting point updip to achieve common midpoints.

(d) Repeat by modifying the model to that shown in fig. 5.44c for reflecting point C.

(e) Shift the reflecting point C as required to achieve common midpoints.

5.13 Velocity analysis usually results in a plot of stacking velocity against travel time. Bauer (private communication) devised a "quick-look" method of determining the interval velocity, assuming horizontal layering and that the stacking velocity is average velocity. The method is shown in fig. 5.45. A box is formed by the two picks between which the interval velocity is to be picked; the diagonal that does not contain the two pick points, when extended to the velocity axis, gives the interval velocity.

(a) Prove that the method is valid and discuss its limitations.

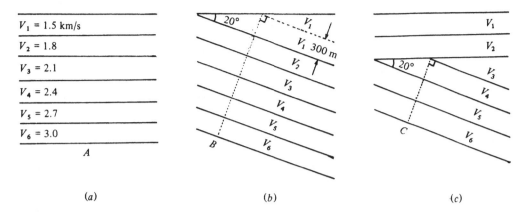

Fig. 5.44 Models of 300-m-thick layers (measured perpendicular to the bedding), each of constant velocity.

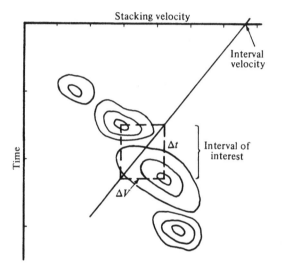

Fig. 5.45 "Quick-look" interval-velocity determination.

(b) This method is useful in seeing the influence of measurement errors; discuss the sensitivity of interval-velocity calculations to (i) error in picking velocity values from this graph; (ii) error in picking times; (iii) picking events very close together, and (iv) picking each event late.

5.14 Figure 5.46 shows data from a well-velocity survey tabulated on a standard calculation form.

(a) Plot time, average velocity, and interval velocity versus depth graphs using a sea-level datum.

(b) How much error in average-velocity and interval-velocity values would result from (i) time-measurement errors of 1 ms and (ii) depth-measurement errors of 1 m?

(c) Determine V_0 and a for a velocity-function fit to these data assuming the functional form $V = V_0 + az$, where V is the interval velocity and z is depth.

5.15 Analysis of an $X^2 - T^2$ survey gives the results in table 5.4. Calculate the interval velocities.

5.16 Determine the velocity by the $X^2 - T^2$ method us-

ing the data given in table 5.5, t_A being for a horizontal reflector and t_B is for a reflector dipping 10° toward the source.

5.17 (a) Given that the trace spacing in fig. 9.46b is 50 m, determine the stacking velocity, depth, and dip at approximately 0.5, 1.0, 1.5, 2.0, 2.3, and 2.4 s.

(b) What problems or ambiguities do you have in picking these events?

(c) How much uncertainty is there in your ability to pick times and how much uncertainty does this introduce into the velocity, depth, and dip calculations?

5.18 (a) In fig. 9.24d, pick stacking velocity versus time pairs and calculate interval velocities; the analysis is for SP 100.

(b) What can you tell about the lithology from this?

(c) If the section that is present in the syncline but missing over the anticline consists of young, poorly consolidated rocks, what values would you expect a velocity function at SP 45 to show?

(d) Note the downdip thinning of the section from about 0.75 to 1.25 s at the left end of the section; suggest the explanation.

5.19 (a) Determine velocity versus depth from fig. 11.8. The direct wave travels through the water and can be used to give source–receiver distances. Assume horizontal reflectors.

(b) Determine the apparent velocities of refractors and correlate refraction with reflection events.

5.20 Given orthogonal dip and strike seismic lines; will velocity analyses at the line intersections yield the same values?

5.21 (a) Because a velocity analysis for a causal wavelet (§15.5.6a) is not made on the wavelet onset, how will this affect the stacking-velocity value?

Table 5.4 $X^2 - T^2$ survey results

i	z (km)	t_i (s)	V_s (km/s)
1	1.20	1.100	2.18
2	2.50	1.786	2.80
3	3.10	1.935	3.20
4	4.10	2.250	3.64

Shothole information - Elevation, Distance & Direction from Well									Company				Well				Elevation (Derrick Floor)	Total Depth	LOCATION: Coordinates	Section, Township, Range	County	Area or Field
1 - 78m, 125m W 2 - 78, 125 W 3 - 78, 125 W 4 - 78, 125 W																	107m	3395m				

Record Number	Shothole Number	D_{gm}	D_s	t_{us}	t_r	T Reading	T Polarity	T Grade	D_{gs}	H	tan i	cos i	T_{gs}	Δsd	$\frac{\Delta sd}{V}$	T_{gd}	T_{gd} Average	D_{gd}	ΔD_{gd}	ΔT_{gd}	V_i Interval Velocity	V_a Average Velocity
1	1	3395	27	026	088	0.990	+	g														
2	2	3395	27	027	088	0.990	+	g														
3	1	3305	25	024	089	0.976	+	g														
4	1	3200	25	024	088	0.958	+	g														
5	1	3100	23	023	088	0.936	+	g														
6	1	2970	23	023	088	0.907	+	g														
7	1	2810	23	024	089	0.868	+	f														
8	1	2690	23	023	088	0.840	+	g														
9	1	2535	23	023	088	0.805	+	f														
10	2	2400	27	027	089	0.769	+	g														
11	2	2270	23	024	089	0.752	+	g														
12	2	2180	23	024	089	0.728	+	g														
13	2	2110	23	024	088	0.709	+	g														
14	2	2045	23	023	089	0.697	+	p														
15	3	1925	27	026	088	0.671	+	g														
16	3	1750	27	026	088	0.649	+	g														
17	3	1670	27	027	088	0.612	+	g														
18	3	1560	23	023	089	0.588	+	g														
19	3	1355	23	023	088	0.546	+	g														
20	3	1265	23	023	088	0.509	+	g														
21	3	1150	23	023	088	0.477	+	f														
22	3	1010	23	023	088	0.426	+	f														
23	3	845	23	023	089	0.373	+	f														
24	4	690	27	027	088	0.310	+	g														
25	4	625	23	024	088	0.288	+	g														
26	4	470	23	024	088	0.226	+	f														
27	4	320	23	024	088	0.165	+	f														
28	4	215	23	024	088	0.129	+	f														

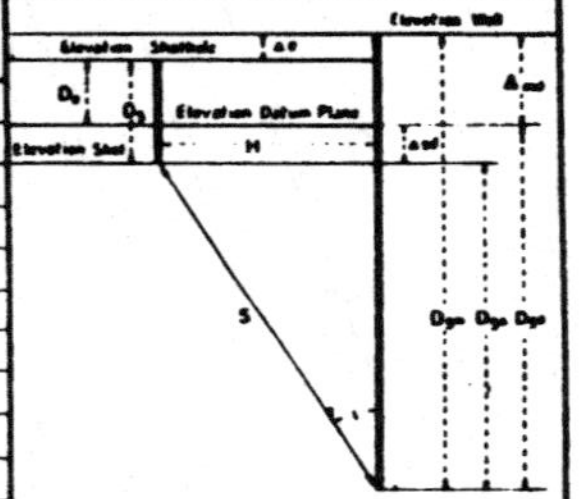

D_{gm} · Geophone depth measured from well elevation
D_{gs} · " " " " shot "
D_{gd} · " " " " datum "
D_s · Depth of shot
D_e · Shothole elevation to datum plane
H · Horizontal distance from well to shotpoint
S · Straight line travel path from shot to well geophone
t_{us} · Uphole time at shotpoint
T · Observed time from shotpoint to well geophone.
t_r · " " " to reference geophone
Δe · Difference in elevation between well & shotpoint
Δsd · " " " " shot & datum plane
Δsd · $D_s - D_e$
D_{gs} · $D_{gm} - D_s \pm \Delta e$, $\tan i = \frac{H}{D_{gs}}$
T_{gs} · $\cos i\ T$ - Vert travel time from shot elev to geophone
T_{gd} · $T_{gs} \pm \frac{\Delta sd}{V}$ " " " " datum plane " "
D_{gd} · $D_{gm} - \Delta e$
V_i · Interval velocity · $\frac{\Delta D_{gd}}{\Delta T_{gd}}$
V_a · Average · $\frac{D_{gd}}{T_{gd}}$

Surveyed by

Date

Weathering Data

Fig. 5.46 Data from a well-velocity survey.

Table 5.5 X^2–T^2 *data*

x (km)	t_A (s)	t_B (s)	x (km)	t_A (s)	t_B (s)	x (km)	t_A (s)	t_B (s)
0.0	0.855	0.906	1.4	1.005	0.977	2.8	1.330	1.202
0.1	0.856	0.902	1.5	1.017	0.991	2.9	1.360	1.234
0.2	0.858	0.898	1.6	1.037	1.004	3.0	1.404	1.253
0.3	0.864	0.898	1.7	1.068	1.019	3.1	1.432	1.272
0.4	0.868	0.899	1.8	1.081	1.037	3.2	1.457	1.296
0.5	0.874	0.902	1.9	1.105	1.058	3.3	1.487	1.304
0.6	0.882	0.903	2.0	1.118	1.066	3.4	1.513	1.334
0.7	0.892	0.909	2.1	1.151	1.083	3.5	1.548	1.356
0.8	0.904	0.916	2.2	1.166	1.102	3.6	1.580	1.377
0.9	0.906	0.922	2.3	1.203	1.121	3.7	1.610	1.407
1.0	0.930	0.932	2.4	1.237	1.127	3.8	1.649	1.415
1.1	0.945	0.943	2.5	1.255	1.158	3.9	1.674	1.438
1.2	0.950	0.950	2.6	1.283	1.177	4.0	1.708	1.459
1.3	0.979	0.965	2.7	1.304	1.195			

(b) What will be the effect of NMO stretch (§9.7.3)?
(c) If a datum (§8.8.2) is used that is appreciably removed from the surface, what effect will this have?

REFERENCES

Al-Chalabi, M. 1979. Velocity determination from seismic reflection data: In *Developments in Geophysical Exploration Methods*, Vol. 1, A. A. Fitch, ed., pp. 1–68. Amsterdam: Elsevier.

Atkins, J. E., and E. F. McBride. 1992. Porosity and packing of Holocene river, dune, and beach sands: *Bull. AAPG*, **76:** 339–55.

Aud, B. W. 1976. History of abnormal pressure determination from seismic data. *OTC Preprints*, paper 2611. Dallas, TX: Offshore Technology Conference.

Biot, M. A. 1956. Theory of propagation of elastic waves in fluid-saturated porous solid. *J. Acoust. Soc. Appl. Phys.*, **26:** 182–5.

Birch, F. 1942, *Handbook of Physical Constants*, GSA Special Paper 36. Geological Society of America.

Birch, F. 1961. The velocity of compressional waves in rocks to 10 kilobars. *J. Geophys. Res.*, **66:** 2199–2223.

Charnock, G., ed. 1990. *Middle East Well Evaluation Review.* Ridgefield, CT: Schlumberger Technical Services.

Christensen, N. I. 1989. Seismic velocities. In *Practical Handbook of Physical Properties of Rocks and Minerals*, R. S. Carmichael, ed., pp. 429–546. Boca Raton, FL: CRC Press.

Dix, C. H. 1955. Seismic velocities from surface measurements. *Geophysics*, **20:** 68–86.

Domenico, S. N. 1974. Effect of water saturation on seismic reflectivity of sand reservoirs encased in shale. *Geophysics*, **39:** 759–69.

Domenico, S. N. 1976. Effect of brine–gas mixture on velocity in an unconsolidated sand reservoir: *Geophysics*, **41:** 882–94.

Domenico, S. N. 1977. Elastic properties of unconsolidated porous sand reservoirs. *Geophysics*, **42:** 1339–68.

Duska, L. 1963. A rapid curved-path method for weathering and drift correction. *Geophysics*, **28:** 925–47.

Dutta, N. C., ed. 1987. *Geopressure:* SEG Geophysics Reprint Series 7. Tulsa: Society of Exploration Geophysicists.

Faust, L. Y. 1951. Seismic velocity as a function of depth and geologic time. *Geophysics*, **16:** 271–88.

Faust, L. Y. 1953. A velocity function including lithologic variation. *Geophysics*, **18:** 271–88.

Gardner, G. H. F., L. W. Gardner, and A. R. Gregory. 1974. Formation velocity and density – The diagnostic basics for stratigraphic traps. *Geophysics*, **39:** 770–80.

Gardner, G. H. F., and M. H. Harris. 1968. Velocity and attenuation of elastic waves in solids. *Transactions of the 9th Annual Log Symposium*, Paper M.

Gassman, F. 1951. Elastic waves through a packing of spheres. *Geophysics*, **16:** 673–85.

Geertsma, J., and D. C. Smit. 1961. Some aspects of elastic wave propagation in fluid saturated porous solids. *Geophysics*, **26:** 1269–81.

Grant, F. S., and G. F. West. 1965. *Interpretation Theory in Applied Geophysics.* New York: McGraw-Hill.

Gregory, A. R. 1976. Fluid saturation effects on dynamic elastic properties of sedimentary rocks. *Geophysics*, **41:** 895–921.

Gregory, A. R. 1977. Aspects of rock physics from laboratory and log data that are important to seismic interpretation. In *Seismic Stratigraphy – Applications to Hydrocarbon Exploration*, C. E. Payton, ed., pp. 15–46. AAPG Memoir 26. Tulsa: American Association of Petroleum Geology.

Gretener, P. E. 1979. *Pore Pressure: Fundamentals, General Ramifications, and Implications for Structural Geology*, AAPG Education Course Note Series 4. Tulsa: American Association of Petroleum Geology.

Hamilton, E. L. 1971. Elastic properties of marine sediments. *J. Geoph. Res.*, **76:** 579–604.

Han, D. 1987. Effects of porosity and clay content on acoustic properties of sandstones and unconsolidated sediments. Ph.D. thesis, Stanford University.

Han, D., A. Nur, and D. Morgan. 1986. Effects of porosity and clay content on wave velocities in sandstones. *Geophysics*, **51:** 2097–2107.

Hicks, W. G., and J. E. Berry. 1956. Application of continuous velocity logs in determination of fluid saturation of reservoir rocks. *Geophysics*, **21:** 739–54.

Hilterman, F. 1990. Is AVO the seismic signature of lithology? A case history of Ship Shoal – South Addition. *The Leading Edge*, **9(6):** 15–22.

Hofer, H., and W. Varga. 1972. Seismogeologic experience in the Beaufort Sea. *Geophysics*, **37**: 605–19.

Jankowsky, W. 1970. Empirical investigation of some factors affecting elastic wave velocities in carbonate rocks. *Geophys. Prosp.*, **18**: 103–18.

Kearey, P., and M. Brooks. 1984. *An Introduction to Geophysical Exploration*. Oxford: Blackwell Scientific.

Keyser, W., L. K. Johnston, R. Reeses, and G. Rodriguez. 1991. Pore pressure prediction from surface seismic. *World Oil* (September): 115–24.

Kokesh, F. P., and R. B. Blizard. 1959. Geometrical factors in sonic logging. *Geophysics*, **24**: 64–76.

Lash, C. C. 1980. Multiple reflections and converted waves found by a deep vertical wave test. *Geophysics*, **45**: 1373–1411.

Lindseth, R. O. 1979. Synthetic sonic logs – A process for stratigraphic interpretation. *Geophysics*, **44**: 3–26.

MacGregor, J. R. 1965. Quantitative determination of reservoir pressures from conductivity logs. *Bull. AAPG*, **49**: 1502–11.

Mareš, S. 1984. *Introduction to Applied Geophysics*. Dordrecht: Reidel.

Meckel, L. D., and A. K. Nath. 1977. Geologic considerations for stratigraphic modeling and interpretation. In *Seismic Stratigraphy – Applications to Hydrocarbon Exploration*, C. E. Payton, ed., pp. 417–38, AAPG Memoir 26. Tulsa: American Association of Petroleum Geology.

Mills, G. F., M. A. Brzostowski, S. Ridgway, and C. A. Barton. 1993. A velocity model building technique for pre-stack depth migration. *First Break*, **11**: 435–43.

Murphy, W. F. 1985. Sonic and ultrasonic velocities: Theory versus experiment. *Geophys. Res. Lett.*, **12**: 85–8.

Musgrave, A. W., and R. H. Bratton. 1967. Practical application of Blondeau weathering solution. In *Seismic Refraction Prospecting*, A. W. Musgrave, ed., pp. 231–46. Tulsa: Society of Exploration Geophysicists.

Nur, A., and G. Simmons. 1969. Stress-induced velocity anisotropy in rocks – An experimental study. *J. Geophys. Res.*, **74**: 66–7.

Nur, A. M., and Z. Wang. 1989. *Seismic and Acoustic Velocities in Reservoir Rocks*, Geophysics Reprint Series 10. Tulsa: Society of Exploration Geophysicists.

Pennebaker, E. S. 1968. Seismic data indicate depth, magnitude of abnormal pressures. *World Oil*, **166(7)**: 73–8.

Pickett, G. R. 1963. Acoustic character logs and their application in formation evaluation. *J. Petrol. Tech.*, **15**: 650–67.

Plumley, W. J. 1980. Abnormally high fluid pressure: Survey of some basic principles. *Bull. AAPG*, **64**: 414–22.

Press, F. 1966. Seismic velocities. In *Handbook of Physical Constants*, rev. ed., S. P. Clark, ed., pp. 195–218, GSA Memoir 97. Geological Society of America.

Rackets, H. M. 1971. A low-noise seismic method for use in permafrost regions. *Geophysics*, **36**: 1150–61.

Reynolds, E. B. 1970. Predicting overpressured zones with seismic data. *World Oil*, **171(5)**: 78–82.

Reynolds, E. B. 1973. The application of seismic techniques to drilling techniques, *SPE Preprint* 4643. Society of Petroleum Engineers.

Robie, R. A., P. M. Bethke, M. S. Toulmin, and J. L. Edwards. 1966. X-ray crystallographic data, densities, and molar volumes of minerals. In *Handbook of Physical Constants*, rev. ed., S. P. Clark, ed., pp. 27–73, GSA Memoir 97. Geological Society of America.

Sharma, P. V. 1986. *Geophysical Methods in Geology*. Amsterdam: Elsevier.

Sheriff, R. E. 1977. Using seismic data to deduce rock properties. In *Developments in Petroleum Geology*, Vol. 1, G. D. Hobson, ed., pp. 243–74. London: Applied Science.

Sheriff, R. E. 1978. *A First Course in Geophysical Exploration and Interpretation*. Boston: International Human Resources Development Corp.

Sheriff, R. E. 1989. *Geophysical Methods*. Englewood Cliffs, NJ: Prentice Hall.

Sheriff, R. E., and L. P. Geldart. 1983. *Exploration Seismology*, Vol. 2. New York: Cambridge University Press.

Shipley, T. H., M. K. Houston, R. T. Buffler, F. J. Shaub, K. J. McMillan, J. W. Ladd, and J. L. Worzel. 1979. Seismic evidences for widespread possible gas hydrate horizons on continental slopes and rises. *Bull. AAPG*, **63**: 2204–13.

Spencer, J. W. 1981. Stress relaxation at low frequencies in fluid saturated rocks: Attenuation and modulus of dispersion. *J. Geophys. Res.*, **88**: 1803–12.

Stulken, E. J. 1941. Seismic velocities in the southeastern San Joaquin Valley of California. *Geophysics*, **6**: 327–55.

Swan, B. G., and A. Becker. 1952. Comparison of velocities obtained by delta-time analysis and well velocity surveys. *Geophysics*, **17**: 575–85.

Tatham, R. H., and M. D. McCormack. 1991. *Multicomponent Seismology in Petroleum Exploration*. Tulsa: Society of Exploration Geophysicists.

Telford, W. M., L. P. Geldart, and R. E. Sheriff. 1990. *Applied Geophysics*, 2d ed. New York: Cambridge University Press.

Timoshenko, S., and J. N. Goodier. 1951. *Theory of Elasticity*, 2d ed. New York: McGraw-Hill.

Timur, A. 1968. Velocity of compressional waves in porous media at permafrost temperatures. *Geophysics*, **33**: 584–95.

Timur, A. 1977. Temperature dependence of compressional and shear wave velocities in rocks. *Geophysics*, **42**: 950–6.

Tucholke, B. E., G. M. Bryan, and J. I. Ewing. 1977. Gas-hydrate horizons in seismic-profiler data from the Western North Atlantic. *Bull. AAPG*, **61**: 698–707.

Wang, Z. 1988. Wave velocities in hydrocarbons and hydrocarbon-saturated rocks, with applications to EOR monitoring. Ph.D. thesis, Stanford University.

Wang, Z., and A. Nur. 1988. Effect of temperature on wave velocities in sands and sandstones with heavy hydrocarbons. *SPE Reserv. Eng.*, **3(1)**: 158–64.

Wang, Z., and A. Nur. 1992a. Aspects of rockphysics in seismic reservoir surveillance. In *Reservoir Geophysics*, R. Sheriff, ed., pp. 295–310. Tulsa: Society of Exploration Geophysicists.

Wang, Z., and A. Nur. 1992b. *Seismic and acoustic velocities in Reservoir Rocks*, Vol. 2: *Theoretical and model studies*, Geophysics Reprint Series 10. Tulsa: Society of Exploration Geophysicists.

Waters, K. H. 1987. *Reflection Seismology*, 3d ed. New York: John Wiley.

Watkins, J. S., L. A. Walters, and R. H. Godson. 1972. Dependence of in-situ compressional wave velocity on porosity in unsaturated rocks. *Geophysics*, **37**: 29–35.

White, J. E. 1965. *Seismic Waves: Radiation, Transmission, and Attenuation*. New York: McGraw-Hill.

White, J. E. 1983. *Underground Sound: Application of seismic waves*. Amsterdam: Elsevier.

Wyllie, M. R. J., A. R. Gregory, and G. H. F. Gardner. 1958. An experimental investigation of factors affecting elastic wave velocities in porous media. *Geophysics*, **23**: 459–93.

Zieglar, D. L., and J. H. Spotts. 1978. Reservoir and source-bed history of Great Valley of California. *Bull. AAPG*, **62**: 813–26.

6
Characteristics of seismic events

Overview

The basic task of interpreting seismic records is that of selecting those events on the record that represent primary reflections (or refractions), translating the arrival times for these into depths and dips, and mapping the reflecting (refracting) horizons. In addition, the interpreter must be alert to events features such as changes in amplitude or character that may yield valuable information about other types of events, such as multiple reflections and diffractions. The characteristics of events are the subject of this chapter.

The features that allow one to recognize and identify an event – coherence, amplitude standout, character, dip, and normal moveout – are discussed in §6.1. Differences in arrival time because of offset provide an especially useful method of distinguishing between reflections, refractions, multiples, and other types of events.

Part of the downgoing wavetrain from a seismic source is reflected at each interface where the acoustic impedance changes. Synthetic seismograms showing the reflections expected from a model aid in understanding seismic reflections. The strongest reflections usually result from unconformities or significant changes in lithology, but reflections also occur from minor lithologic changes. Most reflections are the interference composites from a number of closely spaced interfaces. Occasional reflections are caused by impedance changes associated with phase changes or fluid contacts rather than bedding contacts. Reflections occur not only because of energy returned from the reflecting point, but because of energy returned from the entire Fresnel-zone area. Reflection amplitude, phase, and overall appearance change because of reflector curvature. Where the center of curvature of a synclinal reflector is below the observing level, a buried focus occurs and a reverse branch appears, the reverse branch having convex-upward curvature rather than the concave-upward curvature of the reflector causing it.

Section 6.3 takes up the characteristics of several classes of nonprimary reflection events. Diffractions have more normal moveout than reflections and are curved events on stacked unmigrated sections. They bear certain relationships to the reflections from reflectors whose termination often generates them. The crest of a diffraction gives the location of the diffract-

ing point for simple velocity situations; this point is useful in locating bedding terminations such as occur at faults and salt-dome flanks. Diffractions also provide a mechanism for getting seismic energy into regions that cannot be reached on a geometrical-optics basis.

Multiples are classified as long-path if they show as separate events or short-path if their effect is merely that of changing reflection waveshape. Peg-leg multiples are important factors in changing the waveshape and removing high frequencies with increasing traveltime. Long-path multiples confuse interpretation unless they are recognized for what they are. The characteristics of surface waves conclude §6.3.

Resolution refers to the ability to distinguish between adjacent features. Vertical resolution concerns the minimum separation between interfaces for them to show as separate reflectors; the resolvable limit is about a quarter wavelength. The interpretation of beds thinner than a quarter wavelength has to be based on amplitude rather than time-interval measurements. Where a bed is subdivided into several very thin beds, the reflection amplitude is a measure of the net, rather than gross, thickness of the bed. Horizontal resolution depends on a number of factors. The migration process (§9.12) attempts to increase the horizontal resolution.

The attenuation of reflections with traveltime and frequency is discussed in §6.5.

The shape of the seismic wavelet changes with traveltime because of absorption and peg-leg multiples, and also because of filtering actions in recording and processing. Minimum-phase wavelets are distinguished from zero-phase ones, the Ricker wavelet being the most commonly assumed zero-phase wavelet.

Distinction is made between coherent and incoherent noise, and repeatable and ambient noise. A discussion of the attenuation of noise concludes this chapter.

6.1 Distinguishing features of events

Recognition and identification of seismic events are based upon five characteristics: (a) coherence, (b) amplitude standout, (c) character, (d) dip moveout, and (e) normal moveout.

The first of these characteristics, *coherence,* similarity in appearance from trace to trace (see fig. 6.1), is by far the most important in recognizing an event.

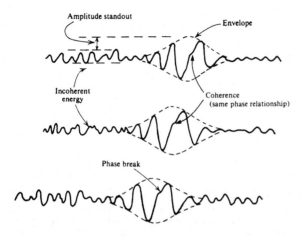

Fig. 6.1. Characteristic of seismic events.

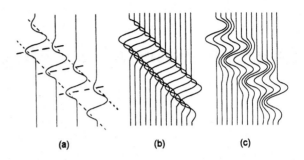

Fig. 6.2 Illustrating spatial aliasing. When traces are spaced too far apart, there is ambiguity in their meaning. Is (b) or (c) the correct interpretation of (a)?

When a wave reaches a spread, it tends to produce approximately the same effect on each geophone. If the wave is strong enough to override other energy arriving at the same time, the traces will look more or less alike during the interval when this wave is arriving. Coherence is a necessary condition for the recognition of any event. Recognition of coherence sets a maximum to the trace spacing. Most events extend for several cycles, so coherence will involve ambiguity if the traces are separated by more than half the shortest apparent wavelength present (see fig. 6.2). The spatial sampling of seismic traces is subject to the sampling theorem (§9.2.2d) just as any sampling is.

Amplitude standout refers to an increase in amplitude such as results from the arrival of coherent energy; it is not always very marked, especially if AGC (see §7.6.3) is used in recording. Coherence and amplitude standout tell us whether or not a strong seismic event is present, but they say nothing about the type of event.

Character (or *signature*) refers to a distinctive appearance of the waveform that identifies a particular event. It involves primarily the shape of the envelope, the number of cycles that show amplitude standout, the dominant frequency, and irregularities in the phase resulting from the interference between components of the event. The reflections observed on seismic sections usually result from the interference between component reflections from a closely spaced series of interfaces, and the appearance of a reflection, that is, its character, depends upon the spacing and magnitude of the individual acoustic impedance contrasts. Usually, these are relatively constant over moderate distances and so the reflection exhibits coherence. Character may help identify reflection events. Reflections are usually fairly short events with little ringing and their frequency components are usually in the range 15–60 Hz, except for shallow high-frequency reflections and very deep reflections, which may have considerable energy even below this range.

Moveout refers to a systematic difference from trace to trace in the arrival time of an event; it is the most distinctive criterion for identifying the nature of events. We distinguish between dip moveout, systematic changes in arrival time because of dip, and normal moveout, systematic changes with source–geophone distance; these have been discussed for reflections in §4.1.1 and 4.1.2. It is fairly easy to separate dip and normal moveouts with split spreads, but not with end-on spreads. With planar reflectors, dip moveout produces a nearly linear alignment, whereas normal moveout is characterized by alignment curvature, but reflector curvature or velocity complications can obscure this distinction. Reflection normal moveouts must fall within certain limits set by the velocity distribution. Reflections often have small dip moveouts but occasionally they have large dip moveouts (as with fault-plane reflections). Events other than primary reflections can also exhibit linear and curved alignments, as will be seen in following sections.

One very powerful technique for distinguishing between reflections, diffractions, reflected refractions, and multiples is to display the data (sometimes after sorting into CMP gathers) after correcting for (a) weathering and elevation (*static corrections*, because the correction is the same for all arrival times on a given trace; see §8.8.2) and (b) normal moveout (dynamic correction). Such corrected records can be made in data processing. Provided the correct normal moveout has been removed, reflections appear as straight lines, whereas diffractions and multiples still have some curvature (fig. 6.3) (because their normal moveouts are larger than those of primary reflections) and refractions and other formerly straight alignments have inverse curvature.

6.2 Reflections

6.2.1 Synthetic seismograms

It is generally assumed that the wavelet reflected from (or transmitted through) an acoustic impedance discontinuity has the same waveshape as the incident wavelet, and that a seismic trace records the succession of such wavelets and thus the succession of

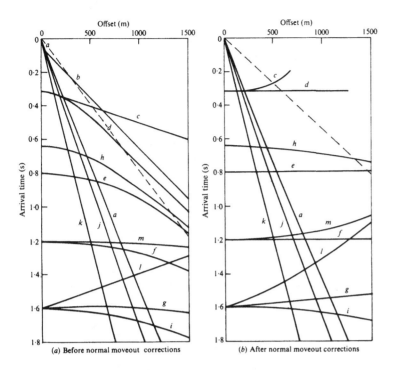

(a) Before normal moveout corrections   (b) After normal moveout corrections

Fig. 6.3 Types of events on a seismic record. Identities of events are as follows: a = direct wave, V = 650 m/s; b = refraction at base of weathering, V_H = 1640 m/s; c = refraction from a flat refractor, V_R = 4920 m/s; d = reflection from the refractor in c, V = 1640 m/s; e = reflection from a flat reflector, $\overline{V}$ = 1970 m/s; f = reflection from a flat reflector, $\overline{V}$ = 2300 m/s; g = reflection from a dipping reflector, $\overline{V}$ = 2630 m/s; h = multiple of d; i = multiple of e; j = ground roll, V_R = 575 m/s; k = air wave, V = 330 m/s; l = reflected refraction from in-line disruption of refractor c; m = reflected refraction from broadside disruption of refractor in c. After proper normal-moveout correction, the primary reflections are straight. Processing usually involves setting to zero all values earlier than some "mute schedule," here indicated by the dashed line; this is called a front-end mute. Consequently, the data that might otherwise appear in the upper-right triangles are usually not seen.

acoustic impedance discontinuities (fig. 6.4). This concept, that a seismic trace is simply the superposition of individual reflections, is basic to the convolutional model (a mathematical calculation that enables us to determine the effect of a filter on a signal; see §9.2.1), the reflection process here being considered as a "filter." Carrying out the calculations for an assumed distribution of physical properties results in a *synthetic seismogram,* one of the commoner types of forward modeling (§10.4.4).

The most common simplification involves a one-dimensional synthetic seismogram where it is assumed that raypaths are vertical and interfaces are horizontal. Reflection and transmission coefficients are thus for normal incidence (eqs. (3.14) and (3.15)). Diffractions and other wave modes are usually ignored, although multiples may be included. Most often, the acoustic impedance values required by these equations are obtained from borehole logs. Often, only sonic logs are available and density is either assumed to be constant or to bear some relation to velocity (such as eq. (5.15)). Small velocity variations are often lumped together into larger steps to reduce the number of interfaces to be considered, and sampling is usually on a regular traveltime interval (rather than on a regular depth interval, as with logs).

Amplitude-changing factors other than those involving reflection coefficients are often ignored. The downgoing seismic waveshape is assumed, often a Ricker

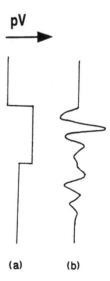

Fig. 6.4 Relation between reflections and acoustic impedance changes causing them. (a) Acoustic impedance changes, and (b) resulting reflections drawn for a minimum-phase wavelet.

wavelet (§6.6.2), but sometimes the waveshape used is the embedded wavelet (determined by wavelet processing, §9.5.9) of the actual data to which the synthetic seismogram is to be compared.

The major use of synthetic seismograms is to compare them with actual seismic data in order to identify reflections with particular interfaces, so that maps can be made on the beds of particular interest. This procedure is also used to distinguish primary reflections from multiples and other events. Seismic sections often involve time or phase shifts (including polarity reversals) of unknown magnitude (which are sometimes time-dependent), so that the ability to match synthetics to actual data adds considerable confidence to an interpretation.

Figure 6.5 illustrates the concept of synthetic seismogram manufacture. The reflectivity model, shown as a stick graph ("stickogram") portraying the magnitude and polarity of discrete reflection coefficients, is often derived from edited log data. Editing involves correction for sonic-log cycle skips and other borehole log errors; editing may involve changing values so that they are consistent with the complete ensemble of logs recorded in a borehole. Conductivity or other logs that show changes in rock properties in a borehole are sometimes used instead of sonic and density logs. The reflectivity model is then convolved with a wavelet, compared with the actual seismic data, and the difference ("error trace") used to modify the model (and sometimes the wavelet); the procedure is iterated until the match is judged to be sufficiently close. As with all types of modeling, the result is not unique, because it is always possible that some differ-

ent model could produce an equally good match.

Figure 6.6 shows a primaries-only synthetic seismogram matched to actual seismic data. Although the match here is considered good, a multitude of differences exist, perhaps because the model or the assumed wavelet was not exactly correct, multiples or density variations were not allowed for, the match is to a common-midpoint section rather than one involving only vertical travel and horizontal bedding, errors in the acquisition or processing, or for other reasons.

The effects of multiples are sometimes incorporated as successive modifications of the propagating waveform (Vetter, 1981), although often they are ignored also. Consider two adjacent interfaces such that the one-way traveltime between them is the sampling interval, Δ, the reflection coefficients for a downgoing wave being R_i and R_{i+1}. If $w(t)_i$ is the downgoing waveform approaching the interface R_i, the upgoing waveform reflected by R_i will be $u(t)_i = R_i w(t)_i$. Neglecting the very slight loss on transmission through R_i, the downgoing wave at R_{i+1} will be $w(t + \Delta)_i$ plus the peg-leg multiples generated between R_i and R_{i+1}, that is,

$$w(t)_{i+1} = w(t + \Delta)_i - R_i R_{i+1} w(t + 3\Delta)_i + (R_i R_{i+1})^2 w(t + 5\Delta)_i - \cdots,$$

the second term on the right being the one-bounce peg-leg multiple, the next the two-bounce, and so on. Thus, we can include the effects of multiples by modifying the downgoing waveform at each step. We need to likewise modify the upgoing (reflection) waveform for the peg-leg multiples that it generates.

Where the match between synthetics and actual

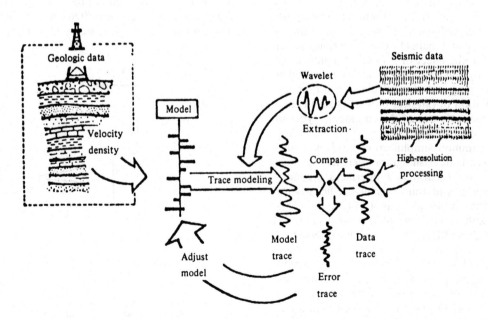

Fig. 6.5   Procedure for manufacturing synthetic seismogram. "Trace modeling" consists of convolving the reflectivity model representing the geologic data with a wavelet that is often extracted from the data to be matched. The difference between the model trace and a trace from the actual data (the error trace) is often used to modify the model. (From Stommel and Graul, 1978.)

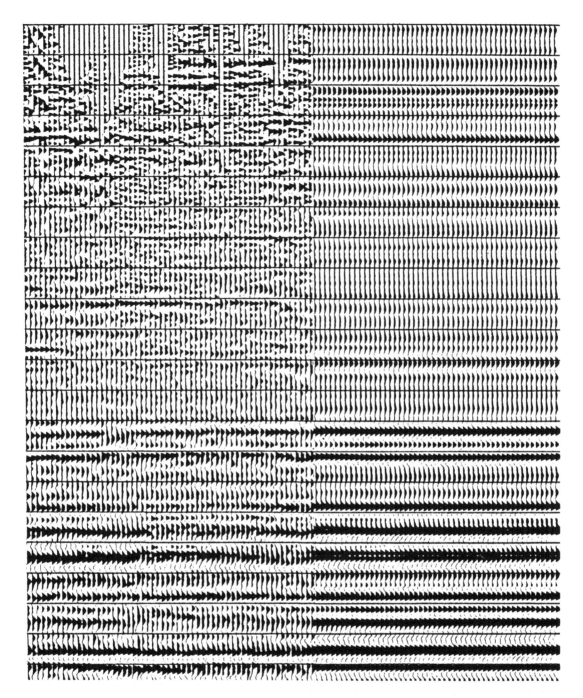

Fig. 6.6 Synthetic seismogram (right half) compared with actual seismic data (left half). The synthetic seismogram manufacture used a sonic log from a well on the seismic line and embedded wavelets extracted from successive traces of the actual data, densities being calculated from Gardner's rule (eq. (5.15)). (Courtesy of Grant Geophysical.)

data is good, the reflecting sequence can be modified according to stratigraphic changes that might occur, so that the effects of such changes on the seismic data can be ascertained. For example, we might assume that a sand unit shales-out in a facies change, that the fill in a stream channel differs from that in the adjacent uneroded formation, that the formation subcropping under an unconformity has changed, or that a small reef grew. Synthetics then give the interpreter a better idea of what to look for in order to locate the hypothesized facies change, channel, subcrop, or reef. This usage of synthetics provides one of the main methods for the stratigraphic interpretation of seismic data (§10.7).

The model for manufacturing one-dimensional synthetic seismograms may actually be two- or three-dimensional. A model might simulate the changes expected along a seismic line connecting two or more wells where the changes between the wells are explained on the basis of facies changes, unconformities, or faults. The synthetic seismogram from such a model is still considered one-dimensional, however, unless nonvertical travel paths are considered.

The synthetic seismogram concept is generalized and discussed further in §10.4.4.

6.2.2 Nature of reflections

Reflecting interfaces in sedimentary rocks usually occur much closer together than the seismic wavelength, and so the observed waveshape is the interference composite of a number of component reflections and (as stated in §6.1) interference is largely responsible for reflection character. Thus, the appearance of a reflection depends upon the spacing and magnitude of the component acoustic impedance contrasts. The composite nature of a reflection is illustrated by fig. 6.7. Although the waveshape assumed in this classic illustration is poorly selected, the point is made clear that observed reflection events do not correspond one-to-one with lithologic changes and that identifying individual cycles with formation tops based on traveltime may lead to errors.

It is often assumed that acoustic impedance contrasts coincide with major lithologic boundaries and indeed this is often true. However, the lithology may change without a change in acoustic impedance and the acoustic impedance may change without any major change in lithology. Hardage (1985), in a vertical seismic profiling study (§13.4), identified some reflections with the acoustic impedance contrasts that cause them (fig. 6.8). Note that reflections B and D occur within fairly massive shale units where the shales differ somewhat in their properties, as evidenced by well logs. In another area, an abundance of reflections on a seismic section (fig. 6.9) was interpreted as indicating appreciable interbedding of sand and shale (low stacking velocity values suggested a siliciclastic section); however, a subsequent stratigraphic hole found hardly any sand; changes in the shale were re-

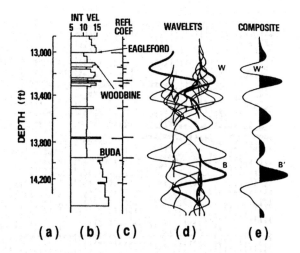

Fig. 6.7 Portion of an East Texas synthetic seismogram. (After Vail, Todd, and Sangree, 1977: 113; reprinted with permission.) (a) Depth in feet, (b) interval velocities blocked from sonic log data, (c) reflection coefficient spikes based on a one-to-one relation between acoustic impedance and interval velocity, (d) wavelets corresponding to the spikes in (c), and (e) composite of the wavelets in (d).

sponsible for the reflections. To be sure, such situations are the "exceptions that prove the rule" and most reflections correspond to distinctive lithologic changes.

A *facies boundary* marks a change in the lithologic or paleontological characteristics of contemporaneous sediments, whereas a *time boundary* marks what was at one time the surface of the solid earth. Although it is commonly stated that facies and time boundaries cross each other, they usually coincide locally and their apparent crossing is the result of the inadequate sampling on which facies boundaries are usually based (see §10.7.3). Because one expects lithologic changes to be associated with facies changes, one might expect reflections to follow the way facies boundaries are drawn. However, overwhelming evidence indicates that reflections coincide with time boundaries (except in occasional unusual circumstances).

Reflections are often associated with unconformities, and often the best and strongest reflections come from unconformity surfaces. The nature of an unconformity reflection, however, changes as the properties of the rocks above or below the unconformity change. Unconformities are often associated with angularities between them and subcropping reflections or onlapping or downlapping reflections; this tends to make unconformity reflections more obvious because an observer's attention is drawn to pattern irregularities. Unconformities have special importance in seismic stratigraphic studies (§10.3.6).

In many areas, especially areas of Tertiary clastic deposition, the nature of deposition changes laterally along time lines and consequently reflections change

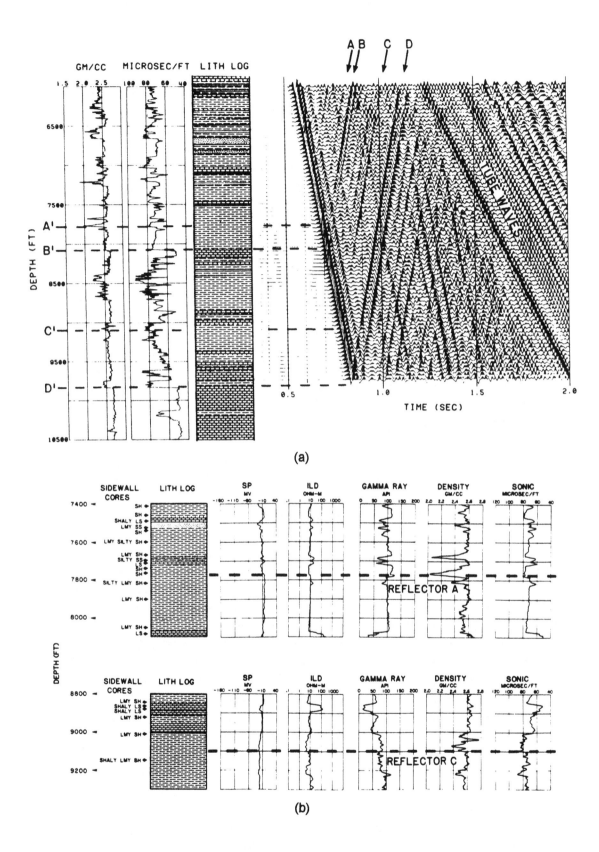

Fig. 6.8 Reflections on a vertical seismic profile related to well logs. The source is at the surface and the geophone at various depths in a borehole. (From Hardage, 1985.) (a) Reflections *A*, *B*, *C*, and *D* correspond with interfaces at the depths *A′*, *B′*, *C′*, and *D′* on the well logs. (b) Detail of response on different logs at interfaces *A′* and *C′*.

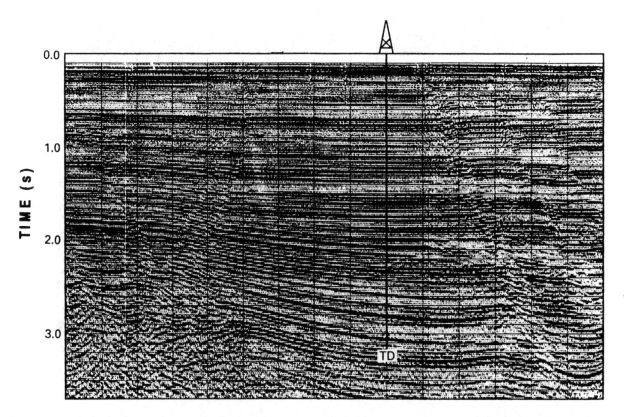

Fig. 6.9 Seismic line in an area of clastic deposition showing abundant reflections from a section that is almost entirely shale.

(Courtesy of Teledyne Exploration Company.)

their character and are discontinuous. Commonly, one can map a given reflection only over a limited area; this leads to the practice called "phantoming," which is discussed in §8.8.3. On the other hand, in some areas, formations extend over very large distances so that individual reflections can be followed over great distances.

Although most reflections mark unconformities and/or time surfaces, fluid contacts within porous permeable rocks provide acoustic impedance contrasts that cut across the bedding. These are responsible for flat spots, one of the most important hydrocarbon accumulation indicators (§10.8). Figure 6.10 shows such a reflection. Acoustic impedance contrasts (hence reflections also) can be caused by chemical or phase changes, such as hydrate reflections (fig. 5.33) and changes due to recrystalization of opal (see fig. 6.11); such situations, however, are relatively rare.

6.2.3 Fresnel zones

When we use rays to represent wave travel, the implication is that a reflection occurs at the reflection point. However, a reflection is made up of energy returning from a fairly large area of the reflector. A *Fresnel zone* is the area from which reflected energy arriving at a detector has phases differing by no more than a half-

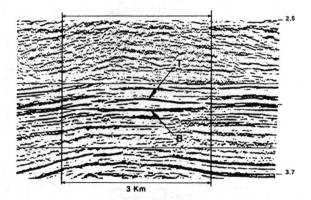

Fig. 6.10 Portion of a West Ekofisk, Norwegian North Sea, seismic section showing a discordant horizontal reflection attributed to a gas–liquid contact. The vertical lines indicate the limit of porosity based on the seismic data; T and B have been interpreted as the top and base of the reservoir. (From d'Heur, 1992: 952; reprinted with permission.)

cycle; thus, this energy interferes more or less constructively.

Consider a source and a coincident detector, S, as in fig. 6.12. SP_0 is perpendicular to a reflecting plane, and R_1, R_2, ... are such that the distances SP_0, SP_1, SP_2, ..., differ by $\lambda/4$. Generally, $h_n \gg R_n \gg \lambda$;

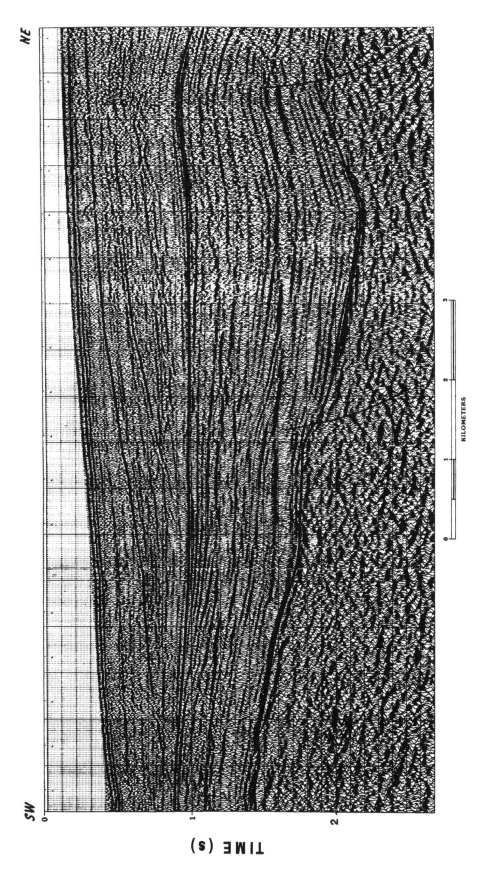

Fig. 6.11 Seismic section showing a horizontal reflector (at ∼ 1.57 s) attributed to pressure-controlled diagenetic change from opal-CT porcellanites above to quartz chert below. A change from opal-A to opal-CT at 1.20 s may also produce reflections. (From Hubbard, Pape, and Roberts, 1985: 85; reprinted with permission.)

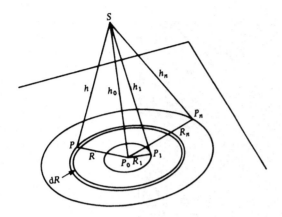

Fig. 6.12   Geometry of Fresnel zones.

hence,

$$R_n \approx (n\lambda h_0/2)^{1/2}, \\ \Delta S \approx \pi\lambda h_0/2, \qquad \Bigg\} \qquad (6.1)$$

where ΔS is the area of each of the annular rings. We shall calculate the energy returning to S from the $(n + 1)$th ring. If we apply eq. (2.129) to a circle with the origin over the center (fig. 6.12), h becomes h_0 and ξ becomes h_n, and the integration with respect to θ merely involves multiplying by 2π. If we now calculate the Laplace transform $\Phi(s)$ for two circles of radii R_n and R_{n+1} and subtract the second from the first, we obtain the effect of the $(n + 1)$th zone:

$$\Phi(S) = \frac{1}{2}ch_0\left(\frac{1}{h_n^2}e^{-2sh_n/V} - \frac{1}{h_{n+1}^2}e^{-2sh_{n+1}/V}\right). \quad (6.2)$$

This solution corresponds to an impulse source $c\,\delta(t)$ (see §2.8.1). Taking the inverse transform (see eqs. (15.180) and (15.187)), we have in the time domain

$$\phi(t) = \frac{1}{2}ch_0\left[\frac{1}{h_n^2}\delta(t - t_n) - \frac{1}{h_{n+1}^2}\delta\left(t - t_n - \frac{T}{2}\right)\right], \quad (6.3)$$

where $t_n = 2h_n/V$, $t_{n+1} = (2h_n + \lambda/2)/V = t_n + T/2$, T being the period.

If the input at the source had been $A\cos\omega t$ instead of an impulse, eq. (6.2) would have the additional factor $As/(s^2 + \omega^2)$ (see eq. (15.183)) and, upon taking the inverse transform, eq. (15.187) would give

$$\phi(t) = \frac{1}{2}h_0A[(1/h_n^2)\text{step}(t - t_n)\cos\omega(t - t_n) \\ - (1/h_{n+1}^2)\text{step}(t - t_n - \tfrac{1}{2}T)\cos\omega(t - t_n - \tfrac{1}{2}T)] \\ = \frac{1}{2}h_0A\{(1/h_n^2)\text{step}(t - t_n) \\ + [1/(h_n + \tfrac{1}{4}\lambda)^2]\text{step}(t - t_n - \tfrac{1}{2}T)\}\cos\omega(t - t_n) \\ \approx (h_0/2h_n^2)A[\text{step}(t - t_n) \\ + (1 - \lambda/2h_n)\text{step}(t - t_n - \tfrac{1}{2}T)]\cos\omega(t - t_n).$$

When $t > (t_n + \tfrac{1}{2}T)$,

$$\phi(t) \approx A[(h_0/h_n^2)(1 - \lambda/4h_n)]\cos\omega(t - t_n) \\ \approx A[h_0/h_n^2]\cos\omega(t - t_n). \quad (6.4)$$

Because t_n and t_{n+1} differ by $\frac{1}{2}T$, contributions from successive zones are alternately plus and minus. Thus, the effect at S is an alternating series, $\phi_T = S_1 - S_2 + S_3 - S_4 + \cdots$, where S_{n+1} is a positive quantity given by A times the factor in the square brackets in eq. (6.4). Because S_n decreases as n increases, the series converges and we may write (Wood, 1961: 34)

$$\phi_T = \frac{1}{2}S_1 + (\tfrac{1}{2}S_1 - S_2 + \tfrac{1}{2}S_3) \\ + (\tfrac{1}{2}S_3 - S_4 + \tfrac{1}{2}S_5) + \cdots. \quad (6.5a)$$

A graph of ϕ_T as a function of the radius is shown in fig. 6.13. In the foregoing equation, the terms in parentheses are approximately zero; hence,

$$\phi_T \approx S_1/2, \quad (6.5b)$$

that is, the major contribution to the reflected signal comes from the first Fresnel zone (the adjective "first" is frequently dropped). The radius of S_1 is

$$R_1 = (\lambda h_0/2)^{1/2} = (V/2)(t/\nu)^{1/2}, \quad (6.6a)$$

where h_0 is the depth, t the arrival time, V the average velocity, and ν the frequency. The outer portion of S_1 makes relatively little net contribution to the final result, as is seen in fig. 6.13, and hence an *effective Fresnel zone* is sometimes taken as a smaller radius $\approx R/\sqrt{2}$.

The first Fresnel zone is often taken as a measure of the horizontal resolution of unmigrated seismic data. For a depth of 3 km and velocity of 3 km/s ($t = 2$ s), the Fresnel-zone radius ranges from 300 to 470 m for frequencies of 50 to 20 Hz (see fig. 6.14). Figure 6.15 shows the response of small segments of a reflecting surface at a depth of 1500 m for a 30-m dominant wavelength for which the Fresnel-zone radius is 150 m. When the reflector dimensions are somewhat smaller than the Fresnel zone, the response is essentially that of a diffracting point.

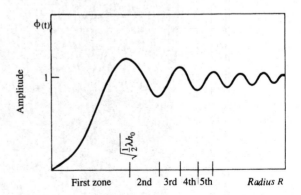

Fig. 6.13   As the radius of a circular reflector increases, the amplitude begins to build up slowly, then rapidly, tapering off toward the edge of the first Fresnel zone. Interference causes the amplitude to oscillate as successive destructive and constructive zones are added. Obliquity causes the oscillation to dampen more rapidly than shown.

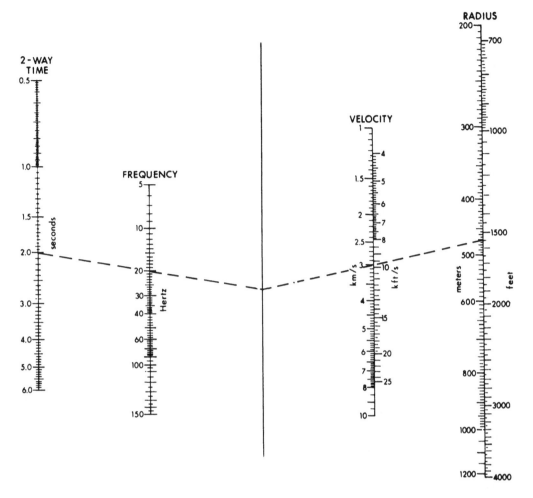

Fig. 6.14 Nomogram for determining Fresnel-zone radii. A straight line connecting the two-way time and the frequency intersects the central line at the same point as a straight line connecting the average velocity and the radius of the zone. For example, a 20-Hz reflection at 2.0 s and a velocity of 3.0 km/s has a Fresnel-zone radius of 470 m.

The foregoing discussion assumes a point source, for which the travel paths from source back to detector differ by a half-cycle for successive Fresnel zones. Fresnel zones are sometimes specified with respect to a plane incident wave rather than a spherical wave, in which case the half-cycle differences between successive Fresnel zones have to be accommodated entirely in the reflector-to-detector portion of the travel path. This results in an enlargement of the Fresnel zone, the radius in this case being

$$R_1 = (\lambda h_0)^{1/2} = \tfrac{1}{2}V(2t/v)^{1/2}. \qquad (6.6b)$$

The amplitude of a reflection will be diminished if the reflectivity anywhere within the Fresnel zone is diminished. For example, if we should be recording immediately over a large vertical step, half of the Fresnel zone would not contribute to a reflection, and the reflection amplitude would only be one-half of that over the reflector remote from the step (see eq. (2.133)). Because seismic traces are spaced much more closely than the dimensions of the Fresnel zone, a specific portion of a reflector will contribute to all of the seismic traces whose Fresnel zones include the portion. We wish the amplitude of a migrated trace to be proportional to the reflectivity, and we do this by summing (actually or effectively) traces around the reflecting point; clearly, if we wish to obtain the correct reflectivity, we must include all observations within the Fresnel-zone radius. We note that migrating seismic line data effectively collapses the Fresnel zone in the in-line direction, but does not shrink its size perpendicular to the seismic line.

6.2.4 Effects of reflector curvature

Geometrical focusing as a result of curvature of a reflector affects the amplitude of a reflection.

In a constant-velocity medium, the wave generated by a point source is spherical with radius of curvature equal to the source-to-wavefront distance. We shall assume that such a wave encounters a spherical reflecting horizon centered directly below the source and with radius $\rho_{\mathscr{G}}$. We can then apply the well-known formula of geometrical optics for reflections from

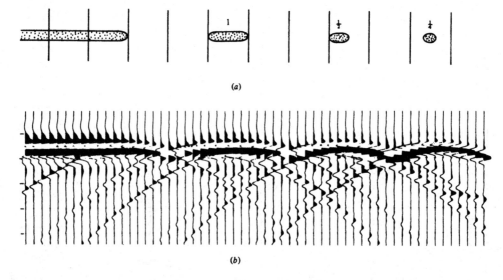

(a)

(b)

Fig. 6.15 Reflection from strips of various widths. (After Neidell and Poggiagliolmi, 1977.) (a) Cross-section of model; spacing of vertical lines equals the Fresnel-zone diameter; (b) seismic section.

curved mirrors:

$$1/u + 1/v = 2/R,$$

where u and v are the object and image distances, respectively, and R is the radius of the mirror. Positive R corresponds to a concave mirror and negative v to a virtual image. The object and image distances are equivalent to the radii of curvature of the incident and reflected waves, ρ_i and ρ_r, the radius of curvature of the reflecting surface ρ_g being positive for a syncline, negative for an anticline. The formula now becomes

$$1/\rho_i + 1/\rho_r = 2/\rho_g. \tag{6.7}$$

If the distance to the reflector is h, $\rho_i = h$ for a point source, and

$$\rho_r = h\rho_g/(2h - \rho_g). \tag{6.8}$$

When the reflected wave reaches the surface, its radius of curvature ρ is

$$\rho = \rho_r - h = 2h(\rho_g - h)/(2h - \rho_g). \tag{6.9}$$

We can express this in terms of curvature, the reciprocal of the radius of curvature, in a normalized form as

$$\frac{h}{\rho} = -\left[\frac{(h/\rho_g) - 1/2}{(h/\rho_g) - 1}\right]. \tag{6.10}$$

Figure 6.16a is a graph of eq. (6.10). Reflector curvature of $\pm\infty$ corresponds to a point diffractor, and zero curvature ($h/\rho_g = 0$) corresponds to a plane reflector for which the convex-upward curvature of the wavefront produces normal moveout. When the reflector center of curvature is at the surface ($\rho_g = h$), the reflected energy concentrates to a point. Curvature greater than this produces a buried-focus as discussed in what follows. Figure 6.16b shows the image points for a diffracting point, anticline, plane and syncline, and fig. 6.17 shows the buried-focus case.

Consider a cone of energy from the source that is reflected from a spherical cap, MN (fig. 6.16b); the reflected energy is spread over a larger area at the surface for the anticline than for the plane, and over a smaller area for the gentle syncline. Thus, reflections should appear stronger over gentle synclines and weaker over anticlines.

When a syncline has a radius of curvature less than its depth, ρ_r is positive and ρ is negative, and the energy passes through a focus below the surface (see fig. 6.17a); this is a *buried-focus* situation. Obviously, the likelihood of a buried focus increases with reflector depth. A reflection involving a buried focus is called a *reverse branch*. The sense of traverse of the reverse branch is reversed from the usual, that is, as the source point travels from left to right, the reflecting point travels from right to left.

Where the source and geophone are not coincident (that is, for offset traces), the reflected wave may focus even where the reflector's center of curvature is not below the surface, as in fig. 6.17b. Thus, long-offset traces may involve buried-focus effects even where short-offset traces do not. Common-midpoint stacking where short- and long-offset traces are combined after normal-moveout correction usually does not allow for this situation correctly.

When the curvature of a syncline is not constant, as in fig. 6.17d, reflections may be obtained from more than one part of the reflector, and the reflected energy has *multiple branches*, most commonly three. The two deeper reflections in fig. 6.18 involve multiple branches; each shows branches from each flank of the syncline and a reverse branch from the curved bottom of the syncline. Three-dimensional multiple-branch effects are discussed in §6.2.5.

Just as light can be focused by passing through a lens, seismic waves can also be focused by curved ve-

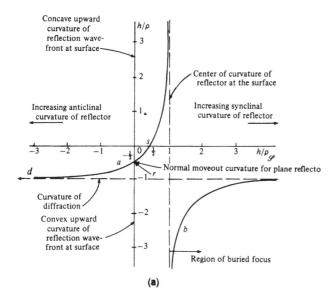

(a)

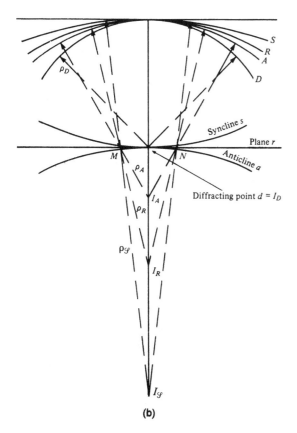

(b)

Fig. 6.16   Effect of changes in reflector curvature. (a) Normalized curvature of wavefront at the surface, h/ρ, as a function of reflector curvature, $h/\rho_{\mathscr{G}}$, for a point source. The letters d, a, r, and s refer to diffracting point, anticline, plane reflector, and syncline, respectively. (b) Effect on wavefront curvature as reflector changes from anticline to syncline. Respective wavefronts for d, a, r, and s are D, A, R, and S; image points I_D, I_A, I_R, and $I_{\mathscr{G}}$; and radii of curvature ρ_D, ρ_A, ρ_R, and $\rho_{\mathscr{G}}$.

locity surfaces that result in seismic rays being bent by refraction; such situations are often very complex. Curvature at the base of the weathering can be especially important because of the large velocity contrast usually associated with this surface. Variations in permafrost thickness and gas accumulations can also cause focusing effects (fig. 6.19).

Fermat's principle explains that a wave will take that raypath for which the traveltime is stationary with respect to minor variations of the raypath, that is, for which the change in the traveltime for an incremental change in raypath is zero. For most situations the raypath involves the minimum traveltime, that is, travel over any neighboring path will take longer; hence Fermat's principle is often called the *principle of least time* or the *brachistochrone principle.* Snell's law, Huygens' principle, and many other laws of geometrical optics can be derived from this principle (see problem 6.3).

An incident wavefront approaching the reflector in a buried-focus situation (fig. 6.17e) encounters the reflector before the wave reaches the reflecting point R, which satisfies Snell's law. Contributions to the reflection from the region surrounding R will thus arrive earlier than the reflection from R itself, that is, the reflection point involves a maximum in a Fermat's principle sense, thus contrasting with the more usual situation where the reflection point involves minimum traveltime. The fact that reflection contributions from the region surrounding the reflection point arrive earlier manifests itself as a change in the waveshape of the reverse branch reflection compared with normal branches. With wavefronts that pass through a focus (as for the reverse branch), this phase shift is π if the reflector is spherical or $\frac{1}{2}\pi$ if it is cylindrical. A $\frac{1}{2}\pi$ phase shift can be seen by comparing the waveshape of the reverse branch for the lower event in fig. 6.18 with other events; the reflectors here are cylindrical. Such a phase shift is rarely useful in identifying buried-focus events, but it will affect calculations of reflector depth where picking is done systematically on the same phase, for example, always picking troughs, and it affects common-midpoint stacking.

6.2.5 Three-dimensional effects

In §4.1.2, we defined the reflecting point as the point at which the angle of incidence equaled the angle of reflection. Seismic data are usually mapped at reflecting points. A line connecting reflecting points is called the *subsurface trace.* There is a subsurface trace for each reflector. Where there is a component of dip perpendicular to the seismic line, the reflecting point lies to the side of the seismic line rather than below it. Such cross-dip effects are often ignored, usually because the cross-dip is not measured, and sometimes such neglect leads to serious errors. In §6.3.1, for example, we assume that the diffracting point is in the vertical plane containing the seismic line; if instead it is from the truncation of a reflector by a fault that is not perpendicular to the line, then the diffracting

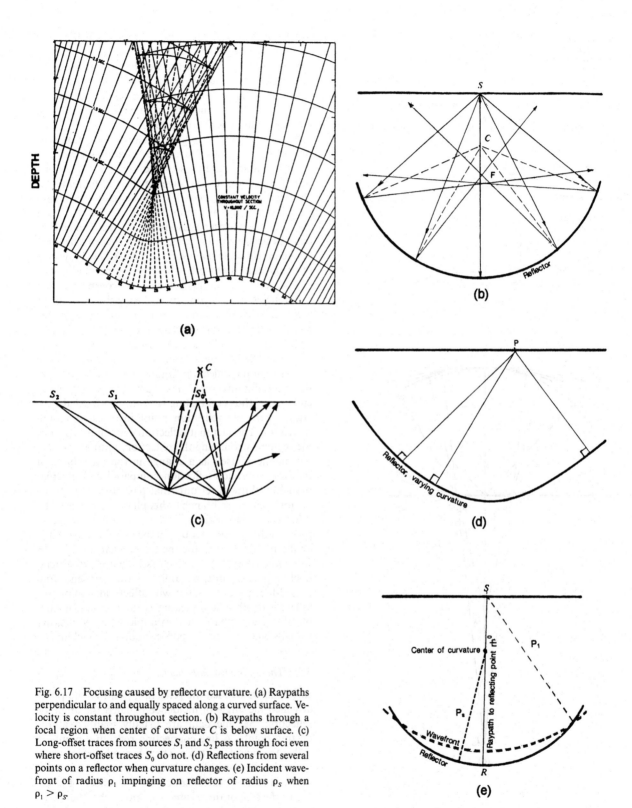

Fig. 6.17 Focusing caused by reflector curvature. (a) Raypaths perpendicular to and equally spaced along a curved surface. Velocity is constant throughout section. (b) Raypaths through a focal region when center of curvature C is below surface. (c) Long-offset traces from sources S_1 and S_2 pass through foci even where short-offset traces S_0 do not. (d) Reflections from several points on a reflector when curvature changes. (e) Incident wavefront of radius ρ_1 impinging on reflector of radius ρ_s when $\rho_1 > \rho_s$.

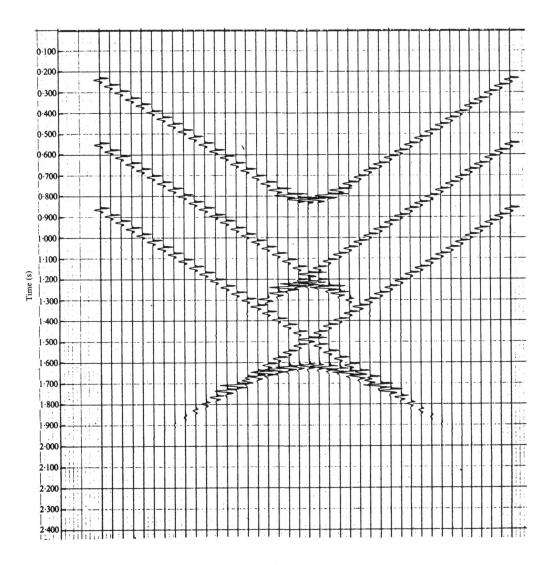

Fig. 6.18 Reflection from cylindrically curved reflectors (which are planar outside the curved region). For all three reflectors, the radius of curvature = 1000 m and V = 2000 m/s. Depths to the bottom of the synclines are 800, 1200, and 1600 m, respectively. The traces are 100 m apart with coincident sources and receivers. (Courtesy of Chevron.)

point may move along the fault as source and/or receivers move and the curvature of the diffraction on the seismic record will be less than that given by eqs. (6.11) and (6.13).

In §6.2.4, we examined reflector curvature effects and in fig. 6.18 showed multiple-branch effects. If a seismic line crosses a syncline other than at right angles (line BB' in fig. 6.20b), the reflection branches may come from opposite sides of the seismic line and the length of the reverse branch may be stretched out and thus show smaller curvature (compare fig. 6.20d with fig. 6.20c). In the extreme situation where the seismic line is parallel to the axis of the syncline, the multiple branches appear as parallel horizontal reflectors (fig. 6.20e). Where the syncline is plunging, the different branches will not be parallel.

The Fresnel-zone concept of §6.2.3 replaces a reflecting "point" with a reflecting "area," the area of the first Fresnel zone. Features to the side of the reflecting point but within the reflecting area will produce effects on the seismic line, as shown in fig. 6.21. Hilterman (1970) showed that such effects can make structures appear to be appreciably larger in area than they actually are (fig. 6.22) when mapping 2-D data.

6.3 Events other than primary reflections

6.3.1 Diffractions

Diffraction phenomena were discussed in §2.8, where it was shown that the reflection from a half-plane and the diffraction from its edge are continuous and indistinguishable on the basis of character. Diffractions from edges that are perpendicular to the seismic line, however, exhibit distinctive moveout. In figs. 6.23a and 6.23b for source positions above the diffracting

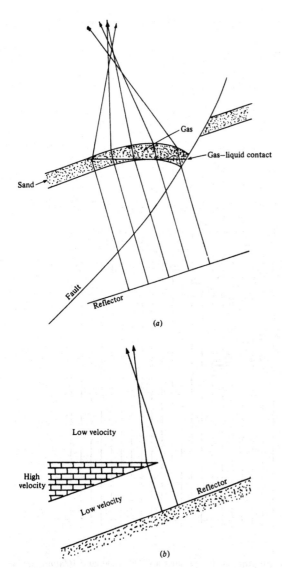

Fig. 6.19 Focusing produced by velocity variations. (a) Low velocity in a gas accumulation focusing raypaths from deeper reflection. (b) High-velocity wedge producing a focus.

edges, the diffraction traveltime curve (for a common-source gather with the source over the diffracting point) for $h \gg x$ is given by

$$t_d = (1/V)[h + (x^2 + h^2)^{1/2}]$$
$$\approx 2h/V + x^2/2Vh = t_0 + 2\,\Delta t_n, \qquad (6.11)$$

where $\Delta t_n = x^2/4Vh$, the normal moveout for a reflection (see eq. (4.7)). However, when the source is not vertically over the diffracting point (fig. 6.23c), the diffraction traveltime curve is given by

$$t_d = (1/V)\{(h^2 + a^2)^{1/2} + [(x - a)^2 + h^2]^{1/2}\}$$
$$\approx (2h/V) + (a^2/2Vh) + (x - a)^2/2Vh$$
$$\approx t_0 + (2a^2 - 2ax + x^2)/2Vh \approx t_0 + 2\,\Delta t_n$$
$$+ a(a - x)/Vh; \qquad (6.12)$$

thus, the diffraction moveout varies with the location of the source. The last term can be either positive or

negative, so the diffraction moveout can be either greater or smaller than given by eq. (6.11). If we consider a coincident source–geophone section, as shown in fig. 6.23d, the diffraction traveltime curve will be

$$t'_d = (2/V)(x^2 + h^2)^{1/2} \approx t_0 + 4\,\Delta t_n. \qquad (6.13)$$

This is the sort of section common-midpoint stacking seeks to simulate. However, a diffraction on a common-midpoint gather (fig. 6.23e) has a traveltime curve

$$t''_d = (1/V)\{[h^2 + (x/2 + b)^2]^{1/2} + [h^2 + (x/2 - b)^2]^{1/2}\}, \qquad (6.14)$$

which differs from that given by eq. (6.13). Obviously, diffractions will be attenuated by common-midpoint stacking. Note that the normal moveout applied in processing usually is that for a deeper reflection (that is, for larger velocity) because the diffraction time t_d is larger than for the reflection terminating at the diffracting point.

The earliest arrival time on a diffraction curve is for the trace that is recorded directly over the diffracting point (except for unusual velocity-distribution situations), but the diffraction will not necessarily have its peak amplitude on this trace.

Consider three half-planes at the same depth but with different dips (fig. 6.24). The diffractions for each of these crest at the location of the edge of the half-plane, have the same curvature, are tangent to the reflections, and the maximum amplitudes of the diffractions occur at this point of tangency. Thus, the diffraction crest locates the diffracting point, the diffraction curvature depends on the depth and the velocity above the diffracting point, and the amplitude distribution along the diffraction depends on the attitude of the half-plane.

A reflector that is bent sharply, as shown in fig. 6.25, could be thought of as the superposition of two dipping half-planes, each terminating at the bend point. Thus, the two diffraction curves would coincide in arrival time and would add together constructively in the region between the respective reflections. In fig. 6.25b, the reflector to the right of $x = 2.1$ km gives rise to $P'B$ and the reflector to the left of $x = 2.1$ km gives AP; diffraction fills in the gap PP' and makes the seismic event continuous without a sharp break in slope.

As another example of diffraction effects, consider the reflection from a reflector with a hole in it, as shown in fig. 6.26. Diffraction tends to fill in the hole.

Figure 6.27 shows the location and amplitude of wave motion a short time after a plane wavefront has passed by the point of a wedge that is a perfect reflector. The reflected wavefront has an associated diffraction BAC and the portion of the wavefront that missed the reflecting wedge also has an associated diffraction FDE. The portion of the diffraction DE represents energy reaching into the shadow zone hidden from the incident wavefront by the reflector. Diffraction pro-

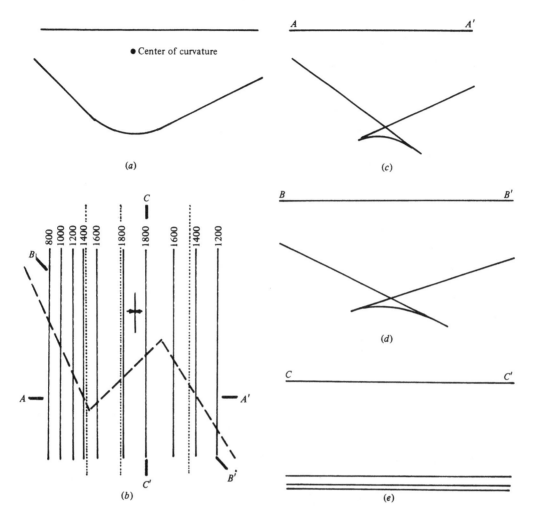

Fig. 6.20 Buried-focus effects on lines at different angles to the strike of a syncline. (a) Syncline cross-section; (b) contour map of reflector showing subsurface traces (dashed line) of BB' and (dotted lines) of CC'; (c) arrival times along line AA'; (d) arrival times along BB'; and (e) arrival times along CC'.

vides a mechanism for getting seismic energy into regions that cannot be reached on the basis of geometrical optics.

The downgoing diffraction FDE in fig. 6.27 might be subsequently reflected by another reflector to give a reflected diffraction, or an upcoming reflection could be diffracted to give a diffracted reflection, and so on. The diffraction curvature for such compound events will differ from that of simple diffraction events that have the same arrival time. Additional compound events are shown in fig. 6.55.

Figure 6.28 shows a vertical step. The top of step A will generate diffraction, D_1, in fig. 6.28b. The bottom of step B will also generate a diffraction, D_1', but the portion of this diffraction to the right of B will travel partially at a lower velocity than the reflection to the left of the step.

The reflection from the left portion of the base of the model will arrive earlier than the reflection from the right portion because more of its travel path is at the higher velocity. Furthermore, each of these re-

flection segments will have an associated diffraction appearing to come from C, even though the base of the model is continuous; such diffractions are called *phantom diffractions* and they result from lateral velocity changes in the section above the reflector. Consider a geophone just to the right of a point directly over the step in fig. 6.28a. The travel path for a reflection from the base of the model is shown by the dashed path, but some energy will travel the dotted path and arrive earlier than the reflection; the latter produces the phantom diffraction (diffraction energy does not necessarily obey the rules of geometrical optics). Phantom diffractions are occasionally seen on seismic records, especially in areas of thrusting where some travel paths pass through thrust plates and others do not.

6.3.2 Multiples

(a) Distinction between types. *Multiples* are events that have undergone more than one reflection. Be-

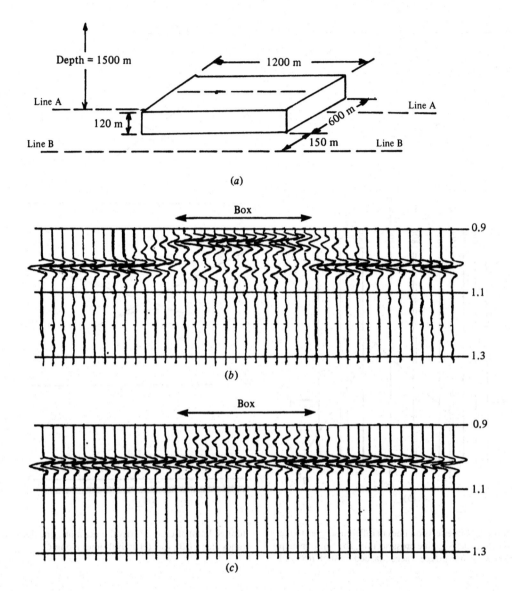

(a)

(b)

(c)

Fig. 6.21 Modeled lines across a box structure. Trace spacing, 85 m, Fresnel-zone radius 280 m for 30 Hz. (Courtesy of Geo-quest.) (a) Model, (b) subsurface trace of line *A* above the top of the box, (c) line *B* above the plane 150 m beyond the box.

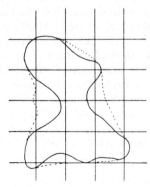

Fig. 6.22 Grid of 2-D seismic lines across a physical model of irregularly shaped gas field (solid curve). Dotted outline shows seismic interpretation; failure to allow for 3-D Fresnel-zone effects led to 40% overestimate of field size. (After Hilterman, 1982.)

cause the amplitude of multiples is proportional to the product of the reflection coefficients for each of the reflectors involved and because *R* is very small for most interfaces, only the largest impedance contrasts will generate multiples that are strong enough to be recognized as distinctive events.

We distinguish between two classes of multiples, which we call long-path and short-path. A *long-path multiple* is one whose travel path is long compared with primary reflections from the same deep interfaces, and hence long-path multiples appear as separate events on a seismic record. A *short-path multiple,* on the other hand, arrives so soon after the associated primary reflection from the same deep interface that it interferes with and adds tail to the primary reflection; hence, its effect is that of changing waveshape rather than producing a separate event. Possible raypaths for

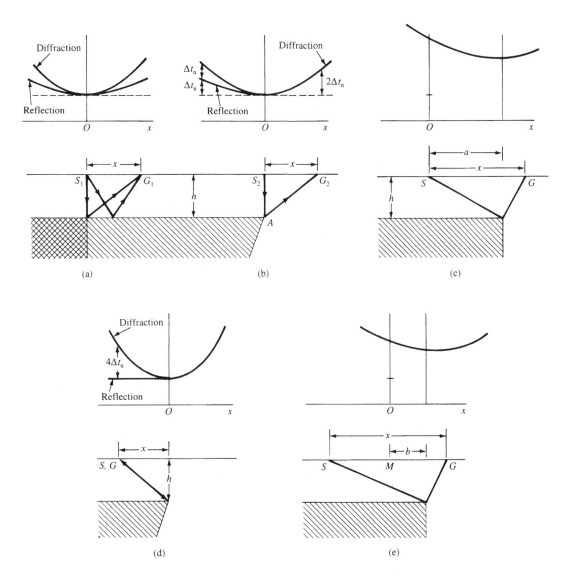

Fig. 6.23 Diffraction traveltime curves. (a and b) Common-source arrivals for diffraction and reflection; Δt_n is the reflection normal moveout; (c) diffraction for source offset from diffract-ing point; (d) source and geophone coincident at different locations; (e) midpoint not over diffracting point.

these two classes are shown in fig. 6.29.

(b) Short-path multiples.

Short-path multiples that have been reflected successively from the top and base of thin reflectors (fig. 6.30a) on their way to or from the principal reflecting interface with which they are associated (often called *peg-leg multiples*) are important in determining the waveforms of the events recorded on a seismogram. These peg-leg multiples delay part of the energy and therefore lengthen the wavelet. The stronger peg-leg multiples often have the same polarity as the primary because successive large impedance contrasts tend to be in opposite directions (otherwise, successive large changes in velocity would cause the velocity to exceed its allowable range). Peg-leg multiples effectively lower the signal frequency as time increases. Figure 6.30b shows how a simple im-

pulse becomes modified as a result of passing through a sequence of interfaces, and their frequency spectra in fig. 6.30c show the loss of high frequency with time. Figure 6.30d shows that the average attenuation due to peg-leg multiples is equivalent roughly to $\eta\lambda = 0.085 \pm 0.055$ dB (see §2.7.2b). This is the same situation that was discussed in relation to synthetic seismograms (§6.2.1).

Ghosts are the special type of multiple illustrated in fig. 6.29. The energy traveling downward from the source has superimposed upon it energy that initially traveled upward and was then reflected downward at the surface or base of the low-velocity layer (LVL) in land surveys or at the surface of the water in marine surveys. A 180° phase shift, equivalent to half a wavelength, occurs at the additional reflection, and hence the effective path difference between the direct wave

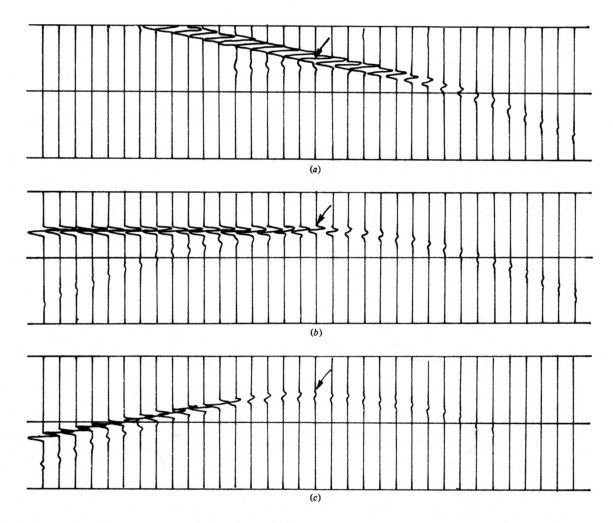

Fig. 6.24 Reflections and diffractions from half-planes terminating at the arrows (coincident sources and geophones). (Courtesy of Chevron.) (a) Termination at the downdip end of the half-plane; (b) the flat half-plane; (c) termination at the updip end of the half-plane.

and the ghost is $\lambda/2 + 2D_s$, where D_s is the source depth below the reflector producing the ghost. Similar ghosting results from buried detectors or a marine streamer towed at depth D_s. The interference between the ghost and the primary depends on the fraction of a wavelength represented by the difference in effective path length. Because the seismic wavelet is made up of a range of frequencies, the interference effect will vary for the different components. Thus, the overall effect on the wavelet shape will vary as D_s is varied. Relatively small changes in D_s can result in large variations in reflection character, creating serious problems for the interpreter. Therefore the depth below the base of the weathering or the surface of the water is maintained as nearly constant as possible.

Ghosts affect directivity as well as waveshape. Figure 6.31 shows a point source at a depth $D_s = c\lambda$; if

the source emits the wave $A \cos(\kappa r - \omega t)$, then the effect at P is

$$\psi_P = A \cos(\kappa r_1 - \omega t) - A \cos(\kappa r_2 - \omega t)$$
$$= A \cos(\kappa r - \omega t - \kappa c \lambda \cos\theta)$$
$$\quad - A \cos(\kappa r - \omega t + \kappa c \lambda \cos\theta).$$

Expanding and noting that $\kappa\lambda = 2\pi$, we get approximately

$$\psi_P = 2A \sin(\kappa r - \omega t) \sin(2\pi c \cos\theta)$$
$$= 2A \sin(2\pi c \cos\theta) \cos(\kappa r - \omega t - \tfrac{1}{2}\pi).$$

$$(6.15)$$

Figure 6.53 is a graph of ψ_P as a function of θ. For r large, the total wave motion lags the original wave motion by 90° and has an amplitude that depends on θ. Because a wavelet contains a spectrum of frequencies, different components will add differently at various

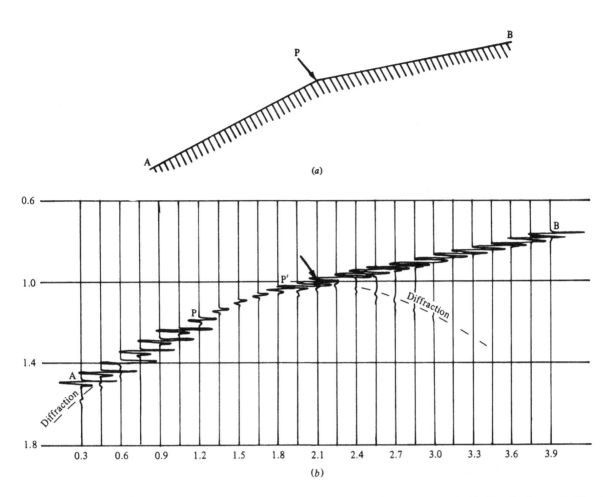

Fig. 6.25   Reflections and diffractions from a sharply bent reflector. Dips are 31° and 11° to the left and right of $x = 2.1$ km.

(Courtesy of Chevron.) (a) Model; (b) reflections and diffractions (dashed curve).

angles of θ, resulting in changes in waveshape.

Ghosts are especially important in marine surveys
because the surface of the water is almost a perfect
reflector and consequently the ghost interference will

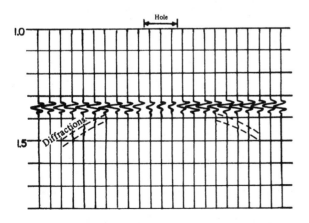

Fig. 6.26   Effect of a hole in a reflector. (Courtesy of Chevron.)

be strong. If D_s is small in comparison with the dominant wavelengths, appreciable signal cancellation will
occur. At depths of 10 to 15 m, interference is constructive for frequencies of 40 to 25 Hz, which is in
the usual seismic range. The same effect occurs with
the upcoming signal from the reflectors we wish to
map. Hence, marine sources and marine detectors are
often operated at such depths.

A particularly troublesome type of multiple produces the coherent noise known as *singing* (also called
ringing or *water reverberation*), which is frequently encountered in marine work (and occasionally on land).
This is due to multiple reflections in the water layer.
The large reflection coefficients at the top and bottom
of this layer result in considerable energy being reflected back and forth repeatedly, the reverberating
energy being reinforced periodically by reflected energy. Depending upon the water depth, certain frequencies are enhanced, and the record looks very
sinusoidal as a result (see fig. 6.32). Not only is the
picking of reflections difficult, but measured travel-

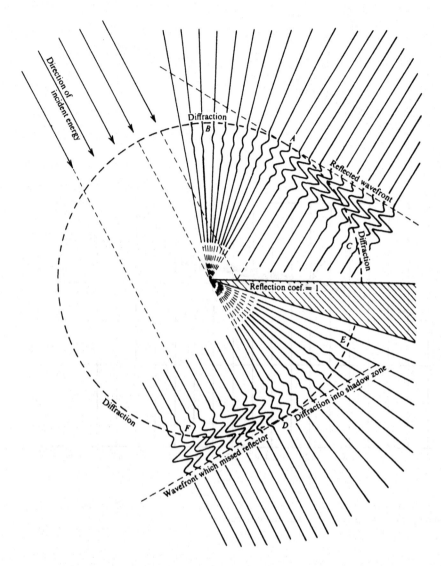

Fig. 6.27   Diffractions from a perfectly reflecting wedge. (Courtesy of Chevron.)

times and dip moveouts will probably be in error. This type of noise and its attenuation are discussed in §9.2.4.

(c) Long-path multiples. The strongest long-path multiples involve reflections at the surface, the sea floor, or (on land) the base of the low-velocity layer (LVL, also called weathered layer; see table 3.1 and §5.3.2), where the reflection coefficient is very large because of the large acoustic-impedance contrast. Because this type of multiple involves at least two reflections at depth, its amplitude depends mainly on the magnitude of reflection coefficients at depth, and multiples of this type will be observed as distinctive events when these coefficients are abnormally high. Because R in eq. (3.14) may be as large as 0.7 at the base of the LVL and perhaps 0.2 for the strongest interfaces at depth, the maximum effective R for such multiples will be of the order of $0.2 \times 0.7 \times 0.2 = 0.03$. This value

is in the range of typical reflection coefficients so that such multiples may have sufficient energy to be confused with primary events. The principal situation where weaker long-path multiples may be observable is where primary energy is nearly absent at the time of arrival of the multiple energy. It is important that multiples be recognized as such so that they will not be interpreted as reflections from deeper horizons.

Because velocity generally increases with depth, multiples usually exhibit more normal moveout than primary reflections with the same traveltime. This is the basis of the attenuation of multiples in common-midpoint processing, which will be discussed in §9.10.4. However, the difference in normal moveout is often not large enough to identify multiples. The attenuation of multiples is also the primary objective of predictive deconvolution (§9.5.7).

The effect of dip on multiples that involve the surface or the base of the LVL can be seen by tracing

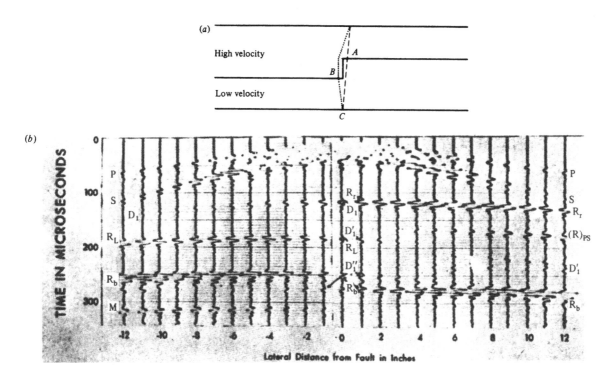

Fig. 6.28   Effects of a step. (From Angona, 1960.) (a) Model; (b) split-spread record with the source over the step: P = direct wave; S = surface wave; R_L and R_r = reflections to left and right of step, respectively; R_b = reflections from the base of the model; D_1 and D_1' = diffractions from top and base of the step; D_1'' = phantom diffractions that continue R_b beyond the center; $(R)_{PS}$ = converted wave; and M = multiple.

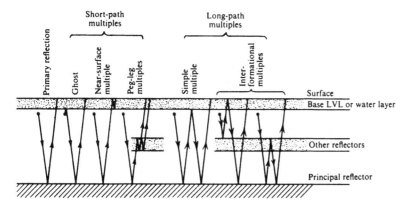

Fig. 6.29   Types of multiples.

rays using the method of images assuming the velocity is constant. In fig. 6.33, we trace a multiple arriving at symmetrically disposed geophones, G_1 and G_{24}. The first image point, I_1, is on the perpendicular from S to AB as far below AB as S is above. We next draw the perpendicular from I_1 to the surface of the ground where the second reflection occurs and place I_2 as far above the surface as I_1 is below. Finally, we locate I_3 on the perpendicular to AB as far below it as I_2 is above. We can now draw the rays from the source S to the geophones (working backwards from the geophones). The dip moveout is the difference between the path lengths I_3G_{24} and I_3G_1; it is about double that

of the primary $(I_1G_{24} - I_1G_1)$. The multiple at the source will appear to come from I_3, which is updip from I_1, the image point for the primary, and I_3S is slightly less than twice I_1S. Hence, we can see that if the reflector dips, the multiple involves a different portion of the reflector than the primary and has a traveltime slightly less than double the traveltime of the primary. The latter fact makes identifying multiples by merely doubling the arrival time of the primary imprecise whenever appreciable dip is present. The arrival time of the multiple will be approximately equal to that of a primary reflection from a bed at the depth of I_1. If the actual dip at I_1 is not double that at

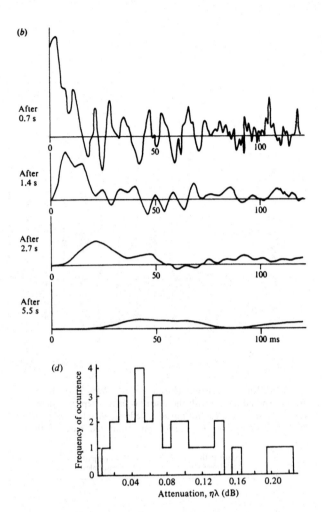

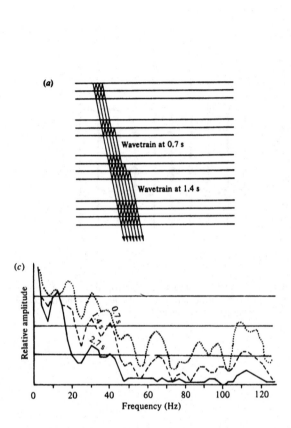

Fig. 6.30   Changes in waveshape produced by peg-leg multiples. (a) Schematic diagram showing peg-leg multiples adding to the wavetrain. (b) Waveshapes after different traveltimes for an actual layered section (after O'Doherty and Anstey, 1971). (c) Frequency spectra of waveshapes shown in part (b). (d) Histogram of attenuation coefficients due to peg-leg multiples based on 31 wells in different basins; η = attenuation constant as in eq. (2.110) (except for the attenuation mechanism), and λ = wavelength (after Schoenberger and Levin, 1978).

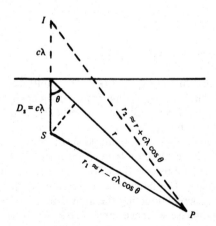

Fig. 6.31   Directivity of source plus ghost. S = source, I = image (effective source of the ghost), and P = observing point.

AB (and one would not in general expect such a dip), then the multiple will appear to have anomalous dip. If the multiple should be misidentified as a primary, one might incorrectly postulate an unconformity or updip thinning, which might lead to erroneous geologic conclusions.

In deep-water marine surveys multiples of the sea floor may be so strong as to virtually obliterate primary reflections. Furthermore, the range of reflection angles may be so great that the effective reflection coefficient varies widely for different offset distances. Figure 6.34 shows the buildup of amplitude observed near the critical angle and how this occurs on different traces for successive multiples. Predictive deconvolution (§9.5.7) is generally ineffective in attenuating such multiples because it assumes that, for any given trace, the reflection coefficients at the reflectors involved are constant. However, if sea-floor dip is small, the reflection angles for the simple reflection at offset x will be

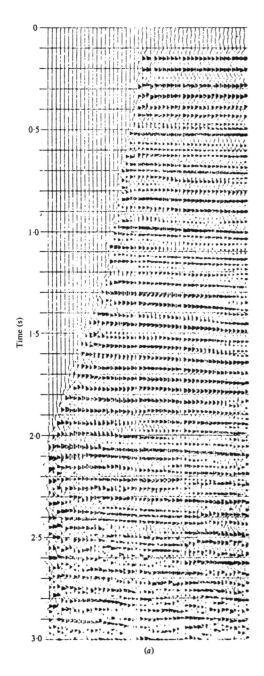

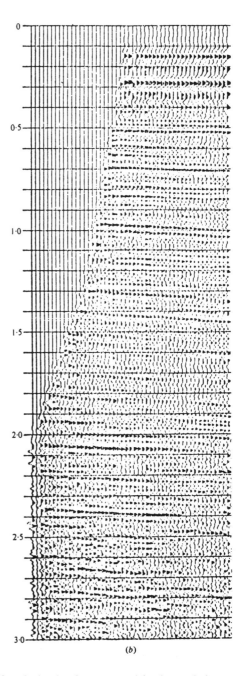

Time (s)

(a)

(b)

Fig. 6.32 Seismic record showing singing. (Courtesy of Haliburton Geophysical Services.) (a) Field record, and (b) the same after singing has been removed by deconvolution processing (§9.2.4).

the same as for the first multiple at offset $2x$ (fig. 6.35) and for the next multiple at offset $3x$, and so on, so that the primary on one trace sometimes can be used to predict and compensate for the multiple seen on another trace at another offset. This provides the basis for radial multiple suppression (§9.5.13).

6.3.3 Refractions

The onset of head waves is often followed by a number of parallel alignments, that is, they seem to involve a long wavetrain consisting of several cycles. As the offset distance increases, the number of cycles increases and the peak energy shifts to later in the wavetrain, an effect called *shingling* (fig. 6.36a). The amount of shingling is greater when the refractor is of limited thickness. Because of this shift of energy, it is often impossible to pick the onset time required for application of head-wave equations such as eqs. (4.36) to (4.48). Most timing of head waves is done on later peaks and troughs, and a correction is applied to obtain the onset time. This process often gives satisfac-

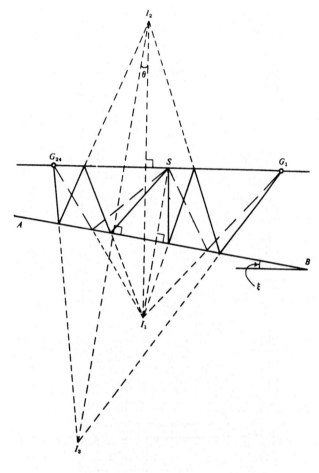

Fig. 6.33 Constructing raypath of a multiple from a dipping bed where velocity is constant.

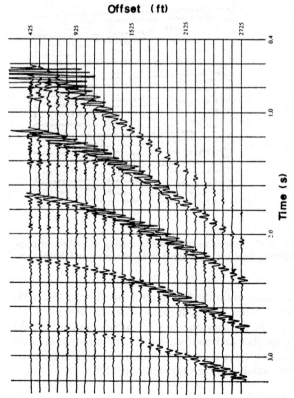

Fig. 6.34 Change of amplitude with offset for sea-floor multiples, offshore eastern Canada. Trace spacing = 100 m and offset of first trace = 425 m. The amplitude buildup occurs near the critical angle (see problem 6.13). (Courtesy of Chevron.)

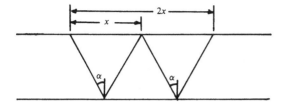

Fig. 6.35 Relation between offset and angle of reflection for primary and multiple reflections from a flat reflector.

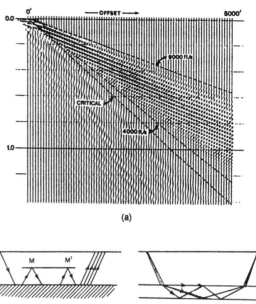

(a)

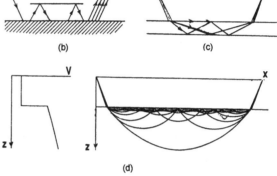

(b)   (c)

(d)

Fig. 6.36 Mechanisms that lengthen the refraction wavetrain. (a) Section showing head-wave shingling. (Courtesy of Geophysical Development Co.) (b) Reflections of head waves from parallel reflectors above the refractor. (c) Repeated reflections within the refractor. (d) Velocity gradient in the refractor.

tory results even though absorption and other mechanisms shift the frequency spectrum lower with increasing distance so that later cycles do not perfectly parallel the wave onset.

Several mechanisms contribute to the shingling effect (figs. 6.36b to 6.36d). Some of the energy that peels off the refractor can be reflected at beds parallel to the refractor and returned to the refractor at the critical angle (such as M and M' in fig. 6.36b) to form delayed head waves. Multiple reflections of this type can peal off the refractor continuously, and for any parallel reflector that is significant in creating multiples, they tend to add in phase (the head wave consequent to a reflection at M having the same distance to travel as one reflecting at M'). The result is to steal energy from the front of the head wave and add cycles at the tail end.

Waves that bounce repeatedly (fig. 6.36c) in layers within the refractor also add tail to the refraction wavetrains. Diving waves resulting from a velocity gradient in the high-velocity refractor (fig. 6.36d) have the similar effect of adding tails. A velocity gradient in the refractor considerably strengthens a head wave, as shown in fig. 6.37.

As the refractor becomes thinner, destructive interference between the head wave and the reflection from the base of the refractor also weakens the head wave, as shown in fig. 6.38. Poisson's ratio also has an effect on head-wave amplitude. Note that the head wave sometimes is phase-shifted, as in fig. 6.38.

Refractions (head waves) are not usually a problem on reflection records. They are generally of low frequency, have straight alignments (prior to normal-moveout correction), and are attenuated by stacking. Head waves are only observed where the offset exceeds the critical distance and, as shown in fig. 4.16 and by eq. (4.39), the critical distance is less than the refractor depth only for $V_2/V_1 > 2.24$. Velocity contrasts of this magnitude are possible below the base of the weathering, for example, where carbonates or evaporites are overlain by sands and shales, but usually head waves from deeper refractors do not appear on enough traces to make their moveout useful in identifying them, and they often disappear in the muting of the first-break region (the upper-right triangular region of fig. 6.3).

6.3.4 Reflected refractions

Where a refractor is terminated, such as shown in figs. 6.39a and 6.39b, the head wave will be reflected backward. It may appear on the later portion of a reflection record some distance from the actual refractor termination. When the refractor termination is nearly perpendicular to the seismic line, the reflected head wave will have a nearly straight alignment with an apparent velocity approximately the negative of the refractor velocity. The head wave will be reflected even though the law of reflection is not satisfied at the refractor termination. The refractor termination may be either against lower- or higher-impedance material so that the reflected head wave may have polarity either opposite to or the same as the head wave. Where the refractor is massive, reflections as in fig. 6.39c may appear much like the reflected head wave (figs. 6.39a and 6.39b). Where the refractor termination is off to the side of the line (fig. 6.39d), the event may have some curvature (pseudo-normal moveout) across the record (see problem 6.15).

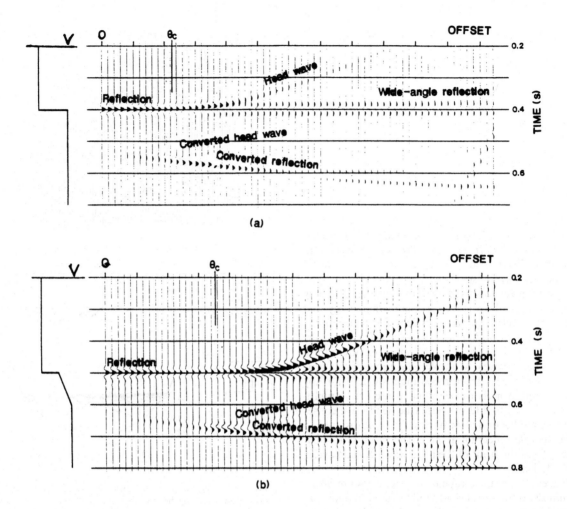

Fig. 6.37 Strengthening of a head wave by a velocity gradient. Reflection normal moveout has been removed so that head waves curve upwards. (Courtesy of Geophysical Development Co.) (a) Velocity step from 2000 to 4000 m/s. (b) Velocity step from 2000 to 3000 m/s, and then gradient increasing the velocity to 4000 m/s. (The lower event from 0.65 to 0.74 s is a converted wave. Note also a phase shift of the wide-angle reflection from that near normal incidence.)

6.3.5 Surface waves

Surface waves (*ground roll*) are usually present on reflection records. For the most part, these are Rayleigh waves with velocities ranging from 100 to 1000 m/s or so. Ground-roll frequencies are usually lower than those of reflections and refractions, often with the energy concentrated below 10 Hz. Ground-roll alignments are straight, just as in the case of refractions, but they have much lower apparent velocities. The envelope of ground roll builds up and decays slowly and often ground roll includes many cycles. Surface-wave energy generally is high enough even in the reflection band to override all but the strongest reflections; however, because of the low velocity, different geophone groups are affected at different times so that only a few groups are affected at any one time. Sometimes there is more than one ground-roll wavetrain, each with a different velocity. Occasionally, where surface waves are exceptionally strong, in-line offsets are used to permit recording desired reflections before the surface waves reach the spread.

Surface-wave effects can be attenuated by the use of arrays (§8.3.5 to 8.3.9 and problem 8.6), by frequency filtering (ground roll can be seen on the 0–6-Hz and slightly on the 6–12-Hz panels of fig. 9.20), and by apparent-velocity filtering (see fig. 9.38).

6.4 Resolution

6.4.1 General

Resolution refers to the minimum separation between two features such that we can tell that there are two features rather than only one. With respect to seismic waves, we may think of (a) how far apart (in space or time) two interfaces must be to show as separate reflectors (*vertical resolution*) or (b) how far apart two features involving a single interface must be separated to show as separate features (*horizontal resolution*). (The word "resolution" is often used loosely to denote the ability to tell that a feature is present.)

Clearly, the ability to see and distinguish features depends on the signal/noise ratio and the knowledge

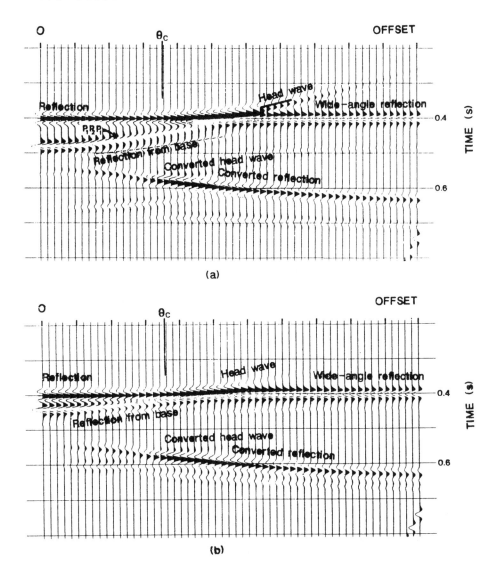

Fig. 6.38 Effect of refractor thickness on head wave. Head wave loses amplitude because of destructive interference with reflection from base of refractor when the refractor is thin. Normal moveout for reflection at top of refractor has been removed so that the head wave and the reflection from the bottom of the refractor curve upwards. (Courtesy of Geophysical Development Co.) (a) Refractor 1.5 wavelengths thick, and (b) refractor 3/4 wavelength thick.

and experience of the interpreter. Where a correct model is used for interpretation, it is possible to exceed conventional resolution limits, that is, if we know a priori exactly what we are looking for in very good data, then subtle differences can be used to locate and identify it.

If seismic wavelets were extremely sharp, resolution would not be a problem. However, real seismic wavelets involve a limited range of frequencies and hence have appreciable breadth (see §6.6.1).

6.4.2 Vertical resolution

Let us first consider resolution in the vertical direction. For two horizontal reflectors a distance Δz apart, the deeper reflection lags behind the shallower by the fraction $2\,\Delta z/\lambda$ of a wavelength. We can tell that there

are two waves when the arrival of the second wave causes a perceptible change in the appearance of the first wave.

Rayleigh (Jenkins and White, 1957: 300) defined the *resolvable (separable)* limit as being when the two events are separated by a half-cycle so interference effects are minimized. Ricker (1953b) and Widess (1973) used slightly different criteria, which resulted in slightly smaller resolvable limits. Kalweit and Wood (1982) discuss resolution criteria.

For a boxcar frequency spectrum (see eq. (15.123)), the wavelet shape is that of a sinc function. The Rayleigh criterion is equivalent to a width of approximately $2/3\nu_u$, where ν_u is the upper frequency limit of the boxcar (see problem 6.18). Thus, we must record higher frequencies if we are to achieve higher resolution (Sheriff, 1977).

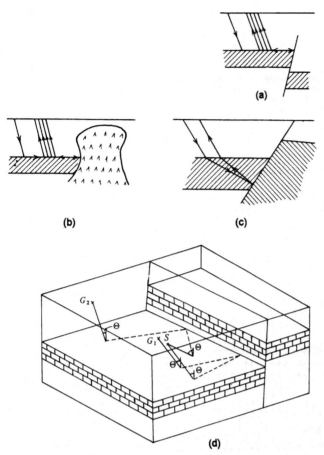

Fig. 6.39 Reflected refractions. (a–c) Refractions reflected from faults or salt domes. (d) Isometric drawing of refractions reflected from the termination of the refractor to the side of the spread; paths are shown from source S to geophones G_1 and G_2; dashed lines indicate head-wave travel in the refractor.

As an illustration of vertical resolution, fig. 6.40b shows the effect of a wedge whose velocity is intermediate between that above and below it. The waveshape clearly indicates more than one reflector when the wedge thickness exceeds $\lambda/4$ (12 ms). Figure 6.40c shows a wedge with a velocity different from that of the surrounding material. The waveshape is nearly constant below a thickness of $\lambda/4$, where the amplitude is at a maximum because of constructive interference (tuning; see §6.4.3). Note that the wedge still produces a significant reflection when it is appreciably thinner than the resolvable limit, and a bed only $\lambda/20$ to $\lambda/30$ in thickness may be detectable although its thickness cannot be determined from the waveshape. For wedge thickness less than $\lambda/4$, the waveshape is the derivative of that for a single interface (see §15.1.7).

Similar resolution considerations apply to structural features. Figure 6.41 shows a series of faults with varying amounts of throw, the fault being quite evident when the throw is $\lambda/4$ or larger. Obviously, the ability to resolve depends on other factors also, such as the signal/noise level and the experience of the interpreter in similar studies. These three examples suggest that the Rayleigh definition of resolvable limit is reasonable.

6.4.3 Tuning and thin-bed effects

When a bed embedded in a medium of different properties is 1/4 wavelength ($\lambda/4$) in thickness, the reflections from the top and base of the bed interfere constructively (as in fig. 6.40c) and the amplitude increases, an effect called *tuning*. Tuning is important in the analysis of hydrocarbon reservoirs and other thin-bed interpretation situations. A *thin bed* is defined as a situation where the aggregate thickness of beds under consideration is less than $\lambda/4$.

Figure 6.42 shows timing and amplitude measurements for the thining-wedge situations shown in fig. 6.40. Where $V_3 > V_2 > V_1$, as in fig. 6.42a, the destructive interference at $\lambda/4$ produces an amplitude minimum. For larger thicknesses, the amplitude gives approximately the correct reflectivity of the interfaces although interference with successive sidelobes of the wavelet causes some oscillation. Peak-to-peak time measurements for thicknesses greater than $\lambda/4$ give approximately the correct thicknesses although the successive sidelobes produce minor errors.

The more common situation encountered in reservoir geophysics (§14.4) is shown in fig. 6.42b, where $V_3 = V_1 \neq V_2$. Then constructive interference at 1/4 wavelength produces a tuning amplitude maximum. Trough-to-peak time measurements give approximately the correct gross thicknesses for thicknesses greater than $\lambda/4$ (although the successive sidelobes produce minor errors) but no information for thicknesses less than $\lambda/4$. Thin-bed thickness information can be obtained from amplitude measurements below $3\lambda/16$ thickness. The amplitude–thickness graph is nearly linear below about $\lambda/8$, but the amplitude is relatively insensitive to thickness in the tuning vicinity.

If one uses zero-phase wavelets and maps the peak and trough that indicate the top and base of the wedge in fig. 6.42b, the arrival times give correct values in the thick-bed situation but not in the thin-bed region. The observed peak and trough can come no closer together than $\lambda/4$, so that for a thin bed, they effectively push each other apart, giving arrival times that are too early for the top and too late for the bottom of the bed. This observation is important in reservoir geophysics studies.

Meckel and Nath (1977) calculated that, for sands embedded in shale, the amplitude would depend on the net sand present provided that the thickness of the entire sequence is less than $\lambda/4$. Mahradi (1983) verified this using physical models (fig. 6.43). For gross thicknesses less than $\lambda/4$, waveshapes are the same and amplitudes (fig. 6.43f) lie on the same curve as in fig. 6.42b, whereas for gross thicknesses greater than $\lambda/4$, waveshapes change and amplitudes no longer lie on this curve. (Note: The measurements for fig. 6.43f

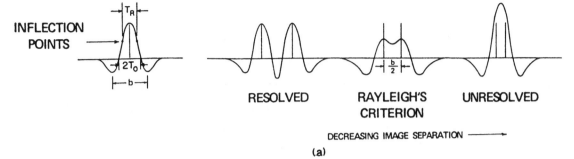

RESOLVED   RAYLEIGH'S   UNRESOLVED
CRITERION

DECREASING IMAGE SEPARATION ⟶

(a)

TWO – WAY LAYER THICKNESS (MILLISECONDS)

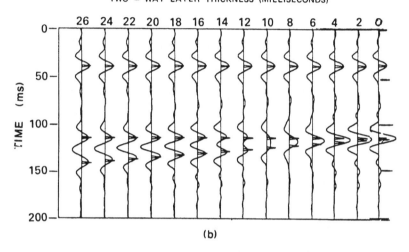

(b)

TWO – WAY LAYER THICKNESS (MILLISECONDS)

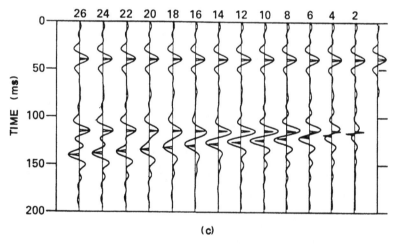

(c)

Fig. 6.40 Reflections illustrating vertical resolution. Zero-phase sinc wavelets; thickness of $\lambda/4$ corresponds to 12 ms. (After Kalweit and Wood, 1982: 1038–9.) (a) Illustrating resolution, (b) reflections from single interface (upper reflection) and wedge of intermediate velocity ($V_3 > V_2 > V_1$ or $V_3 < V_2 < V_1$), and (c) the same as part (b) except the wedge is embedded in a medium of different velocity ($V_3 = V_1 \neq V_2$).

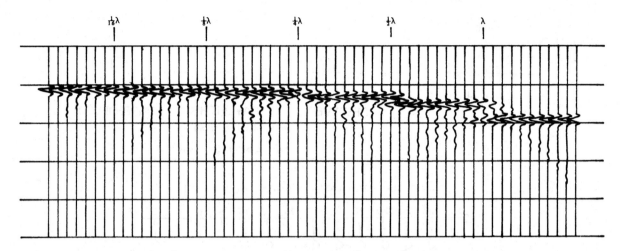

Fig. 6.41 Reflection from a faulted reflector, with the fault throw indicated as fractions of the dominant wavelength.

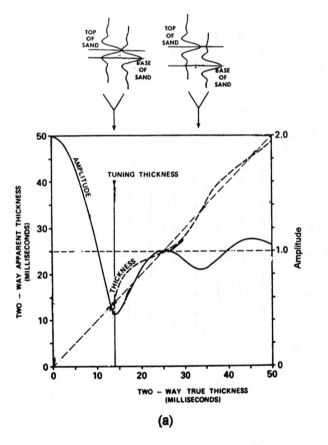

(a)

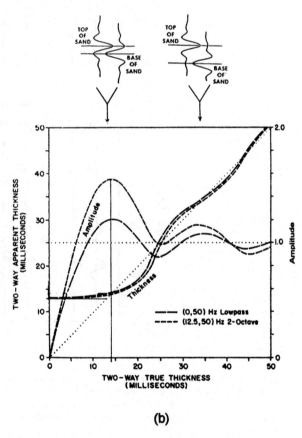

(b)

Fig. 6.42 Amplitude and timing measurements for wedges shown in fig. 6.40. The interference mechanism is shown above the diagrams. Zero-phase sinc wavelets; the horizontal dashed (dotted) lines indicate the amplitude and traveltime as if inter- ference is not involved. (After Kalweit and Wood, 1982: 1043.) (a) Case where $V_3 > V_2 > V_1$ and (b) where $V_3 = V_1 \neq V_2$.

were made at the center of each portion in figs. 6.43a to 6.43e to avoid distortions because of diffractions from the discontinuities.)

6.4.4 Horizontal resolution

The Fresnel zone (§6.2.3) is often taken as limiting horizontal resolution on unmigrated seismic data although other factors such as signal/noise ratio, trace spacing (sampling), three-dimensional effects, and so on, also affect how far apart features have to be to be distinguished as separate features. Note in fig. 6.15 that there is little evidence of reflector shape (that is, that the reflectors are flat) when they are less than one Fresnel zone wide.

Resolution on migrated sections is difficult to quantify because it depends on many factors, especially the presence of noise. Migration (§9.13) can be thought of as collapsing the Fresnel zones, and hence the Fresnel-zone size cannot be used as a criterion for horizontal resolution on migrated sections. Ordinary migration collapses the Fresnel zone only in the direction of the migration so that (unless three-dimensional migration is performed on 3-D data) correction is not made for contributions perpendicular to the line. One of the most important factors is the quality of the unmigrated section; migration rearranges the noise as well as reflections, creating what is sometimes called migration noise.

Actual migration is performed on sampled data (sampled spatially, that is, at discrete geophone locations, as well as at discrete time intervals). Spatial aliasing considerations (§9.1.2b; see also fig. 6.2) limit the angle of approach, which in turn limits the amount of dip that can be migrated.

The sampling theorem dictates that at least two samples per apparent wavelength must be obtained in order to recognize features, even with perfect data. Thus, for example, to recognize a stream channel on a horizon slice (fig. 12.16) generally requires bin sizes no larger than 1/3 or 1/4 the channel width.

Horizontal uncertainty always exceeds vertical uncertainty, often by a factor of at least 2. Schneider (1978) gives an example showing that 5% velocity error smears the position of a discontinuity over a horizontal distance equal to 5% of the depth; local velocities are usually not known better than this.

6.5 Attenuation

6.5.1 Attenuation mechanisms

The amplitudes of events on a seismic record depend upon a multitude of factors (fig. 6.44). Some of these factors (for example, recording/processing) are within our control. The effects of others can be estimated and then compensated for. Still other factors affect data with about the same traveltimes in about the same way and thus do not introduce significant trace-to-trace differences, the main factor on which interpretational decisions are based.

Divergence is usually the major factor causing time-dependent amplitude changes (see §2.7.3). The energy spreads out so that the wave decreases in strength but the total energy in the wavefield does not change. If the medium were homogeneous, the amplitude weakening would be inversely proportional to distance, or Vt; however, because velocity generally increases with depth, raypath curvature makes the wave spread out more and thus makes the decrease in amplitude larger. Newman (1973) showed that, for parallel layering, the amplitude decrease depends approximately on $1/V_{rms}^2 t$, and Hardage (1985) showed that this factor is appropriate for observed data (fig. 6.45).

Absorption (§2.7.2) causes wave energy to disappear by converting it to heat. However, like dispersion, most of the factors affecting the amplitude of waves as they travel through the earth (partitioning at interfaces [chap. 3], interference with other waves such as peg-leg multiples [§6.3.2b], and diffraction or scattering) redistribute the wave energy rather than cause it to disappear. Sometimes compensation for these various factors is approximated by multiplying by an empirical exponential factor.

In general, seismic amplitude decreases exponentially with time, as shown in fig. 2.25. Higher frequencies are attenuated more than lower frequencies so that the spectrum of a seismic wavelet changes with time (fig. 6.46). Hauge (1981) studied cumulative attenuation in a largely clastic section (fig. 6.47) for VSP data. Spencer (1985) concludes that attenuation measurements are not promising as a diagnostic of lithology because of the intrinsic scatter produced by peg-leg multiple interference.

Unlike most of the effects in fig. 6.44, which are generally understood, the basic mechanisms by which elastic-wave energy is transformed into heat are not clearly understood. Toksoz and Johnston (1981) summarized the state of our knowledge about attenuation and absorption. Various absorption mechanisms have been proposed (White, 1965, 1966) but none appears adequate. Internal friction in the form of sliding friction (or sticking and sliding) and viscous losses in the interstitial fluids are probably the most important mechanisms, the latter being more important in high-permeability rocks. Other effects, probably of minor significance in general, are the loss when part of the heat generated during the compressive part of the wave is conducted away, piezoelectric and thermoelectric effects, and the energy used to create new surfaces (of importance only near the source). Many of the postulated mechanisms predict that, in solids, Q should depend upon frequency; however, Q appears to be independent of frequency (that is, η is directly proportional to frequency; see eq. (2.117)). In liquids, Q is inversely proportional to frequency. The loss mechanism in rocks must be regarded as an unsolved problem (Aki and Richards, 1980: 156–7, 169–70).

Often, no distinction is made between "attenuation" and "absorption." Because of difficulties in measuring absorption and also because the quantity of in-

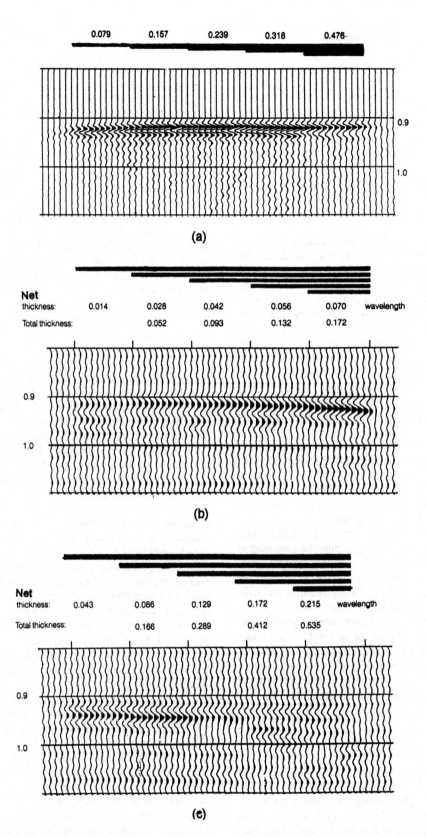

Fig. 6.43 Reflections from interbedded lithologies. Net and gross thicknesses are given in terms of the dominant wavelength. (From Sheriff, 1985, after Mahradi, 1983.) (a) Reflections from plates of varying thicknesses measured as fractions of the dominant wavelength. (b and c) Reflections where lithologies alternate. (d and e) Reflections from beds of different thicknesses. (f) Graph of amplitudes versus net thicknesses, with asterisks indicating points for which gross thicknesses are greater than λ/4; ▲, data from part (a); ● (b); ■ (c); ⊠ (d); ∇ (e); ○, Δ, ⊐, ★ from other models.

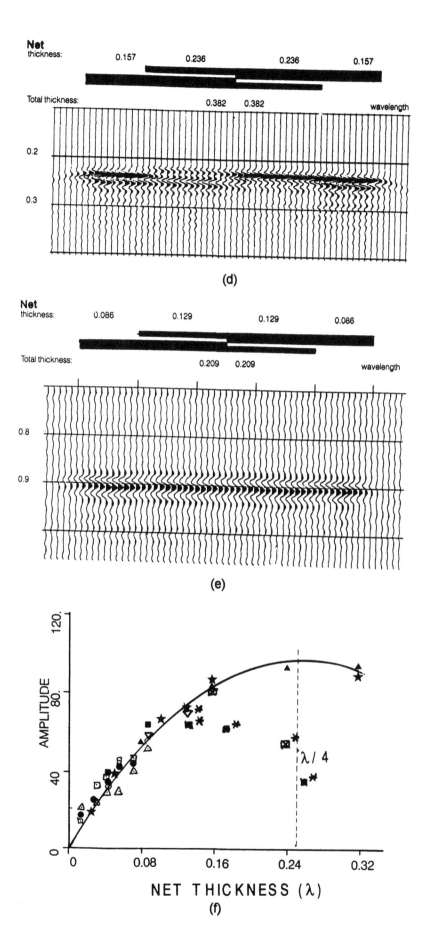

(d)

(e)

(f)

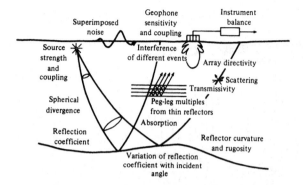

Fig. 6.44 Factors affecting amplitude. (From Sheriff, 1975.)

Table 6.1 *Absorption constants for rocks*

	Q	δ (dB) $= \eta\lambda$
Sedimentary rocks	20–200	0.16–0.02
Sandstone	70–130	0.04–0.02
Shale	20–70	0.16–0.05
Limestone	50–200	0.06–0.02
Chalk	135	0.02
Dolomite	190	0.02
Rocks with gas in pore space	5–50	0.63–0.06
Metamorphic rocks	200–400	0.02–0.01
Igneous rocks	75–300	0.04–0.01

terest is usually the net decrease in wave amplitude, measurements are often made of total attenuation without regard to its cause and the results used to determine a value of η in eq. (2.110) (see fig. 2.25). Although this may be a useful method of treating attenuation, it has no proper mathematical basis because the attenuation due to partitioning, peg-leg multiples, and so on is not a continuous function of distance, as required by eq. (2.110).

6.5.2 Absorption measurements

Attenuation is due to both absorption and a number of more or less predictable factors as described in §6.5.1. In the laboratory, measurements are usually made of absorption because the other factors can be calculated (at least approximately); however, laboratory measurements are invariably made at high frequencies (because of scaling requirements for the model; see Telford, Geldart, and Sheriff, 1990: 384) and so have doubtful significance under actual field conditions (because absorption η increases linearly with frequency; see what follows).

In field measurements of absorption, the effects of partitioning and other significant factors must be allowed for to obtain meaningful absorption values. Difficulties in achieving this have resulted in wide divergence in absorption measurements. Measurements of absorption have been summarized by Attewell and Ramana (1966), Bradley and Fort (1966), and Toksoz and Johnston (1981).

Attewell and Ramana found a best-fit value of $\eta = 0.2$ dB/km for the average of values from 26 authors, and Waters (1987: 33) gives a fairly extensive table of Q values. The ranges of values are summarized in table 6.1. Q values for S-waves appear to be one-half to one-third those for P-waves. Tullos and Reid (1969) report measurements in the first 3 m of Gulf Coast sediments of $\eta = 13$ dB/wavelength ($Q = 0.24$) but 0.15 to 0.36 dB/wavelength for the next 300 m ($Q = 20$ to 9). Often-quoted measurements in the Pierre shale by McDonal et al. (1958) were $\eta = 0.39$ dB/km for P-waves ($Q = 3.5$, $\delta = 0.9$) and $\eta = 3.3$ dB/km for S-waves; the Pierre shale is a massive formation in Colorado about 1200 m thick with a P-wave velocity

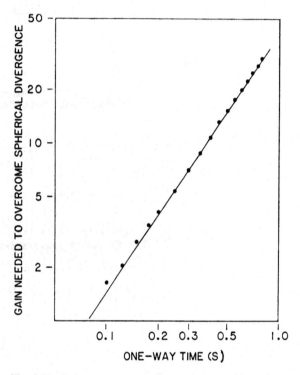

Fig. 6.45 Gain needed to overcome spherical divergence for VSP data. The slope of the line on the log–log plot is nearly $\sqrt{2}$. (From Hardage, 1985: 173.)

of 2330 m/s. Q is generally independent of amplitude for strains less than 10^{-4}, which covers virtually all seismic-wave situations.

Experimental evidence suggests that the absorption coefficient η is approximately proportional to frequency, that is, $\eta\lambda$ is roughly constant for a particular rock. Such an increase of absorption with frequency (§6.5.1) provides one mechanism for the observed loss of high frequencies and the change of waveshape with distance. Peg-leg multiples (§6.3.2b) and possibly other phenomena also produce waveshape changes. In interbedded sections, the loss in amplitude because of peg-leg multiple effects (fig. 6.30d) appears to be comparable to that due to absorption.

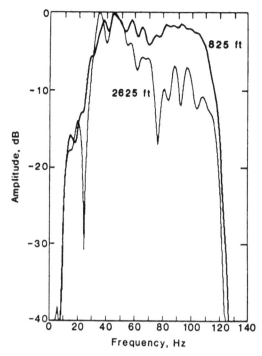

Fig. 6.46 Change in wavelet spectrum from a VSP study. Both curves are normalized with respect to the highest amplitude. (From Balch and Lee, 1984: 16.)

6.6 Shape of the seismic wavelet

6.6.1 Desired wavelet characteristics

An interpreter would like to have seismic sections show simple one-to-one relations to interfaces in the earth and as much detail as possible (maximum resolution), that is, sections where bedding contacts are sharply imaged at their correct locations, with no noise to confuse matters. To achieve short, sharp events requires a broad spectrum with good high-frequency content. To show it at the correct location requires migration and knowledge of the embedded waveshape. To show the contrasts at interfaces, amplitude values must be faithfully preserved.

If we think of a seismic wavelet as resulting from the superposition of many harmonic waves of different frequencies and amplitudes (Fourier synthesis concept), we see that cosine waves with zero phase shift will have maximum constructive interference at $t = 0$, thus producing the maximum possible amplitude there. At certain other values of t, the waves will add up to give smaller peak amplitudes but the broader the band of frequencies included, the farther one has to go from $t = 0$ for these to achieve appreciable amplitude. Higher frequencies in the bandwidth are also necessary to produce a sharp peak. Thus, the desired waveshape is best achieved with a narrow zero-phase wavelet (fig. 6.48a) with minimal sidelobes to interfere with other events.

Figure 6.48 shows how the waveshape changes with the bandwidth characteristics. Note the increased

magnitude of the central peak compared to any other half cycle and the increased sharpness of the central peak as the bandwidth widens (figs. 6.48a to 6.48e). Wavelets do not change very much as bandwidths increase beyond about 2.5 octaves. Wavelets become leggy as the bandwidth slopes become steeper (figs. 6.48f to 6.48h). Two wavelets having the same spectral shape and number of octaves bandwidth but whose spectra are displaced from each other along the frequency scale have the same waveshapes (except for time scaling); the one with the lower frequencies is simply broader in the time domain. Wavelets having the same spectral shape and bandwidth measured in hertz rather than octaves have the same envelope but differing number of cycles within the envelope (figs. 6.48i and 6.48j). In acquisition, we try to achieve higher frequencies and broader bandwidths, but absorption and other mechanisms usually limit energy above about 60 Hz.

Most of the natural mechanisms that affect the shape of real wavelets (§9.2.3) are minimum-phase or nearly so (see Sherwood and Trorey, 1965). A minimum-phase wavelet (§15.5.6a) is *causal* (that is, it is zero for negative times) and has the energy concentrated in the early part of the wavelet. Real wavelets are also causal and the first detectable peak or trough is always delayed from the onset of the wavelet so that the picking and timing of arrival times are always late. Furthermore, as arrival times increase, the increased attenuation of the higher frequencies causes the spectrum to shift toward the low frequencies, so wavelets build up more slowly, and the delays between reflection onsets and their detection increase. Correct compensation for delays is very difficult to achieve.

The embedded wavelet (§9.2.3) after processing is sometimes approximately minimum-phase, but often has a nearly constant-phase spectrum. Most displays in 1994 attempt to achieve zero-phase wavelets (whose phase spectra are identically zero and that are not causal). Antisymmetric wavelets (whose phase spectra are identically 90°; see fig. 6.48b) are also encountered frequently.

The SEG *standard polarity* convention (fig. 6.49) for minimum-phase wavelets is that, for a positive reflection (a reflection from an interface where the acoustic impedance increases), the waveform begins with a downkick, represented by negative numbers; this has a historical basis and is almost universally agreed to. For a zero-phase positive reflection, the wavelet central point of symmetry is a peak represented by positive numbers; a minority use the opposite convention. Displays sometimes show the opposite of the foregoing (*SEG negative polarity* or *reverse polarity*).

6.6.2 Ricker wavelet

The embedded wavelet (§9.2.3) is often converted to a zero-phase equivalent in processing (§9.5.9 and 15.5.6d). The embedded wavelet is made symmetrical and the time scale is shifted (but not always correctly)

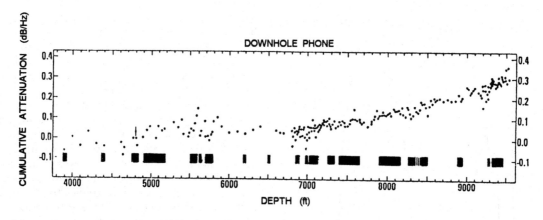

Fig. 6.47 Cumulative attenuation as a function of depth. Silt-sand intervals are shaded. (From Hauge, 1981: 1555.)

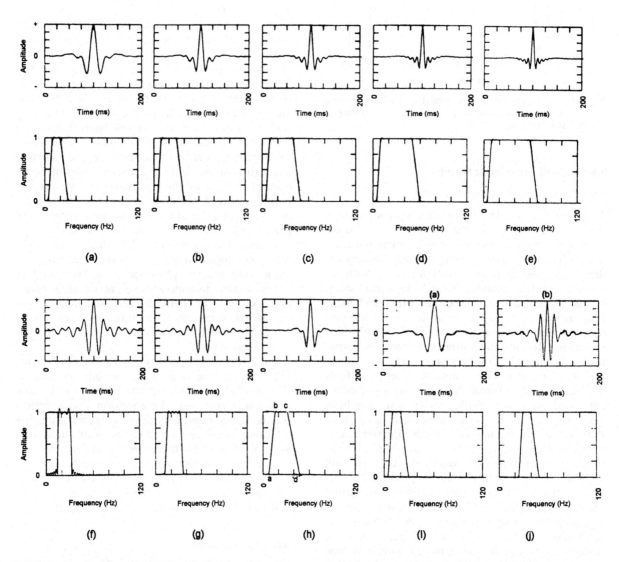

Fig. 6.48 Impulses filtered with various bandpasses. (After Yilmaz, 1987: 23–4.) (a to e) Changing bandwidth by increasing the high-frequency cutoff; bandwidths are approximately 1, 1.5, 2, 2.3, and 2.6 octaves. (f to h) Changing filter slopes; slopes are approximately 120, 60, and 24 dB/octave. (i and j) Shifting bandwidth containing the same frequency interval to higher frequencies.

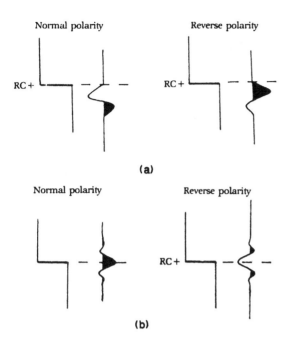

Fig. 6.49 Standard polarity. (a) For a positive reflection, a minimum-phase wavelet begins with a downkick, and (b) the center of a zero-phase wavelet is a peak.

so that the wavelet center indicates the arrival time. Conversion to a zero-phase equivalent does not solve problems with time-variant effects.

The most common zero-phase wavelet is the *Ricker wavelet* (Ricker, 1940, 1944, 1953a), expressed in the time domain (fig. 6.50a) as

$$f(t) = (1 - 2\pi^2 v_M^2 t^2)e^{-(\pi v_M t)^2}, \qquad (6.16)$$

or in the frequency domain (fig. 6.50b) as

$$F(v) = (2/\sqrt{\pi})(v^2/v_M^2)e^{-(v/v_M)^2}, \quad \gamma(v) = 0, \qquad (6.17)$$

where $f(t) \leftrightarrow F(v)$, and v_M is the peak frequency (see problem 6.21). The distance between flanking side lobes in the time domain, T_D (fig. 6.50a), is

$$T_D = (\sqrt{6}/\pi)/v_M.$$

Also, $T_R = T_D/\sqrt{3}$.

6.7 Noise

6.7.1 Types of seismic noise

The reliability of seismic mapping is strongly dependent upon the quality of the records. The quality of seismic data varies tremendously from areas where excellent reflections (or refractions) are obtained to areas in which the most modern equipment, complex field techniques, and sophisticated data processing do not yield usable data (often called *NR areas*, that is, areas of "no reflections"). In between these extremes lie most areas in which useful results are obtained, but

the quantity and quality of the data could be improved with beneficial results.

We use the term *signal* to denote any event on the seismic record from which we wish to obtain information. Everything else is *noise*, including coherent events that interfere with the observation and measurement of signals. The *signal-to-noise ratio*, abbreviated *S/N*, is the ratio of the signal in a specified portion of the record to the total noise in the same portion. Poor records result whenever the signal-to-noise ratio is small; just how small is to some extent a subjective judgment. Nevertheless, when *S/N* is less than unity, the record quality is usually marginal and deteriorates rapidly as the ratio decreases further.

Seismic noise may be either (a) coherent or (b) incoherent. *Coherent noise* can be followed across at least a few traces; *incoherent noise* is dissimilar on all traces, and we cannot predict what a trace will be like from a knowledge of nearby traces. The difference between coherent and incoherent noise is often a matter of scale and if we had geophones more closely spaced incoherent noise would be seen as coherent. Nevertheless, incoherent noise is defined with respect to the records being used without regard for what closer spacing might reveal.

Incoherent noise is often referred to as *random noise* (spatially random), which implies not only nonpredictability but also certain statistical properties; more often than not the noise is not truly random. (It should be noted that spatial randomness and time randomness may be independent; the usual seismic trace is apt to be random in time because we do not know when a reflection will occur on the basis of what the trace has shown previously, with the exception of multiples.

Coherent noise is sometimes subdivided into (a) energy that travels essentially horizontally, and (b) energy that reaches the spread more or less vertically. Another important distinction is between (a) noise that is repeatable, and (b) noise that is not; in other words, whether the same noise is observed at the same time on the same trace when the source is repeated. The three properties – coherence, travel direction, and repeatability – form the basis of most methods of improving record quality.

Coherent noise includes surface waves, reflections, or reflected refractions from near-surface structures such as fault planes or buried stream channels, refractions carried by high-velocity stringers, multiples, and so on (Olhovich, 1964). All of the preceding except multiples travel essentially horizontally and all are repeatable on successive source activations.

Incoherent noise, which is spatially random and also repeatable, is due to scattering from near-surface irregularities and inhomogeneities such as boulders and small-scale faulting; such noise sources are so small and so near the spread that the outputs of two geophones will only be the same when the geophones are placed almost side by side. Nonrepeatable random noise may be due to wind shaking a geophone or caus-

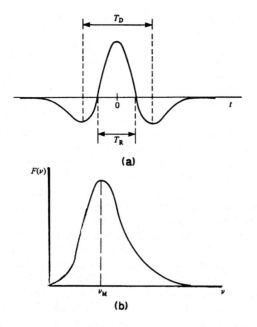

Fig. 6.50 Ricker wavelet. (a) Time-domain representation and (b) frequency-domain representation.

ing the roots of trees to move, generating seismic waves, stones ejected by a shot and falling back to the earth near a geophone, a person walking near a geophone, and so on.

6.7.2 Stacking to attenuate random noise

If we add several random noises together, there will be some cancellation because some will be out of phase with others. Assume that we have n geophones, each of which is responding to coherent signal S but has random noise N_i superimposed on it. A measurement x_i will then be

$$x_i = S + N_i.$$

The average is our best estimate of the signal and we identify the standard deviation σ with the noise, so that

$$S \approx \bar{x} = \frac{1}{n}\sum x_i, \quad N \approx \sigma,$$

$$\sigma^2 = \frac{1}{n}\sum(x_i - \bar{x})^2 = \frac{1}{n}\sum N_j^2.$$

The signal-to-noise ratio, S/N, is thus

$$\frac{S}{N} \approx \frac{\bar{x}}{\sigma} = \frac{\bar{x}}{(1/n^{1/2})(\sum N_i^2)^{1/2}} = \frac{n^{1/2}\bar{x}}{(\sum N_i^2)^{1/2}}. \quad (6.18)$$

As n becomes large, σ approaches a limit that depends on the statistical properties of the noise; hence, for random noise, the signal-to-noise ratio varies as $n^{1/2}$ for n large.

Summing a number of identical traces where there are small random timing differences among them is

equivalent to frequency filtering (fig. 6.51). In this operation, low-frequency components mainly interfere constructively, whereas high-frequency components tend to interfere destructively. This type of summing is apt to occur in the ground mixing of geophones within arrays or sources in source arrays, but it also occurs in vertical stacking and other types of stacking in processing.

6.7.3 Methods of attenuating noise

Because there are many types of noise, various noise-attenuating methods are employed. All are based on differences between properties of the noise and of the signal. Inasmuch as the nature of "signal" is somewhat subjective and the properties of both signal and noise are not completely known, noise attenuation cannot be completely objective.

Noise attenuation begins with the field recording. To the extent that noise has appreciable energy outside the principal frequency range of the signal, it can be attenuated by limiting the frequencies recorded. Very low-frequency components (such as high-energy surface waves rich in low frequencies) may be filtered out during the initial recording provided the low frequencies are sufficiently separated from the reflection frequencies. However, if the spectrum of the noise overlaps the signal spectrum, then frequency filtering is of limited value in improving record quality. The dynamic range of field instruments today is usually sufficiently wide that often the only low-frequency filtering used in the field is that resulting from the limited low-frequency response of the geophones. Likewise, often the only high-frequency filtering employed is that required to prevent aliasing in digitizing.

Cancellation of random noise does not place any restrictions on geophone locations (except that they cannot be so close together that the noise is no longer spatially random). If we connect together, for example, 16 geophones that are spaced far enough apart that the noise is spatially random but still close enough together that reflected energy traveling almost vertically is essentially in phase at all 16 geophones, the sum of the 16 outputs will have a signal-to-noise ratio four times greater than the output when the geophones are placed side by side. If, on the other hand, we are attenuating coherent noise and the 16 geophones are spread evenly over one wavelength of a coherent-noise wavetrain (for example, ground roll), then the coherent noise will be greatly reduced (see §8.3.6 and problem 8.6b). Similar considerations apply to the use of arrays of multiple sources.

The contribution of noise coming from the side of the line has generally been underestimated. We cannot deal properly with data arriving from off to the side of the line unless we record data to the side of the line. Areal arrays (§8.3.8) are sometimes used effectively for attenuating high-angle off-the-line noises. Major

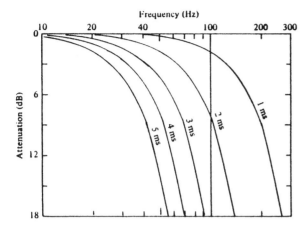

Fig. 6.51 Filter effect of timing errors in stacking. The numbers on the curves are standard deviations of the timing differences among the traces stacked.

noise attenuation results from 3-D recording and processing.

Noise can also be attenuated by adding together traces recorded at different times or different places or both. This forms the basis of several stacking techniques, including vertical stacking, common-midpoint stacking, uphole stacking, and several more complicated methods. The gain in record quality often is large because of a reduction in the level of both random and coherent noise. Provided the static and NMO corrections are accurately made, signal-to-noise improvements for random noise should be about 5 (or 14 dB) for 24-fold stacking.

Vertical stacking involves combining together several records for which both the source and geophone locations remain the same. It is extensively used with weak surface energy sources and many marine sources (see §7.2.4 and 7.4). Vertical stacking usually implies that no trace-to-trace corrections are applied, but that corresponding traces on separate records are merely added to each other. The effect, therefore, is essentially the same as using multiple sources simultaneously. In difficult areas, both multiple sources and vertical stacking may be used. In actual practice, the surface source is moved somewhat (3 to 10 m) between successive recordings. Up to 20 or more separate records may be vertically stacked, but the stacking of many records becomes expensive both in field time and in processing, whereas the incremental improvement becomes small after the first few. Vertical stacking is often done in the field, sometimes in subsequent processing. Marine vertical stacking rarely involves more than four records because, at normal ship speeds, the ship moves so far during the recording that the data are *smeared* when stacked; smearing means that the changes in the reflecting points affect the arrival times so much that the signal may be adversely affected by summing (the effect is similar to using a very large geophone or source array).

The common-midpoint method that is almost universally used is very effective in attenuating several kinds of noise. The summation traces comprise energy from several sources using different geophone and source locations. The field technique will be discussed in §8.3.3 and the processing (which is usually done in a processing center rather than in the field) in §9.10.4.

A number of other noise-attenuating techniques (such as apparent-velocity filtering) are also applied in processing and described in chap. 9. In fact, most of the operations done in seismic processing have the attenuation of noise as their principal objective. Their application has the advantage of trial and error and subjective judgment is usually a factor in deciding which processes to employ and which parameters to vary.

Problems

6.1 In table 6.2, classify different types of events and noise on the basis of commonly observed characteristics.

6.2 A salt dome is roughly a vertical circular cylinder with a flat top of radius 400 m at a depth of 3.2 km. If the average velocity above the top is 3.8 km/s, what is the minimum frequency that will give a recognizable reflection from the dome?

6.3 (a) Use Fermat's principle of least time to derive the law of reflection (§2.7.5). (*Hint:* Express the travel-time for the reflection SMR in fig. 6.52 in terms of the variable x, then set dt/dx equal to zero.)
(b) Repeat part (a) for the refracted path SMQ.
(c) Repeat parts (a) and (b) for reflected and refracted converted S-waves, thus verifying eq. (3.1).

6.4 Redraw fig. 6.16b for a plane wave incident on the reflector, and explain the significance of the changes that this makes.

6.5 (a) Show that the slope of the diffraction event with source S_2 in fig. 6.23b approaches $\pm 1/V$ for large x. (*Hint:* Expand the expression in eq. (6.11) for $x \gg h$.)
(b) What is the slope of the asymptote for fig. 6.23d?

6.6 Assume that fig. 6.34 shows relative amplitudes correctly (divergence having been allowed for). The water depth is 420 m and the velocity below the sea floor 2590 m/s.
(a) If the reflection coefficient is maximum at the critical angle, on what traces would you expect the maximum amplitude for the first, second, third and fourth multiples?
(b) What should be the ratio of the amplitude of the successive multiples on the short-offset trace? How do these calculations compare with observations? What unaccounted-for factors affect this comparison?

6.7 (a) Given that $0 < c < +1$ in eq. (6.15) for the directivity resulting from ghosting, discuss the conditions under which the amplitude of ψ_p is zero.
(b) For a source below the base of the low-velocity layer, compare the amplitude and energy of ghosts generated at the base of the low-velocity layer and at

Table 6.2 *Characteristics of events*

	Incoherent (spacing > 2m)	Predictable trace-to-trace	Predictable from earlier arrivals	Repeatable on successive shots	Low apparent velocity (<25 km/s)	Distinctive apparent velocity	Moveout linear with offset	Attenuated by frequency filtering	Attenuated by geophone arrays	Attenuated by CDP stacking	Attenuated by apparent-velocity filtering	Attenuated by front-end muting	Distinguishable by 3-component recording
Primary reflections, dip <10°													
Multiples													
Primary reflections, dip >25°													
Diffractions													
Head waves													
Reflected refractions													
Ground roll													
Wind noise													
Airwave													
SV-waves (reflected)													
SH-waves (reflected)													

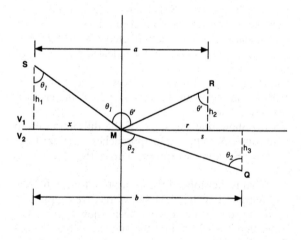

Fig. 6.52 Deriving Snell's law from Fermat's principle.

the surface of the ground, given that $V_H = 1.9$ km/s, $V_W = 0.40$ km/s, and that the densities just below and within the LVL are 2.0 and 1.6 g/cm³, respectively.
(c) Assume that the LVL is $\frac{1}{2}\lambda$ in thickness and that $\eta\lambda = 0.6$ dB for the LVL; now what are the ratios of the ghost amplitudes and energies?

6.8 An air gun is fired at a depth of 10 m. The waveform includes frequencies in the range 10–80 Hz, the amplitudes of the 10- and 80-Hz components being the same near the source. Compare the amplitudes of these components for the wavelet plus ghost at considerable distance from the source in the directions 0°,

30°, 60°, and 90° to the vertical.
6.9 Show that eq. (6.15) gives the directivity diagrams shown in fig. 6.53.
6.10 A multiple reflection is produced by a horizontal bed at a depth of 1.100 km, the average velocity being 2.95 km/s. A primary reflection from a depth of 3.250 km coincides with the multiple.
(a) By how much do arrival times differ at points 200, 400, 800, and 1000 m from the source?
(b) If the shallow bed dips 10°, how much do the arrival times at 400 and 800 m change? What is the apparent dip of the multiple?
6.11 A primary and a multiple each arrive at 0.600 s at $x = 0$; the stacking velocities for them are 1800 and 1500 m/s, respectively. Calculate and plot the residual NMO for offsets of 300 n, where n = 1, 2, . . . , after application of the NMO correction for the primary velocity. What is the shortest offset that will give good multiple suppression for a wavelet with a 50-ms dominant period?
6.12 Pautsch (1927) showed that a horizontal or vertical interface could give identical first-arrival curves (fig. 6.54). Add secondary arrivals and reflections to these diagrams to show how they can distinguish the two cases.
6.13 An unlabeled event can be seen in fig. 6.38b with the projected arrival at $x = 0$ at about 0.60 s; identify it.
6.14 Draw arrival-time curves for the five events shown in fig. 6.55.

(a)

(b)

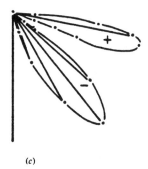
(c)

Fig. 6.53   Directivity of a harmonic source located at various depths below a free surface. (After Waters, 1987.) Depth/wavelength = 0.1, 0.5, and 1.0 in parts (a), (b), and (c), respectively.

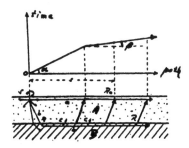

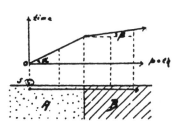

Fig. 6.54   Two models that give identical first-arrival curves. (After Pautsch, 1928.)

6.15 (a) A horizontal refractor at a depth of 1.20 km is being mapped along a N–S line. The overburden velocity is 2.50 km/s and the refractor velocity 4.00 km/s. The refractor is terminated by a linear vertical fault 3.50 km from the sourcepoint. Determine the traveltime curves when the fault strikes (i) E–W, (ii) N–S, (iii) N30°W.

(b) Repeat for the E–W fault for a refractor that dips 10° to the north with the sourcepoint to the south.

(c) What effects will the manner of terminating the refractor have, that is, how will the amplitude of the reflected refraction depend on the dip of the terminating fault?

(d) Most commonly a refractor will terminate against rock of lower acoustic impedance, but the opposite situation can also happen. What differences will this make?

(e) Extend the profile for case (a), part (i), an appreciable distance beyond the fault so as to plot the diffraction from the refractor termination. Assume uniform 2.50-km/s material beyond the refractor termination.

6.16 (a) Determine the traveltime curves for the refraction $SMNPQR$ and the refraction multiple $SMNTUWPQR$ in fig. 6.56.

(b) Determine the traveltime curves when both refractor and reflector dip 8° down to the left, the depths shown in fig. 6.56 now being the slant distances perpendicular to the interface at S.

(c) What happens when the reflector dips 3° to the left and the refractor 5° to the left?

6.17 Explain why waves in fig. 6.40b interfere destructively and in fig. 6.40c constructively when the wedge thickness is $\frac{1}{4}\lambda$.

6.18 (a) A wavelet has a flat frequency spectrum from 0 to v_u, above which the spectrum is zero; show that the Rayleigh criterion gives a resolvable limit t_r, where $t_r = 0.715/v_u$. (*Hint:* Transform a boxcar spectrum $box_{2v_u}(v)$ to the time domain and find the location of the first trough.)

(b) Show that the value of t_r for a wavelet with a flat spectrum extending from v_L to nv_L (that is, m octaves wide, where $n = 2^m$) is given by the solution to the equation

$$nx \cos nx - \sin nx - x \cos x + \sin x = 0,$$

where $x = 2\pi v_L t_r$.

(c) Solve the equation in part (b) for $m = 3, 2$, and 1.5 and compare the relation between t_r and m.

(d) Noting that part (a) involves an infinite number of octaves, how many octaves' bandwidth are required to give nearly the same resolution?

6.19 (a) Approximately what are the dominant frequencies for reflections in fig. 8.5 arriving at the right side of the section at about 0.6, 1.2, and 1.8 s?

(b) If the velocities at those reflectors are 2000, 3000, and 5000 m/s, respectively, what are the resolvable limits (tuning thicknesses)?

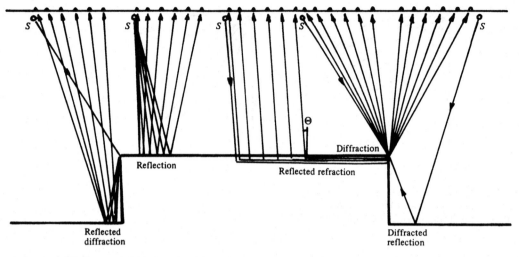

Fig. 6.55   Reflections and diffractions involving a horst.

6.20 Denham's high-frequency limit (§8.3.11) is related both to the loss of high frequencies and to the dynamic range (§7.6.1) of the recording system. Reconcile this limit with a loss by absorption of 0.15 dB/λ (§2.7.2b), spreading, and high-frequency loss because of peg-leg multiples as illustrated in fig. 6.30. Take 84 dB as the dynamic range of the recording system.

6.21 (a) Using the result $e^{-at2} \leftrightarrow (\pi/a)^{1/2}e^{-\omega^2/4a}$, verify that eq. (6.17) for a Ricker wavelet follows from eq. (6.16).

(b) Show that v_M is the peak of the frequency spectrum.

(c) Show that $T_D/T_R = \sqrt{3}$ (see fig. 6.50) and that $T_D v_M = \sqrt{6}/\pi$.

6.22 Select random numbers between ±9 to represent noise N_i and add to each a signal $S = 2$. Sum four values of $S + N_i$ and determine the mean, the standard deviation σ, and the ratio signal/(signal + noise). Repeat for 8, 16, and 32 values. Note how the mean converges toward S as the number of values increases, how σ approaches a limiting value (which depends on the statistical properties of the noise; see §15.2.12), and how the ratio signal/(signal + noise) converges toward 1. (A table of random numbers is given in app. C.)

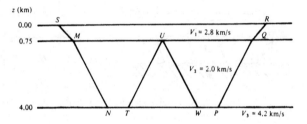

Fig. 6.56   Multiply-reflected refraction.

Technique, Applications, and Case Histories. Boston: International Human Resources Development Corp.

Bradley, J. J., and A. N. Fort. 1966. Internal friction in rocks. In *Handbook of Physical Constants,* S. P. Clark, ed., pp. 175–93, GSA Memoir 97. Boulder, CO: Geological Society of America.

d'Heur, M. 1992. West Ekofisk Field, Norway Central Graben, North Sea. *Bull. AAPG,* **75:** 946–68.

Hardage, B. A. 1985. *Vertical Seismic Profiling: A: Principles,* 2d ed. London: Geophysical Press.

Hauge, P. S. 1981. Measurements of attenuation from vertical seismic profiles. *Geophysics,* **46:** 1546.

Hilterman, F. J. 1970. Three-dimensional seismic modeling. *Geophysics,* **35:** 1020–37.

Hilterman, F. J. 1982. Interpretive lessons from three-dimensional modeling. *Geophysics,* **47:** 784–808.

Hubbard, R. J., J. Pape, and D. G. Roberts. 1985. Depositional sequence mapping as a technique to establish tectonic and stratigraphic framework and evaluate hydrocarbon potential on a passive continental margin. In *Seismic Stratigraphy II,* O. R. Berg and D. G. Woolverton, eds., pp. 79–91, AAPG Memoir 39. Tulsa: American Association of Petroleum Geologists.

Jenkins, F. A., and H. F. White. 1957. *Fundamentals of Optics.* New York: McGraw-Hill.

Kalweit, R. S., and L. C. Wood. 1982. The limits of resolution of zero-phase wavelets. *Geophysics,* **47:** 421–39.

McDonal, F. J., F. A. Angona, R. L. Mills, R. L. Sengbush,

References

Aki, K., and P. G. Richards. 1980. *Quantitative Seismology: Theory and Methods,* Vols. I and II. San Francisco: W. H. Freeman.

Angona, F. A. 1960. Two-dimensional modeling and its application to seismic problems. *Geophysics,* **25:** 468–82.

Attewell, P. B., and Y. V. Ramana. 1966. Wave attenuation and internal friction as functions of frequency in rocks. *Geophysics,* **31:** 1049–56.

Balch, A. H., and M. W. Lee. 1984. *Vertical Seismic Profiling:*

R. G. Van Nostrand, and J. E. White. 1958. Attenuation of shear and compressional waves in Pierre Shale. *Geophysics,* **23:** 421–39.

Mahradi, 1983. Physical modeling studies of thin beds. M.Sc. thesis, University of Houston.

Meckel, L. D., and A. K. Nath. 1977. Geologic considerations for stratigraphic modeling and interpretation. In *Seismic Stratigraphy – Applications to Hydrocarbon Exploration,* C. E. Payton, ed., pp. 417–38, AAPG Memoir 26. Tulsa: American Association of Petroleum Geologists.

Neidell, N. S., and F. Poggiagliolmi. 1977. Stratigraphic modeling and interpretation. In *Seismic Stratigraphy – Applications to Hydrocarbon Exploration,* C. E. Payton, ed., pp. 389–416, AAPG Memoir 26. Tulsa: American Association of Petroleum Geologists.

Newman, P. 1973. Divergence effects in a layered earth. *Geophysics,* **38:** 481–8.

O'Doherty, R. F., and N. A. Anstey. 1971. Reflections on amplitudes. *Geophys. Prosp.,* **19:** 430–58.

Olhovich, V. A. 1964. The causes of noise in seismic reflection and refraction work. *Geophysics,* **29:** 1015–30.

Pautsch, E. 1927. *Methods of Applied Geophysics.* Houston: Minor Printing Co.

Ricker, N. 1940. The form and nature of seismic waves and the structure of seismograms. *Geophysics,* **5:** 348–66.

Ricker, N. 1944. Wavelet functions and their polynomials. *Geophysics,* **9:** 314–23.

Ricker, N. 1953a. The form and laws of propagation of seismic wavelets. *Geophysics,* **18:** 10–40.

Ricker, N. 1953b. Wavelet contraction, wavelet expansion, and the control of seismic resolution. *Geophysics,* **18:** 769–92.

Schneider, W. A. 1978. Integral formulation for migration in two and three dimensions. *Geophysics,* **43:** 49–76.

Schoenberger, M., and F. K. Levin. 1978. Apparent attenuation due to intrabed multiples. *Geophysics,* **43:** 730–7.

Sheriff, R. E. 1975. Factors affecting seismic amplitudes. *Geoph. Prosp.,* **23:** 125–38.

Sheriff, R. E. 1976. Inferring stratigraphy from seismic data. *Bull. AAPG,* **60:** 528–42.

Sheriff, R. E. 1977. Limitations on resolution of seismic reflections and geologic detail derivable from them. In *Seismic Stratigraphy – Applications to Hydrocarbon Exploration,* C. E. Payton, ed., pp. 3–14, AAPG Memoir 26. Tulsa: American Association of Petroleum Geologists.

Sheriff, R. E. 1985. Aspects of seismic resolution. In *Seismic Stratigraphy II,* O. R. Berg and D. G. Woolverton, eds., pp. 1–10, AAPG Memoir 39. Tulsa: American Association of Petroleum Geologists.

Sherwood, J. W. C., and A. W. Trorey. 1965. Minimum-phase and related properties of the response of a horizontally-stratified absorptive earth to plane acoustic waves. *Geophysics,* **30:** 191–7.

Spencer, T. W. 1985. Measurement and interpretation of seismic attenuation. In *Developments in Geophysical Exploration Methods – 6,* A. A. Fitch, ed., pp. 73–110. Amsterdam: Elsevier.

Stommel, H. E., and M. Graul. 1978. Current trends in geophysics. In *Indonesian Petroleum Association Proceedings,* Jakarta, Indonesia.

Telford, W. M., L. P. Geldart, and R. E. Sheriff. 1990. *Applied Geophysics,* 2d ed. Cambridge: Cambridge University Press.

Toksoz, M. N., and D. H. Johnston. 1981. *Seismic Wave Attenuation,* Geophysical Reprint Series 2. Tulsa: Society of Exploration Geophysicists.

Tullos, F. N., and A. C. Reid. 1969. Seismic attenuation of Gulf Coast sediments. *Geophysics,* **34:** 516–28.

Vail, P. R., R. G. Todd, and J. B. Sangree. 1977. Chronostratigraphic significance of seismic reflections. In *Seismic Stratigraphy – Applications to Hydrocarbon Exploration,* C. E. Payton, ed., pp. 99–116, AAPG Memoir 26. Tulsa: American Association of Petroleum Geologists.

Vetter, W. J. 1981. Forward-generated synthetic seismogram for equal-delay layered models. *Geophys. Prosp.,* **29:** 363–73.

Waters, K. H. 1987. *Reflection Seismology,* 3d ed. New York: John Wiley.

White, J. E. 1965. *Seismic Waves – Radiation, Transmission, and Attenuation.* New York: McGraw-Hill.

White, J. 1966. Static friction as a source of seismic attenuation. *Geophysics,* **31:** 333–9.

Widess, M. B. 1973. How thin is a thin bed? *Geophysics,* **38:** 1176–80.

Wood, R. 1961. *Physical Optics.* New York: Dover.

Yilmaz, O. 1987. *Seismic Data Processing.* Tulsa: Society of Exploration Geophysicists.

7
Equipment

Overview

Because the objectives of seismic surveys vary tremendously, from surveys having very shallow objectives to those having deep objectives in difficult areas, the scale of seismic equipment varies. However, the concepts are generally the same and we will not discuss all the variations. Pieuchot (1984) along with Evenden and Stone (1971) provide the principal references for seismic equipment.

Of first-order importance is determining where data are acquired (§7.1). To locate the traces to be combined in common-midpoint stacking, surveying has to be more accurate than formerly required. Land work today often employs electronic distance measuring and Global Positioning System (GPS) measurements. Locating positions at sea where there are no landmarks depends mainly on radiopositioning and satellite observations, with reliance on the Global Positioning System increasing.

Although a multitude of energy sources have been used at times, most large-scale land work uses either explosives or vibrators (§7.2.2 and 7.3.1). The choice of source is usually based on economics. Sources located below the weathering layer usually produce less noise than those on the surface, but the cost of drilling holes is so high in many areas that explosives are not used. Transporting large surface sources prevents their use in some areas; fortunately, drilling is usually easy where the surface is too soft to support surface sources.

Air guns (§7.4.3) are used almost exclusively for large marine surveys. The bubble effect (§7.4.2) is often the limiting constraint for sources immersed in water. Several other types of marine sources are also used, especially for shallow surveys.

Seismic waves are detected almost exclusively by velocity geophones (§7.5.1) on land and pressure hydrophones (§7.5.3) at sea. Exceptions occur in some marsh and transition-zone work. The characteristics of data recorded by geophones and submerged hydrophones differ, and matching records where both are employed are discussed in §7.5.5. The use of very large numbers of recording channels has led to digitization near the geophones and hydrophones, as discussed in §7.5.2 and 7.5.4.

Digital versus analog representation of data is discussed in §7.6.4. Most recording today is digital, but portions of recording systems continue to be analog.

The analog display of data (§7.6.6) is important in communicating with an interpreter. Color displays are increasingly used to facilitate interpretation.

7.1 Determining location

7.1.1 Land surveying

Most types of conventional survey instruments have been used in seismic surveying. Plane table and alidade were extensively used into the 1960s, and transit-theodolites and chaining continue to be used today. Successive source points are often measured with a wire of appropriate length, a theodolite (fig. 7.1) being used to measure elevations, to keep the line straight, and to survey in side features. The theodolite is used to turn both horizontal and vertical angles and to give ranges by stadia measurements.

Electromagnetic distance measuring equipment (fig. 7.2), often called EDM, is widely used today. EDM and GPS (§7.1.5) instruments are often used to run in a base survey and to tie to benchmarks and wellheads even where not used for all source and geophone locations.

Most EDMs use laser diodes that emit a pulse of about 0.9 μm length; the pulse is reflected from the rod, and the round-trip time is measured to give distances to about 5 parts in 10^6 (Spradley, 1984). Horizontal and vertical angles can be measured to about 30 minutes. Laser EDMs have a maximum range of about 2 km.

For long ranges, a microwave EDM transmits a radio beam toward the "rod," which either reflects or retransmits the beam back to the EDM, where the phases of the reflected and transmitted waves are compared. If the two-way range is $(n + k)\lambda$, n being an integer and k a fraction, k is measured to about 1/100 wavelength ($\lambda \approx 10$ m). The ambiguity as to the value of n is resolved by making measurements at several frequencies. A computer calculates the range, which can then be read directly. Because beaming 10-m wavelengths requires large antennas, the 10-m wave modulates a high carrier frequency (10 to 35 GHz) that can be beamed.

EDM instruments can be equipped with digital readouts giving ranges and vertical and horizontal angles, and the data are recorded on a floppy disk, which becomes the surveyor's "notebook." It can then

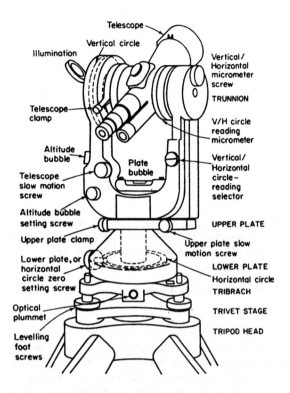

Fig. 7.1   Transit-theodolite. (After Ingham, 1975, vol. 2: 21; reprinted by permission of John Wiley & Sons Ltd.)

be read into a personal computer for automated data reduction.

7.1.2 Marine positioning

Marine seismic navigation involves two aspects: (a) placing the ships at a desired position and (b) determining the actual location afterwards so that the data can be mapped properly. In assessing the accuracy of a navigation method, we distinguish between absolute and relative accuracy. Absolute accuracy is important in tying marine surveys to land surveys and in returning to a certain point later, for example, to locate a well. Relative accuracy is important primarily to ensure accurate relative locations of midpoints. Absolute accuracies of ±70 m are usually sufficient, whereas relative accuracies of ±5 m are desirable. The accuracies obtained in a survey (which are usually very difficult to assess) depend upon the system and equipment used, the configuration of reference (shore) stations, the position of the mobile stations with respect to the reference stations, variations in the propagation of radio waves, instrument malfunctioning, operator error, and so on. Systems capable of giving adequate accuracy under good conditions may not realize such accuracy unless considerable care is exercised at all times (Sheriff 1974). Spradley (1984) discusses seismic marine positioning in detail.

Navigation systems generally measure some of the following: (1) the time between transmission and receipt of a signal, which gives a distance (*range*); (2)

the time difference between two signals, which gives a difference in distances; (3) a difference in frequency because of Doppler shift, which gives a velocity; (4) acceleration by means of oriented accelerometers; and (5) direction with respect to north, usually determined with a gyrocompass. With velocity and acceleration measurements, position is determined by integration.

Navigation systems can also be classified according to the way locations are determined: (1) *piloting,* wherein location is determined with respect to known locations; (2) *dead reckoning,* wherein locations are determined with respect to a known starting point and known course; and (3) *celestial measurements* based on the altitudes of the Sun or stars at known times; measurements with respect to navigation satellites can be included in this class.

Navigation/positioning systems have been undergoing rapid technological development in recent years and this continues in 1994. Prior to the development of atomic clocks, one could not depend on systems at different locations to be operating on the same timing system; the availability of relatively cheap, highly accurate atomic clocks has changed this.

System errors such as minor changes in velocity because of meteorologic or ionospheric uncertainties or minor perturbations in a satellite's orbit limit accuracy; these errors can be removed to a large extent by *translocation,* which involves measurement of appar-

Fig. 7.2   Field survey instruments. The surveyor is replacing a GPS receiver in his left hand with an electronic theodolite and distance meter. The EDM measures distances up to 5 km with accuracy of ±(3 mm + 2 parts per million), the theodolite measures horizontal and vertical angles to 3 seconds of arc. (Courtesy of Leica Inc.)

ent position changes at a fixed station and using these to correct positions of mobile stations. Translocation is now used with systems of different types.

Redundancy of positioning is highly desirable. Where position is determined simultaneously by different systems, the redundancy provides backup in event of failures and a needed check on possible malfunctioning. The overdetermination of position also provides an assessment of the accuracy actually being achieved. The history of marine geophysical surveying includes many instances of work lost in whole or in part because positions could not be relocated. Although the Global Positioning System (§7.1.5) is rapidly rising in prominence, other types of systems are still used and hence are described in the following sections.

Location uncertainty can be reduced significantly by analysis of the data afterwards. Data acquired later in a survey may be used to reduce the uncertainty of data acquired earlier. Much positioning uncertainty is systematic, and analysis of the entire body of data may clarify the nature of errors and permit correcting for them.

7.1.3 Radiopositioning

Radiopositioning (or radionavigation) systems are generally classified according to their frequencies. In general, high-frequency radio systems such as shoran, radar, Autotape, and Hydrodist use frequencies of 3 to 9 GHz (wavelengths of 3 to 10 cm) and achieve good accuracy, but their travel is line-of-sight and so their range is limited to about 40 km (depending on antenna heights). Medium-frequency systems such as Hi Fix, Decca, Toran, and Raydist use frequencies of 1.5 to 3 MHz (wavelengths of 100 to 200 m); they can bend with the Earth's curvature and their range is about 150 km. Low-frequency systems such as Loran-C use frequencies of the order of 100 kHz (wavelength of 3 km) and have a range of 2000 km; the global Omega system operates at the very low frequencies of 10 to 15 kHz (wavelengths 20 to 30 km).

Radiopositioning methods can also be classified depending upon the type of measurement: (a) systems that measure the time required for a radio-frequency pulse to travel between a mobile station and a shore station (examples are radar, shoran, the rho-rho mode of Loran-C), and (b) systems that measure the difference in traveltime (or phase) of signals from two or more shore stations (these include Raydist™, Lorac™, Decca, Pulse-8™, Hi-fix™, Toran™, ANA™, Argo™, the phase mode of Loran, and Omega). Angular measurements are not ordinarily used in radiopositioning because direction cannot be determined with enough precision with antennas of reasonable size.

Radar and shoran are similar in principle. *Radar* depends upon the reflection of pulses by a target, the distance to the target being equal to one-half the product of the two-way traveltime of the reflected pulse and the velocity of the radio waves. *Shoran* differs

from radar in that the target is a shore station that receives the pulse and rebroadcasts it with increased power so that the return pulse is strong. Two or more shore stations are used and the position of the mobile station is found by swinging arcs as in fig. 7.3.

Radar and shoran are high-frequency systems, radar frequencies being in the range 3000 to 10,000 MHz, shoran in the range 225 to 400 MHz. Because such high frequencies are refracted only very slightly by the atmosphere these methods are basically line-of-sight. With normal antenna heights of about 30 m, the range for shoran is roughly 80 km. If the shore stations can be located on hills adjacent to the sea, greater ranges can be obtained. By using very sensitive equipment (directional antennas and preamplifiers), ranges of 250 km can be obtained; this variation is called *extended-range,* or *XR, shoran.* The extension of range beyond the line-of-sight appears to be due to refraction, diffraction, and scattering from the troposphere. In some tropical or subtropical regions, strong temperature gradients in the atmosphere refract the radio waves so that ranges of 300 km or more can be obtained.

The distance between the ship station and each shore station is normally measured within ± 25 m ($\pm$ 0.2 μs), sometimes within ± 5 m. The error in location depends mainly upon the angle between lines joining the shore stations to the mobile station, as shown in fig. 7.3; angles between 30° and 150° are usually considered acceptable.

Several devices utilize the same principles as shoran, but use the higher radar frequencies; they "interrogate" a small *transponder,* a device that emits a signal immediately upon receipt of the interrogating signal. These include RPS™, Miniranger™, and Trisponder™, which use frequencies around 9500 MHz; Autotape™ and Hydrodist™, which use frequencies around 3000 MHz; and Syledis™, which operates at 450 MHz. The effective range of these devices is strictly line-of-sight, but they are extremely portable. Their accuracy is often excellent, of the order of 5 m.

Loran-C involves the broadcast of a coded se-

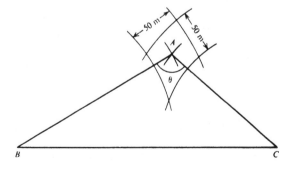

Fig. 7.3 Effect of station angle on errors in range-range position determination. θ = station angle, A = mobile station, B and C = shore stations. Point A can be anywhere within the "parallelogram" formed by the four arcs. (*Note:* Range errors are not to scale.)

quence of pulses of frequency 100 kHz, the broadcast times being controlled very accurately by atomic clocks. The stability of relatively cheap atomic clocks makes it feasible to carry such a standard on a seismic ship, so that the instant of signal transmission can be determined and hence the range to the transmitter. Such range determination is called the *rho* mode, or the *rho-rho* or *rho-rho-rho* modes if ranges are determined to one, two, or three transmitters. Despite the 3-km wavelength, ranges can be determined to 20 to 100 m. Long travel paths may be involved, however, so that minor variations in the speed of the radio waves because of variations in the conductivity of the ground or moisture in the atmosphere can introduce sizeable *propagation errors* (see problem 7.1). To minimize such errors, the system should be calibrated in the local area. The shipboard atomic clock may drift slowly, so that the drift has to be checked every few days.

If two shore stations simultaneously broadcast a radio pulse or coded sequence of pulses, a mobile station can measure the difference in arrival times and so find the difference in distances to the two shore stations. The locus of points with constant difference in distance from two shore stations (A and B in fig. 7.4, for example) is a hyperbola with foci at the two stations; thus, a single measurement determines a hyperbola PQ passing through the location of the mobile station, R. If the difference in arrival times for a second pair of stations (B and C) is measured, the mobile station is located also on the hyperbola VW and hence at the intersection of the two hyperbolas.

This principle forms the basis of the phase-comparison modes of Loran and Omega, long-range radionavigation systems maintained by the U.S. Government. Omega is a worldwide system, but its long wavelength (23 to 30 km) and seasonal and diurnal variations in the ionosphere preclude achieving accuracy greater than about 1 km. Loran-C is available over much of the northern hemisphere, especially in American and European waters. With care, the accuracy of Loran-C phase comparisons may be nearly that of its rho-rho mode. Decca is another system generally comparable to Loran-C, used mainly in Western Europe.

Medium-frequency radiopositioning systems can broadcast continuous waves (CW) from several stations, locations being determined by comparison of phase. *Phase-comparison systems* used in seismic exploration generally operate in the frequency band 1.5 to 4.0 MHz and have ranges up to 650 km.

Referring again to fig. 7.4, shore stations A and B transmit synchronous steady continuous sinusoidal signals that are exactly in phase at M, the midpoint of baseline AB. A mobile station with a phase-comparison meter will show zero phase difference at M and at all points on the perpendicular bisector MN. If $MP = \frac{1}{2}\lambda$ the phase-comparison meter indicates zero phase difference at P also; if the mobile station moves away from P in such a direction that the phase

difference remains constant, it traces out the hyperbola PQ. In general, a point R moving in such a way that

$$RA - RB = n\lambda, \qquad (n = 0, \pm 1, \pm 2, \pm 3, \ldots)$$

traces out the family of hyperbolas shown in the diagram.

The zone between two adjacent zero-phase-difference hyperbolas is called a *lane*. If we start from a known point and maintain a continuous record of the phase difference, we know in which lane the mobile station is located at any given time. By using a second pair of stations (one of which can be located at the same point as one of the first pair of stations) transmitting a different frequency, we obtain a second family of hyperbolas, hence another hyperbolic coordinate of the mobile station. The accuracy of location decreases with increasing lane width as we go farther from the base stations, also as the angle of intersection of the hyperbolas decreases. Location accuracy is of the order of 30 to 100 m. If the continuous count of lanes is lost, however, one could be considerably off location as the phase-difference meters give only the position within a lane and do not indicate which lane. The frequency can be changed periodically with consequent changes in the phase at a given position, which can be used to identify the lane. Atomic clocks are also used with medium-frequency systems to allow their use as range-measuring devices.

Translocation can be used to improve accuracy by removing the effect of propagation variations; this involves using variations in observations at a fixed station to correct those made simultaneously at a mobile station. With translocation and at least one redundant measurement, 1 to 3 m accuracy can be achieved (Musser, 1992).

7.1.4 Transit satellite positioning

Ships can determine their location from Transit satellites in polar orbits 1075 km above the Earth. Each satellite takes about 107 minutes to circle the Earth, being in sight for about 18 minutes horizon to horizon. Each satellite transmits continuous waves of frequencies 150 and 400 MHz. The frequencies measured by a receiver are Doppler-shifted (§7.1.6) because of the relative motion of the satellite with respect to the ship. Because radiowaves travel at the speed of light (V), relativity affects the Doppler-shift equation (see eq. (7.2)); if the velocities of the source and observer are respectively V_S and V_O, the observed frequency v_o is

$$v_o = v_S \left(\frac{V + V_o - V_S}{V - V_o + V_S} \right), \qquad (7.1)$$

where v_S is the source frequency. The difference between the ship's and satellite's longitudes and latitudes at closest approach (see fig. 7.5) are calculated from the Doppler shifts. The satellite transmits information that gives its location every 2 minutes. A small computer on the ship combines this information with the

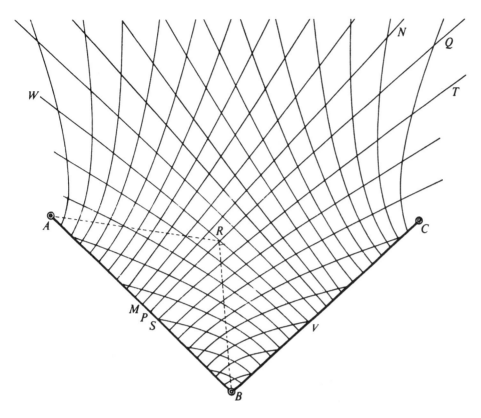

Fig. 7.4 Hyperbolic coordinates for radionavigation system based on measurement of time differences.

Doppler-shift measurements and the speed and course of the ship to give the ship's location.

Each satellite can be observed on four or more orbits each day (except near the equator); hence, with four to five satellites, twenty or more determinations are possible each day. However, the satellites are not uniformly spaced and do not have precisely the same orbital period so that sometimes more than one satellite is visible, whereas at other times, several hours intervene without any satellite being visible. About two-thirds of the "passes" result in satisfactory *fixes* or determinations of position. Satellite fixes may be accurate to within ±50 m, provided the ship's velocity is accurately known (Spradley, 1976). The principal disadvantage of Transit satellite navigation is that it gives no information about position during the interval between fixes.

Commonly, Doppler sonar (§7.1.6), gyrocompass, and satellite navigation are combined (Kronberger and Frye, 1971), or else radionavigation (§7.1.3) and satellite navigation. The satellite gives the periodic updating information needed to maintain Doppler-sonar accuracy or to remove ambiguities or propagation-error effects from radionavigation, whereas the Doppler sonar along with the gyrocompass and/or radio systems give the velocity information needed for an accurate satellite fix. Translocation can be used to improve accuracy of satellite location determination.

7.1.5 Global Positioning System (GPS)

The Global Positioning, or Navstar, System (Dixon, 1992) consists of twenty-one to twenty-four satellites at elevations of 22 200 km; it is operated by the U.S. Government and it permits determination of latitude, longitude, and elevation by trilateration. The system (with 25 satellites in 1994) is extensively used for geophysical positioning in the marine environment and is also used to set base stations on land. Four satellites are to be equally spaced in each of six orbital planes that make 55° angles with the Earth's equatorial plane (fig. 7.6). Each satellite orbits the Earth in about 12 hours.

Each satellite includes four atomic clocks, which provide an ultrastable timing system. Orbital perturbations are observed by stations in Diego Garcia, Hawaii, Kwajelein, and Ascension, and a facility at Colorado Springs, Colorado, keeps the satellites at their proper locations and their timing systems synchronized. Each satellite broadcasts on carrier frequencies of 1575.42 (L1) and 1227.6 (L2) MHz (and another for military use). The 50-Hz information is superimposed on the carriers by biphase phase shifting using +90° to indicate a one and −90° to indicate a zero. The superimposed information includes a "handover word," which permits synchronizing a user's time with a satellite's time, and "almanac" infor-

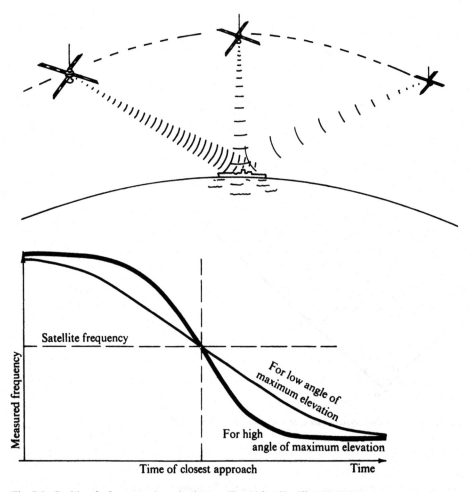

Fig. 7.5   Position fix from transit navigation satellite. (After Sheriff, 1990: 259.)

mation, which gives the satellite's position for the next 18 days, correction factors for tropospheric anomalies, and other information. Broadcasts from the individual satellites are distinguishable because their almanac information begins at different time intervals after the handover word. Two broadcast codes are used, a *P*-code for military use permitting greater accuracy than the civilian-user code.

Several receiver types (Burns, 1992; see also fig. 7.2) are available, including hand-held receivers. The user's position is determined by the simultaneous solution of range information from four satellites. A receiver must search among the transmissions from the satellites visible at any time for the signals that will give the best position information. Three satellites and the user form a tetrahedron, and the most accurate trilateration results when the tetrahedron's volume is maximized; observation of the fourth satellite is required to resolve differences between the satellite and user time systems. If the user knows a priori which satellites will give the best information, the receiver need search only for the signals from them; this speeds up determination of a location.

GPS permits other measurement modes in addition to the system just described. The phase difference between the satellite signals and the receiver reference can be used to give the difference in coordinates of successive stations. The Doppler-frequency shift can be measured to give locations in the same manner as with Transit satellites (§7.1.4). The use of two frequencies permits correction for refraction in the Earth's ionosphere and atmosphere. Translocation (*differential GPS*), in which readings at a fixed second receiver up to 500 km away are utilized, can be used to remove short-term satellite perturbations. Differential GPS geophysical usage can achieve 2 to 5 m accuracy (Jensen, 1992; Musser, 1992).

GPS readings give coordinates with respect to the satellite coordinate system and have to be transformed into local coordinates. The accuracies achieved depend on the modes of use, that is, the measurement duration, whether the receiver is static or in motion, whether locations are required in real time or later for postprocessing, whether absolute locations or only relative locations are required, and especially whether translocation is used. The U.S. Government warns

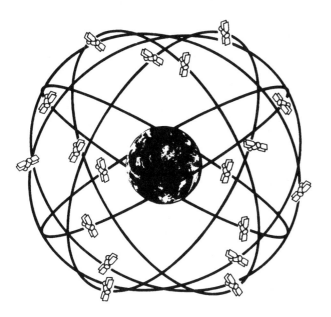

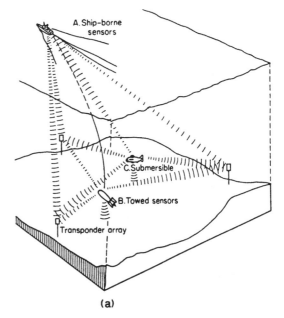

(a)

Fig. 7.6 Global Positioning System schematic showing satellites in orbit about the Earth. (Courtesy of Wild Heerbrugg Ltd.)

that GPS may be degraded for security reasons to give location accuracy of only 50 to 100 m.

7.1.6 Acoustic and inertial positioning

Acoustic or sonar positioning uses both sonar range and frequency-shift measurements. High-frequency acoustic transducers, also called *pingers,* are used in several ways; they emit acoustic (sonar) pulses in the kilohertz range that can be detected by other transducers to give the distances between them. They are sometimes incorporated in the source-detection system (§7.1.7) to locate the sources and streamers with respect to the ship and to each other.

For surveys of restricted areas (fig. 7.7a), anchored pinger transponders can be used. The ship or sonde to be located transmits a sonar pulse and the transponders emit coded responses when they sense the interrogating pulse. Most systems measure the two-way traveltime, though some also measure the phase difference at several sensors on the ship to determine the direction to the transponder (much as moveout gives the apparent direction of a seismic ray). Four or more transponders might be set 1 to 6 km apart where water depths are 20 to 500 m. The range is improved if the transponders are 5 to 10 m above the sea floor. Recoverable transponders having lifetimes of 5 years are available.

The locations of anchored transponders have to be verified, not only because of uncertainties in transponder locations, but also because of local velocity and propagation variations. Verification is usually done by criss-crossing over the area while using some other navigation system. The transponder locations should also be verified periodically because anchored

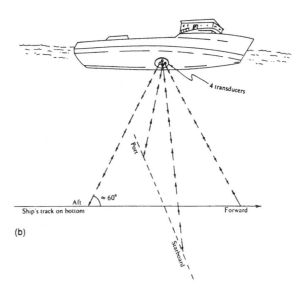

Fig. 7.7 Sonar navigation. (a) Measuring sonar range to bottom-set transducers (from Ingham, 1975; reprinted by permission of John Wiley & Sons Ltd.). (b) Doppler-sonar navigation (from Sheriff, 1990: 90).

transponders sometimes move, especially during storms. Acoustic transponders permit relative positioning of ±5 m, whereas absolute accuracy depends mainly on the method used to position the transponders.

Doppler sonar is a *dead-reckoning* system, that is, it determines position with respect to a starting point by measuring and integrating the ship's velocity. The ship's velocity is measured by projecting sonar beams against the ocean floor in four directions from the ship (fig. 7.7b). These beams are reflected back to the ship but their frequencies undergo a Doppler shift because of motion of the ship with respect to the ocean floor.

The frequency shift in each beam thus gives the component of the ship's velocity in that direction. The Doppler effect relates to the compression of wavefronts ahead of a moving source or as seen by a moving observer. If V is the velocity in the medium and V_s the component of a ship's velocity in the direction of the acoustic beam, the wavelength transmitted will be $(V - V_s)/v_s$ but a stationary observer would see it as V/v_0; hence $v_0 = v_s V/(V - V_s)$. If an observer with component of velocity V_0 is moving toward a stationary source, he would observe $v_0 = v_s(V + V_0)/V$. For a moving source and observer, we have $v_0 = v_s(V + V_0)/(V - V_s)$. When the observer is on the ship moving toward the point of reflection, V_0 becomes V_s, and

$$v_0 = v_s(V + V_s)/(V - V_s). \qquad (7.2)$$

The fore and aft measurements are averaged to minimize the effects of pitching motion of the ship, and starboard and port measurements to minimize the effects of rolling motion. The four beams often actually look in directions 45° to the ship's course, which gives improved sensitivity, rather than in-line and perpendicular to the ship's course. These measurements can be resolved to give the ship's velocity (in conjunction with direction information from a gyrocompass), and the velocity can be integrated to give the ship's position. Small errors in velocity measurement accumulate in the integration, resulting in position uncertainty of the order of 100 m/hr. Accuracy has to be maintained by periodic *updates,* that is, periodic determinations of location by independent measurements. In deep water, scatter of the sonar beams by inhomogeneities in the water dominates and the Doppler shifts give a measure of the velocity with respect to the water rather than the ocean floor, resulting in considerable loss of accuracy. A 300-kHz Doppler-sonar system can usually "see" bottom shallower than about 200 m, whereas a 150-kHz system can see to depths of 400 to 500 m.

Inertial navigation can be accomplished by measuring acceleration in orthogonal directions, integrating once to get velocity and a second time to get location relative to a known starting point. Accelerometers are usually located on a stable platform that is kept horizontal by a leveling feedback system and whose direction in space is maintained by a gyro feedback system. Periodic fixes from an independent navigation system minimize the accumulation of systematic error. The uncertainty with inertial systems in geophysical use increases at a rate of about 200 m/hr.

7.1.7 Locating the streamer

A seismic ship usually tows a long streamer that may extend for 5 km or more behind the ship. Even though the location of the ship is known, the streamer can drift by appreciable amounts. There is usually a tail buoy (fig. 7.8) at the end of the streamer on which a radar reflector is mounted; the direction to this re-

Fig. 7.8   Tail buoy containing a radar corner reflector, a Syledis radiopositioning receiver, and a transmitter for radioing information back to the seismic ship. (Courtesy of Compagnie Générale de Géophysique.)

flector can be measured with the ship's radar. However, it is often impossible in rough seas, to distinguish the tail buoy reflection from water-wave backscatter, particularly when the tail buoy is in the wave troughs. A radio or GPS receiver can be mounted on the tail buoy so that its location is known in the radiopositioning or GPS system being used to locate the seismic ship.

Magnetic compasses, typically 8 to 12 in a 5-km streamer, are included within the streamer (fig. 7.9a) (Proffit, 1991: 30). The readings from these are digitized and sent back to the ship. High-frequency waterbreak detectors are also included within the streamer; these measure the channel wave (§13.3), which travels in the water layer at the speed of sound in water (fig. 13.19), and thus give the distance from the seismic source. High-frequency pingers are sometimes incorporated in the system, particularly when more than one source or more than one streamer are employed (fig. 7.9b), to locate the sources and streamers with respect to the ship and to each other, as in 3-D seismic acquisition (see §12.1). The information from the various sensors is reduced by computer algorithms to give the midpoint bins into which the data fall.

7.2 Impulsive land energy sources

7.2.1 The desired source

An ideal seismic source would generate a wave that (1) contains enough energy that it can be detected readily after traveling great distances, (2) has short duration so that closely spaced interfaces can be resolved, (3) is repeatable, and (4) does not create noise that will

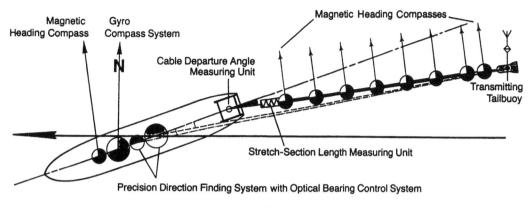

(a)

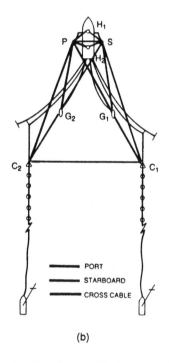

(b)

Fig. 7.9 Locating the streamer. (a) Sensors for locating the streamer with respect to the ship and the seismic line (courtesy of Prakla-Seismos). (b) Acoustic system of pinger transducers for locating sources (G_1 and G_2) and streamers (C_1 and C_2) with respect to pingers P and S, which provide a baseline with respect to the ship. Separate cross-cable pingers at several locations in the streamers measure the streamer separation (such as the distance C_1C_2) (courtesy of Western Atlas International).

interfere with the detection of reflections. A scaling law for sources (Ziolkowski, 1980) indicates that increasing the energy increases both the wavelet amplitude and the time scale, but does not change the waveshape if the source environment is the same (fig. 7.10): This law states that

$$w_B(t) = \alpha w_A(t/\alpha), \tag{7.3}$$

where $w_A(t)$ is the wavelet amplitude generated by source A, and the energy of source B is α^3 times that of source A. Thus, requirements (1) and (2) tend to

contradict; increasing the energy lengthens a wavelet and consequently decreases the resolution. Vertical stacking provides an alternative way of increasing the energy without lengthening the wavelet if the source is repeatable. The problems of interference with nearby sources conceptually can be solved either in processing, if the source characteristics are known, or by source design.

7.2.2 Explosive sources in boreholes

(a) Drilling. When dynamite is being used as the energy source, holes are drilled so that the explosive can be placed below the low-velocity layer. The holes are usually about 8 to 10 cm in diameter and 6 to 30 m in depth, although depths of 80 m or more are used occasionally. Normally, the holes are drilled with a rotary drill, usually mounted on a truck bed, but sometimes on a tractor or amphibious vehicle for working in difficult areas. Some light drills can be divided into units small enough so that they can be carried. Augers are used occasionally. In work in soft marshes, holes are sometimes jetted down with a hydraulic pump. Typical rotary-drilling equipment is shown in figs. 7.11 and 7.12.

Rotary drilling is accomplished with a drill bit at the bottom of a drill pipe, the top of which is turned so as to turn the bit. Fluid is pumped down through the drill pipe, passes out through the bit, and returns to the surface in the annular region around the drill pipe. The functions of the drilling fluid are to bring the cuttings to the surface, to cool the bit, and to plaster the drill hole to prevent the walls from caving and formation fluids from flowing into the hole. The most common drilling fluid is *mud*, which consists of a fine suspension of bentonite, lime, and/or barite in water. Sometimes water alone is used; sometimes air is the circulating fluid. Drag bits are used most commonly in soft formations; these tear out pieces of the earth. Hard rock is usually drilled with roller bits or cone bits, which cause pieces of rock to chip off because of the pressure exerted by the bit teeth. In

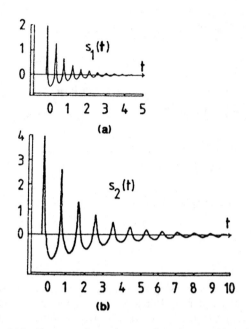

Fig. 7.10 Signature of two air guns where the second has eight times the energy of the first. The polarity is negative SEG standard. (From Ziolkowski, 1984: 13.)

areas of exceptionally hard rock, diamond drill bits are used.

(b) Explosive sources. Explosives were the sole source of energy used in seismic exploration until weight dropping was introduced about 1954. Explosives continue to be an important seismic energy source in land work.

Two types of explosives have been used principally: gelatin dynamite and ammonium nitrate. The former is a mixture of nitroglycerin and nitrocotton (which form an explosive gelatin) and an inert material that binds the mixture together and that can be used to vary the "strength" of the explosive. Ammonium nitrate is cheaper and less dangerous because it is more difficult to detonate than gelatin dynamites. Ammonium nitrate and NCN (nitrocarbonitrite) are the dominant explosives used today (in such forms as Nitramon™). Other types of explosives are also used occasionally.

Explosives are packaged in tins or in tubes of cardboard or plastic about 5 cm in diameter that usually contain 1 to 10 pounds (0.5 to 5 kg) of explosive. The tubes and tins are constructed so that they can be eas-

Fig. 7.11 Shothole drilling. (a) Mayhew 1000 drilling rig (courtesy of Gardner-Denver). (b) Light drill that can be moved by helicopter (courtesy of Prakla-Seismos).

(a)

ily joined together end to end (fig. 7.13a) to obtain various quantities of explosives. Ammonium nitrate sometimes is used in bulk form, the desired quantity being mixed with fuel oil and poured directly into a dry shothole.

The velocity of detonation (that is, the velocity with which the explosion travels away from the point of initiation in an extended body of explosive) is high for the explosives used in seismic work, around 6000 to 7000 m/s; consequently, the seismic pulses generated have very steep fronts in comparison with other energy sources. This high concentration of energy is desirable from the point of view of seismic-wave analysis, but detrimental from the viewpoint of damage to nearby structures.

The U.S. Bureau of Mines, Bulletin 656, "Blasting vibrations and their effects on structures," states that it is the velocity of ground motion rather than displacement or acceleration that correlates best with damage. Damage is minimal if peak velocity does not exceed 12 cm/s but a "safe criterion" is 5 cm/s. This translates into an empirical rule of $x = k\,m^{1/2}$, where $k = 50$ for x in feet and m in pounds, or $k = 23$ for x in meters and m in kilograms. The International Association of Geophysical Contractors sets the following minimum distances:

Pipe lines	200 ft (60 m)
Telephone lines	40 ft (12 m)
Railroad tracks	100 ft (30 m)
Electric lines	80 ft (24 m)
Transmission lines	200 ft (60 m)
Oil wells	200 ft (60 m)
Water wells, cisterns, masonry buildings	300 ft (90 m)

Electric blasting caps are used to initiate an explosion. These consist of small metal cylinders, roughly 0.6 cm in diameter and 4 cm long (see fig. 7.13b). They contain a resistance wire embedded in a powder charge that deflagrates readily. By means of two wires issuing from the end of the cap, a large current is passed through the resistance wire, and the heat generated thereby initiates the deflagration of the powder, which causes the explosion of an adjacent explosive in the cap. The cap has previously been placed inside one of the explosive charges, so that the explosion of the cap detonates the entire charge.

Primers are generally necessary in setting off the explosion in ammonium nitrate explosives. These are cans of more powerful explosives that are used as one of the elements in making up the total charge. A cap is inserted into a "well" in the end of the can of primer to set it off.

(c) Placement and firing of explosives. Charges are usually pushed down a borehole with *loading poles* because the density of the charge may be slightly less than that of the borehole mud. The depth to the top of the charge is measured by how many loading pole lengths are required to push the charge down the borehole. "Wings" fastened to the charge expand and dig into the borehole wall when the charge moves upward, preventing it from rising in the borehole.

The current that causes the blasting cap to explode is derived from a *blaster;* this is basically a device for charging a capacitor to a high voltage by means of either batteries or a hand-operated generator and then discharging the capacitor through the cap at the desired time. Incorporated in the blaster is a device that generates an electrical pulse at the instant that the explosion begins. This time-break pulse fixes the instant of the explosion, $t = 0$. The time-break pulse is transmitted to the recording equipment by a telephone line or radio, where it is recorded along with the seismic data.

The efficiency of an explosion is increased when it is confined and under pressure. A borehole is usually filled with water or mud to "tamp" the explosion. When the hole will not hold water, the borehole is usually filled with sand or loose earth. The explosion generates waste gases that eject the fluid, rocks, and other debris from the borehole (*hole blow*); when this debris falls back to the earth, it causes noise on nearby geo-

(b)

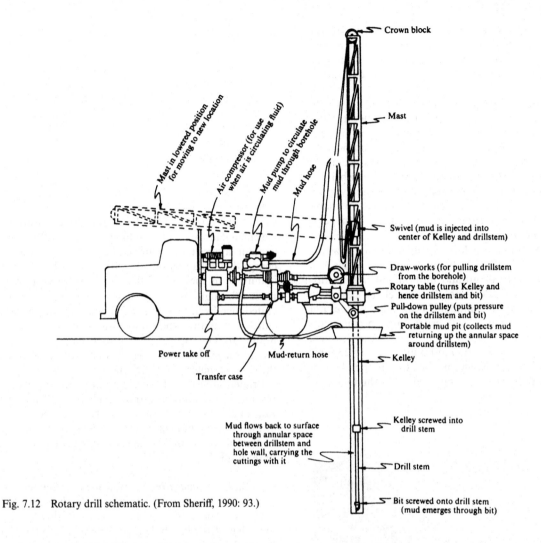

Fig. 7.12   Rotary drill schematic. (From Sheriff, 1990: 93.)

phones. There is often a time delay between the explosion and the hole blow. A device called a "catcher" is sometimes placed over the top of the borehole to prevent the capwires flying into the air and possibly fouling nearby power lines.

(d) Effects of charge size and depth on reflections.

We now give some general rules on the use of charges in boreholes based on experience, the exact processes surrounding an explosion still being unknown.

Shots below the water table are almost always much more effective than those above. Clay is a better shot medium than sand, with more effective energy transmission. The borehole is more apt to be reloadable if the medium is clay; detonation packs the surrounding clay to form a "pot" that is less apt to collapse. A small charge is sometimes used to enlarge the space available ("pot the hole") for large charges where required. Caliche and limestone are often poor shot media.

Hole conditions, especially charge depth, may affect the character of the seismic wavelet, but the changes are usually not extreme when the charge size is roughly the same. Charge depth affects near-surface ghosting and noise generation.

The frequencies of the seismic waves from small explosive charges is generally higher than those from larger charges. Likewise, several small charges separated by a few meters and detonated simultaneously generate higher-frequency waves than a concentrated charge containing the same amount of explosive.

Borehole conditions are different for subsequent shots. The first shot is apt to fracture the borehole walls, force gases and borehole fluid into the surrounding formations, and otherwise change the immediate environment. These changes often cause a few milliseconds delay in the seismic wave generated by subsequent shots, and a "hole-fatigue" correction may be required so that reflection times match. The hole-fatigue correction can be negative (in clays). Subsequent shots in a sand medium may be less effective generators of seismic energy, and a shothole may even "deaden."

(e) Directional charges.

Several techniques are used at times to concentrate the energy traveling downward from an explosion. The detonating front in an explosive usually travels much faster than the seismic wave in the formation, so that the seismic wave originating from the top of a long explosive charge lags behind the wave from the bottom of the charge even when the explosive is detonated at the top (which is the usual method). Explosives with low effective detonating velocity are sometimes used, but these are made in long flexible tubes that are difficult to load. Delay units are sometimes used between several concentrated explosive charges to allow the wave in the formation to catch up with the explosive front; these may consist of delay caps (which introduce a fixed delay between the time the detonating shock initiates them and the time they themselves explode) or helically wound detonating cord (so that the detonating front has to travel a longer distance) (see problem 7.5c). Expendible im-

pact blasters have also been used; these detonate when they are actuated by the shock wave from another explosion.

Consider the effect at a point P (see fig. 7.14a) of the simultaneous explosion of a linear source of length $MN = a\lambda$ and strength σ per unit length, σ being the energy received at P per unit of length per unit of time. Assuming harmonic waves of the form $e^{j(\kappa r - \omega t)}$ (see eq. (2.56)), the resultant at P for a wave generated at all points of MN at $t = 0$ is

$$h(t) = \int_{z_0 - a\lambda/2}^{z_0 + a\lambda/2} \sigma e^{j(\kappa r - \omega t)} \, dz$$

$$= \sigma e^{-j\omega t} \int_{z_0 - a\lambda/2}^{z_0 + a\lambda/2} e^{j\kappa r} \, dz.$$

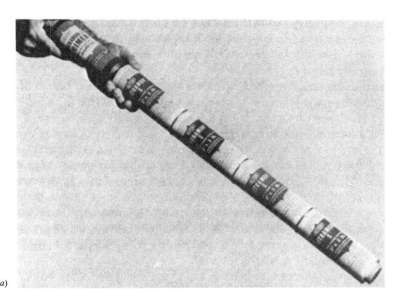

(a)

(b)

Fig. 7.13 Seismic explosives. (Courtesy of Du Pont.) (a) Cans of Nitramon joined end to end, and (b) seismic blasting cap.

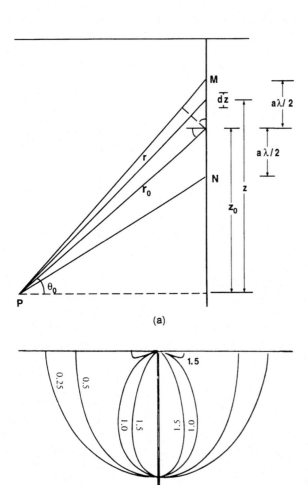

(a)

(b)

Fig. 7.14 Continuous linear source. (a) Geometry for determining directivity, and (b) directivity plot for various values of a.

For $r_0 >> a\lambda$,

$$r \approx r_0 + (z - z_0) \sin \theta_0 = r_0(1 - \sin^2\theta_0) + z \sin \theta_0$$
$$= r_0 \cos^2\theta_0 + z \sin \theta_0.$$

Thus,

$$\begin{aligned}
h(t) &= \sigma e^{-j\omega t} \int_{z_0 - a\lambda/2}^{z_0 + a\lambda/2} e^{j\kappa(r_0 \cos^2\theta_0 + z \sin\theta_0)} \, dz \\
&= \sigma a \lambda e^{j(\kappa r_0 - \omega t)} \frac{\sin(\pi a \sin \theta_0)}{\pi a \sin \theta_0} \\
&= \sigma a \lambda e^{j(\kappa r_0 - \omega t)} \mathrm{sinc}(\pi a \sin \theta_0),
\end{aligned}$$

where $\mathrm{sinc}\, x = (1/x) \sin x$. If the linear source were concentrated at the midpoint of MN, the effect at P would be $\sigma a \lambda e^{j(\kappa r_0 - \omega t)}$; hence, the array response, F (§8.3.6a), is

$$F = |\, \mathrm{sinc}\, (\pi a \sin \theta_0)\,|. \qquad (7.4)$$

The length of the source is apt to be one wavelength or less ($a \leq 1$), and occasionally two wavelengths ($a = 2$). Figure 7.14b shows the directivity response for various values of a. Note that little directivity is achieved when a is small. Shaped charges are sometimes used to concentrate the energy traveling vertically, but they are generally ineffective.

Although fig. 7.14 implies a distributed vertical source, the derivation also applies to a distributed horizontal source.

7.2.3 Large impulsive surface sources

Before magnetic tape recording (in the early 1950s) provided the ability to build up the effective source energy by vertical stacking, explosives provided almost the only seismic energy source. A number of alternative impulsive sources were developed, mostly in the 1960s, but most have now disappeared. The earliest nondynamite source, the *thumper* or *weight dropper* (fig. 7.15) drops a heavy weight (~3 tonnes) from about 3 m; it is still used in some desert areas today. Among the sources that have disappeared are *gas guns* (such as Dinoseis) that used an explosive gas mixture in an expandable chamber held against the surface by the weight of the transport vehicle.

A *land air gun* (fig. 7.16), consisting of a pan containing water and an air gun, is sometimes used. The pan is held against the ground by the vehicle's weight. The air gun is similar to those used in marine work (see §7.4.3). The sudden release of air into the water expands the pan diaphragm and imparts an impulse into the earth. The unit is caught before it can fall back onto the surface, thus preventing a secondary impact.

Explosive charges on or near the surface are sometimes used in areas remote from habitation. In *air shooting* (Poulter, 1950), small "flashless" explosives (so they do not start grass fires) are placed on sticks 1 to 1.5 m high distributed in a source array (fig. 7.17); they are usually connected with detonating cord (such as Primacord™) to initiate the explosions. A very strong air wave is generated that can damage ear drums of anyone nearby. The impact of the air wave on the surface generates the seismic wave.

Explosive detonating cord is sometimes buried 0.3 to 1 m in the earth or laid in shallow water. A vibratory plough (fig. 7.18) can be used to bury the cord; up to 100 m can be used for a shot. The speed of detonation of the cord, about 6.5 km/s, determines the number of caps required to detonate the entire length of cord within the desired time interval. Long cords fired from one end have directional properties that can be utilized in situations where the prevailing dip direction is known. Detonating cord is sometimes laid on the water bottom as a source in shallow water and marine surveys (Aquaseis) and sometimes is used in other ways as an energy source.

The energy from most surface sources is quite weak, and so a number of records are usually verti-

Fig. 7.15   Weight-drop unit. (a) Weight dropping; (b) impressions in the sand left by the impacts (from Sheriff, 1989).

cally stacked together (§6.7.3) to build up the effective strength. The sources usually generate appreciable ground roll and the sources are moved a small distance (3 to 5 m) between source activations to attenuate the ground roll in the vertical stacking.

Although the foregoing are primarily surface sources, gas guns, air guns (Brede et al., 1970), and other devices are sometimes used in boreholes, especially in soft marsh where there is little risk of being unable to recover the equipment from the borehole.

The nature of the surface affects the waveshape that surface sources generate. Figure 7.19 shows the spectra from land air guns on two types of surface.

7.2.4 Small surface sources

Modified versions of the major energy sources are used as sources for engineering, ground water, and other surveys that do not require large energy, but small sources such as those listed in table 7.1 are often more cost effective. Systems that sum the effects of many impacts (see Sosie in §7.3.2) are also used. Gravitational acceleration of weight-drop devices can be supplemented by other means (*enhanced weight drop*). Enhanced weight-drop devices are also used to generate *S*-waves by impact against a baseplate at an angle to the vertical. This generates both *P*- and *S*-waves; the azimuth of the device is then rotated 180° so that the generated *S*-waves have opposite polarity, and the second record is subtracted from the first, thus adding the *S*-wave effects and subtracting the *P*-wave effects (fig. 13.3).

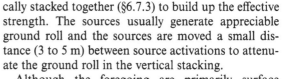

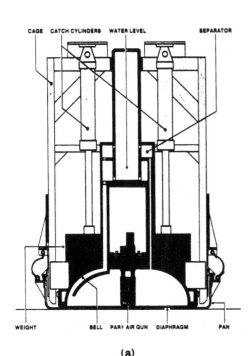

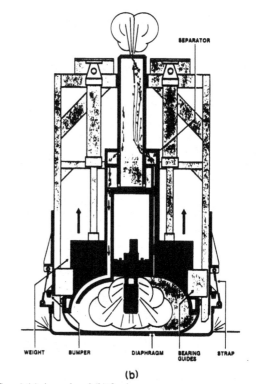

Fig. 7.16   Land air-gun schematic. (Courtesy of Bolt Technology Corp.) (a) Armed and (b) fired.

Fig. 7.17 Air shooting. Charges on small wooden poles are connected by Primacord ready for detonating.

Several of the sources listed in table 7.1 were compared by Miller et al. (1986). The relative energy for single impacts is shown in fig. 7.20. They found about a 48-dB energy decrease between 110 and 1000 Hz. The stronger sources produced good reflection coherence from sand–clay contacts 80 to 200 m deep with 340-Hz low-cut filters.

7.3 Nonimpulsive energy sources

7.3.1 Vibroseis

(a) Introduction. Unlike most energy sources that try to deliver energy to the ground in the shortest time possible, the Vibroseis source delivers energy into the ground for several seconds. A control signal causes a vibrator (usually hydraulic) to exert variable pressure on a steel base plate pressed against the ground by the weight of the vehicle (fig. 7.21b). The pressure $\mathcal{P}$ generally varies according to the relation

$$\mathcal{P}(t) = A(t) \sin 2\pi t[\nu_0 + (d\nu/dt)t], \qquad (7.5)$$

Fig. 7.18 Plough for planting detonating cord. The cord feeds down through pipes behind blades that vibrate as they are pu- illed forward, planting the cord ⅔ to 1 m below the surface. (Courtesy of Primacord Services.)

Table 7.1 *Small seismic sources*

Source	Description of source	Comments	Cost ($)
	Projectiles		
Silenced rifle	Modified rifle with silencing device, fired into ground or water-filled hole		500–5000
Betsy Seisgun™	8-gauge shotgun on portable base	Shotgun shell fired into dirt roadbed (also a downhole source)	5000–15,000
Buffalo gun	Firing rod dropped down 1-m iron pipe to detonate shell below ground surface	4-cm hole augered 1 m into ground; gun placed in hole; hole filled wtih water	< 500
	Impactors (weight drops)		
Sledge hammer	7.3-kg hammer striking steel plate of roughly equivalent weight	Plate set in dirt roadbed with 1–2 hammer blows	< 500
Brutus	136-kg weight raised 1 m by gasoline-powered hydraulic pump and dropped onto steel plate; trailer-mounted	Plate set in dirt roadbed by several weight drops	5000–15,000
Bean Bag™	136-kg weight in soft bag dropped about 3 m	No site preparation	
Soursile™	200-kg weight drop of about 1 m	Both *P*- and *S*-wave generator	
Elastic wave generator™	114-kg weight drop accelerated by elastic bands; reset by electric winch; trailer-mounted	Plate set in dirt roadbed by several weight drops	5000–15,000
Hydrapulse™	Weight drop augmented by compressed gas	No site preparation	
Dynasource	Vacuum-assisted 45-kg weight drop; gasoline-powered, trailer-mounted	Plate set in dirt roadbed by several weight drops	5000–15,000
Primary Source™	Compressed-air accelerated, 136-kg weight drop; truck-mounted; with hydraulic positioning control	Plate set in dirt roadbed by several weight drops; both *P*- and *S*-wave generator	> 15,000
Hydraulic hammer™	Hydraulically accelerated weight drop onto baseplate	No site preparation	
Yumatsu Impactor™	Hydraulically accelerated 200-kg weight drop	No site preparation	
	Electrical		
Piezoelectric	Piezoceramic transducer stack with high-voltage discharge; 1-kV power supply vehicle transported	Impact plate set on ground	5000–15,000
Spark Pak™	Pulse generated by discharging 5–20 kJ of energy between electrodes in salt water; operates off 1-kW generator	50-cm hole dug in ground; garbage bag placed in hole and filled with salt water	> 15,000
	Explosive		
Mini-Primacord	10 cm of 200-grain Primacord inside tubing 1 m below surface; detonated by standard blaster	Small tubing pounded 1 m into ground; Primacord pushed to bottom and tamped with sand	< 500
Explosives	Ammonium nitrate and nitromethane explosive mixed on site; detonated by seismic blasting cap	10-cm hole augered 1 m into ground; explosive and cap placed at bottom; hole packed	< 500
POD (propane–oxygen detonator)	Sleeve is driven against bottom of borehole by explosion		
Omnipulse™	See fig. 13.3	Both *P*- and *S*-wave generator	

Tradenames:	Betsy Seisgun	Mapco	Primary source	Shear-wave Technology
	Bean Bag	Developmental Geophysics	Hydraulic hammer	Prakla-Seismos
	Soursile	Geoméchanique	Yumatsu impactor	Japex Geoscience Institute
	Elastic wave generator	Bison Instruments	Spark Pak	Geomarines Systems
	Hydrapulse	CMI	Omnipulse	Bolt Technology

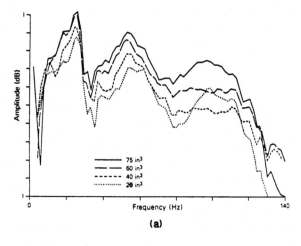

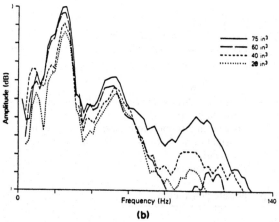

Fig. 7.19 Effect of surface on spectra of wave resulting from a land air gun. (Courtesy of Bolt Technology.) (a) Concrete surface, and (b) clay surface.

(dv/dt) being either positive (*upsweep*) or negative (*downsweep*) and constant in the usual case of a linear "sweep." The amplitude $A(t)$ is usually constant except during the initial and final 0.2 s or so when it increases from zero and decreases to zero (see fig. 9.8). The sweep usually lasts for 7 to 35 s with a frequency varying from about 12 to 60 Hz or vice versa.

Because reflections occur at intervals much smaller than the sweep time, the seismic record is the superposition of many wavetrains and the field records are uninterpretable even to the experienced. Subsequent data processing (see §9.3.4) is necessary to resolve the data; in effect, the processing (cross-correlation with the sweep) compresses each returning wavetrain into a short wavelet, thus removing much of the overlap (see fig. 9.9). The cross-correlation requires that frequencies do not repeat during a sweep; if they did, the cross-correlation would find more than one match for the repeated succession of frequencies, creating fictious events.

(b) Vibrators. A schematic diagram of a vibrator is shown in fig. 7.21a. The base plate is connected by a piston to chambers in the center of a large reacting mass. The chamber is divided into two parts into which oil under high pressure is alternately pumped or exhausted, so that the mass oscillates up and down. The force causing the steel mass to oscillate is equal and opposite to the force exerted by the baseplate on the ground. The vehicle weight acts as a hold-down weight to press the vibrator firmly against the ground; airbags and springs decouple the oscillating system from the vehicle so that it is not unduly shaken. The hold-down weight must exceed the maximum upward instantaneous thrust to prevent the baseplate from leaving the ground; peak thrusts are sometimes as much as 16 tonnes. For more detailed accounts of vibrator design, see Waters (1987: 62ff.) or Geyer (1989).

(c) Field technique. Miller and Pursey (1956) calculated the distribution of energy from a P-wave vibrator in a semi-infinite earth to be 7% P-wave (concentrated in the downward direction), 26% S-wave (beamed at about 30° to the vertical), and 67% surface wave. The near-surface layer causes most of the S-wave energy to be converted into surface-wave energy. Because of the large proportion of surface-wave energy, relatively large source and receiver arrays are generally used. Active geophone groups usually are not located near the vibrators, so that there is an appreciable sourcepoint gap (often exceeding 300 m). By using only groups offset from the vibrator source, the range of amplitudes recorded is much smaller; this simplifies the detection of deep reflections that otherwise would be lost when superimposed on the wavetrains of shallow ones. However, the large source offset and source–receiver arrays also limit the recording of shallow reflections. Several vibrators (usually three or four) and vertical stacking are used in addition to the sweep length to build up the seismic energy. The sources are moved between sweeps to attenuate the ground roll in stacking.

Ideally, the input to the ground is a copy of the pressure applied to the steel plate, but in fact heating, crushing, and compaction of the surface material (pressures are as high as 200 kg/cm^2) result in the input to the ground varying nonlinearly with the pressure exerted by the vibrator. This introduces harmonics not present in the original input signal. The second harmonic is apt to be troublesome because it can correlate with the sweep to superimpose a spurious *correlation ghost* on the record. Correlation ghost problems can be eliminated by using upsweeps so that the ghost arrives before zero time, or by using very long downsweeps so that the ghost does not arrive until after the region of interest. It is generally easier to lock the phase as a downsweep starts but upsweeps are much easier on the equipment, so that there is no clear-cut preference between upsweeps and downsweeps. Standard practice is to move the vibrator a couple of meters between successive sweeps.

Vibroseis sources produce low energy density; as a result, they can be used in cities and other areas where

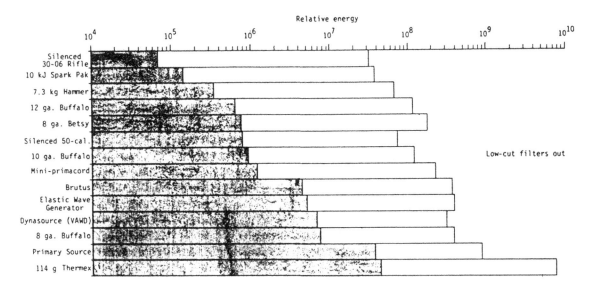

Fig. 7.20 Relative energies for sources listed in table 7.1. Shaded bars represent reflections; open bars, ground roll. (From Miller et al., 1986.)

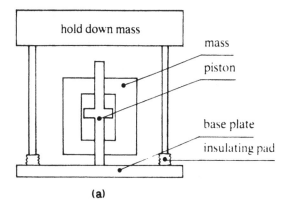

(a)

Fig. 7.21 Vibrators. (a) Basic principle of a vibrator. (b) Photo of a vibrator (courtesy of Conoco).

explosives and other sources would cause extensive damage (Mossman, Heim, and Dalton, 1973). Vibroseis is now used for over half of land seismic exploration.

The large weight of vibrator vehicles restricts their use in some areas. Air-cushion (hovercraft) vehicles can be used across tidal flats and other areas where ground-pressure restrictions apply; to make the vehicle into a seismic vibrator, the air flow is modulated by the sweep.

(d) Phase control. The signal/noise ratio (S/N) is improved more by using several vibrators simultaneously than by vertical stacking. With four vibrators, the S/N with respect to nonsource-generated noise is improved by a factor of 4, whereas stacking four successive sweeps improves the S/N only by $\sqrt{4} = 2$ because the noise will be different on the four sweeps. (S/N also varies as $\sqrt{T}$, where T is the sweep length.) However, the different vibrators must be phase-locked together.

The earth acts as a spring to baseplate motion, creating a resonant system whose properties depend on both the baseplate area and the soil characteristics. Phase-compensating systems are used to keep the injected signals locked onto a reference signal. Each vibrator has its own digitally controlled sweep generator that is triggered by a signal sent from the recording truck; this procedure prevents a noisy radio signal from affecting the sweeps. The output of a strong-motion accelerometer is integrated to give a baseplate velocity, which is then sent to a phase comparator where it is compared against the reference; the output is used to control a phase shifter that keeps constant the phase difference between the reference and the baseplate velocity.

(e) Nonlinear and pseudo-random sweeps. The use of nonlinear sweeps is equivalent to filtering the data (Goupillaud, 1976). The natural attenuation of high frequencies in a wavelet's passage through the earth acts as a filter and is a major factor in limiting resolution. Because the Vibroseis method gives us control over the wavelet's frequency content, we may use a *nonlinear sweep* to increase the high-frequency energy input to compensate for absorption losses. Usually, we cannot significantly decrease the time spent sweeping low frequencies because the total bandwidth needs to be maintained, so increased time spent sweeping higher frequencies means longer sweeps. With a linear sweep, eq. (7.5) shows that it takes double the time to sweep a high-frequency octave than it takes to sweep the next lower octave (because it encompasses twice as many frequencies), hence nonlinear sweeps increase recording time over linear sweeps. The effect of using a nonlinear sweep in improving the high-frequency content and consequently the resolution is shown in fig. 7.22.

Goupillaud (op. cit.) also showed that if a linear sweep is divided into segments and the segments rearranged, the correlation process to extract the information from the recorded data will not change provided there are no discontinuities between the segments. He proposed a random arrangement of the segments that he called a *pseudo-random sweep*. A pseudo-random sweep will not correlate with a sweep using a different sequence. Thus, the use of pseudo-random sweeps makes it possible to record with several sources (at different locations) simultaneously and be able to separate the data in processing. Other coding schemes are also used. It is possible to offset vibrators from each end of a cable and thus double the amount of data acquired in a given recording time. Ground-roll amplitude is also lower with pseudo-random sweeps because the low frequencies that contribute most to ground roll are distributed throughout the sweep.

7.3.2 Sosie

An impactor such as those shown in fig. 7.23 can be used as a source for shallow penetration surveys (up to 1 s) using the Sosie™ method (Barbier and Viallix, 1973). The impactor (*whacker*) strikes the ground 5 to 10 times per second and a recording is made for about 3 minutes (therefore, 900 to 1800 impacts). The impact times can be considered random for seismic frequencies. A sensor on the baseplate records the moment of each impact for use in correlation. Random repetitive firing of other small sources, such as small Vaporchoc™ units (see §7.4.4) in marine surveys, can also be used as Sosie sources.

With the Sosie method, the output of a geophone group is added into its summation register again each time a new impulse is applied to the earth. Because many impulses occur within the recording length, the output of each group is being added to its summation register simultaneously at many places. Source-generated energy adds in phase with respect to the proper arrival times, but adds randomly at other times (see problem 7.6).

7.3.3 Choice of land sources

The choice of seismic source is almost always an economic one. The source (or drill) is usually the largest, heaviest piece of equipment that a seismic crew requires. Hence, if the source can reach its locations, other acquisition operations probably can reach theirs. Explosives in boreholes are probably preferred by most geophysicists except where drilling is difficult or expensive, where legal restrictions exist, or proximity to habitations or structures preclude the use of charges of sufficient size. Some of the large surface sources are so heavy that they require large vehicles for their transport and thus access becomes the major factor. Most equipment can be mounted on a variety of platforms (such as boats or barges, tracked vehicles, marsh buggies, and very shallow-draft air boats pro-

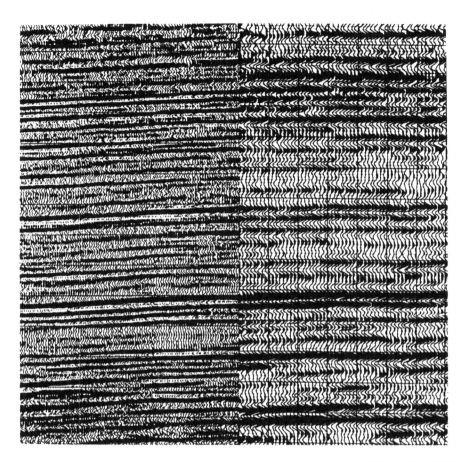

Fig. 7.22   Improvement in resolution resulting from use of non-linear Vibroseis sweep. Left half, nonlinear 10-to-90-Hz sweep; right half, linear 13-to-50-Hz sweep. (From Andrew, 1985.)

pelled by an air propellor) for use in different types of terrain, including marsh, swamp, and shallow-water areas. The local availability of equipment is apt to be a deciding factor on small work projects.

7.4 Marine equipment

7.4.1 General

The ships used for marine seismic acquisition are usually large, 30 to 95 m in length and 5 to 20 m in beam. A typical ship, as shown in fig. 7.24, draws up to 6 m. The ship carries enough fuel, water, and other supplies to operate at sea for 30 to 60 days and accommodates 25 to 60 people. Replenishment of supplies and personnel exchanges can be carried out at sea so that operations can continue steadily. There is usually a platform on which a helicopter can land. The ship is apt to make its own fresh water by reverse osmosis. It is generally equipped with a number of radio and navigation aids for communication and positioning, including a weather facsimile receiver. The ship's speed while acquiring data is usually 5 to 6 knots (2.5 to 3 m/s or 9 to 11 km/hr), but the ship can make about 15 knots (28 km/hr) when not acquiring data. A high degree of automation is employed to sustain efficient continuous operations.

Figure 7.25 shows views aboard seismic ships. The several kilometers of streamers are stored on large reels (fig. 7.25a) from which the streamer can be fed over the stern into the water. Arrays of air guns are sometimes mounted on paravanes that are suspended from the deck ceiling when not in use (fig. 7.25b). The instrument room (fig. 7.25c) often appears much as the control room in any large computing facility.

As on land, early marine seismic acquisition used explosives as sources almost exclusively. Although the efficiency of an explosion in water increases with water depth, it was soon discovered that the waste gases oscillated in the water (*bubble oscillations*), each oscillation effectively producing a new seismic record superimposed on the earlier record so that the resulting record mixture could not be interpreted. To overcome this problem, the practice was to suspend the charge from a floating balloon within about 2 m of the surface, allowing the waste gases to vent to the atmosphere before the bubble could collapse and generate a second seismic wave.

Safety required that only one capped explosive could be aboard a ship at any time; this meant that shots could not be fired at short time intervals, so that the common-midpoint method could not be used effectively in the marine environment. Once nonexplo-

Fig. 7.23   Two impactors used as mini-Sosie source. Geophones on the base plates indicate the times of impacts. (Courtesy of Wacker-Werke.)

Fig. 7.24   A seismic ship towing three streamers (from the two outer paravanes and directly from the ship) and two sources (from the inner paravanes). (Courtesy of Western Geophysical.)

Fig. 7.25 Views on a seismic ship. ((a and b) Courtesy of Prakla-Seismos, (c) courtesy of Seismograph Service.) (a) Streamer reels for two streamers, (b) air-gun subarray, and (c) instrument room.

sive marine sources became available, the use of explosives as marine sources disappeared quickly (fig. 1.21).

7.4.2 Bubble effect

The *bubble effect* (Kramer, Peterson, and Walters, 1968) just described applies to any source that injects high-pressure gases into the water; it is illustrated in fig. 7.26. As long as the pressure in the gas bubble exceeds the hydrostatic pressure of the surrounding water, the net force accelerates the water outward. The net force decreases as the bubble expands and becomes zero when the bubble expansion reduces the gas pressure to the hydrostatic pressure. However, by then, the surrounding water has acquired maximum outward velocity and so continues to move outward while decelerating (because the net force is now directed inward). Eventually, the water comes to rest and the net inward force causes the bubble to collapse. The water converges into the limited bubble volume, thus rapidly compressing the gas to high pressure; this *implosion* is associated with a very rapid increase in the pressure, which starts a new bubble expansion. This process continues as the bubble rises, some energy being dissipated on each oscillation. Eventually, the bubble breaks through the surface of the water, the gases vent to the atmosphere, and the cycle is broken. Seismic waves will be generated by the high pressures associated with each bubble collapse.

The period of the bubble oscillation (see fig. 7.31) depends on the energy stored in the bubble and can

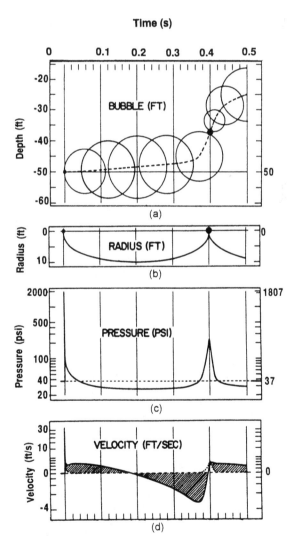

Fig. 7.26 Bubble effect. (From Kramer et al., 1968.) (a) Depth of bubble as function of time, (b) bubble radius, (c) bubble pressure, and (d) velocity of water adjacent to the bubble.

be brought within the seismic range for small sources so that the oscillation contributes to the effectiveness of the source. The bubble energy is roughly proportional to the maximum volume of the bubble as it starts to implode.

The bubble effect is important in determining the source waveshapes of almost all marine sources, even for those designed expressly to minimize the effect by dissipating the bubble energy after the first collapse. The waveshape generated by a small air gun is shown in fig. 7.28a.

7.4.3 Air guns

The most widely used large marine energy source is the *air gun* (fig. 7.27), a device that discharges air under very high pressure into the water (Giles, 1968; Schulze-Gattermann, 1972).

Pressures up to 10 000 psi (70 MPa) are used, although 2000 psi (14 MPa) is most common. The air gun shown in fig. 7.27a is in the armed position, ready for firing. Chambers A and B are filled with high-pressure air that entered A at the top left and passed into B through an axial opening in the "shuttle." The latter is held in the closed position by the air pressure (because flange C is larger than flange D, resulting in a net downward force). To fire the gun, the solenoid at the top opens a valve that allows high-pressure air to reach the underside of flange C. This produces an upward force that is large enough to overcome the force holding the shuttle in the closed position, and consequently the shuttle opens rapidly. This allows the high-pressure air in the lower chamber to rush out through four ports into the water. The bubble of high-pressure air then oscillates in the same manner as a bubble of waste gases resulting from an explosion. However, because the energy is smaller, the oscillating frequency is in the seismic range and therefore has the effect of lengthening the original pulse (rather than generating new pulses as with dynamite). The upward motion of the shuttle is arrested before it strikes the top of chamber A because the upward force falls off rapidly as the air enters the water and the downward force of the air in the upper chamber increases. The shuttle then returns to the armed position, and the lower chamber again fills with air. The explosive release of the air occurs in 1 to 4 ms whereas the entire discharge cycle requires 25 to 40 ms. Other types of air guns operate in essentially the same way. The "sleeve gun" (Harrison and Giacoma, 1984) opens the gun chamber more quickly so that the air is released more rapidly, resulting in increased peak pressure and seismic-wave strength.

The lower chamber in an air gun may be divided into two parts connected by a small orifice (Mayne and Quay, 1971), which results in a delayed discharge of the air in the innermost chamber. The flow of this air into the bubble continues for some time after the initial discharge, retarding the violent collapse of the bubble and diminishing the subsequent bubble effect.

The waveform emitted by a single air gun, shown in fig. 7.28a, oscillates because of the bubble effect. The delayed ghost reflection from the water surface has opposite polarity and comparable amplitude to that from the gun itself; it is primarily responsible for the second half-cycle of the wavelet shape. For the same air pressure, the energy output of an air gun is proportional to the product of volume and pressure. Air-gun size is usually taken as the volume of its lower chamber and gun sizes generally range from 10 to 2000 in.^{3} (0.16 to 33 liters). Usually, several guns (often six or seven) of different sizes are used together in subarrays such as shown in fig. 7.28c. Gun discharges are timed so that their initial impulses interfere constructively, but their subsequent bubble pulses interfere destructively; the wavelet from a subarray is shown in fig. 7.28b.

The action of an air gun is affected by nearby guns unless the gun separation is greater than a wavelength, that is, greater than 10 m for 150 Hz. However, interaction can be calculated and allowed for (Ziolkowski et al., 1982), so that elements can be spaced more closely.

Hydrophones located about 1 m from each gun can measure the pressure field, and this permits calculation of the composite far-field signature. Monitoring the pressure field permits decisions to be made should conditions change, for example, if a gun should fail.

7.4.4 Imploders and other marine sources

Several types of sources create voids in the water into which the water rushes implosively. With the Flexi-choc™, an adjustable-volume chamber is evacuated and the walls of the chamber are held fixed by a mechanical restraint; upon removal of the restraint, the hydrostatic pressure suddenly collapses the chamber. Air is then pumped into the chamber to expand it and to secure the restraint for the next activation. A flexi-choc signature is shown in fig. 7.30f. With the Hy-drosein™, two plates are suddenly driven apart by a pneumatic piston to create a void between them. With the Boomer™, a heavy surge of electrical current through a coil on one plate creates eddy currents that force the plates apart. These imploders are generally weak sources but are suitable for profiling applications (§8.6.3).

The water gun obtains energy from compressed air but does not inject the air into the water. Release of a shuttle (fig. 7.29a) forces the water in the lower chamber out through the ports into the surrounding water at high velocity. The moving slugs of water create near-vacuums behind them, and the implosion as the surrounding water rushing into the voids provides the main seismic energy. Bubble oscillation is minimal because there are no gases to oscillate. A water-gun signature is shown in fig. 7.30g.

Electric arcs (*sparkers*) utilize the discharge of a large capacitor to create a spark between two elec-

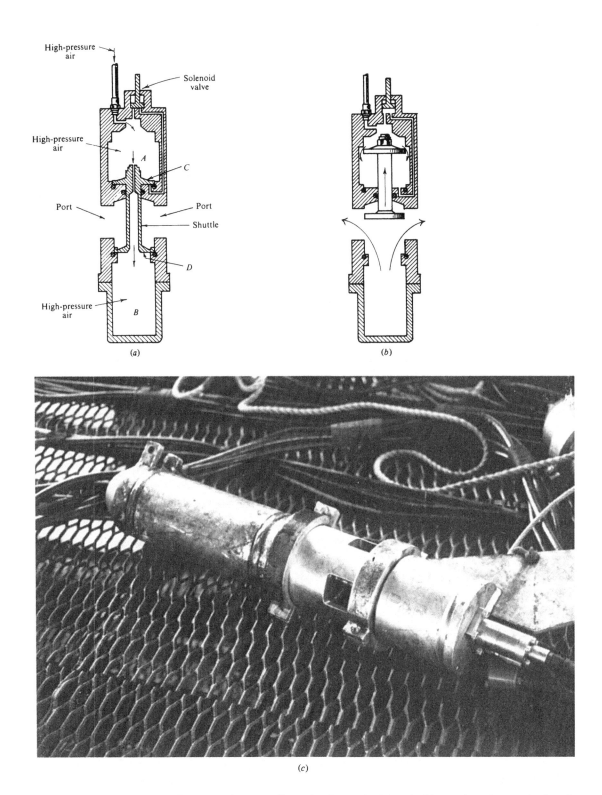

Fig. 7.27 Air gun. (Courtesy of Bolt Associates.) (a) Charged and ready for firing, (b) firing, and (c) photograph of an air gun.

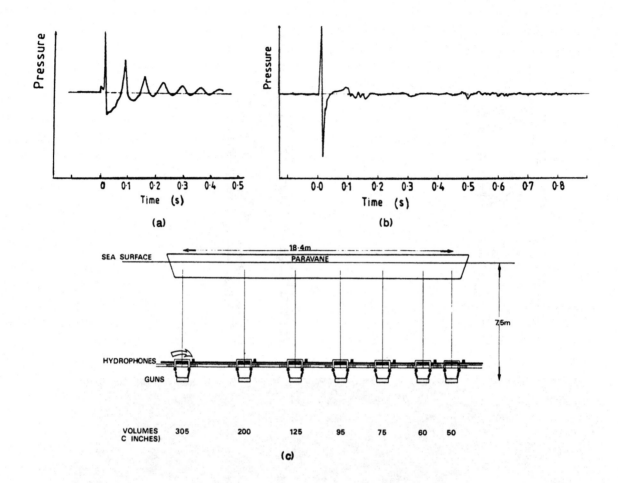

Fig. 7.28 Waveform signatures generated by air guns. ((a and b) From Ziolkowski, 1984: 9, 10, 13; (c) from Parkes et al., 1984: 106.) (a) Single 0.8-L air gun at a distance of 1 m; pressure of 135 bars, gun depth of 7.5 m. (b) Array of seven air guns; polarity is negative SEG standard. (c) Geometry of an air-gun subarray; two to four subarrays separated by 10 to 50 m are often used simultaneously.

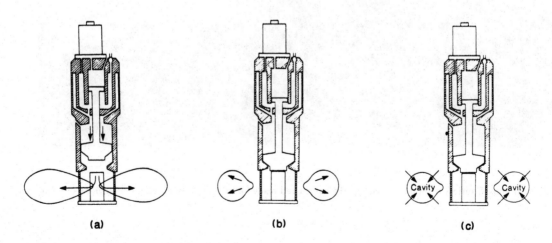

Fig. 7.29 Water gun. (Courtesy of Seismograph Service.) (a) Compressed air forces piston down, ejecting water from lower chamber at high velocity. (b) Because of its high velocity, the water continues moving outward, creating a void behind it. (c) Water rushes into the void, producing a sharp implosion.

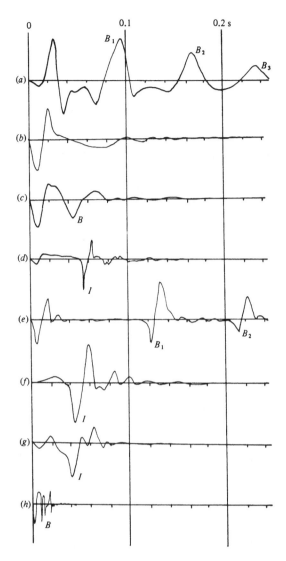

Fig. 7.30 Far-field waveshapes generated by marine seismic sources. (a) Single 120-in.³ air gun, (b) array of air guns of different sizes, (c) sleeve exploder, (d) Vaporchoc™, (e) Maxipulse™, (f) Flexichoc™, (g) water gun, and (h) 5-kJ sparker. Curves show waveshape features, but amplitudes are not to scale. B indicates bubble effects, I implosion. Parts (a) and (g) are from McQuillin, Bacon, and Barclay (1979); (b), (c), and (e) from Wood et al. (1978); (d) from Farriol et al. (1970); (f) from manufacturer's literature; and (h) from Kramer, Peterson, and Walter (1968).

trodes in the water, thus vaporizing the water and effectively causing a small explosion. The penetration of a 5-kJ sparker is generally less than 300 m, but sparker arrays delivering as much as 200 kJ at 50 to 2000 Hz achieve penetration of 1000 m. A variation of the sparker, Wassp™, involves connecting the electrodes by a thin wire that is vaporized in the discharge; this increases the bubble duration and its low-frequency content, and also permits operations in fresh water, where conductivity is too low to permit reliable sparker operation.

Several other types of sources are used for very shallow penetration. These include high-powered piezoelectric transducers made of barium titanate, lead zirconate, or other materials that change their dimensions when subjected to an electric field. Frequencies of 2 to 10 kHz can be generated at about 100 W, achieving 20 to 100 m penetration.

Other kinds of marine sources designed to avoid the bubble effect have nearly disappeared from use. One was the sleeve exploder, or Aquapulse™, which used the explosion of a propane–oxygen mixture in a closed flexible chamber with the waste gases vented to the atmosphere. Another was the Vaporchoc™, or *steam gun,* which injected a bubble of superheated steam into the water. The Flexotir™ involved detonating a small explosive at the center of a steel cage so that the bubble oscillation was damped as the water flowed in and out of openings in the cage. The Maxipulse™ recorded the bubble oscillation and tried to remove it in subsequent processing.

A marine version of the Vibroseis employing several source units has been used; however, because a seismic boat travels an appreciable distance during the time of a Vibroseis sweep, undesirable "smearing" of reflection points results.

7.4.5 Choice of marine sources

The marine-source characteristics most sought after are high peak pressure and low secondary oscillations. These can be determined empirically by firing the source in deep water (so that water-bottom reflections will not confuse the results) and observing the waveform with a calibrated hydrophone 75 to 100 m below the source. Although the experiment seems easy, its implementation is difficult. It is not easy to keep the hydrophone at the desired distance below the source when the ship is moving, and static test conditions are apt to be different from operational conditions. The source waveforms (signatures) for several energy sources are shown in fig. 7.30.

Rayleigh (1917), while studying the sounds emitted by oscillating steam bubbles, related bubble frequency to bubble radius, pressure, and fluid density; and Willis (1941), while studying underwater explosions expressed the relationship in terms of source energy (the Rayleigh–Willis formula):

$$T = 36\rho^{1/2}\mathcal{P}_0{}^{-5/6}E^{1/3}, \tag{7.6}$$

where T is the period of bubble oscillation in seconds, ρ is the fluid density in g/cm³, $\mathcal{P}_0$ is the absolute hydrostatic pressure in pascals (N/m²), and E is the energy in joules. If we assume a density of 1.024 g/cm³ for sea water and replace $\mathcal{P}_0$ with $h + 10$, where h is the depth in meters (10 m is 1 atmosphere), the formula becomes

$$T = 0.017E^{1/3}(h + 10)^{-5/6}. \tag{7.7}$$

Figure 7.31 shows the energies of various sources ver-

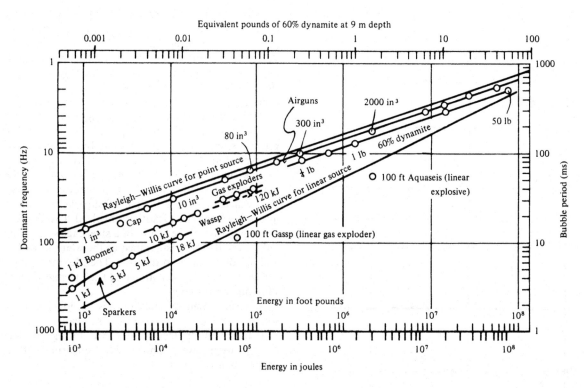

Fig. 7.31 Energy-frequency relationships for marine sources at 9-m depth. (From Kramer et al., 1968.)

sus dominant frequency. In general, large energy involves low frequency and vice versa.

7.5 Detectors

7.5.1 Theory of geophones

(a) General. Seismic energy arriving at the surface of the ground is detected by *geophones* (frequently referred to as *seismometers, detectors, phones,* or *jugs*) or by hydrophones. Although many types of geophones have been used in the past, modern geophones for land work are almost entirely of the moving-coil electromagnetic type. Hydrophones for marsh and marine work (and sometimes for measurements in boreholes) are generally of the piezoelectric type; these will be discussed in §7.5.3.

The moving-coil electromagnetic geophone is shown schematically in fig. 7.32a; and fig. 7.32b shows a cutaway model. The schematic diagram shows a permanent magnet in the form of a cylinder into which a circular slot has been cut, the slot separating the central South Pole from the outer annular North Pole. A coil consisting of a large number of turns of very fine wire is suspended centrally in the slot by means of light leaf springs. The geophone is placed on the ground in firm contact with it in an upright position. When the ground moves vertically, the magnet moves with it, but the coil, because of its inertia, tends to stay fixed. The relative motion between the coil and magnetic field generates a voltage between the terminals of the coil. The geophone output for horizontal motion is essentially zero because the coil is suspended in such a way that it stays nearly fixed relative to the magnet during horizontal motion.

(b) Equations of motion. The theory of geophones has been discussed in several places (Scherbatskoy and Neufeld, 1937; Washburn, 1937; Dennison, 1953; Lamer, 1970; Pieuchot, 1984). We let

z = displacement of the surface = displacement of geophone

z_c = displacement of geophone coil relative to the permanent magnet

m, r, n = mass, radius, number of turns of the coil

i = current in the coil

τ = mechanical damping factor; $\tau(dz_c/dt)$ being the damping force

S = spring constant = $f/\Delta z$, where a force f stretches the spring by Δz

H = strength of permanent magnetic field

$K = 2\pi rnH$

Ki = force on coil due to current

R, L = total resistance and inductance of the coil plus external circuit

A geophone coil in motion is acted upon by three forces: the restoring force of the springs, the force of friction, and the force resulting from the interaction of the permanent magnetic field with the magnetic

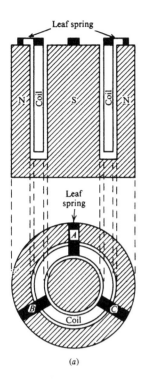

Fig. 7.32 Moving-coil electromagnetic geophone. (a) Schematic and (b) cutaway of digital-grade geophone. (Courtesy of Geo Space.)

field of the current. The first two are retarding (negative) forces, and the last is positive. Newton's second law of motion gives

$$-Sz_c - \tau\frac{dz_c}{dt} + Ki = m\left(\frac{d^2z}{dt^2} + \frac{d^2z_c}{dt^2}\right). \quad (7.8)$$

Faraday's law of induction relates z_c to i:

$$\begin{aligned}\text{emf induced in coil} &= -\frac{d\phi}{dt} = -\frac{d\phi}{dz_c}\frac{dz_c}{dt} \\ &= -2\pi rnH\frac{dz_c}{dt} = -K\frac{dz_c}{dt} \\ &= Ri + L\frac{di}{dt},\end{aligned}$$

where ϕ is the flux through the coil. Solving for z_c,

$$\frac{dz_c}{dt} = -\frac{1}{K}\left(Ri + L\frac{di}{dt}\right).$$

Differentiating eq. (7.8) and substituting for dz_c/dt gives the geophone equation of motion:

$$\begin{aligned}L\frac{d^3i}{dt^3} &+ \left(R + \frac{L\tau}{m}\right)\frac{d^2i}{dt^2} + \left(\frac{SL + \tau R + K^2}{m}\right)\frac{di}{dt} + \left(\frac{SR}{m}\right)i \\ &= K\frac{d^3z}{dt^3}. \quad (7.9)\end{aligned}$$

For a geophone output to be independent of frequency, $L = 0$ (because inductive reactance depends on frequency). Although this cannot be achieved, we assume that L is sufficiently small that we can neglect

it (see problem 7.9), giving

$$\frac{d^2i}{dt^2} + \left(\frac{\tau}{m} + \frac{K^2}{mR}\right)\frac{di}{dt} + \left(\frac{S}{m}\right)i = \left(\frac{K}{R}\right)\frac{d^3z}{dt^3}. \quad (7.10)$$

The term involving (di/dt) represents damping, τ/m giving the mechanical damping and K^2/mR the elecromagnetic damping. If the damping were zero, the system would be simple harmonic with *natural frequency* v_0, where

$$v_0 = \frac{\omega_0}{2\pi} = \left(\frac{1}{2\pi}\right)\left(\frac{S}{m}\right)^{1/2}. \quad (7.11)$$

When the damping is not zero, we write

$$\left.\begin{aligned}\frac{d^2i}{dt^2} &+ 2h\omega_0\frac{di}{dt} + \omega_0^2i = \left(\frac{K}{R}\right)\frac{d^3z}{dt^3}, \\ \text{where} \quad 2h\omega_0 &= \frac{\tau}{m} + \frac{K^2}{mR},\end{aligned}\right\} \quad (7.12)$$

h being the damping factor of eq. (2.111). This is the equation for damped simple harmonic motion, and the solution is given in standard texts (Potter and Goldberg, 1987: 43–7).

(c) Transient response. The transient solution is obtained by setting the right side of eq. (7.12) equal to zero. Let us assume that $i = 0$, $di/dt = u_0$ at $t = 0$; then the solution has the following form, depending

on the value of h:

For $h > 1$ (overdamped),

$$i = \{u_0/[\omega_0(h^2 - 1)^{1/2}]\}e^{-h\omega_0 t} \sinh [\omega_0 t(h^2 - 1)^{1/2}]; \quad (7.13)$$

For $h = 1$ (critically damped),

$$i = u_0 t e^{-\omega_0 t}; \quad (7.14)$$

For $h < 1$ (underdamped),

$$i = \{u_0/[\omega_0(1 - h^2)^{1/2}]\}e^{-h\omega_0 t} \sin [\omega_0 t(1 - h^2)^{1/2}]. \quad (7.15)$$

These solutions are shown in fig. 7.33 in terms of the resonant period T_0; they are transient solutions because i eventually becomes zero owing to the exponential factor. For $h > 1$, the current starts to build up because of the sinh factor, but then decreases as the exponential factor begins to dominate. When $h < 1$, the output is a damped sine wave. For $h = 1$, the *critically damped* case, the output just fails to be oscillatory. When $h < 1$, successive peaks occur at intervals

$$T_0 = 2\pi/\omega_0(1 - h^2)^{1/2}, \quad (7.16)$$

and the ratio of successive peaks is

$$i_n/i_{n+1} = \exp [2\pi h(1 - h^2)^{1/2}]. \quad (7.17)$$

The logarithmic decrement δ (see eq. (2.112)) in nepers (see problem 2.17) is given by

$$\delta = \ln (i_n/i_{n+1}) = 2\pi h(1 - h^2)^{1/2}; \quad (7.18)$$

we can obtain h when $h < 1$ by measuring δ.

(d) Response to a driving force. Imagine a mass suspended by a rubber band from a hand (using an illustration from Pieuchot, 1984). Moving the hand slowly but harmonically moves the suspended mass in unison with it. As the frequency of the hand motion increases (maintaining constant amplitude), motion of the mass will lag behind. As the frequency increases still further, a point will be reached where the suspended mass and the hand are 90° out of phase; this is the natural frequency v_0 (see eq. (7.11)) of the system and the mass is now moving with maximum amplitude. As the hand moves still faster, the motion of the suspended mass decreases and becomes farther out-of-phase with motion of the hand; eventually, at very high frequencies, the phase difference becomes 180° and the amplitude of the motion of the mass approaches zero. Thus, we expect amplitude and phase responses as shown in fig. 7.34. To have the amplitude of the mass relatively independent of the frequency of the hand, two things are necessary: the natural frequency must be well below the frequencies of interest and the phase shift must vary linearly with frequency. To show the latter, we start with the wave $A \cos \omega t$ and add to its phase the quantity $(k\omega + n\pi)$, where k is a constant, and n is integral; the result is the wave $\pm A \cos \omega(t + k)$. Thus, adding the linear phase shift has delayed the wave by k time units and possibly in-

verted it (if n is odd); because these changes do not vary with ω, the waveshape is unchanged.

If a geophone is subjected to a harmonic displacement (Lamer, 1970: 80–1, discusses the response to an input impulse) such that the velocity $dz/dt = V_0 \cos \omega t$, then

$$z = \frac{V_0}{\omega} \sin \omega t; \qquad \frac{dz}{dt} = V_0 \cos \omega t;$$

$$\frac{d^2z}{dt^2} = -\omega V_0 \sin \omega t; \qquad \frac{d^3z}{dt^3} = -\omega^2 V_0 \cos \omega t,$$

and eq. (7.12) becomes

$$\frac{d^2i}{dt^2} + 2h\omega_0 \frac{di}{dt} + \omega_0^2 i = -\frac{\omega^2 K V_0}{R} \cos \omega t. \quad (7.19)$$

The solution of this equation is made up of two parts: a transient solution given by eqs. (7.13) to (7.15) and a solution representing the forced motion of the geophone resulting from the motion of the ground (Potter and Goldberg, 1987: 61–2). The latter is

$$i = (V_0/Z) \cos (\omega t + \gamma) \quad (7.20)$$

where

$$\left. \begin{array}{l} Z = (R\omega_0^2/K\omega^2)\{[1 - (\omega/\omega_0)^2]^2 + (2h\omega/\omega_0)^2\}^{1/2}, \\ \tan \gamma = (2h\omega/\omega_0)/[(\omega/\omega_0)^2 - 1]. \end{array} \right\} \quad (7.21)$$

Thus, the amplitude of i for a given geophone depends upon V_0, ω/ω_0, R, K, and h. When $\omega \to \infty$, $Z \to R/K$ and the amplitude of i becomes $i_\infty = V_0 K/R$.

One of the most important factors of merit of a geophone is the output voltage per unit velocity of the case. We can define the *geophone sensitivity*, Γ (also called the *geophone transduction constant*), by the relation

$$\Gamma = \frac{\text{amplitude of output voltage}}{\text{amplitude of geophone velocity}}. \quad (7.22)$$

Assuming the geophone is connected to an amplifier with essentially infinite input impedance (the usual case), the output voltage is the voltage across R_s, the shunt resistance. Using eqs. (7.20) and (7.21), we get

$$\begin{aligned} \Gamma &= R_s(V_0/Z)/V_0 = R_s/Z \\ &= K(R_s/R)f(\omega/\omega_0), \end{aligned} \quad (7.23)$$

where

$$\begin{aligned} f(\omega/\omega_0) &= 0, & \text{when } \omega = 0, \\ &= 1, & \text{when } \omega = \infty, \\ &= \tfrac{1}{2}h, & \text{when } \omega = \omega_0. \end{aligned}$$

For practical purposes, the geophone sensitivity is determined largely by K and h, that is, by the radius and number of turns of the coil, by the magnetic field strength, and by the damping. Modern geophones have sensitivities of about 0.7 V/cm/s.

Curves of Γ are shown in fig. 7.35a for various values of h. When $h = 0$, the output becomes infinite at

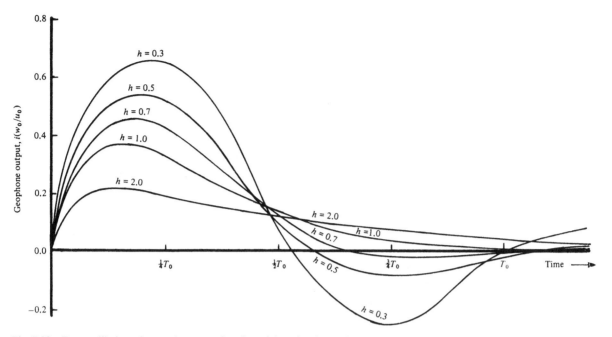

Fig. 7.33   Free oscillation of a geophone as a function of damping factor, h.

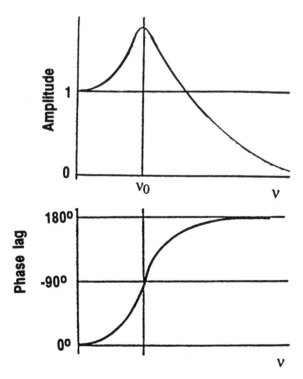

Fig. 7.34   Amplitude and phase for movement of a mass on a spring.

the natural frequency; obviously, this is merely a theoretical result because zero damping can never be achieved. As h increases, the output peak decreases in magnitude and moves toward higher frequencies. When $h \approx 0.7$, the peak disappears and the range of

flat response has its maximum extent. As h increases beyond this value, the low-frequency response falls off. The generally accepted choice of 70% of critical damping for geophones thus results in more or less optimum operating conditions with respect to amplitude distortion in the geophone output. Obviously, the damping of a geophone is a key factor in determining its performance. The damping factor (expressed by h in eq. (7.12)) can be increased by winding the coil on a metal "former" so that eddy currents induced in the former by motion of the coil will oppose the motion; h can be increased to about 0.3 by this means. The damping is usually further increased by a resistance in parallel with the coil (inside the case).

The output of a geophone is shifted in phase with respect to the input, as shown in fig. 7.35b. The phase shift γ (see eq. (7.21)) will change the waveshape (except for a linear phase shift; see the preceding), that is, produce phase distortion, because the seismic signal comprises a range of frequencies.

Figure 7.35a shows that for $h = 0.7$, the distortionless signal band extends from about $1.2\omega_0$ upward; hence, the lower the natural frequency, the wider the distortionless band. The natural frequency of geophones employed in petroleum exploration (ν_0) is usually 7 to 28 Hz for reflection work and 4.5 Hz for refraction. The decrease of sensitivity below the natural frequency (fig. 7.35a) often provides the lower limit to the passband to be recorded.

(e) Other aspects.   Geophone coils are often divided into two parts that are wound in opposite directions and so wired that signals due to motion add, whereas those that result from stray electrical pickup in the

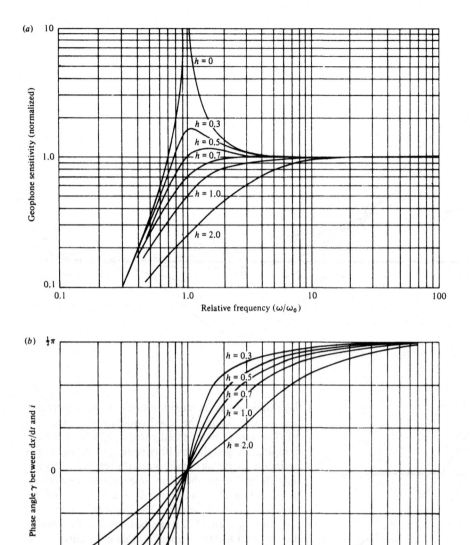

Fig. 7.35   The dependence of geophone responses on damping factor h. (After Dennison, 1953.) (a) Sensitivity and (b) phase response.

coils cancel; this feature is called *hum-bucking*. Geophones also have spurious resonances because of modes of motion other than that intended, but these usually occur at frequencies above the seismic passband.

An entire geophone group is considered to be equivalent to a single geophone located at the center of the group. Because they are generally used in groups, six to nine geophones usually are permanently connected in *strings* with 3 to 6 m of wire separating adjacent phones. Spring-loaded clips at the end of the string provide solid connections to the cable, yet are easy to connect and disconnect. These clips usually are geometrically polarized (for example, one side is wide and the other narrow) to prevent connecting them backwards. Whereas the damping of each geophone will be affected by the presence of the other geophones because of the change in resistance of the circuit, the equation of motion for an array will be the same as for a single geophone except with modified parameters. If there are n parallel branches, each of which contains n geophones in series, the resistance and damping will be the same as for a single geophone (Lamer, 1970: §3.7).

We have assumed that the geophone follows exactly the motion of the surface of the ground, but the geophone is not rigidly fastened to the ground, so the geophone–ground coupling also affects the response. We expect the coupling to affect the response just as source–ground coupling does (fig. 7.19). Lamer (loc. cit.: 88) states that coupling depends on the factor $M/\rho r^3$, where M is the geophone mass (including its case), ρ is the density of the ground involved, and r is the geophone radius. Coupling is better in compacted than in loose soils; the natural frequency of the geophone–ground system is often in the 100 to 200 Hz range, but may be as low as 30 to 40 Hz in swampy ground. The coupling can be improved by increasing the size of the geophone base, but the most common way of improving coupling is to connect it with firmer ground by pushing a spike fastened to the geophone into the ground. Of course, what is most important is to have the geophone planted firmly and vertically.

7.5.2 At-the-geophone digitization

Many geophone channels are necessary to record (1) redundant shallow data and long-offset deep data simultaneously, (2) stations distributed over an area (rather than along a line) for three-dimensional analysis, (3) individual geophones rather than arrays to improve high-frequency response (minor time shifts between different geophones within a group result in attenuation of higher-frequency components; see §8.3.11); and (4) closer geophone group spacing along the line to improve horizontal resolution.

For more than 100 channels or so, cables to carry analog electrical signals from individual channels to the recording instruments become cumbersome. Distortion also occurs in the transmission of analog signals through long cables, especially as they age. Digitizing near the geophone in remote data units (RDUs, also called *distributed systems*) overcomes many of these difficulties. The RDU units (fig. 7.36) are usually individually addressable and battery-powered so that they can be turned on or off remotely. They frequently accommodate one to eight geophone groups and often incorporate solar cells to keep the batteries charged.

The digitized data are sometimes transmitted to the recording unit in multiplexed form over a few pairs of wires or fiber-optic cable or by radio (fig. 7.37); sometimes they are stored on magnetic tape at the RDU to be picked up later. One system transmits a 76-bit burst at 640-kilobit/second rate in a time interval reserved for it, and another uses digitizer–repeater units connected in series, which relay information at a 4-megabit/second rate, adding the digitized output of the geophones at the tail end of the information stream. Another system uses sign-bit recording (as described in §9.3.6) to decrease the amount of data to be transmitted. One system transmits the data by narrow-band radio, which can accept 1536-trace data (4 traces from each of 384 RDU) in real time; another

Fig. 7.36   Battery-powered digitizing unit that filters, amplifies, and digitizes six data channels with 24-bit accuracy. (Courtesy of Input/Output, Inc.)

stores the information temporarily until called upon to transmit it. Radio links are also sometimes combined with conventional cable-linked geophones to span rivers, canyons, and other obstacles across which cables cannot be laid.

7.5.3 Hydrophones

Hydrophones, or marine pressure geophones, are usually of the piezoelectric type (Whitfill, 1970). Synthetic piezoelectric materials, such as barium zirconate, barium titanate, or lead mataniobate, are generally used. A sheet of piezoelectric material develops a voltage difference between opposite faces when subjected to mechanical bending. Thin electroplating on these surfaces allows electrical connection to be made so that this voltage can be measured. Disc hydrophones (fig. 7.38a) are essentially two circular plates of piezoelectric ceramic mounted on the ends of a hollow brass cylinder. Electrical connections are made so that if both bend inward, as they would in response to an increase of pressure outside the unit, the induced voltages add, whereas if the plates bend in the same direction, as they would in response to acceleration, they cancel (fig. 7.38b). This feature is called *acceleration canceling.* Cylindrical hydrophones (fig. 7.38c) are es-

(a)

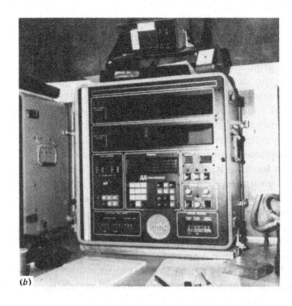

(b)

Fig. 7.37 Opseis™ telemetry system. The system is capable of recording up to 4 lines with 2 spreads/line and 1016 channels/spread. (Courtesy of L. Denham.) (a) Remote telemetry unit serving 4 geophone groups. The seismic traces are digitized and stored in the unit's memory until instructed to transmit them. (b) Program unit of the central recording system. Communication between central and remote units is by horizontally polarized RF waves.

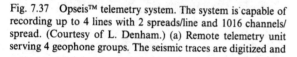

(a)

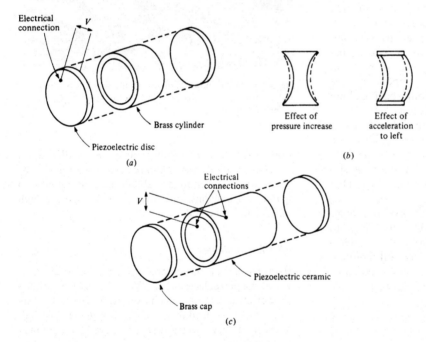

(b)

(c)

Fig. 7.38 Hydrophones. (a) Disc hydrophone, (b) acceleration-canceling feature of a disc hydrophone, and (c) cylindrical hydrophone.

sentially thin hollow piezoelectric ceramic cylinders closed at the ends by brass caps. A change in pressure outside the cylinder induces stresses in the ceramic and hence a voltage difference between the inside and outside of the cylinder.

The sensitivity of each geophone element (75 to 250 μV/Pa) is small so that three to fifty elements are usually combined in series to make up a hydrophone group; these are distributed over 3 to 50 m. Natural frequencies are of the order of 10^4 Hz, well outside the seismic range. Piezoelectric hydrophones have high impedance, so an impedance-matching transformer may be included with each group. Sometimes, charge amplifiers are used instead of transformers. Digitizers may also be located near the hydrophones so that the signal sent to the ship will not be degraded by transmission-line losses.

Hydrophones respond to changes in pressure, that

is, to the acceleration of the fluid medium (see eq. (2.67)). Moreover, the pressure and pressure gradient are proportional to the time derivative of a seismic wave (see eq. (2.68)); this means that for a harmonic wave of amplitude A, the hydrophone output is proportional to $j\omega A$, that is, the output doubles (increases 6 dB) for each octave. Hence, the response of a hydrophone differs from that of a velocity geophone by the factor j resulting in a 90° phase difference (see §15.1.5), and a rise of 6 dB/octave (due to ω). Because pressure is nondirectional, a hydrophone's output is independent of wave direction, whereas reversing the direction of wave travel inverts a geophone's output (see §7.5.5 and 9.5.4).

7.5.4 Streamers

Hydrophones are mounted in a long *streamer* towed behind the seismic ship at a depth between 10 and 20 m. A diagram of a streamer is shown in fig. 7.39. As of 1994, most streamers had 96 to 500 channels and were up to 6000 m in length.

Streamers are constructed in sections 25 to 75 m in length. Groups are usually 6.5 to 50 m in length, containing up to 35 uniformly spaced hydrophones. Data are digitized near the hydrophones with up to 12 channels per digitizer. Transmission to the ship is often by fiber-optic cable at data rates of about 7 megabits/second. Transmission in digital form eliminates distortion produced by leakage and transmission-line effects.

The hydrophones and other sensors, connecting wires, and stress members (to take the strain of towing) are placed inside a neoprene tube 7.5 to 9.0 cm in diameter (fig. 7.40). The tube is then filled with sufficient lighter-than-water liquid to make the streamer neutrally buoyant, that is, so that the average density of the tube and contents equals that of the sea water. A lead-in section perhaps 100 m long is left between the ship's stern and the streamer proper. A depressor paravane pulls the streamer down to the operating depth and a compliant section prevents shocks caused by wave action on the ship from affecting the active sections. Sometimes dead sections are inserted between active sections. The last group is followed by another compliant section and a tail buoy (fig. 7.8) that floats on the surface. The tail buoy with a radar reflector is used to locate the farthest groups in the streamer, the direction from the ship to this buoy being determined by visual or radar sighting. The tail buoy may also include equipment to locate it in the navigation system being used. The tail buoy helps retrieve the streamer if it should be broken accidentally. Depth controllers (one is shown in fig. 7.40b) are fastened to the streamer at 5 to 12 places. These sense the hydrostatic pressure and tilt vanes so that the flow of water over them raises or lowers the streamer to the proper depth; they are ineffective when the streamer is not in motion. The depth that the controllers seek to maintain can be controlled by a signal sent through the streamer; thus, the depth can be changed to accommodate changes in water depth or to allow a ship to pass over the streamer. When not in use, the streamer is stored on a large motor-driven reel (fig. 7.25a) on the stern of the ship.

The streamer contains several (perhaps 10) magnetic compasses and water-break detectors (which respond to high frequencies carried through the water from the seismic source or pinger sources) to help locate the streamer during recording (see §7.1.7). Depth detectors verify the streamer depth.

A marine detection system picks up noises of several kinds (Bedenbecker, Johnston, and Neitzel, 1970): (1) ambient noise due to wave action, shipping, marine life, and so on; (2) locally caused waterborne noise such as that caused by the turbulence generated by motion of the lead-in cable, depressor paravane, depth controllers, and tail buoy through the water, and energy radiated from the ship because of propellers, motors, and other machinery; and (3) mechanically induced noise traveling in the streamer such as results from cable strumming, tail-buoy jerking, and so on. Usually, (3) is dominant except in rough weather when (1) dominates. Towing noise is reduced by (a) making the streamer system as smooth as possible and keeping depth controllers and other deviations from a smooth streamer at least 3 m from the nearest hydrophone, (b) using a lead-in section to increase the distance between the ship and the nearest hydrophone group, and (c) using compliant and stretch sections with nylon rather than steel tensile members to reduce energy transmitted along the streamer. A separate, small, short streamer is sometimes used to record short-offset traces because there is usually an appreciable distance from the ship to the nearest group in the main streamer.

7.5.5 Matching hydrophone and geophone records

Because hydrophones respond to pressure changes and geophones respond to particle velocity, records recorded with hydrophones do not match those recorded with geophones. The mismatch is most often observed in transition-zone work, where land lines recorded with velocity phones are extended into marsh or shallow water where hydrophones are used at shallow depths. The mismatch commonly shows as an apparent phase shift of 90° (fig. 7.41c).

Reflection from the surface of the earth reverses the polarity; this makes an upcoming positive reflection into a negative downgoing reflection or an upcoming compression into a downgoing rarefaction. A buried velocity phone will see the downgoing ghost as having the same polarity as the upgoing reflection (reversed in polarity twice, once by reflection and once because it is traveling in the opposite direction), whereas a buried pressure phone, having no sense as to the direction of wave travel, will see it as having the opposite polarity. Where the ghost reflection follows soon after the primary reflection, the effect is similar to applying

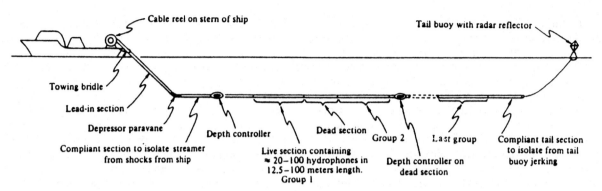

Fig. 7.39   Marine streamer. (From Sheriff, 1990.)

a derivative operator (as was seen in the reflection from the thin wedge of fig. 6.40c; see also §15.1.7). Figure 7.41 shows a minimum-phase wavelet recorded with a pressure hydrophone on the sea floor at various depths. A ghost follows the event delayed by the two-way traveltime in the water, making the waveshape nearly the same as the derivative of the waveshape from a surface geophone. The difference between the response of hydrophones and submerged geophones can be utilized to attenuate ghosting and reverberations (§9.5.4).

7.6 Recording

7.6.1 Amplifier requirements

A seismic amplifier must amplify signals over a range of frequencies and amplitudes. The frequency range is relatively small and designing to cover this range is comparatively simple. However, historically, the large range of amplitudes (fig. 7.42) has presented serious design problems.

The lower limit of signal amplitude is set by the noise level of the amplifier, around 0.2 μV for modern amplifiers; signals weaker than this are lost in system noise. The upper limit, fixed by the geophones, is generally a few tenths of a volt. If we set this at 400 mV, the signal range (or *dynamic range*) that the amplifier must accept is 126 dB. Vertical stacking further increases this range. The newest seismic amplifiers use 24 bits (23 magnitude bits plus one sign bit; §7.6.4) and have a dynamic range of 140 dB.

Accuracy of recording is also important. If we require 0.1% accuracy, we must have four significant figures, or 10 bits (§7.6.4), because $2^{10} \approx 1000$, and we require a gain of 60 dB to achieve this accuracy. If we hope to recover signals below the noise level in subsequent processing, still larger ranges are required. An amplifier with a fixed gain of x dB and dynamic range of y dB could amplify without distortion signals as strong as $(y - x)$ dB, whereas signals as weak as $(60 - x)$ dB would be recorded with the requisite accuracy. Thus, x should be small to record without distortion, but this limits the accuracy; hence, fixed gain often is not desirable, and historically provision has been made to vary the gain during the recording pe-

riod. Modern 24-bit instantaneous floating-point (IFP) amplifiers have sufficient dynamic range to overcome this problem.

7.6.2 Recording instruments

As stated in §7.6.1, recording instruments often have to compress the signal range as well as to record very weak signals. An amplifier is also used to filter the geophone output to enhance the signal relative to noise. Discussions of seismic amplifiers are given by Evenden and Stone (1971) and Pieuchot (1984).

Early recording resulted in paper field records with dynamic ranges of 20 to 26 dB (Pieuchot, loc. cit.); subsequent interpretation had to be done on these. Because the instrumentation could not handle the broad dynamic range of signals actually encountered, automatic gain control (AGC) was used almost universally to compress the signal range. The advent of reproducible recording in the early 1950s expanded the range to 50 or 60 dB, but this range was still inadequate and AGC continued to be used. In order to obtain better amplitude information, efforts were made to record the gain. These sometimes took the form of *ganged gain,* where the gain of all channels was made the same, or of *preprogrammed gain,* where the gain was determined as a function of record time before a record was made. These techniques continued to be used with early digital recording, which had gain ranges of 80 to 90 dB. Digital instruments generally multiplex the data before digitizing so that only a single channel is digitized. Today, digitizing is sufficiently rapid that one does not have to have the gain range in the proper general magnitude ahead of time.

Modern seismic amplifiers generally employ solid-state circuitry, which allows them to be very compact and rugged. Although they are usually carried in a recording truck or other vehicle, they are small and light enough that they can be hand-carried where necessary (fig. 7.43).

Today, most recording instruments output the data in digital format. However, we first describe analog instruments (which are still used somewhat) because they illustrate better the functions that amplifiers are called on to perform.

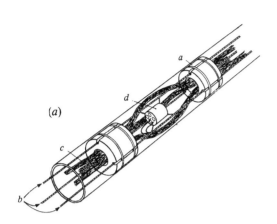

(a)

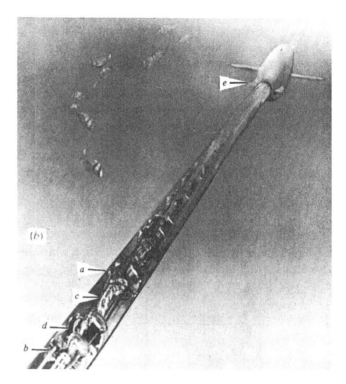

(b)

Fig. 7.40   Seismic streamer. (a) Elements of streamer: *a*, plastic spacers connected by stress members *b*; bundle of conductors *c* to carry data to the ship; *d*, hydrophone sensor. (b) Streamer in the water with depth controller, *e* (courtesy of Seismic Engineering). (c) Connecting streamer sections together (courtesy of Prakla-Seismos).

7.6.3 Analog recording

A block diagram of an analog amplifier is shown in fig. 7.44; the number and arrangement of circuit elements vary, of course. The cable from the geophones may be connected to a balance circuit that permits adjusting the impedance to ground so as to minimize the coupling with nearby power lines, thus reducing pickup of noise at the power-line frequency (*high-line pickup*). The next circuit element usually is a filter to attenuate the low frequencies that arise from strong ground roll and that otherwise might overdrive the first amplification stage and introduce distortion.

Amplifiers are multistage and have very high maximum gain, usually of the order of 10^5 (100 dB), some-

times as much as 10^7 (140 dB); 100 dB means that an input of 5 μV amplitude appears in the output with an amplitude of 0.5 V. Lower amplification can be obtained by means of a multiposition master gain switch. The gain is varied during the recording interval, starting with low amplification during the arrival of strong signals at the early part of the record and ending with the high gain value fixed by the master gain setting. This variation of gain with time (signal compression) can be accomplished with *automatic gain (volume) control*, usually abbreviated AGC or AVC. This is accomplished by a negative feedback loop, a circuit that measures the average output signal level over a short interval and adjusts the gain to keep

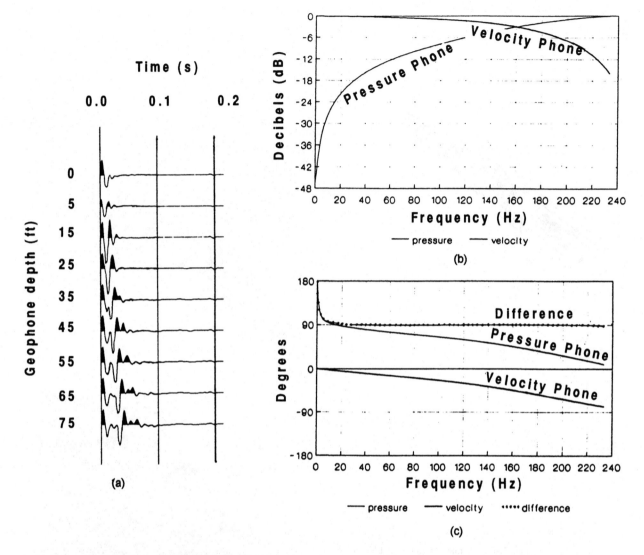

Fig. 7.41 Waveforms observed on the sea floor. ((a) From Barr and Sanders, 1989; (b and c) from Barr et al., 1989.) (a) Waveform observed by a pressure detector in various water depths for a minimum-phase wave. (b) Amplitude spectra for a geophone and a hydrophone at 1.5-m depth. (c) Phase spectra for part (b).

the output more or less constant regardless of the input level. If the time between a change of amplitude and the consequent change of gain is too short, the output amplitude will be nearly constant and reflection events will not stand out; if the time is too long, subsequent reflections will not stand out. In either case, information will be lost. The use of AGC was standard prior to the 1960s, and AGC is still used, especially in making displays. A gain trace is plotted on the record shown in fig. 7.45.

It is important in making corrections for near-surface effects that we be able to observe clearly the *first-breaks,* the first arrivals of energy at the different geophones. (For a geophone near the source, the first arrival travels approximately along the straight line from the source to the geophone; for a distant geophone, the first arrival is a head wave refracted at the base of the low-velocity layer – see the discussion of weathering corrections, §8.8.2.

If we allow the AGC to determine the gain prior to the first arrivals, the low input level (which is entirely noise) will result in very high gain; the output will then be noise amplified to the point where it becomes difficult to observe the exact instant of arrival of the first-breaks. This problem is solved by using *initial suppression* or *presuppression.* A high-frequency oscillator signal (about 3 kHz) is fed into the AGC circuit, which reacts by reducing the gain so that the noise is barely perceptible; the high-frequency signal is subsequently removed by filtering so that it does not appear in the output. With the reduced gain, the relatively strong first-breaks stand out clearly. As soon as the first-breaks have all been recorded, the oscillator signal is removed, usually by a relay triggered by one of

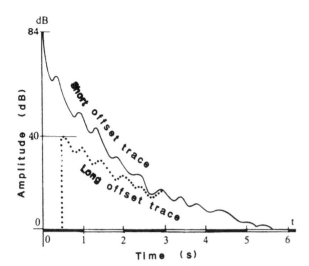

Fig. 7.42 Amplitude of signals from short-offset (solid line) and long-offset (dotted line) groups. (From Pieuchot, 1984: 55.)

the first-breaks. Thereafter, the AGC adjusts the gain in accordance with the seismic signal level (see the gain trace in fig. 7.45).

Seismic amplifiers are intended to reproduce the input with a minimum of distortion, and hence the gain (without filters) should be constant for the entire frequency spectrum of interest. For reflection work, this range is about 10 to 100 Hz, and for refraction work, the range is about 3 to 50 Hz; most amplifiers have flat response for frequencies from about 3 to 200 Hz or more.

Frequency filtering refers to the discrimination against certain frequencies relative to others. Seismic amplifiers have a number of filter circuits that permit us to reduce the range of frequencies that the amplifier passes. Although details vary, most permit the selection of the upper and the lower limits of the passband. Often it is possible to select also the sharpness of the *cutoff* (the rate at which the gain decreases as we leave the passband). Figure 7.46 shows typical filter response curves. The curves are specified by their *cutoff frequencies,* that is, the frequency values at which the gain has dropped by 3 dB (30% of amplitude, 50% of power); the curve labeled "Out" is the response curve of the amplifier without filters.

Seismic amplifiers may include circuitry for *mixing* or *compositing,* that is, combining two or more signals to give a single output. Mixing in effect increases the size of the geophone group and is sometimes used to attenuate low-frequency surface waves. The commonest form, called 50% mixing, is the addition equally of the signals from adjacent geophone groups. Magnetic tape recording now has virtually eliminated the need to mix during recording because we can always mix in playback.

The time-break signal often is superimposed on one of the amplifier outputs (on trace 3 in fig. 7.45), where it appears as a sharp pulse that marks the point $t = 0$

for the record. When explosives are being used, the output of an *uphole geophone* (a geophone placed near the top of the shothole) is also superimposed on one of the outputs (on trace 4 in fig. 7.45); the interval between the time-break and the uphole geophone signal is called the *uphole time* (t_{uh}); it measures the vertical traveltime from the shot to the surface and is important in correcting for near-surface effects (§8.8.2).

For the first 30 years or so of seismic exploration, the outputs of the amplifiers were recorded directly on photographic paper by means of a camera. However, about 1952, recording on magnetic tape began, and today it is nearly universal (see fig. 1.21). The feature that originally led to widespread use of magnetic recording was the ability to record in the field with a minimum of filtering, automatic gain control, mixing, and so on, and then introduce the optimum amounts of these on playback. Later, a more important advantage turned out to be the ability to produce record sections (see §8.8.3), which proved to be powerful aids in interpretation. However, magnetic tape recording did not develop its full potential until the introduction of digital techniques during the 1960s.

Analog magnetic tape recorders usually had heads for recording 26 to 50 channels in parallel. In the early years, direct recording was used; the output from the amplifier went directly to the recording head, the intensity of magnetization of the tape being proportional to the current in the recording head and hence proportional also to the signal strength. Later, direct recording was displaced by frequency modulation and pulse-width modulation techniques because these are more noise-free and can accept a wider range of signal strengths. Today almost all recording is digital.

7.6.4 Digital representation

Digital recording was first introduced into seismic work early in the 1960s and by 1975 was almost universal (see fig. 1.21). Whereas analog devices represent the signal by a voltage (or other quantity) that varies continuously with time, *digital recording* represents the signal by a series of numbers that denote values at regular intervals, usually 2 or 4 ms. Digital recording is capable of higher fidelity than analog recording and permits numerical processing of the data without distortion. However the beginning (geophone response) and end (display) of the recording process continue to be analog.

Before describing digital recording, we shall discuss digital representations. Although we could build equipment to handle data using the scale of 10, which forms the basis of our ordinary arithmetic, it is more practical to operate on the *binary scale* of 2. The binary scale uses only two digits, 0 and 1; hence, only two different conditions are required to represent binary numbers, for example, a switch opened or closed. Binary arithmetic operations are much like decimal ones. The decimal number 20873 is a shorthand way of saying that the quantity is equal to 3 units plus 7 ×

(a)

Fig. 7.43 Portable field recording systems. (a) Recorder for 24-bit 1016-channel distributed system. (Courtesy of Haliburton Geophysical Services.) (b) Portable 24-channel signal-enhancement seismograph incorporating liquid-crystal display (top) and thermal printer. (Courtesy of EG&G Geometrics.)

10 plus 8×10^2 plus 0×10^3 plus 2×10^4. Similarly, the binary number 1011011 is equal to 1 unit plus 1×2 plus 0×2^2 plus 1×2^3 plus 1×2^4 plus 0×2^5 plus 1×2^6, which is the same as the decimal number 91. We can use positive and negative square pulses to represent 1 and 0 or represent them in other ways. Each pulse representing 1 or 0 is called a *bit,* and the series of bits that give the value of a quantity is called a *word.*

7.6.5 Digital instruments

The stages of a digital amplifier are, of course, analog prior to the digitizing unit. These sometimes include (fig. 7.47) a *line filter* to reduce radio-frequency noise and a *notch filter* to remove excessive high-line picked up by the cables where the data are not digitized near the detectors. A *preamplifier* increases the gain by a constant amount before multiplexing and digitizing. A *low-cut filter* is usually provided to supplement geophone filtering in removing excessive ground-roll effects. An *alias filter* is always included to prevent aliasing (see §9.2.2c). A *multiplexer* connects the different channels sequentially to the digitizer; from here onward there is only one channel to the digital tape recorder. The *digitizer* (or *analog-to-digital* (*A/D*) converter) holds signal voltages in a *sample-and-hold circuit* while the voltages are compared with standards in order to measure their values. The measurement output may be in the form of a gain word plus another value. The data are then *formatted* (arranged in the proper manner) and written onto magnetic tape. Today, many systems are distributed so that digitization is done near the geophones or hydrophones (§7.5.2) in small units that handle only a few channels each.

An analog *monitor record* is made of the data written on the tape. The tape values are read by *read-after-write heads, demultiplexed* to sort them into their respective channels, an AGC is usually applied and often filtering, and the resulting voltages are fed to the camera. This is often only an 8-bit procedure because of the very limited dynamic range (about 20 dB) of a paper display.

Binary gain control, where the gain was changed in steps of 2, was used in digital amplifiers beginning in 1968. Sampling near a zero crossing of the seismic trace could produce distortion if accommodating the big gain change required by the amplitude of the next sample took too much time.

In the early 1970s, responses had become fast enough to adjust the amplitude without prior constraints and *instantaneous floating-point* (IFP) amplifiers (fig. 7.48) came into use. These consist of a series of amplifier stages all with the same fixed gain. Gates connect the outputs of the amplifier stages with the system output, only one gate at a time being open. If the output amplitude is above (or below) a preset range, the open gate is closed and an adjacent gate opened to decrease (increase) the gain. Early IFP amplifiers used seven amplifiers each with a gain of 4 and were called *quaternary-gain* amplifiers. Many IFP amplifiers use 15 or more stages, each with a gain of 2, and today 24-bit (sign plus 23 gain stages) systems are coming into use.

(b)

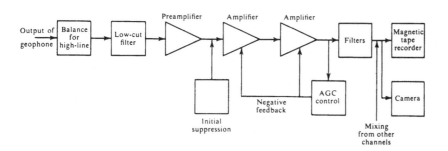

Fig. 7.44 Block diagram of analog seismic amplifier.

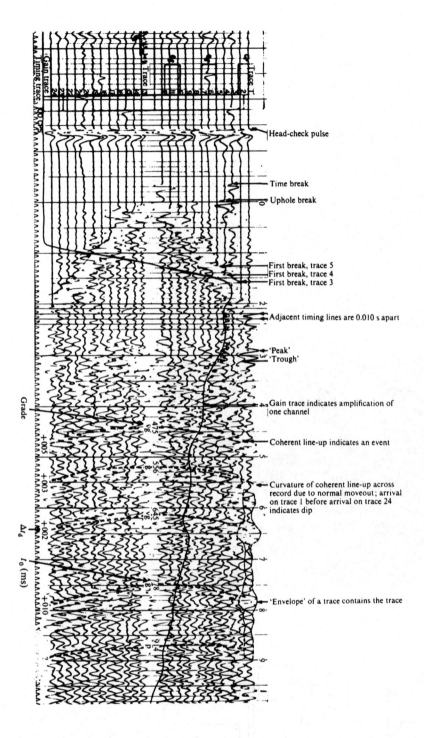

Head-check pulse

Time break

Uphole break

First break, trace 5
First break, trace 4
First break, trace 3

Adjacent timing lines are 0.010 s apart

'Peak'
'Trough'

Gain trace indicates amplification of one channel

Coherent line-up indicates an event

Curvature of coherent line-up across record due to normal moveout; arrival on trace 1 before arrival on trace 24 indicates dip

'Envelope' of a trace contains the trace

Fig. 7.45   Seismic record (playback). Courtesy of Chevron.)

Compact 12–24-channel signal-enhancement seismic recorders are available from several manufacturers; one is shown in fig. 7.43b. *Signal enhancement* means the ability to vertically stack a number of individual records. These recorders include 12–15-bit digitizers and store data on floppy disks. Their frequency range is 3 to 5000 Hz and they can sample data at 0.5 ms. They display data both on graphic displays and

hard copy using thermal printers. They weigh 15 to 25 kg and several are expandable up to 120 channels. Some also include some processing capability facilitating refraction interpretation or application of NMO corrections to reflection data.

Simplified timers are sometimes used in engineering refraction work; they measure only the traveltimes of the first arrivals and display them in digital form.

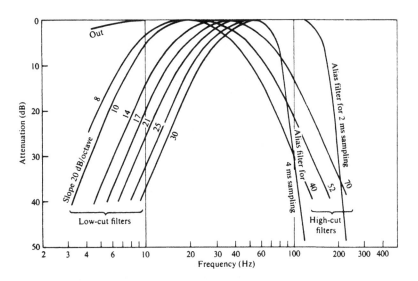

Fig. 7.46   Response of seismic filters in typical 16-bit system.

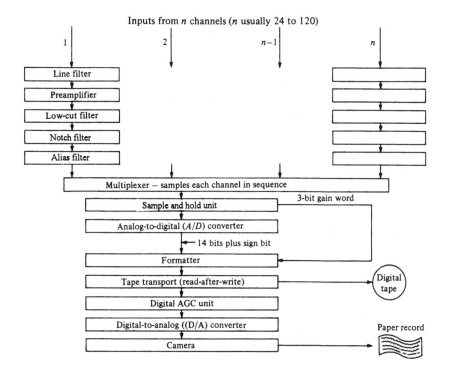

Fig. 7.47   Block diagram of digital recording system.

7.6.6 Display

The data recorded on magnetic tape must be presented in visual form for monitoring and interpretation. The classic camera consists of (1) a series of galvanometers, one for each geophone group, that transforms the electrical signals into intense spots of light moving in accordance with the signals, (2) a device for recording accurate time marks, and (3) a means for recording the positions of the light spots on a moving piece of paper. Historically, this was accomplished mainly by photographic methods. More widely used today are electrostatic cameras wherein the light spot produces an electric charge image and printing powder adheres to the paper wherever it is charged. This dry-write process uses ordinary paper, which is cheaper than photographic paper and also

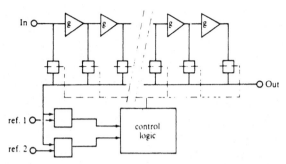

Fig. 7.48   Instantaneous floating-point (IFP) amplifier. (From Pieuchot, 1984: 177.)

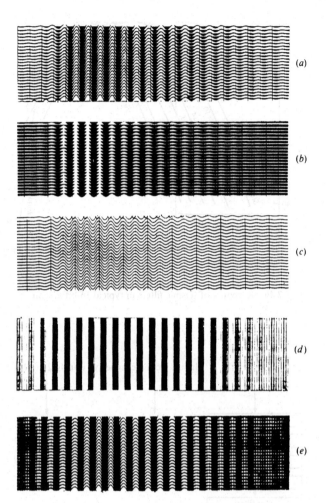

Fig. 7.49   Modes of displaying seismic data. (Courtesy of Geo Space.) (a) Wiggle superimposed on variable area, (b) variable area, (c) wiggle, (d) variable density, and (e) wiggle superimposed on variable density.

dispenses with liquid developer–fixer solutions. Some plotters, especially those in fixed installations, are of the *raster* type wherein a matrix of very fine dots is used to create the image; a very fine beam of light (often from a laser) is swept across the paper, the beam being turned on and off very rapidly to produce the dots. With a raster plotter, the information from the various channels is formatted in a microcomputer and individual galvanometers are no longer used. Ink-jet plotters, where raster dots are sprayed onto the paper, are sometimes used, especially to produce color plots.

Each individual graph representing the average motion of a group of geophones is called a *trace*. A simple graph of amplitude against arrival time is called a *wiggly trace* mode of display (fig. 7.49c). Where part of the area under a wiggly trace curve is blacked in, the display is called *variable-area* (fig. 7.49b); usually positive values (peaks) are blacked in. Sometimes the light intensity is varied instead of the light-spot position to produce *variable-density* mode (fig. 7.49d). Modes are also sometimes superimposed (figs. 7.49a and 7.49e).

Conventional black-and-white, variable-area/wiggle-trace displays have three serious shortcomings: (a) black peaks and white troughs look so different that an interpreter is biased toward the peaks, neglecting the information in the troughs, and it is very difficult to compare the relative amplitudes of a peak and adjacent trough because they look so different; (b) high-amplitude peaks are clipped so that their magnitudes are lost; and (c) horizontal positions are carried sideways by the trace excursions. A color display (e.g., Plate 2) corrects for these shortcomings. By seeing adjacent peaks and troughs with equal clarity, reflections from the top and bottom of a reservoir can often be recognized because their amplitudes vary in unison.

The mode of display and parameter choices greatly affect what an interpreter sees in the data. Among display parameters are horizontal and vertical display scales and trace spacing; width, amplitude, and clip level (maximum amplitude that can be plotted) of wiggly traces; degree of blackness, bias (minimum value, which will be blacked in) and clip level of variable-area traces; and so on. Usually, the effective vertical scale (time scale) is greater than the horizontal scale, that is, sections are horizontally compressed. The vertical scale is, of course, variable with depth when time is plotted linearly, as is usually the case. Scale ratios of approximately 1:1 are most helpful when making a structural interpretation, but considerable vertical exaggeration is often used for a stratigraphic interpretation. Color is sometimes superimposed on sections to display additional information.

Problems

7.1 The velocity of radio waves has the following values (km/s) over various terrains: normal sea water, 299,670; fresh water, 299,250; normal farmland, 299,400; dry sand, 299,900; mountainous terrain, 298,800. If range calculations are based on travel over normal sea water, what are the errors in range per kilometer of path over the various terrains?

7.2 If the error in Shoran time measurements is ±0.1 μs, what is the size of the parallelogram of error in fig. 7.3 when (a) θ = 30° and (b) θ = 150°? Take the velocity of radiowaves as 3×10^5 km/s.

7.3 A satellite is in a stable orbit around the Earth when the gravitational force (*mg*) pulling it earthward equals the centrifugal force mV^2/R, where *g* is the acceleration of gravity, *m* and *V* the satellite's mass and velocity, respectively, and *R* the radius of its orbit about the center of the Earth.

(a) Determine the acceleration of gravity at the orbit of a Transit satellite 1070 km above the Earth, knowing that *g* at the surface of the Earth is 9.81 m/s² and that the gravitational force varies inversely as the square of the distance between the centers of gravity of the masses.

(b) What is the satellite's velocity if its orbit is stable?

(c) How long does it take for one orbit?

(d) How far away is the satellite when it first emerges over the horizon?

(e) What is the maximum time of visibility on a single satellite pass? (Assume the radius of the Earth is 6370 km.)

7.4 Sieck and Self (1977) summarize "acoustic systems," as shown in table 7.2. For each of these calculate the following:

(a) The wavelengths.

(b) The penetration given by Denham's rule (§8.3.11) and reconcile with the stated purposes.

(c) Trade literature claims 30-cm resolution with imploders and 2–5-m resolution with sparkers. How do these figures compare with the resolvable limit (§6.4.2)? (Note that absorption in water is very small, so that effectively absorption does not begin until the sea floor is reached.)

7.5 (a) An explosion initiated at the top of a column of explosives of length $a\lambda_r$ travels down the column with velocity V_c. By comparison with the same amount of explosive concentrated at the center of the column and exploded instantaneously at the same time as the column, show that the array response *F* is

$$F = -\text{sinc}\,[\pi a(\sin\theta_0 - V_r/V_c)],$$

V_r being the velocity in the rock, and θ_0 the same as in fig. 7.14a. Under what circumstances does this result reduce to that of eq. (7.4)?

(b) Calculate *F* for a column 10 m long given that $\lambda_r =$

40 m, V_c = 5.5 km/s, V_r = 2.1 km/s, and θ_0 = 0°, 30°, 60°, 90°.

(c) If the column in (b) is replaced by six charges, each 60 cm long, equally spaced to give a total length of 10 m, the charges being connected by spirals of detonating cord with velocity of detonation 6.2 km/s, what length of detonating cord must be used between adjacent charges to achieve maximum directivity downward?

(d) What are the relative amplitudes (approximately) of the waves generated by the explosives in (c) at angles α_0 = 0°, 30°, 60°, and 90° when λ_r = 40 m?

7.6 Imagine an impulsive source striking the ground at times $n\Delta$ apart, where *n* is a random number between 10 and 20, and Δ is the sampling interval. Given reflections with amplitude +5 at 0, +2 at 5Δ, −1 at 13Δ, +3 at 29Δ, +1 at 33Δ, and −2 at 42Δ, add the reflection sequence as would be done with Sosie recording (§7.3.2) for 10, 20, and 30 impulses to see how the signal builds up as the multiplicity increases.

7.7 How much energy is released (approximately) by the air-gun array in fig. 7.28c when the initial pressure is 2000 psi (14 MPa). (Energy released = work done by the expanding gas = $\int \mathcal{P}\,dV$.) Assume that the change is adiabatic, that is, $\mathcal{P}V^{1.4}$ = constant, the final pressure is 2 atmospheres, and that the guns are far enough apart that they do not interact.

7.8 The dominant period of a marine seismic waveshape is often determined by the source depth, that is, by the second half-cycle being reinforced by the ghost reflected at the surface. Assuming that this is true for the source signatures shown in fig. 7.30, determine their depths.

7.9 If we wish to take it into account approximately the small term $L\,d^3i/dt^3$ in eq. (7.9) (still neglecting other terms involving *L*), show that for a harmonic wave, it can be included in the term involving *h* in eq. (7.12).

7.10 A 96-channel streamer with 25-m groups has the hydrophones spaced uniformly throughout its length. The lead-in and compliant sections together are 200 m in length and the tail section and buoy connection are 150 m. Assume a ship's speed of 5.8 knots (3.0 m/s) and a current perpendicular to the direction of traverse with a speed of 1.9 knots.

(a) What are the perpendicular and in-line components of the distance to the farthest active group with

Table 7.2 *Acoustic systems*

System	Frequency (kHz)	Purpose
Fathometers	12–80	To map water bottom
Water-column bubble detectors	3–12	To locate bubble clusters, fish, etc.
Side-scan sonar	38–250	To map bottom irregularities
Tuned transducers	3.5–7.0	To penetrate 30 m
Imploders	0.8–5.0	To penetrate 120 m and find gas-charged zones
Sparker	0.04–0.15	To map to 1000 m

respect to the traverse direction?

(b) If the velocity to a reflector 2.00 km below the ship is 3.00 km/s and if the reflector dips 20° perpendicular to the traverse direction: (i) By how much will the arrival time be changed for the far trace? (ii) If this should be attributed to a change in velocity rather than cross-dip, what velocity would it imply?

(c) Assume that the amount of *streamer feathering* (drift of the streamer to one side) is ascertained by radar sighting on the tail buoy with an accuracy of only ±3°: (i) How much uncertainty will this produce in locating the far group? (ii) How much change in arrival time will be associated with this uncertainty?

(d) Over what distance will the midpoint traces that are to be stacked when making a CMP stack be distributed?

7.11 Use figs. 7.35 and 7.46 to determine the filter equivalent to a geophone with $v_0 = 10$ Hz and $h = 0.7$ feeding into an amplifier with a 10–70-Hz bandpass filter and a 4-ms alias filter.

7.12 Figure 7.50 illustrates filter characteristics. Evaluate the importance of (a) low-frequency cut, (b) high-frequency cut, (c) bandwidth, and (d) filter slope on: (i) time delay to a point that could be timed reliably; (ii) apparent polarity; and (iii) ringing. The conclusions can be generalized for filters of other design types.

7.13 Figure 7.51 shows waveshape changes produced by the analog filtering in modern digital instruments. What can you conclude about the effects on picking?

7.14 Express the numbers 19 and 10 as binary numbers.

(a) Add the binary numbers together and convert the sum to a decimal number.

(b) Multiply the two binary numbers and convert to decimal.

(Note that mathematical operations are carried out in binary arithmetic in the same way as in decimal arithmetic.)

7.15 Assume a 96-channel seismic system recording with 2-ms sampling and 25-s Vibroseis records. What is the data rate (samples/second) and the number of bits/record? How does the data rate compare with the capacity of a 9-track magnetic tape moving at a 6250-bytes/inch rate, using 4 bytes/sample? How many bits of memory are required to store one channel of data? What is the effect of the header and ancillary information, and parity bits?

References

Andrew, J. A. 1985. The art and science of interpreting stratigraphy from seismic data. In *Seismic Exploration of the Rocky Mountain Region*, R. R. Gries and R. C. Dyer, eds., pp. 95–104. Denver: Rocky Mountain Association of Geologists and the Denver Geophysical Society.

Barbier, M. G., and J. R. Viallix. 1973. Sosie – A new tool for marine seismology. *Geophysics*, **38**: 673–83.

Barr, F. J., and J. I. Sanders. 1989. Attenuation of water-column reverberations using pressure and velocity detectors in a water-bottom cable. Paper read at the 59th Society of Exploration Geophysicists Annual Meeting.

Barr, F. J., R. N. Wright, W. L. Abriel, J. I. Sanders., S. E. Obkirchner, and B. A. Womack. 1989. A dual-sensor bottom-cable 3-D survey in the Gulf of Mexico. Paper read at the 59th Society of Exploration Geophysicists Annual Meeting.

Bedenbecker, J. W., R. C. Johnston, and E. B. Neitzel. 1970. Electroacoustic characteristics of marine streamers. *Geophysics*, **35**: 1054–72.

Brede, E. C., R. C. Johnston, L. R. Sullivan, and H. L. Viger. 1970. A pneumatic seismic energy source for shallow-water/marsh areas. *Geophys. Prosp.*, **18**: 581–99.

Burns, R. F. 1992. GPS receivers – A directory. *Sea Technology* (March): 13–18.

Dennison, A. T. 1953. The design of electromagnetic geophones. *Geophys. Prosp.*, **1**: 3–28.

Dixon, R. C. 1992. Global positioning system. In *Encyclopedia of Earth System Science*, W. A. Nierenberg, ed., pp. 395–407. New York: Academic Press.

Evenden, B. S., and D. R. Stone. 1971. *Seismic Prospecting Instruments, Vol. 2: Instrument Performance and Testing.* Berlin: Gebruder-Borntraeger.

Farriol, R., D. Michon, R. Muniz, and P. Staron. 1970. Study and comparison of marine seismic source signatures. Paper read at the 40th Society of Exploration Geophysicists Annual Meeting.

Geyer, R. L. 1989. *Vibroseis*, Geophysics Reprint Series. Tulsa: Society of Exploration Geophysicists.

Giles, B. F. 1968. Pneumatic acoustic energy source. *Geophys. Prosp.*, **16**: 21–53.

Goupillaud, P. L. 1976. Signal design in the Vibroseis technique. *Geophysics*, **41**: 1291–1304.

Harrison, E. R., and L. M. Giacoma. 1984. A new generation air gun. Paper read at the 54th Society of Exploration Geophysicists Annual Meeting, Atlanta.

Ingham, A. 1975. *Sea Surveying.* New York: John Wiley.

Jensen, M. H. B. 1992. GPS in offshore oil and gas exploration. *The Leading Edge*, **11**(11): 30–4.

Kramer, F. S., R. A. Peterson, and W. C. Walters, eds. 1968. *Seismic Energy Sources – 1968 Handbook.* Pasadena: Bendix United Geophysical.

Kronberger, F. P., and D. W. Frye. 1971. Positioning of marine surveys with an integrated satellite navigation system. *Geophys. Prosp.*, **19**: 487–500.

Lamer, A. 1970. Couplage sol-geophone. *Geophys. Prosp.*, **18**: 300–19.

Mayne, W. H., and R. G. Quay. 1971. Seismic signatures of large air guns. *Geophysics*, **36**: 1162–73.

McQuillin, R., M. Bacon, and W. Barclay. 1979. *An Introduction to Seismic Interpretation.* Houston: Gulf Publishing Co.

Miller, G. F., and H. Pursey. 1956. The field and radiation impedance of mechanical radiators on the free surface of a semi-infinite isotropic solid. *Proc. Royal Soc.*, **A-223**: 321.

Miller, R. O., S. E. Pullan, J. S. Waldner, and F. P. Haeni. 1986. Field comparison of shallow seismic sources. *Geophysics*, **51**: 2067–92.

Mossman, R. W., G. E. Heim, and F. E. Dalton. 1973. Vibroseis applications to engineering work in an urban area. *Geophysics*, **38**: 489–99.

Musser, D. D. 1992. GPS/DGPS in offshore navigation, positioning. *Sea Technology* (March): 61–6.

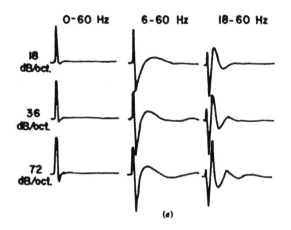

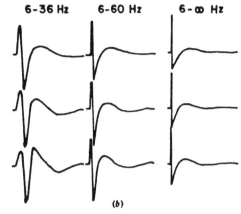

Fig. 7.50 Impulse responses of minimum-phase filters. The respective rows differ in filter slopes and the columns in passbands

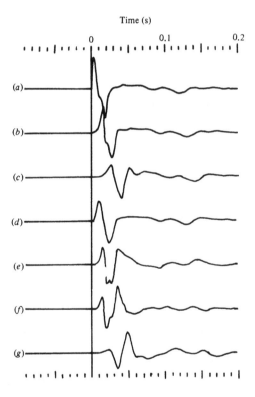

Fig. 7.51 Far-field air-gun signatures through various instrument filters. (a) No extra filtering; (b) out–124 Hz, 18 dB/octave; (c) out–124 Hz, 72 dB/octave; (d) out–62 Hz, 18 dB/octave; (e) 8–124 Hz with slopes of 18 and 72 dB/octave on low- and high-frequency sides, respectively; (f) 18–124 Hz with 18 and 72 dB/octave slopes; and (g) 8–62 Hz with 36 and 72 dB/octave slopes. Timing marks are 10 ms apart.

(specified by 3-dB points). (Courtesy of Grant-Norpac.) Effect of (a) low-cut filtering and (b) high-cut filtering.

Potter, M. C., and J. L. Goldberg. 1987. *Mathematical Methods.* Englewood Cliffs, N.J.: Prentice Hall.

Poulter, T. C. 1950. The Poulter seismic method of geophysical exploration. *Geophysics,* **15:** 181–207.

Proffit, J. M. 1991. A history of innovation in marine seismic data acquisition. *The Leading Edge,* **10(3):** 24–30.

Rayleigh, Lord. 1917. On the pressure developed in a liquid during the collapse of a spherical cavity. *Phil. Mag.,* **34:** 94–8.

Scherbatskoy, S. A., and J. Neufeld. 1937. Fundamental relations in seismometry. *Geophysics,* **2:** 188–212.

Schulze-Gatterman, R. 1972. Physical aspects of the airpulser as a seismic energy source. *Geophys. Prosp.,* **20:** 155–92.

Sheriff, R. E. 1974. Navigation requirements for geophysical exploration. *Geophys. Prosp.,* **22:** 526–33.

Sheriff, R. E. 1989. *Geophysical Methods.* Englewood Cliffs, N.J.: Prentice Hall.

Sheriff, R. E. 1990. *Encyclopedic Dictionary of Exploration Geophysics,* 3d ed. Tulsa: Society of Exploration Geophysicists.

Sieck, H. C., and G. W. Self. 1977. Analysis of high-resolution seismic data. In *Seismic Stratigraphy – Applications to Hydrocarbon Exploration,* C. E. Payton, ed., pp. 353–86, AAPG Memoir 26. Tulsa: American Association of Petroleum Geologists.

Spradley, H. L., 1976. Analysis of position accuracies from satellite systems – 1, 1976 update. In *1976 Offshore Technology Conference Preprints,* paper 2462. Dallas: Offshore Technology Conference.

Spradley, H. L., 1984. *Surveying and Navigation for Geophysical Exploration.* Boston: International Human Resources Development Corp.

Washburn, H. W. 1937. Experimental determination of the transient characteristics of seismograph apparatus. *Geophysics,* **2:** 243–52.

Waters, K. H. 1987. *Reflection Seismology,* 3d ed. New York: John Wiley.

Whitfill, W. A. 1970. The seismic streamer in the marine seismic system. In *1970 Offshore Technology Conference Preprints,* paper 1238. Dallas: Offshore Technology Conference.

Willis, H. F. 1941. Underwater explosions – Time interval between successive explosions. *British Report,* **WA-47:** 21.

Wood, L. C., R. C. Heiser, S. Treitel, and P. L. Riley. 1978. The debubbling of marine source signatures. *Geophysics,* **43:** 715–29.

Parkes, G., A. Ziolkowski, L. Hatton, and T. Haugland. 1984. The signature of an air gun array: Computation from near-field measurements including interactions – Practical considerations. *Geophysics,* **49:** 105–11.

Pieuchot, M. 1984. *Handbook of Geophysical Exploration, Vol. 2: Seismic Instrumentation.* London: Geophysical Press.

Ziolkowski, A. 1980. Source array scaling for wavelet deconvolution. *Geophys. Prosp.*, **28:** 902–18.

Ziolkowski, A. 1984. The Delft airgun experiment. *First Break*, **2(6):** 9–18.

Ziolkowski, A., G. Parkes, L. Hatton, and T. Haugland. 1982. The signature of an air gun array: Computation from near-field measurements including interactions. *Geophysics*, **47:** 1413–21.

8
Reflection field methods

Overview

Field methods for the acquisition of seismic reflection data vary considerably, depending on whether the area is land or marine, on the nature of the geologic problem, and on the accessibility of the area. One of the most important aspects in controlling data costs is avoiding delays such as when some phases of operations have to wait on other phases before work can begin. High-quality field work is essential because nothing done subsequently can remedy defects in the basic data. Evans (1989) and Pritchett (1990) deal with field techniques.

The organizations of field crews who acquire seismic data and procedures for carrying out surveys are described. The common-midpoint (CMP) method is the field method used almost exclusively today. Usually, one wants data to be acquired in the same manner along straight lines so that observed changes in the data may be ascribed to geologic rather than acquisition changes. Practical constraints that restrict acquisition are discussed.

An array of geophones usually feeds each data channel; source arrays are also often used. Arrays have response characteristics that depend on the spectrum and velocity of a wave and the direction from which it comes; these properties are used to attenuate certain types of noise. The selection of field parameters depends on both geologic objectives and noise conditions. Special situations and objectives sometimes require special techniques, such as undershooting, crooked-line, extended resolution, and uphole surveys.

Marine surveys acquire data at a very fast rate and high hourly cost, fundamental facts that distinguish marine operations. Shallow water and obstructions sometimes control acquisition. Special methods may be required in the transition region near a coastline, composing the surf zone, beach, and lagoonal areas inland from the beach; in this zone, environments generally change rapidly.

Corrections have to be made for elevation and weathering variations to prevent them from influencing (distorting and sometimes completely obscuring) the reflection data on which interpretation is based. The corrections calculated by the field crew are the first, and often the most important, corrections. Additional (or residual) corrections are subsequently made in data processing.

Refraction data acquisition is discussed separately in chap. 11, 3-D acquisition in chap. 12, and S-wave, vertical seismic profiling, and crosshole acquisition in chap. 13.

8.1 Basic considerations

8.1.1 Data acquisition

Virtually all seismic acquisition today is performed by geophysical contractors, either for oil- or gas-company clients or on a speculative basis for subsequent sale. The latter probably constitutes 20 to 25% of U.S. acquisitions as of 1994. Acquisition methods have become fairly standardized and, contrary to earlier beliefs, clients generally no longer believe that their own field methods provide a significant competitive edge over their competitor's. Speculative data costs considerably less because costs are distributed over several clients, and lower unit costs permit acquiring more data. Where tracts come up for competitive bidding, companies often feel that they have to buy most available data to avoid the possibility that their competition has an advantage.

The data that result from work done for a sole client belong to the client, who can use the data exclusively, trade the data for data owned by others, or sell the data. The data from speculative surveys belong to the contractor who paid for them or to the group of companies that subsidized the acquisition. The terms of sale usually place restrictions on the purchasers as to who is permitted to see and use the data and also restrictions on future sales by the data owners. In some countries, data go into the public domain after some specified period of time.

Most acquisition in the United States in 1994 is on a turnkey basis, where payment is on a per data-unit basis rather than a time-required basis.

8.1.2 Crew organization

Seismic crews differ greatly in size, ranging from two or three people for a shallow land survey for engineering objectives to more than a hundred people for surveys in jungle areas where many men are required to cut trails and bring in supplies. Consequently, the organization of the crew varies, but those shown in fig. 8.1 are representative for land crews.

A supervisor, or *party chief,* usually a professional

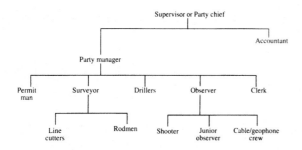

Fig. 8.1 Seismic crew organization.

geophysicist, has the overall responsibility for a field crew. He is often assisted by an administrator or office manager, especially when many personnel are involved.

A *party manager* is usually responsible for field operations. His main responsibility is to obtain maximum production and adequate quality at reasonable cost. Other field personnel report to him; he also hires field helpers. He is responsible for safety, equipment maintenance, maintaining adequate supplies, paying bills, and operation of the field camp where required.

The *surveyor* has the responsibility of locating survey points in their proper places. As the advance man on the ground, he anticipates difficulties and problems that the survey will encounter and seeks to avoid or resolve them. This involves investigating alternatives so that the survey objectives may be achieved at minimum cost. He determines the best access routes for subsequent units. He may be assisted by a *permit man,* who contacts land owners and tenants and secures permissions to conduct the survey. He is also assisted by *rodmen* who help with measurements. In areas of difficult access, he may also supervise brush cutters and bulldozer operators who clear the way.

The *observer* is usually next after the party manager in field authority. He is responsible for the actual field layouts and data acquisition, including operation of the instruments. He is usually assisted by a junior observer and a crew of *jug hustlers* who lay out the cable and geophones.

Other members of a field crew vary depending on the nature of the survey. A crew may have one to four drillers, occasionally more, and assistants to help drill and haul water for the drilling operations, or two to perhaps five operators of surface source units (see §7.2.3 and 7.3.1). A *shooter* is responsible for detonating explosives at the proper time and for cleaning up the shothole area afterwards. Cooks and mechanics may be included where operations are performed out of field camps.

A marine seismic crew usually consists of a party manager, chief observer or instrument engineer, three or four junior observers, two navigation engineers, a chief mechanic, and three or more mechanics. This is supplemented by the ship's crew of captain, mates, engineers, deck hands, and cook and mess/cabin attendants.

8.1.3 Environmental and safety considerations

A seismic crew is not only responsible for any damages from its operations, but also for environmental considerations and safety. Crews have an obligation to minimize the environmental impact of their operations, which should be planned and executed in such a way as to minimize changes in the land (International Association of Geophysical Contractors, 1993). This has not always been the situation; in some areas in former times, the trails that seismic crews cut (often 5 m wide) were apt to be used to "open up the country," but attitudes have changed and seismic operations today should be as unobtrusive as the work permits. Some of the tracks made years ago through forests, swamps, tundra, and deserts are still evident and are cited as arguments against future work. New tracks should be of minimal width and the land should be restored as nearly as possible to its former condition.

Crew and public safety also should be continually on the minds of all crew members. Crews should hold safety meetings periodically to remind crew members of safety concerns. The safety manuals published by the International Association of Geophysical Contractors (1991a, 1991b) should be reviewed by all crew members periodically. Every accident should be studied so that the causes may be rectified in order to avoid similar accidents in the future. Any outdoor work involves dangers from falls, cuts, infections, insect bites, and poisonous plants, and any work with machinery involves hazards. Often, the greatest hazards of all are related to the operation of vehicles.

Public relations is another concern of field work. Courtesy calls should be made on those apt to be affected by the field work or influential in informing the public, even where one has no obligation to do so. Where field work is uncommon, an education program may be required to inform the public as to what seismic operations involve.

8.1.4 Conduct of a field survey

Most seismic field crews today are operated by contractors who acquire the data for client oil companies. Usually, the process begins with a bid request sent out by a client. If experience in the survey area is lacking, prospective contractors scout the area, often accompanied by client personnel, to form opinions about the equipment required and problems likely to be encountered. These matters and any anticipated conflicts are discussed. The contractor estimates costs, suggests modification of specifications, and prepares a bid. The client evaluates bids from the various contractors and selects the best bid. The client and winning contractor meet to resolve changes in the specifications that may have developed in the course of the bidding and a contract for the work results.

The contractor begins equipment preparation and sends an advance group to the field to arrange for office space and personnel accommodations, communications permits, supply, storage, and repair facilities, initiates permitting operations, and recruits local labor. A client representative may participate in some of these activities. Once permitting is complete enough, survey layout is established and the main body of the contractor's equipment and personnel arrive. After some field experimentation, the survey proper gets underway.

The data are prepared for processing and periodically transferred to the processing center. The survey results are also regularly transmitted to the client representative. Unexpected problems will inevitably have arisen during the survey that will have to be resolved between the party chief and the client representative. There probably will be modifications or extensions to the program. Once the field work is concluded, the crew and equipment will be reassigned to the next project, and the party chief will prepare and present a final report to the client.

8.2 Field operations for land surveys

8.2.1 The program

The program of work is usually dictated by the clients, but the conduct of the work is the contractor's responsibility. Acquisition procedures are often developed in meetings between company and contractor geophysicists. A representative of the client company ("birddog") may be attached to the field crew while the work is being done to monitor the work and alter the program in the light of results. Speculative work is done in much the same manner except that the "client" is the same company.

Before beginning a survey, the question should be asked, "Is it probable that the proposed work will provide the required information?" Good practice (Agnich and Dunlap, 1959) is to "shoot the program on paper" before beginning the survey, estimating what the data are likely to show, anticipating problems that may occur, asking what alternatives are available and how data might be obtained that will distinguish between alternative interpretations.

Data migration (§9.12) may require that lines be located elsewhere than directly on top of features in order to measure critical aspects of a structure. Crestal areas may be so extensively faulted that lines across them may be nondefinitive. The structures being sought may be beyond seismic resolving power. Lines may cross features such as faults so obliquely that their evidences are not readily interpretable. Lack of cross control may result in features located below the seismic line being confused by features to the side of the line. Near-surface variations along a proposed line may be so large that the data are difficult to interpret, whereas moving the seismic line a short distance may improve data quality. Obstructions along a proposed line may increase difficulties unnecessarily, whereas moving the line slightly may achieve the same objec-

tives at reduced cost. Where the dip is considerable, merely running a seismic line to a wellhead may not tie the seismic data to the well data. Lines may not extend sufficiently beyond faults and other features to establish the existence of such features unambiguously or to determine fault displacements. In general, lines should extend with full coverage beyond the area of interest to a distance equal to the target depth.

8.2.2 Permitting

Once the seismic program has been decided, it is usually desirable (or necessary) to meet with the owners and/or leasors of the land to be traversed. Permission to enter lands to carry out a survey may involve a payment, sometimes regarded as advance payment "for damages that may be incurred." Even where surface holders do not have the right to prevent entry, it is advantageous to explain the nature of impending operations. Of course, a seismic crew is responsible for damages resulting from their actions whether or not permission is required to carry out the survey.

8.2.3 Laying out the line

Once the preliminary operations have been completed, the survey crew lays out the lines. This is often done by a transit-and-chain survey that determines the positions and elevations of both the sourcepoints and the centers of geophone groups. The chain is often a wire equal in length to the geophone group interval. Successive group centers are laid out along the line using this chain, each center being marked in a conspicuous manner, commonly by means of brightly colored plastic ribbon called *flagging*. The transit is used to keep the line straight and to obtain the elevation of each group center by sighting on a rod carried by the lead chainman. The survey may be tied to points that have been surveyed in with higher precision, perhaps by use of electromagnetic distance measurements (§7.1.1) or GPS (§7.1.5), to avoid accumulating errors, and side shots are made to relate nearby structures, streams, roads, fences, and other features to the line location. Radiopositioning systems (§7.1.3) are sometimes used for horizontal control, especially in marsh and shallow-water areas where elevation control can be obtained from the water level.

A surveyor's field notes should be sufficiently complete that another surveyor can accurately reconstruct the survey from them. With electromagnetic surveying equipment, measurements and survey notes may be recorded on magnetic tapes or floppy disks that can be input into a personal computer after the day's field work. The computer then reduces the survey data, adjusts closure errors, and plots updated maps daily.

One of the surveyor's responsibilities is to plan access routes for the units that follow. In areas of difficult terrain or heavy vegetation, trail-building or trail-clearing crews may be required. These are often under the direct supervision of the surveyor.

8.2.4 Field procedures

When the energy source is explosives, the surveyor is followed by shothole drillers. Depending on the number and depth of holes required and the ease of drilling, a seismic crew may have from 1 to 10 drilling crews. Whenever conditions permit, the drills are truck-mounted. Water trucks are often required to supply the drills with water for drilling. In areas of rough terrain, the drills may be mounted on tractors or portable drilling equipment may be used. In swampy areas, the drills are often mounted on amphibious vehicles. In desert areas, air instead of water or mud may be used as the circulating medium. Where there is hard rock at the surface, percussion drilling is occasionally used; the drill tool is repeatedly dropped onto the rock to break it up. Usually, the drilling crew places the explosive in the holes before leaving the site. Drilling is often a major part of data-acquisition costs.

When surface-energy sources are used, there is of course no shothole drilling. The sources, often consisting of four to five truck-mounted units, move into position and await instructions from the recording crew. Despite the fact that no explosives are involved, terms such as "shot" and "shotpoint" are still sometimes used; often "vibrator point" is used with Vibroseis.

The recording crew can be divided into three units: (1) the source unit responsible for positioning and activating the surface-energy sources or for loading (when required) and firing the explosives; (2) the jug hustlers who lay out the cables, place the geophones in their proper locations, and connect them to the cables, and subsequently pick up the geophones and cables; and (3) the recording unit that does the actual recording of the signals.

After the cables and geophones are laid out and tested, the observer checks that all geophones are connected, that the amplifiers and other units of the recording system are properly adjusted, and that everything is ready for a recording. Finally, he signals the source units via radio or connecting wire (telephone) to activate the sources or to fire the explosive.

When all is ready for a shot (if explosives are being used), the shooter arms his *blaster*, the device used to set off the explosive, by a safety switching arrangement, and advises the observer that he is ready. The observer then presses an "arm" button that causes a "tone" to be transmitted to the shooter and starts the recording system. A signal sent from the recording equipment actually fires the shot. The blaster then transmits back to the recording equipment the shot instant (*time-break*).

When a seismic crew uses surface-energy sources, the source units move into place and a signal from the recorder activates the sources so that the energy is introduced into the ground at the proper time. The energy from each surface source is usually small compared to the energy from a dynamite explosion, so

that many records are made for each sourcepoint and subsequently vertically stacked (§6.7.2) to make a single record. Several source units generally are used and these usually advance a few meters between the component "subshots" that will be combined to make one profile. It is not uncommon to use three or four source trucks and to combine 20 or so component subshots.

After the data are recorded, the observer studies a monitor record to see that the record is free of obvious defects. The monitor record is not used for interpretation, but may be used to determine weathering corrections, discussed in §8.8.2. When finished with the recording at one source location, a *roll-along switch* connects the proper elements for the next record and the source crew moves on. Sometimes this roll-along duty is performed by the instrument software. A computer does some of the checking and recording.

With the standard singlefold recording method (§8.3.2) used before reproducible recording, interpretation had to be done on the paper records obtained in the field, and considerable effort was made to get the best records possible. A geophysicist would examine each record immediately after it was acquired to decide on changes in recording conditions. He would vary explosive size and depth, field layout, and instrument settings in an effort to improve the record. Several shots were generally taken in each borehole, drills sometimes standing by to redrill a hole that might be lost.

The high production and high efficiency needed in order to achieve low cost per kilometer have altered field procedures. With common-midpoint recording, source points are close together, usually 25 to 100 m (75 to 300 ft) apart compared with 400 to 600 m for singlefold recording. The redundancy of coverage lessens the dependence on any individual record, so that occasional missed records can be tolerated. Also, the broad dynamic range of digital recording has removed most of the need to tailor instrument settings to particular local conditions and for filtering in the field. The goal of field recording (and subsequent processing) is generally to have conditions the same for every element, so that changes in the data may be attributed to geologic changes rather than changes in the field conditions.

Cost considerations dictate that the recording operation must not wait on other units. Shotholes may be drilled for the entire line before recording even begins so that the recorder never waits on the drills. Extra cables and geophones are laid out and checked in advance. The roll-along switch makes it possible for the recording unit to be located physically at a place different from where it is located electrically. The recording unit connects to the cable at any convenient location, for example, the intersection of the seismic line and a road. The roll-along switch is adjusted so that the proper geophones are connected. The time between source activations may be only a few minutes and the recording truck may move only once or twice during the day. The shooting unit often walks the line because it needs no equipment except the blaster, and perhaps shovels to fill in the shothole after the shot. The recording unit does not have to traverse the line and so is subject to less abuse. Damages are reduced because less equipment moves along the line. Thus, other benefits accrue besides increased efficiency of recording.

Several points should be noted in the foregoing discussion. Field operations require moving a series of units through the area being surveyed, and balance has to be achieved so that the units do not delay each other, especially so that the recording unit is not delayed. Extra drills or layout personnel or overtime are usually added to achieve the required balance. Crews often work irregular hours, working long days sometimes to make up for time lost because of weather. A variety of transport vehicles are used: trucks where possible, marsh and swamp buggies where the ground is soft, tractors in light forests, boats, jack-up barges, air boats, helicopters, and so on. Generally, the energy source units (drills, vibrators, and so on) are the heaviest units and determine the transport method. In some areas, operations are completely portable, everything, including small drills being carried on men's backs. Transport often represents an important part of a crew's cost and determines how much production can be achieved.

Complete records should be kept so that years later it will be possible to determine field conditions without ambiguity. Most of the routine reporting is done by computer logging, but the field crew should specifically note anything unusual. The most important records are generally those of the surveyor and observer, but drillers and other units should also submit complete reports. All reports should include the date and time of day and should be written as events happen rather than at the end of the day. The daily reports should include tape-reel numbers collated with sourcepoint numbers, specification of source and spread configurations, notes about deviations from surveyed positions, information about all recordings, including repeats, all record settings, size of charge and depth to its top and bottom, any facts that affect the validity of data such as electrical leakage, changes in surface material, excessive noise, reasons for delays in the work, and so on.

8.3 Field layouts

8.3.1 Spread types

By *spread,* we mean the relative locations of the source and the centers of the geophone groups used to record the reflected energy. Several spread types are shown in fig. 8.2 and there are many variations of these. In *split-dip* recording, the source is at the center of a line of regularly spaced geophone groups; for example, if 120 groups are being recorded, the source would be midway between groups 60 and 61. However, the source usually generates considerable noise, and an adjacent

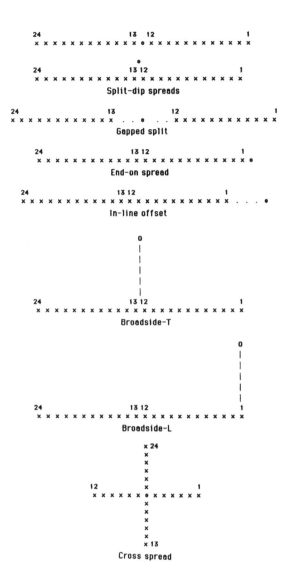

Fig. 8.2 Types of reflection spreads. The symbols **0** and **x** represent source and geophone-group center locations, respectively.

geophone group often yields only a noisy and unuseable trace. The geophone groups nearest the source thus are often not used, producing a gap in the regular geophone-group spacing. The sourcepoint gap may be only a single station or a number of stations (resulting in a *gapped split*) with near-trace offsets of 100 to 700 m.

Asymmetrical spreads are common today. A common spread is the *end-on,* where the source is at one end of regularly spaced geophone groups. This arrangement also often involves in-line offset of the source. Occasionally, the source is offset 500 to 1000 m perpendicular to the seismic line to permit the recording of appreciable data before the arrival of exceptionally strong ground roll; such spreads are called *broadside* spreads and both broadside-T and broadside-L spreads are used, the former having the source opposite the spread center and the latter opposite one end of the spread. With *cross-spreads,* two lines of geophones are laid out roughly at right angles so that three-dimensional dip information may be obtained. Additional spread arrangements used in 3-D recording are discussed in §12.1.2 and 12.1.3.

8.3.2 Singlefold recording

Virtually all routine seismic work consists of *continuous profiling,* that is, the sources and geophone groups are arranged so that there are no gaps in the data other than those due to the discrete sampling because of the geophone-group interval. Prior to the 1960s, each reflecting point was sampled only once to yield singlefold recording. An exception was that the points at the ends of a record (*tie points*) sometimes were sampled again with the adjacent record. Various arrangements of sources and geophone groups are employed to achieve this. Singlefold recording is in contrast to common-midpoint recording where each reflecting point is sampled more than once.

Continuous-coverage split-dip recording is illustrated by fig. 8.3a. Sources are laid out at regular intervals along the line of profiling, often 400 to 540 m apart. A seismic cable that is two source intervals long is used. Provision is made to connect groups of geophones (for example, 24 groups) at regular intervals along the cable (called the *group interval*). Thus, with sourcepoints 400 m apart, 24 groups are distributed along 800 m of cable making the group centers about 35 m apart. With the cable stretched from point O_1 to point O_3, sourcepoint O_2 is used; this gives subsurface control (for flat dip) between A and B. The portion of cable between O_1 and O_2 is then moved between O_3 and O_4 and sourcepoint O_3 is used; this gives subsurface coverage between B and C. The travel path for the last group from sourcepoint O_3 is the reversed path for the first group from sourcepoint O_2 so that the subsurface coverage is continuous along the line. The geophone location at the source is often not recorded.

8.3.3 Common-midpoint method

Common-midpoint (*CMP*) or "roll-along" recording (Mayne, 1962, 1967) is illustrated in fig. 8.4a. We have evenly spaced geophone groups, which we shall number by their sequence along the seismic line rather than by the trace that they represent on the seismic record. Geophone groups 1 to 24 are connected to the amplifier inputs in the recording truck and source A is used. By assuming a horizontal reflector, this gives subsurface coverage from a to g. Geophone groups 3 to 26 are then connected to the amplifier inputs, the change being made by means of the roll-along switch (§8.2.4) rather than by physically moving the seismic cable. Source B is then used, giving subsurface coverage from b to h. Source C is now used with geophones 5 to 28, giving coverage from c to i, and so on down the seismic line. Note that the reflecting point for the energy from source A into geophone group 21 is point

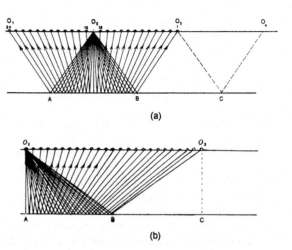

Fig. 8.3   Spreads to give continuous subsurface coverage. (a) Symmetrical split spread where half of the spread is moved forward for successive source locations. (b) End-on spread where sources are located at each end before the entire spread is advanced; the source at O_3 will complete coverage from B to C.

f, which is also the reflecting point for the energy from B into geophone group 19, from C into 17, from D into 15, from E into 13, and from F into 11. After removal of normal moveout, these six traces can be combined (*stacked*) together in a subsequent data-processing operation. In this situation, the reflecting point f is sampled six times and the coverage is called "6-fold" recording (sometimes called 600%). Obviously, the multiplicity tapers off at the ends of the line. Most present-day recording uses at least 12-fold multiplicity, 24- and 48-fold are common, and at times multiplicity exceeds 500.

To help keep track of the many traces involved in CMP acquisition, *stacking charts* are used (Morgan, 1970). A surface stacking chart (fig. 8.4b) has geophone location g as one coordinate and source location s as the other, that is, the trace observed at g from source s is indicated by the location (g, s). A variation of this chart, a subsurface stacking chart (fig. 8.4c), has the trace plotted at $[(g + s)/2, s]$.

Occasionally, one of the regularly spaced locations will not be a suitable place for a source (perhaps because of risk of damage to nearby buildings) and irregularly spaced sourcepoints (or geophone groups) will be used. Thus, if point E could not be used, a source might be located at E' instead and then geophone group 14 (instead of 13) would receive the energy reflected at f. Figure 8.4b shows the surface stacking chart when E' is used instead of E. Note in fig. 8.4b how the six traces that have the common midpoint f line up along a diagonal; points along the opposite diagonal have a common offset, whereas points on a horizontal line have the same source, and points along a vertical line represent traces from a common geophone group. Stacking charts are useful in making static and NMO corrections and ensuring that the

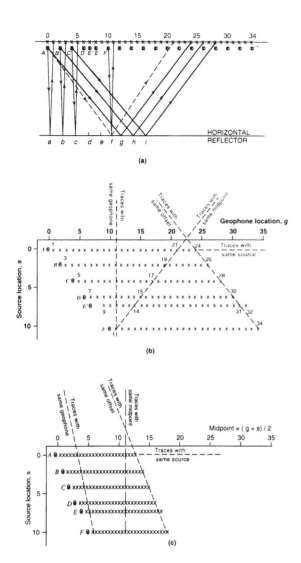

Fig. 8.4 Common-midpoint method. The symbols **0** and **x** represent sources and geophone-group center locations, respectively. (a) Vertical section illustrating CMP recording, (b) surface stacking chart, and (c) subsurface stacking chart.

traces are stacked properly.

Geophones sample the seismic wavefield at discrete locations, just as digitizing samples a seismic trace in time (§7.6.5). This spatial sampling obeys the sampling rules (§9.2.2), and inadequate spatial sampling produces aliasing (that is, creates false dip alignments; see fig. 6.2) just as inadequate temporal sampling creates false frequencies.

Each **x** on the stacking chart in fig. 8.4b represents an observed seismic trace that extends in time. The data can be examined in different directions, as indicated by the dashed lines; this proves useful in the study of noises, such as near-surface irregularities, ghosts, multiples, and converted waves. Displays of the data in different directions are called *gathers* (or sometimes *domains*); thus, a field record is a *common-source gather*, but we can also make a *common-geophone gather*, *common-offset gather*, or *common-*

midpoint gather. The wavefield could also be represented by the samples at the same time for different locations, or as *time slices* (see also §12.3). A three-dimensional representation of data on a single seismic line, oriented in common-midpoint and common-offset directions, is sometimes called *offset space*.

8.3.4 Practical constraints and special methods

(a) Gaps in coverage. As stated earlier, a common goal of field work is to have everything the same at each point along a line, so that an interpreter can attribute a change in the data to a change in the geology rather than changing field conditions. However, uniformity is rare in land recording because access is restricted at some locations, perhaps because of nearby wells or habitations.

Where certain sites cannot be occupied or where the source effort has to be decreased, extra source locations may be used to compensate at least partially for decreased multiplicity or weaker source. The effort at nearby locations may be increased, the seismic source offset to the side, the line direction changed slightly, a dog-leg (jog) introduced in the seismic line, or some other efforts made to partially compensate. Clear notation of the field changes should be included in the field records and subsequently transferred to the seismic sections to alert an interpreter to the changes. Recording condition changes often show on stacked sections by changes in the first-break pattern (see fig. 8.5).

The ends of seismic lines produce differences in multiplicity and data quality (fig. 10.3). To maintain multiplicity closer to the end of the line, extra source locations may be used with land recording. Where end-on shooting is being used with the active spread preceding the source down the line ("pushing the spread"), the source units may proceed through the active spread region, which is held constant, when the end of the line is reached ("shooting through the spread").

(b) Effect of direction of shooting. The direction in which a survey is carried out can affect the data quality. Dangerfield (1992) shows lines run across an area where gas leaking from a reservoir causes distortion (fig. 8.6); by comparison, lines run tangential to the gas area show a remarkable improvement in data quality. O'Connel, Kohli, and Amos (1992) show differences in the quality of vertical sections from marine 3-D data volumes (fig. 8.7), where the acquisition directions differ by 90°. A gather from an east–west line involves raypaths having different amounts of travel in the north–south salt body.

(c) Undershooting. Long in-line or perpendicular offsets are sometimes used where one cannot record over a desired region, perhaps because of structures, river levees, canyons, cliffs, permit problems, and so on. This technique is called *undershooting*. Under-

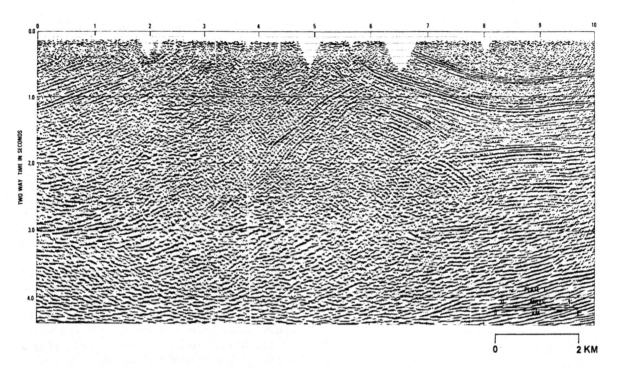

Fig. 8.5   Section in Ardmore Basin, Oklahoma. Lack of access results in the V-shaped gaps seen at the top of the section. Data quality often deteriorates in such regions. The line shows a posi- tive flower structure. (From Harding, Gregory, and Stephens, 1983.)

shooting is used in the marine environment by two boats that travel parallel to each other (fig. 8.8) to obtain data under platforms. Undershooting is also useful where raypaths are so distorted by shallow features of limited extent that sense cannot be made of deep events, as might be the situation in mapping underneath a salt dome, reef, gas leaking from a deep reservoir, or local region of very irregular topography or weathering.

(d) Crooked line methods.   Because many interpretation criteria, such as changes in dip rate, become more difficult to use when line direction changes, efforts are made to keep lines straight. However, sometimes access and/or structural complications make it impossible to locate lines in desired locations. The field recording may be done in the same way as CMP surveying, except that the line is allowed to bend (Lindsey, 1991), and the departures from regularity are accommodated in subsequent processing. The correct source-to-geophone distances (as opposed to distances measured along the line) must be calculated so that the proper amounts of normal moveout can be applied and the correct midpoints actually determined. Usually, a best-fit straight line (or series of straight-line segments) is drawn through the midpoint plot (fig. 8.9), rectangular bins are constructed, and those traces whose midpoints fall within a bin are stacked together. The bins are often perpendicular to the final line, but sometimes bins are oriented in the strike direction. The lateral extent of a bin may be

made smaller as the expected dip increases.

Because the actual midpoint locations are distributed over an area, they contain information about dip perpendicular to the line and in effect produce a series of cross-spreads, from which the true dip can be resolved. Lines are sometimes run crooked intentionally to give cross-dip information.

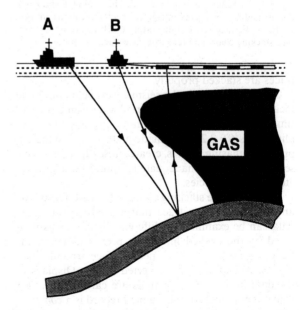

Fig. 8.6   Raypaths for seismic lines across (A) and tangential (B) to a gas-obscured area.

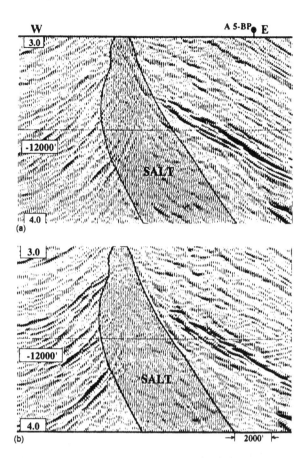

Fig. 8.7   The same east–west line extracted from two migrated 3-D surveys, where the acquisiton lines were oriented respectively east–west (upper) and north–south (lower). Data are better on the latter survey because raypaths did not have to penetrate the north–south salt body (shaded). (From O'Connell, Kohli, and Amos, 1993.)

8.3.5 Array concepts

The term *array* refers either to the pattern of a group of geophones that feed a single channel or to a distribution of sources that are fired simultaneously. It also includes the nearby locations of sources for which the results are combined by vertical stacking. A wave approaching the surface in the vertical direction will affect each geophone of an array simultaneously, so that the outputs of the geophones will combine constructively; on the other hand, a wave traveling horizontally will affect the various geophones at different times, so that there will be some destructive interference. Similarly, the waves traveling vertically downward from an array of sources fired simultaneously will add constructively when they arrive at the geophones, whereas the waves traveling horizontally away from the source array will arrive at a geophone with different phases and will be partially canceled. Thus, arrays provide a means of discriminating between waves arriving from different directions.

8.3.6 Uniform linear arrays

(a) Response to harmonic waves. Arrays are used to discriminate between waves arriving in the vertical and horizontal directions. They are *uniform* and *linear* when the elements are spaced at equal intervals along the seismic line, or *areal* when the elements are distributed over an area. The response of an array is usually illustrated by the *array response,* defined as the ratio of the amplitude of the output of the array to that of the same number of elements concentrated at one location.

Figure 8.10 shows an array of n identical geophones spaced at intervals Δx. We assume that a plane harmonic wave with angle of approach α arrives at the left-hand geophone at time t and that the geophone output is $A \sin \omega t$. The wave arrives at the rth geophone at time $t + r\Delta t$, where $\Delta t = (\Delta x \sin \alpha)/V$; the output of the rth geophone is $A \sin \omega(t - r\Delta t) = A \sin (\omega t - r\gamma)$, where γ is the phase difference between successive geophones, that is,

$$\gamma = \omega \Delta t = 2\pi v(\Delta x \sin \alpha)/V = (2\pi \Delta x/\lambda) \sin \alpha$$
$$= 2\pi \Delta x/\lambda_a,$$

where $\lambda_a = \lambda/\sin \alpha$ is the apparent wavelength in eq. (4.13b). The output of the array of n phones is

$$h(t) = \sum_{r=0}^{n-1} A \sin(\omega t - r\gamma)$$
$$= A \left[\sin \left(\tfrac{1}{2}n\gamma\right)/\sin \left(\tfrac{1}{2}\gamma\right)\right] \sin \left[\omega t - \tfrac{1}{2}(n - 1)\gamma\right]$$

(see problem 15.12c). The array output thus lags behind that of the first geophone; for n odd, the lag is that of the central geophone; for n even, it is the mean of those of the two central geophones. The array response F depends on both n and γ:

$$F = [\text{amplitude of } h(t)/nA] = |\sin \left(\tfrac{1}{2}n\gamma\right)/[n \sin \left(\tfrac{1}{2}\gamma\right)]|$$
$$= |\sin [(n\pi \Delta x \sin \alpha)/\lambda]\{n \sin [(\pi \Delta x \sin \alpha)/\lambda]\}|$$
$$= |\sin [n\pi(\Delta x/\lambda)\sin \alpha]/\{n \sin [\pi(\Delta x/\lambda) \sin \alpha]\}|$$

(8.1)

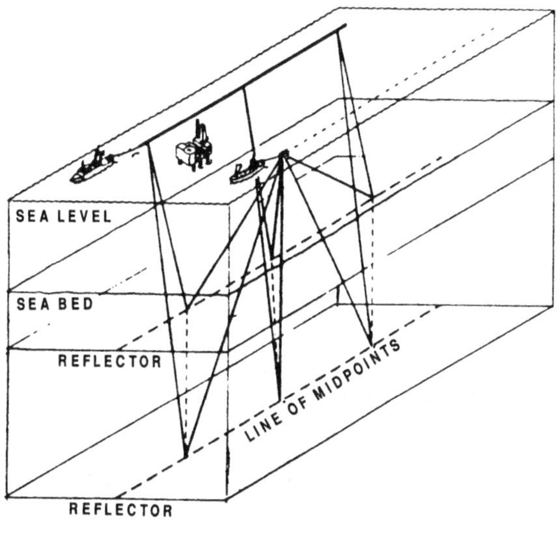

Fig. 8.8   Use of two boats to obtain data underneath a marine platform.

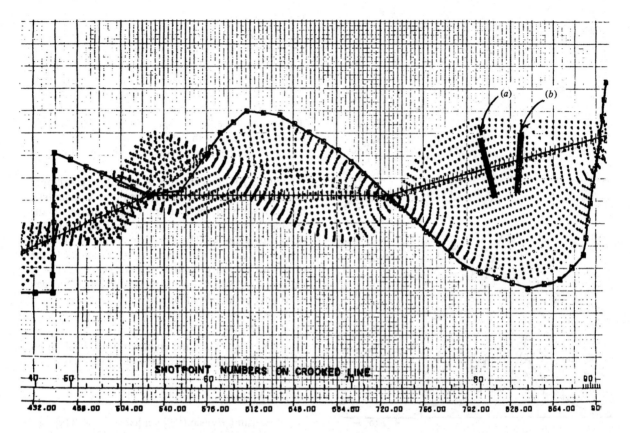

Fig. 8.9 Portion of a computer-drawn midpoint plot for a crooked line. The source locations (indicated by squares) and geophones are laid out along a road, there being one source location to every third geophone group. The midpoints are shown as dots. The line resulting from crooked-line processing has cross-dashes showing the output trace spacing. The solid rectangles show the bins (areas of midpoints combined in stacking) to make a single output trace, one rectangle for projecting perpendicular to the line, the other for projecting along structural strike. (Courtesy of Grant-Norpac.)

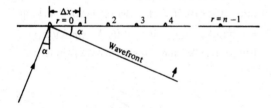

Fig. 8.10 Wavefront approaching a linear array.

(compare with eq. (7.4), which gives the response for a continuous array).

Array response is often plotted using as abscissa Δx, α, λ_a, $V_a = V/\sin\alpha$ (= apparent velocity; see eq. (4.13a)), *apparent dip moveout* $= \Delta t/\Delta x = (\sin\alpha)/V$, and so on, other quantities remaining fixed, or using the dimensionless abscissa, $\Delta x/\lambda_a$ (see fig. 8.11a). The graph usually consists of a series of maxima (lobes) separated by small values. For $\Delta x = \lambda_a$, $F = 1$, giving the first *alias lobe*, and beyond this, the entire pattern repeats. The lobes between the *principal main lobe* ($\alpha = 0$) and the alias lobe are called *side lobes*. For uniform spacing, the position of the first zero, or the width of the principal lobe, depends on $n\,\Delta x$, which

is one geophone spacing greater than the distance between the end geophones, $(n - 1)\,\Delta x$; $n\,\Delta x$ is called the *effective array length*. For nonuniform arrays, the effective array length is taken as the length $n\,\Delta x$ of a uniform array whose principal lobe has the same width at $F = 0.7$. The region between the points where the response is down by 3 dB, that is, where $F = 0.7$, is called the *reject region* (sometimes the reject region is defined with respect to the 6 dB points, that is, $F = 0.5$; occasionally, it is defined by the nulls that separate the side lobes from the main lobe and the principal alias lobe).

The nulls in fig. 8.11 occur when the effective array length is an integral number of wavelengths; wave peaks and troughs are then sampled equally so that they cancel in the sum. An exception to this occurs at the alias lobe, where the wavelength equals the spacing of the individual geophones; each geophone then records the same amplitude so that all add in phase.

Apparent wavelength or apparent velocity is often the variable to be studied and array diagrams are often plotted with a linear wavelength scale, as in fig. 8.12a instead of the reciprocal scale, as in fig. 8.11a, and with a logarithmic vertical scale in decibels. Array response can also be plotted in polar form, as in fig.

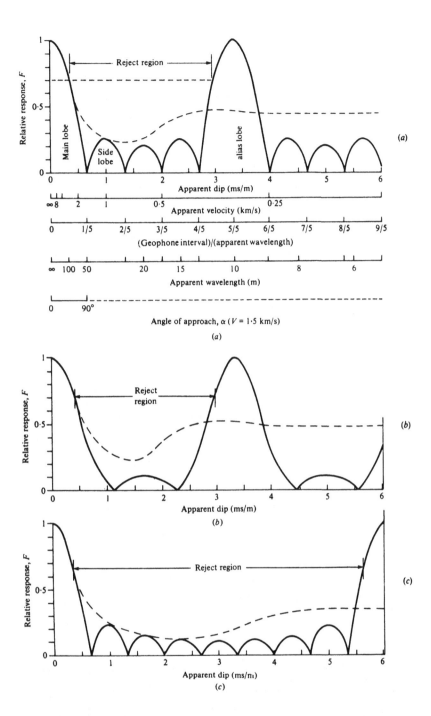

(a)

(b)

(c)

Fig. 8.11 Response of arrays to a 30-Hz signal. The alternative scales shown in part (a) apply to all three arrays. The effective length of the array controls the width of the main lobe, and the element spacing controls the location of the secondary (alias) peak. Weighting increases the attenuation in the reject region. The dashed curves indicate the array response to a bell-shaped frequency spectrum peaked at 30 Hz with a width of 30 Hz. (Courtesy of Chevron.) (a) Five in-line geophones spaced 10 m apart; (b) five geophones spaced 10 m apart and weighted 1, 2, 3, 2, 1 (or nine geophones distributed among the five locations according to these weights); (c) nine geophones spaced 5.5 m apart.

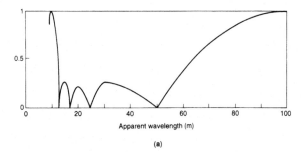

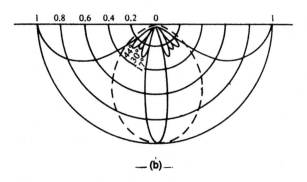

Fig. 8.12 Array directivity plots for five in-line geophones spaced 10 m apart for 30-Hz signal. (a) Plot that is linear in apparent wavelength; (b) polar plot for velocity of 1.5 km/s; solid curve is for 50-m spacing (with nulls at 11.5°, 24°, 37°, and 53°) and dashed curve for 10-m spacing (null at 90°).

8.12b. In this case, the radius vector gives the value of F as a function of the angle α.

The case of a continuous source was discussed in §7.2.2e for a vertical source, but the situation is the same for a horizontal source except for a 90° rotation of fig. 7.14.

(b) Response to transients. Actual seismic wavetrains are almost always relatively short transients involving a spectrum of wavelengths (frequencies) rather than a single harmonic wave as usually assumed by array theory. The effect of changing apparent wavelength is to stretch or compress the array diagram. A transient wavelet can be thought of as a superposition of different apparent wavelength components (the Fourier analysis concept, §15.2), each of which would produce its array response with its peak amplitude equal to the amplitude of the Fourier component, and the effective total response would be the sum of these. This describes the convolution operation (§9.2.1), and the array response to a transient is obtained simply by convolving the harmonic array response with the wavelet spectrum. The effective response for a bell-shaped spectrum is shown by the dashed lines in fig. 8.11. Effective rejection is generally poorer (except in the alias-lobe region) than the rejection for a harmonic wave.

(c) The stack array. The width of an array reject region is proportional to the array length. Anstey (1986)

argues that, with common-midpoint stacking, the entire spread constitutes the effective array length, that is, all the traces in a common-midpoint gather are involved in attenuating ground roll, air waves, and other noises. The *stack array* is a uniform linear array involving the entire common-midpoint gather. This can be achieved in a number of ways with geophones spread uniformly over the entire geophone group interval: (a) with a split spread having sources located midway between group centers and source spacing equal to the geophone-group interval, (b) with an end-on spread having source spacing equal to half the geophone group interval, and (c) in other ways. The NMO correction, sourcepoint gaps, and minor variations of ground-roll properties along the line usually do not lessen the effectiveness of the stack array significantly.

8.3.7 Weighted (tapered) arrays

Arrays where different numbers of elements are located at the successive positions are called *tapered arrays*. Compared with a linear array with the same overall array length, the main lobe and principal alias lobes are broadened, but the response in the "reject region" is generally smaller. The effective array length is less than the actual array length. Figure 8.11b shows the response of a 1, 2, 3, 2, 1 array (the numbers indicating the number of elements bunched at successive locations). Tapering can also be accomplished by varying the outputs of the individual geophones or by varying the spacing of the geophones. Arrays are also sometimes weighted at the ends of the array to attenuate long-wavelength events.

Tapered arrays also result from combinations of source and receiver arrays, where the effective array is the result of convolving (§9.2.1) the source array with the receiver array. The Vibroseis arrangement illustrated in fig. 8.13 provides an example.

8.3.8 Areal arrays

The principal application of linear arrays is in discriminating against coherent noise traveling more or less in a vertical plane through the array. Coherent noise traveling outside this plane can be attenuated by an areal array (Parr and Mayne, 1955; Burg, 1964). Some areal arrays are shown in fig. 8.14. The effective array in a given direction can be found by projecting the geophone positions onto a line in that direction; thus for the diamond array of fig. 8.14a, the effective array in the in-line direction is that of a tapered array 1, 2, 3, 2, 1 with element spacing $\Delta x = a/\sqrt{2}$, whereas at 45° to the line the effective array is 3, 3, 3 (or the same as a three-element uniform array) with $\Delta x = a$.

Where sources are located at different azimuths, as in land 3-D surveying, the differences in array response with direction affect the components within a bin differently and thus introduce undesired differences among the bins. An array such as the windmill

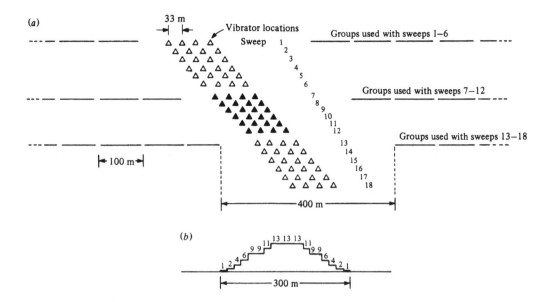

Fig. 8.13   An arrangement used with surface sources such as Vibroseis. Four units 33 m apart follow one another from left to right down the seismic line, operating simultaneously at locations spaced 16.5 m apart. The positions of the vibrators for successive sweeps are displaced vertically to avoid overlap. Records at six successive locations are summed (vertically stacked) to make an output field record. The four central geophone groups (each a linear group 100 m long) are not used for each output record (because of vibrator truck noise). The recording connections are advanced one group after source locations 6 and 12. The source locations used for one output record are shown by solid triangles. (a) Successive locations along the line of source units and active geophone groups; (b) effective array from combining the source and geophone arrays (the result of convolving them); numbers indicate the number of sweeps contributing to each portion, the location being that for source locations 13 to 18.

array of fig. 8.14g whose response is nearly the same in all directions is suitable for this situation.

8.3.9 Practical constraints on arrays

Response diagrams such as those in figs. 8.11 and 8.12 apply equally to arrays of geophones and arrays of sources. They also apply to the summing of traces in vertical stacking or other types of summing, such as is done in data processing. Theoretically, we get the same results by using 1 source and 16 geophones as by using 1 geophone and 16 sources spaced in the same manner and activated simultaneously. However, we use multiple geophones much more than multiple sources because the cost is usually less. In difficult areas, both multiple sources and multiple geophones are used at the same time. With most surface sources, two to four units are used. The records from several successive source locations not very far from each other are often summed to make an *array sum* (vertical stack) and a sizeable effective source array may be achieved in this way (fig. 8.13). Array summing achieves greater attenuation of random noise than using simultaneous multiple sources.

The canceling of coherent noise by using geophone and source arrays presents a more challenging array design problem than does the cancelation of random noise. In the case of random noise, the locations of the elements of the array are unimportant provided no two are so close that the noise is identical for both. For coherent noise, the size, spacing, and orientation of the array must be selected on the basis of the properties of the noise to be canceled (Schoenberger, 1970). If the noise is a long sinusoidal wavetrain, an array consisting of n elements spaced along the direction of travel of the wave at intervals of λ/n, where λ is the apparent wavelength, will provide cancelation (see problem 8.6b). However, actual noise often consists of several types arriving from different directions, each type comprising a range of wavelengths; moreover, the nature of the noise may change from point to point along the line. One sometimes resorts to areal arrays in areas of severe noise problems (although the in-line distribution of elements is almost always the most important aspect). Numerous articles have been written on the subject of arrays; McKay (1954) shows examples of the improvement in record quality for different arrays.

In addition to the difficulties in defining the noise wavelengths to be attenuated, actual field layouts rarely correspond with their theoretical design (see fig. 8.15 and Newman and Mahoney, 1973). Measuring the locations of the individual geophones is not practicable. In heavy brush, one may have to detour when laying out successive geophones, and often one cannot see one geophone from another so that even the orientation of lines of geophones can be very irregular. In rough topography, maintaining an array design might require that geophones be at different

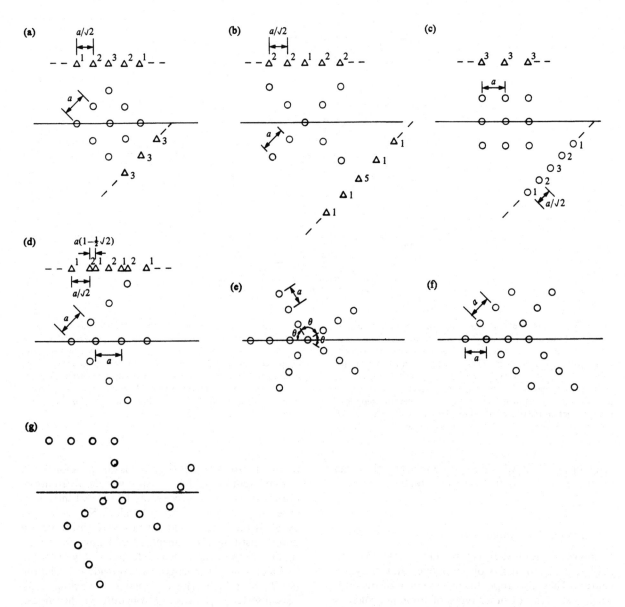

Fig. 8.14 Types of areal arrays. Element locations are indicated by small circles, the effective arrays in different directions are indicated by small triangles with their effective weights. (a) 3×3 diamond; (b) X-array; (c) rectangular array; (d) crow's foot array; (e) odd-arm star; (f) herring-bone array; (g) windmill array.

elevations; this may produce far worse effects than those that the array is intended to eliminate. Similar problems arise where the conditions for planting the geophones vary within a group (Lamer, 1970), perhaps as a result of loose sand, mucky soil, or scattered rock outcrops. The best rules for array design are often (1) to determine the maximum size that can be permitted without discriminating against events with the maximum anticipated dip and (2) to distribute as many geophones as field economy will permit more or less uniformly over an area a little less than the maximum size permitted, maintaining all geophone plants and elevations as nearly constant as possible even if this requires severe distortion of the layout (see also §8.4.2).

Arrays may also be of value in refraction work (Laster and Linville, 1968).

8.3.10 Spatial sampling requirements

A successful seismic survey should be designed with the objectives in mind and with some knowledge of the geology. The subsurface sampling interval should be small enough to avoid aliasing during processing and interpretation. The sampling theorem (§9.2.2) states that signals should be sampled at least twice per wavelength. The highest frequency of interest ν_{max}, the velocity, and the maximum dip (hence, the maximum angle of approach) fix the shortest apparent wavelength and thus determine the maximum permissible subsurface spacing. The limiting value of the sub-surface spacing, D_{max}, is, therefore,

$$D_{max} \leq (\lambda_a)_{min}/2 = \lambda_{min}/(2 \sin \alpha_{max})$$
$$\leq (V/\nu_{max})/(2 \sin \alpha_{max}) = (V/\nu_{max})/[2V(\Delta t/\Delta x)_{max}]$$
$$\leq 1000/[2\nu_{max}(\Delta t/\Delta x)_{max}], \qquad (8.2a)$$

(using successively eqs. (4.13b), (2.4), and (4.13a)), where λ_{min} is the minimum wavelength, $(\lambda_a)_{min}$ is the minimum apparent wavelength (as in eq. (4.13b)), ν_{max} the maximum frequency, α_{max} the maximum angle of approach, $\sin \alpha_{max} = V(\Delta t/\Delta x)_{max}$, with the maximum apparent dip moveout, $(\Delta t/\Delta x)_{max}$, given in millisec-onds per unit distance. It is prudent to allow a margin of safety because it is difficult to determine ν_{max} and $(\Delta t/\Delta x)_{max}$ exactly, and hence we often specify three samples per shortest wavelength (Brown, 1991), that is,

$$D \leq 1000/[3\nu_{max}(\Delta t/\Delta x)_{max}]. \qquad (8.2b)$$

Subsurface sampling intervals computed according to the preceding design considerations generally range from 10 to 100 m. Geological constraints (for ex-ample, the preknowledge that there are no large dips) can permit relaxing the spatial aliasing constraint. However, most data are migrated and migration algo-rithms create noise where spatial sampling is inade-quate; this may provide the limiting constraint. Intelli-gent interpolation (§9.11.2) can be used to relax the spatial aliasing constraint as far as migration is con-cerned.

8.3.11 Extended resolution

Although conventional geophones and recording sys-tems are usually adequate for recording up to 125 Hz (and higher) and normal alias filters (which cut sharply above $1/4\Delta$, where Δ is the sampling rate) per-mit recording up to 250 Hz for 1-ms sampling, the bandwidth of most reflection surveys is only about 10 to 60 Hz. Because both vertical and horizontal resolu-tions (§6.4) are limited by the high-frequency compo-nents, we must expand the passband upward to achieve higher resolution. Techniques for doing this are sometimes called *extended resolution*.

The high-frequency limitations are usually due to (1) limitations in the source, (2) processes within the earth that discriminate against high frequencies, (3) conditions at or near the surface, including array effects, and occasionally (4) recording instruments.

Surface sources are often limited with respect to high frequencies because of mechanical and coupling problems as well as high near-surface attenuation (in comparison with a source in a borehole) resulting from two passes through the weathered layer. Dyna-mite can generate waves relatively rich in high fre-quencies, the high-frequency content increasing as the charge size decreases; hence, the charge size should be the minimum consistent with adequate energy.

High frequencies are attenuated in the earth by ab-sorption (§2.7.2) and peg-leg multiples (§6.3.2b). The practical limiting factor is the amplitude of useful high-frequency reflection energy compared to the noise level. Denham (1981) gives the empirical for-mula $\nu_{max} = 150/t$, where ν_{max} is the maximum useable frequency, and t is the two-way time. According to this, the upper limit should be greater than 60 Hz for events shallower than $t = 2.5$ s. The processes within the earth are largely beyond our control and set the ultimate limit to the resolution that can be achieved.

The greatest single cause of high-frequency loss is often ground mixing owing to the use of geophone arrays (and also source arrays); random time shifts with a standard deviation of 2 ms are equivalent to a 62-Hz filter (fig. 6.51). Thus, although arrays provide one of the best ways of attenuating surface waves and ambient noise, alternative methods should be used if high resolution is desired. Low-frequency filtering and burying the phones should be considered (the best re-sults are obtained by burying the phones below the weathered layer, but burial by only 10 to 50 cm often achieves dramatic improvement). Consideration must also be given to spatial aliasing (§8.3.10), which may dictate group spacing as small as 15 to 20 m. Some-times a hybrid spread is used, group intervals being longer for long-offset groups (see problem 8.13).

If the geophone response is not adequate, acceler-ometers (which have a frequency response increasing 6 dB/octave relative to velocity geophones (§7.5.3)) can be used even though maintenance costs increase. When low-cut filtering is used to attenuate severe ground roll, a passband of at least 2 octaves should be retained in order to achieve good wavelet shape.

Significant improvement in high-frequency re-sponse may be obtainable in some areas without marked increase in acquisition costs, although pro-cessing costs may increase. Extending the frequency spectrum upward usually improves both shallow and deep resolution.

8.4 Selection of field parameters

8.4.1 Noise analysis

Systematic investigation of coherent noise often be-gins with shooting a *noise profile* (also called a *micro-spread* or *walkaway*). This is a small-scale profile with a single geophone per trace, the geophones being spaced as closely as 1 to 3 m over a total spread length of the order of 300 m or more. If the weathering or elevation is variable, corrections should be made for each trace. The corrected data, often in the form of a record section, such as shown in fig. 8.16a, are studied to determine the nature of the coherent events on the records, the frequencies and apparent velocities of the coherent noise, *windows* between noise trains where reflection data would not be overridden by such noise, and so on. Once we have some indications of the types of noises present, we can design arrays or other field techniques to attenuate the noise and then field-test

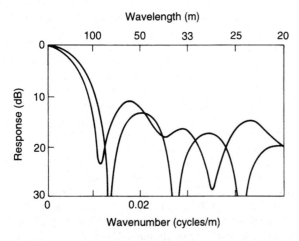

Fig. 8.15   The theoretical response of a 12-element in-line array (heavier curve) and as it was actually laid out on the ground.

our techniques to see if the desired effect is achieved.

The data from noise analyses are sometimes transformed (see §9.1.3 and 15.2.3) into the frequency-wavenumber domain and displayed as ν–κ plots (often called f–k plots), such as shown in fig. 8.16b. Radial lines on such a plot show a constant proportionality between ν and κ_a, that is, constant apparent velocity (by eq. (2.4)). Such ν–κ plots are helpful in seeing the characteristics of different types of events and determining the best ways of attenuating the portions regarded as noise-dominated in order to improve the signal-to-noise ratio (for example, with arrays, frequency filtering, apparent velocity filtering (§9.9), and so on).

8.4.2 Determining field parameters

Field parameters should be determined in a logical way (Anstey, 1970; Sheriff, 1978), although existing equipment (number of channels and geophones, how cables and geophones are wired, and so on) usually prejudices decisions considerably.

1. The maximum offset, the distance from source to the farthest group, should be comparable to the depth of the deepest zone of interest. This usually results in large enough normal-moveout differences to distinguish primary reflections from multiples and other coherent noise, but offsets should not be so large that reflection coefficients change appreciably, that conversion to shear waves becomes serious, and that the approximations of the CMP method become invalid. If data quality in the deepest zone of interest is sufficiently good, the maximum offset may be increased up to the value of the basement depth.

2. The minimum offset distance should be no greater than the depth of the shallowest section of interest. Getting sufficiently far from source-generated noise sometimes dictates a greater distance, but this may cause a loss of useful shallow data.

3. The maximum array length is determined by the minimum apparent velocity of reflections. The minimum apparent velocity usually occurs at the maximum offset. The shortest apparent wavelength (highest frequency) at this minimum apparent velocity should be just within the main lobe of the array's directivity pattern (fig. 8.11).

4. The minimum useful in-line geophone spacing within arrays is usually determined by the ambient noise, sometimes by source-generated noise. Ambient-noise characteristics can be determined experimentally by recording individual geophones spaced 0.5 to 1 m apart to determine the minimum geophone spacing for which the noise appears to be still incoherent. This minimum spacing is often 2 to 5 m, the smaller value being where noise is mostly generated locally (such as noise caused by the wind blowing grass, shrubs, or trees), and the larger value where noise is mainly caused by distant sources (such as microseisms, surf noise, traffic noise, and so on). If more geophones are available, an areal distribution of phones at this minimum spacing will be more effective than crowding the phones closer together in-line. Areal arrays are rarely required to attenuate noise coming from the side of the line. Air-wave attenuation may require closer geophone spacing.

5. Group interval should be no more than double the desired horizontal resolution, thus providing subsurface spacing equal to the desired resolution. However, the group interval should not exceed the maximum permissible array length indicated by rule 3.

6. The minimum number of channels required is determined by the combination of spread length and group interval decisions already reached.

7. The minimum charge size or source effort is determined by the ambient noise late on the record. Random noise should not affect repeated records until after a depth below the deepest section of interest; if this is not the case, the source effort should be increased. Figure 8.17 shows the difference between 6- and 12-fold multiplicity for a particular prospect; the 12-fold data are sufficiently superior to warrant the additional effort. However, charge size and source effort are more often too large rather than too small.

It should be remembered that line length, line orientation, and line spacing are also field parameters. The deterioration of data quality near the ends of a seismic line can be seen in fig. 10.3. Figure 8.18a shows another example (in addition to figs. 8.6 and 8.7) of differences in data quality as a result of line orientation, and fig. 8.18b illustrates how line orientation and spacing might affect the interpretation of a faulted domal structure.

Figure 8.19 summarizes how the field parameters depend on the nature of the exploration problem. Special circumstances may require variations from customary guidelines. For example, mapping a zone of

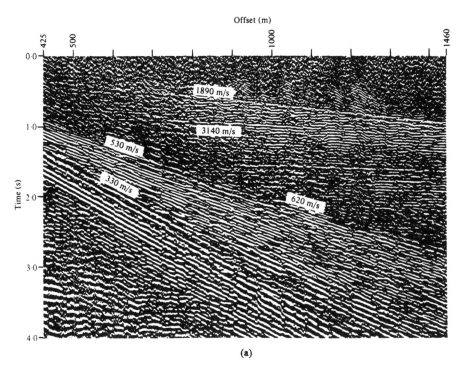

(a)

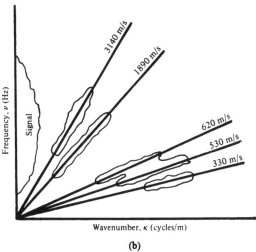

(b)

Fig. 8.16 Noise analysis or walkaway. Geophones are spaced 1.5 m apart; the offset to first geophone is 425 m. Identification of events: 1890 m/s is the refraction from base of weathering; 530 and 630 m/s are ground-roll modes; 330 m/s is the air wave; and 3140 m/s is a refraction event. (After Sheriff, 1991.) (a) Noise analysis and (b) frequency wavenumber sketch. A horizontal slice of the data diagrammed in part (b) can be taken by frequency bandpass filtering, a vertical slice by the use of arrays, and a pie-shaped wedge by apparent-velocity filtering (see also fig 9.36c).

maximum interest may require recording in an offset window where desired reflections are between the wavetrains following the first arrivals and those caused by surface waves.

Although one is usually constrained somewhat by the equipment available, it may be possible to use geophones with higher natural frequency where ground roll is especially strong or accelerometers where high-frequency response is inadequate. Field parameter choices with marine surveys are apt to be limited by the equipment available.

8.4.3 Field testing

Sustaining efficient field operations with many channels and changes in surface conditions, as generally occur on land, is challenging. Recording equipment today includes considerable computing capacity, which is used to carry out a variety of tests, keep records of what was done, and assist in planning field design changes that may be required during the survey. The system can generate a variety of control plots that monitor operations as they proceed.

The recording equipment executes and analyzes instrument tests to verify that the instruments are functioning properly. It checks the entire system, including geophones and vibrators. Simplified processing can be carried out, including transformations between time, frequency, and f–k domains, autocorrelations and crosscorrelations, frequency and velocity filtering, deconvolution, velocity analysis, muting, amplitude scaling, refraction statics, stacking, and other operations.

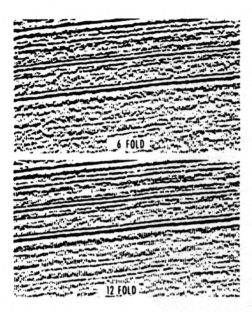

Fig. 8.17 An example of the results of 6- and 12-fold multiplicity.

Recording equipment also does much of the bookkeeping required. It can plot and update stacking charts as data are acquired. This capability can be used to show how stacking would be affected by rearranged source–receiver locations, which may be required because of access problems.

The recording equipment and portable workstations can also be used to generate plots to assist in field design. Software is available for the design and display of in-line and areal arrays, correlating Vibroseis records, performing spectral analyses, calculating and displaying frequency–wavenumber ($f–k$) information, plotting changes of amplitude with time or offset, and other analysis tools. This capability can be especially helpful while experimenting to determine optimum recording parameters.

8.5 Defining the near surface

8.5.1 Uphole surveys

An *uphole survey* is one of the best methods of investigating the near surface and finding the thickness (D_W) and velocity (V_W) of the low-velocity layer (LVL) and the velocity of the subweathering layer (V_H). An uphole survey requires a borehole deeper than the base of the LVL. Usually shots are fired at various depths in the hole, as shown in fig. 8.20a, beginning at the bottom and continuing until the shot is just below the surface of the ground. Usually, a complete spread of geophones and an uphole geophone are used. Arrival times are plotted against source depth for the uphole geophone and for several distant geophones, including two spaced at least 200 m apart, as shown in fig. 8.20b. Sometimes the procedure is reversed with a source on the surface recorded with a

downhole cable containing a number of sensors spaced perhaps 3 m apart.

The plot for the uphole geophone changes abruptly where the source enters the LVL; the slope of the portion below the LVL gives V_H, the break in slope gives D_W, and the slope above the break gives V_W. For distant geophones, because the path length changes very little, the traveltime plot is almost vertical as long as the source is in the high-velocity medium. However, when the source enters the LVL, there is an abrupt change in slope and the traveltime increases rapidly as the path length in the LVL increases. The refraction velocity at the base of the LVL (V_{H^*}) is obtained by dividing the time interval between the vertical portions of the curves for two widely separated geophones (for example Δt_{17} in fig. 8.20a) into the distance between the geophones. The refraction velocity V_{H^*} may be different from the velocity given by the slope of the deeper portion of the uphole geophone curve, V_H, because the highest-velocity portions of the subweathering carry the refraction first-break energy, whereas lower-velocity portions also contribute to the uphole measurement; however, often $V_{H^*} \approx V_H$ and these differences are ignored. The values of V_{H^*} and V_H may also differ because the time interval t_{17} is usually more accurate than that measured by the uphole geophone.

8.5.2 Near-surface refraction

The contact between the near-surface, low-velocity (LVL) layer (§5.3.2) and the underlying rocks (subweathering) is usually fairly sharp and has a large velocity contrast. Consequently, the first energy to reach most geophones is head-wave energy involving the base of the LVL. Refraction first-break interpretation techniques (§11.3 to 11.5) are suitable for determining the vertical traveltime through the LVL and thus for correcting reflection traveltimes. The interpretation problem is often framed in terms of determining the LVL thickness from which the LVL traveltime can be determined, but the LVL velocity may also vary significantly, so that measuring traveltimes avoids having to decide whether observed changes are caused by LVL thickness or velocity changes. When using surface sources, there may be some ambiguity about partitioning traveltimes between those under the source and those under the geophones where the LVL correction varies, but the relative point-to-point corrections, which are more important, are apt to be correct.

Errors in the assumptions tend to cause LVL correction errors to accumulate along the line, creating overall time-shift errors, but these usually do not cause serious problems. The most serious errors of this kind often have a wavelength about the same as the spread length.

The first-break pattern can suggest changes in the LVL, as illustrated in fig. 8.21 for buried sources. Subsequent processing also often employs refraction statics analysis (§9.6.4) to refine statics determinations.

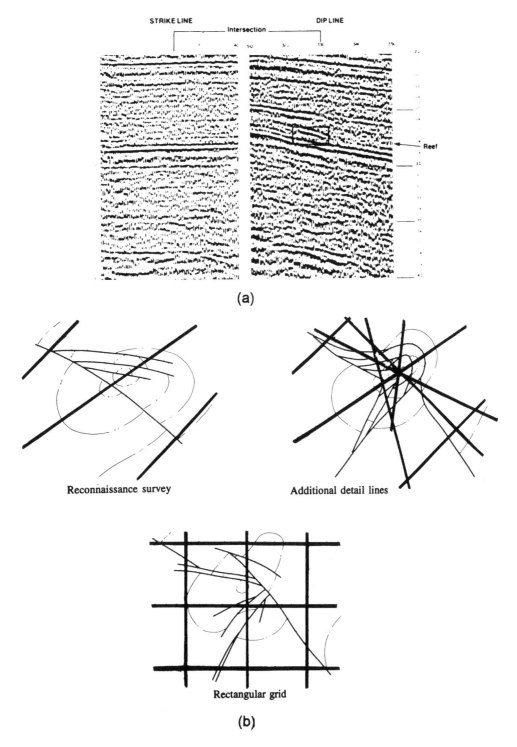

STRIKE LINE   DIP LINE
├── Intersection ──┤

Reef

(a)

Reconnaissance survey   Additional detail lines

Rectangular grid

(b)

Fig. 8.18 Line orientation and spacing effects. (a) Intersecting strike and dip land lines; the small reef seen on the dip line is not evident on the strike line; and (b) maps resulting from different layouts of lines over a faulted dome are mapped differently.

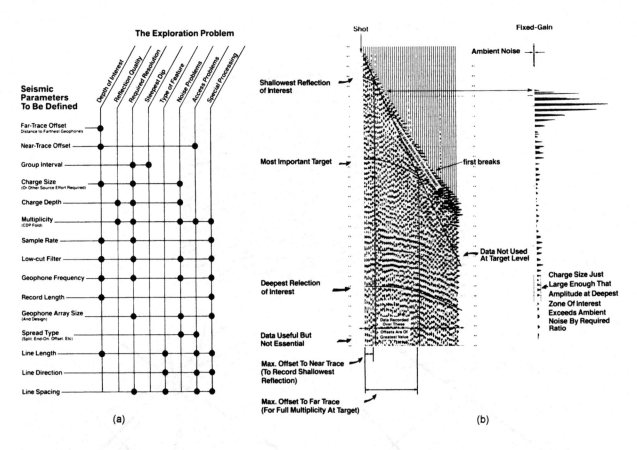

Fig. 8.19 Determination of survey parameters. (a) Aspects of the exploration problem that define field parameters, and (b) determination of parameters from a seismic record.

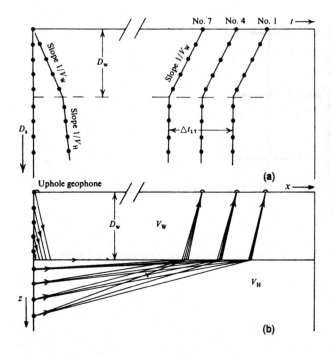

Fig. 8.20 Uphole survey. (a) Traveltime versus depth of source, and (b) vertical section showing raypaths.

8.6 Marine methods

8.6.1 Conventional marine operations

Conventional marine operations involve one ship that tows both the sources and hydrophone streamer. The ship has to be sufficiently large to tow so much equipment, and hence efficiency in acquiring data becomes of high priority. With so much towed behind it, the ship must maintain its forward motion so as not to lose control over the tow and tangle the towed elements. With such a long tow, changes in direction must occur gradually, and hence it takes appreciable time to change from one line to another (usually more than 1 hour). The tows also restrict operations to water depths greater than 10 m and to areas that are relatively free of obstructions. These, then, are the major factors governing marine data acquisition. Today, most 2-D marine data acquisition employs ships that are larger than actually required because they are used mainly for 3-D acquisition (§12.1.2), which involves towing more equipment than required for 2-D acquisition.

When on station, the streamer, sources, and other towed equipment are unreeled from the stern of the ship; this typically requires 4 hours or more, de-

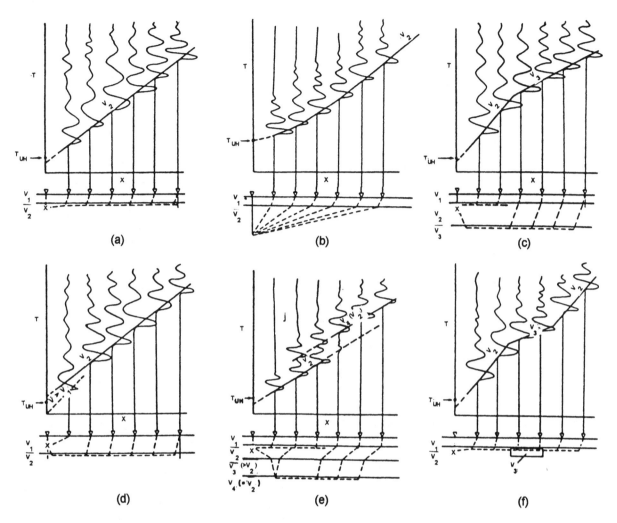

Fig. 8.21   First-break patterns from a buried source for various LVL conditions. (a) "Normal" pattern from source slightly below the base LVL, (b) from source deep within the subweather-ing, (c) from double-weathering layers, (d) source in the LVL, (e) low-velocity layer below the first high-speed refractor, and (f) short high-velocity "stringer" within or near the base LVL.

pending on difficulties in ballasting the streamer so that it is neutrally buoyant. Navigation of the ship during recording is done by a navigation computer, which steers the ship and, at the proper times, issues "on-location" commands to start the recording system and fire the sources. The helmsman keeps overall watch so that he can take over steering of the ship if something unforseen should occur. The recording of the various sensors is largely automatic, the chief function of the observers, navigators, and others being to see that everything is functioning properly.

Marine seismic lines are usually fairly long to minimize the fraction of time spent getting into position. Minor failures during the acquisition of a line of data, such as the failure of one or two hydrophone groups or air guns in an array or of a missed source location, are tolerated because of the huge cost in time required to remedy the situation. Normal operations proceed at about 6 knots (11 km/hr or 3 m/s) on an around-the-clock basis. Hence, about 250 km of CMP data

could be recorded in a day if all the time were spent acquiring data. Of course, this much production is never achieved because of the time spent going to the survey area, moving from line to line, repairing equipment, waiting on good weather, and other factors. A ship typically collects data about 60% of the time. With sources activated every 10 to 15 s, source locations are spaced about 30 to 45 m apart. Records (sometimes called *subshots*) may be vertically stacked as the data are acquired or recorded individually for combining later. At regular intervals, a few records are given a cursory examination to detect fairly obvious defects, but detailed monitoring is not possible at the pace of operations. Nevertheless, the relatively constant water environment surrounding sources and hydrophones and the general absence of a weathering layer usually result in uniformly good data quality.

A variation occasionally employed is to tow the streamer at an angle with the water surface so that ghosting effects (§6.3.2b) differ for the different hydrophone groups; these effects are then attenuated when the data are stacked. Another variation used to reduce ghost effects is to tow two streamers at different depths. Sometimes a short ministreamer is towed at shallower depth to record near-offset data, which may be missing with the regular streamers that are towed deeper.

A tremendous amount of data is generated by a marine survey with data obtained from a large number of sensors. Essentially, all the data are recorded automatically on magnetic tape to be read and digested later by computer programs. A modern 3-D seismic ship, for example, towing multiple sources and streamers with perhaps 500 channels each, generates a huge volume of data with subshots perhaps every 15 to 20 s. With redundant navigation systems and source–streamer locating systems, such as shown in fig. 7.9, the volume of navigation data may equal the volume of seismic data.

8.6.2 Shallow-water and obstructed operations

For operation in shallow water, drag cables carrying hydrophones or gimbel-mounted geophones can be pulled along the sea floor. Cables can be laid on the bottom in areas where obstacles restrict operations, the cables being connected to recording equipment on an anchored recording ship or barge, with a separate mobile boat used for the source.

8.6.3 Profiling methods

Environmental and engineering objectives usually do not require appreciable penetration of seismic energy and hence do not require the strong sources nor elaborate detection systems used in hydrocarbon studies. We use the general term *profiling* for such methods. Profiling involves smaller equipment and does not require large, dedicated ships. Equipment is usually installed on boats of opportunity (leased locally).

Profiling almost always involves a trade-off between the use of small, high-frequency sources to achieve high resolution and larger sources to achieve greater penetration. Profiling covers a broad range of operational sophistication, ranging from the use of single detectors that simply record reflections from the sea floor and a thin veneer of sediments below it to the use of streamers employing the common-midpoint method to achieve penetrations of the order of 1000 m. Thus, there is no sharp distinction between profiling and full-scale marine operations. Profiling systems are summarized in table 7.2.

Profiling systems were entirely analog for many years. The simpler systems employed a single source and single hydrophone group whose output was merely amplified and plotted. One simple strip recorder used a wire wrapped around a cylinder in a helical manner; the electrical current from the detection system passed through the wire and completed its circuit through a metal plate tangent to the cylinder. A

record of the current was made on electrosensitive paper passing slowly (in the x-direction) between the cylinder and the plate. The source was fired when the wire touched the top of the paper, and the plot point moved in the y-direction across the paper as the cylinder rotated. Reflections with arrival times greater than the cylinder rotation time were sometimes written on top of the recording of shallower data, an effect called *paging* (see fig. 8.23).

Today, both small air guns (§7.4.3) and other sources such as described in §7.4.4 are used. Sea-floor multiples usually are quite strong, because systems with only a single detector cannot discriminate between events based on their normal moveout or directions of arrival, and data arriving after the sea-floor multiple are often so contaminated that they are unuseable. Data are sometimes recorded digitally so that they can be processed. The major amplitude loss in travel through the water is due to geometrical divergence, there being little absorption and loss of high frequencies until waves enter the sediments; towing systems close to the sea floor reduce losses because of divergence.

Profiling is extensively used in engineering studies to determine the nature of sediments (fig. 8.22), for example, to find sand layers that can support structures erected on pilings. It is used to locate pipelines buried in the mud. An important application is in surveying for hazards that might cause trouble in drilling, such as foundation problems, gas seeps, shallow gas accumulations, buried channels, and faults. Because it is relatively inexpensive, profiling is extensively used in oceanographic studies (fig. 8.23).

Profiler operations often use a number of different kinds of sensors simultaneously (figs. 8.24 and 14.3), including magnetometers and sniffers (to sample the seawater for analysis). One profiling variation is *side-scan sonar*; transducers in a towfish radiate short sonar pulses that scan the sea floor to the side of the fish and record energy that is back-scattered from the sea floor (fig. 8.25). The result is essentially a map of the sea-floor relief showing reflectors such as rocks, ledges, metal objects, and sand ripples as dark areas, and depressions and other features as light areas. The useful coverage is out to about three times the height of the towfish (often about 20 m above the sea floor), with a gap of no coverage underneath the towfish. Data can be corrected for divergence, slant range, and variations in boat speed. The *synthetic-aperture* concept is sometimes used to sharpen the in-line footprint; the directivity of a radiation pattern is inversely proportional to the antenna size, and the successive images obtained as the towfish moves through the water can be combined to effectively give a source array and thus improve directivity.

8.7 Transition-zone operations

A *transition zone* is a region where environments change rapidly. Usually, we try to minimize acquisition changes to avoid misinterpreting their effects as geologic features, but changes are necessary in transition zones. The most common transition zone is the region near a coastline where a survey area may include shallow water, surf zone, beach dunes, mud flats, lagoons, marsh, and swamp. These different environments are apt to require quite different kinds of equipment.

Energy sources may be those ordinarily used on land (dynamite in boreholes and Vibroseis, for example) and at sea (air guns). Explosive detonating cord might be used where the water is not deep enough for the use of air guns. Velocity geophones of both conventional land types and gimble-mounted in bay cables and hydrophones in both marsh cases and streamers might be used in different parts of the area in conjunction with several kinds of sources. Telemetry might be used to get information from the detectors to the recording unit.

Means of moving about the region may include trucks, small boats, jack-up barges (to raise them above wave action), rubber boats, pontoons, and amphibious craft such as hovercraft and marsh buggies, and parts of the area may have to be portable. Locations and elevations might be determined partly by marine positioning and partly by conventional land survey methods.

8.8 Data reduction

8.8.1 Field processing

The first step in either field or processing office is usually to check the inventory of data to make certain that all data are in hand. Checks for data consistency may also be made.

Field processing usually includes vertical stacking and sometimes other processing. *Vertical stacking* usually implies combining records that were obtained without any major movement of sources or detectors and without any differential static or normal-moveout corrections made to them. Especially with surface sources, four to six source locations within the geophone-group spacing distance are occupied to give the effect of a source array (as in fig. 8.13). Occasionally, vertical stacking also includes combining records from sources at several depths in a borehole after time shifting according to differences in uphole time. Instrument software carries out testing and analyzing the test results.

Field recording instruments often have the capability of doing appreciable processing, but usually field personnel do not have time for doing anything except almost automatic routine processing. The correlation of Vibroseis records, the summation of Sosie records, and demultiplexing to trace-sequential format are examples of such routine processes done in the field. The processing capabilities of field instruments are utilized during field experimental periods and sometimes at night when survey results are needed quickly. Marine operations may involve more extensive processing and interpretation.

Field records of all kinds must be completely identified. "Observer's sheets" are used to record field data such as (a) company and prospect names; (b) line, profile, and magnetic tape numbers; (c) information about source and active geophone locations; (d) number and spacing of geophones and geophone groups; (e) source pattern or size and depth of explosive charges; (f) instrument settings; (g) time of the recording; (h) history of any processing done; and especially (i) anything unusual, such as changes in source locations because of access problems. Computers may automatically record much of this data. In earlier times, the appropriate information was written on the back of paper records immediately after recording.

8.8.2 Elevation and weathering corrections

The first calculation of corrections for elevation and weathering variations is carried out in the field. These *field statics* are based on survey, uphole, and firstbreak information, and are subsequently used in processing as the first estimate. Marsden (1993) reviews statics corrections.

Variations in the elevation of the surface affect traveltimes and it is necessary to correct for such variations as well as for changes in the low-velocity layer (LVL). Usually, a *reference datum* is selected and corrections are calculated so that in effect the source and geophones are located on the datum surface, it being assumed that conditions are uniform with no LVL material below the datum level.

Many methods exist for correcting for near-surface effects. These schemes are usually based on (1) uphole times, (2) head waves from the base of the LVL (refraction statics), or (3) the smoothing of reflections. Methods (2) and (3) are employed in automatic statics processing (§9.6), often on an iterative basis; these generally involve some statistical smoothing.

In the following we shall assume that V_W and V_H, the velocities in the LVL and in the layer just below it, respectively, are known, having been determined from an uphole survey or the refraction first-breaks. With explosives in boreholes as the source, we also assume that the source is below the base of the LVL; if this is not true, some modifications have to be made in the following equations (see problem 8.18). Similar modifications have to be made when surface sources are used.

Figure 8.26 illustrates a method of obtaining the corrections for t_0, the source arrival time. E_d is the elevation of the datum, E_s the elevation of the surface at the source, E_g the elevation of a geophone group, D_s the depth of the source below the surface, and t_{uh} the uphole time, the time required for energy to travel vertically upward from the source to a geophone on the surface (in practice, the uphole geophone is within a couple of meters of the shothole). The deviation of reflection paths from the vertical is usually small

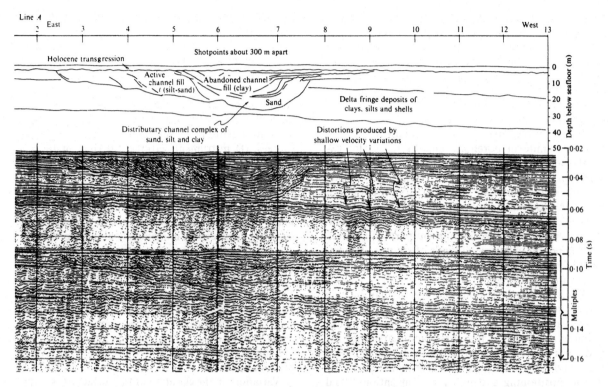

Fig. 8.22   Profiler record showing subbottom deposits. (After King, 1973.)

enough that we can regard the paths as vertical; therefore, the time required for the seismic wave to travel from the source down to the datum is Δt_s, where

$$\Delta t_s = (E_s - D_s - E_d)/V_H. \qquad (8.3)$$

Similarly, the time for the wave to travel up from the datum to a geophone on the surface at B is Δt_g, where

$$\Delta t_g = \Delta t_s + t_{uh}. \qquad (8.4)$$

The correction, Δt_0, for the traveltime at the source is then

$$\begin{aligned} \Delta t_0 &= \Delta t_s + \Delta t_g = 2\Delta t_s + t_{uh} \\ &= 2(E_s - D_s - E_d)/V_H + t_{uh}. \end{aligned} \qquad (8.5)$$

Subtraction of Δt_0 from the arrival time t_0 is equivalent to placing the source and the geophone group at the source location on the datum plane, thereby eliminating the effect of the low-velocity layer if the source is beneath the LVL. At times, the source may be so far below the datum plane that Δt_0 will be negative.

When eq. (4.11) is used to calculate dip, the dip moveout must be corrected for elevation and weathering. The values needed in eq. (4.11) are the differences in arrival time at A' and C' in fig. 8.26. The correction to be applied, Δt_c, called the *differential weathering correction,* is the difference in arrival times at opposite ends of a split-dip spread for a reflection from a horizontal bed. The raypaths from the source B down to a horizontal bed and back to geophones at A and C have identical traveltimes except for the portions $A'A$ and $C'C$ from the datum to the surface. Assuming as

before that $A'A$ and $C'C$ are vertical, for Δt_c we get the expression

$$\begin{aligned} \Delta t_c &= (\Delta t_g)_C - (\Delta t_g)_A \\ &= (\Delta t_s + t_{uh})_C - (\Delta t_s + t_{uh})_A, \end{aligned} \qquad (8.6)$$

where we assume that A and C are source locations so that the quantities in parentheses are known. If we take the positive direction of dip to be down from A toward C, then Δt_c must be subtracted algebraically from the observed difference in arrival times at A and C to obtain that for A' and C'.

The following calculation illustrates the effect of the correction. We take as datum a horizontal plane 200 m above sea level; V_H is 2075 m/s. Data for three successive sources, A, B, and C, such as those in fig. 8.26, are shown in table 8.1.

Let us suppose that a reflection on a split profile from source B gives the following data: $t_0 = 2.421$ s, $t_A = 2.419$ s, and $t_C = 2.431$ s. Then, the corrected value of t_0 is $2.421 - 0.074 = 2.347$ s, and the corrected Δt_d is

$$\begin{aligned} \Delta t_d &= 2.431 - 2.419 - (-0.009) = 0.012 + 0.009 \\ &= 21 \text{ ms.} \end{aligned}$$

Suppose that (for another reflection) $t_0 = 1.392$ s, $t_A = 1.401$ s, and $t_C = 1.395$ s; then the corrected value of t_0 is 1.318 s and the corrected Δt_d is

$$\begin{aligned} \Delta t_d &= 1.395 - 1.401 - (-0.009) \\ &= -0.006 + 0.009 = +3 \text{ ms.} \end{aligned}$$

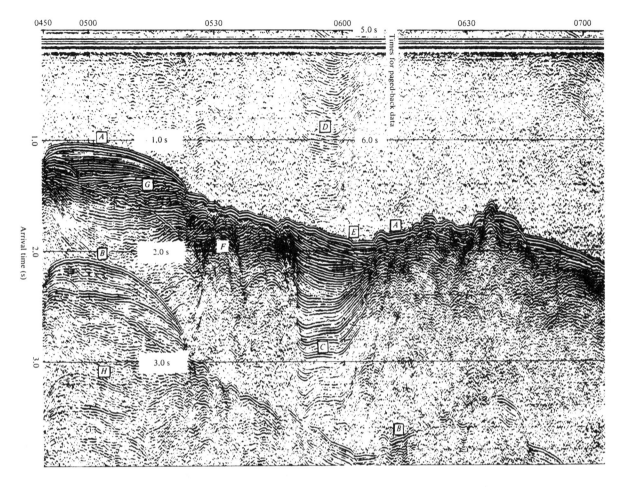

Fig. 8.23   Profiler record, offshore Japan. The water-bottom reflection, *A*, with traveltime 1.0 to 2.0 s indicates water depths of
750 to 1500 m. The ship traveled 8.5 km between the 30-min
marks at the top of the record. Most primary reflections are
obscured by the first and second water-bottom multiples, *B* and

H. More than a kilometer of sediments are indicated by the reflections near *C;* multiples of these appear paged back at *D. E*
indicates a fault scarp on the ocean floor. *F* are diffractions,
probably from sea-floor relief slightly offset to the line. Note the
onlap and thinning above *G.* (Courtesy of Teledyne.)

Table 8.1 *Corrections to reflection traveltimes*

	Source *C*	Source *B*	Source *A*	
Surface elevation, E_s (m)	248	244	257	measured
Depth of source, D_s (m)	15	13	20	
Uphole time, t_{uh} (ms)	48	44	53	
Source-to-datum time, Δt_s (ms)	16	15	18	calculated
Datum-to-geophone, Δt_g (ms)	64	59	71	
Correction time, Δt_0 (ms)	80	74	89	
Differential corr., Δt_c (ms)		−9		

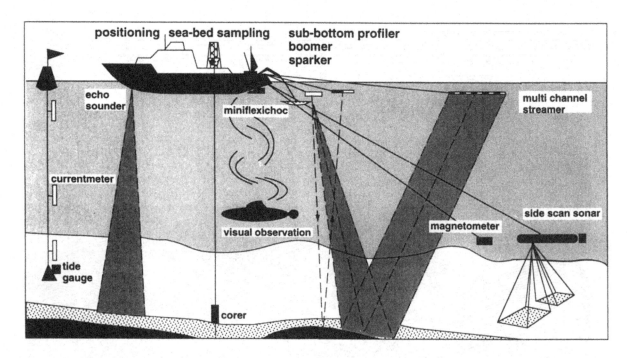

Fig. 8.24 Profiling operations employing several types of sensors. (Courtesy of Bureau d'Études Industrielles et de Coopération de l'Institut du Pétrole.)

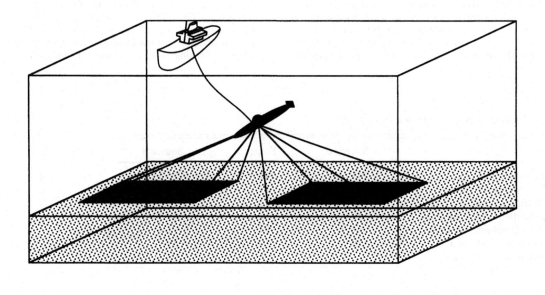

(a)

Fig. 8.25 Side-scan sonar. (a) Schematic diagram; (b) portion of a record showing reflections from sea-floor relief and gap underneath the towfish (courtesy of CGG); (c) recorder, tow fish, and towing cable (courtesy of EG&G Marine Instruments).

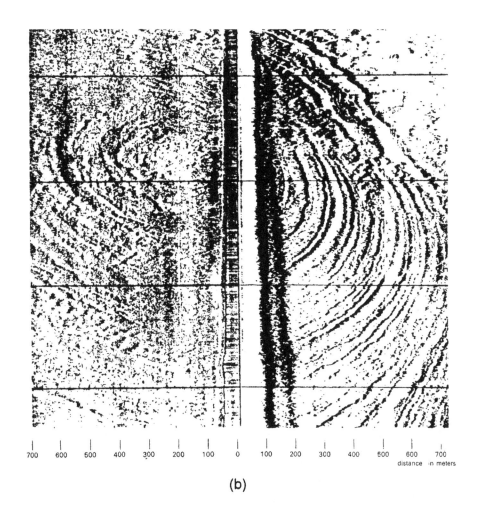

700 600 500 400 300 200 100 0 100 200 300 400 500 600 700

distance in meters

(b)

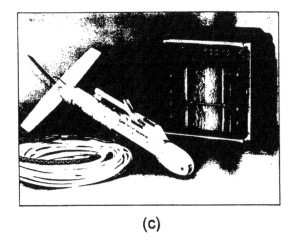

(c)

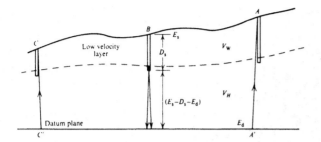

Fig. 8.26 Calculation of weathering corrections.

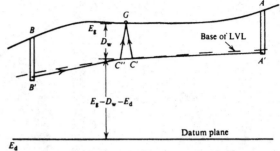

Fig. 8.27 Datum correction for geophones in between buried sources.

Thus, the correction can change the apparent direction of dip as well as the dip magnitude. Accurate corrections are essential.

Corrections are often required for geophones in between source locations where uphole data are not available. The first-breaks can be used for this purpose. In fig. 8.27, G is a geophone intermediate between adjacent sources A and B for which we have first-break traveltimes. Let t_{AG} and t_{BG} be the first-break times for paths $A'C'G$ and $B'C''G$. Almost always GC' and GC'' are within 20° of the vertical and $C'C''$ is therefore small. Thus, we can write the approximate relation

$$t_{AG} + t_{BG} \approx (A'B'/V_H) + 2t_W \approx (AB/V_H) + 2t_W,$$

t_W being the traveltime through the weathered layer at G. Thus,

$$t_W \approx \tfrac{1}{2}[t_{AG} + t_{BG} - (AB/V_H)]. \qquad (8.7)$$

Subtracting t_W from the arrival times in effect places the geophone at the base of the LVL; to correct to datum, we must subtract the additional amount, $(E_g - E_d - D_W)/V_H$, where E_g is the mean elevation of the geophone group, D_W being found by multiplying t_W by V_W.

Occasionally, special refraction profiles are acquired to obtain data for making corrections for intermediate geophones. These profiles are usually of the standard type using small charges placed near the surface or a nondynamite source on the surface; these are interpreted using standard methods such as the GRM (§11.3.3) or Wyrobek's (see §11.4.4) to find the depth and traveltime to the base of the LVL. Alternatively, a source may be placed just below the LVL, as shown in fig. 8.28; in this event, we must modify eq. (4.36) because the source is at the base rather than the top of the upper layer. Thus,

$$t = \frac{x - D_W \tan \theta}{V_H} + \frac{D_W}{V_W \cos \theta} = \frac{x}{V_H} + \frac{D_W \cos \theta}{V_W}. \qquad (8.8)$$

Most near-surface correction methods require a knowledge of V_H and sometimes of V_W as well. The former can be determined by (1) an uphole survey, (2) a special refraction survey, as previously described, or (3) analysis of the first-breaks for distant geophone groups (because these are equivalent to a refraction profile such as that shown in fig. 8.28). The weathering velocity V_W can be found by (1) measuring the slope of a plot of the first-breaks for geophones near the source (correcting distances for obliquity), (2) dividing D_s by t_{uh} for a source placed near the base of the LVL, (3) an uphole survey, or (4) firing a cap at the surface and measuring the velocity of the direct wave.

8.8.3 Picking reflections and preparing cross-sections

Although the following procedures are those used prior to the era of computer-processed record sections, they are also used occasionally today, and the concepts that they involve continue to be central to today's techniques. When the record quality is poor, almost any alignment may be mistakenly identified as a reflection. The best criterion in such cases is often the geological picture that results. If this picture does not make sense, we should reexamine the geophysical data with more skepticism. Naturally, "making sense" does not mean that the result must fit our preconceived ideas, but rather that it must be geologically plausible.

When the interpreter decided that an event was a legitimate reflection, he marked and "timed" it. When working with individual split-dip records, the arrival times at the center of the record and at the two outside traces (or the difference between these outside times, Δt_d), usually corrected for weathering and elevation, were recorded.

Besides timing reflections, the interpreter assigned

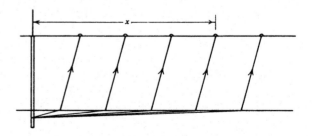

Fig. 8.28 Refraction weathering profile.

a *grade* to each, for example, VG, G, P, ? (for very good, good, poor, questionable). These grades referred to the certainty that the event was a primary reflection and the accuracy of measurement of the arrival times. Sometimes a two-letter grading system was used to separate the grading of certainty from the grading of timing accuracy.

The process of identifying events on a seismic record and selecting and timing the reflections is referred to as *picking* the records. Figure 7.45 shows a picked record.

The next stage after picking individual records was to prepare a composite representation, usually by plotting on a sheet of graph paper. The source locations were marked at the top on a horizontal line indicating the datum plane. The reflection events were plotted, and sometimes also the surface elevations, the depth of the base of the LVL, and the first-break arrival times.

Cross-sections are called *time sections* if the vertical scale is linear with time or *depth sections* if linear with depth. A cross-section is *unmigrated* when reflections are plotted vertically below the source. A *migrated section* (such as fig. 8.29) is one for which we assume that the seismic line is normal to strike so that the dip moveout indicates true dip, and we attempt to plot in their actual locations the segments of reflecting horizons that produced the recorded events. The scale on hand-migrated sections was usually linear with depth.

Unmigrated sections give a distorted picture of the subsurface, the distortion increasing with the amount of dip; they are used for interpretation only in areas of gentle dip or when an accurate structural picture is not required. Figure 8.30 shows the effect of lack of migration on simple structures. The anticline at the left would appear on an unmigrated section as the dashed line $R'ST'$ and the syncline would appear as the dashed line $T'UV'$. Failure to migrate decreases the curvature of an anticline and increases that of a syncline. Failure to migrate also positions a fault incorrectly.

Several methods of migration have been used. The simplest is to assume a constant velocity down to the reflector and swing an arc whose radius is half the arrival time t_0 multiplied by the average velocity. The radius O_1R in fig. 8.30 makes an angle with the vertical equal to the reflector dip (calculated using eq. (4.11)) and a straight-line segment equal to half the spread length is drawn at R perpendicular to the radius (tangent to the arc) to represent the reflecting segment (see also fig. 1.3b). This method of migrating positions the reflector segments incorrectly and gives them the wrong dip, compared with correctly allowing for a velocity gradient (see problem 8.20e).

Migration was sometimes carried out with simple plotting machines. Some of these assumed a functional form for the velocity, such as one described by Rockwell (1967), which used eq. (4.35) assuming a linear increase of velocity.

Probably the commonest method of hand-migrating reflections used a wavefront chart, a graph showing wavefronts and raypaths for an assumed vertical distribution of velocity. Figure 8.31 is a simplified version of such a chart. It shows the location of a

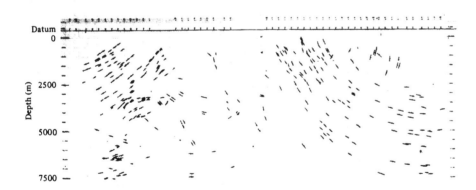

Fig. 8.29   Hand-migrated cross-section. (Courtesy of Chevron.)

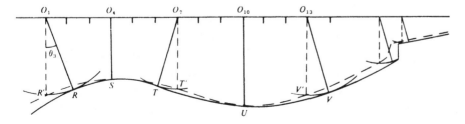

Fig. 8.30   Effect of lack of migration.

wavefront at different times after the source instant, the successive positions being labeled with the two-way traveltime. Raypaths are also shown; these are found by applying Snell's law at each change in velocity. Raypaths are labeled in terms of dip moveout, $\Delta t_d/\Delta x$. Dip moveout can be measured from corrected record sections such as stacked sections, as the difference in arrival time, Δt_d, between points a distance Δx apart. To plot a cross-section, the wavefront chart often was placed under transparent graph paper with the chart origin at the appropriate source location. A reflection with certain values of t_0 and $\Delta t_d/\Delta x$ was plotted by interpolating between the wavefronts and rays (actual wavefront charts have more closely spaced wavefronts and rays than those shown in fig. 8.31 so that interpolation was more accurate) and drawing a straight line of length equal to half the spread length tangent to the wavefront at the point (t_0, $\Delta t_d/\Delta x$). The reflection denoted by the symbol –o– in fig. 8.31 corresponds to $t_0 = 2.350$ s and $\Delta t_d/\Delta x = 110$ ms/km.

With an asymmetric spread, we correct the measured moveout for the difference in normal moveout between the two ends of the spread to find the dip moveout, $\Delta t_d/\Delta x$. With in-line offsets, we find $\Delta t_d/\Delta x$ by applying normal-moveout corrections (and correcting the arrival time to find t_0 when the offsets are large). On record sections (see what follows), we may measure the dip over any distance over which the horizon is dipping uniformly.

Identifying a reflection seen on one record as representing the same interface as a reflection on another record is called *correlation*. It was based partly on similarity of character and partly upon the dip and agreement in arrival times (*time-tie*). The travel paths for reflections on the outside traces of adjacent profiles that provide continuous subsurface coverage usually were the same except that they were traversed in opposite directions; hence, the traveltimes should be the same provided that adequate weathering and elevation corrections have been applied. For example, in fig. 8.3, trace 1 on the profile from source O_2 (with path O_2BO_3) should have the same corrected traveltime as trace 24 on the profile from source O_3.

If the record quality is too poor to provide continuous reflections or if the continuous reflections that are present are not in the part of the section where information is required, we draw *phantoms*, lines on the section so that they are parallel to adjacent dip symbols that the interpreter considers valid. Where data conflict or are absent, the phantom is drawn in the manner that seems most reasonable to the interpreter on the basis of whatever evidence may be available.

Record sections can be regarded as a form of time section obtained by placing successive records side by side. Record sections are usually corrected for elevation, weathering, and normal moveout. They display a large amount of data in compact form. A potential disadvantage with record sections is that they are so

graphic that people who lack understanding of the significance of the data tend to attach incorrect meanings to them, as though the section were a "photograph" of the rock formations in the ground.

Problems

8.1 Land cables are built in sections that are identical. The connections to the plugs at each end of the section effectively rotate the system by the number of groups to be connected to that section. Thus, if pins 1, 2, and 3 at one end of the section are connected to takeouts for 3 groups in a section, at the other end of the section they will connect with pins 4, 5, and 6 of the next section. There are more sets of wires than channels being used at any one time, for example, perhaps 96 independent pairs of wires for use with 48 channels. However, occasionally so many sections and geophones are laid out that a distant group of phones is connected to the same channel as a nearer group; this mistake is called *rollover*. Sketch a possible arrangement for the connections in one section with takeouts for three channels and explain how rollover could appear on the seismic records.

8.2 Assume a reflector 2.0 km beneath the midpoint and a dip of 20° with constant overburden velocity; how much does the reflecting point move between members of the common-midpoint set for offsets of 0, 0.5, 1.0, 1.5, and 2.0 km?

8.3 (a) Draw surface and subsurface stacking charts for a 24-trace split spread where the source is midway between the geophone-group centers and the centers of geophone groups 12 and 13 are separated by three times the normal geophone-group spacing. Assume a source interval double the geophone-group interval and that source and geophone locations beyond point A cannot be occupied. The source, however, can move on to location A, producing asymmetric spreads ("shooting through the spread").

(b) Assume that geophones but not sources can be located over a distance of 10 geophone-group intervals centered at point B. What is the effect on multiplicity? If the data quality deteriorate markedly when multiplicity falls, what can be done to alleviate the deterioration?

8.4 For an airwave with a velocity of 330 m/s and two geophones separated by 5 m, at what frequency is maximum attenuation achieved?

8.5 Reflections in the zone of interest have apparent velocities around 6.5 km/s, whereas the velocity just below the uniform LVL is 2.1 km/s. If we wish to avoid cancelation below 80 Hz when using an array, what is the maximum in-line array length?

8.6 (a) Under what conditions is the response of a linear array of n evenly spaced geophones zero for a wave traveling horizontally (such as ground roll)?

(b) If n geophones are distributed uniformly over one full wavelength, show that the response is $F = 1/n$.

(c) What is the response of the array in part (a) when

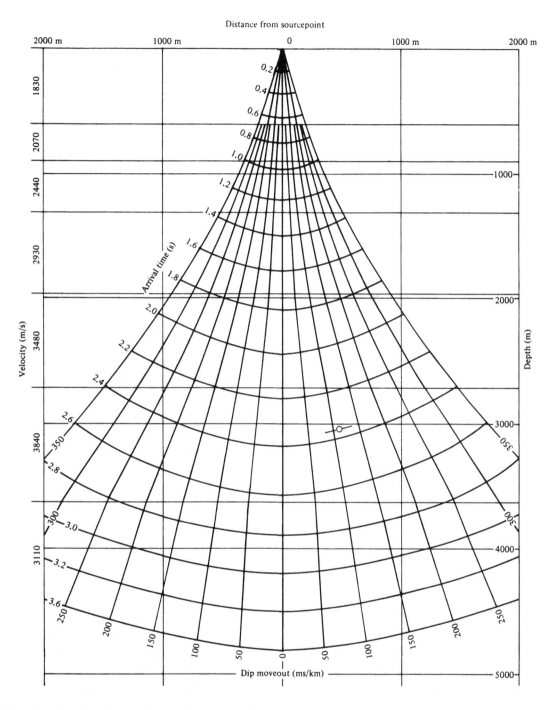

Distance from sourcepoint

Fig. 8.31   Simplified wavefront chart for horizontal velocity layering.

the waves arrive perpendicular to the line of geo-phones?

(d) What is the response of the array in part (a) when the waves arrive at 45° to the line and $n = 8$.

(e) Repeat part (d) for $n = 16$.

8.7 Show that eqs. (8.1) and (7.4) are consistent.

8.8 Array tapering is sometimes achieved by (1) dou-bling elements at some locations, (2) weighting equally spaced elements either (a) within the element or (b) in

a mixing box in the field, or (3) using unequal spacing of elements. What arguments would you use for or against each of these approaches?

8.9 Figure 8.32 shows the directivity effect of a group length typical of end-on marine shooting. How will the curves change (a) with arrival time, (b) as offset increases, and (c) for greater stacking velocity? Is it better to proceed in the updip or downdip directions?

8.10 (a) The tapered array [1, 2, 3, 3, 2, 1] = [1, 1, 1,

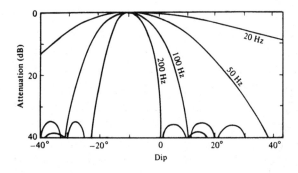

Fig. 8.32   Response to apparent dip of a tapered array with effective length of 50 m. Reflection arrival time is 1.0 s, stacking velocity is 1.5 km/s, offset is 300 m. (After Savit and Siems, 1977.)

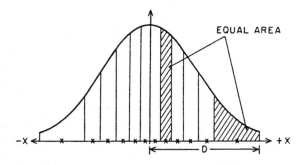

Fig. 8.33   Approximating a cosine array with unequally spaced geophones. (From Evans, 1989.)

1] ∗ [1, 1, 1]. Use this fact to sketch the array response. (b) This approximates a cosine array, which could be achieved more exactly by spacing elements unequally as shown in fig. 8.33. How could three strings, each having four equally spaced geophones, be used to approximate a cosine array more exactly than in part (a)?

8.11 (a) The noise test shown in fig. 8.34 used 36 geophones spaced 10 m apart and dynamite shots various in-line distances away. The event A_1–A_2 indicates ground roll. What is its velocity, dominant frequency, and wavelength? What length should a geophone group have to attenuate it?

(b) What are the velocity, dominant frequency, and wavelength of other wavetrains? Will the geophone group length determined in part (a) attenuate them?

(c) What causes the alignment B_1–B_2?

8.12 Assume that you wish to map objectives 3 to 5 km deep in an area with topography ranging from flat to gentle hills (surface gradients usually less than 3 m/ 100 m). Dips at objective depth may be up to 30°. The velocity at the base of the low-velocity layer is 2 km/s, that at objective depth is 4 km/s, and at the basement (8 km) probably about 6 km/s. Five surface-source units and 48-channel recording equipment are available. Both ground roll ($V_R \approx 800$ m/s) and air-waves may be problems, but the low-velocity layer (about 10 m thick with a velocity about 600 m/s) is probably fairly uniform. The area is fairly noisy and moderate effort will probably be required to achieve adequate data quality. Propose field methods and explain the bases for your proposals.

8.13 Seismic field work is usually carried out in a uniform manner with group intervals everywhere the same and the spread either all on one side or symmetric about the sourcepoint. The layout dimensions tend to be determined by what equipment is at hand or by habit rather than the nature of the problem to be solved; for example, the length of geophone strings (several geophones for a single group being permanently wired together) may dictate the geophone interval and the available equipment the number of channels and thus the effective spread length. Sometimes

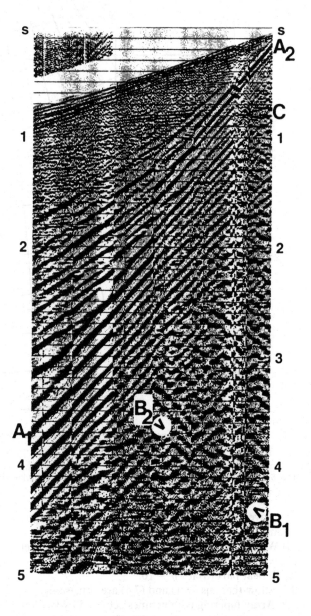

Fig. 8.34   Walk-away noise test. (From Yilmaz, 1987: 35.)

hybrid spread (§8.3.11) arrangements are used to make fuller use of equipment. Assume that you have more channels available than the number given by following the guidelines of §8.4.2; what circumstances might lead you to use the extra channels to (a) extend the spread length beyond that given by guideline (1); (b) fill in the space between the source and the minimum offset given by (2); (c) interleave additional groups somewhere in the middle of your spread; (d) lay out a partial spread on the other side of the source-point where an end-on arrangement is being used; and (e) lay out a short cross arm? If you have almost but not quite enough channels to use a split arrangement compared to an end-on, what are the advantages and disadvantages of using the split with: (f) longer group intervals than given by guideline (5); (g) shortening the maximum offset; or (h) increasing the minimum offset?

8.14 Signal and noise characteristics determined from a previous dynamite survey are shown in fig. 8.35. The principal objective is at 3000 m with a stacking velocity of 3000 m/s.

(a) What frequencies should be covered by a linear sweep?

(b) If a downsweep of 8 s is used, at what time will the correlation ghost appear? Will it interfere with the objective?

(c) If a single vibrator sweep of 8 s yields an S/N ratio $= 0.2$ at the objective depth for a 20-to-60-s sweep, how many sweeps per vibrator point will be required to give $S/N = 2.0$?

(d) Assume that recording continues for an additional time of 6 s (listen time) beyond the sweep time and that it takes 10 s for moveup between sweep points; how long will be required for four vibrators to record one vibrator point?

8.15 You wish to survey a 70 × 70 km area where anticlinal structures with their long axes north–south are expected, the minimum size structure economically viable being 2 km across. Maximum dip expected is 15° and reflectors are listed in table 8.2. The Recent reflector will be useful in making statics corrections. A noise test gave a prestack $S/N = 0.5$.

(a) Determine line spacing and orientation.

(b) What multiplicity is required to give $S/N = 3.0$ after stack?

(c) What spread geometry should be used, that is, near- and far-channel offsets, spacing required to avoid aliasing for 15 to 40 Hz, and minimum number of channels required?

(d) How long will the survey require, assuming 270 km/month for 24-fold?

(e) Answer part (d) for 210 km/month for 48-fold.

8.16 (a) Take the wavelet that has values at successive 4-ms intervals of 0, . . . , 0, 8, 7, −8, −6, 0, 4, 2, 0, . . . , 0 (with 10 zeros at each end) and add random noise (a random number table is given in app. C) in the range from +10 to −10, that is, with signal/noise ≈ 1. Do this five times for different noise values and

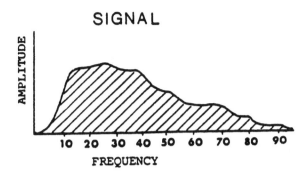

SIGNAL

NOISE

← ground roll

Fig. 8.35 Signal and noise spectra from a particular area. (From Evans, 1989.)

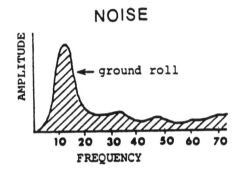

Table 8.2 *Reflection data*

Age	Depth (m)	$\bar{V}$ (m/s)	t_0 (s)	V_i (m/s)
Recent	300	2000	0.333	2000
Eocene	3000	3000	2.222	3140
Cretaceous	5000	3300	3.000	4137

plot the results, shading positive values as in a variable-area display.

(b) Repeat for noise ranging from + 20 to −20 (signal/noise ≈ $\frac{1}{2}$). What can you conclude about the extraction of signal buried in noise?

(c) Sum the five waveforms in part (a) and also in part (b) to show how stacking enhances coherent signal versus random noise.

(d) Replace the elements in the wavelets in parts (a) and (b) with +1 or −1 as the values are positive or negative (sign-bit expression) and repeat part (c).

8.17 Uphole surveys in five different (unrelated) areas give the uphole-time versus depth information in table 8.3. Explain the possible velocity layering for each case. How reliably are velocities and depth of weathering defined?

8.18 The correction methods discussed in §8.8.2 assume that the source is below the base of the LVL. What changes are required in the equations of this section if this is not the case?

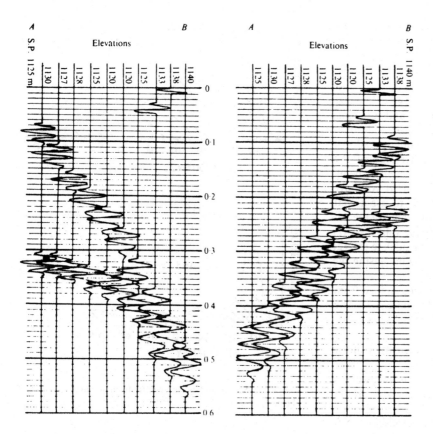

Fig. 8.36   Reflection first-breaks.

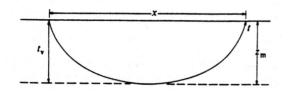

Fig. 8.37   Illustrating Blondeau weathering correction.

8.19 Figure 8.36 shows the first arrivals at geophone stations 100 m apart from shots 25 m deep at each end of the spread. (There are actually 11 geophone stations with the shotpoints being at the first and eleventh stations; however, the geophone group at each shotpoint is not recorded because of hole noise.) The uphole geophone is recorded on the third trace from the right. The weathering velocity is 500 m/s.

(a) Estimate the subweathering velocity V_H by averaging the slopes of lines approximating the first-breaks. The valley midway between the shotpoints produces a change in the first-break slopes, as if two refractors are involved, which is not the case. How can we be sure of the latter? (*Hint:* Plot an elevation profile as an aid in determining the best-fit line through the first breaks to give V_H.)

(b) Determine the weathering thicknesses at the two shotpoints from the uphole times.

(c) What corrections Δt_0 should be applied to reflection times at the two shotpoints for a datum of 1100 m?

(d) Calculate the weathering thickness and the time correction for each geophone station.

(e) Plot corrected reflection arrival times in an X^2–T^2 plot and determine the depth, dip, and average velocity to the reflector giving the reflection at 0.30 and 0.21 s.

8.20 The arrival time of a reflection at the sourcepoint is 1.200 s and the difference in arrival times at geophones 1000 m on opposite sides of the source is 0.150 s, near-surface corrections having been applied. Assume the line is perpendicular to the strike. Determine the reflector depth, dip, and horizontal location with respect to the sourcepoint assuming: (a) no dip moveout and the average velocity associated with a vertical traveltime is $\overline{V} = 2630$ m/s; (b) the observed dip moveout and $\overline{V}$; (c) straight-ray travel at the angle of approach and $V_H = 1830$ m/s; (d) straight-ray travel at the local velocity above the reflector, 3840 m/s; (e) the migrated position determined from the wavefront chart of fig. 8.31.

8.21 The Blondeau method of making weathering corrections starts with a curve of emergent arrival times versus offsets, t versus x in fig. 8.37 plotted on log–log graph paper. The slope gives $B = (1 - 1/n)$,

Table 8.3 *Uphole surveys in five areas*

Depth (m)	Area A (s)	Area B (s)	Area C (s)	Area D (s)	Area E (s)
5	0.012	0.011		0.012	0.008
8			0.020	0.016	
10	0.025	0.023	0.024	0.018	
12		0.024	0.027		0.020
15	0.030		0.031	0.022	
18		0.028	0.034	0.030	0.030
21	0.034		0.036	0.033	0.031
25	0.036	0.032		0.035	0.032
30	0.039	0.035	0.039		
35		0.037		0.039	0.036
40	0.046		0.044	0.042	
50	0.051	0.044	0.048	0.047	0.043

where $1/n$ is the exponent in eq. (5.18). To find t_v, the vertical traveltime to a given depth z_m, we find F, a tabulated function of B (see Musgrave and Bratton, 1967: 233). Then, $x = Fz_m$ so that we can get t from the x–t curve. Finally, $t_v = x/F$.

Verify this procedure by deriving the following relations:

(a) $z = z_m \sin^n i$, where i is the angle of incidence measured with respect to the vertical at depth z.

(b) $x = Fz_m$, where $F = 2n\int_0^{\pi/2} \sin^n i \, di$ = function of n, hence of B also.

(c) $t = (G/a)(x/F)^B$, where $G = 2n\int_0^{\pi/2} \sin^{n-2} i \, di$.

(d) $dx/dt = x/Bt$ is the horizontal component of apparent velocity at the point of emergence.

(e) dx/dt, the horizontal component of apparent velocity of the wavefront at any point of the trajectory, is $V_m = x/Bt$.

(f) $t_v = \int_0^{z_m} dz/V = t/F$.

References

Agnich, F. J., and R. C. Dunlap. 1959. Standards of performance in petroleum exploration. *Geophysics*, **24**: 916–24.

Anstey, N. A. 1970. *Seismic Prospecting Instruments, Vol. 1: Signal Characteristics and Instrument Specifications*. Berlin: Gebruder-Borntraeger.

Anstey, N. A. 1986. Whatever happened to ground roll? *The Leading Edge*, **5(3)**: 40–5.

Brown, A. R. 1991. *Interpretation of 3-Dimensional Seismic Data*, 3d ed., AAPG Memoir 42. Tulsa: American Association of Petroleum Geologists.

Brown, A. R. 1992. Technologies of reservoir geophysics. In *Reservoir Geophysics*, R. E. Sheriff, ed., pp. 51–70. Tulsa: Society of Exploration Geophysicists.

Burg, J. B. 1964. Three-dimensional filtering with an array of seismometers. *Geophysics*, **29**: 693–713.

Dangerfield, J. A. 1992. Ekofisk Field development: Making images of a gas-obscured reservoir. In *Reservoir Geophysics*, R. E. Sheriff, ed., pp. 98–109. Tulsa: Society of Exploration Geophysicists.

Denham, L. R. 1981. Extending the resolution of seismic reflection exploration. *J. Can. Soc. Expl. Geophys.*

Evans, B. 1989. *Handbook for Seismic Data Acquisition in Oil Exploration*. Perth, West Australia: Curtin University of Technology.

Harding, T. P., R. F. Gregory, and L. H. Stephens. 1983. Convergent wrench fault and positive flower structure, Ardmore Basin, Oklahoma. In *Seismic Expressions of Structural Styles*, A. W. Bally, ed., AAPG Studies in Geology 15. Tulsa: American Association of Petroleum Geologists.

International Association of Geophysical Contractors. 1991a. *Land Geophysical Operations Safety Manual*, 7th ed. Houston: IAGC.

International Association of Geophysical Contractors. 1991b. *Marine Geophysical Operations Safety Manual*, 7th ed. Houston: IAGC.

International Association of Geophysical Contractors. 1993. *Environmental Guidelines for Worldwide Geophysical Operations*. Houston: IAGC.

King, V. L. 1973. Sea bed geology from sparker profiles, Vermillion Block 321, offshore Louisiana, 1973 Offshore Technology Conference Preprints, paper 1802. Dallas: Offshore Technology Conference.

Lamer, A. 1970. Couplage sul-geophone. *Geophys. Prosp.*, **18**: 300–19.

Laster, J., and A. F. Linville. 1968. Preferential excitation of refractive interfaces by use of a source array. *Geophysics*, **33**: 49–64.

Lindsey, J. P. 1991. Crooked lines and taboo places. *The Leading Edge*, **10(11)**: 74–7.

Marsden, D. 1993. Static corrections – A review. *The Leading Edge*, **12(1)**: 43–9 and **12(2)**: 115–20.

Mayne, W. H. 1962. Common-reflection-point horizontal data-stacking techniques. *Geophysics*, **27**: 927–38.

Mayne, W. H. 1967. Practical considerations in the use of common-reflection-point techniques. *Geophysics*, **32**: 225–9.

McKay, A. E. 1954. Review of pattern shooting. *Geophysics*, **19**: 420–37.

Morgan, N. A. 1970. Wavelet maps – A new analysis tool for reflection seismograms. *Geophysics*, **35**: 447–60.

Musgrave, A. W., and R. H. Bratton. 1967. Practical application of Blondeau weathering solution. In *Seismic Refraction Prospecting*, A. W. Musgrave, ed., pp. 231–46. Tulsa: Society of Exploration Geophysicists.

Newman, P., and J. T. Mahoney. 1973. Patterns – With a pinch of salt. *Geophys. Prosp.*, **21**: 197–219.

O'Connell, J. K., M. Kohli, and S. Amos. 1993. Bullwinkle: A unique 3-D experiment. *Geophysics,* **58**: 167–76.

Parr, J. O., and W. H. Mayne. 1955. A new method of pattern shooting. *Geophysics,* **20**: 539–64.

Pritchett, W. C. 1990. *Acquiring Better Seismic Data.* London: Chapman and Hall.

Rockwell, D. W. 1967. A general wavefront method. In *Seismic Refraction Prospecting,* A. W. Musgrave, ed., pp. 363–415. Tulsa: Society of Exploration Geophysicists.

Savit, C. H., and L. E. Siems. 1977. A 500-channel streamer system, 1977 Offshore Technology Conference Preprints, paper 2833. Dallas: Offshore Technology Conference.

Schoenberger, M. 1970. Optimization and implementation of marine seismic arrays. *Geophysics,* **35**: 1038–53.

Sheriff, R. E. 1978. *A First Course in Geophysical Exploration and Interpretation.* Boston: International Human Resources Development Corp.

Sheriff, R. E. 1991. *Encyclopedic Dictionary of Exploration Geophysics,* 3d ed. Tulsa: Society of Exploration Geophysicists.

Yilmaz, O. 1987. *Seismic Data Processing.* Tulsa: Society of Exploration Geophysicists.

9
Data processing

Overview

The "digital revolution" changed seismic exploration beginning in the early 1960s, about 20 years before the widespread use of CDs in recorded music. The changes were of the same kind: much less noise, no deterioration of signal by repeated processing (playing), and the ability to reshape the information content into more easily understood forms.

Radar, one of the technological advances of World War II, was used in detecting ships and aircraft. However, noise frequently interfered with its application and considerable theoretical work devoted to the detection of signals in the presence of noise led to the development of a new field of mathematics, information theory. At first, this theory was very difficult to understand because it was formulated in complex mathematical expressions and employed an unfamiliar vocabulary. However, the development of digital computer technology considerably simplified the understanding of the basic concepts, and the number of applications has expanded greatly.

Early in the 1950s, a research group at the Massachusetts Institute of Technology studied the application of information theory to seismic exploration (Flinn, Robinson, and Treitel, 1967). These studies combined with the new digital technology changed seismic exploration considerably. Today, most seismic data are recorded in digital form and subjected to data processing before being interpreted.

The basic concepts are expressed in a number of books and papers (Lee, 1960; Robinson and Treitel, 1964, 1980; Silverman, 1967; Anstey, 1970; Finetti, Nicolich, and Sancin, 1971; Kanasewich, 1987). Probably the two best current books on seismic data processing in general are Hatton, Worthington, and Makin (1986) and Yilmaz (1987).

In chap. 15, we give the mathematical concepts most important in data processing, and in this chapter, we show how these concepts are applied without being concerned with mathematical proofs. This chapter discusses mainly functions in digital form, whereas chap. 15 treats both continuous and digital functions. Equation numbers give the nearly identical equations in chap. 15.

Usually, we think of seismic data as the variation with time (measured from the source instant) of the amplitudes of various geophone outputs. When we take this viewpoint, we are thinking in the *time domain*, that is, time is the independent variable. We also sometimes find it convenient to regard a seismic wave as the result of the superposition of many sinusoidal waves differing in frequency, amplitude, and phase; the relative amplitudes and phases are regarded as functions of frequency and we are thinking in the *frequency domain*. The frequency-domain approach is illustrated by electrical systems that are specified by their effects on the amplitudes and phases of sinusoidal signals of different frequencies. For example, graphs of filter characteristics usually show amplitude ratios or phase shifts as ordinates with frequency as the abscissa.

Three types of mathematical operations constitute the heart of most data processing: Fourier transforms, convolution, and correlation. Fourier transforms (§9.1.3) convert from the time domain to the frequency domain and vice versa, and they and other types of transforms can be used to convert into and out of other domains also. Sometimes we transform in two dimensions to other domains, such as the f–k (§9.9), τ–p (§9.11.1), or other domains. The essential aspect of transforms is that in principle no information is lost in the procedure, although in actual application, very minor degradation occurs because of approximations, truncation, and so on. Transforms provide alternate ways of doing things that are sometimes advantageous.

Convolution (§9.2) is the operation of replacing each element of an input with a scaled output function; it is the mathematical equivalent to filtering, such as occurs naturally in the passage of seismic waves through the earth, in passing electrical signals through circuits, and so on. The limitations on sampling and signal reconstitution, that there be no frequency components above half the sampling frequency, are explained using convolution concepts. Sometimes undesirable filtering can be undone by deconvolution.

Correlation (§9.3) is a method of measuring the similarity between two data sets. A common application is determining the time shift that will maximize the similarity. Correlation is also the means for extracting short signals of known waveshape from long wavetrains, as is used in Vibroseis processing. If a data set is correlated with itself (autocorrelation), a measure of the repetition in the data is obtained.

Phase (§9.4) plays an especially important role in seismic exploration. The wavelet injected into the earth by most impulsive sources is nearly minimum phase but the wavelet injected by the Vibroseis is far from minimum phase. A zero-phase output provides the best resolution and hence is the desired wavelet for interpretation, and the phase of seismic data is changed in processing.

The objective of most data processing is enhancing the signal with respect to the noise (§9.5). Deconvolution is probably the most important process that aims to improve vertical resolution. Improvements resulting from discrimination on the basis of frequency are employed in various deconvolution techniques: deterministic inverse filtering, recursive filtering, least-squares (Wiener) filtering, wavelet processing, and so on. Deconvolution is, of course, not restricted to just one dimension.

Compensation for near-surface time delays is the objective of statics corrections (§9.6). The first approximation to statics is generally based on measurements in the field, but this is usually supplemented by use of a surface-consistent statics correction procedure. However, surface-consistent statics can leave long-wavelength errors and refraction statics analysis is often employed to remove such errors.

Normal moveout is the discriminant employed in velocity analysis (§9.7). Both the velocity spectrum and velocity panels as used in velocity analysis generally yield velocity information at locations a kilometer or so apart. Horizontal-velocity analysis provides a continuous examination of the stacking velocity for specific reflecting horizons.

The preservation of amplitude is important because of its significance in interpretation (§9.8). Amplitudes vary for a number of reasons. Surface-consistent amplitude analysis is similar in concept to surface-consistent statics correction.

Apparent-velocity (or 2-D, dip, or f–k) filtering (§9.9) provides a very powerful tool for discriminating against coherent noise trains. Stacking techniques (§9.10) also aid considerably in discriminating against noises of various types. DMO processing removes much of the reflection-point smear inherent in CMP stacking when reflectors dip. DMO is sometimes applied after NMO corrections and sometimes before. Muting of noise trains can also be used effectively to discriminate against coherent noise trains. Weighted stacks can provide considerably more discrimination against certain types of noises than simple CMP stacking.

Special processing techniques (§9.11) such as slant (τ–p) stacking, intelligent interpolation, and automatic picking can be used under special circumstances. Attribute analysis and other methods that rearrange the information content of seismic data sometimes provide viewpoints that give interpretive insight into features of various kinds.

Migration, the repositioning of data elements to represent the subsurface location of features, is often the last major step in processing. Migration is generally based on the exploding reflector model, where reflectors are replaced by distributed sources with mag-

nitude proportional to the reflectivity, the sources are exploded simultaneously, and the waves travel only upward (at $\frac{1}{2}$ the velocity) to generate the recorded section. Migration involves using the wave equation to backtrack the wavefield to locate the reflectors. Most methods employ variations of three methods: an integral solution (Kirchhoff or diffraction-stack migration), a solution in the frequency domain (Stolt or Gadzag migration), or a finite-difference solution in the time domain. The major problem is how to handle (in fact, how to determine) velocity variation, especially laterally varying velocity. Depth migration is used to take into account lateral velocity variations.

Typical data-processing procedures are described in §9.13. Processing decisions are often made at workstations, and to some degree, processing can be tailored to the special requirements of specific data. Often practical compromises have to be exercised.

This chapter concludes with a brief introduction to generalized inversion.

9.1 Transforms

9.1.1 Integral transforms

Fourier transforms are an example of a class of operations called integral transforms, which are used to transform a function into a related function of different variables. The transform is accomplished by multiplying the original function by a "kernel," which is a function of both sets of variables, and then eliminating the first set of variables by integrating the product with respect to these variables between definite limits. The kernel must be such that the process can be reversed by integration using a different kernel (sometimes the reciprocal of the first) to obtain a function of the first variables.

The original function and its transform are said to be in *domains* specified by the variables; thus, if $f(t)$ is in the time domain, its transform $F(\nu)$ is in the frequency domain (note that the dimensions of t and ν are reciprocals).

Although in general the most widely used transform is the Laplace transform, the Fourier transform (including the special forms known as z-transforms and Hilbert transforms) is most important in seismic work. The most common transforms are listed in table 9.1.

A crucial factor regarding transforms is that (theoretically) no information is lost in transforming. Thus, we can start with a waveform in the time domain, transform it into the frequency domain, perform various operations on the transform, and then transform the result back to the time domain to obtain the original function with modifications equivalent to the frequency-domain operations. For example, if we remove all frequencies below 60 Hz in the frequency domain, the time-domain result will be the original wavelet minus these frequencies. Transforms enable us to do part of our processing in one domain and part in another, taking advantage of the fact that some pro-

Table 9.1 *Common transforms used in seismic work*

Original domain	Transform type	Transform domain	Kernel
time (t)	Fourier	Frequency	$e^{\pm j2\pi\nu t}$
1-D space (x)	Fourier	Wavenumber	$e^{\pm j\kappa x}$
2-D space (x, y)	2-D Fourier	Wavenumber–wavenumber	$e^{\pm j(\kappa_x x + \kappa_y y)}$
time (t)	Laplace	s-domain	e^{-st} (cf. §15.3)
$x - t$	Radon (slant stack)	$\tau - p$	(cf. §9.1.5)

cesses can be done more easily in one domain than in another. In actual practice the integrals must be calculated by using series and a small amount of information is lost in transforming because of the truncation of these series and round-off errors, but we can make these errors as small as desired by carrying the calculations far enough.

9.1.2 Fourier analysis and synthesis

Fourier analysis in our context involves transforming functions from the time domain to the frequency domain and Fourier syntheses the inverse process of transforming from the frequency domain to the time domain. This distinction is artificial, however, and analysis and synthesis can be interchanged without making any difference in the result.

If we have a reasonably "well-behaved" (§15.2.1) periodic function $g(t)$ of period T (that is, if $g(t)$ repeats itself every time t increases by T), then the function can be represented by a *Fourier series*,

$$g(t) = \tfrac{1}{2}a_0 + \sum_{n=1}^{\infty} (a_n \cos 2\pi\nu_n t + b_n \sin 2\pi\nu_n t),$$
$$(9.1; 15.93)$$

$$= \tfrac{1}{2}c_0 + \sum_{n=1}^{\infty} c_n \cos (2\pi\nu_n t - \gamma_n), \quad (9.2; 15.102)$$

$$= \sum_{n=-\infty}^{\infty} \alpha_n e^{j2\pi\nu_n t}, \qquad (9.3; 15.105)$$

where

$$\nu_n = n/T = n\nu_0 = n\omega_0/2\pi = \omega_n/2\pi,$$

$$a_n = (2/T) \int_{-\frac{1}{2}T}^{\frac{1}{2}T} g(t) \cos 2\pi\nu_n t \, dt,$$

$$b_n = (2/T) \int_{-\frac{1}{2}T}^{\frac{1}{2}T} g(t) \sin 2\pi\nu_n t \, dt,$$

$$(9.4; 15.99, 15.100)$$

$$c_n = (2/T) \int_{-\frac{1}{2}T}^{\frac{1}{2}T} g(t) \cos (2\pi\nu_n t - \gamma_n) \, dt,$$

$$\gamma_0 = 0, \gamma_n = \tan^{-1}(b_n/a_n) \quad (n\neq0),$$

$$(9.5; 15.103)$$

$$\alpha_{\pm n} = (1/T) \int_{-\frac{1}{2}T}^{\frac{1}{2}T} g(t) e^{\mp j2\pi\nu_n t} \, dt \quad (9.6; 15.106)$$

(Subscripts indicate discrete sets, as with ν_n, a_n, and so on, whereas functional notation, such as $g(t)$, indicates

a continuous variable. Equation numbers also show equivalents in chap. 15.)

Equation (9.1) expresses $g(t)$ in terms of cosine and sine curves of amplitudes a_n and b_n, eq. (9.2) in terms of cosine curves of amplitude c_n that have been phase shifted by γ_n, and eq. (9.3) in terms of complex exponentials. All three forms are equivalent, that is,

$$c_n = a_n \cos \gamma_n + b_n \sin \gamma_n, \, = (a_n^2 + b_n^2)^{1/2},$$
$$(9.7, 15.103)$$

$$a_n = \alpha_n + \alpha_{-n}; \, b_n = j(\alpha_n - \alpha_{-n}); \qquad (9.8)$$

$$\alpha_{\pm n} = \tfrac{1}{2}(a_n \mp jb_n). \qquad (9.9; 15.106)$$

Note that

$$\tfrac{1}{2}c_0 = \tfrac{1}{2}a_0 = \alpha_0 = \text{average values of } g(t).$$
$$(9.10; 15.101; 15.103)$$

With seismic data $g(t)$ is usually a voltage or equivalent with a zero mean; in this case, eq. (9.10) means that there is no dc component ($a_0 = b_0 = c_0 = \alpha_0 = 0$). Equations (9.1) and (9.2) show that $g(t)$ can be regarded as the sum of an infinite number of harmonic (cosine and sine) waves of frequencies ν_n having amplitudes a_n, b_n, or c_n and phases γ_n. These equations thus represent the analysis of $g(t)$ into component harmonic waves.

9.1.3 Fourier transforms

As period T becomes larger, it takes longer for $g(t)$ to repeat; in the limit when T becomes infinite, $g(t)$ no longer repeats. In this case, we get in place of eqs. (9.3) and (9.6) (see §15.2.3)

$$g(t) = \int_{-\infty}^{\infty} G(\nu) e^{j2\pi\nu t} \, d\nu,$$
$$(9.11; 15.109)$$

$$G(\nu) = \int_{-\infty}^{\infty} g(t) e^{-j2\pi\nu t} \, dt.$$
$$(9.12; 15.108)$$

The function $G(\nu)$ is the *Fourier transform* of $g(t)$ and $g(t)$ is the *inverse Fourier transform* of $G(\nu)$. Using the symbol $\leftrightarrow$ to denote equivalent expressions in different domains, we write

$$g(t) \leftrightarrow G(\nu)$$

We also refer to $g(t)$ and $G(\nu)$ as a *transform pair*.

Equations (9.11) and (9.12) can be written in several ways. In general, $G(v)$ is complex:

$$G(v) = A(v)e^{j\gamma(v)}, \quad (9.13; 15.113a)$$

where $A(v)$ and $\gamma(v)$ are real, and $A(v)$ is also positive. We call $A(v)$ the *amplitude spectrum* (often called the frequency spectrum – see §15.2.1) and $\gamma(v)$ the *phase spectrum* of $g(t)$. Substitution in eq. (9.11) gives

$$g(t) = \int_{-\infty}^{\infty} A(v)e^{j[2\pi vt + \gamma(v)]}\, dv. \quad (9.14)$$

For actual waveforms, $g(t)$ is real and hence from eq. (9.14) and problem 15.12a, we get

$$g(t) = \int_{-\infty}^{\infty} A(v) \cos[2\pi vt + \gamma(v)]\, dv. \quad (9.15)$$

Because $G(v)$ is complex, we may separate it into real and imaginary parts,

$$G(v) = R(v) + jX(v); \quad (9.16; 15.113a)$$

when $g(t)$ is real,

$$\left.\begin{array}{l} R(v) = \displaystyle\int_{-\infty}^{\infty} g(t) \cos 2\pi vt\, dt, \\[2ex] -X(v) = \displaystyle\int_{-\infty}^{\infty} g(t) \sin 2\pi vt\, dt. \end{array}\right\} \quad (9.17; 15.112)$$

The integrals $R(v)$ and $-X(v)$ are called the *cosine* and *sine transforms,* respectively. When $g(t)$ is real, $R(v)$ and $X(v)$ are respectively even (symmetrical, that is, $R(v) = R(-v)$) and odd (antisymmetrical, that is, $X(v) = -X(-v)$) functions. When $g(t)$ is real and even, $X(v) = 0$; when $g(t)$ is real and odd, $R(v) = 0$ (see problems 15.17b to 15.17d).

We have regarded the Fourier transform as a transformation of the independent variable, $t \leftrightarrow v$. The variables have reciprocal dimensions: t is measured in time units (for example, seconds), v in reciprocal time units (for example, per second). The same mathematics clearly applies to any other dimension and its reciprocal such as distance (for example, meters) and reciprocal distance (for example, per meter).

9.1.4 Multidimensional Fourier transforms

In §9.1.3, we described the Fourier transform as involving a change from one independent variable to another, $t \leftrightarrow v$. The transformation can be applied to any number of variables, but most often only two are involved; for example, a seismic section is usually a plot of time versus distance (for example, trace spacing). In this case, we can generalize eqs. (9.11) and (9.12) to the two-dimensional transform:

$$g(x, t) = (1/2\pi) \int_{-\infty}^{+\infty} \int_{-\infty}^{+\infty} G(\kappa, v)e^{j(\kappa x + 2\pi vt)}\, d\kappa\, dv, \quad (9.18; 15.117)$$

$$G(\kappa, v) = \int_{-\infty}^{+\infty} \int_{-\infty}^{+\infty} g(x, t)e^{-j(\kappa x + 2\pi vt)}\, dx\, dt, \quad (9.19; 15.116)$$

the $1/2\pi$ factor enters because $\kappa = 2\pi/\lambda$, whereas $v = 1/T$ (see §2.1.1). When we regard κ and v as the independent variables, we are in the *f–k domain.*

9.1.5 Radon (τ–p) transforms

The Radon transform, also known as a slant stack or projection, is a line integral of some property of the medium along a specified line (usually a straight line). The transform can be defined in many ways, the following one being that given by Stewart (1991). The Radon transform of a property such as amplitude, $g(x, y)$ along the line RS in fig. 9.1, is the integral

$$g(r, \phi) \leftrightarrow G(\ell, \theta) = \int_{R}^{S} g(r, \phi)\, ds. \quad (9.20)$$

Line RS has the equation $x \cos \theta + y \sin \theta = \ell$, so the integral can be written

$$g(x, y) \leftrightarrow G(\ell, \theta)$$
$$= \int_{R}^{S}\int_{R}^{S} g(x, y)\, \delta(x \cos \theta + y \sin \theta - \ell)\, dx\, dy, \quad (9.21)$$

where $\delta(t) = 0$ if $t \neq 0$, $\delta(t) = 1$ if $t = 0$ (see §9.2.1 and 15.2.5); using the delta function ensures that the only contributions to the integrals are at points on line RS where $x \cos \theta + y \sin \theta - \ell = 0$.

The Radon transform changes (maps) $g(x, y)$ from the x–y domain onto a cylinder in the ℓ–θ domain, the cylinder being given by the equations $0 \leq \theta \leq 2\pi$, $0 \leq \ell \leq +\infty$ or onto the half-cylinder $0 \leq \theta \leq \pi$, $-\infty \leq \ell \leq +\infty$.

The Radon transform is closely related to the τ–p transform, the only difference being in the equation used to specify RS. For the τ–p transform, we define RS by the equation $y = \tau - px$ (that is, $\tau = \ell/\sin \theta$, $p = \cot \theta$) and eq. (9.21) becomes

$$g(x, y) \leftrightarrow G(\tau, p) = \int_{R}^{S}\int_{R}^{S} g(x, y)\, \delta(y + px - \tau)\, dx\, dy. \quad (9.22)$$

The theory of the Radon transform will be developed further in §13.5 on tomography; applications of the τ–p transform will be discussed in §9.11.1.

9.1.6 Implementation of transforms

In §9.1.1, we described the Fourier transform in terms of continuous variables and integrals from $-\infty$ to $+\infty$.

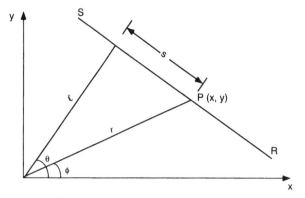

Fig. 9.1   Illustrating the Radon transform.

In practice, we deal with a finite number of discrete samples and the operation is done with a fast Fourier transform, which requires 2^k samples, where k is an integer; we add (pad) a string of zeros to satisfy this condition. A long string of zeros added at the end of our waveform also helps avoid wraparound, the aliasing that occurs because our waveform repeats every 2^k samples. Fourier transforms do not perform well at abrupt discontinuities (§15.2.7). The two numerical integrations (as in eqs. (9.11) and (9.12)) required to transform from the time domain to the frequency domain and back again involve roundoff errors, but these can be made as small as desired by using sufficient terms (at the cost of increased computer time).

The Fourier transform involves fitting cosine curves to our waveform, so we must include a sufficiently long portion in our analysis if the frequency-domain representation is to have meaning. A time series extending from 0 to $n-1$ consists of only n independent pieces of information, and its representation in the frequency domain, therefore, can have only n independent values. The time domain involves only real values, whereas values are complex in the frequency domain, that is, each frequency requires two independent values (real and imaginary, or amplitude and phase) to express it. Hence, there can be independent values for only $n/2$ frequencies.

Transforms provide equivalent ways of representing the same information content in different domains. We sometimes refer to transform operations as *mapping* from one domain to another, and we utilize mapping in a number of different ways. For example, in addition to the transforms listed in §9.1.1, we map between the domains listed in table 9.2.

Another way of looking at Fourier synthesis of a seismic wavelet is that each component, being infinite in length, spreads its energy over all of space, but that other components interfere destructively with it except where the wavelet is located. This concept provides a useful way of thinking in a number of situations. For example, DMO (§9.10.2) and migration algorithms (§9.13) spread the energy of each trace out over the operator shape, and we rely on the spread-

out energy from other traces to provide cancellation except where the migrated data lie. As another example, we can think of a reflector as the aggregate of many closely spaced diffracting points and the reflection as the interference composite of their diffractions. An alias (§9.2.2) results when not enough components are included to produce the required destructive interference.

9.2 Convolution

9.2.1 The convolution operation

Convolution is the time-domain operation of replacing each element of an input function with an output function scaled according to the magnitude of the input element, and then superimposing the outputs. Most systems with which we deal are linear and time-invariant, or nearly so (§15.4). The output of a *linear system* is directly proportional to the input and is independent of the time when the input occurred, that is, it is *time-invariant*. Let us assume that we feed into the system data sampled at regular intervals, Δ, for example, a digital seismic trace. The output of the system can be calculated if we know the *impulse response* of the system, that is, the response of the system when the input is a *unit impulse* (§15.2.5). We write the unit impulse δ_t or $\delta(t)$, depending on whether we are dealing with sampled data or continuous functions. The impulse response of the system will be zero prior to $t = 0$ and then have the values $f_0, f_1, f_2, \ldots$ at successive sampling intervals. We represent the process diagramatically thus:

$$\delta_t \rightarrow \boxed{\text{system}} \rightarrow f_t = [f_0, f_1, f_2, \ldots].$$

Writing δ_{t-n} for a unit impulse that occurs at $t = n$ rather than at $t = 0$, we can illustrate linear and time-invariant systems as follows:

Linear:

$$k\delta_t \rightarrow \boxed{\text{system}} \rightarrow kf_t = [kf_0, kf_1, kf_2, \ldots];$$

Time-invariant:

$$\delta_{t-n} \rightarrow \boxed{\text{system}} \rightarrow f_{t-n} = \underbrace{[0,0,0, \ldots, 0,}_{n \text{ zeros}} f_0, f_1, f_2, \ldots]$$

In the last bracket on the right, the first output different from zero is f_0 and occurs at the instant $t = n\Delta$ (which we write as $t = n$ by counting in units of Δ).

Obviously, any input that consists of a series of sampled values can be represented by a series of unit impulses multiplied by appropriate amplitude factors. We can then use the above two properties to find the output for each impulse, and by superimposing these, we get the output for the arbitrary input.

We shall illustrate convolution by considering the output for a filter whose impulse response f_t is $[f_0, f_1, f_2] = [1, -1, \frac{1}{2}]$. When the input g_t is $[g_0, g_1, g_2] = [1, \frac{1}{2}, -\frac{1}{2}]$, we apply to the input the series of impulses $[\delta_t, \frac{1}{2}\delta_{t-1}, -\frac{1}{2}\delta_{t-2}]$ (the last two subscripts mean-

Table 9.2 *Mapping between domains*

Offset-CMP traces	⇐NMO⇒	Common source–receiver traces
Common-midpoint domain	⇐DMO⇒	Common reflection-point domain
Unmigrated domain	⇐Migration⇒	Migrated domain
Location–time section (or map)	⇐T–D conversion⇒	Location–depth section (or map)
Offset-time (gather)	⇐Velocity analysis⇒	Stacking-velocity time at zero offset

ing that the impulses are delayed by one and two sampling intervals, respectively) and obtain the following outputs:

$$\delta_t \to [1, -1, \tfrac{1}{2}],$$
$$\tfrac{1}{2}\delta_{t-1} \to [0, \tfrac{1}{2}, -\tfrac{1}{2}, \tfrac{1}{4}],$$
$$-\tfrac{1}{2}\delta_{t-2} \to [0, 0, -\tfrac{1}{2}, \tfrac{1}{2}, -\tfrac{1}{4}].$$

Summing, we find the output

$$[\delta_t + \tfrac{1}{2}\delta_{t-1} - \tfrac{1}{2}\delta_{t-2}] \to [1, -\tfrac{1}{2}, -\tfrac{1}{2}, \tfrac{3}{4}, -\tfrac{1}{4}].$$

Convolution is illustrated in fig. 9.2. This operation is equivalent to replacing each element of the one set by an appropriately scaled version of the other set and then summing elements that occur at the same times. If we call the output h_t and denote the operation of taking the convolution by an asterisk, we can express this as

$$\begin{aligned} h_t &= f_t * g_t \\ &= \sum_k f_k g_{t-1} \\ &= [f_0 g_0, f_0 g_1 + f_1 g_0, f_0 g_2 + f_1 g_1 + f_2 g_0, \ldots]. \end{aligned}$$
$$(9.23; 15.211)$$

(We use the sum sign to mean summation over all appropriate values of the summation index.) Note that we would have obtained the same result if we had input f_t into a filter whose impulse response is g_t; in other words, convolution is commutative:

$$h_t = f_t * g_t = g_t * f_t = \sum_k f_k g_{t-k} = \sum_k g_k f_{t-k}. \quad (9.24)$$

Whereas we have been expressing convolution as an operation on sampled data, we can also convolve a sample set with a continuous function:

$$h(t) = f_t * g(t) = \sum_k f_k g(t - k). \quad (9.25)$$

Each term in the summation represents the function $g(t)$ displaced and scaled (displaced to the right k units and multiplied by f_k). A special case of eq. (9.25) is that of convolving a continuous function $g(t)$ with a unit impulse located at $t = n$:

$$\delta_{t-n} * g(t) = \sum_k \delta_{k-n} g(t - k) = g(t - n) \quad (9.26)$$

(because δ_{k-n} is zero except for $k = n$, where $\delta_{k-n} = 1$). Hence, convolving $g(t)$ with a time-shifted unit impulse displaces the function by the same amount and in the same direction as the unit impulse is displaced.

One can also convolve two continuous functions. In this case, the summation becomes an integration:

$$h(t) = f(t) * g(t) = \int_{-\infty}^{\infty} f(\tau)\, g(t - \tau)\, d\tau.$$
$$(9.27; 15.145; 15.196)$$

The *convolution theorem* states that the Fourier transform of the convolution of two functions is equal to the product of the transforms of the individual functions; we can state the theorem as follows:

$$\begin{aligned} f_t &\leftrightarrow F(\nu) = |F(\nu)|\, e^{j\gamma_f(\nu)}, \\ g_t &\leftrightarrow G(\nu) = |G(\nu)|\, e^{j\gamma_g(\nu)}, \\ f_t * g_t &\leftrightarrow F(\nu)G(\nu) = \{|F(\nu)|\, e^{j\gamma_f(\nu)}\}\{|G(\nu)|\, e^{j\gamma_g(\nu)}\}, \\ &\leftrightarrow |F(\nu)||G(\nu)|\, e^{j[\gamma_f(\nu)+\gamma_g(\nu)]} \quad (9.28; 15.145) \end{aligned}$$

where $|F(\nu)|$ and $|G(\nu)|$ are the amplitude spectra, and $\gamma_f(\nu)$ and $\gamma_g(\nu)$ the phase spectra. This means that if two sets of data are convolved in the time domain, the effect in the frequency domain is to multiply their amplitude spectra and to add their phase spectra. Equation (9.28) thus provides an alternative way to carry out a convolution operation: (1) transform the function to be convolved, say, g_t, and the convolution (or filter) operator f_t into the frequency domain; (2) multiply their amplitude spectra at each frequency value and add their phase spectra; and (3) transform the result back into the time domain.

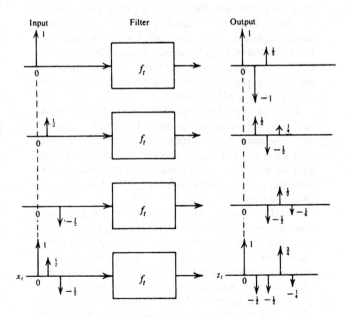

Fig. 9.2   Filtering as an example of convolution.

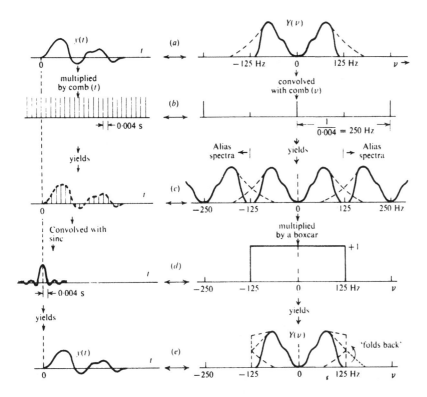

Fig. 9.3 Sampling and reconstituting a waveform. The left column shows a time-domain sequence, and the right shows the equivalent frequency-domain sequence.

Because of certain symmetry properties of the Fourier transform, the reciprocal relationship also holds, that is, multiplication in the time domain is equivalent to convolution in the frequency domain (see §15.2.8):

$$2\pi f_t g_t \leftrightarrow F(\nu) * G(\nu). \quad (9.29;\ 15.146)$$

9.2.2 Sampling, interpolating, and aliasing

(a) Sampling. In analog-to-digital conversion, we replace the continuous signal with a series of values at fixed intervals. It would appear that we are losing information by discarding the data between the sampling instants. The transform relationship in eqs. (9.28) and (9.29) can be used to understand sampling and the situations in which information is not lost (see also §15.5.1).

We make use of the *comb* or *sampling function;* this consists of an infinite set of regularly spaced unit impulses (fig. 9.3b). The transform of a comb is also a comb:

$$\text{comb}(t) \leftrightarrow k_t \text{ comb}(\nu),$$
$$(9.30;\ 15.155)$$

where $k_1 = 2\pi/\Delta$, Δ being the sampling interval (see §15.2.6). If the comb in the time domain has elements every 4 ms, the transform has elements every $1/(0.004\ \text{s}) = 250$ Hz. We shall also make use of the *boxcar* (fig. 9.3d), a function that has unity value between the values $\pm\nu_0$ and is zero everywhere else. The transform of a boxcar, $\text{box}_{2\nu_0}(\nu)$, is a *sinc function:*

$$\text{box}_{2\nu_0}(\nu) \leftrightarrow k_2 \text{ sinc } 2\pi\nu_0 t = k_2 \frac{\sin 2\pi\nu_0 t}{2\pi\nu_0 t},$$
$$(9.31;\ 15.152)$$

where k_2 depends upon the area of the boxcar (see eq. (15.123)).

Figure 9.3a shows a continuous function $y(t)$ and its transform $Y(\nu)$:

$$y(t) \leftrightarrow Y(\nu)$$

The amplitude spectrum, $|Y(\nu)|$, is symmetric about zero for real functions, negative frequencies having the same amplitude values as positive frequencies. (Negative frequencies result from the use of Euler's formula when we combine Fourier series in sine–cosine form into the complex exponential form.)

The sampled data that represent $y(t)$ can be found by multiplying the continuous function by the comb (hence the name "sampling function"). If we are sampling every 4 ms, we use a comb with elements every 4 ms. According to eqs. (9.29) and (9.30),

$$2\pi \text{ comb}(t)y(t) \leftrightarrow k_1 \text{ comb}(\nu) * Y(\nu).$$

Convolution is equivalent to replacing each data element (each impulse in comb (ν) in this instance) with the other function, $Y(\nu)$ (see eq. (9.23)). This is illustrated in fig. 9.3c. Note that the frequency spectrum of the sampled function differs from the spectrum of the continuous function in this example by the repetition of the spectrum.

We can recover the transform of the original function by multiplying the transform of the sampled function by a boxcar. The equivalent time-domain operation (see eqs. (9.28) and (9.31)) is to convolve the sample data with the sinc function; as shown in fig. 9.3e, this restores the original function in every detail. The sinc function thus provides the precise "operator" for interpolating between sample values.

(b) Interpolating. Interpolating between sample values is necessary in many processes, for example, to time shift a trace by less than a multiple of the sampling interval for static or normal-moveout correction. In the simplest form, we might try to obtain a value in between two samples by linear interpolation; fig. 9.4, which shows the amplitude spectrum of a linear operator, indicates that this is unsatisfactory for frequencies greater than a sixth of the sampling frequency (about 40 Hz for the 250 Hz sampling in this case). An 8-point truncated sinc function, an interpolator of reasonable length, gives good results up to 3/8 of the sampling frequency (about 95 Hz), which is adequate for most operations. Interpolating is also discussed in §9.11.2 and 15.1.7 (see eq. (15.74)).

(c) Aliasing; sampling theorem. When the complete sinc function is used for interpolating, no information whatsoever is lost in the process of sampling and interpolating. However, if the continuous function had had a spectrum (shown dashed in fig. 9.3a), which included frequency components higher than 125 Hz (in this example), then the time-domain multiplication by the sampling function would have produced an overlap of frequency spectra and no longer would we be able to recover the original spectrum from the spectrum of the sampled data nor could we recover the original waveform. Whether the original waveform is recoverable depends, therefore, upon whether the original waveform contains frequencies higher than half of the sampling frequency.

The relationships demonstrated in the foregoing are

summarized by the *sampling theorem;* no information is lost by regular sampling provided that the sampling frequency is greater than twice the highest frequency component in the waveform being sampled (see also §15.5.1). This is equivalent to saying that there must be more than two samples per cycle for the highest frequency. The sampling theorem thus determines the minimum sampling we can use. Because this minimum sampling allows complete recovery of the waveform, we can further conclude that nothing is gained by using a finer sampling. Thus, sampling rates of 2 and 4 ms permit us to record faithfully provided none of the signal spectrum lies above 250 and 125 Hz, respectively. In actual practice, the limits are half to two-thirds of these frequencies because of analog aliasing filters (see what follows).

Half the sampling frequency is called the *Nyquist frequency, v_N*, that is,

$$v_N = 1/2\Delta \qquad (9.32)$$

Any frequency present in the signal that is greater than the Nyquist frequency by the amount Δv will be indistinguishable from the lower frequency $v_N - \Delta v$. In fig. 9.5, we see that a sampling rate of 4 ms (that is, 250 samples per second) will allow perfect recording of a 75-Hz signal, but 175- and 250-Hz signals will appear as (that is, will *alias* as) 75 and 0 Hz (0 Hz is the same as a direct current), respectively. Alias signals that fall within the frequency band in which we are primarily interested will appear to be legitimate signals. To avoid this, *alias filters* (see fig. 7.46) are used before sampling to remove frequency components higher than the Nyquist frequency. This must be done before sampling because afterwards the alias signals cannot be recognized. Actual analog alias filters often have slopes of 72 dB/octave and cutoff frequencies one- to two-thirds of an octave below the Nyquist frequency in order to prevent aliasing (see problem 9.5). Digital filters can be made with almost arbitrarily steep slopes; some instruments sample with a frequency high enough that there is little energy to alias and then apply a very steep digital alias filter before resampling to the desired sampling frequency.

Aliasing is an inherent property of all systems that sample, whether the sampling is done in time, in space, or in any other domain such as frequency. For example, trace-to-trace miscorrelation because the trace spacing is larger than half the apparent wavelength constitutes aliasing (spatial aliasing; see § 9.2.2d). Alias filtering must be done before sampled data are resampled at a lower rate, as is often done in processing to reduce processing time. The term "aliasing" is also used to denote other types of errors in sampling, for example, errors because the sampling is not regular and errors in correlating sedimentary cycles well-to-well where the well spacing is too large.

Signals with frequencies higher than the Nyquist frequency can be recovered uniquely if all the frequencies are known to lie within a narrow passband (see problem 9.4). Thus, 4-ms sampling would be adequate

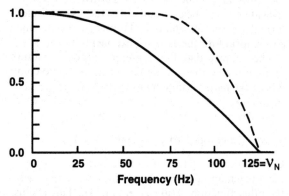

Fig. 9.4 Amplitude spectra of interpolators for the midpoint between two samples of data sampled every 4 ms. The solid line is the linear interpolator; the dashed line is the 8-point truncated sinc function. (From Hatton, Worthington, and Makin, 1986: 38.)

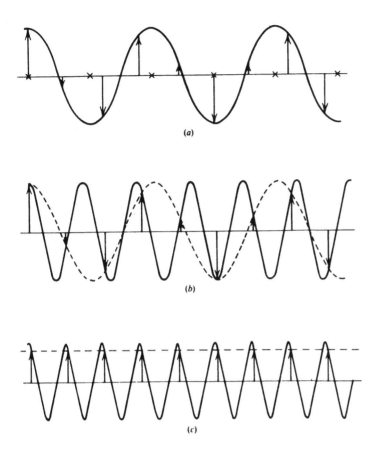

Fig. 9.5 Sampling and aliasing. Different frequencies sampled at 4-ms intervals (250 times per second). (a) 75-Hz signal, (b) 175-Hz signal yields the same sample values as 75 Hz, and (c) 250-Hz signal yields samples of constant value (0 Hz).

to define signals uniquely if the only frequencies present were, for example, between 1 and 1.1 kHz.

(d) Spatial sampling. Spatial sampling occurs when we use geophones to sample a wave at different points in space. Figure 9.5 represents a harmonic wave recorded at a fixed point as a function of time and the sampling is done at fixed time intervals Δt. However, the figure could equally well represent wave motion observed along a line of geophones at a given instant of time, the sampling being done at fixed intervals of distance Δx. In time sampling, we take $1/\Delta t$ samples per unit time, whereas in space sampling, we take $1/\Delta x$ samples per unit distance. In time sampling, we had a Nyquist frequency, ν_N of eq. (9.32), which gave the maximum number of waves per two unit time intervals, and aliasing occurred for components that have frequencies higher than ν_N. In space sampling, we have a Nyquist wave number, $\kappa_N/2\pi$:

$$\kappa_N/2\pi = 1/\lambda_N = 1/2\Delta x, \qquad (9.33)$$

which gives the maximum number of waves per two unit lengths. Aliasing occurs when a component has more waves per unit length than this, that is, when $\kappa > \kappa_N$ (or $\lambda < \lambda_N$). Figure 6.2c shows an alias alignment.

9.2.3 Filtering by the earth

We can think of the earth as a filter of seismic energy. We might consider the wave resulting from an explosion as an impulse $k\delta_0$, that is, the wave motion at the source of the explosion is zero both before and after the explosion and differs from zero only in an extremely short interval (essentially at $t = 0$), and during this infinitesimal interval, the motion is very large. Ideally, the signal that we record would be simply $k\delta_t$ convolved with the impulse response of the earth (assuming the earth is a linear system – see §15.4). The result would be zero except for sharp pulses corresponding to the arrivals of different reflections. If this were so, we could easily determine from the recorded data the complete solution to the seismic problem. However, in practice, we get back not only primary reflections, but also multiples, diffractions, surface waves, scattered waves from near-surface irregularities, reflected refractions, and so on, all modified by filtering because of absorption and other causes, and with random noise always superimposed.

We can regard the recorded seismic wave as the result of successive convolutions of the initial impulse δ_t with a number of impulse responses representing various factors modifying the wave during its passage through the earth. The major zones affecting the signal are

    (a) the zone near the source where the stresses and absorption of energy (especially at higher frequencies) often are extreme; we write s_t for the impulse response of this zone;

    (b) the sequence of reflectors with impulse response e_t; this is the signal that seismic reflection work is intended to find;

    (c) the near-surface zone, which has a disproportionate effect in modifying the signal; its impulse response is n_t;

    (d) additional modifying effects because of absorption, wave conversion, multiples and diffractions, and so on; the combination of these is represented by p_t.

Combining these effects, we obtain for the recorded waveform g_t,

$$g_t = k\,\delta_t * s_t * e_t * n_t * p_t = (k\delta_t * s_t * n_t * p_t) * e_t. \tag{9.34}$$

This equation expresses the *convolutional model*, that is, a seismic trace can be thought of as a series of convolutions. The convolutional model is central to most data processing. Equation (9.34) is also written

$$g_t = e_t * w_t, \tag{9.35}$$

where w_t includes $(k\,\delta_t * s_t * n_t * p_t)$; w_t is called the *embedded* (or *equivalent*) wavelet; it is the wavelet that would be reflected from a single isolated interface. The convolutional model often includes additive noise r_t (usually, but not necessarily, assumed to be random). The convolutional model of a noisy seismic trace is

$$g_t = e_t * w_t + r_t. \tag{9.36}$$

When we use a Vibroseis source, the input to the earth is a long wavetrain v_t and the resulting seismic trace g_t' is

$$g_t' = v_t * s_t' * e_t * n_t * p_t \tag{9.37}$$

(where we write s_t' rather than s_t because the filtering processes near the vibrator may be different from those near an impulsive source owing to the different magnitude of the stresses involved).

9.2.4 Water reverberation and deconvolution

Let us examine the effect of multiples resulting from reflection at the top and bottom of a water layer (Backus, 1959). We write $n\Delta$ for the round-trip traveltime from top to bottom and back, n being an integer. We assume that the reflection coefficients at the surface and bottom of the water layer are such that the ratio of the reflected to incident amplitudes are -1 and $+R$, respectively, the minus sign denoting phase reversal at the water–air interface. We assume also that the amplitude of a wave returning directly to a hydrophone after reflection at a certain horizon (without a "bounce" round trip between top and bottom

of the water layer) is unity and that its traveltime is t. A wave that is reflected at the same horizon and suffers a bounce within the water layer either before or after its travel down to the reflector will arrive at time $t + n\Delta$ with the amplitude $-R$. Because there are two raypaths with the same traveltime for a single-bounce wave, one that bounced before traveling downward and one that bounced after returning from depth, we have in effect a wave arriving at time $t + n\Delta$ with the amplitude $-2R$. There will be three waves that suffer two bounces: one that bounces twice before going downward to the reflector, one that bounces twice upon return to the surface, and one that bounces once before and once after its travel downwards; each of these are of amplitude R^2, so that their sum is a wave of amplitude $3R^2$ arriving at time $t + 2n\Delta$. Continuing thus, we see that a hydrophone will detect successive signals of amplitudes 1, $-2R$, $3R^2$, $-4R^3$, $5R^4$, ..., arriving at intervals of $n\Delta$. We can therefore write the impulse response of the water layer for various water depths, $z = \frac{1}{2}nV\Delta$, where V is the velocity in water:

$$
\begin{aligned}
f_t &= [1,\ -2R,\ 3R^2,\ -4R^3,\ 5R^4,\ \dots\,], & (n = 1)\\
&= [1,\ 0,\ -2R,\ 0,\ 3R^2,\ 0,\ -4R^3,\ \dots\,], & (n = 2)\\
&= [1,\ 0,\ 0,\ -2R,\ 0,\ 0,\ 3R^2,\ \dots\,], & (n = 3)
\end{aligned}
\tag{9.38}
$$

and so on. Thus, the water layer acts as a filter.

If we transform this to the frequency domain, we find a large peak (the size of the peak increasing with increasing R) at the frequency $1/2n\Delta$ and at multiples of this frequency. These are the frequencies that are reinforced at this water depth (that is, the frequencies for which interference is constructive). The result of passing a wavetrain through a water layer is the same as multiplying the amplitude spectrum of the waveform without the water layer by the spectrum of the impulse response of the water layer. Whenever the reflection coefficient is large (and hence R is large) and the frequency $1/2n\Delta$ (or one of its harmonics) lies within the seismic spectrum, the seismic record will appear very sinusoidal with hardly any variation in amplitude throughout the recording period (see fig. 6.32). Because of the overriding oscillations, it will be difficult to interpret the primary reflections.

A filter i_t having the property that

$$f_t * i_t = \delta_t \tag{9.39}$$

is called the *inverse filter* of f_t. If we pass the reverberatory output from the hydrophones through the inverse water-layer filter (in a data-processing center), we will remove the effect of the water-layer filter. The inverse of the water-layer filter is a simple filter (the Backus filter) with only three nonzero terms:

$$
\begin{aligned}
i_t &= [1,\ 2R,\ R^2, & (n = 1)\\
&= [1,\ 0,\ 2R,\ 0,\ R^2], & (n = 2)\\
&= [1,\ 0,\ 0,\ 2R,\ 0,\ 0,\ R^2] & (n = 3)
\end{aligned}
\tag{9.40}
$$

and so on (see problem 9.7a and §15.5.5). Figure 6.32b shows the result of applying such a filter to the data shown in fig. 6.32a.

The process of convolving with an inverse filter is called *deconvolution* and is one of the most important operations in seismic data processing (Middleton and Whittlesey, 1968). Whereas we have illustrated deconvolution as removing the singing effect of a water layer, we could also deconvolve for other filters whose effects we wish to remove if we know enough about the filters and the signal (see §9.5 and 15.7).

9.2.5 Multidimensional convolution

In the foregoing sections, we have assumed only one independent variable, time. We have assumed that each seismic trace is being processed by itself and that the only data available are the succession of samples of that trace, the actual situation for many processes. There is, however, no need to restrict ourselves to a succession of time samples for only one trace, and some processes involve convolution with two or even more variables. The convolution operation, eq. (9.23), can be written for two variables, t and w, as

$$h_{t,w} = f_{t,w} * g_{t,w} = \sum_k \sum_m f_{k,m} g_{t-k,w-m}$$

$$(9.41; 15.164)$$

This equation states that the convolution result is the superposition of values at nearby times and locations (if t and w respectively indicate time and location) after each has been weighted by the filter $f_{t,w}$. As with one-dimensional convolution, the operation is commutative. It is also possible to carry out the operation by transforming $f_{t,w}$ and $g_{t,w}$ using a two-dimensional Fourier transform (see §9.1.4, 15.2.4), multiplying the two-dimensional amplitude spectra and adding the phase spectra, and then transforming back to give $h_{t,w}$. If t is time and w is distance, then the transformed domain is the frequency–wavenumber domain.

9.3 Correlation

9.3.1 Cross-correlation

The *cross-correlation* function is a measure of the similarity between two data sets. One data set is displaced varying amounts relative to the other and corresponding values of the two sets are multiplied together and the products summed to give the value of cross-correlation. Wherever the two sets are nearly the same, the products will usually be positive and hence the cross-correlation is large; wherever the sets are unlike, some of the products will be positive and some negative and hence the sum will be small. If the cross-correlation function should have a large negative value, it means that the two data sets would be similar if one were inverted (that is, they are similar except that they are out of phase). The two data sets might be dissimilar when lined up in one fashion and yet be similar when one set is shifted with respect to the

other; thus, the cross-correlation is a function of the relative shift between the sets. By convention, we call a shift positive if it involves moving the second function to the left with respect to the first function.

We express the cross-correlation of two data sets, x_t and y_t, as

$$\phi_{xy}(\tau) = \sum_k x_k y_{k+\tau}, \qquad (9.42; 15.147)$$

where τ is the displacement of y_t relative to x_t. (Note that $\phi_{xy}(\tau)$ is a data set rather than a continuous function, because x and y are data sets.) Let us illustrate cross-correlation by correlating the two functions, $x_t = [1, -1, \frac{1}{2}]$ and $y_t = [1, \frac{1}{2}, -\frac{1}{2}]$, shown in fig. 9.6. Diagram (c) shows the two functions in their normal positions. Diagram (a) shows y_t shifted two units to the right; corresponding coordinates are multiplied and summed as shown below the diagram to give $\phi_{xy}(-2\Delta)$. Diagrams (a) to (e) show y_t shifted by varying amounts while (f) shows the graph of $\phi_{xy}(\tau)$. The cross-correlation has its maximum value (the functions are most similar) when y_t is shifted one unit to the left ($\tau = +\Delta$). Obviously we get the same results if we shift x_t one space to the right. In other words,

$$\phi_{xy}(\tau) = \phi_{yx}(-\tau). \qquad (9.43; 15.157)$$

The similarity between eq. (9.42) and the convolution eq. (9.23) should be noted. We may rewrite eq. (9.42) in the form

$$\phi_{xy}(\tau) = \phi_{yx}(-\tau) = \sum_k y_k x_{k-\tau} = \sum_k y_k x_{-(\tau-k)}$$
$$= y_\tau * x_{-\tau} = x_{-\tau} * y_\tau. \qquad (9.44; 15.158)$$

Hence, cross-correlation can be performed by reversing the first data set and convolving.

If two data sets are cross-correlated in the time domain, the effect in the frequency domain is the same as multiplying the complex spectrum of the second data set by the conjugate of the complex spectrum of the first set. Because forming the complex conjugate involves only reversing the sign of the phase, cross-correlation is equivalent to multiplying the amplitude spectra and subtracting the phase spectra. In mathematical terms,

$$x_t \leftrightarrow X(\nu) = |X(\nu)| e^{j\gamma_x(\nu)},$$
$$y_t \leftrightarrow Y(\nu) = |Y(\nu)| e^{j\gamma_y(\nu)},$$
$$x_{-t} \leftrightarrow \overline{X(\nu)} = |X(\nu)| e^{-j\gamma_x(\nu)},$$
$$\phi_{xy}(\tau) \leftrightarrow \overline{X(\nu)} Y(\nu) = |X(\nu)||Y(\nu)| e^{-j[\gamma_x(\nu)-\gamma_y(\nu)]}.$$

$$(9.45; 15.147)$$

We note that changing the sign of a phase spectrum is equivalent to reversing the trace in the time domain. Anstey (1964) gives a particularly clear explanation of correlation.

9.3.2 Autocorrelation

The special case where a data set is being correlated with itself is called *autocorrelation*. In this case, eq. (9.42) becomes

$$\phi_{xx}(\tau) = \sum_k x_k x_{k+\tau}. \qquad (9.46)$$

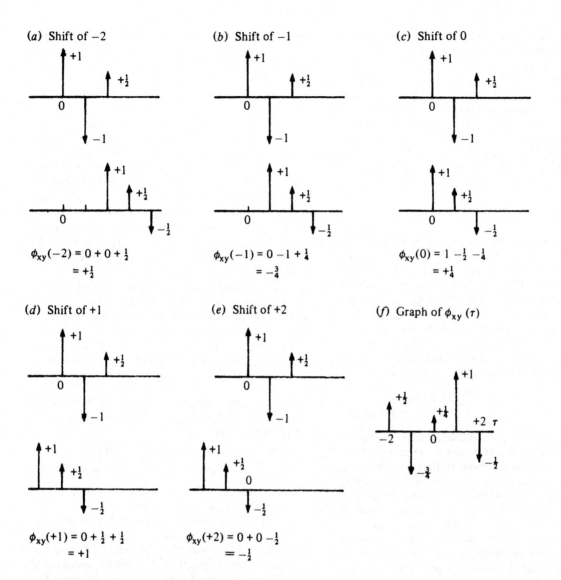

Fig. 9.6   Calculating the cross-correlation of two functions.

Autocorrelation functions are symmetrical because a time shift to the right is the same as a shift to the left; from eq. (9.43),

$$\phi_{xx}(\tau) = \phi_{xx}(-\tau). \qquad (9.47)$$

The autocorrelation has its peak value at the zero time shift (that is, a data set is most like itself before it is time shifted). If the autocorrelation should have a large value at some time shift $\Delta t \neq 0$, it indicates that the set tends to be periodic with the period Δt. Hence, the autocorrelation function may be thought of as a measure of the repetitiveness of a function.

We can express the preceding concepts in integral form applicable to continuous functions. Equations (9.42) and (9.46) now take the forms

$$\phi_{xy}(\tau) = \int_{-\infty}^{\infty} x(t)y(t+\tau)\ dt,$$

$$(9.48;\ 15.147)$$

$$\phi_{xx}(\tau) = \int_{-\infty}^{\infty} x(t)x(t+\tau)\ dt.$$

$$(9.49;\ 15.161)$$

9.3.3 Normalized correlation

The autocorrelation value at zero shift is called the *energy* of the trace:

$$\phi_{xx}(0) = \sum_{k} x_k^2. \qquad (9.50;\ 15.162)$$

(This terminology is justified on the basis that x_t is usually a voltage, current, or velocity, and hence x_t^2 is proportional to energy). For the autocorrelation function, eq. (9.45) becomes

$$\phi_{xx}(\tau) \leftrightarrow |X(\nu)|^2. \qquad (9.51;\ 15.161)$$

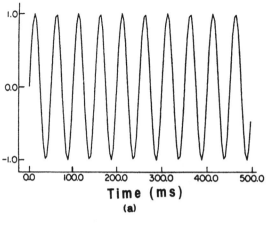

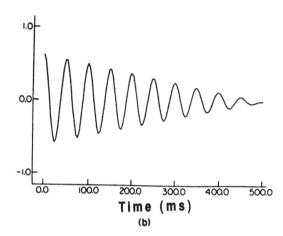

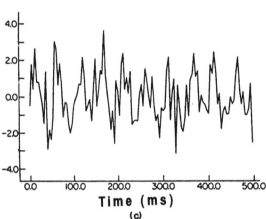

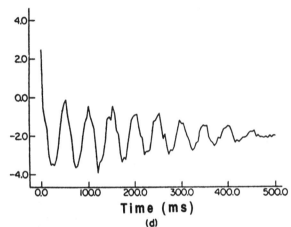

Fig. 9.7   Effect of random noise on an autocorrelation. (From Hatton et al., 1986: 27–8.) (a) Sinusoidal signal $\frac{1}{2}$ s long, (b) half of autocorrelation of part (a); (c) sinusoid with random noise superimposed, and (d) half of autocorrelation of part (c).

In continuous-function notation,

$$\phi_{xx}(0) = \int_{-\infty}^{\infty} |x(t)|^2\, dt = \int_{-\infty}^{\infty} |X(\nu)|^2\, d\nu.$$

$$(9.52;\ 15.162)$$

Because the zero-shift value of the autocorrelation function is the energy of the trace, $|x(t)|^2$ is the energy per unit of time or the *power* of the trace and $|X(\nu)|^2$ is the energy per increment of frequency, usually called the *energy density* or *spectral density*.

We often normalize the autocorrelation function by dividing by the energy:

$$\phi_{xx}(\tau)_{\text{norm}} = \frac{\phi_{xx}(\tau)}{\phi_{xx}(0)}.\qquad (9.53)$$

The cross-correlation function is normalized in a similar manner by dividing by the geometric mean of the energy of the two traces:

$$\phi_{xy}(\tau)_{\text{norm}} = \frac{\phi_{xy}(\tau)}{[\phi_{xx}(0)\phi_{yy}(0)]^{1/2}}.\qquad (9.54)$$

Normalized correlation values must lie between ± 1. A value of $+1$ indicates perfect copy; a value of -1 indicates perfect copy if one of the traces is inverted.

Purely random noise affects only the zero-shift value of an autocorrelation. A periodic signal of frequency ν buried in noise increases the autocorrelation value at the frequency ν and hence shows up much more clearly in autocorrelations than in a time series (fig. 9.7).

9.3.4 Vibroseis analysis

The signal g_t', which our geophones record when we use a Vibroseis source (fig. 9.8d), bears little resemblance to e_t, the impulse response of the earth. To obtain a meaningful record, the data are correlated with the Vibroseis sweep (control) signal v_t. The recorded signal g_t' is

$$g_t' = v_t * e_t',$$

where we let $e_t' = s_t' * e_t * n_t * p_t$ in eq. (9.37). Using

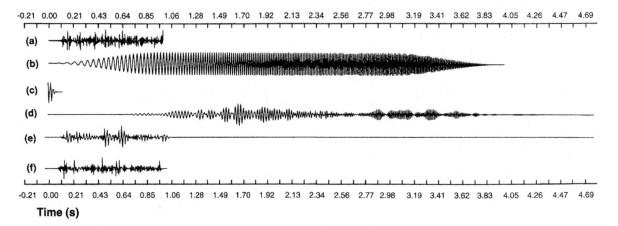

-0.21 0.00 0.21 0.43 0.64 0.85 1.06 1.28 1.49 1.70 1.92 2.13 2.34 2.56 2.77 2.98 3.19 3.41 3.62 3.83 4.05 4.26 4.47 4.69

Time (s)

Fig. 9.8 Composition of Vibroseis signal and its deconvolution. (From Yilmaz, 1987: 143.) (a) A reflectivity sequence; (b) 10-to-120-Hz sweep; (c) assumed minimum-phase near-surface filter; (d) assumed effect of $(e_t * v_t * s'_t)$ in eq. (9.37), that is, the convolution of (a), (b), and (c), what we would expect to observe with Vibroseis (ignoring $n_t * p_t$); (e) cross-correlation of (b) with (d); and (f) 10-to-120-Hz zero-phase filtered version of (a).

eq. (9.44), we find for the cross-correlation of the sweep and the recorded signal

$$\phi_{vg}(t) = g'_t * v_{-t} = (v_t * e'_t) * v_{-t}$$
$$= e'_t * (v_t * v_{-t})$$
$$= e'_t * \phi_{vv}(t). \qquad (9.55)$$

(The next to the last step is possible because convolution is commutative.) Hence, the overall effect is that of convolving the earth function with the autocorrelation of the Vibroseis sweep signal. The autocorrelation function, $\phi_{vv}(t)$, is quite sharp and has sizeable values only over a very narrow range of time shifts. Therefore, the overlap produced by the passage of a long sweep through the earth has been eliminated almost entirely (fig. 9.8e). This is shown also in fig. 9.9, where parts (b) and (c) are the same Vibroseis record before and after cross-correlation.

9.3.5 Multichannel coherence

The cross-correlation function can only be used as a measure of the coherence between two traces. As a coherence measure for a large number of traces we could make use of the fact that when we stack several channels together, the resulting amplitude is generally large where the individual channels are similar (*coherent*) so that they stack in phase, and small where they are unlike (incoherent).

If we let g_{ti} be the amplitude of the individual channel i at time t, then the amplitude of the stack at time t will be $\sum g_{ti}$ and the square of this will be the energy. The average amplitude of the stacked trace over a time window from t to $t + m\Delta$ is given by

$$C_A = \frac{\sum\limits_{t}^{t+m\Delta} \left| \sum\limits_{i=1}^{N} g_{ti} \right|}{1 + m\Delta}. \qquad (9.56)$$

Sometimes the energy of the stacked trace is used instead of the amplitude:

$$C_E = \frac{\sum\limits_{t}^{t+m\Delta} \left(\sum\limits_{i=1}^{N} g_{ti} \right)^2}{1 + m\Delta}. \qquad (9.57)$$

Both of these depend on whether the event within the time gate is strong or weak. Some normalized measure of coherence would be a better indication of the phase agreement of traces than the foregoing. The ratio of the energy of the stack compared to the sum of the energies of the individual components could be used as a coherence measure:

$$E_T = \frac{\left(\sum\limits_{i=1}^{N} g_{ti} \right)^2}{N \sum\limits_{i=1}^{N} (g_{ti})^2}. \qquad (9.58)$$

We expect a coherent event to extend over a time interval; hence, a more meaningful quantity than E_T is the *semblance*, S_T (Neidell and Taner, 1971), which denotes the ratio of the total energy of the stack within a gate of length $(1 + m\Delta)$ to the sum of the energy of the component traces within the same time gate. Using the same terminology as before, we can write

$$S_T = \frac{\sum\limits_{t}^{t+m\Delta} \left(\sum\limits_{i=1}^{N} g_{ti} \right)^2}{\sum\limits_{t}^{t+m\Delta} \sum\limits_{i=1}^{N} (g_{ti})^2}. \qquad (9.59)$$

The semblance will not only tend to be large when a coherent event is present, but the magnitude of the semblance will also be sensitive to the amplitude of the event. Thus, strong events will exhibit large semblance and weak events will exhibit moderate values of semblance, whereas incoherent data will have very low semblance.

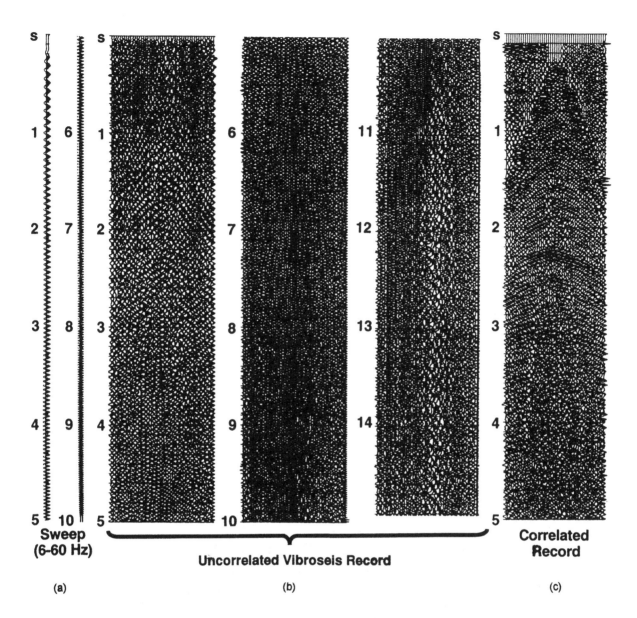

Fig. 9.9 Vibroseis field record and record after cross-correlating with sweep. (From Yilmaz, 1987: 21.) (a) 6-to-60-Hz sweep; (b) uncorrelated record (in three segments); and (c) correlated record.

Semblance and other coherence measures are used to determine the values of parameters that will "optimize" a stack. The semblance is calculated for various combinations of time shifts between the component channels, and the optimum time shifts are taken to be those that maximize the semblance. Semblance, therefore, can be used to determine static corrections or normal-moveout corrections.

9.3.6 Sign-bit recording

In a conventional Vibroseis field recording, such as shown in fig. 9.9, the information about any individual reflection is distributed over the time duration of the sweep. The information content at any time is small compared to the noise and can be thought of as a bias superimposed on stationary random noise (§15.2.12). The probability that any individual sample will be positive or negative will depend on the magnitude of this information bias compared to the noise level. If we record only the sign of each sample (ignoring completely the magnitude of each sample) and then sum many samples, we will in effect measure the time variations of this probability and thus the magnitude of the signal bias. This is called *sign-bit recording*.

The situation can be thought of as the testing of two mutually exclusive hypotheses: signal present (T) or signal absent (F) (Cochran, 1973). The signal is a quantity r, for example, a displacement at time t, which can be measured and compared with a preselected threshold value r_0. When $r > r_0$, we say that hypothesis T is true. However, this decision is not necessarily correct because there is a finite probability $P(r|T)$ that a signal is present for any value of r and conversely a finite probability that the signal is absent for any value of r. The probability of a correct decision Q_T is

$$Q_T = \int_{r_0}^{\infty} P(r \mid T) \, dr \qquad (9.60)$$

and the probability of a "false-alarm" error Q_F is given by

$$Q_F = \int_{r_0}^{\infty} P(r \mid F) \, dr \qquad (9.61)$$

The probability of missing the signal is

$$Q_m = \int_{-\infty}^{r_0} P(r \mid T) \, dr. \qquad (9.62)$$

Because the signal must occur for $-\infty < r < +\infty$ and the probability of a certainty is unity, we must have

$$Q_T + Q_m = 1, \qquad Q_T = 1 - Q_m. \qquad (9.63)$$

Clearly, we should select r_0 to make Q_T a maximum and Q_m a minimum. The choice of r_0 depends on what is known a priori about the signal and noise.

Sign-bit detection consists of measuring the coincidences in polarity of a group of n traces. We can specify a sign-bit semblance (§9.3.5) as follows. Let x_{ik} be the kth sample in the observation window on the ith trace. The product $(x_{ik}x_{jk})$ will be positive when and only when a coincidence of polarity occurs. The number of coincidences among the n traces at $t = k$ is

$$\sum_{i=1}^{n-1} \text{sgn}(x_{ik}x_{jk})u(x_{ik}x_{jk}), \qquad (j = i + 1),$$

where $\text{sgn}(y) = y/|y| = \pm 1$, the factor $u(x_{ik}x_{jk})$ (§15.2.5) being included to eliminate noncoincidences (that is, when $\text{sgn}(x_{ik}x_{jk}) = -1$). To normalize the sum, we divide by the number of coincidences when all traces have the same polarity, that is, by $[(n - 1) + (n - 2) + \cdots + 3 + 2 + 1] = \frac{1}{2}n(n - 1)$. Summing over the observation window of width m, we get

$$\hat{S} = \frac{2}{mn(n - 1)} \sum_{k=1}^{m} [\sum_{i=1}^{n-1} \text{sgn}(x_{ik}x_{jk})u(x_{ik}x_{jk})],$$
$$(j = i + 1), \qquad (9.64)$$

If we wish to include coincidences when two traces are both zero, we note that $\text{sgn}(0)$ and $u(0)$ are both undefined, so we can assign arbitrary values to them; when one of the traces is zero, we say that $\text{sgn}(x_{ik}x_{jk})u(x_{ik}x_{jk}) = 0$. When both traces are zero at $t = r$, the product is still zero except when the traces coincide in polarity at $t = r - 1$; in this case, the product is $+1$.

Figure 9.10 is a synthetic example to show that, in the presence of noise, the sign-bit method gives results comparable to conventional Vibroseis. Trace 1 is a synthetic Vibroseis trace resulting from 10 irregularly spaced reflectors of different reflectivity. Random noise three times the rms amplitude of trace 1 has been added to it to give trace 2. Trace 3 is a sign-bit representation of trace 2; it flips back and forth between its only two states, positive and negative. Trace 4 is the result of correlating trace 2 with the sweep, that is, it is a conventional Vibroseis output, and trace 5 the result of correlating trace 2 with a sign-bit version of the sweep. Trace 6 is the sum of 20 traces manufactured like trace 2 except with different noise, and trace 7 the sum of 20 traces, each manufactured like trace 3. Traces 8 and 9 represent respectively a conventional correlated Vibroseis trace and a correlated sign-bit trace.

With sign-bit recording, only one bit is recorded per sample so that the volume of data is greatly reduced and instrumentation can be simplified. Instrumentation fidelity requirements are also greatly relaxed; geophone nonlinearity, for example, becomes less important because only the sign and not the magnitude of the output is measured. Geocor™ uses sampling boxes in the field with 16 geophone channels connected to each box. The sampling box determines whether each channel output is positive or negative at the sampling instant and records this information as one of the bits in a 16-bit word that is then relayed to the recording truck. A 16-to-1 saving in the number of cable channels is thus achieved. Field systems employing 1024 channels are now in use.

9.4 Phase considerations

The Fourier synthesis of wavetrains according to eq. (9.2) involves adding together cosine waves of different frequencies and different phases. If the same components are added together with different phase relations, different waveforms result. Changing the waveform changes the location of a particular peak or trough, and hence measurements of arrival times are affected by variations in the phase spectra. Because seismic exploration involves primarily determining the arrival times of events, preservation of proper phase relationships during data processing is essential.

Out of all possible causal wavelets with the same amplitude spectrum, that wavelet whose energy builds up fastest is called the *minimum-delay* wavelet; its phase is always less than the other wavelets with the same amplitude spectrum and hence it is also called *minimum-phase*. The simplest wavelet (except for an impulse) is a data set that contains only two elements, the set (a, b). The amplitude spectrum of this data set is identical with that of the set (b, a), but no other data set has the same spectrum. If $a > b$, energy is concentrated earlier in the wavelet in the set (a, b) than in the set (b, a), and hence (a, b) is minimum-phase (or minimum-delay). Larger wavelets can be expressed as the successive time-domain convolution of two-element wavelets (see §15.5.6a); a large wavelet is

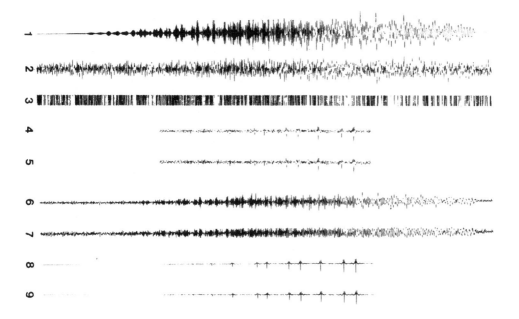

Fig. 9.10 Comparison of conventional Vibroseis and sign-bit recordings. (Courtesy of Geophysical Systems.) (Trace 1) Uncorrelated record from 10 reflectors; (2) uncorrelated record with noise three times the rms amplitude of (1) added; (3) sign-bit equivalent of 2; (4) trace 2 correlated with the sweep, that is, a correlated Vibroseis trace; (5) trace 3 correlated with the sign-bit version of the sweep, that is, a correlated sign-bit trace; (6) sum of 20 traces manufactured as for trace 2, where each has different random noise; (7) sum of 20 traces manufactured as for 3; (8) correlated trace 6; (9) correlated trace 7. (*Note:* Traces 4, 5, 8, and 9 have been time shifted.)

minimum-phase if all of its component wavelets are minimum-phase. Minimum-phase can also be defined in other ways, for example, by the location of roots in the z-domain (§15.5.6a). Most seismic sources generate waves that are nearly minimum-phase and the impulse response of many of the natural filtering processes in the earth are minimum-phase.

Minimum-phase does not necessarily mean that the first half-cycle is the largest, however. In the presence of interfering events and noise, it is often difficult to tell whether a reflected wavelet has the same or opposite sign as the downgoing wavelet, that is, whether the reflection coefficient is positive or negative. It is also difficult to tell the onset time of a reflection, and it is this time that is needed in determining reflector depth.

The embedded wavelet can be changed in processing to a zero-phase wavelet (see §15.5.6d) to facilitate interpretation. A *zero-phase wavelet* has its phase spectrum identically zero, that is, $\gamma(\nu) = 0$ for all ν. Such a wavelet is symmetrical about a central peak (or trough), which has higher amplitude than any other peaks or troughs. (An autocorrelation function is zero-phase.) We shift the time scale so that the amplitude maximum gives the arrival time. Such a wavelet is anticipatory because half of the wavelet precedes the arrival time.

Some filtering processes require that assumption be made about the phase of the signal; generally minimum-phase is assumed (Sherwood and Trorey, 1965). Deconvolution based upon autocorrelation information has to assume the phase because the phase

information of the waveform was lost when its autocorrelation was formed. This can be seen from eq. (9.51), where we note that the autocorrelation function, $\phi_{xx}(t)$, has the transform, $|X(\nu)|^2$ with zero phase for all values of frequency. Thus, all of the phase information present in $X(\nu)$ has been lost in the autocorrelation. However, most signals and natural filtering processes are represented by real, causal functions (see §15.5.6a). The Hilbert transform technique (§15.2.13) can be used to determine the phase information if the function is real and causal.

Causal wavelets that are not minimum-phase can be made minimum-phase by applying an exponential gain (taper) that cuts down the size of the latter part of the wavelet. Many actual wavelets contain only a few nonminimum-phase roots and can be made minimum-phase by using an exponential multiplier with a very gentle slope, perhaps 0.995^t, where t is the time in milliseconds. A criterion for determining how much taper to apply is discussed in §15.6.

Correlated Vibroseis records are sometimes erroneously called "zero-phase" because the effective wavelet, $\phi_{vv}(t)$ in eq. (9.55), is zero-phase. However, the earth response e_t is usually nearly minimum-phase and thus the result, $\phi_{vg}(t)$, is mixed-phase.

It is sometimes believed that zero-phase filtering of minimum-phase signals does not alter the phase because it involves subtracting zero from the phase spectrum; however, this is only true if the amplitude spectrum is unaltered (see fig. 9.11 and Hatton et al., 1986: §2.6.5).

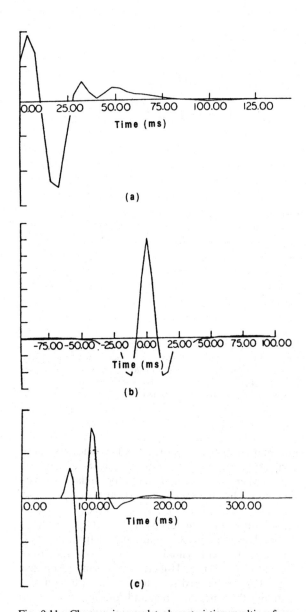

Fig. 9.11 Changes in wavelet characteristics resulting from zero-phase filtering. (From Hatton et al., 1986: 24.) (a) Minimum-phase wavelet (10–24 Hz); (b) zero-phase wavelet with same amplitude spectrum; and (c) convolution of (a) with (b).

9.5 Deconvolution and frequency filtering

9.5.1 General

In §9.2.4, we defined deconvolution as convolving with an inverse filter. Equations (9.34) to (9.36) expressed the seismic trace as a convolution of an earth reflectivity function e_t and a series of distorting filters, whose combined expression is the equivalent wavelet w_t. The ultimate objective of deconvolution is to extract the reflectivity function from the seismic trace and thus improve the vertical resolution and recognition of events. Sometimes deconvolution only tries to undo the effects of some prior filtering.

Deconvolution to extract the reflectivity function e_t

is nonunique unless additional information is available or additional assumptions are made. Equation (9.36) contains three unknowns, w_t, e_t, and r_t, but only one known, g_t. The most common additional assumptions are that w_t is minimum-phase and/or e_t has a flat (white) spectrum (at least over some limited bandpass). Additional constraints sometimes used include imposing a maximum length to w_t or use of a multichannel procedure.

Deconvolution operations are sometimes cascaded, a deconvolution to remove one type of distortion being followed by a different kind to remove another type of distortion. Some of the types of deconvolution are described in the following sections. Webster (1978) gives a bibliography of the literature on deconvolution up to that date.

9.5.2 Deterministic inverse filtering

Where the nature of the distorting filter is known, the inverse can sometimes be found in direct deterministic manner. In §9.2.4, we gave an example of water reverberation wherein we used a model of a water layer to derive the distorting filter, f_t. We also expressed the inverse filter, i_t, by eq. (9.39) but this equation cannot be used to find the sequence i_t (except on a trial-and-error basis). However, we can transform eq. (9.39) into the z-domain (see §15.5.3), carry out the division,

$$I(z) = 1/F(z) = \sum_k i_k z^k \qquad (9.65)$$

(provided the division does not "blow up"), and then transform back to get i_t. Even where we know the nature of the filter, we may have to determine part of the solution by trial and error or other techniques; in any case, we still have to determine the value of R in eq. (9.40).

Deterministic (or semideterministic) solutions are also used to remove the filtering effects of recording and processing systems. The source waveshape is sometimes recorded (though not always with proper ghost effects) and used in a deterministic source-signature correction. In marine work, the source waveform is often assumed to be constant; the waveshape may be recorded in deep water with a hydrophone suspended below the source. Another procedure is to monitor each energy release using hydrophones near the source and calculate changes in the signature from such monitor records; this procedure corrects for source-to-source variations before stacking. Still another procedure uses the sea-floor reflection to determine the source waveshape, assuming that the sea floor is a simple, sharp reflector.

9.5.3 Deghosting and recursive filtering

With a source below the base of the weathering (fig. 9.12), ghost energy is reflected at the base of the weathering (see §6.3.2b), where the coefficient (approaching from below) is $-R$ (ignoring ghost energy reflected at the free surface). An impulse followed by

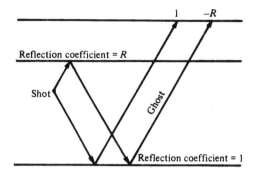

Fig. 9.12   Reflection plus ghost.

its ghost constitutes a filtering action; the transform can be written as

$$F(z) = 1 - Rz^n, \qquad (9.66)$$

where z^n represents the delay associated with the two-way traveltime from the source to the reflector producing the ghost. The inverse filter is an infinite series,

$$\begin{aligned} F^{-1}(z) &= 1/(1 - Rz^n) \\ &= 1 + Rz^n + (Rz^n)^2 + (Rz^n)^3 + \cdots + \end{aligned}$$
$$(9.67)$$

Because R is less than unity, this series converges and so could be used as a satisfactory deghosting filter. An input that includes ghost effects, g_t, can be convolved with this inverse filter to give a deghosted output h_t; in z-transform form,

$$\begin{aligned} H(z) &= G(z)F^{-1}(z) = G(z)/(1 - Rz^n), \\ H(z)(1 - Rz^n) &= G(z), \\ H(z) &= G(z) + Rz^n H(z). \end{aligned} \qquad (9.68)$$

The last term on the right represents a delay of $H(z)$ by n units; this last equation can be written in time notation as

$$h_t = g_t + Rh_{t-n} . \qquad (9.69)$$

Thus, we can determine an output value by adding to the input a proportionate amount of a previous output value. If the ghost delay is the sample interval, $n = 1$ and

$$h_0 = g_0,$$

because there was no output prior to zero;

$$\begin{aligned} h_1 &= g_1 + Rh_0, \\ h_2 &= g_2 + Rh_1, \\ h_3 &= g_3 + Rh_2, \cdots \end{aligned}$$

Filtering that involves feeding back part of the output is called *feedback* or *recursive filtering* (see §15.7.2). Recursive filtering allows us to carry out complete deghosting without using many terms and thus is economical in computing. In more general terms, we can express a filter $F(z)$ as a quotient of polynomials (see §15.5.3 and 15.7.2),

$$H(z) = G(z)F(z) = G(z)[N(z)/D(z)],$$

and write

$$G(z)N(z) = H(z)D(z), \qquad F(z) = N(z)/D(z).$$
$$(9.70)$$

The right side is the "recursive" part where previous output values are used in deriving future output values.

9.5.4 Deghosting by combining geophone and hydrophone records

If an upgoing compression gives a positive kick on a velocity geophone, a downgoing compression will give a negative kick, whereas both waves will give a kick of the same kind on a hydrophone (§7.5.5). This observation can be used to attenuate near-surface reverberation when geophones and hydrophones are combined in sea-floor cables. The 180° reversal of phase upon reflection at a free surface causes a hydrophone to see a rarefaction and thus a polarity reversal of the ghost with respect to the primary impulse, whereas a velocity phone sees also the reversal of wave direction and hence sees the ghost and primary with the same polarity. If R is the sea-floor reflection coefficient and Γ_H and Γ_G are the hydrophone and geophone transduction constants (§7.5.1d), respectively, the resulting impulse responses (fig. 9.13a; see also §9.2.1 and 15.4.1) have opposite polarities and amplitude ratios of $\Gamma_H(1 + R)/\Gamma_G(1 - R)$. This scale factor can be determined by locating a source directly above the sea-floor cable and recording the respective signal amplitudes. Barr and Sanders (1989) found that values for R ranged from 0.20 to 0.25 offshore Louisiana. Although fig. 9.13 assumes normal incidence, R does not change very much until the critical angle is approached.

Figure 9.13b shows the result of convolution of fig. 9.13a with a minimum-phase wavelet in both time and frequency domains. Notches in the geophone and hydrophone spectra (caused by interference as waves reverberate within the water layer) occur at the lobe peaks in each other's spectra, resulting in notchless spectra when added together. Over the seismic range, the hydrophone spectrum rises 6 dB/octave, whereas the geophone spectrum is flat, and the phase characteristics differ by 90°; these are the conditions associated with a derivative operator (this can be seen by differentiating the wave in eq. (2.56) with respect to t). Thus, summing geophone and hydrophone outputs provides a way of removing reverberation.

9.5.5 Least-squares (Wiener) filtering

Sometimes we wish to determine the filter that will do the best job of converting an input into a desired output. The filter that most nearly accomplishes this objective in the least-squares error sense is called the *least-squares filter* or the *Wiener filter*, occasionally the *optimum filter* (Robinson and Treitel, 1967).

Let the input data set be g_t, the filter that we have

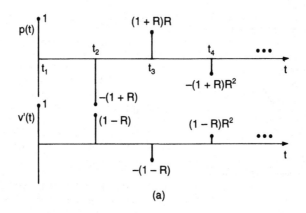

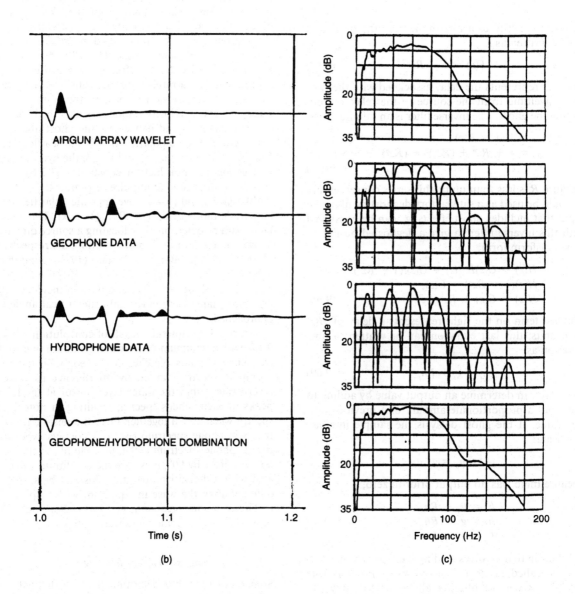

Fig. 9.13 Reverberation seen by a pressure hydrophone and a velocity geophone on the sea floor. (Courtesy of Haliburton Geophysical Services Inc.) (a) Impluse responses of a hy-drophone (above) and a geophone (below). (b) Waveforms and (c) spectra for the source array, the individual elements, and their combination.

to determine be f_t, and the desired output set be h_t. The actual result of passing g_t through this filter is $g_t * f_t$ and the "error" or difference between the actual and the desired outputs is $h_t - g_t * f_t$. With the least-squares method (§15.1.6), we add together the squares of the errors, find the partial derivatives of the sum with respect to the variables f_i (the elements of f_t), and set these derivatives equal to zero. This gives the following simultaneous equations, where g_t and h_t are known:

$$\frac{\partial}{\partial f_i} \sum_t (h_t - g_t * f_t)^2 = 0, \qquad i = 0, 1, 2, \ldots, n,$$

(9.71)

or

$$\sum_t (h_t - g_t * f_t) \frac{\partial}{\partial f_i} (g_t * f_t) = 0, \qquad i = 0, 1, \ldots, n.$$

(9.72)

One equation is obtained for each of the $n + 1$ elements in f_t. Writing the convolution as a sum using eq. (9.24) gives

$$\sum_t \left(h_t - \sum_k g_k f_{t-k} \right) \frac{\partial}{\partial f_i} \left(\sum_k g_k f_{t-k} \right) = 0.$$

The only terms in the convolution that involve f_i are those containing g_{t-i}. Hence,

$$\sum_t (h_t - \sum_k g_k f_{t-k}) g_{t-i} = 0,$$
$$\sum_t h_t g_{t-i} = \sum_t \sum_k g_k g_{t-i} f_{t-k}.$$

The left side is $\phi_{gh}(i)$ according to eqs. (9.42) and (9.43). We let $j = t - k$ and sum over j instead of over k:

$$\phi_{gh}(i) = \sum_t \sum_j g_{t-j} g_{t-i} f_j = \sum_j f_j \sum_t g_{t-j} g_{t-i}$$

upon interchanging the order of summation. The last factor is $\phi_{gg}(i - j)$ according to eqs. (9.46) and (9.47). Hence, we arrive at the *normal equations:*

$$\sum_{j=0}^{n} \phi_{gg}(i-j) f_j = \phi_{gh}(i), \qquad i = 0, 1, 2, \ldots, n$$

(9.73a)

In matrix form, this is

$$\begin{vmatrix} \phi_{gg}(0) & \phi_{gg}(-1) & \cdots & \phi_{gg}(-n) \\ \phi_{gg}(1) & \phi_{gg}(0) & \cdots & \phi_{gg}(1-n) \\ \cdots & \cdots & \cdots & \cdots \\ \phi_{gg}(n) & \phi_{gg}(n-1) & \cdots & \phi_{gg}(0) \end{vmatrix} \begin{vmatrix} f_0 \\ f_1 \\ \cdots \\ f_n \end{vmatrix} = \begin{vmatrix} \phi_{gh}(0) \\ \phi_{gh}(1) \\ \cdots \\ \phi_{gh}(n) \end{vmatrix}.$$

(9.73b)

(See also problem 15.11.) The normal equations for least-squares filtering also have an integral expression for continuous functions,

$$\int_{-\infty}^{\infty} \phi_{gg}(\tau - t) f(t) \, dt = \phi_{gh}(\tau).$$

(9.74; 15.273)

These equations can be used to *cross-equalize* traces, that is, to make traces as nearly alike as possible. Suppose we have a group of traces to be stacked, such as the components of a common-midpoint stack. After the normal-moveout corrections have been made, the traces may still differ from each other because they have passed through different portions of the near surface. The normal equations can be used to find the filters that will make all the traces as nearly as possible like some *pilot trace*, such as the sum of the traces. This procedure will improve the trace-to-trace coherence before the stack and hence improve the quality of the stacked result.

The least-squares method is also called "minimizing the L_2 norm." Sometimes other differences are minimized, for example, the absolute error, giving instead of eq. (9.71)

$$(\partial / \partial f_j)(\sum |h_t - g_t * f_t|) = 0,$$

(9.75)

which is called "minimizing the L_1 norm."

Much geophysical processing is done as matrix operations (see §15.1.3). These may become unstable if some eigenvalues have the value zero (because division by zero is then involved; see §15.1.3c). The addition of a small amount of white noise (perhaps as little as 0.0001% but typically 0.5 to 2%) makes the eigenvalues nonzero and stabilizes calculations.

Practical Wiener filtering involves specifying the desired output, the filter length, the filter's zero-time position (or, equivalently, the amount of anticipation component), and the amount of white noise to be added in the filter design. Truncation and tapering of the input wavelet's autocorrelation are generally required, recognizing that estimates based on the data are contaminated with noise.

Figure 9.14a shows a minimum-phase wavelet after being passed by a band-pass filter and figs. 9.14b and 9.14c are its frequency-domain representation. The wavelet is no longer minimum-phase after the band-pass filtering. Figure 9.14d is its autocorrelation, which is the same as that of the minimum-phase desired wavelet having the same amplitude spectrum. Figure 9.14e is the minimum-phase Wiener operator to convert fig. 9.14a into the desired minimum-phase wavelet. The amplitude spectrum of fig. 9.14e, shown in fig. 9.14f, is simply the inverse of fig. 9.14b, but its phase spectrum (fig. 9.14f) differs from fig. 9.14b because it is minimum-phase. The result of convolving fig. 9.14e with fig. 9.14a is shown in fig. 9.14h; the wavelet has been shortened but not made into a spike; its amplitude spectrum is white (flat) but it is mixed phase.

9.5.6 Whitening

(a) Spiking deconvolution. The normal equations are used to accomplish spiking deconvolution. The desired output is the earth's impulse response, e_t. We assume that e_t is random, that is, knowledge of the shallow reflections does not help in predicting the ar-

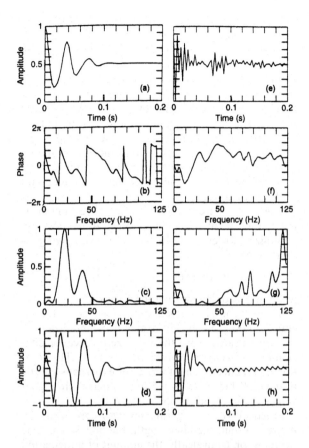

Fig. 9.14 Band-pass-filtered minimum-phase wavelet and its spectra before deconvolution and after Wiener deconvolution. (After Yilmaz, 1987: 101.) (a) Wavelet; (b and c) phase and amplitude spectra of (a); (d) autocorrelation of (a); (e) inverse operator calculated from (d) assuming minimum phase; (f and g) phase and amplitude spectra of (e); (h) result of convolving inverse operator (e) with (a). A spike is not achieved because the band-pass filtering destroyed the minimum-phase of (a).

rival times and amplitudes of deeper reflections. Consequently, the autocorrelation of e_t is negligibly small except for zero shift, and we can write

$$\phi_{ee}(t) = k\,\delta_t. \tag{9.76}$$

The geophone input g_t is regarded as the convolution of e_t with various filters (see eq. (9.34)) (the most important of which results from near-surface effects), the overall effect being represented by the single equivalent filter w_t (see eq. (9.35)):

$$g_t = e_t * w_t.$$

Because e_t (assumed to be minimum-phase) is the desired output h_t; h_{-t} eq. (9.44) enables us to write

$$\phi_{gh}(t) = g_t * h_{-t}$$
$$= e_t * (e_{-t} * w_{-t}) = (e_t * e_{-t}) * w_{-g} = k\,\delta_t * w_{-t}$$
$$= kw_0 \tag{9.77a}$$

There can be no output from filter w_t until after there has been an input to the filter; this is equivalent to saying that g_t is causal. Hence, $w_t = 0$ for $t < 0$. Thus,

$$\phi_{gh}(t) = 0 \qquad \text{for } t < 0. \tag{9.77b}$$

Therefore, if we concern ourselves only with positive values of t, we have the values required to solve eq. (9.73) for the spiking deconvolution filter; thus, eq. (9.73b) becomes

$$\begin{vmatrix} \phi_{gg}(0) & \phi_{gg}(-1) & \cdots & \phi_{gg}(-n) \\ \phi_{gg}(1) & \phi_{gg}(0) & \cdots & \phi_{gg}(1-n) \\ \cdots & \cdots & \ddots & \cdots \\ \phi_{gg}(n) & \phi_{gg}(n-1) & \cdots & \phi_{gg}(0) \end{vmatrix} \begin{vmatrix} f_0 \\ f_1 \\ \cdots \\ f_n \end{vmatrix} = \begin{vmatrix} k\omega_0 \\ 0 \\ \cdots \\ 0 \end{vmatrix}. \tag{9.78}$$

The constant kw_0 is usually set equal to 1 when we are only concerned with the relative values of f_i; ignoring scale factors is common and proper in many processing methods.

Note that much of the noise is random, but presumably the assumption of minimum-phase discriminates against the noise.

(b) Spiking deconvolution in frequency domain (whitening).

The ability to transform into the frequency or other domains not only provides alternative computing methods, but also provides insights as to what deconvolution methods imply. Thus, thinking of response e_i as random implies equal probabilities that the amplitudes at all frequencies will be equal and reminds us of white light, and we call spiking deconvolution *whitening.* It is equivalent to finding an inverse filter whose transform is $I(\nu)$, where

$$I(\nu) = 1/G(\nu) \tag{9.79}$$

(see fig. 9.15a), $G(\nu)$ being the transform of the input, so that the product $I(\nu)G(\nu)$ is constant. In applying the inverse filter, the phase has to be known. Often we start with ϕ_{gg}, assume that $G(\nu)$ is minimum-phase so that the wavelet can be determined uniquely from the spectrum of ϕ_{gg} (see §15.5.6c), and then invert $G(\nu)$ to obtain $I(\nu)$ (note that $I(\nu)$ is also minimum-phase; see problem 15.32a).

Equation (9.79) applies to any frequency value, for example, $I(\nu_1) = 1/G(\nu_1)$. If $G(\nu_1)$ should be small, then $I(\nu_1)$ will be large. Thus, a whitening filter emphasizes the weak frequency components, resulting in improvements to the extent that they are attenuated signals. Above some frequency ν_u noise dominates, so whitening is performed only over a limited band-pass. If, as in fig. 9.15a, signal $G(\nu)$ should be especially weak over a narrow band near frequency ν_1 (sometimes referred to as a notch in the signal spectrum), the inverse filter will magnify noise at this frequency, sometimes with disastrous results. To prevent excessive magnification of noise, white noise, ε, is sometimes added when the filter is being designed, the magnitude of ε being small compared to the average of $G(\nu)$. This does not substantially change the filter (fig. 9.15b) at most frequencies but makes it smaller at the notch frequency ν_1 (that is, $1/[G(\nu) + \varepsilon] \approx 1/G(\nu)$ except when $G(\nu)$ is very small). The white noise is added only for filter-design purposes and a "white noise added" comment on a seismic section thus indicates less noise gen-

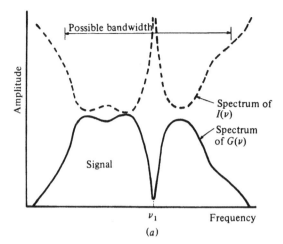

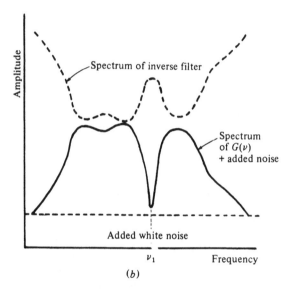

Fig. 9.15   Spectra of signal before deconvolution (solid curves) and of inverse filter used to achieve whitening (dashed curves). (a) Without addition of white noise; the bandwidth is usually specified without knowledge of the exact spectrum; (b) with white noise added for inverse-filter design purposes.

erated by the whitening deconvolution. Equation (9.78) now becomes

$$
\begin{vmatrix}
(1+\varepsilon)\phi_{gg}(0) & \phi_{gg}(-1) & \cdots & \phi_{gg}(-n) \\
\phi_{gg}(1) & (1+\varepsilon)\phi_{gg}(0) & \cdots & \phi_{gg}(1-n) \\
\cdots & & \ddots & \cdots \\
\phi_{gg}(n) & \phi_{gg}(n-1) & \cdots & (1+\varepsilon)\phi_{gg}(0)
\end{vmatrix}
$$

$$
\times
\begin{vmatrix} f_0 \\ f_1 \\ \cdots \\ f_n \end{vmatrix}
=
\begin{vmatrix} 1 \\ 0 \\ \cdots \\ 0 \end{vmatrix}.
\tag{9.80}
$$

Note that adding white noise changes the minimum-phase characteristic of a wavelet (see §15.5.6a).

The least-squares inverse filter is designed from autocorrelation values, eq. (9.73). The central peak of the autocorrelation function represents shifts of less than half the dominant period; it contains most of the information about average wavelet shape (fig. 9.16a). On the other hand, peaks and troughs for greater lags represent repetition of information, such as produced by multiples. If our desired wavelet h_t is to be essentially the same as the early part of the wavelet involved in g_t but is to die out rapidly but smoothly (since sharp changes induce ringing – see §15.2.7), we can take for ϕ_{gh} the early part of ϕ_{gg} multiplied by a suitable taper that truncates it after perhaps one cycle (fig. 9.16b), giving us everything needed to solve eqs. (9.73).

(c) Delayed spike. Spiking deconvolution often magnifies high-frequency noise, especially when the embedded wavelet is not minimum-phase. A better result can often be obtained by delaying the spike of the desired wavelet by m samples:

$$
\begin{vmatrix}
\phi_{gg}(0) & \phi_{gg}(-1) & \cdots & \phi_{gg}(-n) \\
\phi_{gg}(1) & \phi_{gg}(0) & \cdots & \phi_{gg}(1-n) \\
\cdots & \cdots & \ddots & \cdots \\
\phi_{gg}(m) & \phi_{gg}(m-1) & \cdots & \phi_{gg}(m-n) \\
\phi_{gg}(m+1) & \phi_{gg}(m) & \cdots & \phi_{gg}(1+m-n) \\
\cdots & \cdots & \ddots & \cdots \\
\phi_{gg}(n) & \phi_{gg}(n-1) & \cdots & \phi_{gg}(0)
\end{vmatrix}
$$

$$
\times
\begin{vmatrix} f_0 \\ f_1 \\ \cdots \\ f_m \\ f_{m+1} \\ \cdots \\ f_n \end{vmatrix}
=
\begin{vmatrix} 0 \\ 0 \\ \cdots \\ 1 \\ 0 \\ \cdots \\ 0 \end{vmatrix}.
\tag{9.81}
$$

A better result can often be obtained by shifting the shaping-filter operator, that is, making it two-sided (noncausal), so that it has coefficients for both negative and positive time values. Such a filter has both *anticipation components* (which act on future input values) and *memory components* (which act on past values). Two-sided filters are sometimes called *shaping filters* and they can be used to produce a zero-phase embedded wavelet. A particular wavelet, such as a recorded air-gun waveshape, can be used as the desired wavelet in what is sometimes called *signature processing*. Such applications of shaping filters are examples of wavelet processing (§9.5.9).

The effect of a one-sided filter using only memory components is shown in fig. 9.17. Figure 9.17b is a smoothed version of the zero-phase autocorrelation of fig. 9.17a and fig. 9.17c is a one-sided version. Figure 9.17d is a minimum-phase inverse operator to shorten fig. 9.17b, and fig. 9.17e is half of its autocorrelation. Figure 9.17f is the zero-phase wavelet corresponding with fig. 9.17e, but the result of applying it to fig. 9.17b gives fig. 9.17g; memory components alone cannot achieve the desired zero-phase result (although the multiple has been attenuated).

9.5.7 Predictive (gapped) deconvolution

Predictive deconvolution (Peacock and Treitel, 1969) attempts to remove multiple effects, which can be predicted from knowledge of the arrival time of the primaries involving the same reflectors. Predictive deconvolution operators often do not begin to exert an effect until after some time L (called the *prediction lag*), which is usually the two-way traveltime to the first multiple-generating reflector. We use the portion of ϕ_{gg} after the time L as ϕ_{gh} in eq. (9.73a) so that the filter predicts the multiples (fig. 9.16c), that is, we write

$$\sum_{j=0}^{n} \phi_{gg}(i - j)f_j = \phi_{gg}(L + i); \qquad (9.82a)$$

or expressing it in matrix form, eq. (9.82a) becomes

$$
\begin{vmatrix}
\phi_{gg}(0) & \phi_{gg}(-1) & \cdots & \phi_{gg}(-n) \\
\phi_{gg}(1) & \phi_{gg}(0) & \cdots & \phi_{gg}(1 - n) \\
\cdots & & \cdots & \cdots \\
\phi_{gg}(L) & \phi_{gg}(L - 1) & \cdots & \phi_{gg}(L - n) \\
\phi_{gg}(L + 1) & \phi_{gg}(L) & & \phi_{gg}(1 + L - n) \\
\cdots & & \cdots & \cdots \\
\phi_{gg}(n) & \phi_{gg}(n - 1) & \cdots & \phi_{gg}(0)
\end{vmatrix}
$$

$$
\times
\begin{vmatrix}
f_0 \\
f_1 \\
\cdots \\
f_L \\
f_{L+1} \\
\cdots \\
f_n
\end{vmatrix}
=
\begin{vmatrix}
\phi_{gg}(-L) \\
\phi_{gg}(-L + 1) \\
\cdots \\
\phi_{gg}(0) \\
\phi_{gg}(1) \\
\cdots \\
\phi_{gg}(n - L)
\end{vmatrix}
\qquad (9.82b)
$$

This gives a *prediction filter* of length $n + 1$ and lag L. We can subtract the predicted trace from the observed trace to give the *prediction error*, which is the trace with the predicted multiples removed:

$$h_t = g_t - g_{t-L} * f_t. \qquad (9.83)$$

Where the first multiple-generating reflector is deep, as with marine data in deep water, the deconvolution operator may be set to zero over portions of its length (corresponding to the zeros in the inverse filters in eq. (9.40)) to make the computation more economical; this is called *gapped deconvolution* (Kunetz and Fourmann, 1968).

Corrections may be made for variations in the wavelet shape from location to location. One procedure is to record the initial waveform. Another is to sum the autocorrelations of all the traces from a single source activation, estimate the waveshape for this autocorrelation sum assuming that the waveform is

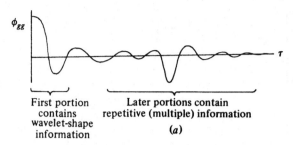

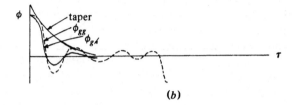

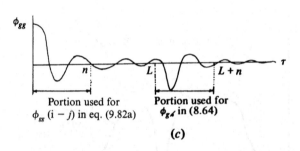

Fig. 9.16   Determining autocorrelation values to be used in deconvolution. (a) Autocorrelation of a trace; (b) tapering of ϕ_{gg} to use for ϕ_{gh} in eqs. (9.73); and (c) prediction of multiple effects for predictive-deconvolution filter design.

minimum-phase, and then apply a Wiener filter to convert this wavelet into a desired constant waveshape. This sort of procedure may be used to correct for variations in wavelet shape produced by different factors, such as changes in the source, detectors, recording instruments, or near-surface conditions.

9.5.8 Other types of deconvolution

Homomorphic or cepstral deconvolution, Kalman filtering, and other techniques are occasionally employed.

Homomorphic deconvolution involves a transformation from a space where functions are convolved (the time domain) to one where they are added (the *cepstrum domain;* see §15.6). The transformation of a time-domain function g_t to a cepstrum-domain function $\hat{g}(\zeta)$ is accomplished in three steps,

$$
\left.
\begin{aligned}
g_t &\leftrightarrow G(z), \\
\ln\{G(z)\} &= \hat{G}(z), \\
\hat{G}(z) &\leftrightarrow \hat{g}(\zeta).
\end{aligned}
\right\}
\qquad (9.84; 15.235)
$$

The cepstrum-domain equivalent of eq. (9.35) is

$$\hat{g}(\zeta) = \hat{w}(\zeta) + \hat{e}(\zeta), \qquad (9.85)$$

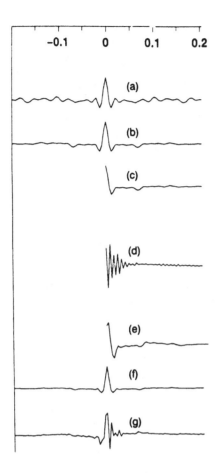

Fig. 9.17 Wavelet processing. (From Yilmaz, 1987: 107.) (a) Autocorrelation; (b) same after smoothing; (c) one-sided version of (b); (d) spiking–deconvolution operator calculated from (c); (e) minimum-phase inverse of (d), which is often assumed to be the embedded wavelet; (f) zero-phase equivalent of (e) having the same amplitude spectrum; and (g) shaping filter to convert (e) into (f).

thus the contributions of the wavelet, $\hat{w}(\zeta)$, and of the reflectivity, $\hat{e}(\zeta)$, add. The wavelet is usually slowly varying and its cepstrum lies mainly at low ζ values, whereas that of the reflection coefficients is mostly spread out over larger values. Thus, low-pass and high-pass filtering in the cepstrum domain (called *liftering*) achieves a large measure of separation. An inverse transformation of the high-pass (reflectivity) portion back to the time domain then completes the *homomorphic deconvolution*.

The cepstrum of a minimum-phase function is one-sided, that is, $\hat{g}(\zeta) = 0$ for $\zeta < 0$ if $G(z)$ is minimum-phase; this fact is sometimes used in separating minimum- and maximum-phase elements. To take advantage of the one-sided aspect in the cepstral domain, one sometimes forces $G(z)$ to be minimum-phase by applying exponential gain, that is, by using $G'(z) = G(z)k^z$, where $k \leqslant 1$ (Stoffa, Buhl, and Bryan, 1974). The reflectivity $E(z)$ is then found by liftering the high-pass portion of $\hat{g}'(\zeta)$, transforming back to find $E'(z)$, and then applying the inverse exponential weighting $E(z) = E'(z) k^{-z}$.

Otis and Smith (1977) use spatial averaging in the cepstrum domain as a way of determining source wavelet shape. They assumed that the source wavelet is stationary and the earth's response is spatially non-stationary, so that phase contributions in the earth's response at different locations disappear in the averaging.

Entropy is a measure of the chaos or lack of order in a system. Primary reflections are nonpredictable from preceding data and thus lack order. *Maximum-entropy deconvolution* attempts to extract such reflections by separating orderly (for example, equivalent wavelet) from disorderly (for example, signal) elements. Maximum-entropy deconvolution is discussed further in §15.7.6d.

Most of the foregoing techniques assume *stationarity,* that is, that the statistics of the waveshape do not change with time. However, we know that higher frequencies are attenuated more rapidly than lower frequencies and that peg-leg multiples and other factors cause the downgoing wavetrain to lengthen with time. One should deconvolve more effectively if the change in waveshape with time were accounted for (see §9.5.11). Kalman filtering (Crump, 1974) and other types of *adaptive filtering* attempt to take changes with time into account by continuously updating the statistics on which the filters are based. The most common mode of *time-variant deconvolution* (often abbreviated TV decon; see Clarke, 1968) involves designing one operator based on an autocorrelation of the early portion of the data and another based on an autocorrelation of the late portion of the data, each data window being 1 s or longer so as to give adequate statistics. The early and late operators are then applied at any given time in inverse proportion to the difference in time to the centers of the design windows, that is, the early operator is gradually *ramped* out as the late operator is ramped in. Sometimes the design windows overlap, and sometimes more than two design windows are used.

9.5.9 Wavelet processing

The convolutional model and Wiener filtering make it possible to replace a known embedded wavelet with a more desirable wavelet, within limitations imposed by the signal-to-noise level. The wavelet embedded in the data results from the convolution of many filters, some of which are space- and time-variant, and the wavelet usually contains some mixed-phase components. Most wavelet-estimating techniques average over a number of traces and sizeable time windows. The desired wavelet is almost always either minimum- or zero-phase, the output to the final display almost always being zero-phase nowadays.

A variety of different processes that involve determining, assuming, or operating on the effective wavelet shape go under the name *wavelet processing.* Some of these (1) attempt to make the wavelet shape

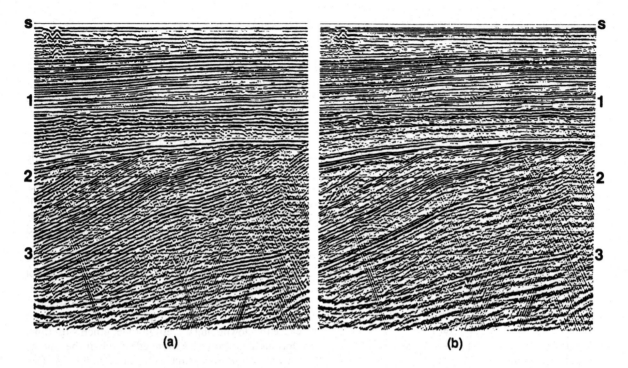

Fig. 9.18   Effect of deconvolution. (From Yilmaz, 1987: 85.) (a) Undeconvolved stacked section, and
(b) stacked section of deconvolved gathers.

everywhere the same, (2) some change the effective wavelet to some "more desirable" shape, and (3) some endeavor to separate the earth's reflectivity from wavelet-shape effects.

Wavelet processing, which attempts to make the wavelet shape everywhere the same, should be done as a prestack process so that all the component traces to be stacked have the same effective wavelet shape. Low-frequency components are more likely to be stacked in phase than high-frequency components, so that stacking often acts as a filter attenuating higher frequencies; this type of wavelet processing decreases this filtering action. Sometimes the source wavelet is actually recorded for every energy release in marine recording and then used in deterministic wavelet processing. More commonly, the wavelet is determined from the autocorrelation function by summing the autocorrelations of all traces recorded from the same source, assuming that the only common element is the source wavelet, so that the autocorrelation sum is simply the autocorrelation of the source wavelet. Examples of wavelet processing are shown in figs. 9.18 and 9.19.

The second type of wavelet processing, changing to some more desirable waveform, is used to correct for filtering actions (especially phase shifts) associated with instrumentation, so as to change hydrophone-recorded data to look more like geophone-recorded data or to produce a better match between lines recorded with different recording instruments. Sometimes the effective wavelets associated with certain source types have been measured and "cataloged,"

and the catalog wavelet is used in wavelet processing. Sometimes the effective wavelet is determined from the sea-floor reflection.

The third type of wavelet processing attempts to remove wavelet-shape effects and leave the earth's reflectivity function, that is, to separate w_t and e_t in eq. (9.35) such as in the liftering example in §9.5.8. Such processing is usually applied after other processing has removed as much of the noise as possible. Wavelet-processing techniques usually improve the high-frequency response and, consequently, the resolution. They often precede trace inversion (§5.4.5).

An example of wavelet processing of actual seismic data is shown in fig. 9.19. The shortening of the wavelet reduces ringing and permits seeing more stratigraphic detail.

9.5.10 Frequency filtering

Reflection signals often dominate over noise only within a limited frequency band. The filter should pass frequencies where the signal dominates and not pass those where the noise dominates in order to optimize the signal-to-noise ratio. Filter panels, such as shown in fig. 9.20, which display a portion of record section filtered by a succession of narrow band-pass filters, are often used to determine the optimum band-pass limits.

9.5.11 Time-variant processing

The frequency spectrum of seismic reflections usually becomes lower with increasing arrival time as the

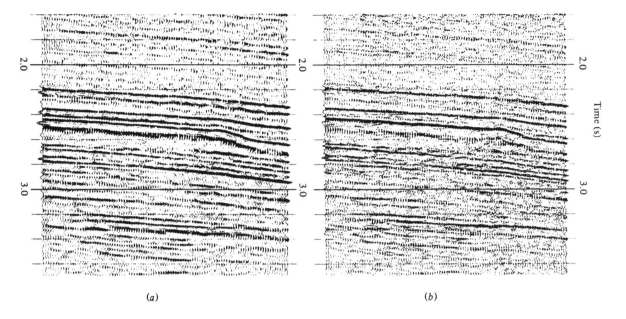

(a)   (b)

Fig. 9.19 Wavelet processing. (Courtesy of Grant Geophysical.) (a) Portion of migrated seismic section, and (b) the section after processing to broaden the bandwidth of embedded wavelet and make it zero-phase.

FIELD DATA
FILTER PANEL

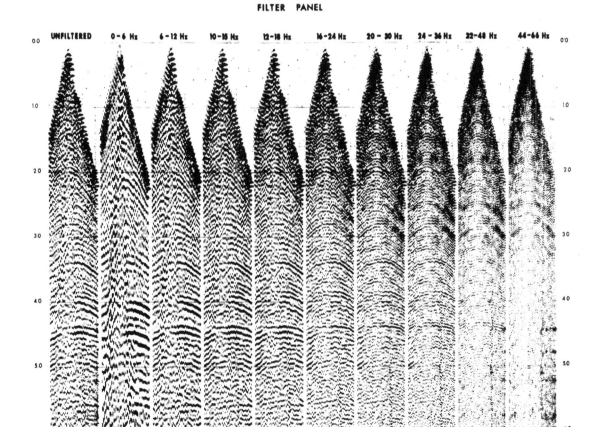

Fig. 9.20 Filter panel. (Courtesy of Grant Geophysical.)

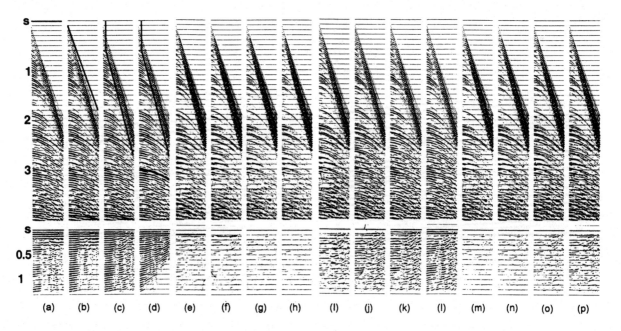

Fig. 9.21   Tests of deconvolution parameters. A gather and its autocorrelation (below) are shown under various circumstances. (From Yilmaz, 1987: 132–4.) (a) The input gather; (b to d) varying autocorrelation windows (between the heavy lines); (e to h) deconvolution-operator lengths of 40, 80, 160, and 240 ms; (i to l) prediction lags of 12, 32, 64, and 128 ms; (m to p) percent prewhitening of 1, 4, 16, and 32%. Best choices are (c), (g), (j or k), and (m).

higher-frequency components are attenuated faster by absorption, peg-leg multiple, and other natural filtering processes (see §6.5.1 and 6.3.2b). Hence, we often wish to shift the passband toward lower frequencies for later portions of the records, that is, we wish to accomplish *time-variant* (*TV*) *filtering*. Decisions as to the time-variant filter parameters are often based on filter panels such as shown in fig. 9.20, the deepest coherent energy in any passband being taken as the point where noise begins to dominate over signal.

Any discontinuous change, such as an abrupt change in band-pass parameters, will produce undesirable effects on the seismic section, however, including Gibbs' phenomena (§15.2.7). Changes are therefore usually distributed over a *merge zone*. For example, filter A might be used down to time t_A and filter B below time t_B ($t_A < t_B$); the merge zone then is between times t_A and t_B. A *linear ramp* may be used in the merge zone; the data in this zone may be filtered with both filters A and B and the data at $t_A + \Delta t$ within the merge zone will be the sum of the results of applying these two filters, where the results are weighted according to the position within the zone, that is, the weights would be $(t_B - t_A - \Delta t)/(t_B - t_A)$ and $\Delta t/(t_B - t_A)$, respectively. More than two filters and hence more than one merge zone might be used.

In addition to filtering, other processes such as deconvolution and statics correction are sometimes applied in the *time-variant* mode following similar procedures. Changes in filter parameters or in the parameters in other processes should not be made in the region where mapping is to be done lest the effects

of changing parameters be misinterpreted as having structural or stratigraphic significance.

9.5.12 Choosing deconvolution parameters

Yilmaz (1987: 109–31) discusses the selection of deconvolution time gate, operator length, prediction lag, and percent prewhitening for both wavelet shaping and multiple-suppression purposes (fig. 9.21); he shows examples of the effects of varying these parameters. The best time gate should exclude the early part of a record, which contains energy corresponding to guided waves, and also exclude the deeper part of the record, where ambient noise dominates (choice (c) in fig. 9.21b–d).

Yilmaz observes that the operator length generally should be chosen to include the first energy packet in the autocorrelation; too long an operator may suppress genuine reflections whereas one that is too short may produce overshoot and ripples. The 40-ms operator (e) in fig. 9.21e–h leaves some residual energy attributed to the basic wavelet and reverberating wavetrain, whereas operators longer than 160 ms (h) have little effect on the results. Deconvolution generally assumes minimum-phase and zero offset, so it is generally less effective in suppressing multiples where mixed-phase or nonzero-offset traces are involved.

As the prediction lag increases, results become less spikey, producing a band-limiting effect that supresses high frequencies, but increasing the prediction lag is not equivalent to using spiking deconvolution followed by band limiting. Increasing the lag tends to produce a wavelet with a duration equal to the prediction lag and makes deconvolution less effective in broadening the spectrum. Common prediction lags for predictive deconvolution are the autocorrelation's first or second zero crossings (choices (j) or (k) in fig. 9.21i–l), or unity for spiking deconvolution.

Prewhitening preserves a spiky character but adds a low-amplitude, high-frequency tail. The spectrum becomes less broadband as the percent prewhitening increases. Use of spiking deconvolution with prewhitening is similar to spiking deconvolution without prewhitening followed by band-pass filtering. Prewhitening is similar to adding random noise. Usually, only 0.1 to 1% prewhitening (choice (m) in fig. 9.21m–p) is required for stability.

9.5.13 Multichannel deconvolution

Most of the foregoing discussions imply that the design of the deconvolution operator is based on data from the same trace as that to which it is to be applied. One of the wavelet-processing methods described was based on the sum of a number of autocorrelations and then the application was to all of the components of the sum. Occasionally, other multichannel schemes are utilized.

The radial multiple-suppression method given by Taner (1980) involves using data from one trace as the basis for designing the operator to be applied to another trace. For flat reflectors, the angle of incidence is the same for the first multiple as it is for a primary at half the offset distance (see fig. 6.35), for the second multiple as for the primary at one-third the offset distance, and so on. The reflectivity for the sea-floor reflection changes so much (because of the changing angle of incidence) that predictive deconvolution applied to the same trace as that used in the operator design does not work well. Taner achieved better multiple attenuation by designing operators on traces where the angle of incidence for primaries was more nearly the same as for the multiples to be attenuated.

9.6 Automatic statics determination

9.6.1 Interrelation of statics and normal-moveout corrections

Statics corrections can be determined most easily after normal-moveout corrections have been optimized, but (as will be seen) normal-moveout determination is best when statics corrections are optimum. Because one of these determinations must precede the other, the calculations are often repeated with more refined inputs. Corrections for elevation differences or corrections based on uphole or first-break information from the field monitor records and estimated velocity are usually made before the first automatic statics determination. This is then followed by normal-moveout determination using these statics values. The values determined from the first statics and normal-moveout determinations are applied and then a second statics determination is made. The cycle of refining parameter values may be repeated several

times to obtain an optimum solution. Marsden (1993) reviews statics corrections.

9.6.2 The surface-consistent model

Automatic statics determination is often based on a *surface-consistent model* that associates a delay R_i with the geophone group at location i and a delay S_j with the source at location j. All data received by geophone group i will be delayed by R_i, possibly because the geophone group is at a higher elevation or because there is a thicker or slower low-velocity layer underneath it. All data from source j will be delayed by S_j, possibly because there was a delay between the source firing signal and the actual energy release, from the source being in (or on) a medium with lower velocity than other sources, from a higher source elevation or shallower shothole, and so on. Following the method of Taner, Koehler, and Alhilali (1974), we refer subscripts i and j to a common origin and make the station increments equal (as in the surface stacking chart, fig. 8.4b); hence offset distance is proportional to $j - i$. If there is structure along the line, a delay L_k may be associated with the location k (that is, L_k is some sort of average of time shifts because of structure at different depths below k). For flat reflectors, $k = \frac{1}{2}(i + j)$, that is, it is referenced to the midpoint location, and, if the dip is gentle, k is nearly constant for common-midpoint traces. If the normal-moveout correction is only approximate, some residual normal moveout M_k will remain, and this residual normal moveout will vary as the square of the offset distance. Because the residual normal moveout varies with arrival time, the delay associated with M_k will be some sort of average, as was L_k. Thus, for the surface-consistent model, the total time shift for a trace, t_{ij}, will be given by

$$t_{ij} = R_i + S_j + L_k + M_k(j - i)^2. \quad (9.86)$$

(The surface-consistent model is not restricted to determining time shifts for statics correction. In surface-consistent amplitude adjustment, for example, we assume an attenuation associated with each geophone and an attenuation associated with each source; the subsequent analysis follows basically the same procedure about to be outlined for surface-consistent statics. Likewise, wavelet extraction, deconvolution, and other operations are sometimes based on surface-consistent models. Note that "surface-consistent" does not necessarily require that the static shift, attenuation, and so on are the same when a geophone is located at point P as when the source is located at P.)

Although we may not know the amount of time shift to be associated with any trace, cross-correlation affords a means of determining $t_{ij} - t_{mn}$, the time shift of one trace relative to another, that produces the optimum alignment of the two traces:

$$t_{ij} - t_{mn} = R_i - R_m + S_j - S_n + L_{i+j} - L_{m+n}$$
$$+ M_{i+j}(j - i)^2 - M_{m+n}(n - m)^2.$$
$$(9.87)$$

(We use the subscript $i + j$ for L rather than $k = \frac{1}{2}(i + j)$ to assure that the subscripts are integers; the subscript magnitudes are not important because they are merely index values.) The shift that maximizes the cross-correlation produces the optimum alignment (match) of the two traces, and the magnitude of the cross-correlation indicates quantitatively how much improvement such a shift produces. With CMP data, we have many combinations of traces that have some of the unknowns R_i, S_j, L_{i+j}, or M_{i+j} in common. Because we can cross-correlate any two traces, we have more relative shift data than we have unknowns, that is, we have an "overdetermined" set of equations to be satisfied. However, we also have uncertainty in our measurements, that is, opposite sides of eq. (9.87) differ by some "error." The solution for R_i, S_j, L_{i+j}, and M_{i+j} is usually by the least-squares method, sometimes in an iterative manner.

The least-squares problem is to minimize the sum of the squares of the errors:

$$E = \sum e_p^2 = \sum \Big[t_{ij} - t_{mn} - R_i + R_m - S_j + S_n$$
$$- L_{i+j} + L_{m+n} - M_{i+j}(j - i)^2$$
$$+ M_{m+n}(n - m)^2 \Big]^2$$

$$= \text{minimum.} \qquad (9.88)$$

We wish to solve eq. (9.88) for the best set of R_i, S_j, L_{i+j}, and M_{i+j}. The least-squares solution is found by setting

$$\left. \begin{array}{ll} \partial E/\partial R_i = 0, & \partial E/\partial S_j = 0, \\ \partial E/\partial L_{i+j} = 0, & \partial E/\partial M_{i+j} = 0. \end{array} \right\} \qquad (9.89)$$

This results in many equations because there are as many R_i as there are geophone-group locations, as many S_j as there are source locations, and so on.

Often, the component traces are correlated with a pilot trace rather than with each other. One may, for example, select one of the better gathers to begin the analysis, equalize their rms values within a correlation window, and sum them (after removing normal moveout based on a first guess) to yield a first pilot trace. The pilot trace may be refined by applying the shifts determined by cross-correlation, and then iterating the procedure. One may then analyze adjacent gathers using this pilot trace, modifying the pilot trace as one proceeds. Band-pass filtering is sometimes applied before the correlation process. The correlation window may change laterally to follow the structure.

We do not want to produce an overall time shift. One way is to require that $\sum R_i = 0$ and $\sum S_j = 0$. Taner et al. (1974) achieved this by adding extra equations and modifying eq. (9.88), writing

$$\sum e_p^2 + \lambda \left(\sum_i R_i^2 + \sum_j S_j^2 + \sum_{i+j} L_{i+j}^2 + \sum_{i+j} M_{i+j}^2 \right)$$
$$= \text{minimum,} \qquad \lambda > 0, \qquad (9.90)$$

λ being a weighting factor expressing the relative emphasis to be given to the latter part of the equation (for example, see Claerbout, 1976: 112–14). Equation (9.90) has a unique solution for any λ. It is often solved in an iterative manner to achieve any desired degree of accuracy. One can assign different values of λ to the different component equations that went into eq. (9.90) (or to different terms in eq. (9.90)), for example, if one believes some of them yield better values and so should be weighted more heavily than others.

Additional constraints are often applied to some of the variables in eq. (9.88). One can remove part of the ambiguity between L_{i+j} and R_i or S_j by limiting L_{i+j} to small values. One may postulate a relation between R_i and S_j, for example, that the receiver and source statics for the same location should be similar, especially when using surface sources.

Taner et al. (loc. cit.) showed that solutions to eq. (9.90) have five arbitrary constants that represent intrinsic indeterminacies, some of which correspond to (a) an overall time shift of the section, that is, all events are too shallow or too deep; (b) an overall tilt of the section, which may create fictitious structure; and (c) masking of real structure, that is, making structure show up as a statics correction or vice versa.

The maximum shift allowed should be a little larger than possible source and geophone statics combined. This will sometimes cause cycle-jump problems, especially where the data are somewhat ringy, as the match between traces may be almost equally good when the traces are displaced by an additional cycle. Usually, the minimum-trace shift is the preferred one. One might think that cascading a series of small shifts in an iterative solution would give the same result as one large shift, but this is usually not the case.

The cross-correlation concept implies that the best match of traces results in the best match of primary reflections. Sometimes other types of energy are so strong that the calculated corrections optimize nonreflection rather than reflection alignments. Because cross-correlation is performed over a window, the best solution may be to narrow the window so that obvious noise is excluded. However, if the window is made too narrow, an alignment will be determined even if there are no reflections. The window should include as many primary reflections as possible while excluding nonprimary energy.

Equations (9.87), (9.88), and (9.90) and other equations that express additional constraints can be written in matrix form (§15.1.6); for example, eq. (9.87) with weighting becomes

$$\mathcal{H}\mathcal{A} - \mathcal{Y} = \varepsilon \qquad (9.91)$$

where

$\mathcal{H}$ = matrix of coefficients and weightings, λ,

$\mathcal{A}$ = matrix of unknowns, R_i, S_j, L_{i+j}, and M_{i+j},

$\mathcal{Y}$ = matrix of time shifts, $t_{ij} - t_{mn}$,

ε = matrix of error terms.

The least-squares solution is

$$\mathcal{A} = (\mathcal{H}^T \mathcal{H})^{-1} \mathcal{H}^T \mathcal{Y} \qquad (9.92; 15.57)$$

Surface-consistent statics sometimes considerably improves marine as well as land data. Modifications have to be made because the characteristics of hydrophone groups are apt to be consistent with respect to their locations in the streamer rather than locations along the line.

9.6.3 Maximizing the power of the stacked trace

Another approach assumes that the optimum static corrections are those that maximize the power of the stacked trace. A time-shift relation similar to eq. (9.87) provides the starting point, with the R_i, S_j, L_k, and M_k quantities being regarded as independent variables, x_n. Appropriate traces are stacked and the square of the amplitude (proportional to the power P) determined. The amount by which the power changes for changes in each variable, that is, $(\partial P/\partial x_n)\Delta x_n$, is determined for each variable, and Δx_n is selected so that P increases. This is the method of *steepest ascent* and similar methods are used in many data-processing methods. In practice, two problems are encountered: (1) how to find the correct maximum if there are several maxima, and (2) how to get to the maximum with the fewest calculations.

To solve the first problem, one assumes that the first estimate is on the slope of the correct maximum (seismic data are semiperiodic and adjacent maxima usually represent cycle jumps). Sometimes a search is made for other maxima so that one can determine which is the largest. Another technique is to make a first solution after filtering out higher frequencies so that the maxima are broader and fewer; the first solution is then used as the starting point for solving the problem with the unfiltered data.

The ideal solution to the second problem is to climb toward the maximum in relatively few steps without overshooting the top by very much. The step size is often related to $\partial P/\partial x_n$. Another technique is to calculate the curvature (or second derivative) to estimate how far away is the maximum. To minimize calculations, problems are often subdivided, limiting the number of variables being considered at one time.

Figures 9.22 to 9.24 illustrate the improvement in data quality that can result from application of automatic statics. Marked improvement is often achieved.

9.6.4 Refraction statics

Although surface-consistent statics corrections, which are based on differences among traces within the spread length, generally accommodate trace-to-trace variations, they may accumulate small errors and do a poor job at handling long-wavelength statics variations (variations of the order of the spread length or larger). Refraction statics corrections, which are based on first-break refraction arrival times, provide a means of dealing with such long-wavelength variations.

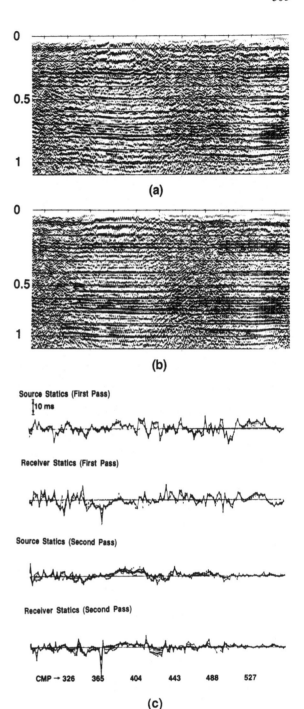

(a)

(b)

Source Statics (First Pass)

Receiver Statics (First Pass)

Source Statics (Second Pass)

Receiver Statics (Second Pass)

CMP → 326   365   404   443   488   527

(c)

Fig. 9.22 Quality improvement resulting from surface-consistent statics. (From Yilmaz, 1987: 222, 224.) (a) Stack with only field statics applied; (b) stack after two residual statics passes; and (c) diagnostics after the first and second passes.

The refraction first-breaks are picked automatically, usually after an approximate refractor velocity has been used to produce reduced refraction profiles. With CMP data, there is usually appreciable redundancy, and a logic is used to throw out those traces that do not appear to involve travel along the base of

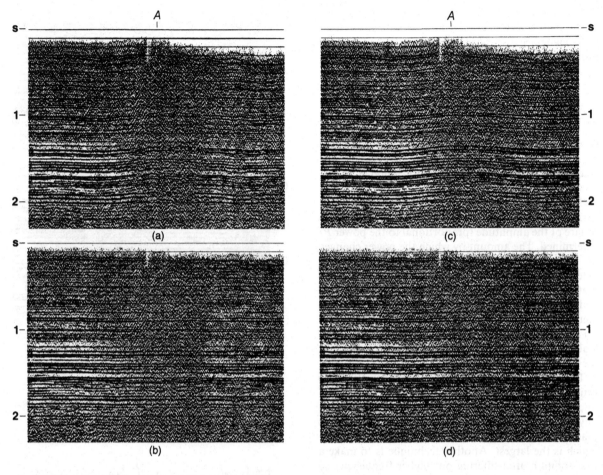

Fig. 9.23 Improvement resulting from applying both refraction and surface-consistent statics. (From Yilmaz, 1987: 229–33.) (a) Stack with only field statics applied; (b) stack after surface-consistent statics; (c) stack after refraction statics; and (d) stack after both refraction and surface-consistent statics.

the LVL. Then the remaining data are averaged for each location and analyzed by refraction calculation methods, which may be as simple as those discussed in §8.8.2 or more elaborate, such as the plus-or-minus method or generalized reciprocal methods discussed in §11.5.2 and 11.3.3.

9.7 Velocity analysis (velocity spectrum)

9.7.1 Conventional velocity analysis

The variation of normal moveout with velocity and arrival time has already been discussed in connection with eq. (4.7). Several techniques utilize the variation of normal moveout with record time to find velocity (Garotta and Michon, 1967; Cook and Taner, 1969; Schneider and Backus, 1968; Taner and Koehler, 1969). Most assume a stacking velocity (V_s) as discussed in §5.4.4a and apply the normal moveouts appropriate for the offsets of the traces being examined as a function of arrival time, and then measure the coherence (degree of match) among the traces available to be stacked. Several measures of coherence can be used; some of these were discussed in §9.3.5 (see eqs. (9.56) to (9.59)). Another stacking velocity is then

assumed and the calculation repeated, and so on, until the coherence has been determined as a function of both stacking velocity and arrival time. (Sometimes normal moveout is the variable rather than stacking velocity.)

Velocity analysis is usually done on common-midpoint gathers where the assumption of hyperbolic alignment is often reasonable. Where dips are large, a common reflecting point is not achieved and DMO (§9.10.2) or equivalent processing may be required.

A velocity-analysis display is shown in fig. 9.25. This is a good analysis because the data involved in fig. 9.25a are good. Peaks on the peak amplitude trace (fig. 9.25b) correspond to events. The locations of the highs yield the velocities (or normal moveouts) that have to be assumed to optimize the stack (hence the name stacking velocity), but these may not all be primary reflections. Velocity analyses are also commonly displayed as contour plots (fig. 9.26) rather than as in fig. 9.25.

Other events as well as primaries give rise to peaks, and hence the results have to be interpreted to determine the best values to be used to stack the data (see §9.7.3). In many areas where the velocity increases

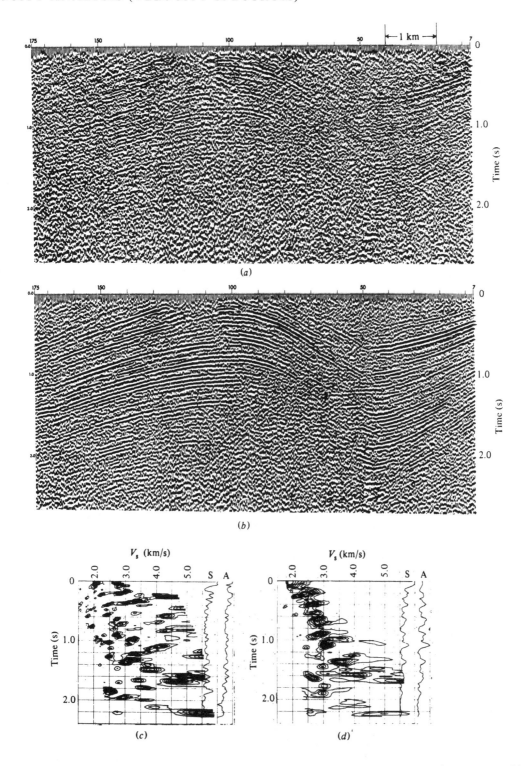

Fig. 9.24 Improvement resulting from use of surface-consistent statics. (Courtesy of Grant Geophysical.) (a) Section using only the field-determined statics; (b) section using also statics determined by a surface-consistent program; (c) velocity analysis using field statics; and (d) velocity analysis after application of surface-consistent statics.

more or less monotonically with depth, the peaks associated with the highest reasonable stacking velocities are assumed to represent primary reflections and peaks associated with lower velocities are attributed to multiples of various sorts. In other areas, the relationships are not as obvious, and even where the velocity relationships are generally regular, difficulties may be encountered.

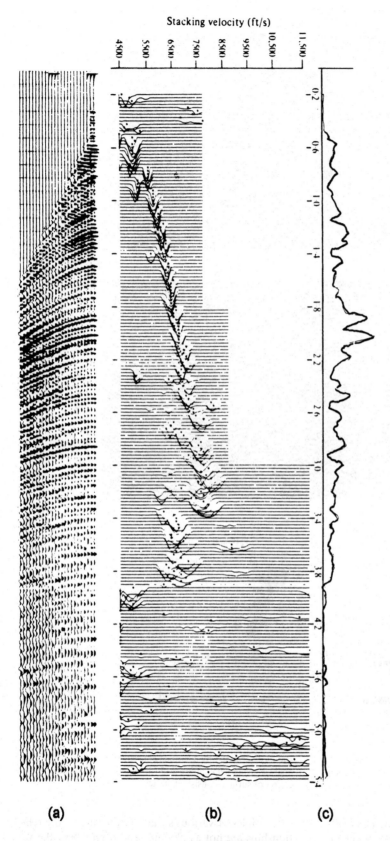

Stacking velocity (ft/s)

(a) (b) (c)

Fig. 9.25 Velocity analysis. (Courtesy of Petty-Ray Geophysical.) (a) Common-midpoint gather showing the data involved in
the analysis, (b) amplitude of the stacked trace as a function of
stacking velocity at 100-ms intervals, and (c) maximum amplitude achievable on stacked traces. The low velocities below 2.7
s are probably multiples and there are few primary reflections
below 3.3 s.

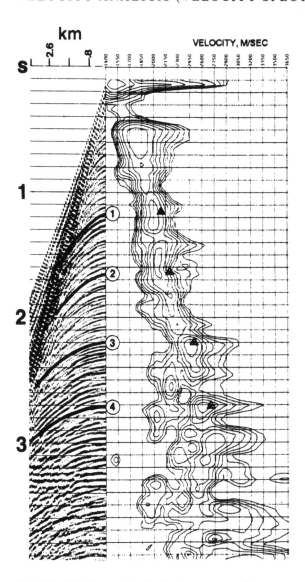

Fig. 9.26 Velocity analysis displayed as contours of a measure of coherence (semblance in this case). (From Yilmaz, 1987: 168.)

A compromise has to be made between using the small amount of data appropriate to a specific spot, in which case the velocity analysis is apt to be nondefinitive, and using more data but distributed over a larger area, in which case velocity may be defined better but the velocity measurements are then averages over a sizeable region. The compromise is often to use data for three to five adjacent midpoints. Measurements are also usually based on all the data within a window, which is often 50 to 100 ms long, in order to increase the amount of data and hence improve the velocity definition.

9.7.2 Velocity panels

Velocity panels (fig. 9.27) provide another display from which stacking velocity can be determined. A set of data is plotted several times, each plot being based on a different stacking velocity. The central two panels, figs. 9.27e and 9.27f, utilize an approximate velocity function; the panels to the left use velocities successively lower by some velocity increment and those to the right utilize higher velocities. Such a set of velocity panels shows whether increasing or decreasing the velocity will enhance individual events. Because stacking velocity is not necessarily single-valued (see fig. 9.28), different events might require different velocities to be optimized. A velocity panel is often run as a check on the interpretation of velocity analyses of the type shown in figs. 9.25 and 9.26. Velocity panels are often made of sections (or portions of sections) stacked with different velocities as well as of common-midpoint gathers.

9.7.3 Picking velocity analyses

Velocity analysis involves a considerable number of calculations and hence is fairly expensive to execute; therefore, too few analyses are often run, sometimes only every ½ to 5 km along the line. Where only a limited number of velocity analyses are to be run, their locations should be selected judiciously, based on the best available geologic information, so that analyses are not wasted in noisy areas and so that changes in geology are adequately sampled. Where the number of traces in a CMP gather is large, only every other trace may be used in order to reduce the cost.

Velocity analyses are ordinarily picked by an interpreter. Picking involves selecting the time–velocity values to be used in subsequent processing. The velocity-analysis interpreter often has in mind only achieving a good stack, and stacking can often tolerate appreciable velocity errors. Velocity interpretation is time-consuming and hence expensive and has significant potential for error, especially when the picker knows little about the local geology and hence does not factor this into the interpretation. It is not uncommon for analyses to be picked as stand-alone operations and consequently successive analyses may not even be picked consistently. These errors are becoming less frequent today where velocity interpretation is done at a workstation where adjacent analyses already interpreted can be displayed alongside the new analysis as a guide for picking consistently. A plot (fig. 9.29) of the interval velocities (calculated by the Dix equation, eq. (5.25)) that a particular interpretation implies is often helpful in interpreting velocity analyses.

The interpreter is ordinarily guided by a set of simple rules (Cochran, 1973: 1048–9):

1. an increase in stacking velocity V_s with increasing depth is more probable than a decrease;
2. successive reflections are ordinarily separated by more than 100 ms in two-way time;
3. an interval velocity greater than 6700 m/s (22,000 ft/s) or less than 1430 m/s (4700 ft/s) is unlikely;

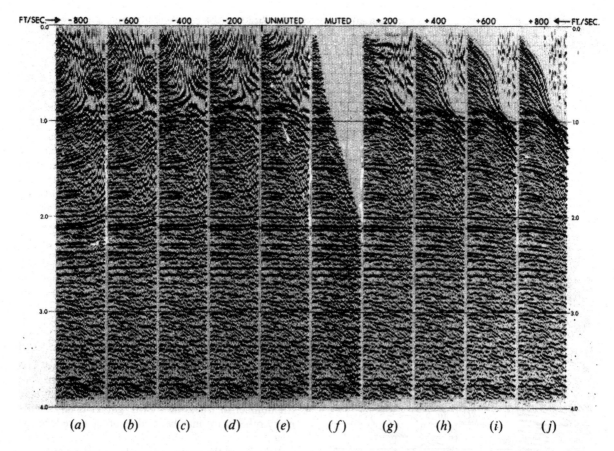

Fig. 9.27 Velocity panel of a CMP gather. Panels (e) and (f) employ the velocity resulting from a velocity analysis with a mute applied in panel (f). Panels (a) to (d) show results where the stacking velocity is decreased from that in (e) by $n\, \Delta V_s$, where n is respectively 4, 3, 2, and 1, and ΔV_s is sometimes as much as 200 ft/s. Panels (g) to (i) show results where the stacking velocity is increased by $n\, \Delta V_s$. (Courtesy of Grant Geophysical.)

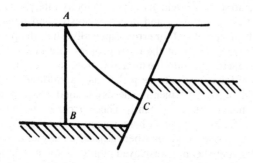

Fig. 9.28 Multivalued stacking velocity values. Reflections B and C arriving at the same time may have different stacking velocities.

4. the difference in interval velocity for successive layers should exceed 2% (see problem 5.13 for a quick approximate method of determining interval velocity);
5. any event at about twice the t_0 of a previous event and with approximately the same V_s is probably a multiple and should not be used.

Computer picking based on similar rules is sometimes used. For example, a possible pick must satisfy rule 2 and then must pass rules 3, 4, and 5; if more than one pick passes these tests, they are tested against rule 1; if more than one pick is still possible, that with a velocity nearest the V_s of the preceding pick is selected. Multiples are apt to have velocities that are low and diffractions and sideswipe events (for example, diffractions from faults nearly parallel to the line or reflections for which the line makes a small angle with the strike direction) are apt to have velocities that are unreasonably high.

Because the amount of normal moveout applied varies with arrival time, frequencies are lowered as offsets increase (fig. 9.30); this is called *normal-moveout stretch* and it affects velocity-analysis picks. Long-offset traces are muted (§9.10.3) to avoid excessive stretch effects; clearly the amount of mute applied affects the measured velocities. In the usual case, where velocity varies with depth, the alignment of events is actually some other curve rather than a hyperbola. However, the errors in assuming a hyperbolic alignment are usually small.

The accuracy and resolution of stacking velocity values clearly depend on acquisition factors such as the spread length, the multiplicity (fold), the recorded bandwidth, the signal-to-noise ratio, and the lack of

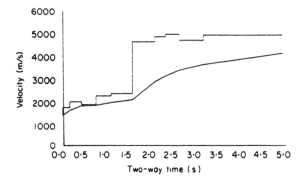

Fig. 9.29 Interval-velocity bar graph produced from a velocity analysis.

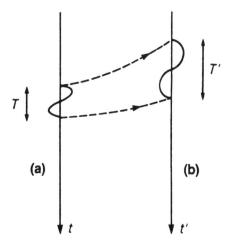

Fig. 9.30 Normal-moveout stretching. (From Yilmaz, 1987: 162.) (a) A signal with a period T, which after applying NMO (b) has period $T' > T$.

near- or far-offset traces or irregular spacing in the field. They also depend on processing parameters such as muting and the weighting of input traces, location and length of the time gate, sampling intervals, and the coherency measure used. Gathers are often decimated for velocity analysis, perhaps reducing the number of input traces by 1/4 and/or subsampling in time, to reduce analysis costs; in addition, coherence is usually checked only for stacking velocities within a window centered on the expected velocities. Figure 9.31 shows the effects of analyzing combined adjacent gathers, subsampling, and insufficient offsets, and Yilmaz (1987: 173–82) discusses the effects of other factors on velocity determinations.

9.7.4 Uses and limitations of velocity analyses

The precision of reading values from hard-copy velocity analyses is usually ± 10 ms in t_0 and ± 50 m/s in V_s, but the accuracy is often less than this. Velocity values have to be interpolated for intermediate times and in-

tervening locations. Values for times between picks are often interpolated linearly, and then the values for traces between analyses are interpolated from these, a process called *bilinear interpolation;* this procedure may introduce errors where analyses are inadequately spaced, of poor quality, or picked in a nonsystematic manner. A plot showing interpolated values (fig. 9.32) provides valuable control by making the consequences of velocity assumptions clear.

Velocity analyses should be plotted at the same vertical scale as the seismic section so that they can be overlaid on the section to make it easier to identify stacking-velocity picks with specific events. The same events should be picked on successive analyses.

Analyses should be continuously compared to neighboring analyses to check that variations make geologic sense; comparing analyses along a line allows an interpreter to assess the uncertainty in individual picks and smooth out noise effects. Where data are good, systematic changes may indicate stratigraphic changes. Generally, as many events should be picked as possible. While picking just a few events may suffice for stacking purposes, picking many events often discloses important interpretation clues.

Whereas velocity analyses are generally interpreted as if reflectors were horizontal and the seismic line were perpendicular to strike, stacking velocity depends on both quantities. Levin (1971) showed the dependence of stacking-velocity measurements on dip ξ and trace Ξ (the angle between the strike and the line) for constant-velocity overburden:

$$V_s = V(1 - \sin^2\zeta \cos^2\Xi)^{1/2} \qquad (9.93)$$

This relationship is shown in fig. 9.33.

Although the objective of velocity analyses is to achieve good stacked data, the velocity values also have interpretational importance (§10.5). With seismic data that are not unduly distorted by structural complexities, approximate interval velocities can be obtained from stacking velocities by simple relationships (see eq. (5.23) and problem 5.13); however, interval-velocity values determined in this way should be routinely checked for reasonableness.

9.7.5 Horizon velocity analysis

The determination of stacking velocities continuously along a seismic line is called *horizontal velocity analysis.* Such analyses are often made for only a single or a few reflections. Generally, horizons are picked by either an autopicker or manually, and analyses are made over a narrow time window about the reflections. The analysis is essentially the same as for a conventional velocity analysis. Coherency is measured within the window as the assumed velocity is varied and the selected velocity is that which maximizes the coherency. Figure 9.34 shows horizontal velocity analyses along five horizons, and fig. 9.35 shows the improvement in data quality for a reflection below major lateral velocity changes, caused by salt diapirism in

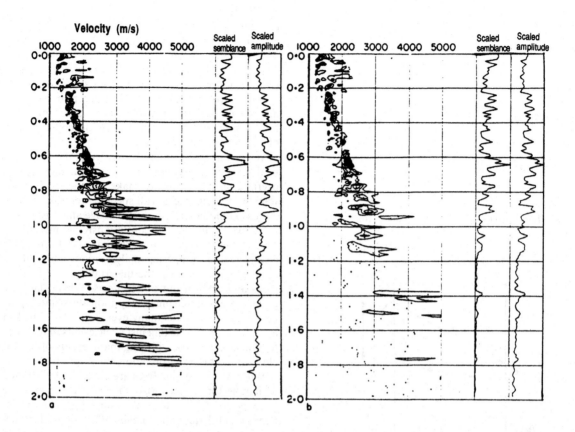

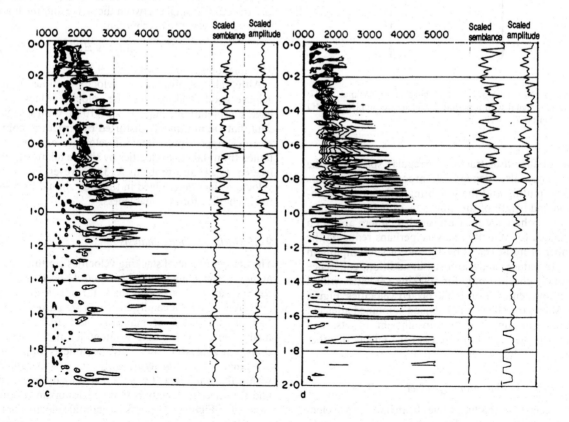

Fig. 9.31 Effects of velocity-analysis parameters. (From Hatton et al., 1986: 68–9.) (a) Analysis based on two adjacent CMP, (b) based on eight adjacent CMP, (c) analysis using only every third trace, and (d) analysis using only near offsets.

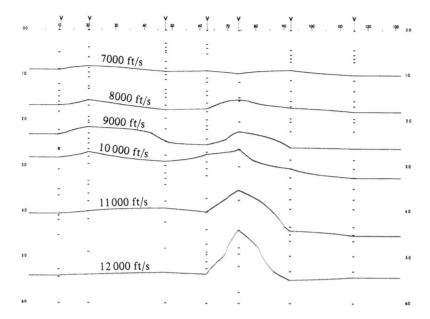

Fig. 9.32 Stacking velocity along a seismic line. Values are interpolated by the computer from input picks indicated by the dashes. This line is also shown in fig. 10.33. (Courtesy of Grant Geophysical.)

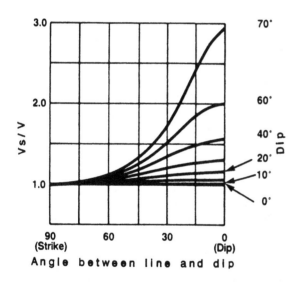

Fig. 9.33 Increase in stacking velocity with dip ξ and the angle between the strike Ξ and the line direction. (From Levin, 1971.)

this instance. Changes in the velocity in the interval between horizontal velocity analyses on adjacent parallel horizons are sometimes used as an interpretation tool to sense stratigraphic changes.

9.8 Preservation of amplitude information

The amplitude of a reflection depends on the acoustic impedance contrast at the reflecting interface. However, other factors, such as those listed in fig. 6.44, often obscure the acoustic-impedance-contrast information. The effects of spherical divergence and

raypath curvature can be calculated and corrected for. The gain of the recording instruments normally is known. Array directivity rarely has a significant effect on the amplitude of nondipping events and so its effects are generally ignored. Corrections for offset-dependent amplitude effects are also usually ignored. Migration can correct for reflector curvature effects.

Remaining effects are mostly of two kinds: (1) those associated with energy losses because of absorption, scattering, transmissivity losses, and peg-leg multiples, and (2) those that vary with source strength and source coupling, geophone sensitivity and geophone coupling, and offset. The effects in the first group are difficult to determine but they usually do not vary appreciably along a line and so may not obscure lateral variations. The high multiplicity of CMP data permits determining the second group of effects in a surface-consistent amplitude-correction program (actually, so that the effects are additive rather than multiplicative, the log of the amplitude rather than the amplitude) similar to automatic statics correction (§9.6.2; see Taner and Koehler, 1981).

A correction for frequency-dependent absorption and peg-leg multiples (a *Q-correction*) is sometimes made:

$$A(t) = A(0)e^{\pi v t / Q}, \qquad (9.94)$$

where $A(0)$ refers to some reference time. Because Q is usually known only approximately, it is often taken as $0.01V$, where V is the velocity in ft/s.

One processing routine adjusts amplitude in several steps. After first correcting for amplitude adjustments made in recording, a time-dependent spherical-divergence correction based on assumed velocity is

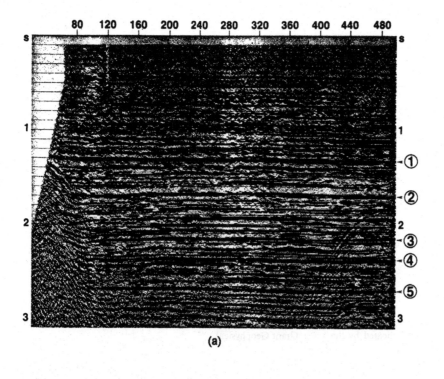

(a)

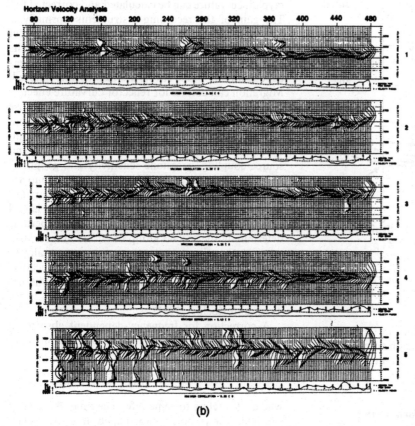

(b)

Fig. 9.34   A stacked section with horizon-velocity analyses of five horizons. (From Yilmaz, 1987: 184.)

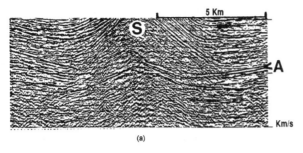

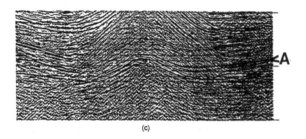

Fig. 9.35 Portion of a section across a salt dome prior to migration. (From Yilmaz, 1987: 185.) (a) Conventionally processed, (b) horizon-velocity analysis (HVA) along a base salt reflector A (center), and (c) processed utilizing HVA velocities.

applied. Such a correction makes the range of amplitude values smaller and therefore easier to handle. These corrections constitute the "preliminary gain recovery/adjustment" shown in the "editing" phase of fig. 9.62. Surface-consistent amplitude analysis and/or correction is then done during one or more of the processing passes in the "main processing" phase. After velocity has been determined, the spherical divergence correction is changed to depend on $V_s^2 t$, which allows approximately for raypath curvature, using some arbitrary time as a reference value. An additional arbitrary exponential gain can be applied to make the range of amplitude values smaller for display purposes. This correction may be based on mean absolute or rms amplitude averages over time windows a few hundred milliseconds in length and also averaged over many traces. Sometimes the previous step-by-step amplitude adjustment is simply replaced by an arbitrary gain function.

Amplitudes are sometimes adjusted so that their rms averages over a time window (perhaps 200 ms in length) are equal, this step being called *equalization*.

It should be noted that, because reflection amplitude varies with incidence angle (or with offset; see §3.4), CMP stacking does not result in normal-incidence amplitudes even if the amplitudes of all of the component traces should be correctly preserved (see also §9.10.5).

9.9 Apparent-velocity (2-D) filtering

Apparent-velocity filtering, also called *dip, fan, moveout,* or *pie-slice filtering* (Fail and Grau, 1963; Treitel, Shanks, and Frasier, 1967) for reasons that will become obvious, depends upon the apparent velocity (defined by eq. (4.13a)) of a wave as it approaches a recording spread. Equations (2.4) and (4.13) can be combined to give

$$V_a = \omega / \kappa_a = 2\pi\nu / \kappa_a \qquad (9.95)$$

For a fixed apparent velocity V_a, the plot of frequency ν versus apparent wavenumber κ_a is a straight line. For a seismic spread along the x-axis, κ_a is positive or negative according as V_a is in the positive or negative directions. For a vertically traveling signal, $\kappa_a = 0$ and $V_a = \infty$ and the ν–κ_a representation is along the ν-axis. For most reflection signals, $V_a > V_{min}$, some minimum apparent velocity, and hence the reflections lie within a relatively narrow wedge containing the ν-axis, as shown in fig. 9.36a. Coherent noise generally has a lower V_a than reflections (fig. 9.37) and therefore separates from them in the ν–κ_a plot, usually called an *f–k plot* (frequency vs. wavenumber plot).

We can use two-dimensional transforms (§9.1.4 and 15.2.4) to define an apparent-velocity filter,

$$\begin{aligned} F(\nu_s, \kappa_a) &= 1, \qquad |\kappa_a| < 2\pi\nu / V_m, \\ &= 0, \qquad |\kappa_a| > 2\pi\nu / V_m, \end{aligned} \qquad (9.96)$$

that will pass the signal but reject the noise (as shown in fig. 9.36c). Such a filter that passes a narrow wedge in the ν–κ_a domain is a "pie-slice" filter. Of course, neither signal, noise, nor filter need be symmetric about the ν-axis. For example, there are hardly any coherent alignments dipping to the left in fig. 8.16a, and so fig. 8.16b if extended to the left of the ν-axis would be essentially blank. Apparent-velocity filters can also be designed to remove a noise wedge rather than pass a signal wedge; such a filter is called a "butterfly" filter.

Just as frequencies above the Nyquist frequency may alias back into the passband unless excluded by alias filters before the sampling, so spatial sampling involves *wrap-around* aliasing (fig. 9.36b) of data for wavenumber values exceeding the Nyquist wavenumber (see eq. (9.33)). The only way to prevent aliasing is to filter before sampling, which is not possible with respect to spatial sampling, or to move the Nyquist points farther out by sampling more closely.

The filter in the space–time domain (x, t) equivalent to the filter given by eq. (9.96) is obtained by taking the two-dimensional inverse Fourier transform (see eq. (15.117))

$$f(x, t) = (1/2\pi) \int_{-\kappa_N}^{+\kappa_N} \int_{-\nu_N}^{+\nu_N} F(\nu, \kappa_a) e^{j(\kappa_a x + 2\pi\nu t)} \, d\kappa_a \, d\nu$$

or

$$f(x, t) = (1/2\pi) \int_{-\kappa_N}^{+\kappa_N} \int_{-\nu_N}^{+\nu_N} \cos(\kappa_a x + 2\pi\nu t) \, d\kappa_a \, d\nu, \qquad (9.97)$$

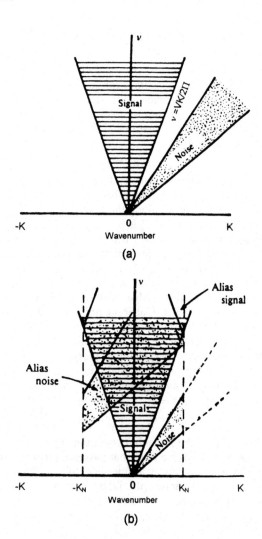

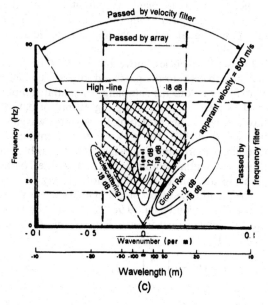

Fig. 9.36 A seismic gather in the frequency–wavenumber domain. (After Sheriff, 1991.) (a) Signal, generally near the ν-axis, and noise tend to separate; (b) illustrating wrap-around aliasing where κ_N is the Nyquist wavenumber; and (c) filtering effects of frequency, array, and velocity filters.

because $f(x, t)$ must be real. The convolution of $f(x, t)$ with the input (signal + noise), $g(x, t)$, gives the output $h(x, t)$,

$$h(x, t) = g(x, t) * f(x, t)$$
$$= \int_{-\infty}^{\infty} \int_{-\infty}^{\infty} g(\sigma, \tau)\, f(x - \sigma, t - \tau)\, d\sigma\, d\tau.$$

$$(9.98, 15.164)$$

This equation can also be written in digital form:

$$h_{x,t} = \sum_{m} \sum_{n} g_{m,n} f_{x-m,t-n}, \qquad (9.99)$$

where the space-sample interval is usually the trace spacing in the x-direction and the time-sample interval in the t-direction.

Instead of transforming the 2-D filter to the time domain and calculating $g(x, t) * f(x, t)$ as we did in eq. (9.41), we can transform $g(x, t)$ to the (ν, κ_a) domain, multiply $G(\nu, \kappa_a)$ by $F(\nu, \kappa_a)$ and use the two-dimensional convolution theorem (eq. (15.165)) to obtain $h(x, t)$.

The use of 2-D filtering to attenuate noise trains such as severe ground roll on common-source gathers is illustrated in fig. 9.38. Using 2-D filtering reduces the amount of muting required so that more reflection data can be used in velocity analysis and in stacking, providing better stacking-velocity definition and better attenuation of multiples in stacking.

Figure 9.39 shows that 2-D filtering can be effective in attenuating surface multiples where there is a steady increase of velocity with depth. 2-D filtering may also be applied after stacking (fig. 9.40).

9.10 Stacking

9.10.1 Gathers

Common-midpoint stacking is the most important data-processing application in improving data quality. The principles involved have already been discussed along with the field procedures used to acquire the data. The component data are sometimes displayed as gathers. A common-midpoint gather (see figs. 9.25 to 9.27) has the traces for the same midpoint arranged side by side, and a common-offset gather has the traces for which the source-to-geophone distance is the same arranged side by side. Gathers are displayed either before or after normal-moveout correction. The traces within a common-midpoint gather are summed to yield a single stacked trace.

9.10.2 DMO (dip-moveout) correction

The result of stacking CMP traces after normal-moveout correction is assumed to be the trace that would be recorded by a coincident source and geophone located at the midpoint. However, the reflection point is displaced updip (§4.1.4), and in fig. 4.9b, the reflecting point is R, not P; this results in an offset change Δx given by eq. (4.22a) and a decrease in the

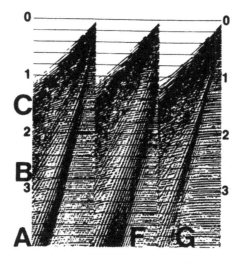

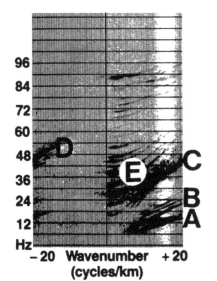
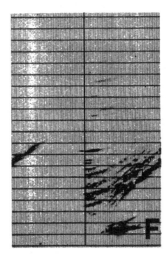
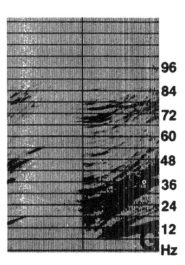

Fig. 9.37   Three common-source gathers (above) and their $f-k$ spectra (below). A, B, and C are high-amplitude, dispersive, coherent noise trains; D is the wrap-around of C, and E consists of reflection events. As the spatial extent of the noise train becomes wider, its $f-k$ equivalent becomes narrower; compare F and G. (From Yilmaz, 1987: 70.)

zero-offset traveltime Δt given by eq. (4.22b). Both effects are proportional to the square of the offset, so stacking produces smearing unless proper DMO corrections are applied. Also, the velocities determined in velocity analyses are dip-dependent unless a DMO correction has been applied. Dip also causes peg-leg multiples to divide into two sets, one with apparent stacking velocity higher than the zero-dip stacking velocity, the other lower (Levin and Shah, 1977), so that stacking alters the character of events that include appreciable peg-leg energy.

Unlike the classical dip moveout, which is simply the effect of dip on traveltime for a common-source record (gather), DMO processing creates common-reflection-point gathers. It effectively moves a reflection seen on an offset trace to the location of the coincident source–receiver trace that would have the same reflecting point (fig. 9.41). It thus involves shifting both time and location. The result is that the reflection moveout no longer depends on dip, reflection-point smear of dipping reflections is eliminated, and events with various dips have the same stacking velocity. It is often carried out as a convolution in the common-offset domains.

Levin (1971) showed that the reflecting point moved updip (fig. 9.42a) from that for the coincident source–geophone trace by $\Delta = (h^2/D) \cos \xi \sin \xi$. To avoid reflection-point smearing, offset traces should be gathered at a point a distance $\Gamma = (-h^2/D) \sin \xi$ updip. However, such a gather is not hyperbolic but has the

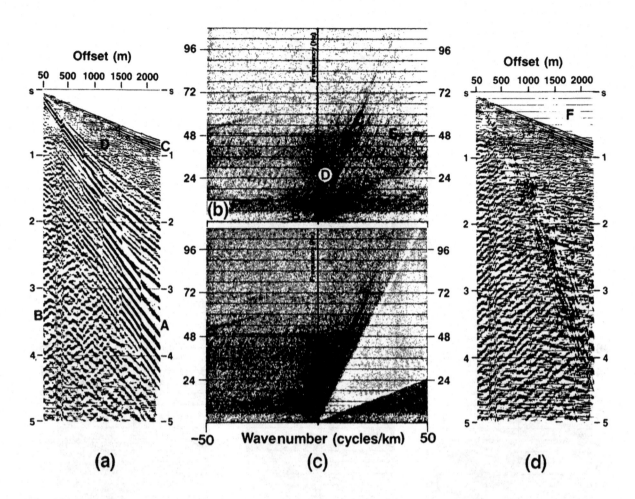

(a)   **(c)**   **(d)**

Fig. 9.38   Velocity filtering of a gather. (From Yilmaz, 1987: 71.) (a) Unfiltered gather; (b) f–k spectra of (a); (c) the velocity filter eliminating wedge from (b); and (d) the transform of (c) to the time domain showing how noises A and E have been eliminated but B and D retained.

shape of the DMO ellipse,

$$x^2 - \frac{Vt_0}{2\sin\xi}\,x - h^2 = 0. \qquad (9.100)$$

The DMO correction makes this gather hyperbolic.

Because DMO involves considerable computer time, Hale (1991: 2–9) gives an empirical rule that DMO correction is required whenever it exceeds one-half the dominant period. By using eqs. (4.11) and (4.22b), the rule is that DMO processing should be carried out whenever

$$(4s^2/V^2t_n)\sin^2\xi = (\Delta t_0/\Delta x)^2(s^2v_d/t_n) \geq 1, \qquad (9.101)$$

where t_0 is the zero-offset time, t_n is the NMO-corrected time, $2s$ is the offset, v_d is the dominant frequency, and ξ is the dip.

Corrections for DMO can be made in various ways, including prestack partial migration (Yilmaz and Claerbout, 1980), time-domain finite-difference methods or offset continuation (Bolondi, Loinger, and Rocca, 1982), Fourier-domain implementation (Hale, 1984), and integral (Kirchhoff) methods (Hosken and

Deregowski, 1985). DMO is usually applied after velocity-dependent NMO, but Gardner's DMO (Forel and Gardner, 1988) applies velocity-dependent DMO prior to velocity-dependent NMO. For further information, the reader is referred to Hale (1991: chaps. 3–4) or Bancroft (1991), who discuss several methods and give references to original sources.

For 3-D surveys, the 2-D ellipse in fig. 4.9b becomes an elliptical "bowl." Raypaths (assuming the velocity is constant) lie in a plane containing the source and receiver, and this plane intersects the bowl along an ellipse similar to that given by eq. (4.19). Thus, 3-D DMO is essentially the same as 2-D unless the azimuth changes. Under these circumstances, if $\Delta\phi$ is the azimuth angle, eq. (9.101) becomes

$$\left|\frac{dt_0}{dx}\frac{dt_0}{dy}\frac{s^2v_d}{t_0}\Delta\phi\right| > \frac{1}{2} \qquad (9.102)$$

A diffraction in location-offset space is called a Cheops pyramid (fig. 9.42b); it is not a hyperboloid. Application of NMO changes the Cheops pyramid into a saddle-shaped surface (fig. 9.42c); DMO makes it into a cylindrical hyperboloid (fig. 9.42d).

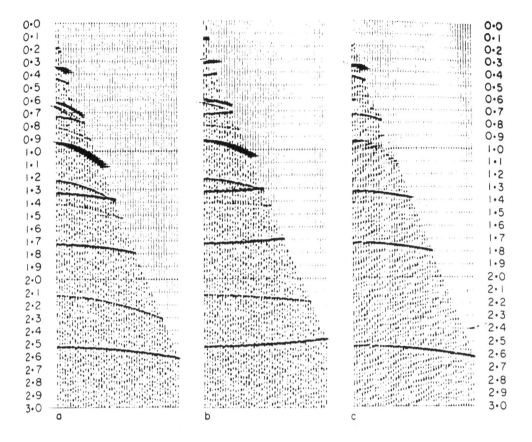

Fig. 9.39 Use of *f–k* filtering to attenuate multiples. (From Hatton et al., 1986: 98.) (a) Gather; (b) gather with approximate NMO applied; this gather is then *f–k* filtered; and (c) gather after filtering followed by removing the approximate NMO.

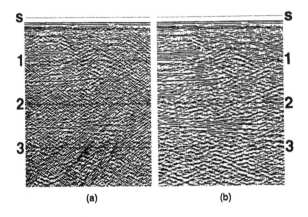

Fig. 9.40 Velocity-filtered stacked section. (From Yilmaz, 1987: 76.) (a) CMP stack contaminated by coherent noise, and (b) filtered after stacking.

9.10.3 Muting

First-breaks and the refraction wavetrains that follow them are usually so strong that they have to be excluded from the stack to avoid degrading the quality of shallow reflections (see fig. 9.43). This is done by *muting,* which involves arbitrarily assigning values of zero to traces during the mute interval. Also, the re-

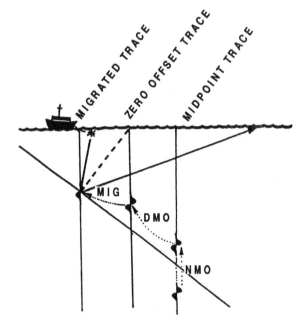

Fig. 9.41 NMO corrects for the time delay on an offset trace assuming zero dip; DMO moves the data to the correct zero-offset trace for a dipping reflection; migration further moves it to the subsurface location. (After Deregowski, 1986: 13.)

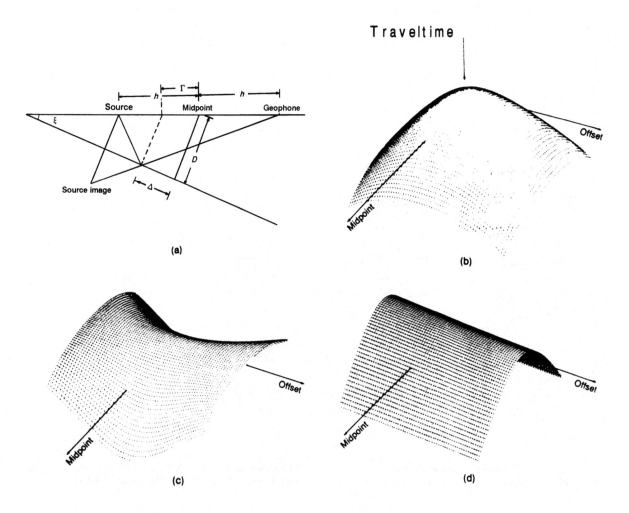

Traveltime

Fig. 9.42 DMO. (From Sheriff, 1991.) (a) Terms involved in reconstruction of the reflecting point assuming constant velocity, (b) a diffraction in location-offset space is not a hyperboloid, (c) NMO correction makes (b) into a saddle-shaped surface, and (d) DMO correction along with NMO yields a cylindrical hyperboloid.

flection waveshape on longer-offset traces is stretched because of rapid changes in the normal moveout (fig. 9.30) and directivity effects of geophone arrays. Stretching effectively changes the frequency spectrum of the wavelet, resulting in attenuation of higher frequencies in subsequent stacking. Therefore, long-offset traces usually are muted before the stretching reaches 25%. Figure 9.27 also shows the effect of muting.

A consequence of muting is that the multiplicity of a stack increases by steps, the shallowest data often being a twofold stack, slightly deeper data being a fourfold stack, and so on until the full multiplicity of the stack is achieved after the muted events have passed beyond the most distant geophones. To avoid amplitude discontinuities associated with changes in the multiplicity, the amplitude is usually divided by the number of nonzero traces that have been added.

Sometimes an *inner mute* (*tail mute*) is also applied, setting short-offset traces to zero as air waves or ground roll strikes the geophones. Traces near a shotpoint may become very noisy as time after the shot increases, perhaps because of *hole noise* (noise produced by oscillation and venting of gases generated by the shot and/or ejection of material from the borehole). Traces near surface sources may likewise become noisy as time increases.

Occasionally, a wedge of data across the gather (such as a portion dominated by ground roll) will also be muted (*surgical mute*), although it is more common to use apparent-velocity filtering (§9.9) in such situations.

9.10.4 Common-midpoint stacking

Combining a sequence of common-midpoint gathers after NMO correction yields a *common-midpoint stack*. Multiples spend more of their traveltimes in the shallower part of the earth than do primaries with the same traveltimes, and hence usually have smaller stacking velocities than the primaries and so do not align on the NMO-corrected gather. Thus, stacking severely attenuates most multiples. Common-

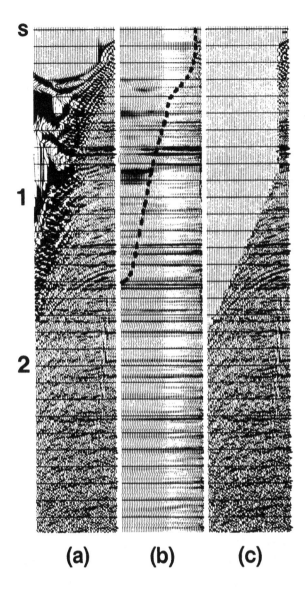

(a)   (b)   (c)

Fig. 9.43 Dependence of reflection quality on mute selection. (From Yilmaz 1987: 164.) (a) A CMP gather. (b) The stacked trace resulting from varying the mute; the right trace is the same as the inside trace of the gather; the next the result of stacking the two inside traces; the next stacking the three inside traces; and so on. The best mute includes as much data as possible without degrading reflection quality. (c) Muted gather.

midpoint gathers are sometimes apparent-velocity (f–k) filtered (§9.9) to remove coherent noise trains before stacking.

Common-midpoint stacking ordinarily assumes that all traces in the gather being stacked have equal validity and thus should be given equal weight. The output amplitude is divided by the number of live traces entering the stack, that is, adjustments are made for muted and occasional missing or dead traces.

A CMP stacked section is often regarded as a zero-offset section, especially when migrating the data.

Even where DMO has been applied to convert CMP traces to common-reflecting-point traces, CMP and zero-offset sections differ in important regards. Noises on the two types of sections are generally markedly different, especially multiple noise.

Amplitude-variation-with-offset (AVO) differences cause reflection events to have different amplitude relations to each other than in the zero-offset case, a point usually neglected in inversion, and the assumption of hyperbolic stacking may have also changed the amplitudes of different events in different ways (see Yilmaz, 1987: 244, 251).

9.10.5 Weighted stacking

In certain situations, unequal weighting (producing a *weighted stack*) of the traces in a gather may yield results that are better than the CMP stack. Offset-dependent weighting is sometimes used. The difference in NMO between primaries and multiples depends, for example, on the square of the offset distances so that better multiple attenuation may be achieved by weighting the long-offset traces more heavily than the short-offset traces (fig. 9.44). Most weighting is empirical, often varying linearly with offset, the weights usually varying from 0.5 to 1.5. More complicated weighting schemes are sometimes used. Where the relations between stacking velocity and time are known accurately for primaries and for one type of multiples, use of a stacking velocity different from either can maximize attenuation of these multiples compared with the primaries even though it does not maximize the primaries; this is the basis of "optimum wide-band horizontal stacking" (Schneider, Prince, and Giles, 1965). However, because various types of multiples have different stacking velocities, this type of stacking rarely produces optimum results.

One goal of CMP stacking is to produce the reflection amplitude appropriate for normal incidence. However, amplitudes vary with incident angle (§3.4), that is, with offset, and especially so where the interstitial fluid changes. One scheme (Denham, Palmeira, and Farrell, 1985) fits amplitude-offset measurements with a best-fit curve and then gives the stacked trace the zero-offset amplitude value. Such processing may be especially appropriate as a prelude to one-dimensional inversion, which assumes normal incidence.

Weighting is also sometimes done to enhance coherence, weights being based on a coherence measurement (§9.3.5) such as semblance. Enhancement of certain dips can be achieved in this way.

Several iterative or adaptive weighting schemes have been used (Naess and Bruland, 1985) for various types of noise problems. Estimates of the signal and noise amplitudes are usually required. Weighting (Naess, 1979) can be used to suppress abnormal amplitudes. Muting (§9.10.3) is a type of weighted stack where noisy traces are given weights of zero compared

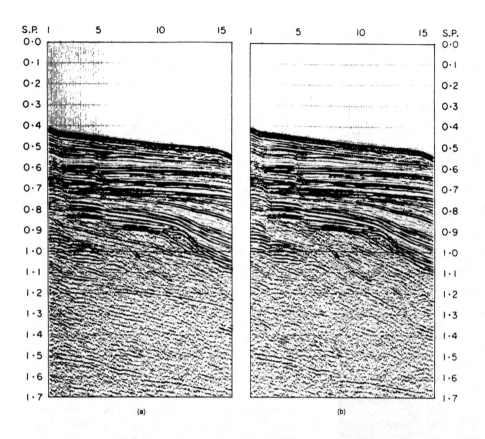

Fig. 9.44 Weighted stacking to attenuate multiples. (From Hatton et al., 1986: 97.) (a) Unweighted stack with strong multi-ples from the sea floor (e.g., ~ 0.85 s at left) and interbeds (e.g., ~ 1.2 s); (b) weighted stack.

with weights of one for unmuted traces. Simply elimi-nating noisy traces is another form of weighting. Sometimes noisy traces are replaced with estimates based on interpolation rather than being simply elimi-nated; this is equivalent to changing the weighting of the traces adjacent to the noisy trace. Diversity stack-ing (§9.10.6) is another form of discriminating against noise that occasionally affects acquisition in a nonsys-tematic manner.

9.10.6 Diversity stacking

Much data processing is far less exotic than is sug-gested by the mathematical relationships expressed in the foregoing pages. Some of these processes involve merely excluding certain elements of the data, such as the muting operation that has already been discussed. It is almost always better to throw away noisy data than to include it (often on the theory that its adverse effects will be averaged out). A very powerful pro-cessing technique, which is not used as much as it should be, is to simply look at the data and delete por-tions that appear to be mainly noise.

Diversity stacking is another technique used to achieve improvements by excluding noise. Records in high-noise areas, such as in cities, often show bursts of large-amplitude noise, whereas other portions of the records are relatively little distorted by noise. Under such circumstances, amplitude can be used as a dis-criminant to determine which portions are to be ex-cluded. This often takes the form of merely excluding data where the amplitude exceeds some threshold, or perhaps some form of inverse weighting might be used. Such noise bursts are often randomly located on repeated recordings so that sufficient vertical stacking after the weighting tends to produce records free from the high-amplitude noises.

9.10.7 Simplan stacking

Most sources are effectively points and hence seismic waves are spherical or nearly so. An alternative to CMP stacking of component spherical-wave records is to simulate sections that would have been generated by plane or cylindrical waves; such sections are called "Simplan" sections (Taner, 1976).

Simplan utilizes reciprocity (§4.3.4) and superposit-ion (§2.1.4). The sum of the outputs of a geophone for a number of in-line point sources simulates the output from a line source, that is, a cylindrical wave. Figure 9.45 shows a split-spread record and the Simplan trace that results from simple stacking without mak-ing any time shifts for normal-moveout correction. Only those traces of the gather that lie within the first

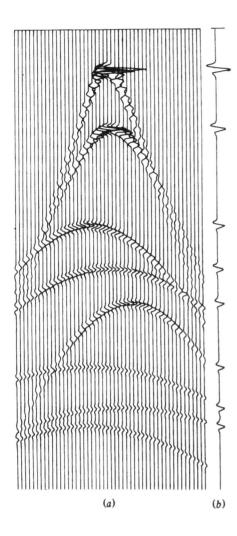

(a)   (b)

Fig. 9.45 Synthetic common-source gather and Simplan trace.
(Courtesy of Grant Geophysical.) (a) Gather showing reflections symmetrical about the trace $x = -2h \sin \xi$, where ξ is the
dip, and h the distance to the reflector, as in fig. 4.2. (b) The
Simplan trace that results from summing all the traces; in effect,
only the first Fresnel zone contributes.

Fresnel zone make an appreciable contribution to the
Simplan trace. Even moderate dip has little effect on
the size of the zone, so dip has little effect on the Simplan trace. The first Fresnel zone also includes more
traces as arrival time increases, so that the rate of amplitude decay on the Simplan trace is less than on
the traces of the gather (the Simplan trace undergoes
cylindrical divergence rather than the spherical divergence of the component traces). The traces from geophones closely spaced can be used in the same way
as the traces from sources closely spaced. Customary
group spacing and range of offset distances are usually sufficient to avoid undesirable end effects.

Split-spread and Simplan records can be simulated
from end-on records. Note (fig. 9.46a) that the trace
at (r_{k+i}, s_k) on the surface diagram is the same as the
trace at (r_k, s_{k+i}) by reciprocity. Thus, end-on records
can be used to produce a split-spread record for twice

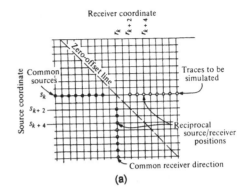

(a)

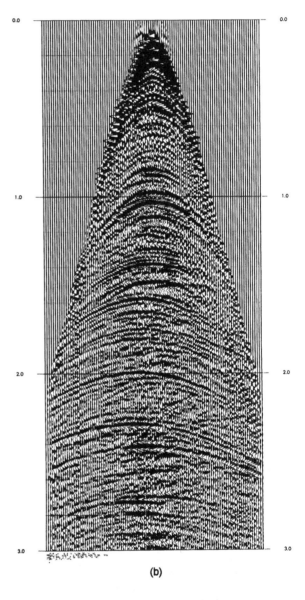

(b)

Fig. 9.46 Simulating split-spread record from end-on records.
(a) Reciprocal relations between traces on a surface stacking
chart; traces on one side of the zero-offset line have identical
raypaths to traces symmetrically disposed on the opposite side
of the line; (b) 96-trace split-spread record simulated from 48-
trace end-on records (courtesy of Grant Geophysical).

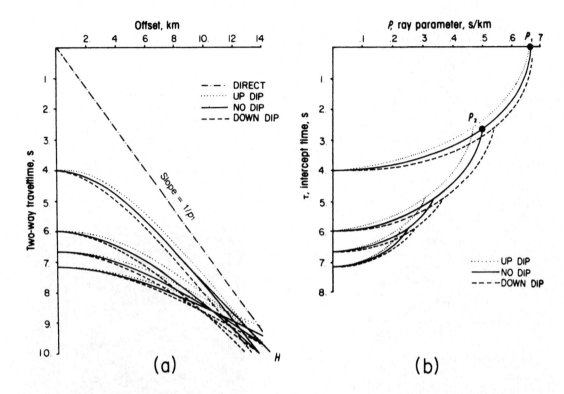

Fig. 9.47 τ–p mapping. Reflection hyperbolas in time domain map into ellipses in the τ–p domain and straight lines (direct wave and head waves) into points. (From Sheriff, 1991.) (a) An end-on seismic record $f(x, t)$, where x is the source–receiver distance and t is the arrival time; solid lines indicate no dip, dotted and dashed lines indicate up-dip and down-dip directions, respectively. (b) τ–p domain showing points P_1 for the direct arrival and P_2 for the head wave H. The dotted and dashed lines show changes if the profile is in up-dip or down-dip direction.

the number of channels, using the common-source and common-receiver traces, respectively, for the two halves of the split. Figure 9.46b shows a 96-trace split-spread record simulated from 48-trace end-on records. The stack of these 96 traces yields one Simplan trace.

Simplan sections contain all primaries, multiples, and diffractions without amplitude bias or waveform distortion, whereas CMP stacking emphasizes primary reflections compared to multiples and diffractions.

9.11 Other processing techniques

9.11.1 τ–p transform processing (slant stacking)

The τ–p transform or *slant stack* is a form of Radon transform (see §9.15 and eq. (9.22)). When applied to seismic records, the slant stack maps the amplitude $g(t, x)$ from the t–x domain to the τ–p domain (fig. 9.47), the integral in eq. (9.22) becoming a summation. Both reflection and refraction data can be slant stacked. The inverse transformation can be carried out by filtered backprojection, as in §13.5.2 (see eq. (13.12)).

As in the case of other transforms, the slant stack is used because certain operations can be carried out more easily and efficiently in the τ–p domain than in

the t–x domain. Several of these applications are listed in Yilmaz (1987: 429). Stoffa et al. (1981) applied slant stacking to obtain semblance (§9.3.5) and eliminate spatial aliasing (§9.2.2d). Clayton and McMechan (1981) applied the technique to refraction data to produce velocity–depth models. Gardner and Lu (1991) have collected together papers dealing with slant stacking.

9.11.2 Intelligent interpolation

Intelligent interpolation is an interpolation process that mimics the interpreter's ability to jump correlate using seismic character. It is often based on cross-correlation, sometimes on recognition of trace attributes (§9.11.4). It is used to interpolate between data spaced four or five times farther apart than spatial alias considerations (§9.2.2d) permit if aliasing during migration is to be avoided. However, intelligent interpolation does not alter the resolution of the resulting data, which is determined by the original sampling rather than that after interpolation. Intelligent interpolation is also used to permit cheaper 3-D acquisition (§12.1.2 and 12.1.3) to compensate for relaxed line-spacing requirements and to fill in undersampled grid loops of 2-D coverage to create pseudo-3-D surveys.

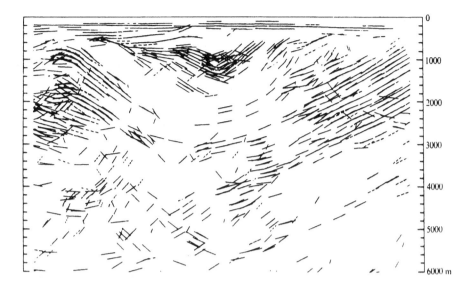

Fig. 9.48   Automatically picked migration section. (From Paturet, 1971.)

9.11.3 Automatic picking

Conceptually, events can be picked and graded automatically using coherence measures as criteria (Paulson and Merdler, 1968; Bois and la Porte, 1970; Garotta, 1971). Whenever coherence exceeds a threshold value, an event can be picked, the arrival time, NMO, and dip moveout being determined corresponding to the maximum coherence. Grades can be assigned based on coherence values, the distance over which coherence can be maintained being included as a factor. The picks can be automatically migrated and plotted, as shown in fig. 9.48. Automatic picking can be expanded to include intersecting lines. The picks can be posted on a map and contoured automatically. Thus, conceptually, the output of processing could be contoured depth maps of reflecting horizons, and much of the work usually thought of as interpretation could be automated. However, in the process, many decisions have to be made. Criteria have to be specified for determining which events are primary reflections and which multiples are for deciding what to do when events interfere or terminate, and so on; the process breaks down or produces meaningless results if each of these decisions has not been anticipated and specified correctly in advance.

Although automatic picking was never used very much with 2-D data, its equivalent, horizon tracking (see §12.4), is extensively used with 3-D data. Improvements in data quality and the areal density of sampling are largely responsible for this success. However, horizon tracking still has to be monitored carefully to produce reliable results.

9.11.4 Complex-trace analysis

Let us assume a seismic trace of the form

$$g(t) = A(t) \cos 2\pi v t, \qquad (9.103)$$

where $A(t)$ varies slowly with respect to $\cos 2\pi v t$; $A(t)$ is the *envelope* of $g(t)$, often called the *envelope amplitude*. For $A(t)$ constant, the Hilbert transform (§15.2.13) of $g(t)$ is given by

$$g(t) \leftrightarrow g_\perp(t) = -A(t) \sin 2\pi v t \qquad (9.104)$$

(see problem 15.23a). Thus, we can form a complex signal, $h(t)$, where

$$h(t) = g(t) + jg_\perp(t) = A(t)e^{-j2\pi v t}, \qquad (9.105)$$

$h(t)$ being known as the *analytical* or *complex trace* (Bracewell, 1965), $g_\perp(t)$ as the *quadrature trace* of $g(t)$ (see fig. 9.49). If v is not constant but varies slowly, we define the *instantaneous frequency,* $v_i(t)$, as the time derivative of the *instantaneous phase,* $\gamma(t)$; thus,

$$2\pi v_i(t) = \frac{d\gamma(t)}{dt} = \frac{d}{dt}(2\pi v t) \qquad (9.106)$$

The quantities $A(t)$, $\gamma(t)$, $v_i(t)$, and other measurements derived from the seismic data are called *attributes.*

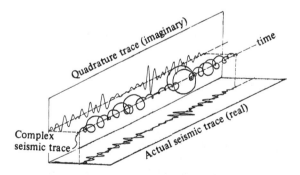

Fig. 9.49   The complex trace shown as a helix of variable amplitude in the direction of the time axis. Projection onto the real plane gives the actual seismic trace and onto the imaginary plane the quadrature trace.

To find $A(t)$, $\gamma(t)$, and $v_i(t)$, we obtain $h(t)$, either by eq. (15.176), that is,

$$g_\perp(t) = g(t) * (1/\pi t) = (1/\pi) \sum_{n=-\infty}^{\infty} g_{t-n}(e^{jn\pi} - 1)/n \tag{9.107}$$

for digital functions (see problem 15.23c), or by using eq. (15.177), that is, we calculate the transform of $g(t)$, set the result equal to zero for negative frequencies, multiply by 2, and then inverse transform to get $h(t)$. Because $A(t)$ is real and $|e^{j2\pi vt}| = 1$, we see that

$$\left.\begin{aligned} A(t) &= |h(t)|, \\ \gamma(t) &= 2\pi v t = \tan^{-1}[g_\perp(t)/g(t)], \\ v_i(t) &= \frac{1}{2\pi}\frac{d}{dt}[\gamma(t)]. \end{aligned}\right\} \tag{9.108}$$

Complex-trace analysis can be used in convolution, correlation, semblance, and other types of calculations (Taner, Koehler, and Sheriff, 1979), sometimes facilitating the calculations.

Attributes sometimes reveal features that are not as obvious otherwise, especially lateral changes along the bedding, such as those associated with stratigraphic changes or hydrocarbon accumulations (§10.7 and 10.8); see Taner and Sheriff (1977). Phase plots facilitate picking weak coherent events, and lateral discontinuities in phase facilitate picking reflection terminations as at faults, pinchouts, and so on. Instantaneous frequency patterns tend to characterize the interference patterns resulting from closely spaced reflectors and thus aid in correlating from line to line or across faults.

9.12 Processes to reposition data

9.12.1 Introduction

Seismic data prior to migration are oriented with respect to the observation points. *Migration* involves repositioning data elements to make their locations appropriate to the locations of the associated reflectors or diffracting points. The need to migrate seismic data to obtain a structural picture was recognized at the beginning of seismic exploration and the very first seismic reflection data in 1921 were migrated (fig. 1.3b).

Consider the constant-velocity situation shown in fig. 9.50 A reflection from a reflector with dip ξ at point C underneath E is observed at A and is plotted at C' on an unmigrated section. Clearly,

$$\tan \xi_a = \sin \xi, \tag{9.109}$$

where ξ_a is the apparent dip on the unmigrated section. The reflector lies updip from its apparent location, $\xi > \xi_a$, and a segment of reflection $C'D'$ is shortened to CD by migration. Equation (9.109) is called the *migrator's equation.*

Migration ordinarily assumes a coincident source-receiver section and is generally carried out after stacking. This usually gives good results where dips

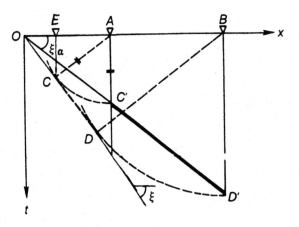

Fig. 9.50 Migration principle. Migration of segment $C'D'$ into CD increases the dip from ξ_a to ξ.

are small and where events with different dips do not interfere on the unmigrated section. Migration before stacking almost always gives better or at least equivalent results, but is expensive because then many more data have to be migrated. DMO removes much of the need for prestack migration, so that today prestack migration is mainly associated with depth migration in areas where the velocity distribution is complex.

Although the objective of migration is to obtain a picture of reflectors at their correct locations in depth, the velocity required for time-to-depth conversion is usually not known accurately and the result of migration is usually a migrated time section, which is a vertically stretched version of the depth domain provided velocity varies in the vertical direction only. The process "depth migration" (§9.12.5) attempts to account for changes in velocity in the horizontal direction as well. Another limitation on migration is the *migration aperture,* the range of data included in the migration of each point; the aperture is often less than ideal because of the volume of data to be processed.

Migration generally is based on the premise that all data elements represent either primary reflections or diffractions. The migration of noise, including energy that does not travel along simple reflection paths, produces meaningless results. Migration requires a knowledge of the velocity distribution; changes in velocity bend raypaths and thus affect migration. Although migration can be extended to three dimensions with ordinary 2-D seismic lines, we usually assume that the cross-dip is zero, which results in two-dimensional migration. Ignoring cross-dip sometimes results in undermigration, but an undermigrated section is at least easier to interpret than one not migrated at all. Moreover, cross-dip information is often not available, two-dimensional migration is appreciably more economical, and the results are often adequate.

The simplest approach to migration is to determine the direction of approach of energy and track the raypath backwards to the reflecting point at half the traveltime, or to find the common tangent to wave-fronts for half the traveltime; these methods were extensively used in hand-migrating data. Computer methods generally involve solutions of the scalar wave equation, eq. (2.28). We replace the time with half the traveltime, that is, in effect we start with the energy originating at each reflector, as if each reflector were covered by elementary point sources as postulated by Huygens' principle, all actuated at the instant $t = 0$ (the "exploding-reflector" model). We regard $\psi(x, z, c)$ as a vertical section showing the wave motion at the point (x, z) at time $t = c$, that is, an unmigrated seismic section corresponds to $\psi(x, 0, t)$, whereas a migrated seismic section corresponds to $\psi(x, z, 0)$. There are various ways of solving for $\psi(x, z, 0)$, including (a) integral methods based on Kirchhoff's equation (§9.12.2), where the integration is over those elements in unmigrated space that contribute to an element in migrated space, (b) methods based on a solution in the frequency–wavenumber domain (§9.12.3), and (c) finite-difference solutions in the time domain (§9.12.4), which accomplishes backward-tracing of seismic waves in a downward-continuation manner.

The methods discussed in the next sections accomplish full-waveform migration; they involve large numbers of calculations and so are restricted to computer implementation.

9.12.2 Kirchhoff (diffraction-stack) migration

Diffraction-stack migration is based on a concept of Hagedoorn (1954). We assume constant velocity V and convert arrival times to distances by multiplying by $\frac{1}{2}V$. Figure 9.51a relates a diffraction PMR and a reflection MN seen on an unmigrated section. A reflector PQ with dip ξ passes through P at a depth z_0, S_0P is perpendicular to the reflector. Arcs are swung with centers S_0, S_1, S_e, and so on and with radii equal to the distances to the reflector. Hagedoorn called the unmigrated diffraction curve PMR a *curve of maximum convexity,* because no other event from the depth z_0 can have greater curvature (see fig. 9.51a). The diffraction curve is a hyperbola with apex at P and the unmigrated reflection is tangent to it at M (see problem 9.27).

The concept for carrying out migration as a manual operation is to plot a diffraction curve for each depth and slide it along the unmigrated section (keeping the top lined up with zero depth) until a segment of a reflection is tangent to one of the curves; on the corresponding migrated section, the reflector is located at the crest of the diffraction curve tangent to the wavefront that passes through the point of tangency of the reflection to the diffraction curve (fig. 9.51b). The principle is the same if the velocity is not constant and if the sections, wavefronts, and diffraction curves are plotted in time rather than in depth.

To carry out diffraction-stack migration, diffraction curves are calculated for each point on the section.

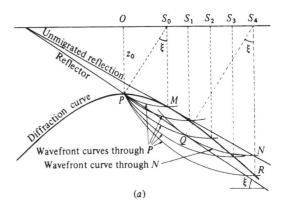

(a)

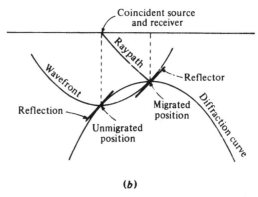

(b)

Fig. 9.51 Wavefront and diffraction curves intersecting at the unmigrated and migrated positions. (a) Unmigrated reflection MN migrates into reflector PQ; (b) relation between wavefront and diffraction curves (from Hagedoorn, 1954).

The data on the unmigrated section lying along each diffraction curve are summed to give the amplitude at the respective point on the migrated section. If there is indeed energy involving the point at the crest of the diffraction curve, then the addition will produce the value appropriate to the energy involving that point; if only noise is present, positive and negative values will be equally probable along the diffraction curve, so the sum will be very small.

In effect, diffraction-stack migration treats each element of an unmigrated reflection as a portion of a diffraction, that is, a reflector is thought of as a sequence of closely spaced diffracting points (fig. 9.52). The relationship between points shown in fig. 9.51b suggests that the data at each point could be distributed along the wavefront through that point (wavefront smearing), and when the distributed data for all points are superimposed, they will reinforce where reflectors exist but otherwise positive and negative values will be equally probable so the sum will be small. Bursts of noise will not have neighboring elements to cancel their effects and hence will be smeared out along wavefronts on a migrated section to become "smiles" (fig. 9.53).

Migration by the method of wavefront smearing produces results identical to diffraction-stack migra-

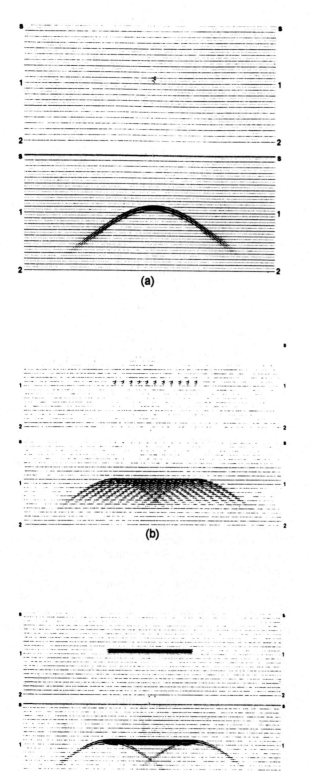

Fig. 9.52 Migrating reflections as diffractions. (From Yilmaz, 1987: 257, 258.) (a) A diffraction (below) migrates into a point (above). (b) With a sequence of diffracting points the diffractions tend to merge to form the reflector. (c) If closely enough spaced, only the reflector and diffractions at its ends are evident.

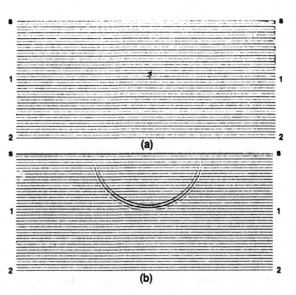

Fig. 9.53 A burst of noise on an unmigrated section (a) migrates into a wavefront (smile) (b). (From Yilmaz, 1987: 258.)

tion, the only difference being in that operations are performed in a different sequence. The "common-tangent" method of migration (Sheriff, 1978) is in effect wavefront smearing.

A more elegant formulation of diffraction-stack migration is based on the Kirchhoff integral (see Schneider, 1978). This approach makes it clear that this technique is an integral solution to the wave equation, as opposed to a finite-difference solution or a Fourier-transform solution of the wave equation (usually called "frequency-domain migration").

Amplitudes are adjusted for obliquity and divergence before summing along the diffraction curves in Kirchhoff migration. The former factor gives the cosine of the angle between the direction of travel and the vertical, and the divergence factor corrects for $1/r^2$ for 2-D migration or $1/r$ for 3-D migration. In addition, a wavelet-shaping factor corrects amplitudes by the inverse square root of the frequency and phase by 45° for 2-D migration, or by the inverse of the frequency and 90° for 3-D migration. The reasons for the wavelet-shaping factor are explained in Schneider (1978) and Berryhill (1979). If we consider collapsing diffraction hyperbolas as wave propagation in spherical coordinates, the near-field terms are generally neglected.

How far down a diffraction hyperbola integration (summation) should extend is the aperture-definition problem. In general, the collapse of diffraction hyperbolas to points is inversely related to the aperture width. Inadequate aperture widths effectively discriminate against steep dips and aperture width can be used as a dip filter. A general rule is that the aperture should exceed twice the horizontal distance of migration of the steepest dips. Clearly, aperture width should increase with depth because diffractions flatten with depth. Use of a wide aperture also has a detrimental effect in tending to organize horizontal noise where reflections are weak.

9.12.3 Migration in the frequency–wavenumber domain

Equation (9.109) provides the basis for frequency-domain migration. If velocity is constant, lines in x, t space that have the same slope (same apparent velocity or same apparent wavenumber) transform by the 2-D Fourier transform into a single line in κ_x, ω space (fig. 9.54); the separate parallel lines in x, t space are distinguished by different phases in κ_x, ω space. Frequency-domain migration changes the slope of lines in κ_x, ω space according to eq. (9.109); the inverse transform then gives a migrated section in x, t space. Thus, conceptually, frequency-domain migration becomes a very simple operation (Robinson, 1983). (Similar migration, called *slant-stack migration*, can be done in τ, p space using the Radon transform; see Hubral, 1980.) The problems with frequency-domain (and slant-stack) migration come about because of the assumption that the velocity is constant (see what follows).

Stolt (1978) introduced the Fourier-transform migration method, sometimes called *Stolt migration.* This method starts with eq. (2.30) in two dimensions, the x-axis being along the profile direction and the z-axis positive vertically downward. Thus,

$$\frac{\partial^2 \psi}{\partial t^2} = V^2 \left(\frac{\partial^2 \psi}{\partial x^2} + \frac{\partial^2 \psi}{\partial z^2} \right). \qquad (9.110)$$

We use eq. (15.11b) to take the three-dimensional transform of $\psi(x, z, t)$ and obtain

$$\psi(x, z, t) \leftrightarrow \Psi(\kappa_x, \kappa_z, \omega) = \int_{-\infty}^{\infty} \int_{-\infty}^{\infty} \int_{-\infty}^{\infty} \psi(x, z, t)$$
$$\exp\left[-j(\kappa_x x + \kappa_z z + \omega t)\right] dx\, dz\, dt.$$

Equation (15.141) now gives

$$\frac{\partial^2 \psi}{\partial t^2} \leftrightarrow (j\omega)^2 \Psi(\kappa_x, \kappa_z, \omega),$$

$$\frac{\partial^2 \psi}{\partial x^2} \leftrightarrow (jk_x)^2 \Psi(\kappa_x, \kappa_z, \omega),$$

$$\frac{\partial^2 \psi}{\partial z^2} \leftrightarrow (jk_z)^2 \Psi(\kappa_x, \kappa_z, \omega).$$

Substituting in eq. (9.110), we get

$$\omega^2 - V^2(\kappa_x^2 + \kappa_z^2) = 0. \qquad (9.111)$$

Returning to eq. (9.110), we take the two-dimensional transform of $\psi(x, z, t)$ with respect to x and z, obtaining

$$\Psi_{xz}(\kappa_x, \kappa_z, t) = \int_{-\infty}^{\infty} \int_{-\infty}^{\infty} \psi(x, z, t)$$
$$\exp\left[-j(\kappa_x x + \kappa_z z)\right] dx\, dz. \qquad (9.112)$$

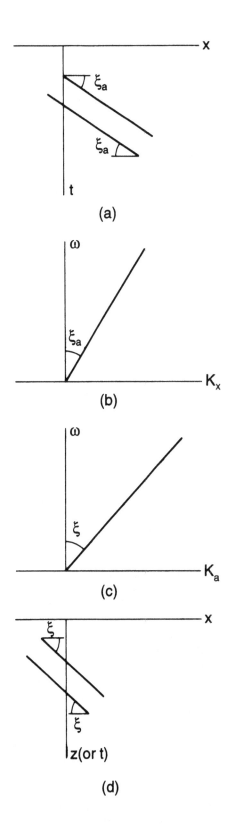

Fig. 9.54 Frequency-domain migration where velocity is constant. (a) Two dipping events on an unmigrated section; (b) these events map into the same line in κ_x–ω domain but with different phases; (c) changing the line slope from ξ_a to ξ according to eq. (1.109); (d) transforming back to x, t, or x, z domains produces a migrated section.

If we restrict the solution to harmonic waves, we can write

$$\Psi_{xz}(\kappa_x, \kappa_z, t) = \Psi_{xz}(\kappa_x, \kappa_z, 0)e^{-j\omega t} \quad (9.113)$$

(to verify this relation, substitute $t = 0$); the first factor on the right is the required solution to our problem.

To find $\Psi_{xz}(\kappa_x, \kappa_z, 0)$, we start by calculating the transform of the recorded data with respect to x and t:

$$\Psi_{xt}(\kappa_x, 0, \omega) = \int_{-\infty}^{\infty} \int_{-\infty}^{\infty} \psi(x, 0, t)$$
$$\exp[-j(\kappa_x x + \omega t)] \, dx \, dt; \quad (9.114)$$

inverting the transform, we have

$$\psi(x, 0, t) = (1/2\pi)^2 \int_{-\infty}^{\infty} \int_{-\infty}^{\infty} \Psi_{xt}(\kappa_x, 0, \omega)$$
$$\exp[j(\kappa_x x + \omega t)] \, d\kappa_x \, d\omega. \quad (9.115)$$

From eqs. (9.112) and (9.113), we obtain

$$\psi(x, z, t) = (1/2\pi)^2 \int_{-\infty}^{\infty} \int_{-\infty}^{\infty} \Psi_{xz}(\kappa_x, \kappa_z, 0) \, e^{-j\omega t}$$
$$\exp[j(\kappa_x x + \kappa_z z)] \, d\kappa_x \, d\kappa_z;$$

hence,

$$\psi(x, 0, t) = (1/2\pi)^2 \int_{-\infty}^{\infty} \int_{-\infty}^{\infty} \Psi_{xz}(\kappa_x, \kappa_z, 0)$$
$$\exp[j(\kappa_x x + \omega t)] \, d\kappa_x \, d\kappa_z. \quad (9.116)$$

Comparing eqs. (9.115) and (9.116), we see that

$$\Psi_{xt}(\kappa_x, 0, \omega) \, d\omega = \Psi_{xz}(\kappa_x, \kappa_z, 0) \, d\kappa_z,$$

or

$$\Psi_{xz}(\kappa_x, \kappa_z, 0) = \Psi_{xt}(\kappa_x, 0, \omega)\frac{\partial\omega}{\partial\kappa_z}$$
$$= V\Psi_{xt}(\kappa_x, 0, \omega) \, [1 + (\kappa_x/\kappa_z)^2]^{-1/2}$$

from eq. (9.111), ω being equal to $V(\kappa_x^2 + \kappa_z^2)^{1/2}$. Because $\Psi_{xt}(\kappa_z, 0, \omega)$ is known from eq. (9.114), we can calculate $\Psi_{xz}(\kappa_x, \kappa_z, 0)$, then invert it to get the solution, $\psi(x, z, 0)$.

This migration is exact for constant velocity, but practical migration must allow for change of velocity, which is usually a function of the depth z. To get a section with effectively constant velocity, we stretch the time scale, even though the velocity as a function of time differs for reflections with different amounts of dip. Where velocity variations with depth involve serious problems, migration can be carried out by transforming with respect to t only and then downward continuation can be carried out in the space–frequency (x–ω) domain. One could also transform into the wavenumber–time κ_x–t domain. Although the velocity field of an unmigrated section may be multivalued, stretching is sufficiently tolerant that results are often satisfactory. The stretch factor also changes the effective aperture. Lateral variation of velocity can sometimes be handled similarly by some sort of stretching. Gazdag and Squazzero (1984) handle lateral velocity variation by migrating with a number of laterally constant velocities and then interpolating to get the migrated wavefield.

Phase-shifting migration (Gazdag, 1978), also called *Gazdag migration,* carries out migration in the frequency domain by using a pure phase-shifting operator $e^{-j\kappa_z z}$ at each z-step in downward continuation (§9.12.4). The parameter of the maximum dip to be migrated in Gazdag migration can be used to discriminate against dipping coherent noise, but choosing too small a maximum can unintentionally filter out real dips and result in smearing the data. Use of too-small migration steps tends to produce discontinuities at the step boundaries. Typically, the step size is something less than the dominant period, often 20 to 30 ms. Gazdag migration is generally more sensitive to velocity error than finite-difference migration (§9.12.4).

9.12.4 Finite-difference method of wave-equation migration

The concept underlying time-domain migration is downward continuation of the seismic wavefield. Continuation is a familiar process with gravity and magnetic fields (see Telford, Geldart, and Sheriff, 1990: 34, 108). It utilizes the continuity property of fields, one expression of which is that we can determine the field over any arbitrary surface if we know the field completely over one surface, provided the field satisfies Laplace's equation, $\nabla^2\phi = 0$. Evidences of subsurface features spread out as distance from the features increases, or, conversely, the evidences converge on the location of the features as they are approached. If we let $t' = jVt$ and assume harmonic waves, we can write the scalar wave equation (eq. (9.110)) as

$$\frac{\partial^2\psi}{\partial x^2} + \frac{\partial^2\psi}{\partial z^2} - \frac{1}{V^2}\frac{\partial^2\psi}{\partial t^2} = 0 = \frac{\partial^2\psi}{\partial x^2} + \frac{\partial^2\psi}{\partial z^2} + \frac{\partial^2\psi}{\partial t'^2}$$
$$= \nabla^2\psi(x, z, t') = 0, \quad (9.117)$$

thus expressing the wave equation in the form of Laplace's equation. We know the wavefield at the surface of the earth, $z = 0$, so our problem is to continue this field downward to determine what geophones would see if they were buried at arbitrary depths. We downward-continue in a series of steps, effectively lowering the geophones gradually downward through the earth. At each geophone depth, we expect to get a clear picture of reflectors that lie immediately below the geophones, so we retain this portion of the downward-continued record section from each continuation step and combine the upper portion of these to give our complete migrated section. Figure 9.55 illustrates the downward continuation of a diffraction.

We let t be the one-way traveltime (half the arrival time for coincident source–detector data). A plane wave approaching the surface at the angle θ is given by

$$\psi(x, z, t) = A \exp\{j\omega[t - (x/V)\sin\theta$$
$$- (z/V)\cos\theta]\}. \quad (9.118)$$

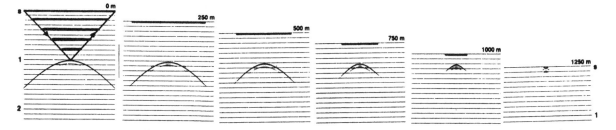

Fig. 9.55 Simulation of downward continuation showing a diffraction hyperbola collapsing to a point. Note that the effective aperture (heavy line) shrinks in the process. (From Yilmaz, 1987: 261.)

If we restrict ourselves to small angles of θ, we can approximate $\sin \theta \approx \theta$ and $\cos \theta \approx 1 - \frac{1}{2}\theta^2$, so that eq. (9.118) becomes

$$\psi(x, z, t) = A \exp [j\omega(t - x\theta/V - z/V + z\theta^2/2V)]. \tag{9.119}$$

We now define a new time scale, $t^* = t - z/V$. This change means that our coordinate system effectively rides along on an upcoming wavefront (see problem 9.28). We now have

$$\psi^*(x, z, t^*) = A \exp [j\omega(t^* - x\theta/V + z\theta^2/2V)]. \tag{9.120}$$

and

$$\frac{\partial \psi}{\partial t} = \frac{\partial \psi^*}{\partial t^*} \frac{\partial t^*}{\partial t} = \frac{\partial \psi^*}{\partial t^*}; \quad \frac{\partial^2 \psi}{\partial t^2} = \frac{\partial^2 \psi^*}{\partial t^{*2}}$$

$$\frac{\partial \psi}{\partial x} = \frac{\partial \psi^*}{\partial x}; \qquad \frac{\partial^2 \psi}{\partial x^2} = \frac{\partial^2 \psi^*}{\partial x^2}$$

$$\frac{\partial \psi}{\partial z} = \frac{\partial \psi^*}{\partial z} + \frac{\partial \psi^*}{\partial t^*} \frac{\partial t^*}{\partial z} = \frac{\partial \psi^*}{\partial z} - \frac{1}{V} \frac{\partial \psi^*}{\partial t^*}$$

$$\frac{\partial^2 \psi}{\partial z^2} = \frac{\partial^2 \psi^*}{\partial z^2} - \frac{2}{V} \frac{\partial^2 \psi^*}{\partial z \partial t^*} + \frac{1}{V^2} \frac{\partial^2 \psi^*}{\partial t^{*2}}$$

Substituting into the wave equation (9.110) gives a new wave equation,

$$\frac{\partial^2 \psi^*}{\partial x^2} + \frac{\partial^2 \psi^*}{\partial z^2} - \left(\frac{2}{V}\right) \frac{\partial^2 \psi^*}{\partial z \partial t^*} = 0 \tag{9.121}$$

For waves that are traveling nearly vertically, the change in ψ^* with respect to z is small in our moving coordinate system, so we neglect $\partial^2 \psi^*/\partial z^2$. This is called the "15° approximation," and neglecting this term means that we shall not be able to migrate steep dips well. We thus get

$$\frac{\partial^2 \psi^*}{\partial x^2} - \left(\frac{2}{V}\right) \frac{\partial^2 \psi^*}{\partial z \partial t^*} = 0. \tag{9.122}$$

(A "45° approximation" gives the equation

$$\frac{\partial^3 \psi^*}{\partial x^2 \partial t^*} - \frac{1}{2} V \frac{\partial^3 \psi^*}{\partial x^2 \partial z} - \left(\frac{2}{V}\right) \frac{\partial^3 \psi^*}{\partial x \partial t^{*2}} = 0; \tag{9.123}$$

see Claerbout, 1976: 198.)

If we think of ψ^* as a three-dimensional array (fig. 9.56a), that is, having values at discrete intervals of

Δx, Δz, and Δt^*, the plane $z = 0$ represents the unmigrated time section (which is our starting point) and the diagonal plane $t = t^* - z/V$ represents the migrated time section. (The projection of this diagonal plane onto the $t^* = 0$ plane would also give a migrated depth section.) We approximate derivatives by finite differences:

$$\frac{\partial^2 \psi^*}{\partial x^2}$$
$$\approx \frac{\psi^*(x, z, t^*) - 2\psi^*(x - \Delta x, z, t^*) + \psi^*(x - 2\Delta x, z, t^*)}{(\Delta x)^2}$$

$$\frac{\partial^2 \psi^*}{\partial z \partial t^*} \approx \{\psi^*(x, z, t^*) - \psi^*(x, z - \Delta z, t^*)$$
$$- \psi^*(x, z, t^* - \Delta t^*)$$
$$+ \psi^*(x, z - \Delta z, t^* - \Delta t^*)\}/\Delta z \, \Delta t^*.$$

Equation (9.122) now becomes

$$\psi^*(x, z, t^*) = \frac{2\Delta z \, \Delta t^* (\Delta x)^2}{2(\Delta x)^2 - V \Delta z \, \Delta t^*} \left[\frac{\psi^*(x, z - \Delta z, t^*)}{\Delta z \, \Delta t^*}\right.$$
$$+ \frac{\psi^*(x, z, t^* - \Delta t^*)}{\Delta z \, \Delta t^*}$$
$$- \frac{V\psi^*(x - \Delta x, z, t^*)}{(\Delta x)^2}$$
$$- \frac{\psi^*(x, z - \Delta z, t^* - \Delta t^*)}{\Delta z \, \Delta t^*}$$
$$\left. + \frac{V\psi^*(x - 2\Delta x, z, t^*)}{2(\Delta x)^2}\right]. \tag{9.124}$$

This is a relation among six elements of the array, as shown in fig. 9.56b:

$$\psi^*(x, z, t^*) = a_1 \psi^*(x, z - \Delta z, t^*)$$
$$+ a_2 \psi^*(x, z, t^* - \Delta t^*)$$
$$+ a_3 \psi^*(x, z - \Delta z, t^* - \Delta t^*)$$
$$+ V[a_4 \psi^*(x - \Delta x, z, t^*)$$
$$+ a_5 \psi^*(x - 2\Delta x, z, t^*)].$$

This relation can be used to extend the three-dimensional array in the $+x$-, $+z$-, $+t^*$-directions. Our problem is getting enough of the array to begin with so that we can extend it. For negative z, $\psi^* = 0$ because this refers to wave values in the air above the surface of the ground. If our data begin at $x = 0$, we need values of ψ^* for $x = -\Delta x$ and $x = -2\Delta x$, which

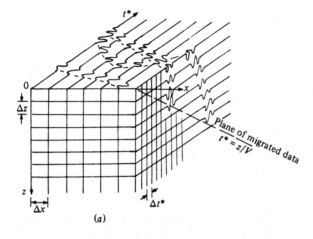

(a)

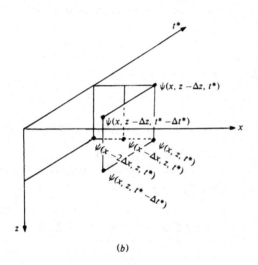

(b)

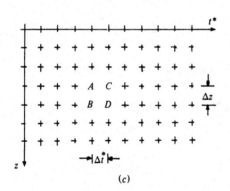

(c)

Fig. 9.56   Relationship of elements in (x, z, t^*)-space. (a) Seismic traces on the top surface, $z = 0$, show the unmigrated section; those at successive layers show what would be recorded by geophones buried at depth z. (b) Elements entering into the time-domain calculation of $\psi^*(x, z, t^*)$. (c) Table of values of $\psi^*(z, t^*)$.

are not available to us; we guess these starting values, and it turns out that the solution is not affected very much if these are in error, so that a stable solution can be achieved.

Various alternatives to the foregoing method of approximating the derivatives are available, some of which lead to highly stable algorithms and permit calculations using a fairly coarse grid, which of course makes the calculation much more economical. One approach is to take the Fourier transform with respect to x:

$$\psi^*(x, z, t^*) \leftrightarrow \Psi_x^*(\kappa_x, z, t^*)$$

so that eq. (9.122) becomes

$$\kappa_x^2 \Psi_x^* - \left(\frac{2}{V}\right)\frac{\partial^2 \Psi_x^*}{\partial z \, \partial t^*} = 0. \qquad (9.125)$$

We assume that we have a table of values of Ψ_x^* for discrete values of z, t^*, as illustrated in fig. 9.56c, and we consider the portion of the table centered between the values A, B, C, and D. The approximate value of Ψ_x^* at this point is $\frac{1}{4}(A + B + C + D)$. We can approximate $\partial^2 \Psi_x^*/\partial z \partial t^*$ by the expression

$$\left(\frac{D - C}{\Delta z} - \frac{B - A}{\Delta z}\right)\frac{1}{\Delta t^*} = \frac{A - B - C + D}{\Delta z \, \Delta t^*},$$

that is, as the difference of differences. We now write eq. (9.125) in the form

$$\varepsilon\begin{Vmatrix}1 & 1 \\ 1 & 1\end{Vmatrix} - \begin{Vmatrix}1 & -1 \\ -1 & 1\end{Vmatrix} = \begin{Vmatrix}\varepsilon - 1 & \varepsilon + 1 \\ \varepsilon + 1 & \varepsilon - 1\end{Vmatrix} = 0,$$

where $\varepsilon = \frac{1}{8}V\kappa_x^2\Delta z \, \Delta t^*$. This box of four compartments is laid over the table of Ψ_x^* values, and each is multiplied by the overlying factor $(\varepsilon \pm 1)$, and the sum is set equal to zero. If we know three of the four values, we can calculate the fourth. Again, some values have to be guessed to get started.

The upper surface in fig. 9.56a, the unmigrated time section observed at the surface of the earth, thus provides the basis for calculating the time section that would be observed at $z = \Delta z$, and so on. In effect, we are calculating the output of geophones buried at $z = z_1$ as the superposition of filtered outputs of the geophones in the layer $z = z_1 - \Delta z$ and previously calculated values in the layer z_1. The filters are mainly phase shifting to allow for the difference in traveltime from a geophone directly above the buried geophone to those at adjacent locations.

In continuing the wave field from $z = z_1$ to $z = z_1 + \Delta z$, we use the velocity of the layer between z_1 and $z_1 + \Delta z$, so that accommodating vertical variations of velocity is fairly easy and straightforward in time-domain migration. Velocity V can also be made a function of x, but often this is not done. A common way of accounting for lateral velocity variations is to migrate with different velocities and then merge the

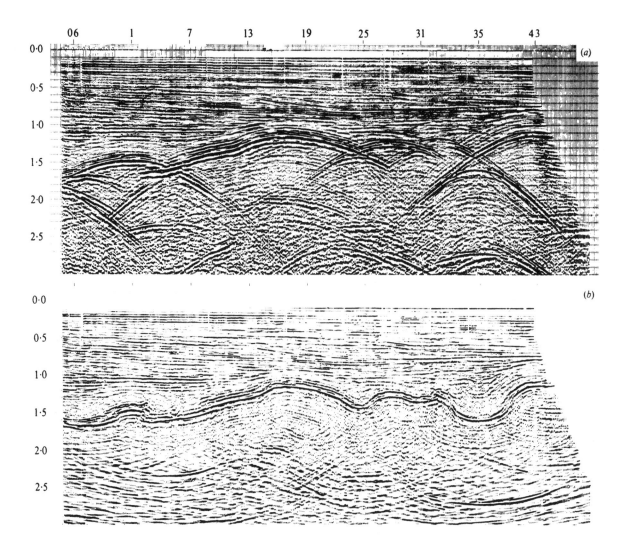

Fig. 9.57 Migration removes the confusion produced by multiple-branch reflections. The deeper data, which are probably multiples, out-of-plane diffractions, and other types of noise are smeared out because they are migrated as if they were primary reflections. (Courtesy of AGIP.) (a) Unmigrated section; (b) migrated.

two sections, that is, $\psi = k\psi_1 + (1 - k)\psi_2$, where k varies linearly over the merge region, the same type of procedure as is used to accomplish "time-variant" filtering.

Because downward continuation proceeds in steps, the size of the step has to be selected. Steps that are too large cause undermigration, kinks in reflection continuity, and dispersion, whereas small steps increase computation time and costs. The undermigration increases with increasingly steep dips. The dispersive noise is an effect of approximating differential operators with difference operators. The step size is usually selected somewhat smaller than the dominant period, that is, between 20 and 40 ms.

9.12.5 Depth migration

Migrated time sections may be simply stretched according to a vertical velocity function to give sections where the vertical scale is linear in depth rather than in time (fig. 9.59b). However, where velocity varies appreciably in the horizontal direction, raypath bending introduces additional complications that depth migration (Judson et al., 1980; Schultz and Sherwood, 1980; Larner et al., 1981) attempts to accommodate.

Hubral (1977) observed that the apex of a diffraction curve is where the *image ray*, a ray that approaches the surface at right angles, emerges. Therefore, if we follow the image ray as it refracts according to Snell's law down through the earth, it will lead to the correct position of the diffracting point even if velocity surfaces are not horizontal. This concept is the heart of *depth migration*, migration that accommodates horizontal changes in velocity. Conventional migration collapses diffractions to the image-ray positions, so an additional step is needed to move elements to their correct subsurface locations. Larner et al. (1981) develop a two-dimensional velocity model, $V(x, z)$, and then ray trace image rays to locate features. This procedure is illustrated in fig. 9.59c and 9.59d.

The velocity model defines the major velocity surfaces where significant raypath bending occurs; key horizons on a conventionally migrated time section are mapped assuming that these are the major velocity interfaces. Clearly, defining the velocity model adequately is the key to successful depth migration. Specifying velocities is a very difficult task because choices are not obvious. Robinson (1983) says, "it is at this point that the true inverse problem is being solved." Detailed knowledge of the velocity distribution is often not available, especially in the structurally complex areas where depth migration is most needed. However, even though velocity errors create depth and location errors in the final product, the improved structural clarity often makes the procedure worthwhile and an appreciable amount of depth migration is being done today.

Subsalt imaging is important in several areas to locate hydrocarbons trapped beneath salt. Appreciable raypath bending occurs at the large contrast between the salt and sediments and the surfaces of the salt may be quite irregular. Migration is usually done in steps: conventional migration first defines the top of the salt, then the base-of-salt reflection is defined using the salt velocity, and finally migration is completed with sediment velocities. Subsalt imaging provides a severe test of migration accuracy and requires very reliable data, which are usually 3-D data, and processing, often prestack migration.

9.12.6 Hybrid migration

The limitations of different migration methods (§9.12.7) can sometimes be overcome by flipping back and forth between methods. For example, the Stolt algorithm can be used to improve the performance of finite-difference migration, which has difficulty with steep dips. A frequency-domain migration can be used with a constant low velocity to perform a partial migration, and follow this with a residual finite-difference migration. Residual migration may not work well when too much of the migration is left to the second migration step, as is apt to be the case with steep velocity gradients.

Frequency-space finite-difference migration (Yilmaz, 1987: 309–11) is sometimes carried out. This method, sometimes called *omega-x migration,* can handle both steep dips and lateral-velocity variations. Kjartansson (1979) implemented a 45° version, which can be further modified to accommodate even steeper dips.

9.12.7 Relative merits of different migration methods

In practical implementation, each migration method involves approximations and limitations that affect data with different characteristics in different ways, so that one method may migrate better for one data set but another method might be superior for a different data set, or one implementation of a method may yield better results than another implementation of the same method. Among the characteristics of the methods are the following:

Diffraction-stack migration:

Migrates steep dips

Allows weighting and muting according to dip or coherency

Aperture can be varied explicitly

Usually not adapted to accommodate lateral-velocity variation

Stolt frequency–wavenumber migration:

Migrates dips up to spatial-aliasing limitations

Difficult to accommodate lateral velocity variations

Can be applied to specific limited areas

Is often the most economical method

Finite-difference migration:

Migrates dips up to 60° ($\approx$ 30° with 15° version)

Produces less migration noise

Is effective in low signal-to-noise areas

Can accommodate lateral velocity variations

Frequency–space migration:

Migrates dips up to spatial-aliasing limitations

Often the easiest method for depth migration (§9.13.8) because velocity as a function of space is explicit

One important point should be made: migration (by almost any method) almost always produces a result that is closer to the correct picture than failure to migrate, even where the fundamental assumptions are grossly violated.

9.12.8 Resolution of migrated sections

In §6.4.4, it was pointed out that the horizontal resolution of seismic data is limited by noise and migration considerations rather than Fresnel-zone size, as with unmigrated data. One of the basic assumptions underlying migration is that the data show only the reflected wavefield; data of any other type (including multiples) will be migrated as if they were primary reflected energy and will be rearranged and superimposed on the migrated images of reflectors. Noise of limited extent that might be quite recognizable as noise in the unmigrated data may smear out and degrade the sharpness of reflector images. For example, a noise burst on a single trace will become a complete wavefront (sometimes called a "smile") on a migrated section (fig. 9.53). Data with a dip component perpendicular to the line will not be migrated correctly, thus degrading horizontal resolution. Spatial-aliasing con-

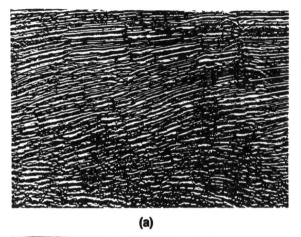

(a)

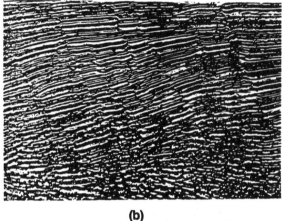

(b)

Fig. 9.58   Migration collapses diffractions and makes faulting clearer. (Courtesy of Grant Geophysical.) (a) Unmigrated section; (b) migrated section.

siderations limit the amount of dip that can be migrated. The use of a limited migration aperture and the approximations involved in the migration algorithms further restrict resolution. With good data and good migration algorithms, features are sometimes defined very sharply; for example, faults can sometimes be located to within the trace-spacing distance. Usually, however, the net effect of the various limiting factors limits the precision to much less than this. Figures 9.57 and 9.58 illustrate the improved clarity that migration can produce.

9.12.9   Other migration considerations

Migrating data before stacking provides an alternative to DMO processing (§9.10.2) in avoiding the smearing of reflecting points for dipping reflections during stacking. Various schemes (Sattlegger and Stiller, 1974; Sattlegger et al., 1980; Schultz and Sherwood, 1980; Jain and Wren, 1980) are sometimes employed to obtain the benefit of migration before stack without incurring excessive cost. Several stacks may be made, each with data involving only a limited range of offsets

so that the smear produced by stacking is minor for each of them and then these partial stacks are migrated and the migrated results stacked. Thus, the full stack and full noise attenuation may be achieved without significant reflection-point smear.

Turning-wave migration is another aspect of migration that is important today. A *turning wave* is a diving wave (§4.3.5) that has been reflected after reaching its maximum depth, that is, on its upward travel toward emergence at the surface. Stacking may destroy such arrivals, whose normal moveout may be negative. Turning waves are especially used to locate steeply dipping reservoirs truncating against the overhanging flanks of salt domes.

The migration of unmigrated reflection-time maps (Kleyn, 1977) often provides an excellent way of achieving correct depth-contour maps. Usually, the unmigrated map surface (the consequence of picking a reflection event, posting, and contouring) and also intermediate velocity surfaces are approximated by a grid of small planar elements (fig. 9.60). Rays are then traced, the starting direction at the surface being given by the attitude of the elements on the unmigrated map (which give the angle of approach at the surface), the rays being bent in accordance with Snell's law whenever they encounter an intermediate velocity interface. The consequent depth map (fig. 9.61) then has data posted in an irregular manner, so these are usually regridded and contoured by the computer.

9.13   Data-processing procedures

9.13.1   Typical processing sequence

(a) Initial steps.   Seismic data processing often follows a basic sequence (fig. 9.62) that is varied to tailor to specific needs of data.

A small group of people is usually responsible for processing. They prepare specific instructions, choose processing parameters, and monitor the quality of results. This group needs to know the objectives of the processing in order to make optimal decisions. Processing that is optimal for one set of objectives may not be optimal for another set.

The first step after field tapes are received at a processing center is to verify the arrangement of data on the magnetic tape. This involves *dumping* (displaying the tape's magnetic pattern on a printout) the first few (possibly 10) records and comparing with what is expected. With Vibroseis data, format verification includes a check on the Vibroseis sweep length and spectrum.

(b) Editing.   Editing follows format verification. The data are rearranged or *demultiplexed;* field data are usually time-sequential, that is, the first sample for each channel is recorded before the second sample for any channel, whereas most processing requires trace-sequential data, that is, all the data for the first channel before the data for the second channel.

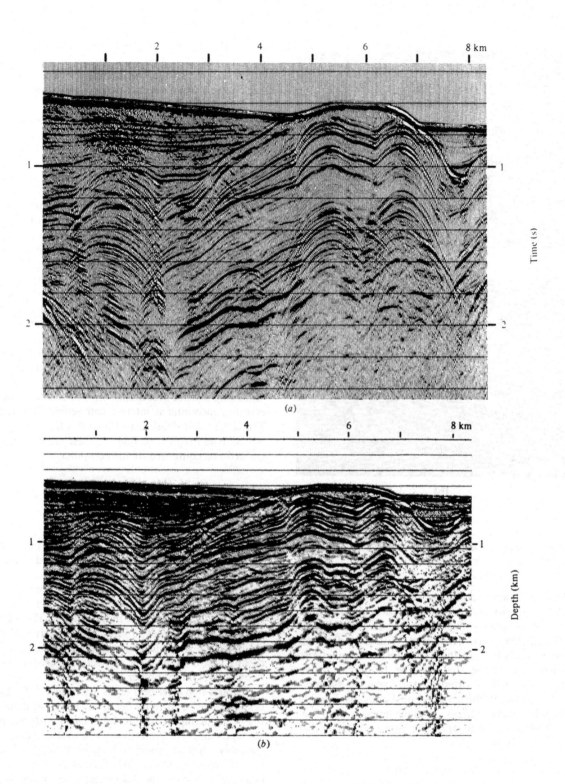

Fig. 9.59 Depth migration. (From Hatton, Larner, and Gibson, 1981.) (a) Unmigrated time section, (b) finite-difference migrated section stretched vertically to a linear depth scale, (c) ray tracing through velocity layers, and (d) migration stretched along the raypaths in part (c).

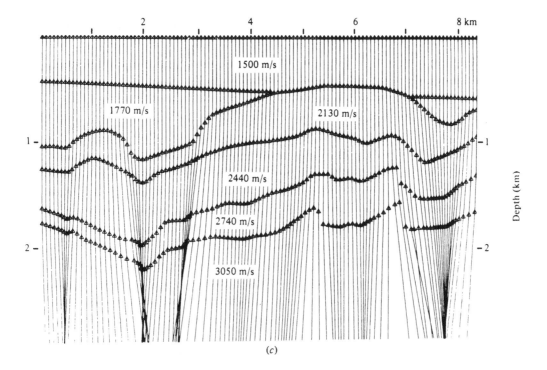

(c)

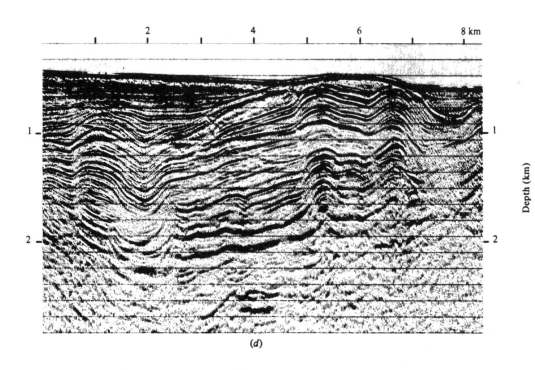

(d)

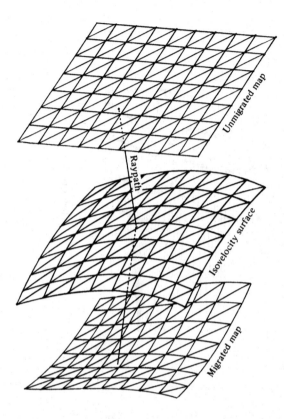

Fig. 9.60 Map migration. The unmigrated map (top) is subdivided into elements and a raypath is traced downward, the direction of the raypath being determined by the values at the corners of the elements. Isovelocity surfaces are likewise subdivided into elements, and when a raypath strikes one of the isovelocity elements, it is bent according to Snell's law. The (x, y, z) coordinates when the traveltime is satisfied are contoured to give a migrated map.

Traces may be summed or arrays formed, especially with marine data recorded with 240 to 250 channel streamers, to reduce the data for further processing.

Editing may involve detecting dead or exceptionally noisy traces. "Bad" data may be zeroed out or replaced with interpolated values. Anomalously high amplitudes, which are probably noise, may be reduced to zero or to the level of the surrounding data.

Outputs after editing usually include: (1) a plot of each file, so that one can see what data need further editing and what types of noise attenuation are required; (2) a plot of the shortest-offset trace from each record (near-trace plot), to give a quick look at the geologic structure and for use in making decisions as to where velocity analyses are to be made; (3) near-trace autocorrelation plots, which indicate multiple problems and aid in making deconvolution decisions; (4) a trace-sequential tape that will be used for subsequent processing. The field tape is then stored in a tape library.

If the input waveform has been recorded, a deterministic source-signature correction (§9.5.2) may be applied to make the effective waveform the same for all records. If the source is Vibroseis or some other source that continues for some time, the equivalent of a source-signature correction is accomplished by correlation with the source waveform.

Field data encompass a very wide range of amplitude values that have been recorded in encoded form, so the field gain is decoded and a first approximation of a spherical-divergence correction may be applied to give a smaller range of values for input to subsequent processes. Sometimes data are also vertically stacked or resampled to make less data for future processing. Resampling should be preceded by alias filtering.

(c) Parameter determination. The object of the next sequence of processes is to determine processing parameters that are data-dependent, such as static time shifts, amplitude adjustments, normal-moveout values, and frequency content. Field-geometry information is input so the computer can determine which data involve common geophones or have other factors in common, and so the offset for each trace is known for normal-moveout determinations.

If the near-trace plot shows strong coherent noise trains from surface waves, shallow refractions or other horizontally traveling energy, apparent-velocity filtering may be applied to remove them. Statistical wavelet contraction may be applied. The analysis of the statistical characteristics of the data for wavelet contraction and for statics and velocity analysis should be based on data that do not include known bad traces and zones of nonreflection energy, such as the first-breaks region. A wide window is usually chosen so as to provide appreciable statistics, but the window should exclude zones where primary reflections are not expected, such as below the basement. Surface-consistent deconvolution may be used.

The statics-analysis program looks for systematic variations such as would be expected if time shifts were associated with particular source activations, particular geophones, and so on. Preliminary statics, as determined in the field office from first-break information and from the elevation of geophone stations, is usually input before the statics analysis, so that the statics analysis determines residual statics errors. The results of this analysis are output on a control plot. Surface-consistent statics may be supplemented by refraction-based statics (§9.6.4).

An analysis similar to the time-shift analysis of the statics program is carried out for amplitude, to determine systematic amplitude effects such as might be associated with a weak source, poor geophone plant, and so on, and a control plot is output. Data are then re-sorted to CMP gathers. 2-D filtering may be applied to the gathers to improve the subsequent velocity analysis.

Velocity analyses are usually run 1 to 2 km apart at locations selected because they are relatively free of structural complications. The locations for analyses are based on the near-trace output from the editing pass. A first guess as to velocity is input prior to velocity analysis so that the velocity analysis determines re-

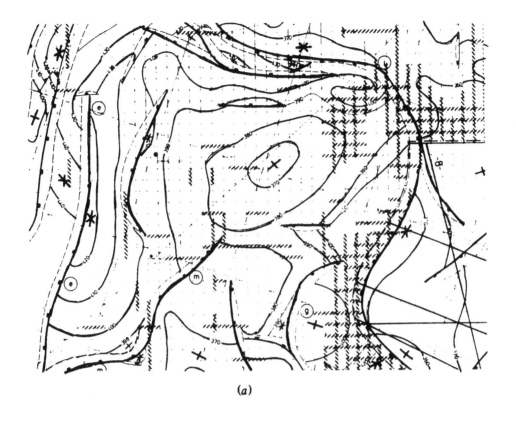

(a)

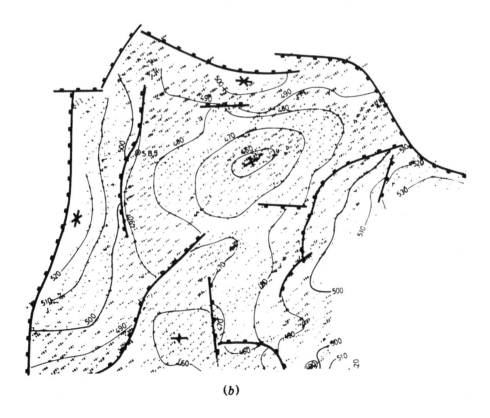

(b)

Fig. 9.61   May migration. (Courtesy of Prakla-Seismos.) (a) Unmigrated map; and (b) migrated map.

sidual normal moveout. If the near-trace plots show dip, this information is input because velocity determination depends on dip. Compromises have to be made as to how much data should be included in an analysis; more data yield better statistics but then the results do not apply at specific points, so the compromise is between determining a more accurate average and a less accurate value that applies to a specific location. Outputs from velocity analysis may include (1) a velocity spectral plot, such as in fig. 9.25 or 9.26, which shows the coherence achieved when various stacking velocities are assumed; (2) velocity panels, as shown in fig. 9.27, which show the data stacked according to the input-velocity information and also according to velocities slightly smaller and larger than the input velocity; such panels allow one to see if certain events require different stacking velocity than other events, because stacking velocity is not necessarily a single-valued function, and also how sensitive the stack is to velocity assumptions; and (3) a graph such as in fig. 9.32 showing how velocity determinations at different locations along the lines relate to each other. Velocity analysis is sometimes done both before and after DMO. Instead of DMO, data are sometimes migrated before stacking. Multiple attenuation by filtering after Radon transforming is sometimes done after DMO.

The data may also be filtered by a sequence of narrow-band-pass filters yielding a filter panel (fig. 9.20) that is used to determine subsequent filtering parameters. Autocorrelations and spectral plots of various sorts may also be output. Preliminary stack sections may be made to show the effectiveness of processing parameters and as an aid in diagnosing additional problems.

(d) Principal processing pass. The main processing pass usually begins with the tape from the editing pass, and the sequence of operations is almost the same as in the parameter determination pass, the differences being in the values that are input for statics, normal moveout, and so on. The correct spherical divergence based on the actual velocities replaces preliminary gain assumptions. The main processing pass or portions of it may be repeated using more refined values, especially statics and normal-moveout values.

The final output from the main processing pass will be one or more stacked sections that may differ in the processes applied or in the choices of parameters, especially display parameters such as amplitude, polarity, and filtering choices.

(e) Other processing. The stacked data may then be used as input to various other processes, such as migration (see §9.12) and attribute analysis. Velocity information is required for migration; the velocity may be based on stacking velocity values where dips are not extreme, but in general, the optimum velocity for migration differs from the optimum velocity for stacking. Some empirically determined percentage of the stacking velocity may be used as the migration velocity.

Either the stacked or the migrated data may be further analyzed for amplitude, frequency content, apparent polarity, inverted to acoustic impedance or synthetic-velocity traces, and so on, and displayed in various ways. The data may also be used in iterative modeling or other types of processing.

9.13.2 Interactive processing and workstations

Seismic data processing today is a usually a sequence of batch processes that is often interrupted to insert parameter decisions based on previous results. We would like to make optimum decisions as to parameter values and processes to be executed and to have processing flow along smoothly and rapidly. Historically, the time required for many processes has been so great that an operator becomes bored waiting at a terminal for the response to decisions; the cost of tying up computers or computer memories while waiting for an operator to make decisions has militated against interactive processing. However, workstations that have enough memory and computing power (especially when tied to other computers) to yield results rapidly are currently changing this situation. Indeed, as computers become faster, memory cheaper, and data storage larger, seismic processing may become a geologic interpretation activity with real-time visual feedback to guide parameter decisions.

One of the most useful applications of workstations is simply checking input parameters and instructions for batch processing to make sure that these are complete and consistent before involving an expensive computer in their execution, and also displaying the quality-control and parameter-determination outputs (such as those listed in fig. 9.62). The results of processing samples of the data in different ways can help visualize the results of complete processing. Workstations can prompt one regarding overlooked decisions and then communicate the parameters and instructions directly to a large computer for the actual processing.

Velocity analysis is an especially critical process where many decisions have to be made. Historically, this has been done for occasional individual gathers without an easy way to check their consistency either internally or with respect to adjacent analyses. Being able to view at a workstation adjacent velocity analyses and the interval velocities implied by the operator's picking allows more consistency and more complete picking. It also encourages introducing geologic interpretation into the picking decisions.

9.14 Generalized inversion

In a direct problem, we calculate the effects produced by a model. In an inverse problem, we try to derive the model from observation of its effects. This is basically the function of interpretation, determining the distribution of the physical properties of the earth (the

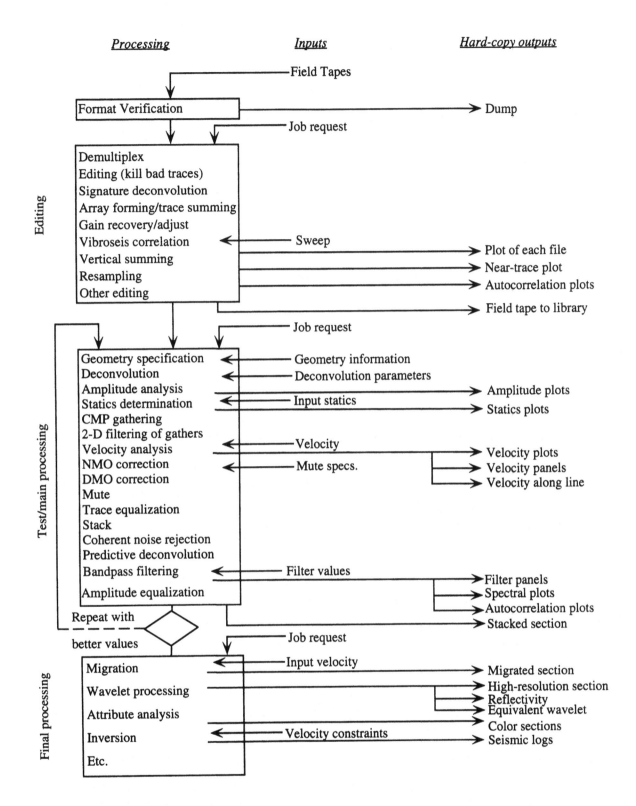

Fig. 9.62 Typical processing flow chart. Some of the processing shown is optional, sometimes the sequence is changed or other processing is added.

parameters of the problem) from seismic observations. We generally use "inversion" for the algorithmic aspects of interpretation. We have encountered one-dimensional inversion (§5.4.5) and automatic picking (§9.11.3), and we discuss tomographic methods in §13.5. This discussion draws heavily from Hatton et al. (1986) and Russell (1988).

Inversion is nonunique, that is, a number of different models can lead to the same set of observations. This is true partly because measurements are incomplete and also because they involve uncertainties. Often an infinite number of models can lead to the same set of measurements within the uncertainties involved, and this may lead one to question the value of inversion. However, constraints usually limit physical property values so that the set of possible solutions exists only within narrow bounds. Much of the non-uniqueness often can be removed by careful model construction.

Model construction involves, among other things, determining the model complexity. The selection of too fine a grid, for example, might permit including an anomalous layer that is too thin to be detected (and also lead to excessive calculation time), whereas too coarse a grid may involve so much smoothing that the results lack utility.

Generalized inversion attempts to determine the spatial distribution of physical properties that could produce observed seismic data, usually in an iterative manner. Inversion is usually treated as a linear problem, that is, measurements are assumed to bear a linear relation to the parameters. However, many problems are intrinsically nonlinear; such problems are usually solved by a sequence of linear approximations. A starting model (which may be as simple as a uniform half-space) is given, the *error field* (the difference between the direct-modeling solution and the observed data) is used to perturb the model in such a way as to make the error field smaller, the new error field is then used to further perturb the model, and so on. The objective is to minimize the error field; iterations are stopped when the error field becomes smaller than some threshold quantity or when successive iterations fail to change the error field more than some threshold amount.

We must measure the goodness of fit in some suitable way. The best estimate to a set of numbers may be the median. The most common error measure is the *least-squares fit* (norm), which measures the square root of the mean of the squares of the individual errors; least squares implies a Gaussian distribution of errors. A *least absolute deviation fit* averages the errors without regard for their sign; it corresponds to the maximum-likelihood estimate when the errors have a Laplace (double exponential) distribution. These are special cases of L_p fits that minimize

$$\Psi = \sum_i |w_i e_i|^p, \qquad (9.126)$$

where the errors are $e_i = y_i - m_i$, y_i being the data points, m_i the values calculated from the model, and

w_i are weighting factors. If $p = 1$, this yields the least-absolute deviation fit and, if $p = 2$, the least-squares fit. For $p = \infty$, we get the *minimax* or *Chebychev fit*. The criterion of goodness of fit is not necessarily technical; for example, it can be economic.

Models are often formulated as matrices representing a set of equations to be solved simultaneously. However, the matrices are usually much too large to be solved by conventional matrix methods.

Usually, one is searching for an extremum (in this case, the minimum error), but the problem is usually phrased in terms of searching for a maximum or *hill climbing* (rather than valley searching). Hill climbing is sometimes performed by methods such as the *method of steepest ascent*. The maximum is discovered by taking an increment along the steepest gradient to arrive at the next approximation, the step length often being proportional to the magnitude of the gradient. Provision can be made to speed up convergence and prevent oscillation about the maximum. Convergence (or at least the speed of convergence) is apt to depend on the starting model as well as the data and the algorithm used. Convergence assumes that the function is continuous and that the initial estimate is already on the lower slopes of the correct maximum (see Lines and Treitel, 1984).

However, there may be many "hills" (or "valleys"), and we want to climb the highest hill. Iteration may result in "climbing the wrong hill" and thus converging on a suboptimal solution. We wish to perform global, rather than local, optimization, that is, we want to search widely through model space in order to find the highest hill. We wish for something in between a global random search and a local climb. Sometimes filtering to retain only the low-frequency components in early iteration stages helps get one to the vicinity of the global solution.

The simplest method of determining model parameters is trial and error, but there is no assurance that enough trials have been performed to arrive at an adequate determination of the parameters. We might use a random method, such as the trial-and-error *Monte Carlo method,* which steps out in random directions for random distances in its search to determine the highest hill.

Another method is *simulated annealing* (see Vasudevan, Wilson, and Laidlaw, 1991); it uses algorithms based on an analogy between optimization and the growth of long-range order, such as the growth of large crystals in a slowly cooling melt. It is usually implemented by a "drunkard's walk" through model space, where steps begin in random fashion, but progressively become biased toward the uphill direction. Simulated annealing has three components: an "energy function" that defines the problem in terms of a set of parameters and constraints (including interactions between parameters), an "order function" that measures coherence, and a "temperature" that regulates the system's energy and order; high temperature implies high energy and low order.

Another class of methods are called *genetic algorithms* (see Smith, Scales, and Fischer, 1992; Stoffa and Sen, 1992); they begin with a loose analogy between optimization and a biological system composed of a relatively few organisms that react in a relatively complex way. Algorithms try to evolve a population of trial answers in a way mimicking biological evolution (the survival of the fittest). Each model has a "fitness" associated with it; the goal is to find the most fit of all possible models. A genetic algorithm consists of a set of operations that we apply to a model population to produce a new population whose average fitness exceeds that of the predecessors. The characteristics of models are specified by "chromozone strings." One type of genetic algorithm selects parents in a somewhat random manner but weighted by their fitness ("selection"); the chromozones for a "child" are somewhat randomly selected from the two parents ("crossover"). The child then joins the population, and the least fit member of the population (which might be the child itself) is eliminated. At random times, a "mutation," a random change in a member's chromozones, occurs; this permits introducing into the species chromozone elements not present in the original population.

Problems

9.1 Derive the expression for the amplitude of frequency components, eq. (9.6), from the equation for Fourier analysis, eq. (9.3). (*Hint:* Multiply both sides of eq. (9.3) by $e^{-j2\pi v_n t}$ and integrate over one cycle, for example, from $-T/2$ to $T/2$. Before substituting the values at the limits, use Euler's theorem, $e^{\pm jx} = \cos x \pm j \sin x$ (see problem 15.12a) to express the complex exponentials in terms of sines and cosines.)

9.2 The techniques and concepts of convolution, aliasing, z-transforms, and so on can be applied to other than the time–frequency domain. Express the source and group patterns of fig. 8.13a as functions of x (horizontal coordinate) and convolve the two to verify the effective pattern shown in fig. 8.13b.

9.3 (a) Because an impulse $\delta(t)$ is zero except for $t = 0$, where it equals $+1$ (see §15.2.5), we can apply the Fourier transform equation (9.12) and find that $\delta(t) \leftrightarrow +1$; show that

$$\delta(t - t_0) \leftrightarrow e^{-j2\pi v(t - t_0)}$$

(b) Show that $\delta(t) * g(t) = g(t)$ and $\delta_t * g_t = g_t$.
(c) Show that a boxcar of height h and extending from $-v_0$ to $+v_0$ in the frequency domain has the transform

$$h \, \text{box}_{2v_0} \leftrightarrow A \, \text{sinc} \, 2\pi v_0 t = A(\sin 2\pi v_0 t)/2\pi v_0 t,$$

where $A = 2hv_0$, the area of the boxcar.

9.4 Show that 4-ms sampling is sufficient to reproduce exactly a signal whose spectrum is confined to the range v_0 to $v_0 + 125$ Hz. (*Hint:* Modify fig. 9.3 to fit this signal.)

9.5 The standard alias filter such as shown in fig. 7.46 has its 3-dB point at about half the Nyquist frequency and a very steep slope so that noise above the Nyquist frequency is highly attenuated relative to the passband of the system.

(a) Assuming an initially flat spectrum, alias filtering with a 125-Hz, 72-dB/octave filter, and subsequent resampling from 2 to 4 ms (without additional alias filtering), graph the resulting alias noise versus frequency.
(b) Some believe standard alias filters are unnecessarily severe. Assume a 90-Hz, 72-dB/octave filter and 4-ms sampling and graph the alias noise versus frequency.

9.6 The sampling theorem applies to uniform sampling. Suppose the spacing between samples alternates from a to $3a$, that is, the impulse response of the sampling is $[\dots, \delta, \delta, 0, 0, \delta, \delta, 0, 0, \dots]$. What is the effect on aliasing?

9.7 (a) Verify the equation for the inverse filter for water reverberation, eq. (9.40), by convolving it with the equation for water reverberation, eq. (9.38).
(b) The spectrum of the water-layer filter is shown in fig. 9.63 for $n = 1$; the large peaks occur at the "singing frequency." Sketch the amplitude spectrum of the inverse filter. (*Hint:* Time-domain convolution such as shown in eq. (9.39) corresponds to frequency-domain multiplication, eq. (9.28); the frequency spectrum of the unit impulse is $+1$, that is, it is flat.)
(c) Verify your sketch of the water-reverberation inverse filter by transforming

$$[1, 2R, R^2] \leftrightarrow 1 + 2Re^{-j2\pi v\Delta} + R^2 e^{-j4\pi v\Delta},$$

and calculating the value of the spectrum for $v = 0$, v_1, and $2v_1$.

9.8 Four causal wavelets are given by $a_t = [2, -1]$, $b_t = [4, 1]$, $c_t = [6, -7, 2]$, and $d_t = [4, 9, 2]$. Calculate the cross- and autocorrelations ϕ_{ab}, ϕ_{ac}, ϕ_{ca}, ϕ_{aa}, and ϕ_{cc} in both the time domain and in the frequency domain.

9.9 Fill in the values in table 9.3.

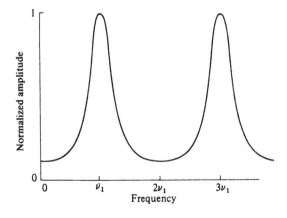

Fig. 9.63 Spectrum of a water-layer filter for a water-bottom reflection coefficient of 0.5. If z is the water depth and V is the water velocity, $v_1 = V/4z$.

Table 9.3 *Digital notation and operations*

	$t = -3$	$t = -2$	$t = -1$	$t = 0$	$t = +1$	$t = +2$	$t = +3$	$t = +4$	$t = +5$
$a_t = [\overset{\downarrow}{2}, 1, -2, 1]$									
$b_t = -2a_t$									
$c_t = -3a_{t-2}$									
$d_t = 4a_{-t}$									
$e_t = \pi a_{3-t}$									
$f_t = [-1, 1]$.									
$g_t = a_t * f_t$									
δ_{t+2}									
δ_{2-t}									
$\phi_{ff}(t)$									
$\phi_{fa}(t)$									

9.10 Show that the semblance, $S_T = 1$, where S_T is given by eq. (9.59), when the N values of g_{ti} are identical for all values of t in the interval Δt.

9.11 (a) Convolve $[2, 5, -2, 1]$ with $[6, -1, -1]$.

(b) Cross-correlate $[2, 5, -2, 1]$ with $[6, -1, -1]$. For what shift are these functions most nearly alike?

(c) Convolve $[2, 5, -2, 1]$ with $[-1, -1, 6]$. Compare with the answer in part (b) and explain.

(d) Autocorrelate $[6, -1, -1]$ and $[3, -5, -2]$. The autocorrelation of a function is not unique to that function (see §15.5.6c), for example, other wavelets having the same autocorrelations as the preceding are $[-2, -5, 3]$ and $[-1, -1, 6]$. . . . Which of the four is the minimum-delay wavelet?

(e) What is the normalized autocorrelation of $[6, -1, -1]$? What is the normalized cross-correlation in part (b)? What do you conclude from the magnitude of the largest value of this normalized cross-correlation?

9.12 (a) Of the four causal wavelets given in problem 9.8, which are minimum-phase?

(b) Find $a_t * b_t$ and $a_t * c_t$ by calculating in the time domain.

(c) Repeat part (b) except using transforms.

(d) Find $a_t * b_t * c_t$.

(e) Does the maximum value of a minimum-phase function have to come at $t = 0$?

(f) Can a minimum-phase wavelet be zero at $t = 0$?

9.13 Using the wavelets $W_1(z) = (2 - z)^2 (3 - z)^2$ and $W_2(z) = (4 - z^2)(9 - z^2)$, calculate the composite wavelets: $W_1(z) + W_2(z)$, $W_1(z) + zW_2(z)$, $zW_1(z) + W_2(z)$, and $z^{-1}W_1(z) + W_2(z)$. Plot the composite wavelets in the time domain. The results illustrate the effects of phase shifts (note that all of the composite wavelets have the same frequency spectrum but different phases because multiplication by z shifts the phase (see §15.5.6a).

9.14 The following wavelet is approximately minimum-phase: $[11, 14, 5, -10, -12, -6, 3, 5, 2, 0, -1, -1, 0]$ (fig. 9.64a), the sampling interval being 2 ms. Use $V_{sd} = 2.0$ km/s for the velocity in sand, $V_{sh} = 1.5$ km/s for the velocity in shale, and reflection coefficients (scaled up and rounded off) shale-to-sand = $+0.1$ = sand-to-limestone.

(a) Determine the reflection waveshape for sands 0, 2,

4, 6, 8, and 10 m thick encased in shale. (A thickness of 6 m is approximately a quarter wavelength.)

(b) Repeat for the sand overlain by shale and underlain by limestone.

(c) Determine the waveshape for two sands each 6 m thick and separated by 4.5 m of shale, the sequence being encased in shale; this illustrates a "tuned" situation (§6.4.3).

(d) Repeat parts (a) and (b) with the wavelet $[6, 11, 14, 14, 10, 5, -2, -10, -11, -12, -10, -6, 0, 3, 4, 5, 4, 3, 1, 0]$ (fig. 9.64b), a minimum-phase wavelet stretched out so that it has about half the dominant frequency of the former wavelet. Comparison with the results of parts (a) and (b) illustrates the effect of frequency on the resolution.

(e) Repeat parts (a) and (b) using the zero-phase wavelet $[1, 1, -1, -4, -6, -4, 10, 17, 10, -4, -6, -4, -1, 1, 1]$ (fig. 9.64c), which has the same frequency spectrum as the wavelet used in part (a).

9.15 The wavelet $[-0.9505, -0.0120, 0.9915]$ is not minimum-phase. How would you make it minimum-

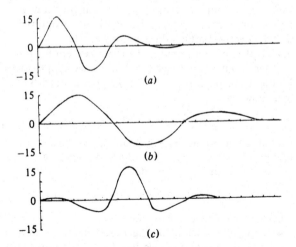

Fig. 9.64 Determining a composite reflection. (a) Minimum-phase wavelet, (b) minimum-phase wavelet of lower frequency, and (c) zero-phase wavelet corresponding to part (a) (but time-shifted and with the polarity reversed).

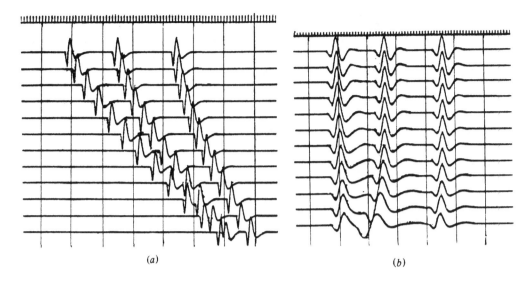

Fig. 9.65   End-on records from a model with four horizontal layers of velocities: 1490, 1895, 2215, and 2440 m/s. (a) Before NMO correction, and (b) after NMO correction.

phase, at the same time changing it as little as possible? Give two methods.

9.16 Show that the result of passing a minimum-phase signal through a zero-phase filter is mixed phase.

9.17 In §9.5, several deconvolution methods were described. List the assumptions of the different methods, such as invariant wavelet and randomness of the reflectivity or of the noise, that a source wavelet is the same as a wavelet recorded near the source or measured from a sea-floor reflection by a group offset a few hundred meters, and so on.

9.18 Assuming the signature of an air-gun array is a unit impulse, find the inverse filter for the recorded wavelet: $[-12, -4, +3, +1]$. How many terms should the filter include to get 1% accuracy?

9.19 A source is located 7 m below the base of the LVL. Given that $V_H = 2.0$ km/s, $V_W = 0.3$ km/s, $\rho_H = 2.3$ g/cm³, $\rho_W = 1.8$ g/cm³, $\Delta = 4$ ms and that the reflected signal is $[6, -7, -2.8, 5.6, -1.6]$, find the original wavelet using: (a) the inverse filter of eq. (9.67); (b) eq. (9.69).

9.20 The ghost reflection from the surface acts as a notch filter for receivers planted on the sea floor.
(a) Graph the notch frequency versus water depth.
(b) If air-gun sources are fired at a depth of 10 m, what effect will the ghost have on the spectrum?

9.21 Show that

$$\sum_t g_{t-j} g_{t-i} = \phi_{gg}(i - j)$$

(see the derivation of eq. (9.73a)).

9.22 Consider wavelets $A = [1, -2, 3]$ and $B = [3, -2, 1]$.
(a) Plot cumulative energy as function of time.
(b) Calculate the three-element Wiener inverse filters assuming the desired output is (i) $[1,0,0,0]$ and (ii)

$[0,1,0,0]$; then apply the inverse operators to each of the wavelets.
(c) Add white noise $\varepsilon = 0.01$ and $\varepsilon = 0.1$ and repeat.

9.23 In fig. 9.32, the hump in the 8- to 12 000-ft/s contours might cause some concern. If it is known that this is the same section as that shown in fig. 9.29, what would the hump imply? How would you modify the hump to do a better job of stacking?

9.24 The horizon velocity analysis for horizon A shown in fig. 9.35 indicates higher stacking velocities on opposite sides of the salt dome and low stacking velocities below the salt dome. Why? Does this have geological significance? The streamer was about 4 km long. Note that only the portion of the section from 2 to 4 s is shown, the upper 2 s of section having been cut off.

9.25 On a north–south line, the noise arriving from the south is confined to the band $V_s \leq 6$ km/s and the noise arriving from the north is in the band $V_s \leq 3$ km/s.
(a) Given that $\Delta x = 50$ m, sketch an f–k plot such as fig. 9.36b.
(b) Repeat for $\Delta x = 25$ m.
(c) Calculate $f(x, t)$ for parts (a) and (b) (see eqs. (9.96) and (9.97)).

9.26 Given the wavelet $[10, -8, 0, 9, -11, 6, 0, -7, 12, -5, 0, 0]$, calculate the following:
(a) The quadrature function, $g_\perp(t)$.
(b) $\gamma(t)$ (add multiples of π to obtain a monotonically increasing function).
(c) $h(t)$, $A(t)$, and $v(t)$ at $n = 4$. Take $\Delta = 4$ ms.

9.27 Prove the following:
(a) The equation of a diffraction curve (curve of maximum convexity; see fig. 9.51) is

$$z^2 - x^2 = z_0^2,$$

where O is the origin and $z_0 = OP$.

(b) The unmigrated reflection is tangent to the diffraction curve.

(c) The coordinates of P and the slope of the wavefront at P (hence, the dip also) can be obtained from the recorded data.

9.28 Show that the coordinate system (x, z, t^*) in eq. (9.120) in effect "rides along on an upcoming wavefront."

9.29 Derive the finite-difference migration equation, eq. (9.124).

9.30 Interpret the faulting in figs. 9.58a and 9.58b to see how much improvement migration makes.

9.31 (a) The operator $f_t = [-1, +1]$ is called the "derivative operator"; explain why.

(b) What is the integral operator?

9.32 Figure 9.65 shows three reflections before and after NMO removal. Explain (a) the broadening of the wavelets by the NMO corrections; (b) why the reflections do not have straight alignments after NMO correction.

References

Anstey, N. A. 1964. Correlation techniques – A review. *Geophys. Prosp.*, **12**: 355–82.

Anstey, N. A. 1970. *Seismic Prospecting Instruments, Vol. 1: Signal Characteristics and Instrument Specifications.* Berlin: Gebrüder Borntraeger.

Backus, M. M. 1959. Water reverberations – Their nature and elimination. *Geophysics*, **24**: 233–61.

Bancroft, J. C. 1991. *A Practical Understanding of Migration and Dip Moveout*, SEG Course Note Series. Tulsa: Society of Exploration Geophysicists.

Barr, F. J., and J. I. Sanders. 1989. Attenuation of water-column reverberations using pressure and velocity detectors in a water-bottom cable. Paper read at the 59th SEG Annual Meeting.

Berryhill, J. R. 1979. Wave-equation datuming. *Geophysics*, **44**: 1329–39.

Bois, P., and M. la Porte. 1970. Pointe automatique. *Geophys. Prosp.*, **18**: 489–504.

Bolondi, F., F. Loinger, and F. Rocca. 1982. Offset continuation of seismic sections. *Geophys. Prosp.*, **30**: 813–28.

Bracewell, R. 1965. *The Fourier Transform and Its Applications.* New York: McGraw-Hill.

Chun, J. H., and C. A. Jacewitz. 1981. Fundamentals of frequency-domain migration. *Geophysics*, **46**: 717–33.

Claerbout, J. F. 1976. *Fundamentals of Geophysical Data Processing.* New York: McGraw-Hill.

Clarke, G. K. C. 1968. Time-varying deconvolution filters. *Geophysics*, **33**: 936–44.

Clayton, R. W., and G. W. McMechan. 1981. Inversion of refraction data by wavefield continuation. *Geophysics*, **46**: 860–8.

Cochran, M. D. 1973. Seismic signal detection using sign bits. *Geophysics*, **38**: 1042–52

Cook, E. E., and M. T. Taner. 1969, Velocity spectra and their use in stratigraphic and lithologic differentiation. *Geophys. Prosp.*, **17**: 433–48.

Crump, M. D. 1974. A Kalman filter approach to the deconvolution of some signals. *Geophysics*, **39**: 1–13.

Denham, L. R., R. A. R. Palmeira, and R. C. Farrell. 1985. The zero-offset stack. Paper read at the 55th SEG Annual Meeting.

Deregowski, S. M. 1986. What is DMO (dip moveout)? *First Break*, **4**(7): 7–24.

Fail, J. P., and G. Grau. 1963. Les filtres en eventail. *Geophys. Prosp.*, **11**: 131–63.

Finetti, I., R. Nicolich, and S. Sancin. 1971. Review of the basic theoretical assumptions in seismic digital filtering. *Geophys. Prosp.*, **19**: 292–320.

Flinn, E. A., E. A. Robinson, and S. Treitel. 1967. Special issue on the MIT Geophysical Analysis Group reports. *Geophysics*, **32**: 411–535.

Forel, D., and G. H. F. Gardner. 1988. A three-dimensional perspective on two-dimensional dip moveout. *Geophysics*, **53**: 604–10.

Gardner, G. H. F., and L. Lu. 1991. *Slant-Stack Processing*, Geophysical Reprint Series 14. Tulsa: Society of Exploration Geophysicists.

Garotta, R. 1971. Selection of seismic picking based upon the dip moveout and amplitude of each event. *Geophys. Prosp.*, **19**: 357–70.

Garotta, R., and D. Michon. 1967. Continuous analysis of the velocity function and of the normal-moveout corrections. *Geophys. Prosp.*, **15**: 584–97.

Gazdag, J. 1978. Wave-equation migration by phase shift. *Geophysics*, **43**: 1342–51.

Gazdag, J., and P. Squazzero. 1948. Migration of seismic data by phase shift plus interpolation. *Geophysics*, **49**: 124–31.

Hagedoorn, J. G. 1954. A process of seismic reflection interpretation. *Geophys. Prosp*, **2**: 85–127.

Hale, D. 1984. Dip movement by Fourier transform. *Geophysics*, **49**: 741–57.

Hale, D. 1991. *Dip Moveout Processing*, SEG Course Note Series 4. Tulsa: Society of Exploration Geophysicists.

Hatton, L., K. Larner, and B. S. Gibson. 1981. Migration of seismic data from inhomogeneous media. *Geophysics*, **46**: 751–67.

Hatton, L., M. H. Worthington, and J. Makin. 1986. *Seismic Data Processing: Theory and Practice.* London: Blackwell Scientific.

Hosken, J. W. J., and S. M. Deregowski. 1985. Migration strategy. *Geophys. Prosp.*, **33**: 1–33.

Hubral, P. 1977. Time migration, some ray theoretical aspects. *Geophys. Prosp.*, **25**: 728–45.

Hubral, P. 1980. Slant-stack migration. In *Festschrift Theodor Krey*, pp. 72–8. Hanover, Germany: Prakla-Seismos.

Jain, S., and A. E. Wren. 1980. Migration before stack: Procedure and significance. *Geophysics*, **45**: 204–12.

Judson, D. R., J. Lin, P. S. Schultz, and J. W. C. Sherwood. 1980. Depth migration after stack. *Geophysics*, **45**: 361–75.

Kanasewich, E. R. 1987. *Time Sequence Analysis in Geophysics.* University of Alberta Press.

Kjartansson, E. 1979. Modeling and migration by the monchromatic 45-degree equation, Stanford Exploration Project Report No. 15, Stanford University, Stanford, California.

Kleyn, A. H. 1977. On the migration of reflection time contour maps. *Geophys. Prosp.*, **25**: 125–40.

Kunetz, G., and J. M. Fourmann. 1968. Efficient deconvolution of marine seismic records. *Geophysics*, **33**: 412–23.

Larner, K. L., L. Hatton, B. S. Gibson, and I. C. Hsu. 1981. Depth migration of imaged time sections. *Geophysics*, **46**: 734–50.

Lee, Y. W. 1960. *Statistical Theory of Communications*. New York: John Wiley.

Levin, F. K. 1971. Apparent velocity from dipping interface reflections. *Geophysics*, **36**: 510–16.

Levin, F. K., and P. M. Shaw. 1977. Peg-leg multiples and dipping reflectors. *Geophysics*, **42**: 957–81.

Lines, L. R., and S. Treitel. 1984. A review of least-squares inversion and its application to geophysical problems. *Geophys. Prosp.*, **32**: 159–86.

Lines, L. R., and S. Treitel. 1985. Inversion with a grain of salt. *Geophysics*, **50**: 99–109.

Marsden, D. 1993. Static corrections – A review. *The Leading Edge*, **12(2)**: 43–9, 115–20.

Middleton, D., and J. R. B. Whittlesey. 1968. Seismic models and deterministic operators for marine reverberation. *Geophysics*, **33**: 557–83.

Millahn, K. O. 1980. In-seam seismics: Position and development. *Prakla-Seismos Report*, **80(2 and 3)**: 19–30.

Naess, O. E. 1979. Attenuation of diffraction noise through very long arrays. *Expanded Abstracts, 49th Annual International SEG Meeting*, p. 32.

Naess, O. E., and L. Bruland. 1985. Stacking methods other than simple summation. In *Developments in Geophysical Exploration Methods – 6*, A. E. Fitch, ed., pp. 189ff. Amsterdam: Elsevier.

Neidell, N. S., and M. T. Taner. 1971. Semblance and other coherency measures for multichannel data. *Geophysics*, **36**: 482–97.

Otis, R. M., and R. B. Smith. 1977. Homomorphic deconvolution by log spectral averaging. *Geophysics*, **42**: 1146–57.

Paturet, D. 1971. Different methods of time–depth conversion with and without migration. *Geophys. Prosp.*, **19**: 27–41.

Paulson, K. V., and S. C. Merdler. 1968. Automatic seismic reflection picking. *Geophysics*, **33**: 431–40.

Peacock, K. L., and S. Treitel. 1969. Predictive deconvolution – Theory and practice. *Geophysics*, **34**: 155–69.

Robinson, E. A. 1983. *Migration of Geophysical Data*. Boston: International Human Resources Development Corp.

Robinson, E. A., and S. Treitel. 1964. Predictive decomposition of time series with applications to seismic exploration. *Geophysics*, **32**: 418–84.

Robinson, E. A., and S. Treitel. 1967. Principles of digital Wiener filtering. *Geophys. Prosp.*, **15**: 311–33.

Robinson, E. A., and S. Treitel. 1980. *Geophysical Signal Analysis*. Englewood Cliffs, N.J.: Prentice Hall.

Russell, B. H. 1988. *Introduction to Seismic Inversion Methods*, SEG Course Note Series 2. Tulsa: Society of Exploration Geophysicists.

Satlegger, J. W., and P. K. Stiller. 1974. Section migration before stack, after stack, or in between. *Geophys. Prosp.*, **22**: 297–314.

Satlegger, J. W., P. K. Stiller, J. A. Echterhoff, and M. K. Hentschke. 1980. Common-offset-plane migration. *Geophys. Prosp.*, **28**: 859–71.

Schneider, W. A. 1978. Integral formulation for migration in two dimensions and three dimensions. *Geophysics*, **43**: 49–76.

Schneider, W. A., and M. M. Backus. 1968. Dynamic correlation analysis. *Geophysics*, **33**: 105–26.

Schneider, W. A., E. R. Prince, and B. F. Giles. 1965. A new data-processing technique for multiple attenuation exploiting differential normal moveout. *Geophysics*, **30**: 348–62.

Schultz, P. S., and J. W. C. Sherwood. 1980. Depth migration before stack. *Geophysics*, **45**: 376–93.

Sheriff, R. E. 1978. *A First Course in Geophysical Exploration and Interpretation*. Boston: International Human Resources Development Corp.

Sheriff, R. E. 1991. *Encyclopedic Dictionary of Exploration Geophysics*, 3d Ed. Tulsa: Society of Exploration Geophysicists.

Sherwood, J. W. C., and A. W. Trorey. 1965. Minimum-phase and related properties of the response of a horizontally stratified absorptive earth to plane seismic waves. *Geophysics*, **30**: 191–7.

Silverman, D. 1967. The digital processing of seismic data. *Geophysics*, **32**: 988–1002.

Smith, M. L., J. A. Scales, and T. L. Fischer. 1992. Global search and genetic algorithms. *Leading Edge*, **11**: 22–6.

Stewart, R. R. 1991. *Exploration Seismic Tomography Fundamentals*, SEG Course Note Series 3. Tulsa: Society of Exploration Geophysicists.

Stoffa, P. L., P. Buhl, and G. M. Bryan. 1974. The application of homomorphic deconvolution to shallow-water marine seismology. *Geophysics*, **39**: 401–26.

Stoffa, P. L., P. Buhl, J. B. Diebold, and F. Wenzel. 1981. Direct mapping of seismic data to the domain of intercept time and ray parameter – A plane-wave decomposition. *Geophysics*, **46**: 255–67.

Stoffa, P. L., and M. K. Sen. 1992. Nonlinear multiparameter optimization using genetic algorithms: Inversion of plane-wave seismograms. *Geophysics*, **56**: 1794–810.

Stolt, R. H. 1978. Migration by Fourier transform. *Geophysics*, **43**: 23–48.

Taner, M. T. 1976. Simplan, simulated plane-wave exploration. Paper read at the 46th Annual SEG Meeting (abstract in *Geophysics*, **42**: 186–7).

Taner, M. T. 1980. Long-period sea-floor multiples and their suppression. *Geophys. Prosp.*, **28**: 30–48.

Taner, M. T., and F. Koehler. 1969. Velocity spectra: Digital computer derivation and applications of velocity functions. *Geophysics*, **34**: 859–81.

Taner, M. T., and F. Koehler, 1981. Surface-consistent corrections. *Geophysics*, **46**: 17–22.

Taner, M. T., F. Koehler, and K. A. Alhilali. 1974. Estimation and correction of near-surface time anomalies. *Geophysics*, **39**: 441–63.

Taner, M. T., F. Koehler, and R. E. Sheriff. 1979. Complex seismic trace analysis. *Geophysics*, **44**: 1041–63.

Taner, M. T., and R. E. Sheriff. 1977. Application of amplitude, frequency, and other attributes to stratigraphic and hydrocarbon determination. In *Seismic Stratigraphy – Applications to Hydrocarbon Exploration*, C. E. Payton, ed., pp. 301–28.

AAPG Memoir 26. Tulsa: American Association of Petroleum Geologists.

Telford, W. M., L. P. Geldart, and R. E. Sheriff. 1990. *Applied Geophysics,* 2d ed. Cambridge, U.K.: Cambridge University Press.

Treitel, S., J. L. Shanks, and C. W. Frasier. 1967. Some aspects of fan filtering. *Geophysics,* **32:** 789–900.

Vasudevan, K., W. G. Wilson, and W. G. Laidlaw. 1991. Simulated annealing statics computation using an order-based energy function. *Geophysics,* **56:** 1831–9.

Webster, G. M. 1978. *Deconvolution.* Tulsa: Society of Exploration Geophysicists.

Yilmaz, O. 1987. *Seismic Data Processing.* Tulsa: Society of Exploration Geophysicists.

Yilmaz, O., and J. F. Claerbout. 1980. Prestack partial migration. *Geophysics,* **45:** 1753–79.

10
Geologic interpretation of reflection data

Overview

Interpretation, as we use the word in this chapter, involves determining the geologic significance of seismic data. This necessarily involves geologic terminology and we follow the usage in Bates and Jackson (1987). Interpretation also sometimes includes data reduction, selecting events believed to be primary reflections, and locating the reflectors with which they are associated. Indeed, a number of decisions have to be made in data processing, acquisition, and even in the initial planning of a survey that prejudice the geologic conclusions and thus could be legitimately included as part of interpretation.

There are few books that concentrate on the geologic interpretation of seismic data; we cite Fitch (1976); Anstey (1977, 1980a, 1980b); Sheriff (1980); McQuillin, Bacon, and Barclay (1984); Badley and Anstey (1984); Badley (1985); Gries and Dyer (1985); and Brown (1991). Numerous scientific sections and their structural/stratigraphic interpretations are shown in Bally (1983–4, 1987–9). A number of interpretation papers are published in *The Leading Edge,* and the Society of Exploration Geophysicists periodically publishes reprints of these (SEG, 1989–92).

It is rare that the correctness or incorrectness of an interpretation can be ascertained because the actual geology is rarely ever known in adequate detail. The test of a good interpretation is consistency rather than correctness (Anstey, 1973). Not only must a good interpretation be consistent with all the seismic data, it also must be consistent with all that is known about the area, including gravity and magnetic data, well information and surface geology, as well as geologic and physical concepts.

One can usually be consistent and still have a choice of interpretations, the more so when data are sparse. The interpreter should explore various possibilities, but usually only one interpretation is wanted, that which offers the greatest possibilities for significant profitable hydrocarbon accumulation (assuming this is the objective). An interpreter must be optimistic, that is, he must find the good possibilities. The optimistic interpretation is often preferable to the "most probable," because the former will probably cause additional work to be done to test (and perhaps modify) the interpretation, whereas a nonoptimistic interpretation may result in abandoning the area. Management is usually tolerant of optimistic interpretations that are disproven by subsequent work, but failing to

recognize a possibility is an "unforgivable sin." It should be noted that "success" or "failure," that is, finding or failing to find hydrocarbons in commercial quantity, is often a poor test of interpretation because many factors critical to commercial accumulations cannot be predicted from seismic data. It may also be noted than an optimistic interpretation is not always what is sought. In engineering and coal seismic work, for example, one is apt to want to know the worst possible case so that these eventualities can be investigated further.

Seismic data are usually interpreted by geophysicists or geologists. The ideal interpreter combines training in both fields. He fully understands the processes involved in the generation and transmission of seismic waves, the effects of the recording equipment and data processing, and the physical significance of the seismic data. His geologic experience helps him assimilate the mass of data, some of it conflicting, and arrive at the most plausible geologic picture. Unfortunately, not all interpreters have the requisite knowledge and experience in both geology and geophysics, and often the next best alternative is to have a geophysicist–geologist team working in close cooperation.

Deducing geologic significance from the aggregate of many minor observations tests the ingenuity of an interpreter and his in-depth understanding of physical principles. For example, downdip thinning of reflection intervals might result from a normal increase of velocity with depth as well as from sediment thinning, and flow of salt or shale may cause illusory structure on deeper horizons. Geometric focusing produced by reflector curvature can produce various effects, especially if the data have not been migrated correctly, and energy that comes from a source located off to one side of the line can interfere with the patterns of other reflection events to produce effects that might be interpreted erroneously, unless their true nature is recognized. Improper processing likewise can create opportunities for misinterpreting data (Tucker and Yorsten, 1973).

Inasmuch as our interpretation objective is usually locating hydrocarbon accumulations, this chapter begins with a summary of concepts about the generation and migration of hydrocarbons and the types of situations that trap hydrocarbon accumulations.

A section on interpretation procedures (§10.2) includes both discussions of the philosophy of seismic

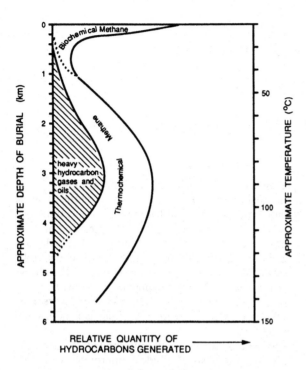

Fig. 10.1 The relation between liquid and gaseous hydrocarbons generated by temperature and depth of burial, assuming a geothermal gradient of 22°C/km. (From Batzie and Wang, 1992.)

interpretation, questions to be answered by the interpretation, and cautions to help avoid erroneous conclusions. Deducing the geologic history of the area is one interpretation objective. Well data have to be related to seismic data so that the interpretation is consistent with both.

Structural maps are commonly the foremost interpretation objective. The tectonic setting usually governs which types of structures are present and how structural features relate to each other, so a review of structural styles precedes discussions of the evidences of various geological features. Among the features examined are faults, folded and flow structures, reefs, unconformities, channels, and stratigraphic traps. Workstations (§10.2.7) are playing increasing roles in interpretation.

Modeling provides a major tool in interpretation. Direct modeling, the making of a synthetic seismogram to show what should be expected from a geologic model, helps in understanding what seismic features should be looked for as evidences of sought-for geologic anomalies. Inverse modeling, the making of synthetic acoustic-impedance or sonic logs from seismic data, aids in seeing the geologic significance of seismic waveshape variations near well control, especially in locating nearby stratigraphic changes suggested by well data.

Lateral variations in velocity can produce the illusion of unreal structural features. Velocity changes affect structural features much more frequently than usually realized. Unrecognized three-dimensional as-

pects can also lead to misinterpretation. However, as 3-D work becomes more common (see fig. 1.21), the importance of 3-D aspects is now being recognized more and more (see also chap. 12).

Stratigraphic interpretation involves delineating seismic sequences, which represent different depositional units, recognizing seismic facies characteristics, which suggest the depositional environment, and analyzing reflection character variations to locate both stratigraphic changes and hydrocarbon accumulations. Three-D work is especially important in recognizing stratigraphic features by their distinctive shapes. Hydrocarbon accumulations are sometimes indicated by amplitude, velocity, frequency, or waveshape changes (§10.8). The variation of amplitude with angle (or with offset) is one of the newer hydrocarbon indicators (HCI).

New roles are being played by seismic methods in the delineation, description, and surveillance of reservoirs. These are discussed further in §14.4 and 14.5.

Seismic data are also proving useful in crustal studies, as aids in indicating and delineating deeper features than those usually associated with sedimentary basin exploration.

10.1 Basic geologic concepts

10.1.1 Generation and migration of hydrocarbons

The interpretation of seismic data in geologic terms is the objective and end product of seismic work. However, before discussing this most important and critical phase of interpretation, we shall review briefly some basic geologic concepts that are fundamental in petroleum exploration.

Petroleum is a result of the decomposition of plant or animal matter in areas that are slowly subsiding. These areas are usually in the sea or along its margins in coastal lagoons or marshes, occasionally in lakes or inland swamps. Sediments are deposited along with the organic matter, and the rate of deposition of the sediments must be sufficiently rapid that at least part of the organic matter is preserved by burial before being destroyed by decay. Restricted circulation and reducing (rather than oxidizing) conditions favorable for hydrocarbon preservation are found in the deeper portions of both marine and lacustrine waters. As time goes on and an area continues to sink slowly (because of the weight of sediments deposited or because of regional tectonic forces), the organic material is buried deeper and hence is exposed to higher temperatures and pressures. Eventually, chemical changes result in the generation of petroleum, a complex, highly variable mixture of hydrocarbons, including both liquids and gases (part of the gas being in solution because of the high pressure). Temperature in the earth generally increases at a rate of 20–55°C/km, in some places (for example, Sumatra) by as much as 100°C/km. The habitat of liquid petroleum generation (fig. 10.1) is generally 65–150°C, which is usually in the 1.5–3-km range. At depths of 3–6 km, reservoirs predominantly contain gas rather than oil, and at still greater depths, the temperature is apt to be so high as to cause gas to decompose.

Sedimentary rocks are porous, generally being deposited with about 45% porosity. As sediments are piled on top, the weight of the overburden compacts the rocks and the porosity becomes less (fig. 5.3). Some of the water that filled the interstices in the rock (*interstitial water*) escapes during the compaction process until the pressure of the water equals that of the hydrostatic head corresponding to its depth of burial. If the formation water cannot escape, it becomes overpressured (see §5.3.4).

Petroleum collects in the pore spaces in the source rock or in a rock adjacent to the source rock, intermingled with the remaining water that was buried with the sediments. When a significant fraction of the pores is interconnected so that fluids can pass through the rock, the rock is *permeable*. Permeability permits the gas, oil, and water to separate partially because of their different densities. The oil and gas tend to rise, and they will eventually reach the surface of the earth and be dissipated unless they encounter a barrier (called a *trap*) that stops the upward fluid migration. Fracturing sometimes plays an important role in the movement of fluids to the boreholes from which they can be produced.

10.1.2 Types of traps

The essential characteristic of a trap is a porous, permeable bed (*B* in fig. 10.2a) overlain by an impermeable bed (*A*), which prevents fluid from escaping. Oil and gas can collect in the reservoir of an anticline until the anticline is filled to the *spill point*. Whereas fig. 10.2a is two-dimensional, similar conditions must hold for the third dimension, the structure forming an inverted bowl. The spill point is the highest point at which oil or gas can escape from the anticline; the contour through the spill point is the *closing contour*, and the vertical distance between the spill point and the highest point on the anticline is the *amount of closure*. In fig. 10.2b the closing contour is the −2085-m contour and the closure is 30 m. The quantity of oil that can be trapped in the structure depends upon the amount of closure, the area within the closing contour, and the thickness and porosity of the reservoir beds.

Figure 10.2a might be the cross-section of an anticline or a dome, and the trap is called an *anticlinal trap*. Other structural situations can also provide traps. Figure 10.2c shows *fault traps* in which permeable beds, overlain by impermeable beds, are faulted against impermeable beds. A trap exists if there is also closure in the direction parallel to the fault, for example, because of folding, as shown by the contours in fig. 10.2d. Figure 10.2e shows possible traps associated with thrust faulting.

Figure 10.2f shows a *stratigraphic trap* in which a permeable bed grades into an impermeable bed, as might result where a sand grades into a shale. Sometimes permeable beds gradually thin and eventually pinch out to form *pinchout traps*. (Stratigraphic traps of various types are also shown in fig. 10.42.) Closure must also exist at right angles to the diagram, possibly because of folding or faulting. Many traps involve both stratigraphic and structural aspects.

Figure 10.2g shows *unconformity traps*, which may result from permeable beds onlapping an uncomformity or beds truncated by erosion at an unconformity (see also fig. 10.42). If the permeable beds are overlain by impermeable ones and if there is closure at right angles to the diagram, hydrocarbons can be trapped at the unconformity.

Figure 10.2h shows a limestone *reef* that grew upwards on a slowly subsiding platform. The reef was originally composed of coral or other marine animals with calcareous shells that grow prolifically under the proper conditions of water temperature and depth. As the reef subsides, sediments are deposited around it. Eventually the reef stops growing, perhaps because of a change in the water temperature or the rate of subsidence, or because there is so much sediment suspended in the water that sufficient light to support reef growth no longer exists, and the reef may be buried. The reef material is often highly porous and covered by impermeable sediments. Reefs sometime produce arching in overlying sediments because of differential-compaction effects, the reef being generally less compactible than the sediments on either side of it. The reef may form a trap for hydrocarbons generated in the reef itself or flowing into it from other beds.

Figure 10.2i represents a *salt dome* formed when a mass of salt flows upwards under the pressure resulting from the weight of the overlying sediments. Below 1–1.5 km, salt is apt to be buoyant compared to denser surrounding sediments, which tend to subside as the basin subsides, whereas the salt tends to remain at roughly the same depth. Eventually, the salt becomes cut off from the underlying mother salt and may take on a teardrop shape, overhanging deeper sediments. The salt dome bows up sedimentary beds, produces faulting, and affects the nature of the beds being deposited. Consequently, traps may be produced over or around the sides of the dome and sometimes within cavities in caprock over the salt, the trapping sometimes resulting from dip reversal, faulting, unconformities, or stratigraphic changes. Because salt is impermeable, hydrocarbons also may be trapped beneath the salt.

The primary objective of a seismic survey for hydrocarbons is usually to locate structures such as those shown in fig. 10.2. However, many structures that provide excellent traps do not contain oil or gas in economic quantities. We also try to derive from the seismic data as much information as possible about the geologic history of the area and about the nature of the rocks in an effort to form an opinion about the probability of encountering petroleum in the structures that we map.

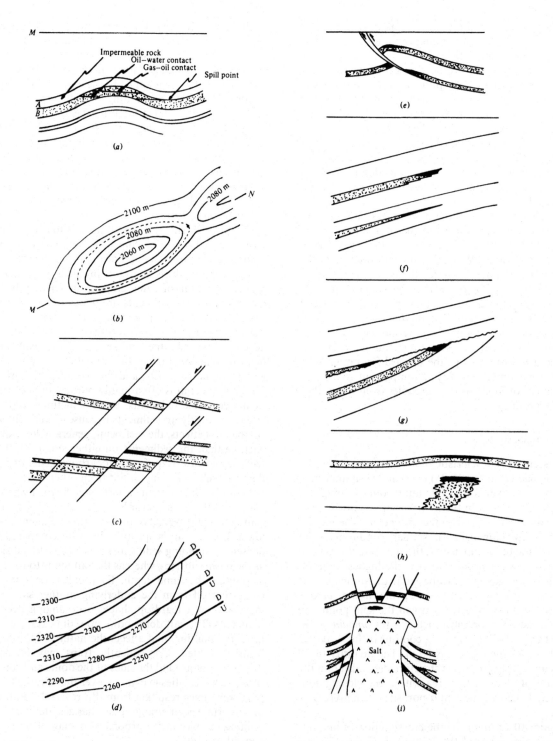

Fig. 10.2 Sedimentary structures that produce hydrocarbon traps. Permeable beds are dotted in the cross sections; hydrocarbon accumulations are in black. (a) Vertical section through anticline along line *MN* in (b); (b) map of the top of the permeable bed in (a) with the spill-point contour dashed; (c) vertical section through fault traps; (d) map of the middle permeable bed in (c); (e) possible traps associated with thrust faulting; (f) stratigraphic traps produced by lithologic change and pinchout; (g) unconformity traps; (h) a trap in a reef and in draping above the reef; and (i) possible traps associated with a salt dome.

10.2 Interpretation procedures

10.2.1 Fundamental geophysical assumptions

Seismic interpretation generally assumes (1) that the coherent events seen on seismic records or on processed seismic sections are reflections from acoustic-impedence contrasts in the earth, and (2) that these contrasts are associated with bedding that represents the geologic structure. Thus, mapping the arrival times of coherent events is related to the geologic structure, and by allowing for velocity and migration effects, we obtain a map showing the geologic structure. We also assume (3) that seismic detail (waveshape, amplitude, and so on) is related to geologic detail, that is, to the stratigraphy and the nature of the interstitial fluids; we examine this assumption further in §10.7 and 10.8.

10.2.2 Collection and examination of data

(a) Introduction. The interpreter gathers together all the data relevant to the interpretation, including geologic, well data, and so on. The relevant seismic data usually include seismic sections, a base map, and velocity and other data from the field or generated in processing. Sometimes the interpretation is done concurrently with the field and processing work, so that the interpreter receives additional data while carrying out the interpretation, and he may be able to feed back conclusions from the preliminary interpretation so that field or processing procedures can be changed or additional work can be carried out, in order to prove or disprove points that are not resolved.

Alternative ways of interpreting data are almost always possible. This "inherent ambiguity" exists with almost any data, although ambiguity in seismic interpretation is less than with most geophysical and geologic data. Ambiguity arises because data are incomplete and/or inaccurate, and the best way to reduce ambiguity is to add more data. The added data might be more seismic data, but it also might be information from surface geology, wells, gravity measurements, and so on. The regional geologic setting and concepts about the tectonic stresses to which the region has been subjected should also be used as a check on seismic information.

As an example, one sometimes encounters disruptions in a seismic reflection. If we explain this as caused by faulting, then we must determine what else the fault did. Where did it cut shallower beds, or did it die out? Where did it cut deeper beds, or was the fault displacement absorbed by flowage in mobile salt or shale sediments, or did the fault sole out into the bedding plane? Where is the fault on parallel and intersecting lines, or did it die out laterally? Is the fault a normal fault indicating extension or a reverse fault indicating compression? An interpretation cannot be regarded as complete until such questions have been answered as completely as possible. A fault that dies out both shallower and deeper is difficult to justify

(though occasionally this is the correct interpretation). Faults that have not produced effects on nearby lines may also be difficult to justify.

(b) Examining sections. One of the first tasks of an interpreter is to examine the data for evidences of mislocation (do sections tie properly?) or improper acquisition or processing. Such an examination, although not conclusive, often uncovers gross errors. Unmigrated seismic traces at the intersections of seismic lines ought to copy. When they do not, mislocation or mislabeling of one or both of the lines is a possible explanation, but differences in acquisition or processing techniques also provide possible explanations. The title blocks of the sections should be examined to see what differences exist. Different-size arrays, different mixes of offset distances, or different processing procedures may have resulted in noise contributions being different. Lack of full multiplicity at the ends of lines or where source or geophone spacing is irregular (possibly because of access problems) may have affected data quality. Features that line up vertically on unmigrated sections are especially suspect because geologic features are usually not vertical, whereas the effects of statics errors often are. Occasionally, files are mixed up and data are assembled incorrectly. The various data elements should be consistent; if the velocity were assumed to vary, are the assumptions consistent with the structure and the character of the sections? Are certain data that show on sections made as intermediate steps in the processing missing or changed on the final sections? Unexplained differences or departures from what is geologically reasonable should be investigated, so that geologic significance is not attributed to errors in the data.

Figure 10.3 shows both ends of a typical seismic section, including the data block. The data block is often subdivided into parts listing information about line identification, data acquisition, and processing. A generalized line direction and horizontal scale may be given. Where the horizontal scale is not indicated, it can be determined by counting the number of traces per centimeter, the trace spacing being half the geophone group spacing (assuming no horizontal compositing). The locations at which velocity analyses were run are usually indicated, often with the results of the analyses tabulated as time–velocity pairs. These should be examined for consistency along the line. The locations of changes in line direction or abrupt surface changes (such as elevation differences or changes in water depth) should be noted for their possible effects on reflection quality or attitude. Irregularities in coverage are common in land data because of surface or access problems; these often show as irregularities in the first-break patterns and they may also affect reflection quality and the apparent attitude of reflections. The multiplicity involved in each trace is sometimes shown by encoding at the bottom of the section, providing a key to irregularities of coverage.

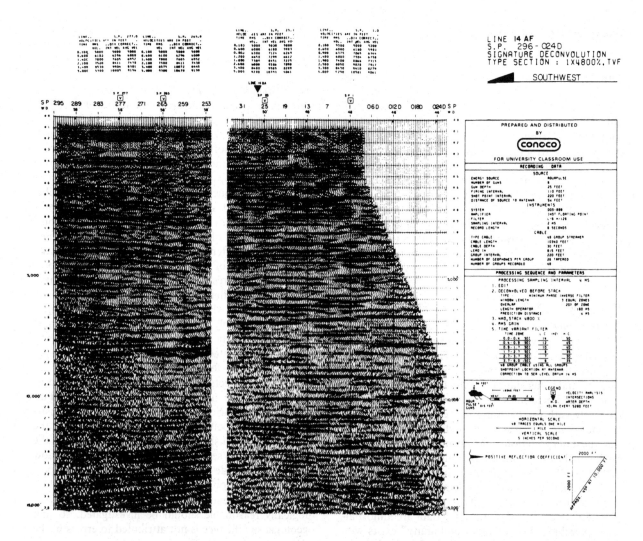

Fig. 10.3   Left and right ends of a seismic section. The data were recorded with a 48-channel streamer with numbered source points every 220 ft, individual traces being spaced 110 ft (33.5 m) apart (this information can be used to get the horizontal scale). Full 48-fold CMP multiplicity is not achieved for the 47 traces at each end of the line, which may explain some of the data-quality deterioration evident in these regions; the lack of full multiplicity affects multiple attenuation, so that multiples become more likely in those regions. The locations of velocity analyses are indicated by a "V" at the top of the section, and the tabulated data give stacking (labeled RMS), interval, and average velocities (see eqs. (4.26), (5.25), and (5.20)) assuming horizontal velocity layering. The solid triangle above SP 27 indicates an intersecting seismic line; TVF stands for time-variant filter, WD for water depth, RMS gain means that the RMS amplitude of each trace has been brought to the same value; LC and HC indicate the frequencies for which the low-cut and high-cut filters introduce 3-dB attenuation. (Courtesy of Conoco.)

The vertical scale shown on the section is usually linear in time. Depth equivalents are sometimes given, but these are only intended to be approximate; depth determinations can be made more accurately by measuring the arrival times and converting these to depths using the appropriate velocity relations. Where processing or display parameters are changed with arrival time, the locations of changes should be noted so that changes in quality produced by the changes in processing are not interpreted as geologic changes. If changes have been made in the midst of the objective section, or if the horizon being mapped varies in depth so as to cross the zone of changed parameters, special care has to be employed to avoid possible misinterpretations.

(c) Interpretation approaches. Interpretation involves building a model of the prospect area in the interpreter's mind. Some interpreters hang all their data on their walls, thus surrounding themselves with the data so that they can look from one section to another to see interrelationships better. They develop much of their interpretation sitting back and pondering about which ideas are feasible, rather than being busy all the time timing events, transferring data to base maps, and drawing contours. Imagination is required in interpretation and it takes time to develop an interpretation that leads to new discoveries.

Two basically different approaches are made to seismic interpretation: the one focuses on objectives; the other gradually builds up a complete interpretation.

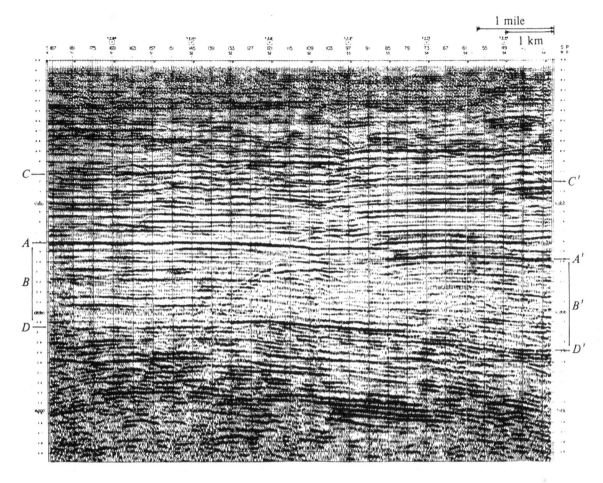

Fig. 10.4 Portion of a line in the U.S. Gulf Coast area. (Courtesy of Conoco.)

Often only a few reflectors are considered to be of interest because it is already known that the only prospective reservoirs are in one part of the section. Such areas usually contain wells in which formations of interest are identified and then related to seismic sections, either by synthetic seismograms or, more simply, by using velocity to relate depth in the wells to arrival time on the seismic sections. The associated events on the seismic sections are followed throughout the region to give structural relationships, looking for faults that displace the reflections and watching for reflection character changes that may indicate thickness changes, sand pinchouts, patch-reef growths, other stratigraphic variations, or hydrocarbon accumulations.

The alternative approach is to develop a more complete interpretation of an entire section. An interpreter generally starts with the most obvious features, usually the strongest reflection event or the reflection event that possesses the most distinctive character, and follows this feature as long as it remains reliable. When the feature being followed deteriorates or changes character so that it is not clear what is happening, it is dropped rather than "pushed" or extrapolated beyond its region of reliability, and the interpreter returns to it subsequently after he has

developed other features of reasonable reliability. By following features in the order of their reliability, overall interpretation is developed before attention is concentrated on the objective reflections. This type of approach usually leads to a better interpretation, but it is much more time-consuming. An interpretation of the section shown in fig. 10.4, for example, might start with mapping of the strong reflector AA', distinguished in part by the following zone of poor reflections BB'. Reflector AA' is cut by a normal fault, and following this fault onto intersecting seismic lines might be the next step. Attention might then shift to mapping a shallower horizon, perhaps CC', but here it will be difficult to be sure one is staying with the same reflector. Perhaps one would try to map the base of the BB' zone next; this will also involve uncertainties, and to resolve some of these it may be necessary to go back and revise the parts interpreted earlier. The petroleum objectives are about the middle of the BB' portion, so a map will have to be made here, but this map can be made more reliably with the aid of more certain events already mapped.

The interpretation of seismic sections is sometimes done from the top downward. Shallow features such as channels, surface relief, local pockets of gas, and irregular coverage because of surface obstacles may produce deeper artifacts, so that early recognition of shallow features may prevent misinterpretation of the deeper features. However, sometimes deep structure such as basement faulting has prejudiced subsequent structure, so that it is better to proceed from the bottom upward. There is no automatically "best way" to proceed with an interpretation.

10.2.3 Picking reflections

Making a seismic structure map generally consists of four operations: (1) selecting which events, and what point (peak, trough, zero crossing) on each event, are to be mapped (*picking*); (2) measuring the arrival time of each pick (*timing*) and converting the values to depths if a depth map is to be made rather than a time map; (3) writing the values on a base map (*posting*); and (4) connecting the posted values to represent the structure (*contouring*). Other relevant information such as well data, regional trends, anticlinal and synclinal axes, the location of gravity highs and lows, interruptions because of faulting or indications as to the relative reliability of posted values (*grading*) will also be posted on the base maps.

Reflections are usually identified with rock units by correlating with well-log data, often using a synthetic seismogram (§6.2.1) or vertical seismic profile (§13.4) as the relating device. Clearly, the amplitude and

phase spectra of the wavelets embedded in both the seismic data and the synthetic seismogram must match. A match is often determined by varying the spectra (especially the phase spectrum of the synthetic seismogram). The match over any appreciable portion of a seismic trace is hardly ever good enough that one can be absolutely confident that it is correct. A seismologist is apt to attribute matching problems to errors in the seismic data, unaccounted-for noise being the most probable source. However, the problems may also lie in the well-log data as these also involve measurement uncertainties.

Picking reflections is usually based on following the same phase from trace to trace, usually a peak or a trough. Where the reflection character changes becaue component beds thicken or thin or change their lithologic makeup, routine following of the same phase may introduce errors, and the interpreter should think about what he is trying to map and take appropriate liberties in interpreting the data. Where peaks (or troughs) split into two events (for example, in figs. 10.5d to 10.5f), which way to follow the event depends on what the interpreter thinks is happening, for example, whether there was erosional topography at an uncomformity so that the cause of local thickening

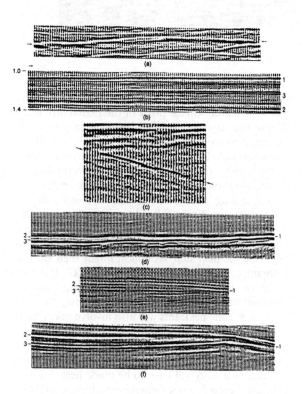

Fig. 10.5   Problems in reflection picking. (From Fitch, 1976: 5, 11, 12.) (a) Variations in character due to noise; signal/noise ratio = 1.5; (b) convergence of reflectors attributed to differential subsidence; (c) dipping reflector terminating at an unconformity; (d, e, f) reflection 1 splitting into reflections 2 and 3.

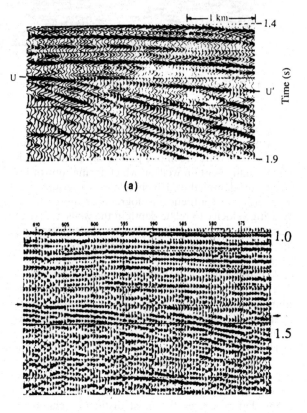

Fig. 10.6   Changing character of unconformity reflections. (a) Polarity reversal as different beds subcrop at unconformity (courtesy of Grant Geophysical). (b) Weak and variable reflection at an angular unconformity (from Fitch, 1976: 68).

lies below the unconformity surface or whether the thickness of the basal members above the unconformity may be changing. The polarity of unconformity reflections may even reverse as different beds subcrop (fig. 10.6) or onlap the unconformity.

10.2.4 Mapping reflecting horizons

The horizons that we draw on seismic sections provide us with a two-dimensional picture only. A three-dimensional picture is necessary to determine whether closure exists, the area within the closing contour, the location of the highest point on the structure, and so on. To obtain three-dimensional information, we usually run lines in different directions. Most reflection surveys are carried out along a more-or-less rectangular grid of lines, often with common midpoints at the intersections of lines to facilitate correlating reflections on the intersecting profiles.

Events picked on one section are compared with those on intersecting sections in order to identify the same horizons; identification is made on the basis of character and arrival times. The horizons are now "carried" along the cross-lines, and, ultimately, along all lines in the prospect to the extent that the quality of the data permits.

When a horizon can be carried all the way around a closed loop, we should end up with the same arrival time with which we started. This *closing of loops* provides an important check on reliability. When a loop fails to close within a reasonable error (which depends mainly upon the record quality and the accuracy of the weathering corrections), the cause of the misclosure (mistie) should be investigated carefully. Migrated sections have to be tied by finding the same reflection on the intersecting sections; such tie points will be displaced from the midpoint by the amount of the migration on each of the lines. Often, misclosure is due to an error in correlating along the line or from line to line, possibly because of inaccurate corrections, a change in reflection character, or error in correlating across faults. When the dip is different on the two sides of a fault or the throw varies along the fault, an incorrect correlation across the fault may result in misclosures (but not necessarily). After the sources of misclosure have been carefully examined and the final misclosure reduced to an acceptable level, the remaining misclosure is distributed around the loop.

Interpretation is usually done on migrated sections whenever these are available. Migrating usually improves the signal/noise ratio, sharpens up fault evidences, and clarifies features whose evidences are events having different dips, such as pinchouts. However, because migration is based on only the components of dip in the in-line direction (except for 3-D data), dipping reflections will generally not time tie at line intersections. It may be necessary to locate the events being mapped on the unmigrated sections to ascertain that they tie.

Fault planes should be picked as accurately as pos-

sible on seismic lines as an aid to identifying fault cuts seen on different lines involving the same fault. Fault planes are often located by the termination of reflections that have only small dips, so fault planes often can be time tied at line intersections with sufficient confidence to determine fault strike (fig. 10.29).

After horizons have been carried on the sections, maps are prepared. For example, we might map a shallow horizon, an intermediate horizon at roughly the depth at which we expect to encounter oil, if any is present, and a deep horizon. We map on a *base map*, which shows the locations of the seismic lines (usually be means of small circles representing midpoints) and other features such as oil wells, rivers, shorelines, roads, land and political boundaries, and so on. Values representing the depth of the horizon below the datum plane are posted on the map, usually at the midpoints (although, strictly speaking, they should be posted at the reflecting points on the respective horizons). Whereas interpreters often time data at regular midpoint intervals, determining and posting the intersections of contour values with the picked horizon eliminates the need to interpolate along the lines while doing the contouring. Faults that have been identified on the record sections are drawn on the map and the depth values are then contoured.

Where unmigrated data are contoured, the unmigrated map can be digitized and migrated using map-migration algorithms.

The amount of contour smoothing to be done depends on the assessment of how much noise there is in the data. One does not want to either smooth through real geologic features or contour the noise. The amount of noise clearly sets the minimum contour interval that should be used. Depicting small structural features believed to be real generally sets the contour interval, but larger contour intervals are often used because small features may not be significant even though real. Dashed contours intermediate between integral contours or using different contour intervals for different parts of a map often conveys the impression of structural relief better than use of a smaller uniform contour interval.

After the structural information has been extracted, the next step is to work out as much as possible of the geologic history of the area. Fundamental in this connection is the determination of the ages of the different horizons, preferably according to the geologic time scale, but at least relative to one another. Often, seismic lines pass close enough to wells to permit correlating the seismic horizons with geologic horizons in the wells. Refraction velocities (if available) may help identify certain horizons. The strongest, most obvious, and easiest identifiable reflections are often associated with unconformities. Occasionally, a particular reflection has a distinctive character that persists over large areas, permitting not only it to be identified, but also other events by their relation to it. Notable examples of persistent identifiable reflections are the low-frequency reflections sometimes associ-

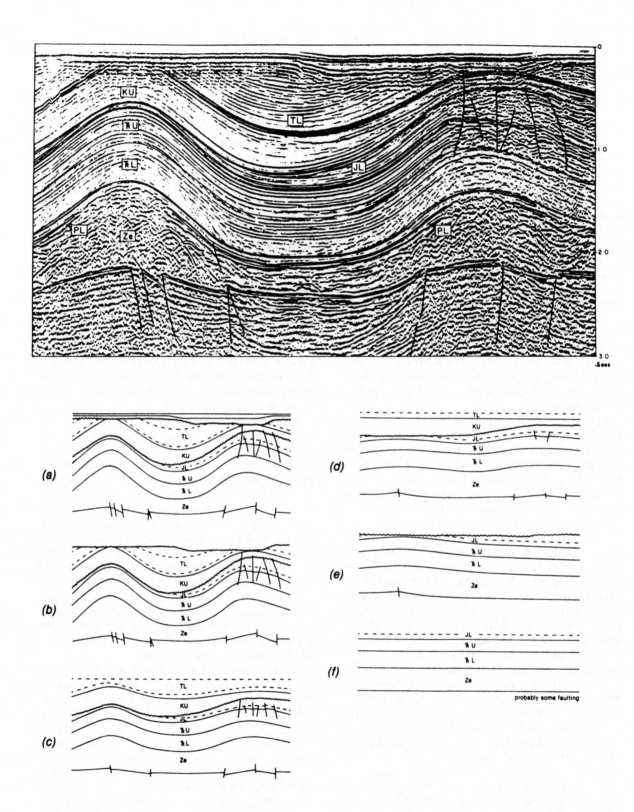

Fig. 10.7 Flattening reflections on a seismic section to work out geologic history. (From Taylor, 1981.) (a) Major interpreted horizons in the above seismic section; (b to f) restoration of horizons to their presumed attitude as successive portions are removed. The nearly constant thickness of the Triassic section (ℝ)

and lowermost part of the Jurassic indicates little salt movement had occurred. The left dome began growing in the Jurassic (J), as evidenced by erosion of the top Jurassic; the right dome began growing in the Cretaceous (K). The major dome growth took place during the Lower Tertiary (TL).

ated with massive basement and the prominent reflection from the top of the Ellenburger, a limestone
encountered in Northern Texas.

10.2.5 Deducing geologic history

Seismic sections often subdivide naturally into units.
The boundaries between units are often the better reflectors, and the units often have angular relations to
each other that indicate features of geologic history:
periods of tectonism, unconformities, transgressions,
and so on. The boundaries between units generally indicate a gap in geologic time and often the unit
boundaries separate sediments deposited in different
kinds of environments. Velocity and other seismic
measurements, such as of amplitude or instantaneous
frequency, and their variations in the direction of the
bedding yield additional information. Lithology and/
or stratigrapic situations are usually inferred from
many evidences that are individually weak but that,
taken together, make a coherent pattern.

Isopach maps, which show the thickness of sediments between two horizons, are useful in studying
paleostructure and structural growth. Ideally, only
one rock unit should be encompassed by the interval
between horizons, but often the only horizons that
can be mapped reliably are separated by more than
one rock unit, so that the resulting isopach map may
show more than one period of movement or more
than one depositional trend. The interval between horizons is often measured in terms of two-way traveltime rather than thickness in meters or feet, it being
implied that velocity-variation effects are minor compared to thickness variations. Isopach maps are often
prepared by overlaying maps of two horizons and subtracting the contour values wherever the contours on
one map cross the contours on the other. The differences are recorded on a blank map and then contoured. If isopach contours show a trend toward increased thickness in a certain direction, it may suggest
that the region was tilted downward in this direction
during the period of deposition or that the source of
the sediments is in this direction. Uniform thickness
of a folded competent bed indicates that the folding
came after the deposition, whereas deposition probably was contemporaneous with the growth of an anticline if the thickness increases away from the crest.
Growth during deposition is usually more favorable
for petroleum accumulation because it is more likely
for reservoir sands to be deposited on the flanks of
structures with even slight relief.

Paleosections (palinspastic sections) can be made by
time shifting traces to flatten some distinctive horizon
that can be assumed to have been deposited horizontally; the objective is to show relationships as they ex-
isted at the time of deposition of this horizon (see figs.
10.7 and 10.8). In practice, such flattening is often
done in the interpreter's mind rather than by actually
manipulating the data because of the cost of reprocessing, but in areas of even moderate complica-

tions actual flattening can be worthwhile. Obviously,
migration before the flattening is necessary and the
flattened horizon should be selected judiciously. Interpretation workstations make it easy to flatten a horizon by simple subtraction. Some also permit removal
of fault throws along the fault planes to avoid creating
artifacts of faults vertically underneath fault restora-
tions. Compaction effects and changes in velocity
since the deposition of the flattened horizon should be
allowed for, but usually the information required to
do this is lacking.

The unraveling of the geologic history of the area is
important in answering questions such as the following: (a) How would the paleotopography have affected
the stratigraphy and the lithology deposited? (b) Was
the trap formed prior to, during, or subsequent to the
generation of the oil and gas? (c) Has the trap been
tilted sufficiently to allow any trapped oil to escape?
(d) Did displacement of part of a structure by faulting
occur before or after possible emplacement of oil? Although the seismic data rarely give unambiguous answers to such questions, often clues can be obtained
that, when combined with other information such as
surface geology and well data, permit the interpreter
to make intelligent guesses that improve the probability of finding oil. Alertness to such clues is the "art"
of seismic interpretation and often the distinction between an "oil finder" and a routine interpreter.

10.2.6 Integrating well data into an interpretation

Wells drilled in the area provide geologic information
that must be consistent with the interpretation. Borehole logs (fig. 10.9) are interpreted to determine formation tops, lithology, depositional environment, the
location of faults (fig. 10.10) and unconformities with
an indication of the amount of section missing, and
so on. Well logs plotted linearly in time at the seismic
section scale aid in correlating (fig. 10.11). Although
a borehole provides an opportunity for actual measurements, the results available to a seismic interpreter
are usually interpretations of measurements. When a
disagreement between well and seismic data appears,
both well and seismic data should be reexamined in
order to resolve the problem.

Synthetic seismograms made from well-log data
provide a means of identifying reflections with formation tops provided the wavelet embedded in the seismic data is close to that used in manufacturing the
synthetic. Vertical seismic profiles (§13.4) provide an
even better way of accomplishing this (Hardage,
1985).

Well information usually has to be projected into
seismic data, the wells not having sampled the same
subsurface locations as the seismic data. Even where
a wellhead is located on a seismic line, dip may result
in the seismic work seeing a different portion of the
subsurface than the well. The projection of well information involves an interpretation so that the data are
projected up- or downdip by the correct amount

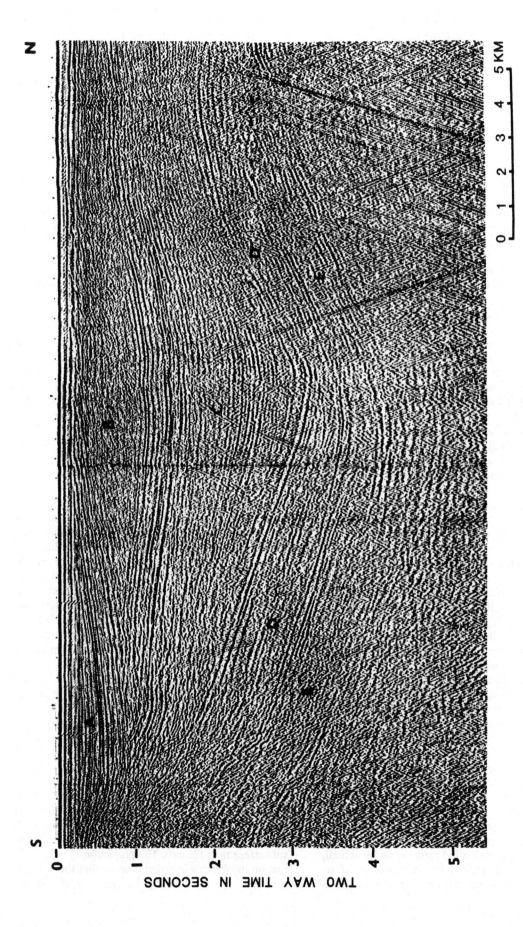

Fig. 10.8 Section across part of the St. George's Channel (Irish Sea) basin. Fairly uniform Permo-Triassic sequences (*D, E*) were deposited with some possible thinning over the basement arch at the right end. The fault at the left end did not have any significant effects. An asymmetric basin developed in the Jurassic (*B, C*), and sometime later uplift and erosion occurred. Finally, faulting developed at the left, producing a Tertiary basin (*A*). (From Dimitropoulis and Donato, 1983.)

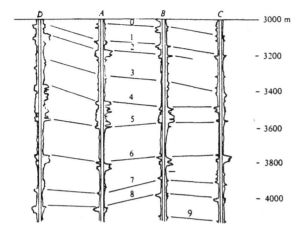

Fig. 10.9 Correlations between well logs within the same field. For each well, the left curve indicates the SP (spontaneous potential) response and the right curve resistivity. Several correlation lines have been drawn and numbered. Some intervals are thinner in one well than in another; the interpreter must decide whether intervals are thinner because section is missing (as a result of faulting or an unconformity), because of stratigraphic variations, or because of miscorrelation. Part of the 3–4 section (60 m) is faulted out of well C, 40 m of the 6–7 section is faulted out of well B, and horizon 5 marks an unconformity, explaining the thicker 4–5 section in A. Obviously, other interpretations are possible and a seismic interpreter should not regard well information as infallible.

(which may vary with depth in both magnitude and direction), making appropriate allowance for faults or other features that intervene. The seismic section is usually plotted in time, whereas the well data are usually in depth, so that the choice of an appropriate velocity for converting the one into the other has to be made.

Even when location problems are not present, the well information results only from the rock within a few centimeters of the borehole (which may have been altered by the drilling process), whereas seismic data include contributions from a large Fresnel-zone region. Well and seismic data may be plotted with respect to different data and time shifts may have been introduced into the seismic data in recording or processing. Furthermore, most reflections are the interference composites of several reflections, and multiples from shallow reflectors or other noise may also affect the interference. In consequence, relating interfaces to specific seismic events is not easy and is often done incorrectly (see also §6.2.1 and 13.4.2).

10.2.7 Workstations

Much seismic interpretation today is done interactively at workstations and the use of workstations is expected to increase rapidly (Brown, 1992). A *workstation* consists of a computer terminal, usually with two fairly large high-resolution screens, upon which data can be displayed, often in color. The display is

driven by a dedicated computer that can be controlled by instructions entered by means of a keyboard or a mouse selecting from a menu displayed on one of the screens. The workstation is sometimes connected to a larger computer or computer network that gives it access to a larger data bank and that can be called upon to perform computing tasks that are too large to be handled efficiently by the workstation computer.

The interpreter can cause portions of the data stored in the system to be displayed quickly (within a few seconds) on the screens and the screen displays can be subdivided so that different portions of the data can be displayed adjacent to each other. Rapid access to arbitrarily chosen portions of the stored data

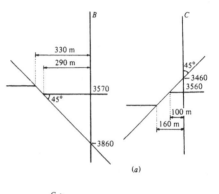

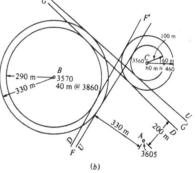

Fig. 10.10 Construction to aid in mapping faults from well picks. Assume that it has been determined that the same normal fault cuts well B at 3860 m with 40 m missing, well C at 3460 m with 60 m missing, and is not seen in well A. From regional considerations, we expect the fault to dip about 45°. Our mapping horizon is at 3605 m in well A, 3570 m in B, and 3560 in C. (a) Section through wells B and C neglecting dip of the horizon. The mapping horizon should encounter the downthrown side of the fault 290 m (3860 − 3570) from well B and the upthrown side 330 m from the well. Likewise the fault's upthrown and downthrown sides should lie 100 and 160 m from well C. (b) Map view showing location of wells and circles to which fault should be tangent; the fault may strike NE (trace *FF′*) or SE (trace *GG′*). If the fault strikes SE, well A is downthrown and we expect the fault at 3600 + 200 = 3805 m in A; if the fault strikes SE, well A is upthrown and we expect the fault at 3605 − 330 = 3275 m. Growth along the fault, dip, and variations of the fault angle introduce uncertainty into the construction. In this instance, it is believed that A is not faulted at 3805 m, but the fault may die out upward before 3275 m, so the lack of fault evidence in A does not prove that A is not upthrown. Hence fault trace *FF′* is preferred to fault trace *GG′*.

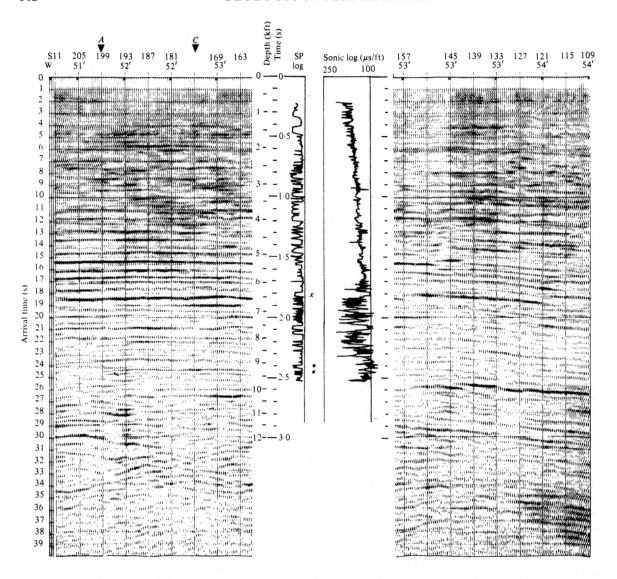

Fig. 10.11 Relating well logs to seismic data. The well is located near the seismic line, the seismic data are plotted at "true" amplitude and are migrated, and the well logs are plotted at the same time scale as the seismic section. Some of the sands as seen on the SP-log (indicated by excursions to the left; see Telford, Geldart, and Sheriff, 1990: fig. 11.12) seem to relate to specific reflections, assuming the log section is slightly high with respect to the seismic section. The sands marked with a dot are productive in the well. The sand marked x is nearly the tuning thickness $\lambda/4$ (§6.4.3), which partially explains why it produces a prominent reflection. (Courtesy of Conoco.)

provides one of the major advantages of workstations, as it makes it easy to check portions of the data against each other to verify their consistency. An interpreter never has time to verify all aspects of the consistency of his interpretation and many more can be verified when it is easier and quicker to do so. Workstations thus provide more consistent interpretations than are made without them.

An interpretation showing picked horizons, faults, well-log ties, and other features can be superimposed on the data. The seismic (and other) data can be manipulated to time shift or color encode them. Algorithms can be applied to carry out simple data-processing operations (§9.13) or so that attributes

(§9.11.4) may be displayed. The facility of being able to see the effects of varying processing parameters encourages trial-and-error experimentation in order to optimize sought-for features, and the ability to display easily various kinds of attributes permits seeing the data from various viewpoints to lessen the likelihood of missing significant features. Color displays considerably broaden the dynamic range of data and thus contribute to the ability to see nonobvious features.

Workstations permit the easy flattening of picked horizons to aid in seeing the attitudes of bedding at the time the picked horizon was deposited, and thus working out the history of structural changes. This is apt to be especially important in areas that have been involved in structural inversion, uplifts becoming depressions, normal faults undergoing reverse movement, and so on, as a result of changes in tectonic stresses.

10.2.8 Drawing conclusions from reflection data

In mapping seismic data, one looks for *leads,* the possibilities of hydrocarbon traps that require more work to define them completely. Although structural leads and local amplitude anomalies may be fairly evident, the interpreter should also be alert for subtle clues, perhaps to channels, that may indicate stratigraphic accumulations. Information from wells in the area may help locate the parts of the section where stratigraphic trapping is most probable. Dip, reflection character, or amplitude variations may indicate stratigraphic or porosity changes. Careful study of the maps, sections, and records plus broad experience and ample imagination will at times disclose accumulations.

Whereas we imply that locating hydrocarbon accumulations is the goal, this is not always the situation. Whatever the goal, the interpreter should be alert to clues that suggest aspects other than those being primarily sought. Interpretation for stratigraphic features (§10.7) should be incorporated with structural interpretation (and vice versa). Much seismic work today is done to aid in field development and production (White and Sengbush, 1987; Sheriff, 1992).

Seismic interpretations are usually cut short before all the meaning that can be found has been extracted, because decisions have to be made, or because the interpreter is needed on another project. The interpreter should try to anticipate the questions that may be asked of the interpretation and prioritize the work so that at any stage of the interpretation, the best answers can be given, recognizing that the answers may change based on further study.

When an interpretation is concluded, a report (see app. G) is usually prepared, often both for submission in writing and for oral presentation. In some ways, this is the most difficult and most important task of the interpreter. He must present his findings in such a way that the appropriate course of action is defined as clearly as possible. The important aspects should not be obscured by presenting a mass of details nor should they be distorted by presenting carefully selected but nonrepresentative maps and sections. Evidences to support significant conclusions should be given. Alternate interpretations should be presented and an estimate given of the reliability of the results and conclusions. Finally, the interpreter should recommend what further action should be undertaken.

10.2.9 Display techniques; color

Display techniques strongly affect the ease with which features can be seen. As one truism states, "You cannot interpret what you do not see."

The optimum display for one interpretation may not be optimum for another. A regional interpretation needs a synoptic view and reduced sections so that the features over a large area can be seen relative to each other. The mapping of a prospect requires larger sections in order to see detail to resolve faults and structures. Locating stratigraphic traps associated with an unconformity requires a full-waveform display of the unconformity reflection on a fairly large scale. Evidences of hydrocarbon accumulations may require displays at very low amplitude (so that "bright spots" become evident), displays with reversed polarity, and displays of attributes such as frequency and velocity. A display of velocity and other attributes may also assist in lithologic identification. Stratigraphic variations may be more evident if appreciable vertical exaggeration is employed, whereas structural interpretation is usually easier if horizontal and vertical scales are nearly the same. The varying requirements in interpretation call for a variety of displays of seismic data. Sheriff and Farrell (1976) show a section displayed with various plotting parameters. Feagin (1981) discusses plotting parameters.

A good black-and-white wiggle-trace display has about 24 dB dynamic range at best, a range that can be considerably enlarged by the use of color (Brown, 1991; Russell, 1992). Color is, however, somewhat subjective because different individuals perceive and distinguish between colors differently. The "brighter" colors of red and orange especially stand out and are most often used to indicate what is perceived to be good – high amplitude, amplitude increasing with offset, low velocities, and so on. Color assignments to numerical values are often done in spectral sequence – red (largest values), orange, yellow, green, and blue. Some advocate juxtaposing contrasting (almost complementary) colors, a practice that emphasizes small contrasts. Color schemes can emphasize significant features, but they can also sometimes emphasize irrelevant features and obscure significant ones (for example, the relatively common use of blue printing on a black background makes it almost invisible to many people). Color can be used to hide as well as to illuminate features.

One of the problems with interpretation is that there is often too much data to be examined and comprehended. Superimposed displays may help in seeing the interrelationship of various data aspects. Color overlays provide one way of adding an additional variable to a display, and color overlays are employed to show amplitude, velocity, frequency, and other aspects (Balch, 1971; Taner and Sheriff, 1977; and Taner, Koehler, and Sheriff, 1979).

Most often interpretations are based on migrated sections; features on these should be checked against unmigrated sections to guard against possible migration errors and to effect ties to intersecting seismic lines. Where an appreciable range of depths is of interest, sections plotted so that the vertical scale is linear with depth rather than with time are useful, especially in working out structural problems. The velocities used in the time-to-depth conversion should be checked, especially if the velocities vary horizontally. Where several outputs from data processing are available, for example, where there are outputs employing different filtering or special processing, these should be examined to see what differences they produce. Different displays may prove better for mapping different horizons, but waveshapes and time delays may change with processing parameters and these changes have to be taken into account in the mapping.

The various products used to control processing should be examined so that the interpreter understands more clearly exactly what was done in the processing and how the decisions made there affect the final product. Velocity analyses should be especially studied for consistency in picking and to yield clues regarding lithology, high-pressure zones, and so on. Where velocity data exist only in tabular forms, one might wish to redisplay them graphically to facilitate understanding the significance of variations, especially where velocity varies along the seismic line.

10.3 Evidences of geologic features

10.3.1 Concepts from structural geology

(a) Structural style and plate-tectonic setting. Many areas have been subjected to fairly simple stress fields that have determined the types and orientations of structural features present, that is, the *structural style*. The stress field may have changed with time, so that the structural style may be different for different portions of the section, or later patterns may be superimposed on older, different patterns. Because the interpretation of features on seismic sections often involves some ambiguity, knowledge of the structural style appropriate to the region can aid in selecting the most probable interpretation and in making the interpretation consistent with everything known about the area, not merely with the seismic data alone. It is also important in programming data acquisition in the most economical manner.

Structural style depends on the tectonic setting (Lowell, 1985), especially the location with respect to plate boundaries and the types of boundaries. The types of plate boundaries (see fig. 10.12) relate to the relative movement of the plates involved: (1) a pull-apart zone where plates are separating (*divergent boundary*), (2) a collision or subduction zone where they are coming together (*convergent boundary*), and (3) a strike-slip zone or *transform boundary* where they are sliding past each other. The latter may involve transform faults, a kind of fault that accommodates the changes where different portions of pull-apart or subduction zones are offset, or the changes between boundaries of different types. Subsidence and isostatic adjustment are important factors influencing structural style away from plate boundaries, for example, at the "passive margins" of continents.

First-order or primary structural features relieve

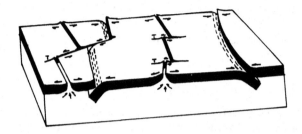

Fig. 10.12 Plate-tectonic model. Magma upwells in rift zones where plates move apart; in convergence (subduction) zones, one plate plunges under another and eventually gets so hot that it melts. Transform faults (*T*) link offsets in rift and/or convergence zones; transform faults may involve plates sliding by each other over only part of their length, for example, only between the active rifts, other portions having been active before the plates grew at the rift zones. (After Isaacs, Oliver, and Sykes, 1968.)

the main stresses produced by plate movements; they also generate secondary stresses that produce secondary structural features, and these generate tertiary features, and so on.

(b) Types of structural style. The major structural styles as classified by Harding and Lowell (1979) are listed in table 10.1. Their first-order distinction is between styles that involve the basement and those that are detached from it.

The beginning of sea-floor spreading seems to be an uplift resulting from heating by upwelling magma. This may produce three grabens radiating from a triple junction (fig. 10.13); two of these generally take over and form the rift zone that subsequently forms a new ocean. (Local uplifts, such as salt domes, develop similar faulting patterns.) Sea-floor spreading (fig. 10.14) produces many more-or-less parallel normal faults trending perpendicular to the direction of extension, the faults becoming older as distance from the spreading centers increases. The faults are often steep and planar and may dip in either direction. They may be *en echelon* and throw may shift from one to another. Except near the spreading center, the faults are often inactive. They are sometimes reactivated by later tectonism, which m   change the stress pattern and even reverse the throw direction. The fault blocks are often rotated, occasionally producing a high edge that may become the venue for reef development. In intraplate areas, spreading may have occurred for only a short time.

Compression produces high-angle reverse faults and basement thrusts at convergent boundaries. Secondary compression is sometimes produced in other settings.

A complex array of features can be produced at convergent boundaries. In the subduction of one plate under another, as in fig. 10.15, portions of the oceanic plate may be scraped off in a bulldozerlike action to produce a melange wedge with thrust faulting. The melange includes sediments derived from the continental plate as well as rocks originally remote from the subduction zone. Thrusting and other effects will also be seen in the forward portions of the continental plate and the affected zone may be very wide. The materials carried down by the subducting plate may melt and produce a volcanic arc parallel to the subducting edge but an appreciable distance from the trench. Back-arc basins and other features may also develop. Occasionally, pieces of the oceanic arc or continental blocks riding on the subducting plate may adhere to the continental plate and the subducting edge may jump to a new location, so that a variety of complications are possible.

Wrench-fault assemblages are commonly associated with transform boundaries (fig. 10.12). Although the predominant motion is strike-slip, vertical component of throw is often the most evident aspect. Features are usually confined to a relatively narrow linear zone along the principal strike-slip direction. Some associated secondary features are illustrated in fig. 10.16. These secondary features are often fairly straight and arranged *en echelon*. Fault traces are generally straight and steepen with depth. The main stresses may have components of extension and compression perpendicular to the main strike-slip motion, and irregularities along strike-slip faults also produce extensional or compressional features. Flower structures may occur (fig. 10.17).

Basement warps are often solitary, very gentle features, sometimes with associated normal faulting. They may be of basin size (Williston Basin) or occur as regional arches or local domes. They tend to persist over long periods of time and hence localize trunca-

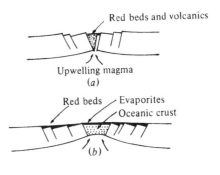

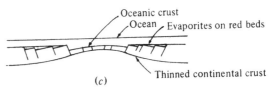

Fig. 10.14 Structure associated with rifting and sea-floor spreading. (a) Early phase of rifting: some crustal thinning, normal faulting mainly down to a central graben or half-graben; infilling sediments are mainly continental. (b) Transition from rifting to drifting: new oceanic crust results in isostatic subsidence, possibly restricted circulation and evaporite deposition. (c) Drifting and growth of an ocean, with further subsidence as the oceanic crust gradually cools. Bounding faults are rarely symmetrical and the fault slip tends to shift from one fault to another along strike.

tion, uncomformities, and various types of stratigraphic traps.

Thrusts tend to exist as a set of subparallel salients on the overriding plate at subduction zones. Thrust zones may be very broad and of somewhat different forms (fig. 10.18). Thrusts generally parallel the bedding in incompetent rocks and cut across the bedding in competent rocks; anticlines often overlie where thrusts cut across competent members (figs. 10.18e and 10.18f). Abnormal pressure zones (§5.3.4) probably provide the décollement (detachment) zones along which the gliding takes place, the plate thrust in effect "floating" into place (Gretener, 1979). Where massive carbonates form a significant part of the section, thrust sheets often involve repeated slabs (figs. 10.18a to 10.18.c); where ductile formations predominate, hanging-wall folds are common (fig. 10.18d). Dips often decrease with depth. The shortening associated with thrusting and folding may be distributed among features in the strike direction by perpendicular tear faults (fig. 10.19a) or by transfer zones in which parallel features grow or decrease in magnitude (fig. 10.19b).

In folding, the length and volume of beds tend to remain constant; however, often both cannot be conserved at the same time, especially with intense folding, and flow and/or faulting occurs in some beds (fig. 10.20). Folding cannot persist to great depths but must give way to flowage or faulting mechanisms.

Normal faults detached from basement occur adja-

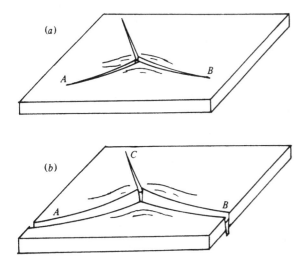

Fig. 10.13 Rupturing produced by an uplift. (a) A localized uplift tends to produce three sets of faults that produce three grabens. (b) As stresses continue, two of these (*A*, *B*) tend to become pull-apart zones; the third (c) may become a *failed arm*, or *alaucogen*.

Table 10.1 *Structural styles and plate-tectonic habitats*

Structural style	Characteristics	Dominant deformational stress	Plate-tectonic habitat	Typical profile
Pull-apart zones	Fairly high-angle normal faults dipping 60–70° in either direction Rotated fault blocks	Extension	Divergent boundaries (1) at spreading centers (2) aborted rifts Intraplate rifts Transform boundaries with component of divergence Secondary at convergent boundaries: (1) Trench outer slope (2) Arc massif (3) Stable flank of foreland and fore-arc basin (4) Back-arc marginal seas	
Compressive faults and basement thrusts	High-angle reverse faults, upward imbricating of faults	Compression	Convergent boundaries (1) Foreland basins (mostly) (2) Orogenic belt cores (3) Trench inner slopes and outer highs Transform boundaries with component of convergence	
Wrench-fault assemblages	Strike-slip faulting is primary, secondary features at about 30° angle to main trend Fairly narrow trend Faults generally steepen with depth	Couple	Transform boundaries Convergent boundaries at an angle: (1) Foreland basins (2) Orogenic belts (3) Arc massifs Divergent boundaries with offset spreading centers	
Basement warps	Gentle structure: domes, arches, sags	Isostatic adjustment Heat flow	Plate interiors Passive boundaries Other areas	

BASEMENT-INVOLVED STYLES

Table 10.1 *continued*

	Structural style	Characteristics	Dominant deformational stress	Plate-tectonic habitat	Typical profile
BASEMENT-DETACHED STYLES	Thrust assemblages	Faults sole out at décollement in incompetent rocks	Compression	Convergent boundaries (1) Inner slopes of trenches and outer highs (2) Mobile flank of forelands (orogenic belts) Transform boundaries with component of convergence	
	Growth faults and other normal fault assemblages	Downthrown toward basin or toward center of uplift Dip often lessens with depth (for growth faults) Often contemporaneous with deposition	Extension	Passive boundaries Secondary to uplifts (folds, salt domes)	
	Salt structures	Pillows, domes, salt walls	Plastic flow Solution	Divergent boundaries (rifts provide venue for salt deposition)	
	Shale structures		Plastic flow (often involving overpressuring produced by rapid burial)	Passive boundaries	
	Drape features		Differential compaction	Subsiding basins Over reefs	
	Volcanic plugs		Igneous intrusions		

Source: After Harding and Lowell, 1979.

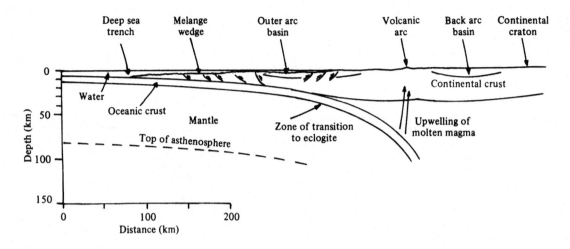

Fig. 10.15   Tectonic features associated with subduction of an oceanic plate under a continental plate.

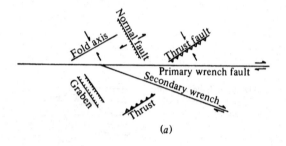

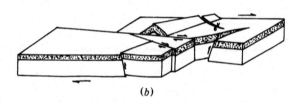

(a)   (b)

Fig. 10.16   Features secondary to a wrench fault (which often make angles of about 30° or 60° to the primary fault). (a) Map view (for symbols, see app. H); (b) isometric diagram showing features on the map.

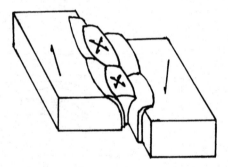

Fig. 10.17   Flower structure resulting from strike-slip motion that involves a component of compression. Motion involving a component of expansion produces negative flower structures. Flower structures are usually not symmetrical. (After Lowell, 1972.)

cent to subsiding basins, especially at the passive margins of continents. They are generally downthrown toward the basins and grow contemporaneously with deposition as basins subside, as evidenced by thickening into faults on the downthrown side (fig. 10.21). The fault planes generally decrease in dip with depth, generally being concave upward. Rotation of the downthrown block in growth faulting produces *roll-*

over into the fault (sometimes called *reverse drag*), the consequent dip reversal possibly making hydrocarbon traps. Growth faults are not only curved in cross-section but also in plan view, generally being concave toward the depression. Growth faults frequently die out upward. The throw on growth faults often increases with depth. They sometimes sole-out in the bedding, the movement parallel to the bedding sometimes producing *toe structures.* Toe structures also result from downslope sliding (fig. 10.22).

Normal faults are often secondary elements of other structural styles; for example, normal faults occur on the crests of folds and above diapirs. They are often associated with shale or salt flowage. Normal faults are probably the most common structural feature because rocks are especially weak under tension.

Salt and shale structures can be of a variety of types involving flow and withdrawal structures, and sometimes collapse features resulting from salt removal by solution. In fig. 10.23, the system of growth faults developed just beyond the underlying shelf edge, where presumably shale under abnormal pressure acted as a fluid and flowed to the left and up to form shale diapirs. Salt and shale often provide the detachment surfaces associated with detached faulting (as with thrusting) and folding. Buoyant salt structures are es-

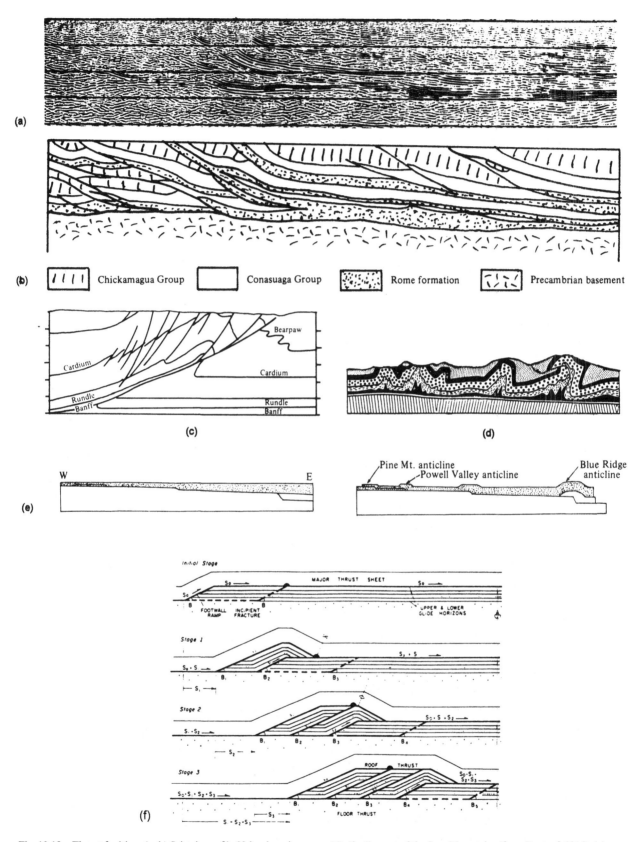

(a)

(b)

| | | | Chickamagua Group   Conasuaga Group   Rome formation   Precambrian basement

(c)   (d)

(e)

W   E

Pine Mt. anticline
Powell Valley anticline   Blue Ridge anticline

(f)

Initial Stage

MAJOR THRUST SHEET S_0
S_0
S_0
B   B   FOOTWALL   INCIPIENT   UPPER & LOWER
RAMP   FRACTURE   GLIDE HORIZONS

Stage 1
$S_1 \cdot S$
$S_1 \cdot S$
B_1 B_2 B_3
S_1

Stage 2
$S \cdot S_2$
$S_0 \cdot S \cdot S_2$
B_1 B_2 B_3 B_4
S_2

Stage 3
$S_0 \cdot S_1 \cdot S_2 \cdot S_3$
ROOF THRUST
$S_0 \cdot S_1 \cdot S_2 \cdot S_3$
B_1 B_2 B_3 B_4 B_5
S_3
FLOOR THRUST
$S \cdot S_2 \cdot S_3$

Fig. 10.18   Thrust faulting. (a, b) Seismic profile 32 km long in Valley-Ridge Province of east Tennessee and its interpretation (from Harris and Milici, 1977); (c) thrusting in Canadian Rockies producing the Turner Valley structure (from Gallup, 1951); (d) décollement of the Jura Mountains (from Buxtorf, 1916); (e) two stages of southern Appalachian décollement with anticlines developing over ramps; (f) development of duplex structure (after Boyer and Elliott, 1982).

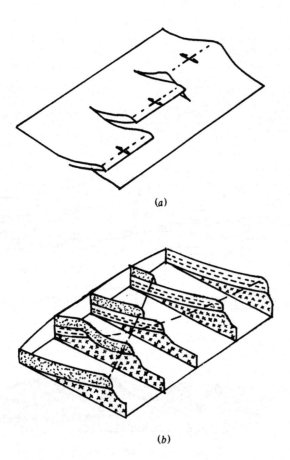

(a)

(b)

Fig. 10.19 Transfer of throw from one thrust fault to another. (After Dahlstrom, 1970.) (a) Tear faults separate thrusts; and (b) faults grow or die laterally.

pecially common along passive continental margins, the basinward progression often being a sequence of salt-withdrawal structures, pillows, nonpiercement and piercement domes, and finally, a "wall" of salt (figs. 10.24 and 10.25); the salt often has moved considerable distances basinward as deposition continues and as basin subsidence moves seaward.

Depositional and structural features are often interrelated. Reefs may grow over shelf edges or on the upturned edges of rotated fault blocks. Differential compaction may produce drape over reefs and shelf edges. Progradation beyond a shelf edge may produce growth faults. The weight of sediments deposited by a river delta system may produce subsidence because of isostatic adjustment, affecting faulting patterns and causing salt and shale movements.

10.3.2 Balancing sections

Preservation of volume and bed length were mentioned earlier. The total mass cannot change during faulting or flowage and the length of competent beds also cannot change. Mass can be lost or gained locally because of erosion or increased deposition or other

factors that move mass into or out of the system being considered. Interpreters often do not check to see that their interpretations are consistent with this principle because calculations are tedious and often require more extensive data than are available. Modeling programs can ascertain whether discrepancies exist. Jones (1988) discusses the balancing of seismic sections.

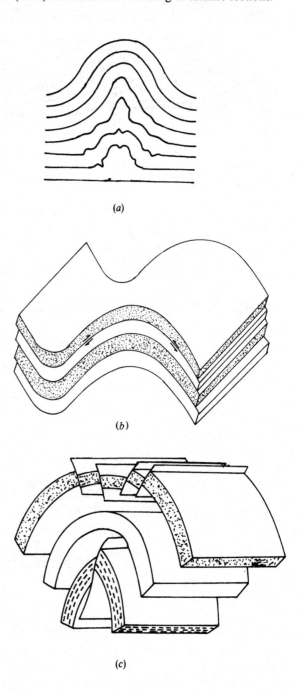

(a)

(b)

(c)

Fig. 10.20 Mechanisms that maintain bed length and volume in fold interpretation. (a) "Concentric" folding with flow or severe distortion as the mechanism (after Goguel, 1962); (b) "similar" folding with bedding-plane shear and thickness variations; (c) combination of folding and faulting; the more competent members mainly fault (after Hobbs, Weams, and Williams, 1976).

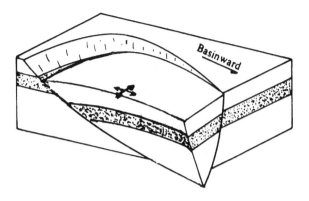

Fig. 10.21 Growth faulting. Sediments, especially poorly consolidated ones, slide toward the basin along a concave-upward
(*listric*) fault plane. Rotation of the down-dropped block contemporaneous with deposition results in thickening into the
fault producing a rollover anticlinal axis parallel with the fault
plane. The fault is arcuate in plan view with the throw gradually
diminishing away from the center of the fault. Seismic data
showing a fault of this kind are seen in fig. 10.29.

10.3.3 Faulting

(a) Introduction. Faults constitute one of the more
important hydrocarbon trapping mechanisms (Downey, 1990) and recognizing them and determining
their precise locations are often the keys to success.

Fault nomenclature is illustrated in fig. 10.26a. The
fault quantities that can be measured on good-quality
seismic sections are usually the vertical throw and the
apparent fault-plane dip; horizontal components of
fault slip can rarely be determined.

Ideally reflection events terminate sharply as the
point of reflection reaches the fault plane and then
they resume again in displaced positions on the other
side of the fault. In addition, ideally, the reflection has
a sufficiently distinctive character that the two portions on opposite sides of the fault can be recognized
and the fault throw determined. In practice, diffractions usually prolong events, so that the locations of
fault planes are not clearly evident, although often
faults do show clearly as reasonably sharp reflection

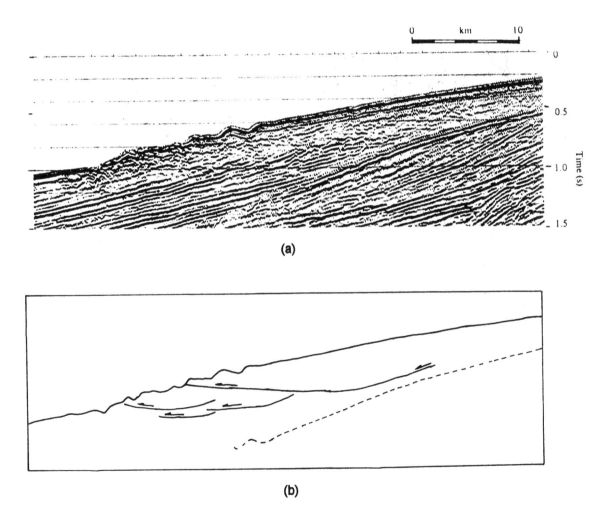

Fig. 10.22 Seismic section showing downslope sliding and
slumping, with toe structures. (Courtesy of Exxon.) (a) CMP
section; the marks at the top are 2 km apart so that the vertical
exaggeration is about 11×, the sea-floor slope being less than
1°; (b) interpretation of fault glide planes and sea-floor multiple (dashed).

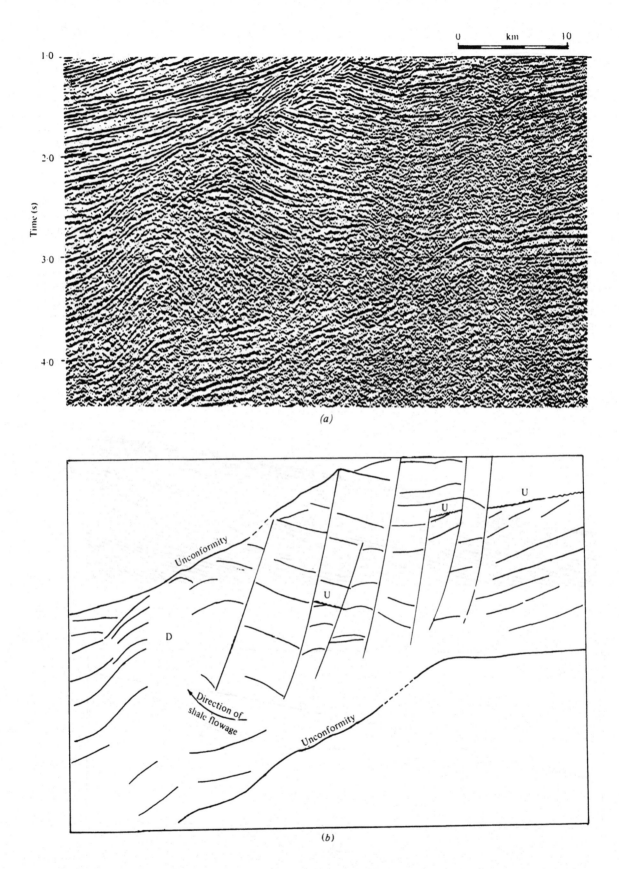

Fig. 10.23 Shale flowage. (Courtesy of Exxon.) (a) CMP section with vertical exaggeration of 5× to 8×, decreasing with depth; (b) interpretation indicating the attitude of reflections (which have not been correlated except for the angular unconformity U). With depth the faults die out in flowage of the overpressured shale into shale diapirs D.

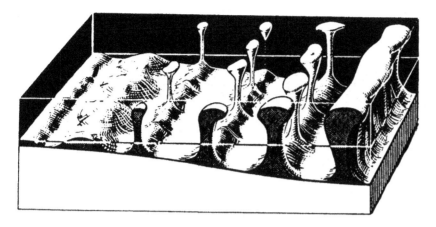

Fig. 10.24 Salt structures grow basinward as salt increases in thickness, either because more salt is deposited or because salt is pushed basinward as sediments are deposited further landward. (From Trusheim, 1960.)

terminations on migrated lines that are roughly perpendicular to the faults. Moreover, although sometimes the same reflection can be identified unequivocally on the two sides of a fault, in many cases, we can make only tentative correlations across faults. Campbell (1965) and Sheriff (1982) discuss criteria for detecting faults on seismic sections.

(b) Characteristics of faulting. Faults are produced by unbalanced stresses that exceed the strength of rocks, the type of fault depending largely on whether the vertical or horizontal stresses are the larger (fig. 10.26b). Normal faults result when the maximum compressive stress is vertical and the minimum horizontal, often producing fault-plane dip of the order of 50°–60°. When the maximum compressive stress is horizontal, thrusts result, often with a fault plane dip of 30°–40°. Where the maximum and minimum stresses are both horizontal, wrench faults result, the faulting often being at about 30° to the maximum stress direction.

Because velocity ordinarily increases with depth, the same vertical distance is represented by less time as depth increases; in consequence a postdepositional fault with constant attitude and throw usually appears to dip less with depth, that is, has concave-upward curvature on a time section (fig. 10.27). Furthermore, constant throw is represented by fewer wavelengths as depth and velocity increase and so constant-throw faults appear to die out with depth on a time section. Postfaulting compaction (fig. 10.27) with increased depth of burial also produces concave-upward curvature. Thus, fault traces are rarely straight on seismic sections.

The locations of faults often are determined by underlying features. An underlying uplift places the overlying sediments under tension as distances are stretched to accommodate the drape over the structure. Graben faulting forms to relieve the stress, but if the uplift is three-dimensional rather than two-dimensional, radial faulting is also required (fig. 10.13). Normal faulting commonly accommodates stretching above a hinge line or at a shelf edge where the basin side of the shelf edge subsides more rapidly (fig. 10.23). The location of faulting can be the key to underlying features, and conversely the underlying features can aid in connecting sometimes confusing faulting evidences into a probable pattern.

(c) Fault example. The two record sections in fig. 10.28 join at their north and west ends at right angles. The reflection band consisting of four strong legs marked Σ can be correlated readily across the normal faults. In fig. 10.28a, this event is down-thrown to the south by about 75 ms (about 2 cycles, ignoring rollover, §10.3.1b) at its 1.6 s arrival time; at a velocity of 2500 m/s, this represents a vertical throw of about 95 m. The exceptionally strong event near 2.3 s (marked χ) indicates a throw of 120 ms (about 3 cycles, the dominant frequency having become slightly lower); at a velocity of 3000 m/s, this represents 180 m of throw, so the fault appears to be growing rapidly with depth.

Although the evidence suggests that the fault is a simple break in the shallow section, at greater depths there may be a fault zone of some width with subsidiary faults (shown dashed in the figure); perhaps a small piece of the event χ can be seen between the faults. Because data quality deteriorates in the region under a fault, details often cannot be seen clearly. If the deeper correlations are correct, the downthrown event Ω at 3.5 s is found around 2.9 s on the upthrown side; by assuming a velocity of 3500 m/s, this corresponds to a throw of 1050 m.

The basis for correlation across the fault for the Σ event is reflection character; for χ, it is strong amplitude; and for other events, the intervals between reflections, systematic growth with depth, time ties around loops, and so on. Sometimes the displacement of an unconformity or other recognizable feature indicates the amount of throw. Often, however, the throw cannot be determined clearly from the seismic data.

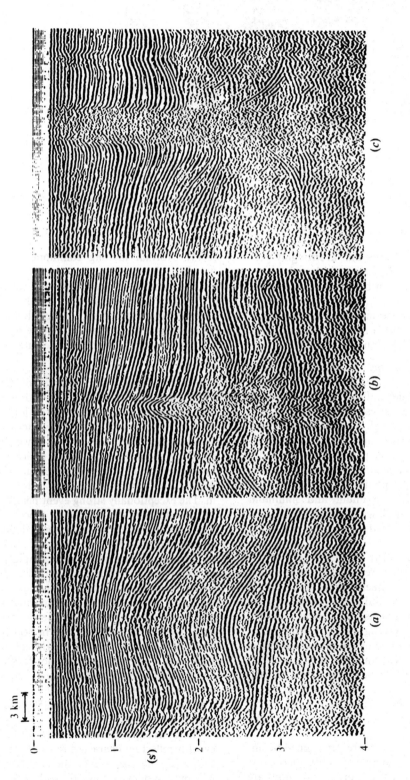

Fig. 10.25 Portions of an unmigrated section in the North Sea. (Courtesy of Grant Geophysical.) (a) Salt swell or pillow; (b) salt dome that has pierced some of the sediments; (c) salt dome that has pierced to the sea floor.

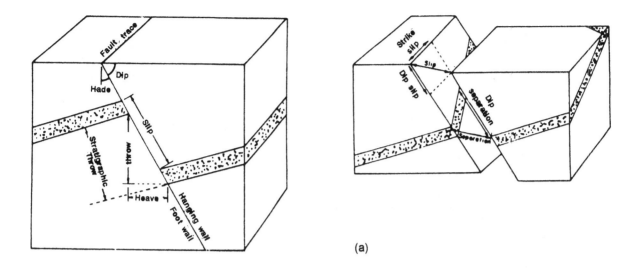

(a)

FAULT TYPE	RELATED TERMS	STRESS DIRECTION		CHARACTERISTICS
		MINIMUM	MAXIMUM	
NORMAL	TENSION FAULT GRAVITY FAULT SLIP FAULT LISTRIC FAULT (CURVED FAULT PLANE)	HORIZONTAL (Tension)	VERTICAL (Gravity)	Dip usually 75° to 40°
REVERSE	THRUST FAULT LOW ANGLE (dip < 45°) HIGH ANGLE (dip > 45°)	VERTICAL	HORIZONTAL (Compression)	Fault plane may disappear along bedding
STRIKE - SLIP	TRANSCURRENT FAULT TEAR FAULT WRENCH FAULT RIGHT LATERAL (Dextral) LEFT LATERAL (Sinistral)	HORIZONTAL	HORIZONTAL	Fault trace often 30° to maximum stress
ROTATIONAL	SCISSORS FAULT HINGE FAULT			Throw varies along fault strike; may vary from normal throw to reverse.
TRANSFORM	DEXTRAL SINISTRAL	HORIZONTAL		Associated with separation or collision of plates New material fills rifts between separating plates or one plate rides up on another if plates collide.

(b)

Fig. 10.26 Fault nomenclature. (From Sheriff, 1991: 112–15.) (a) Fault terms; (b) fault types.

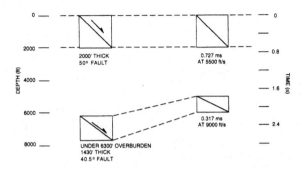

Fig. 10.27 Compaction and increase of seismic velocity with depth decrease the dip of fault planes with depth, tending to produce concave-upwards curvature. The left column shows the effect on shale of adding 6300 ft of overburden, thus shrinking vertical dimensions (with no renewed faulting); the right column indicates how this might appear on a seismic time section because of the consequent velocity increase. (From Sheriff, 1982: 54.)

If the data in figs. 10.28a and 10.28b are transformed into depth sections, we get fig. 10.28c. The components of fault-plane dip are around 55° and 48°. Note that the fault that is nearly straight on a depth section is concave upwards on a time section because of the increase in velocity with depth. If the fault surface is actually concave upwards, the curvature will be accentuated on a time section. Where the fault was most active (indicated by the most rapid growth in fault throw), the fault surface is most curved.

The fault has not completely died out by the north end of the line and hence the fault trace should appear on the intersecting line (fig. 10.28b). As picked on the E–W section, the fault offsets the event Σ at 1.6 s by only about 25 m, indicating that the fault is dying out rapidly toward the east. The fault plane has nearly as much dip in the E–W section so that the strike of the fault plane near the intersection of the two lines is NE–SW and the fault plane dips to the southeast. The true dip of the fault plane is about 62° (the apparent dip on sections is always less than the true dip unless the line is perpendicular to the strike of the fault). Fault indications are not evident below about 2 s on the E–W section so that the fault appears to have died out at depth toward the east. In poorly consolidated sediments, such rapid dying out of faults is common. In this instance, we are dealing with a radial fault from a deep salt-cored diapir located just south and slightly west of these lines; such radial faults often die out rapidly with distance from the uplift.

(d) Evidences for faulting. A number of the more common faulting evidences can be seen in the foregoing example. Several diffractions can be identified along the fault trace in fig. 10.28a between 1.9 and 2.5 s. If we had been dealing with migrated sections these diffractions would have been nearly collapsed (but not completely because the fault is not perpen-

dicular to the lines). Terminations of events and offset of reflections (and nonreflection zones) across the fault are other important faulting evidences.

Different reflection dips are often seen on the two sides of the fault. Some of these dip changes are real, involving slight rotation of the section as the fault moved along a slightly curved fault plane, drag, and other real phenomena. On the other hand, some (especially those seen through the fault plane) are distortions resulting from raypath bending (refraction) in passing through the fault plane because of local velocity changes at the fault. Although the upthrown sediments are most apt to have the higher velocity at any given level, the polarity and magnitude of velocity contrasts vary down the fault plane as units are juxtaposed against different units, so that the nature of the distortion varies from one place to another. In fact, the distortions may be so great and may change so rapidly as to cause marked deterioration of data quality below the fault, sometimes so great that reflections are almost entirely absent (a "shadow zone") below the fault. This is especially apt to be true for CMP sections because raypaths for the components stacked together crossed the fault at different places.

Occasionally, the fault plane itself generates a reflection, but generally the fault plane is a highly variable reflector because of the rapid changes in velocity contrast along the fault plane. Also, faulting is often distributed over a zone and involves many fracture surfaces. Furthermore, most reflection recording and processing discriminate against fault-plane reflections because of the use of arrays and of stacking velocities that do not optimize such events. In addition, fault-plane reflections on unmigrated data are usually displaced an appreciable distance from the fault and often the traveltimes to them are so great (because of the long slant paths) that they are not recorded and processed. Many of the earlier evidences for faulting can be seen for the growth fault in fig. 10.29 (of the type illustrated in fig. 10.21); other faults are also present in fig. 10.29.

The increased detail made possible by 3-D methods often shows not only more and smaller faults, but also many that are short along strike and disconnected (fig. 12.10). The continuity of stratigraphic features across faults seen on 3-D horizon slices (fig. 12.14) sometimes provides convincing evidence that fault throws have been picked correctly.

10.3.4 Folded and flow structures

When subjected to stress, rocks may fault, fold, or flow, depending on the magnitude and duration of the stresses, the strength of the rocks, the nature of adjacent rocks, and so on. The folding of rocks into anticlines and domes provides many of the traps in which oil and gas are found.

Figure 10.30 shows a migrated seismic section across an anticline. Some portions such as *A*, which are composed of the more competent rocks (for ex-

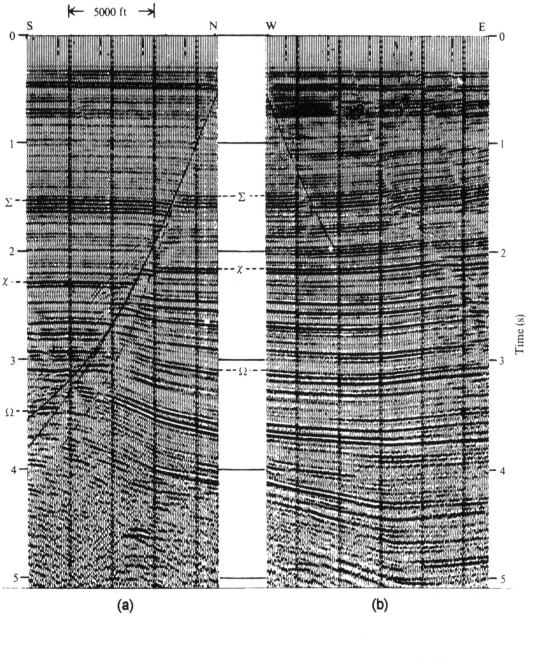

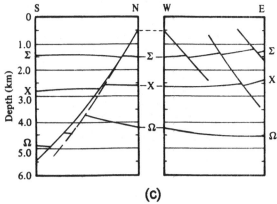

Fig. 10.28 Intersecting unmigrated sections showing faulting. (Courtesy of Haliburton Geophysical Services.) (a) North–south section, (b) east–west section, and (c) depth sections.

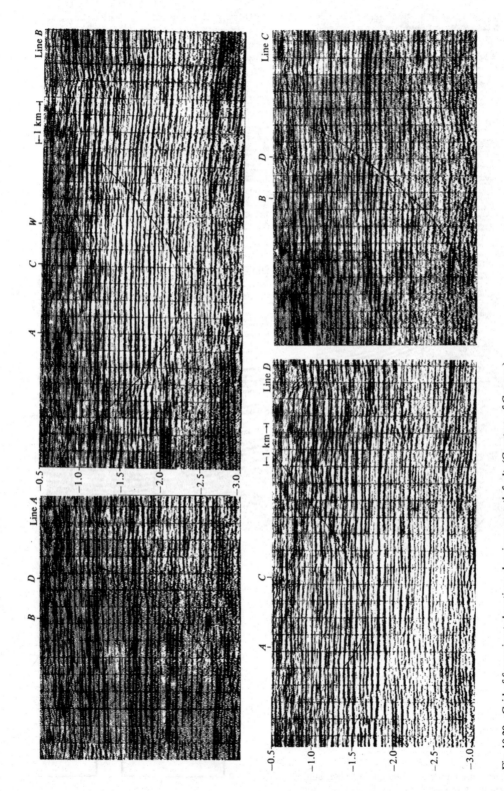

Fig. 10.29   Grid of four migrated sections showing a growth fault. (Courtesy of Conoco.)

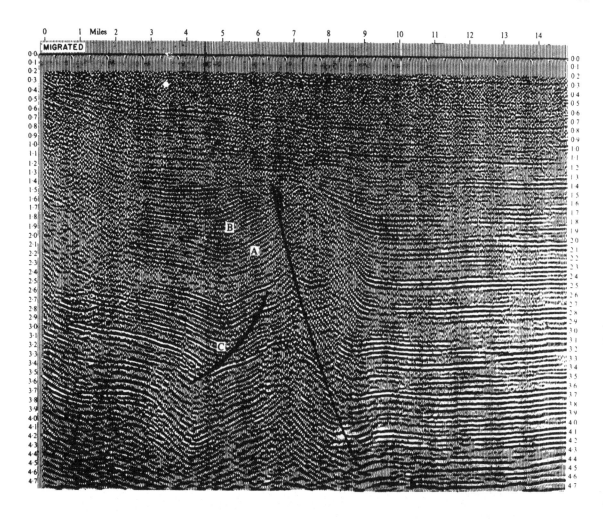

Fig. 10.30   Migrated seismic section showing anticlinal structure in the Central Valley of California. (Courtesy of Getty Oil.)

ample, limestones and consolidated sandstones), tend to maintain their thickness as they fold. Other portions such as *B*, which contain less competent rocks (often shales and evaporites), tend to flow and slip along the bedding, resulting in marked variations in thickness within short distances. Geometry places limits on the amount of folding that is possible and folded structures almost always involve faulting (fig. 10.20c). Note at *C* in fig. 10.30 how a fault is involved with the folding and in fig. 10.31 how the force fold is associated with the underlying fault and inverted structure.

Arching causes extension; often the sediments break along normal faults and produce graben-type features on the top. Folding must disappear by faulting or flowage at some depth. Anticlinal curvature tends to make seismic reflections weaker as well as increase the likelihood of faulting and flowage, so that data quality commonly deteriorates over anticlines.

Salt flow often produces anticlines and domes. In many parts of the world, thick salt deposits have been buried fairly rapidly beneath relatively unconsolidated sediment. The sediments compact with depth and so increase their density, whereas the salt density remains

nearly constant. Thus, below some critical depth the salt is less dense than the overlying sediments. Salt behaves like a very viscous fluid under sufficient pressure, and buoyancy may result in the salt flowing upward to form a salt dome, arching the overlying sediments and sometimes piercing through them (fig. 10.25). Piercement does not necessarily imply uplift, however, because subsidence of the sediments surrounding a salt plug accomplishes nearly the same structural result. Often the velocity in "uplifted" rocks is nearly the same as that in laterally adjacent nonuplifted rocks, implying that neither was ever buried deeper; if they had been, they would have irreversibly lost porosity and attained a higher velocity.

Grabens and radial normal faults (whose throw decreases away from the dome) often result from arching of the overlying sediments (fig. 10.13), to relieve the stretching that accompanies the arching. Salt domes tend to form along zones of weakness in the sediments, such as a large regional fault. The side of a salt dome may itself be thought of as a fault.

Figure 10.32 shows a seismic section across a salt dome. Shallow salt domes are apt to be so evident that they can scarcely be misidentified. Because of the

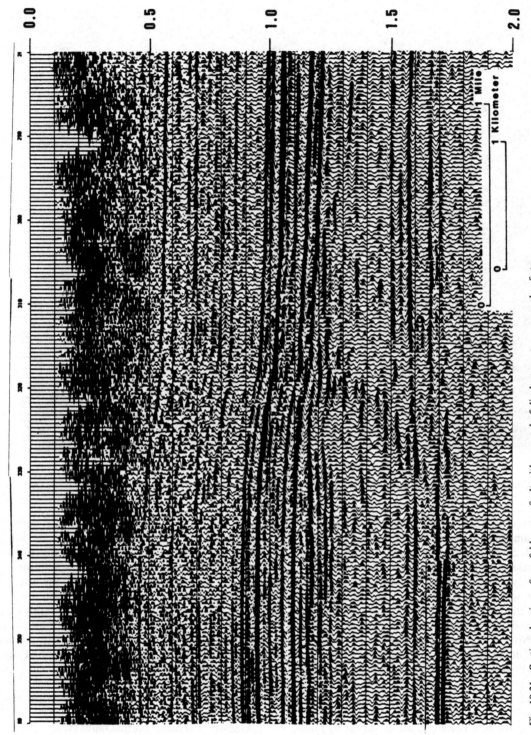

Fig. 10.31   Section showing a force fold over a fault with reversal of displacement, an example of structural inversion, Central Montana. (From Plawman, 1983.)

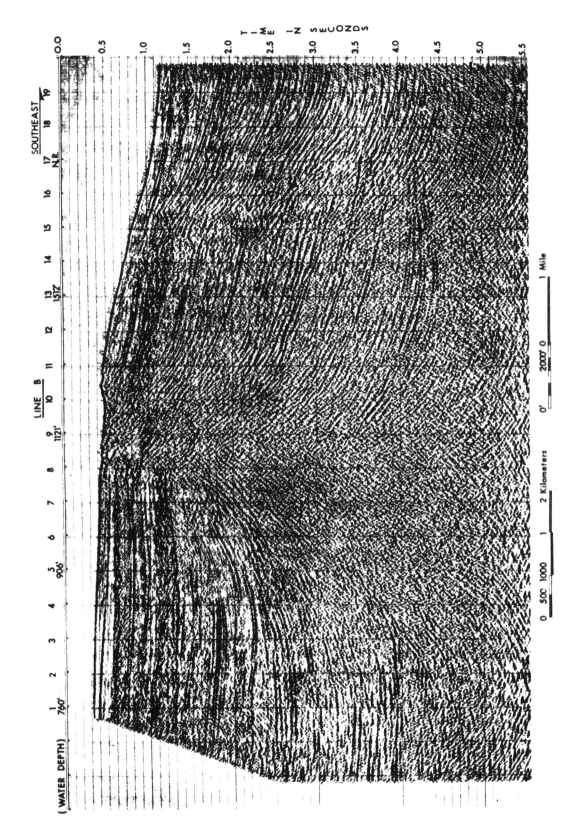

Fig. 10.32 Section across salt ridge located at the shelf break. (From Wanslow, 1983.)

large impedance contrast, the top of the salt dome (or caprock on top of the dome) may be a strong reflector. Steep dips may be seen in the sediments adjacent to the salt dome as a result of these having been dragged up with the salt as it flowed upward. The sediments often show rapid thinning toward the dome. The salt itself is devoid of primary reflections, although multiples often obscure this feature, especially if AGC is used.

Defining the flank of a salt dome precisely is often economically important and seismically difficult. Frequently, oil accumulation is in a narrow belt adjacent to the dome flank, but because the flank is often nearly vertical or even overhung, it usually does not give a recognizable reflection with conventional processing. Fortunately, velocities are often only slightly affected by the growth of the dome (except for the velocity in the salt and caprock) so that the steep dips of the sediments adjacent to the flanks can be migrated fairly accurately and the flank outlined by the terminations of often strong, steeply dipping reflections. Even reflections from overhung flanks are sometimes used (*turning waves*, diving waves that reflect on their upward leg). Proximity surveys (§13.7.1) using wells in or near the salt are also used to define the flanks more accurately. Nevertheless, there remains much art and experience in defining salt-dome flanks.

Although the weight of sediments prograding toward a salt mass may push the salt seaward for large distances, the salt in a salt dome generally has come from the immediately surrounding region. The removal of the salt from under the sediments around the dome has allowed them to subside, producing a rim syncline. The seismic data over such synclines are often very good and aid in mapping the adjacent dome by indicating the volume of salt involved, when movement took place (by sediment thickening), and so on. Such synclines may also help provide closure on neighboring areas where the sediments continue to be supported by residual salt.

Figure 10.25 is a portion of a section in the North Sea (the horizontal scale has been compressed so as to display a long line on a short section, producing considerable vertical exaggeration). This line shows deep salt swells that have not pierced through the overlying sediments (fig. 10.25a), salt that has pierced through some of the sedimentary section (fig. 10.25b), and also salt that has pierced all the way to the sea floor (fig. 10.25c). The reflection from the base of the salt is generally continuous except for some suggestions of faulting. Distortion is produced by the variable salt thickness. Because the salt velocity is greater than that of the adjacent sediments, the base-of-salt event appears to be pulled up where the salt is thicker. In areas where the salt velocity is lower than that of the surrounding sediments, flat reflectors beneath the salt may appear to be depressed where the overlying salt is thicker.

Figure 10.33 shows a salt uplift at a shelf edge. Most of the salt movement occurred prior to the unconfor-

mity U_2, although the right side continued to subside somewhat even after U_1, producing the monocline in the shallower sediments. Note the graben faulting seen most prominently around 2.0 s.

Occasionally, substances other than salt form flow structures. Poorly consolidated shale may flow (fig. 10.23), forming structures that strongly resemble salt domes on reflection sections; also, at times, shale flows along with salt, producing a salt dome with a sheath of shale. Magma also sometimes flows into a sedimentary section to produce structural uplifts, including piercement domes. Shale often flows upward on the upthrown side of growth faults.

10.3.5 Reefs

The term "reef" as used by petroleum geologists comprises a wide variety of types, including both extensive barrier reefs that cover large areas and small isolated pinnacle reefs. It includes carbonate structures built directly by organisms, aggregates comprising limestone and other related carbonate rocks, as well as banks of interstratified carbonate (and sometimes, also noncarbonate) sediments. Reef dimensions range from a few tens of meters to several kilometers, large reefs being tens of kilometers in length, a few kilometers wide and 200–400 m or more in vertical extent. Some reefs grow at the boundary between different environments, such as the shelf-margin and barrier types shown in fig. 10.34a, whereas others, such as patch and pinnacle reefs, are surrounded by the same environment.

We shall describe a model reef so that we may develop general criteria by which reefs can be recognized in seismic data, keeping in mind that deviations from the model may result in large variations from these criteria. Our model reef forms in tropical to subtropical waters far enough from river mouths that the water contains little suspended sediment. The area is generally tectonically quiet, characterized by flat-lying bedding that is more-or-less uniform over a large area. The uniformity of the section makes it possible to attribute significance to subtle changes produced by the reef that might go unnoticed in more tectonically active areas. The reef is the result of the buildup of marine organisms living in the zone of wave action where the water temperature is suitable for sustaining active growth and the water sufficiently clear to permit significant penetration of sunlight. The site of the reef is usually a topographic high that provides the proper depth. Although the topographic high may be due to a structure in the underlying beds or basement, such as a tilted fault block, more often it is provided by a previous reef; as a result, reefs tend to grow vertically, sometimes achieving thicknesses of 400 m or more and thereby accentuating their effects on the seismic data. In order for the reef to grow upward, the base must subside as the reef builds upward, maintaining its top in the wave zone as the sea transgresses. The reef may provide a barrier between a lagoonal area

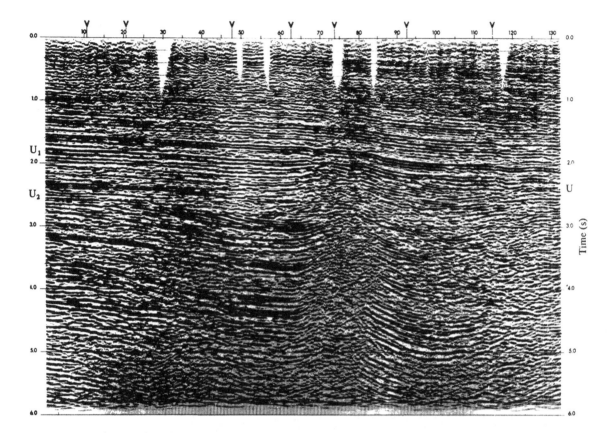

Fig. 10.33 Migrated section across a salt dome. Note evidences for unconformities (U_1, U_2). Stacking velocity data for this line were plotted in fig. 9.32. (Courtesy of Grant Geophysical.)

(the *backreef*) and the ocean basin (the *forereef*), so that sedimentation (and consequently the reflection pattern) may be different on opposite sides of the reef. The surrounding basin may be *starved* (that is, not have sufficient sediments available to keep it filled at the rate at which it is subsiding); at times, only one side, more often the ocean side of the reef, may be starved. Alternatively, the reef may not be a barrier to movement of the sediments and in this case it will be surrounded by the same sediments. Erosion of the reef often provides detritus for deposition adjacent to the reef, resulting in *foreset beds* with dips up to 20°, but usually with smaller dips, often of only 1–2°. The primary reef may possess considerable porosity, which makes it a good potential reservoir rock, but other organisms such as sponges penetrate into and replace much of the reef rock, altering the porosity in the process. Usually, the actual *biohermal* portion of the reef (the portion produced by reef-building organisms) cannot be distinguished from other portions except by the examination of samples, and the entire complex associated with the reef is called the "reef," only perhaps 15% of which is biohermal. Eventually, the environment for the reef organisms changes so that they can no longer continue to live and build the reef; this might come about because of changes in the water temperature, a change to more turbid water, possibly

an increase in the rate of subsidence so that the organic buildup cannot keep pace (called *drowning* of the reef), or various combinations of circumstances. Subsequently, the reef may become buried by deep-water shales that may provide both an impermeable cap to the porous reef and sufficient hydrocarbons that the reef becomes a petroleum reservoir. Additional sediments may continue to be deposited, their weight compacting the sediments that surround the reef more than they compact the relatively rigid reef; thus, the overlying sediments that were deposited flat may develop a *drape* over the reef. The interior of the reef may be more porous and less rigid than the edges so that some differential compaction may occur over the reef itself.

Based on the foregoing model, we develop the reef criteria illustrated in fig. 10.35. The top of the reef may be outlined by reflections (fig. 10.35a), perhaps with some onlapping reflections, but the reef interior is apt to be a reflection void (fig. 10.35b). Occasionally, diffractions from beds terminating at the reef (fig. 10.35c) and the abrupt termination of reflections from the surrounding sediments (fig. 10.35d) can be used to locate reef flanks, but the precise outline of reefs is usually difficult to determine. If the reef provided a barrier to sedimentation, the pattern of reflections may differ on opposite sides of the reef (fig. 10.35e).

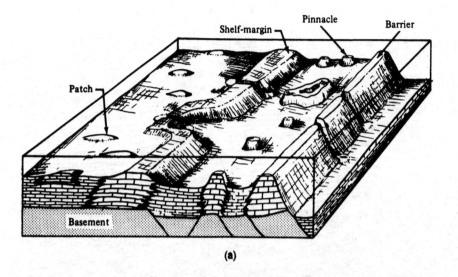

(a)

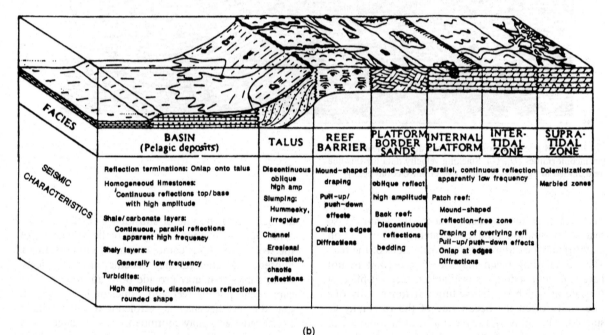

(b)

Fig. 10.34   Carbonate features. (a) Types of reefs (from Bubb and Hatlelid, 1977); (b) seismic-facies characteristics of carbonate environments (from Fontaine et al., 1987).

Overlying reflections may show small relief (usually only a few milliseconds in magnitude) because of differential compaction, the effect decreasing with distance above the reef (fig. 10.35f). A velocity difference between the reef materials and the surrounding sediments may cause the traveltime to flat-lying reflections below the reef to vary (Davis, 1972), and this velocity difference may produce pseudo-structure on reflecting horizons below the reef. Usually, the velocity in the reef limestone is greater than that in surrounding shales, so that the reef may be indicated by time thinning between reflections above and below the reef and by a pseudo-high under the reef (fig. 10.35g); the mag-

nitude of such an anomaly is small, usually less than 20 ms. Sometimes, however, the reef may be surrounded by evaporites or other rocks with higher velocity than that of the porous reef limestone so that the time anomaly is reversed (fig. 10.35h). The hinge-line or high that localized the reef development may also be detectable (figs. 10.35i and 10.35j).

Barrier reefs, which separate different environments, are often fairly evident, but small patch reefs may be so subtle that seismic mapping is feasible only in good record areas. Of importance is geologic information about the nature of the sediments and the environment of deposition, so that one knows before-

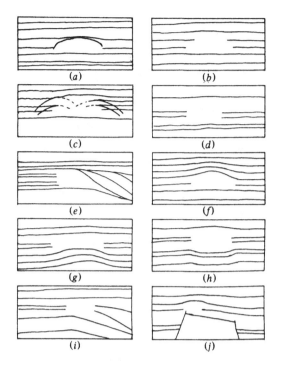

Fig. 10.35 Criteria for reef identification. (After Bubb and Hatlelid, 1977.) (a) Reef outlined by reflections; (b) reflection void; (c) diffractions from reef edges; (d) abrupt termination of reflections; (e) change in reflection pattern on opposite sides of reef; (f) differential compaction over the reef (isopach thinning); (g) velocity anomaly underneath the reef where $V_{reef} > V_{surrounding}$, or (h) where $V_{reef} < V_{surrounding}$. (i) Reef location on a hingeline or shelf edge, and (j), on a structural uplift.

hand in what portion of the section reefs are more likely to occur. Subtle features in the part of a seismic section where reefs are expected may be interpreted as reefs whereas similar features elsewhere are ignored. Kuhme (1987) discusses reef interpretation.

Similarities between reefs and salt features cause problems at times. The lagoonal areas behind and around reefs often provide conditions for evaporite deposition and salt is frequently present in the same portion where reefs are expected. The amount of salt may not be thick enough to produce flow features but differential solution of salt beds followed by the collapse of the overlying sediments into the void thus created may produce seismic features that are similar in many ways to those that indicate reefs.

A seismic line across a barrier reef is shown in fig. 10.36; note the change in reflection pattern across the reef, the differential compaction and velocity uplift evidences, and the change in regional attitude of reflections beneath the reef that indicates a weak hingeline. A line across patch reefs is shown in fig. 10.37. Patch reefs are usually much smaller in vertical extent than these and, consequently, often difficult to locate.

Typical seismic facies characteristics of different carbonate depositional settings are illustrated in fig. 10.34b.

10.3.6 Unconformities

Unconformities represent a missing sequence of rock, a time period during which rocks were being eroded

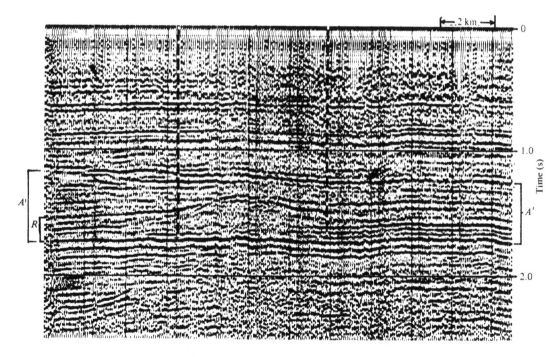

Fig. 10.36 Section across Horseshoe Atoll in West Texas. *R* denotes portion of the section that contains the reef (just left of center). The backreef area of flat-lying, strong, continuous reflections (*A'*) is to the right; the forereef showing an entirely different reflection pattern (*A*) is to the left. (Courtesy of Conoco.)

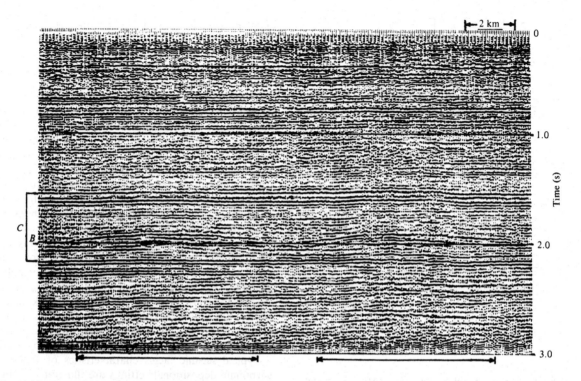

Fig. 10.37 Two patch reefs in the Etosha Basin of Namibia. *C* denotes the carbonate portion of the section; *B* is the base of the reefs. The region of reefs is indicated approximately by the arrows below the section. The reef to the left has about 210 m (85 ms) thickness, the one to the right 300 m (120 ms). (Courtesy of Etosha Petroleum.)

away, or at least not deposited. Conditions probably changed during the hiatus, so that the nature of the sediments above the unconformity are often different from those below and an acoustic-impedance contrast is likely to exist at the unconformity. Hence, unconformities are usually good reflectors. They frequently involve some angularity between the bedding below and above, and this also tends to make them stand out as reflectors. The result is that unconformities are often among the easiest and most distinctive reflectors to map. On the other hand, the rocks that an unconformity separates often vary from one location to another so that the contrast at the interface changes and hence the unconformity reflection varies in amplitude (see fig. 10.38) and sometimes even in polarity (see fig. 10.6a). There may be large regions over which the beds above and below the unconformity parallel the unconformity so that there is no angularity to distinguish the unconformity reflection from other reflections. In such regions, the unconformity has to be mapped by correlating it along the bedding with places where the unconformity can be identified by angularities on the seismic section or where it can be identified from well or other types of data.

Rather prominent unconformities can be seen in figs. 9.59 and 10.38, mainly evidenced by angularities and fairly strong reflections from the unconformities themselves.

Various types of hydrocarbon traps are associated with unconformities – both (1) pinchouts and truncations of reservoir beds below the unconformity where the unconformity constitutes a seal, and also (2) stratigraphic variations in the sediments laid down on the unconformity (see also fig. 10.42). Unconformities are involved in most stratigraphic traps. Streams flowing across the unconformity surface may have eroded valleys into the surface, and the stream deposits may constitute the reservoir or sometimes the seal.

10.3.7 Channels

Ancient stream channels are involved with a number of oil and gas accumulations (fig. 10.39a). The relief associated with large river valleys may be sufficient to give structural evidences, but most often the seismic effects are slight. The velocity of the sediments infilling a channel may differ from that of the sediments into which the channel is cut so as to distort underlying reflections, especially where the channels are deep (fig. 10.39b). Figure 10.40 represents a shallow seismic time slice from a 3-D data set (see also §12.3); it shows the pattern of a meandering stream. This suggests that evidences of very minor features are contained in seismic data if we can but find an economical, feasible way of extracting them.

Profiler surveys in deep water often reveal channel cuts and fill (fig. 10.41), indicating that channels are important in the deep marine environment as well as on land. Turbidity currents at places clearly have eroded deep channels and have sometimes built up extensive levee systems under deep marine conditions. (A turbidity current is a density current in water

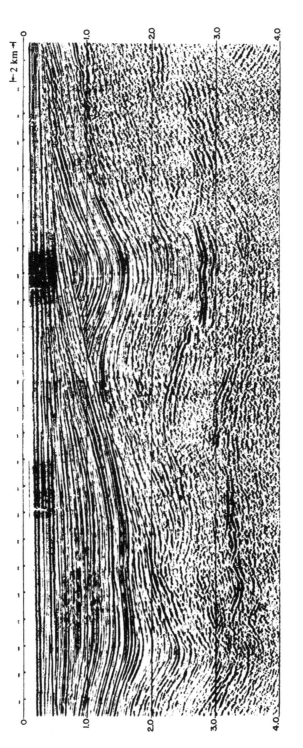

Fig. 10.38 Migrated section from offshore Oregon showing folding and faulting below an unconformity. (Courtesy of Western Geophysical.)

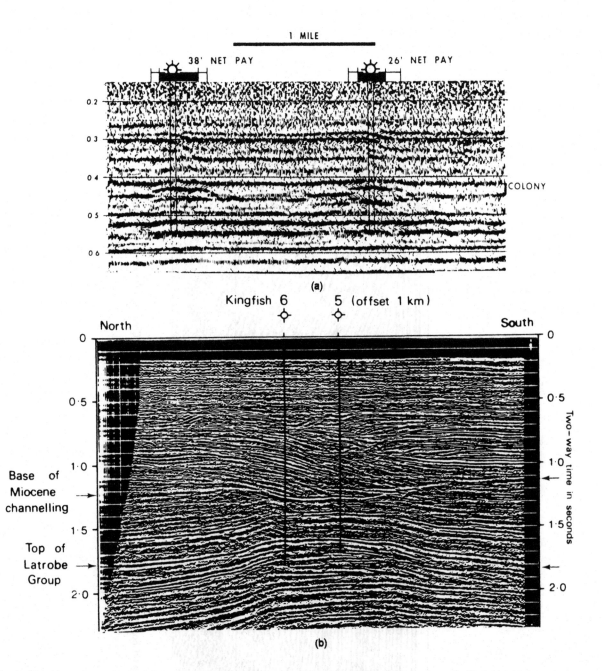

Fig. 10.39 Channels on seismic sections. (From McQuillin, Bacon, and Barclay, 1984: 183, 245.) (a) Gas-filled Colony channel sands; their amplitudes correlate with the net pay thicknesses; (b) seismic line across the East Kingfish field, Bass Straits, Australia; Miocene channels filled with higher velocity sediments produce significant velocity anomalies at the deeper Latrobe producing horizon.

caused by different amounts of solids in suspension; they are important in submarine erosion and deposition and in hydrocarbon accumulations.) Some channels undoubtedly result from the sea level lowering and some from marine erosional processes. Brown and Fisher (1980) developed a "destructive shelf" concept relating the erosion of channels into shelves to the lack of new material available for deposition; this concept has helped discover accumulations in fans on the slopes offshore Brazil and South Africa.

10.3.8 Stratigraphic traps

Rittenhouse (1972) gives a classification scheme for stratigraphic traps, summarized in table 10.2. His first-order division depends on whether the traps are or are not adjacent to unconformities. Illustrations of some of these are given in fig. 10.42.

A large portion of the hydrocarbon accumulations remaining to be found probably involve stratigraphic traps. Marr (1971), Lyons and Dobrin (1972), and Dobrin (1977) discuss seismic evidences of strati-

Table 10.2 *Classification of stratigraphic traps*

I. Not adjacent to unconformities
  A. Facies-change traps involving current-transported reservoir rock
    (1) Eolian (dunes or sheets)
    (2) Alluvial fan
    (3) Alluvial valley (braided stream, channel fill, point bar)
    (4) Deltaic (distributary mouth or finger bars, sheet, channel fill)
    (5) Nondeltaic coastal (beach, barrier bar, spit, tidal delta or flat)
    (6) Shallow marine (tidal bar, sand belt, washover, shelf edge, shallow turbidite or winnowing)
    (7) Deep marine (marine fan, deep turbidite or winnowing)
  B. Noncurrent-transported reservoir rock
    (1) Gravity (slump)
    (2) Biogenic carbonate (shelf-margin reef, patch reef, algal buildup or blanket)
  C. Diagenetic traps
    (1) Change from nonreservoir to reservoir
      (a) Replacement and leached (dolomitized)
      (b) Leached
      (c) Brecciated
      (d) Fractured
    (2) Change from reservoir to nonreservoir
      (a) Compaction (physical or chemical)
      (b) Cementation
II. Adjacent to unconformities
  A. Traps below unconformities
    (1) Seals above unconformity
      (a) Subcrop at unconformity
      (b) Topography (valley flank or shoulder, dip-slope, escarpment, valley, beveled)
    (2) Seal below unconformity
      (a) Mineral cement
      (b) Tar seal
      (c) Weathering product
  B. Traps above unconformities
    (1) Reservoir location controlled by unconformity topography
      (a) On two sides (valley, canyon, fill)
      (b) On one side (lake or coastal cliff, valley side, flank of hill or structure)
    (2) Transgressive

Source: After Rittenhouse, 1972.

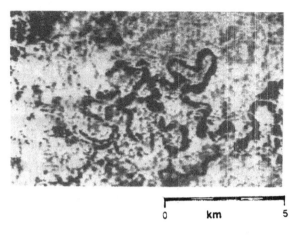

Fig. 10.40 Time slice (§12.3) through a 3-D data set showing reflectivity variations along (mostly) the same reflector. The pattern shows a meandering stream channel. (From Brown, Dahm, and Graebner, 1981.)

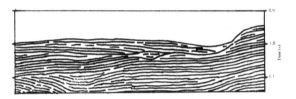

Fig. 10.41 Tracing of a profiler section showing channel cut and fill. A present-day channel can be seen in the sea-floor reflection at about 750 m depth, with indications of a natural levee to the left of it. Earlier channels and levees are also indicated.

graphic traps drawn from published case histories. They paint a discouraging picture; most stratigraphic accumulations have been found by searching for something else, that is, by serendipity. Sheriff (1980) includes the following quote:

Stratigraphic case histories had one important moral. While the discovery of stratigraphic accumulations was not generally attributed to a sound exploration program, the genius lay in being alert when a surprise occurred. Often the surprise occurs in the record from a borehole; some portion differs from what we expected in such a way as to suggest the possibility of a stratigraphic trap nearby. But where? This is where reflection-character analysis comes into its power; it can help us locate the nearby accumulation that the unexpected in a well suggests. It can help us search for strati-

graphic traps directly rather than relying on luck and statistics.

In the last few years, the search for stratigraphic traps has changed significantly. Patterns recognizable in the seismic data (§10.7) sometimes indicate the environment of deposition and thus narrow the choices of rock types, and this has led to a number of discoveries where the stratigraphy has been accurately predicted from seismic patterns. Usually, seismic stratigraphic techniques are combined with well data and geologic insight. Further improvements in techniques will expand considerably the circumstances under which this can be done successfully.

10.3.9 Integration with other geophysical data

Although it should go without saying that an interpreter should utilize borehole and other geologic data in his interpretation, he should also utilize other types of geophysical data, especially gravity and magnetic data. Where available, these data should be examined to see if they suggest anything not otherwise evident and whether the gravity and magnetic fields are consistent with the mapped features. The nature of a diapir is not always evident from examination of seismic data alone, and other data may reduce the ambi-

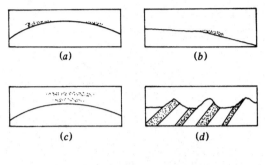

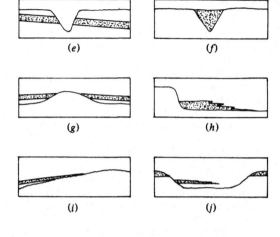

Fig. 10.42 Some types of stratigraphic traps. Reservoir rocks are dotted, impermeable rocks are clear. (After Rittenhouse, 1972.) (a) Accumulation of sands on flanks of growing structure resulting from winnowing and lateral transport of sand; (b) sand body formed at edge of shelf resulting from lowering of sea level; (c) accumulation of sand over growing structure resulting from winnowing; (d) reservoir beds subcropping at an unconformity; (e) trapping against impermeable sediments in valley fill; (f) reservoir sediments in valley fill; (g) trapping against hill or other topography; (h) accumulation against cliff; (i) reservoir sands onlapping an unconformity; and (j) accumulation subcropping at sides of valley fill.

guities. A gravity model can be constructed from a seismic structural interpretation by assigning density values to various portions of the section; the gravity field calculated from this model then can be compared with the measured gravity field. Sheriff (1989: 102, 106–9) shows an example of comparing seismic and gravity expressions of a salt dome. Such a comparison may reveal areas of disagreement that call for a reexamination of the seismic interpretation. The depth of basement, which may not be evident from seismic data, may be indicated by magnetic data. Refraction velocities may help in the identification of the nature of certain reflectors. Especially where seismic record quality is poor, such as in areas of karst or volcanics on the surface, magnetotelluric soundings may be useful in reducing interpretational ambiguities.

10.4 Modeling

10.4.1 Introduction

Interpretation of seismic data invariably involves a conceptual "model" of the portion of the earth involved in seismic measurements. The model is a simplification of the actual earth in which the only elements included are those expected to be most important in affecting the measurements. For example, the identification of stacking velocity with rms velocity is based on a model in which velocity does not vary in the horizontal direction, and statics corrections are based on a model in which travel through the weathering is vertical regardless of raypath direction below the weathering. A model may be an actual physical model, mathematical expressions, or merely a rather vague mental picture.

Modeling is often subdivided into two types, forward and inverse. *Forward* or *direct modeling* involves computing the effects of a model and *inverse modeling*

involves calculating a possible model from observation of effects. Inverse modeling in a sense includes the entire interpretation process and invariably involves uncertainty and ambiguity. Often "modeling" without a preceding adjective implies forward modeling. Modeling is important as an aid to understand how various types of possible features might appear in seismic data (Edwards, 1988; Fagin, 1991; Noah, Hofland, and Lemke, 1992).

In forward modeling, expected values are calculated from the model and compared with actual measurements, differences ("errors") being attributed to either inaccuracies in the model or factors not accounted for. Modeling is usually iterative; the model is altered in an effort to account for errors, errors from the altered model are calculated, and so on, until the errors have been reduced to what is considered acceptable. Adequate agreement, however, does not "prove" that the model corresponds to the actual earth; a different model might also provide adequate agreement.

10.4.2 Physical modeling

Many geologic phenomena are too complicated to be amenable to theoretical treatment; hence, modeling sometimes involves experiments with miniature physical models (fig. 10.43). However, models must be geometrically, kinematically, and dynamically similar to the systems being modeled (Hubbert, 1937) if the results are to be useful. Geometric similarity is achieved by making angles in the model equal to those in the system and corresponding lengths proportional. If λ is the ratio of lengths, the ratios of areas and volumes are proportional to λ^2 and λ^3, respectively. Kinematic similarity concerns the ratio of times, τ, required to effect similar changes in position or shape. The ratios of velocities and accelerations will be λ/τ and λ/τ^2, and angular velocity and angular acceleration ratios will

Fig. 10.43 Tank for seismic modeling at the University of Houston. Seismic models with horizontal dimensions of 30–60 cm and vertical dimensions of 5–10 cm are made from layers of resins or other materials to simulate three-dimensional layered structures. They are immersed in the water-filled tank for *P*-wave simulation, and sources and receivers are moved over them to obtain seismic data, motions being controlled by a computer to simulate various field-recording arrangements. Directional transducers are also held directly against solid models to study *S*-waves and wave conversion.

be $1/\tau$ and $1/\tau^2$. Dynamic similarity concerns the ratio of mass distributions, μ; this fixes density ratios, μ/λ^3. Forces acting on corresponding mass elements must be such that the motions and changes in shape produced are geometrically and kinematically similar; force ratio is $\mu\lambda/\tau^2$. Dimensionless quantities (like Poisson's ratio) must have the same numerical value. Thus, there are only three independent values, λ, τ, and μ.

We might, for example, wish to represent 1 km by 10 cm in a model; hence, the ratio of model to actual distance is $\lambda = 10^{-4}$. In practice, seismic velocity is restricted by available materials, and the ratio of model to actual velocity can range only by a very small amount, that is, $\lambda/\tau \approx 1$. Because λ has already been selected, this restricts τ. If the model material has the same velocity as the earth, $\tau = 10^{-4}$ and we must use frequencies 10^4 times what is used in the earth (because frequency ratios depend on $1/\tau$). The density of available model materials is probably about the same as that of earth materials; because the density ratio $\mu/\lambda^3 \approx 1$, this determines the mass ratio $\mu \approx 10^{-12}$. If we wish to model several types of things simultaneously, such as various modes of wave propagation, at-tenuation, and so on, we must be sure that the relevant physical properties are consistent with our model ratios λ, τ, and μ. Examples of physical modeling are shown in figs. 6.21, 6.28, and 6.43.

10.4.3 Computer modeling

More commonly, modeling is done by computer and several examples have been given (for example, figs. 2.32, 3.3, 6.18, 6.23 to 6.25, 6.37, 6.38, 6.40, and 6.41). Many types of algorithms are used in computer modeling, ranging from simply convolving a wavelet with a sequence of reflection coefficients, tracing rays through models where the raypaths bend in accordance with Snell's law, to full-waveform methods based on relations such as Kirchhoff's equation (2.42), or wave-equation methods such as used in migration (§9.13.4) (§9.13.4; Gazdag, 1981).

Synthetic seismograms may help in determining how stratigraphic changes might affect seismic records, and raypath modeling (§10.4.5) in determining the distortions that complicated velocity distributions produce. Where the algorithms and models are good, the resemblance to actual seismograms is good. Modeling is an invaluable pedagogical tool (see Hilterman, 1970), but it invariably involves assumptions and approximations that should not be forgotten when drawing conclusions.

10.4.4 Synthetic seismograms

The one-dimensional synthetic seismogram was introduced in §6.2.1; it is simply a wavelet convolved with the reflectivity assuming zero offset and horizontal layering, the velocity and density values most often being those measured from borehole logs in a well. Its most common use is in identifying reflections on CMP sections with specific interfaces in the earth, but a perfect match should not be expected. CMP sections tend to average amplitudes over a range of offsets and involve dip effects that are not generally allowed for with synthetic seismograms, as well as including noises of various kinds. Synthetic seismograms can be generalized to allow for multiples and other types of events that may be superimposed on the primaries-only synthetic. Horizontal changes may be made in the reflectivity so that a sequence of one-dimensional synthetic traces simulates a common-midpoint line recorded over the changing reflectivity.

Two-dimensional synthetic seismograms are not limited to vertical travel nor to zero offset. They permit modeling diffractions and the dependence on offset of arrival times, waveshape, and amplitude. Mode conversion may be allowed for. A variety of methods are used, some permitting dipping interfaces and some employing simplifications (such as the scalar form of the wave equation allowing for *P*-waves only). Trorey (1977) approximated reflectors by a series of semiinfinite plane strips and based his method on the Kirchhoff equation (2.42). Other methods utilize wave-equation methods of the type used in migration processing (§9.12.3 and 9.12.4). A common model is the *exploding-reflector model;* each reflecting interface is assumed to be a distributed source detonated at time $t = 0$, the source density being proportional to the reflectivity at the interface; seismic waves are radiated upward at half the actual velocity (to give the traveltime for two-way travel). The record received at the surface simulates a common-midpoint section in many (but not all) regards. Waves may be tracked by some of the methods. More elaborate methods allow for mode conversion, the variation of reflectivity with incident angle, surface waves, head waves, and so on.

10.4.5 Ray-trace modeling

Where velocity varies in other than a very simple way, the tracing of raypaths through a model obeying Snell's law at each velocity change is one way of developing an understanding of how a seismic section relates to a portion of the earth where velocity complications exist. Horizontal changes in velocity especially can distort structural pictures (§10.5) and make it difficult to appreciate the significance of structural evidences.

Taner, Cook, and Neidell (1970) carried out ray tracing for several models, one of which is shown in fig. 10.44. An illustration of ray-trace modeling for a thrust-fault situation is shown in fig. 10.45. Downward ray tracing for offset traces is not feasible except on a trial-and-error basis because initially we do not know the starting direction. Where source and detector are coincident, as is assumed on common-midpoint sections, the raypath to and from a reflector must be coincident and so must strike the reflector at right angles. This makes it easy to trace rays upward. If increments along the reflectors are constant, then the density of emergent raypaths at the surface may give a qualitative indication of amplitude variations. Note the buried-focus effects in fig. 10.44d. Ray-trace modeling is also used in depth migration (see fig. 9.59c).

Ray-trace modeling is useful in seeing how stacking-velocity measurements get distorted by velocity complications (May and Covey, 1981). Taner et al. (loc. cit.) also traced rays to obtain simulated gathers on which to make velocity analyses (fig. 10.46). They did this iteratively on a trial-and-error basis because the reflection-point shifts updip with offset in a manner not easily predicted. The stacking velocities determined for fig. 10.46 do not bear any simple relation to rms velocities, as was pointed out in §5.4.4a, and, in fact, the curve of arrival time versus offset is not even a hyperbola, so that the values of stacking velocity obtained from best-fit hyperbolas depend on the mix of offset data used in the calculation. Most ray tracing does not allow for mode conversion.

10.5 Lateral variations in velocity

10.5.1 Gradual changes

Often, velocity variations in the horizontal direction are sufficiently gradual that their effects can be treated as a second-order correction. This situation is especially common in Tertiary basins filled mainly with clastic sediments that have not been subject to uplift. The horizontal variations often result from gradual changes in the lithology, for example, as distance from the source of the sediments increases. Sometimes, the vertical velocity function is changed slightly from location to location and the horizontal gradient is otherwise ignored in plotting data. A common variation of this technique is to map reflectors using a single function for the area and then to add location-dependent depth corrections to the mapped values.

Lateral effects also result from changes in the thickness of a water layer on top of the sediments. The change in velocity with depth effectively begins at the sea floor, variations within the water layer usually being insignificant. The velocity of the sediments is not affected greatly by the amount of water overburden; the difference between overburden and interstitial pressures is usually the factor determining velocity (see §5.1.2) and, because the water layer increases both pressures by the same amount, it does not

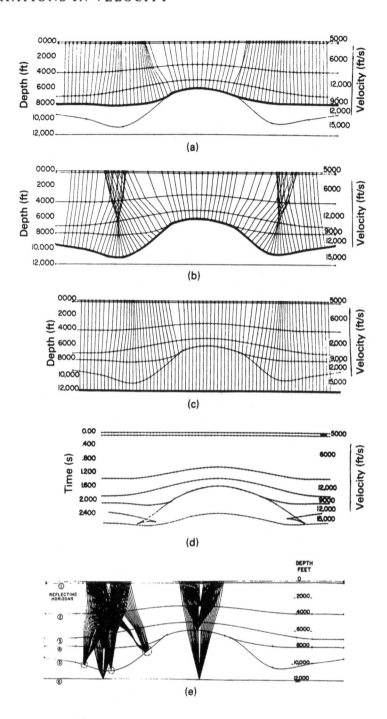

Fig. 10.44 Ray-trace modeling over a salt swell. (From Taner, Cook, and Neidell, 1970.) (a to c): Normal-incident rays (the exploding-reflector model) to three reflectors; (d) the conse-quent time section; (e) gathers for modeling offset traces; note the triplication of reflecting points for the left gathers and the reflection-point smear for the steeply dipping one.

change the differential pressure or velocity. Of course, the average velocity down to a reflector is affected by inclusion of more travel path at water velocity. Figure 10.47 shows a seismic line that goes from shallow to deep water; much of the apparent dip is a velocity effect rather than real dip (compare figs. 10.46a and 10.46b). The apparent dip can be corrected by changing the velocity function with water depth when making depth calculations.

Lateral velocity changes also affect the horizontal positions of features (fig. 9.59). This is illustrated for a diffracting point and a simple two-layer model in fig. 10.48 (see also problem 10.14). The crest of a diffraction usually locates the diffracting point, but lateral changes of velocity shift the crest of the diffraction. If we consider more complicated models, for example, two dipping layers with different strikes, serious distortions exist that would be very difficult to unravel

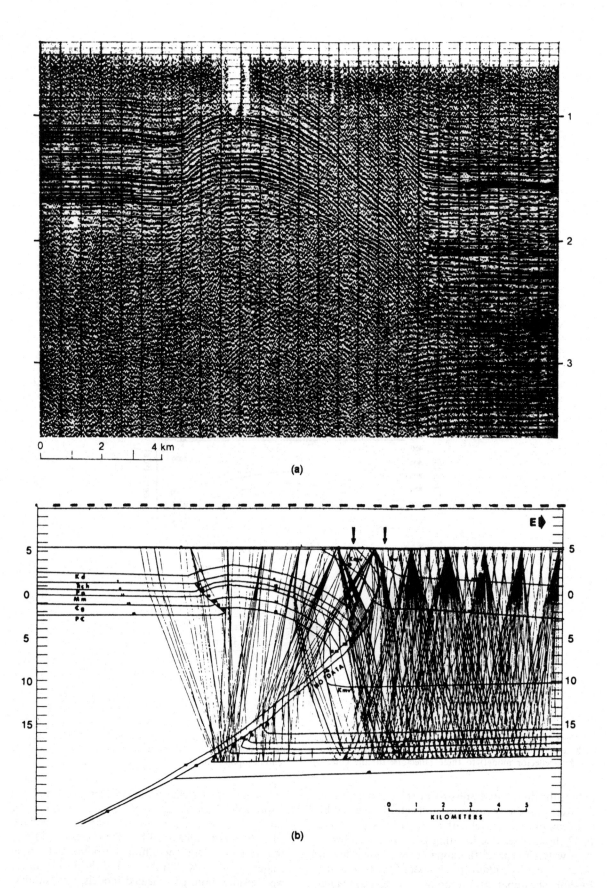

Fig. 10.45 Section across a thrust fault in the Big Horn Basin, Wyoming. (After Stone, 1985.) (a) Seismic section, (b) ray trac-ing through a depth model, and (c) seismic section predicted from the model.

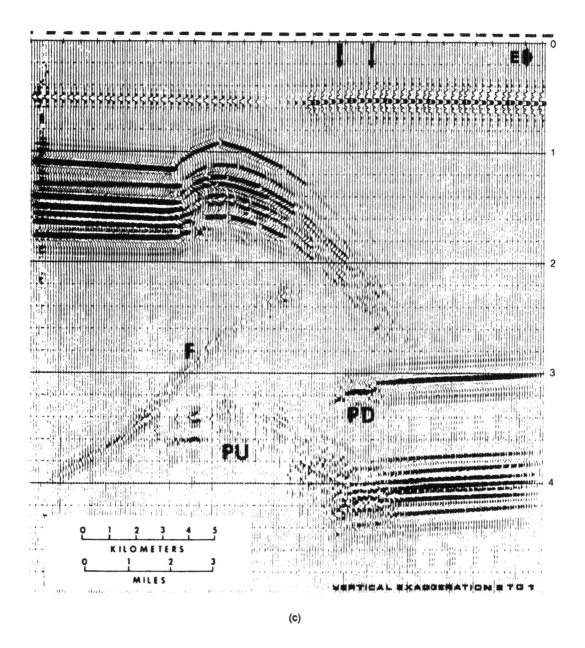

(c)

from seismic data. Such a situation could easily result where the section and the sea floor dip in different directions.

Correction for gradual velocity changes usually hinges on being able to determine the velocity changes with sufficient reliability. Often, velocity has to be determined from the seismic data themselves (see §9.7), but velocity analyses (although they may be adequate for use in stacking) often have appreciable uncertainty that may make them unsuitable for such corrections without smoothing. Displays of the type of fig. 9.32 are often especially helpful. Velocity variations are usually reasonably systematic with structure, although in attenuated fashion, that is, velocity relief is usually not as great as the structural relief. Usually, data should be smoothed with the geology in mind.

10.5.2 Sudden changes

Where lateral changes in velocity are more sudden, correction may not be simple. Consider the effects of the sea-floor relief in fig. 10.49. The velocities of the sediments immediately below the canyon probably are markedly different from those of their lateral equivalents because of the differences in overburden, but at large depths, the effects of the canyon probably vanish. Furthermore, the sediments below the bottom of the canyon may be in fluid-pressure equilibrium with their lateral equivalents, thus having fluid pressure appropriate to the uneroded thickness whereas their overburden pressure is less because of the erosion, so that they are overpressured. A "correct" method of removing the velocity effect is not evident, and usually

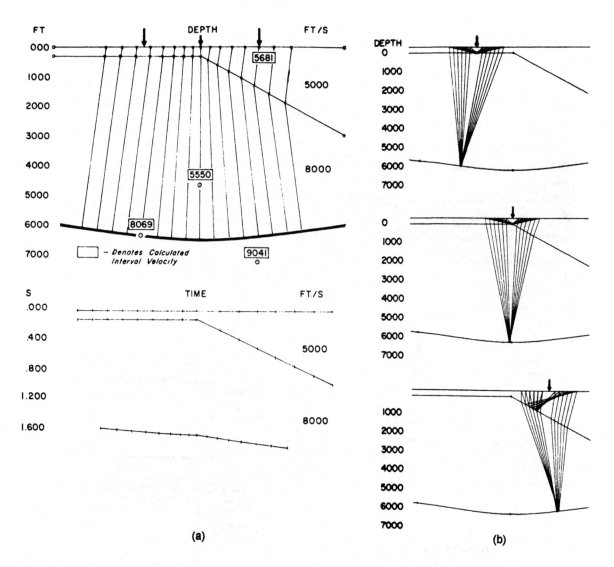

(a)

(b)

Fig. 10.46 Use of ray-trace modeling for velocity analysis. (From Taner, Cook, and Neidell, 1970.) (a) Depth model with normal-incident raypaths from synclinal reflector; the dots show where the reflection would be plotted using the velocities (in ft/s in boxes) determined from gathers at three locations; (b) raypaths for gathers at the locations indicated by arrows.

the method adopted is the empirical one that produces the most sensible results. In areas of purely erosional relief, such as that in fig. 10.49, this may not be too difficult, but where structural complications accompany, and perhaps cause, the sea-floor relief, objective criteria may be lacking.

The flow of salt into lenses and domes may produce velocity anomalies in the section below them. Salt velocity, about 4.5 km/s, may be either higher or lower than that of the laterally adjacent sediments, and so the velocity anomalies resulting from a salt lens may be either a pull-up or a push-down. A pull-up is most common because lower-velocity clastic rocks are the most common lateral equivalent, but lime-rich sediments, anhydrite, or other high-velocity rocks may produce push-down, and in some areas (such as Mississippi) both can occur. Figure 10.50 shows a salt pillow with a consequent pull-up. Similar velocity effects

can result from other situations, such as reefing (see figs. 10.35g and 10.35h), where either pull-up or push-down can occur, depending on how the velocity in the reef (which depends in part on the reef's porosity) compares with the velocity of the lateral equivalents.

Velocity complications may be very drastic in regions of compressional or thrust tectonics. Figure 10.51 shows velocity effects resulting from the overthrusting of high-velocity sediments in the Rocky Mountain thrust belt. The complications involve not only velocity pull-up, but also fictitious faulting evidences, phantom diffractions, and other effects. The usual solutions to such extreme complications are to trace rays through a model of the section in an effort to achieve reasonable agreement with what is observed (see §10.4.5), but this approach is subject to the uncertainties of modeling, mainly lack of information as to how to construct the model and determine ap-

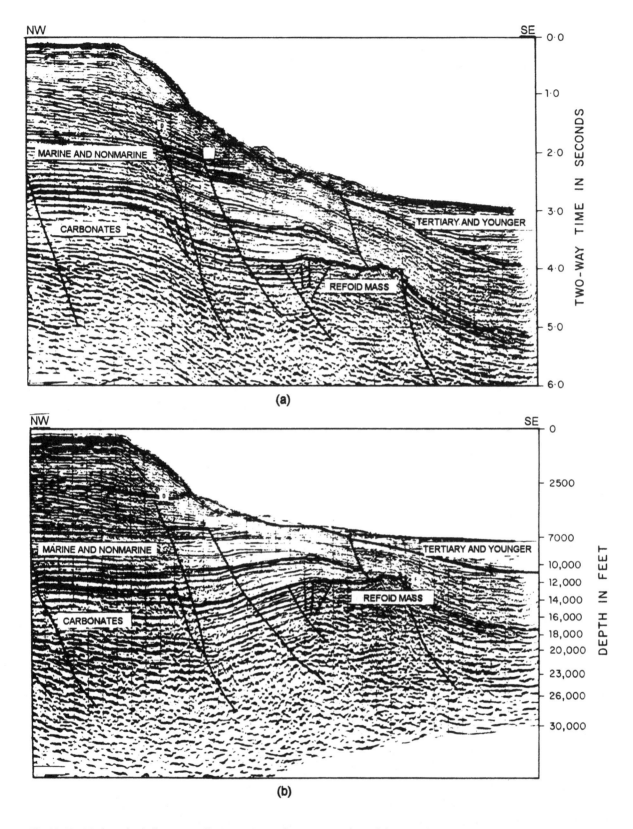

Fig. 10.47 Marine seismic line perpendicular to the continental slope. Variation of the water depth creates false dip. (From Morgan and Dowdall, 1983.) (a) Seismic time section, and (b) migrated depth section. Note the reversal of dips under the dipping sea floor.

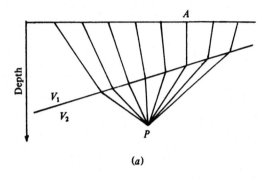

(a)

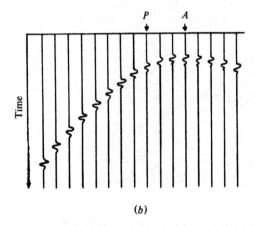

(b)

Fig. 10.48 Distortion of diffraction arrival-time curve when $V_2 > V_1$. The earliest arrival of the diffraction is at A rather than at P over the diffracting point. (From Larner et al., 1981.) (a) Depth model, and (b) time section.

propriate velocities; such information is apt to be lacking where needed most.

10.6 Three-dimensional interpretation of 2-D data

Reflecting points lie updip from the points where the reflections are observed. Migration accommodates the components of dip in the in-line direction (although not always correctly), so that most of the problems in three-dimensional mapping result because the component of dip perpendicular to the line is unknown or not taken into account properly. The subsurface trace (line of reflecting points) lies in the direction updip from the seismic line and this should be allowed for in mapping data oriented along seismic lines (see Sheriff, 1978, chap. 21), that is, data ought to be posted on maps at the reflecting points on the reflector being mapped rather than at the source points on the surface. Where data are limited to a grid of a few lines, the first step in mapping is to determine reflecting points where cross-information permits this to be done, as at line intersections (fig. 10.52). Reflecting points in between the points where such determinations can be made then often can be inferred with adequate accuracy. An alternative that is sometimes

feasible is to map unmigrated data and then migrate the maps to produce a correct structural map (see fig. 9.61).

The techniques of acquiring data specifically for three-dimensional (3-D) analysis are discussed in §12.1, of 3-D data processing in §12.2, of 3-D display in §12.3, and of 3-D interpretation in §12.4 and 12.5. As of 1994, 3-D is one of the fastest growing areas of geophysics. Most 3-D work has been (and is still) devoted to detailing fields after hydrocarbons have been discovered in order to optimize field development and exploitation (Brown, 1991; Sheriff, 1992), and 3-D techniques have been very cost-effective. There seems to be unanimous agreement that 3-D surveys result in clearer and more accurate pictures of geological detail and that their costs are more than repaid by the elimination of unnecessary wells and by enabling the recovery of more hydrocarbons through the discovery of isolated pools that otherwise might be missed. 3-D techniques are now also being used for exploration in a number of areas.

In areas where 2-D lines are particularly dense and of good quality, interpolation is sometimes used to yield a uniform data grid that is then processed as 3-D data. Clearly, the 2-D data must be compatible, that is, must involve the same frequencies, same embedded wavelet, and so on. The result, called a *2½-D survey,* has some of the benefits of, but is usually markedly inferior to, a true 3-D survey.

10.7 Stratigraphic interpretation

10.7.1 Introduction

Extracting nonstructural information from seismic data is called *seismic stratigraphy* or *seismic-facies analysis. Facies* refers to the sum total of features that characterize the environment in which a sediment was deposited. Facies involves, among other things, sedimentary structure, the form of bedding, original attitude, and the shape, thickness, thickness variations, and continuity of sedimentary units. Our interest here is in inferring stratigraphy rather than in locating stratigraphic traps (discussed in §10.3.8), though the implications for stratigraphic traps are obvious. Books on seismic stratigraphy include Sheriff (1980) and Hardage (1987); many case histories showing applications in various circumstances are given by Halbouty (1982), Berg and Woolverton (1985), Van Wagoner et al. (1990), and the three volumes edited by Bally (1987–9). The "classic" reference is Payton (1977) and one of the best references is Wilgus et al. (1988).

Depositional patterns such as progradation, pinchouts, and channels can sometimes be seen in seismic data (for example, in figs. 10.40 and 10.53) although many stratigraphic features are too small to be resolvable (Sheriff, 1977) or too gradual to see. Depositional patterns are sometimes associated with depositional energy (which determines the degree of separation of fine particles from coarse), lithology, porosity, and

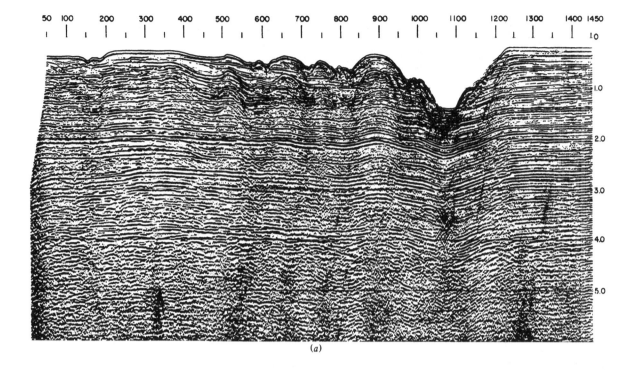

(a)

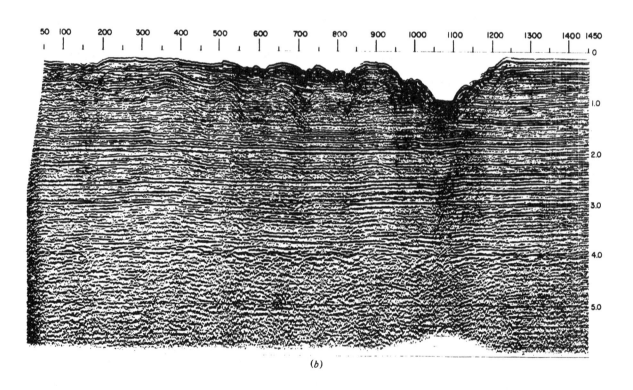

(b)

Fig. 10.49   Velocity effects of sea-floor canyon. (Courtesy of
Grant Geophysical.) (a) Before correction for velocity, and (b)
after correction.

Fig. 10.50   Two uplifts in the Mediterranean Sea. The left uplift is a salt pillow with the salt from 5.0 to 5.3 s (at the left end) and the right one is piercement salt (located slightly to the side of the line). The apparent uplift below the salt is probably all velocity anomaly. (Couirtesy of C.G.G.)

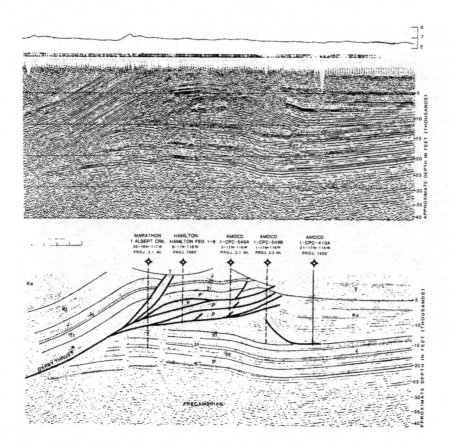

Fig. 10.51   Thrust faulting in Wyoming overthrust belt showing velocity pull-up under the thrusts. (From Williams and Dixon, 1983.)

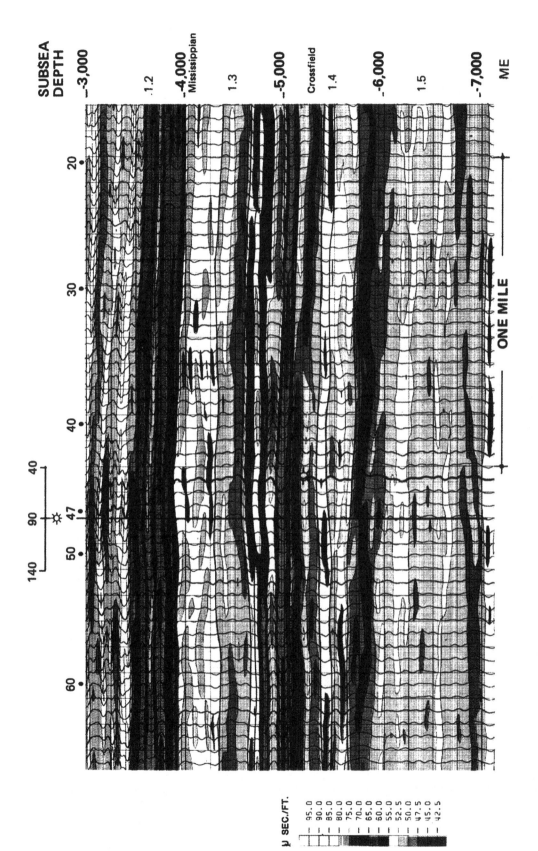

Plate 1 Seismic-log display (as transit time) of a portion of a seismic section in Alberta. Production comes from local porosity development in the Crossfield formation, which is associated with local velocity lowering. Low-frequency components and "ground truth" as to velocities associated with hydrocarbon accumulation were available from nearby well control. (Courtesy of Technica.)

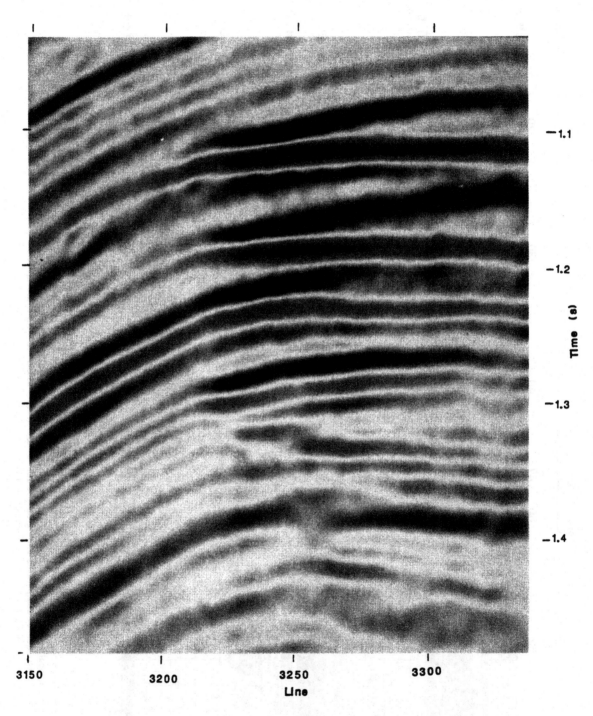

Plate 2 Bicolor display showing stacked hydrocarbon accumulations. Note high amplitudes (bright spots), polarity reversals, flat spots, time sags due to the increased time to traverse the hydrocarbon accumulations. (From Brown, 1991: 163.)

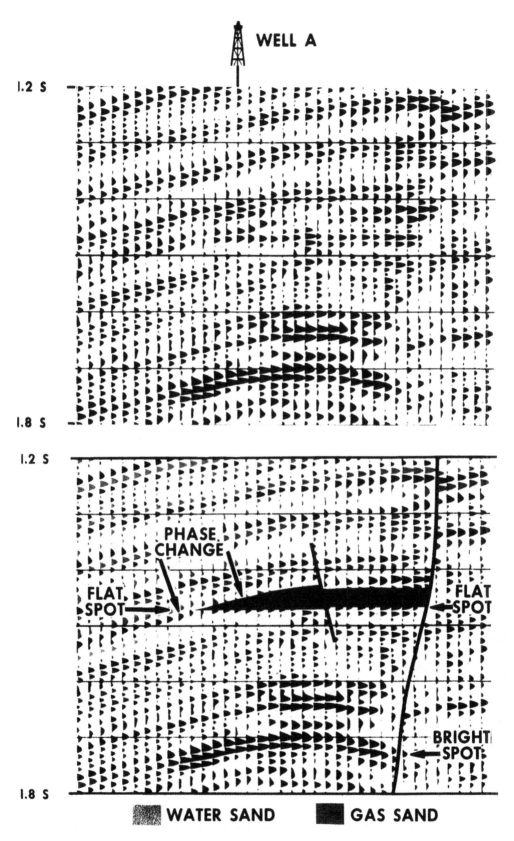

Plate 3   Hydrocarbon indicators shown on dual-polarity variable-area section. (From Brown, 1991: 139.)
(a) Variable-area section; (b) interpretation.

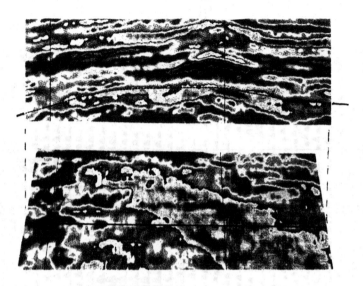

Plate 4 A vertical section (top) from a 3-D volume inverted to seismic-log form and a horizon slice (bottom) over a hydrocarbon accumulation. (Courtesy of CGG.)

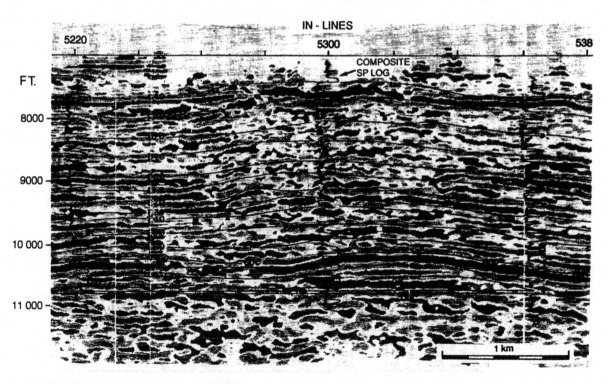

Plate 5 Fault slice, Nun-River Field, Nigeria. This section 75 m from a complexly curved fault plane shows bright spots and flat spots to indicate hydrocarbons trapped against the fault. (From Bouvier et al., 1989.)

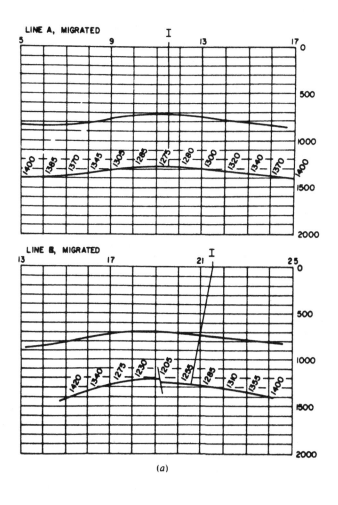

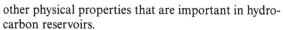

(a)

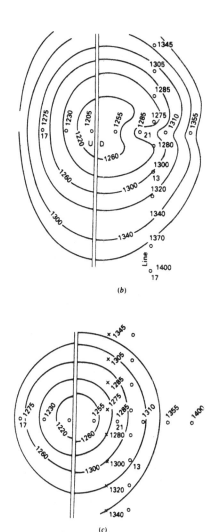

(b)

(c)

Fig. 10.52 3-D effects in interpreting 2-D data. (a) 2-D migrated, intersecting seismic lines with one horizon timed at the source-point locations; (b) contours of data plotted at source points show fictious structure; and (c) when data are plotted at reflecting points, contours indicate simple structure.

other physical properties that are important in hydrocarbon reservoirs.

Seismic stratigraphy is often divided into several subareas:

1. Seismic-sequence analysis, separating out time-depositional units based on detecting unconformities or changes in seismic patterns;
2. Seismic-facies analysis, determining depositional environment from seismic-reflection characteristics;
3. Reflection-character analysis, examining the lateral variation of individual reflection events, or series of events, to locate where stratigraphic changes occur and identify their nature; the primary tool for this is modeling by both synthetic seismograms and seismic logs, as already discussed in §6.2.1, 5.4.5, and 10.4.4.

4. Detection of hydrocarbon indicators, which will be discussed in §10.8.

10.7.2 Sequence stratigraphy

Seismic stratigraphy has evolved into sequence stratigraphy as depositional features that were first seen in seismic data have been recognized in outcrops and in well-log and paleontological data.

The concept underlying *sequence stratigraphy* is that sediment deposition is controlled by four factors:

1. Subsidence of the crust because of tectonic and/or isostatic reasons; this creates the space (*accommodation*) available to receive sediments. Thermal expansion and/or tectonic forces cause rocks to rise so that their higher portions are exposed and eroded. Upon cooling, rocks become denser and sink to restore isostatic equilibrium. The cooling

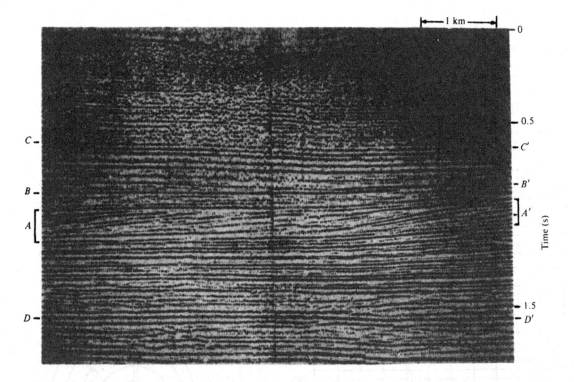

Fig. 10.53 Section showing prograding unit *AA'* with toplap where reflections pinchout at the top of the unit and downlap where reflections converge at the base. Note also the channel at

is a very slow process because of the extremely low thermal conductivity of rocks (Jurassic oceanic basement is still losing its heat of formation). Isostatic adjustment is also a very slow process, although not as slow as conductive heat loss (isostatic rebound from Pleistocene continental glaciation is still going on). The filling up of the accommodation space adds to the load, producing further subsidence and additional accommodation; sediment volume ultimately may be up to three times the original accommodation.

2. Sediment inflow, which provides the sediments for infilling the accommodation.

3. *Eustasy,* rises and falls of absolute sea level. The combination of subsidence and eustasy determines relative sea level, which in turn fixes the amount of accommodation. Eustatic changes are generally rapid compared with subsidence and they predominate in determining the locale of sediment deposition. The depositional patterns left in sedimentary rocks thus mainly document the eustatic variations. Sea-level changes occur on several time scales, and the more rapid sea level rises and falls (*paracycles*) produce changes within the sequence of rock being deposited; these changes may produce seismic reflections. Seismic reflections thus indicate the

surface of the solid earth at the time of sediment deposition (§10.7.3).

4. Climate, which mainly determines the nature of the sediments being deposited. For example, carbonates are apt to dominate in warm climates.

The interplay of these factors, especially those of factors 2 and 3, determines depositional patterns. The effects can be seen in rock outcrop, well-log, paleontologic, and seismic data.

The heat injected by thermal/tectonic events is conducted away very slowly, producing gradual subsidence. These long-term subsidence effects occur at a rate that is relatively constant compared with the more rapid changes resulting from eustasy (fig. 10.54). The net effect on sea level is the superposition of the tectonic (thermal) and eustatic effects. Where absolute sea level is rising, it will increase the rate of growth of accommodation, but where absolute sea level is falling the net effect will depend on the relative rates of the tectonic subsidence and the eustatic fall. Where net sea-level fall results, we have a "type-1" situation and a significant type-1 unconformity involving subareal erosion; where the effects of the two causes nearly cancel each other, a "type-2" situation results.

Clearly, relative (rather than absolute) sea-level changes are what affects deposition, that is, subsidence will produce the same effects as sea-level rise. We expect effects to be local if caused by local subsidence but widespread if caused by eustatic changes.

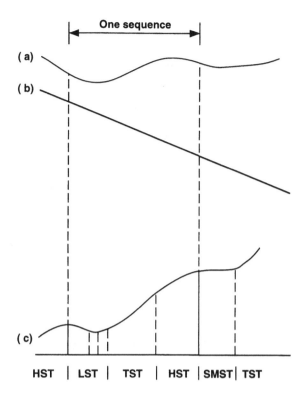

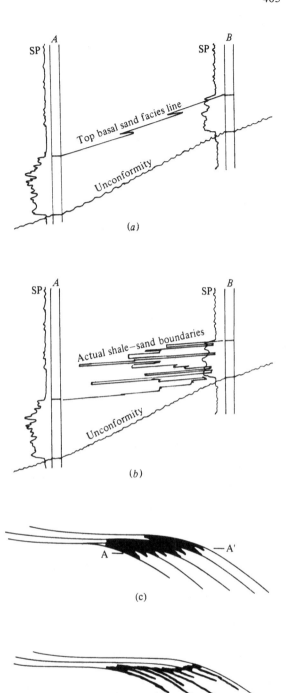

Fig. 10.54 Changes in relative sea level. (a) Assumed eustatic variation, which is rapid compared with tectonic subsidence (b). Tectonic subsidence is nearly linear over this short time span, although logarithmic over a long time. The sum of these two (c) gives the relative change of sea level that provides the accommodation space for sediments. The weight of deposited sediments will produce further subsidence by isostatic adjustment, also a long-term process. System tracts (§10.7.5) are indicated at the bottom.

Many of the effects of sea-level changes appear to be contemporaneous in widely separated basins.

10.7.3 Time significance of reflections

Implied in seismic-sequence analysis is the concept that the attitude of seismic reflections is that of depositional time surfaces rather than facies surfaces. A *time surface* indicates a surface that at one time was the surface of the solid earth. The passage of major storms, floods, and other short-term events redistribute sediments within very short periods of time along time surfaces, whereas the long periods between such events often do not leave a record because the new sediments brought in are rearranged by the next major event. Stratal surfaces thus follow time surfaces. Because the thicknesses of stratal units are generally small, much smaller than the seismic resolving power, they only produce very minor reflection contributions, but these tend to interfere in essentially the same way over a widespread area because the stratal surfaces are generally parallel over a wide area and change very slowly laterally. The interference produces the coherent lineups of reflection events. The fact that seis-

Fig. 10.55 The nature of facies surfaces. (Data for a and b from Vail, Todd, and Sangree, 1977b.) (a) Facies surface based on data from two wells 17 km apart; the *SP*-log curves distinguish the sand from surrounding shale. (b) Redrawing of the facies surface based on intervening well-control points; the major portions parallel stratal or time surfaces. Seismic data show reflections parallel to the time surfaces onlapping the unconformity. (c) Classical picture of sand-rich sediments in a prograding/aggrading system suggests a reflection along the facies boundary AA', which does not show. (d) Occasional major storms and other catastrophic events rework the sand-rich sediments and spread them along time surfaces, which is the attitude of reflections.

mic reflections parallel time surfaces is well established by many observations but it is somewhat contradictory to intuitive feelings that reflections should be due to changes in rock nature, such as from sand to shale along facies surfaces. Facies surfaces are often based on fairly widespread control (for example, on well control) so that the detailed information as to how to draw the facies surface is not available. The major portions of correctly drawn facies surfaces parallel time surfaces (figs. 10.55 and 10.56).

10.7.4 Depositional models

The angularities between reflections where one of them terminates (fig. 10.58) are the principal seismic evidences of seismic stratigraphy. Where the data quality and resolution are sufficient, we observe that reflections group themselves naturally into packages bounded by systematic reflection angularities; these packages correlate with sediment packages, groups of sediments deposited contemporaneously between eustatic events.

Two types of angularities occur at the bottom of reflection packages: onlap and downlap. *Onlap* is indicated by reflections that are horizontal or dip away from their terminations and *downlap* by reflections that dip toward their terminations. Except for dip direction, onlap and downlap often look very similar in seismic data. Subsequent rotation, because of differential subsidence or other reasons, may have changed the dips so that present attitudes no longer have the original implications, that onlap indicates locations that are proximal (deposition close to the source of sediments, that is, on the landward side of a sediment package) and downlap locations that are distal (deposition distant from the sediment source).

Three types of angularities are seen at the top of reflection packages: toplap, erosional truncation, and apparent truncation. *Erosional truncation* indicates that the sediment package formerly extended higher than it does today but that portions were removed by erosion, whereas *toplap* indicates deposition near sea level and that the sediment package never extended significantly higher in the section. With good seismic data quality, toplap sometimes can be distinguished from erosional truncation because there were changes in the depositional environment near toplap and consequently reflections are changeable in attitude and character, whereas no such changes occurred at erosional truncation terminations. *Apparent truncation* (like downlap) is a consequence of sediment *starvation*, that is, insufficient sediments were available to continue a resolvable reflection. Sometimes reflections terminate because of starvation other than at sediment-package boundaries, as in divergent patterns associated with differential subsidence. Over large portions of sediment-package boundaries, reflections parallel the boundaries, a situation called *concordance*.

Diagrams from Vail, Mitchum, and Thompson

(1977) indicate the depositional patterns expected from sea-level rises and falls (fig. 10.57). A relative rise of sea level can be produced by either an absolute sea-level rise or by land subsidence. The primary evidence in seismic data for a sea-level rise is a *coastal onlap* pattern, the progressive termination of reflections in the landward direction. Whereas a sea-level rise is usually associated with a *transgression* (coastline moving landward) over an unconformity (fig. 10.57a), it can also be associated with coastal *regression* (coastline moving seaward; see fig. 10.57b) provided the influx of sediments is large enough. Coastal onlap is seen in both situations. The primary evidence for a *stillstand* of sea level (fig. 10.57c) is *toplap*. A fall of sea level (fig. 10.57d) would expose previously deposited sediments to erosion, so erosional truncation is the primary evidence for a sea-level fall. Where sediment packages are thick enough and noise sufficiently low, reflections showing these features can be seen in seismic data and used to determine the sea-level changes.

The technique used by Vail et al. (loc. cit.) is to first mark reflection angularities on a seismic section and then draw the unconformity boundaries that join onlaps (heavy solid lines in fig. 10.58) and downlaps (*downlap surfaces,* indicated by dashed lines in fig. 10.58). A line connecting onlaps marks a relative sea-level fall and indicates an unconformity (and a *sequence boundary*); this boundary is then continued through regions where angularities are not evident (the *correlative conformity*). The procedure continues with mapping the boundaries over a grid of lines, constructing maps of structural relief and isopach thickness for the intervals between the boundaries, subdividing the sequences according to seismic-facies evidences (§10.7.6), relating them to adjacent sequences, and finally attributing stratigraphic significance to them. This procedure is illustrated by the examples shown in fig. 10.59. (Some people use the downlap surface of maximum flooding, which generally separates transgressive and highstand tracts (§10.7.5), as the sequence boundary.)

The package of sediments between the sequence boundaries is a *sequence,* that is, a three-dimensional set of facies deposited contemporaneously and linked by depositional processes and environments. For example, fluvial deposition on land, deltaic deposition near the coastline, a strand plain off to the side, reefs near the shelf edge, and turbidite fans at the base of the slope might all be contemporaneous and parts of the same sequence. Downlap surfaces occur within sequences.

Vail et al. (loc. cit.) believe that many sea-level changes were contemporaneous worldwide and they developed a eustatic-level chart showing the worldwide pattern (Haq, Hardenbol, and Vail, 1988). By correlating local coastal onlap charts with the master eustatic-level chart, one can sometimes date reflection events with rather high precision.

In the early days of seismic stratigraphy, the magni-

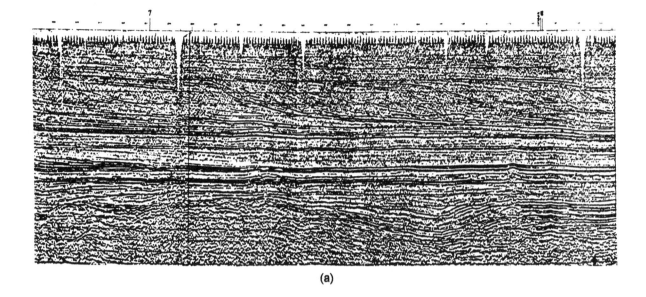

(a)

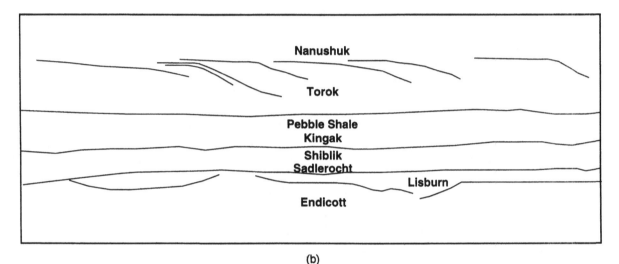

(b)

Fig. 10.56 Line on the North Slope of Alaska. The top of the Torok formation in contact with transgressive sediments (and then nonmarine Nanushuk sediments) is a facies surface. As the sea level has varied, the deltas have stepped successively higher in the section, thus jumping from one time surface to another. (USGS data.) (a) Seismic line and (b) interpretation of the top Torok formation.

tudes of sea-level falls and rises were determined simply by measuring the vertical distances between onlap points; this led to estimates that were unreasonably large. However, onlap can occur both well below sea level (*marine onlap*) and above sea level, and recognition of this removed the early objections. Periodic sea-level changes are now widely accepted. They appear to occur with several periodicities (table 10.3). Wilgus et al. (1988) (and especially Greenlee and Moore, 1988) discuss calculating sea-level changes.

10.7.5 System tracts

Depositional patterns depend on whether accommodation is growing or decreasing as well as on the avail-

ability of sediments. Accommodation is dominated by the rate of eustatic change. Let us conceptually model one simple eustatic cycle of sea-level fall and rise, starting with a sufficiently rapid eustatic fall that, when superimposed on tectonic subsidence, produces a net fall of relative sea level (as in fig. 10.54). The sea will regress and expose more area to aerial erosion. If the sea level falls below the shelf edge (fig. 10.60a), a *type-1 unconformity* will result and sediments will be deposited directly in deep water rather than on the shelf (*sediment bypassing*); the result is deposition of a *lowstand* or *basin-floor fan* (or a *slope fan*). Coarse sediments will probably be included in the lowstand fan. The steep slopes will result in instability of sediments resting at the angle of repose, producing slumps

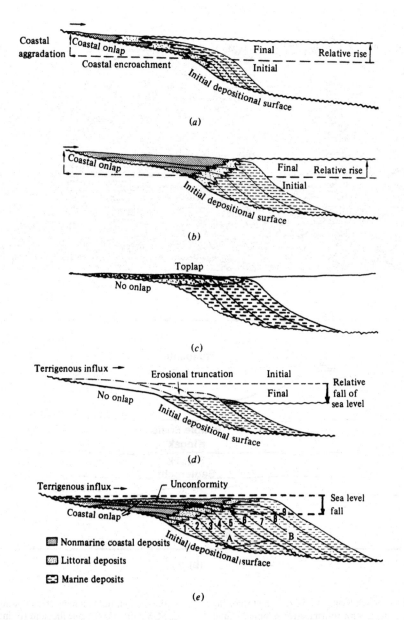

Fig. 10.57 Patterns associated with relative sea-level changes. (After Vail, Todd, and Sangree, 1977b.) (a) Relative sea-level rise produces a transgression if terrigenous influx is low and (b) a regression if terrigenous influx overwhelms the effects of the rise. (c) Progradation associated with stillstand of sea level. (d) Grad-ual sea-level fall produces downward shift in pattern but tops of patterns are eroded. (e) Rapid sea-level fall produces major seaward shift in the locale of onlap; the pattern indicates a grad-ual rise, then a sudden fall between units 5 and 6, followed by another gradual rise.

Table 10.3 *Eustatic cycles*

Order of cyclicity	Period	Magnitude	Cause	Terminology
1	≈50 Ma		Plate tectonics and continental breakup	
2	3–50 Ma	≈100 m	Tectonics and eustasy	Supercycles
3	½–3 Ma	≈50 m	Glacial eustasy	Cycles
4	80–500 ka		Orbital (Milankovitch) cycles	
5	30–80 ka		Climatic changes	Paracycles
6	10–30 ka			

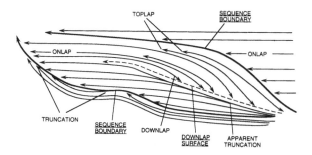

Fig. 10.58 Terminology for reflection terminations and their relation to sequence boundaries and downlap surfaces. (After Mitchum, Vail, and Thompson, 1977.)

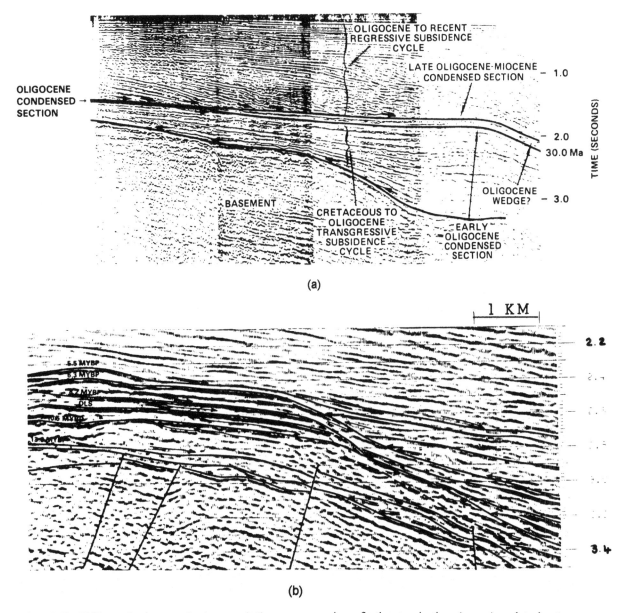

Fig. 10.59 Picking reflection terminations to indicate sequences. (a) Section on east coast of New Zealand (from Loutit et al., 1988); and (b) section from Midland Basin, Texas, showing reflection terminations (arrows) used to locate sequence boundaries (from Sarg, 1988).

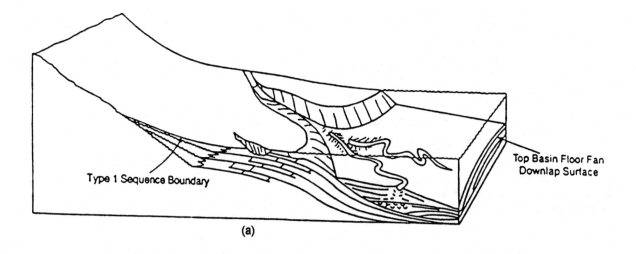

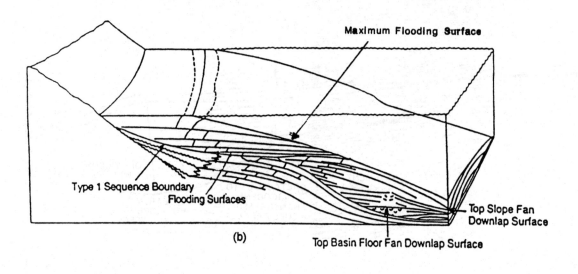

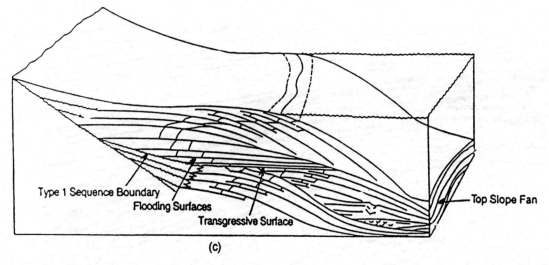

Fig. 10.60 Schematic of system tracts on a passive margin. (After Posamentier, Jervey, and Vail, 1988.) (a) Sea-level low-stand producing basin-floor and slope mounds, level-channel complexes, and (not shown) lowstand prograding wedges; (b) transgressive tract topped with a maximum flooding surface; and (c) highstand tract of aggrading followed by prograding deposition.

and slides, a condition that will encourage turbidity flows. At this time, rivers will cut (incise) valleys into the exposed shelf.

As the rate of eustatic fall becomes about the same as the rate of subsidence, relative sea level will remain roughly constant and a *lowstand* will result. Incoming sediment volume will decrease because gradients of the land subject to erosion will no longer be increasing; turbidity flows will continue and tend to build up on top of the lowstand fan or on the slope; the result is deposition of a *slope fan*. As sea level begins to rise, sediments will begin to *aggrade* (build upward) and *prograde* (build outward) to produce a *lowstand wedge*. Some valley filling may also occur. The overall package of sediments involving lowstand fan, slope fan, and lowstand wedge is called a *lowstand system tract*.

If the eustatic fall had been smaller it might have merely balanced out the effects of subsidence on sea level and a sea-level stillstand and would have resulted. The sediments would first prograde and then aggrade as relative sea level begins to rise slowly. This would produce a *shelf-margin system tract*. A *type-2 unconformity* separates a shelf-margin tract from the underlying highstand system tract (see what follows).

As a eustatic rise becomes greater, it will have the same effect on relative sea level as the tectonic subsidence; thus, relative sea level will rise rapidly and accommodation will increase rapidly. The coastline will transgress over the shelf (fig. 10.60b) producing *marine flooding*. Because the shelf can now accommodate more sediments, few sediments will be transported far from the coast, resulting in relatively thin deep-water deposits called a *condensed section*. The condensed section is often rich in both numbers of fossil specimens and species, and usually provides the best paleontological age dating. The package of sediments deposited during the rapid sea-level rise is called a *transgressive system tract*.

As the rate of sea-level rise slows down, becomes static, and begins to fall, sediments first aggrade and then prograde (fig. 10.60c). The package of sediments is called a *highstand system tract*. A eustatic fall at the end of the highstand system tract marks the top of the sequence that began with the preceding sea-level fall.

Of course, eustatic variations will not usually be a simple cycle, as assumed in the foregoing. Small, more rapid oscillations superimposed on larger oscillations result in *parasequences*. The sequences will also not always occur in the foregoing order and the local setting and tectonic situations will affect the patterns that develop. Nevertheless, the system-tract concepts are central to sequence stratigraphy.

10.7.6 Seismic-facies analysis

Seismic facies (§10.7.1) concerns the distinctive characteristics that make one group of reflections look different from adjacent reflections; inferences as to the depositional environment are drawn from seismic facies. Analysis and classification schemes are given by

Roksandic (1978), Sangree and Widmier (1979), and Brown and Fisher (1980); table 10.4 concerns facies classifications. Classifications are sometimes based on reflection terminations (fig. 10.58), reflection characteristics (abundance, continuity, amplitude, amplitude consistency, and so on), stratal patterns (fig. 10.61a), and the external shapes of sequences (fig. 10.61b).

Parallel reflections suggest uniform deposition on a stable or uniformly subsiding surface, whereas divergent reflections indicate variation in the rate of deposition from one area to another or else gradual tilting. Chaotic reflections suggest either relatively high depositional energy, variability of conditions during deposition, or disruption after deposition, such as can be produced by slumping or sliding or turbidity-current flow. A reflection-free interval suggests uniform lithology such as a relatively homogeneous marine shale, salt, or massive carbonates; however, distinguishing reflection-free patterns from multiples and noise that obscures reflections may be difficult.

Reflection terminations such as onlap (fig. 10.58) and downlap (sometimes called offlap), already described in §10.7.4, give a genetic context: onlap is the the landward edge of a unit, whereas downlap results from inadequate sediment supply (starvation) and thus is the seaward edge of a unit.

Oblique progradational patterns (fig. 10.62a) are characterized by toplap angularity (also sometimes called offlap) and reflection-character variability. The tops of oblique patterns indicate periods during which sea level was not changing markedly (stillstands) and deposition near the wave base, with consequent high depositional energy. Thus, the tops of oblique patterns often contain relatively clean sands. Sigmoid progradational patterns, on the other hand, are characterized by gentle S-shaped reflections of rather uniform character, the tops of the reflections exhibiting concordance with the top of the sequence unit. These indicate relative sea-level rise and usually consist of fine-grained sediments, sometimes calcareous.

The three-dimensional shape of units provides the principal basis for classification in basin settings (fig. 10.62b). Units that drape over preexisting topography are generally low-energy fine-grained pelagic units. Those with mounded tops or chaotic reflections are generally variable-to-high-energy deposits.

High-reflection continuity suggests continuous strata deposited in an environment that was relatively quiet and uniform over a widespread area, such as marine shales interbedded with silts and calcareous shales. Fluvial sediments with interbedded clays and coals sometimes produce strong reflections.

The lateral equivalents of units sometimes provide the key to identification. Thus, a low-reflection-amplitude facies representing prodelta shales may grade landward into a facies of high continuity and amplitude resulting from interbedded silts and/or sands, whereas a low-reflection-amplitude sand facies may grade landward into a nonmarine, low-continuity, variable-amplitude facies. The prodelta

Table 10.4 *Seismic-facies classification*

Regional setting	Basis of distinction	Subdivisions	Interpretation	Other characteristics
Shelf	Reflection character Unit shape: widespread sheet or gentle wedge Reflections generally parallel or divergent	High continuity, high amplitude	Generally marine – alternating neritic shale/limestone, interbedded high/low energy deposits, or shallow marine clastics transported mainly by wave action	Possibly cut by submarine canyons Distinguish on basis of location compared to other facies
		Variable continuity, low amplitude, occasional high amplitude	Fluvial or nearshore clastics, fluvial/wave-transport processes (delta platform), or low-energy turbidity current or wave transport	Distinguish on basis of location compared to other facies Shale-prone if seaward of unit above Sand-prone if seaward of unit below
		Low continuity, variable amplitude	Nonmarine clastics, fluvial or marginal-marine	Occasional high amplitude and high continuity from coal members
	Mounded shape	Variable continuity and amplitude	Delta complex	Internal reflections gently sigmoid to divergent Occasional high amplitudes
		Local reflection void	Reef	See fig. 10.35
Self margin – prograded slope	Internal reflection pattern	Oblique, fan-shaped or overlapping fans	Adequate sediment supply Shelf margin – deltaic High energy deposits in updip portions Occasionally due to strong currents in deep water	Moderate continuity and amplitude, reflections variable Foreset (clinoform) dips to 10° (averaging 4–5°), steeper dips are calcareous Often fan-shaped (including multiple fans)
		Sigmoid, elongate lens/fan	Low sediment supply Low depositional energy	High continuity, high to moderate amplitude, uniform reflections

Table 10.4 *Seismic-facies classification*

Regional setting	Basis of distinction	Subdivisions		Interpretation	Other characteristics
Basin slope, basin floor	Overall unit shape	Drape	Sheet drape	Deep marine hemipelagic; mainly clay Low energy	High continuity, low amplitude Drapes over preexisting topography
		Mounded	Contourite	Deep Low energy	Variable continuity and amplitude
			Fan-shaped	Variable energy, slump/turbidity currents	Discontinuous, variable amplitude At mouth of submarine canyons Composition depends on what was eroded up above
		Fill	Slope front fill	Low energy Deep marine clay and silt	Variable continuity and amplitude Fan-shaped to extensive along slope
			Onlapping fill	Low-velocity turbidity currents	High continuity, variable amplitude
			Mounded onlap fill or chaotic fill	High or variable-energy turbidites	Overall mound in a topographic low, gouge common at base Discontinuous, variable amplitude
			Canyon fill	Variable superimposed strata Coarse turbidites to hemipelagic	Variable continuity and amplitude

Source: After Sangree and Widmier, 1979.

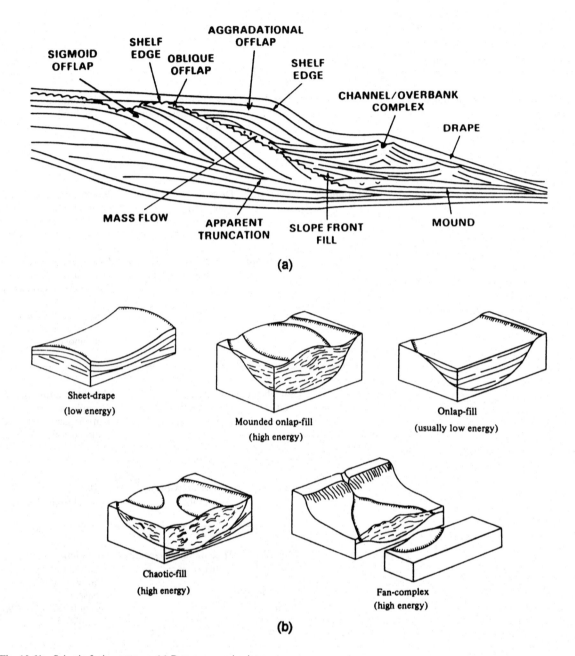

Fig. 10.61   Seismic-facies patterns. (a) Patterns on seismic sections, and (b) three-dimensional shapes of basinal sequences (from Sangree and Widmier, 1979).

shale may grade basinward into a prograded-slope facies, whereas the sand may grade basinward into high-continuity, high-amplitude marine facies.

Examples of seismic-facies classifications are shown in figs. 10.62 and 10.63.

10.7.7 Reflection-character analysis

Reflection-character analysis involves study of the trace-to-trace changes in the waveshape of one or more reflections with the objective of locating and determining the nature of changes in the stratigraphy or fluid in the pore spaces. Special displays may be used

to make it easier to see the changes, such as enlarged displays of the portion of the section being studied, displays of attribute measurements (Taner and Sheriff, 1977; Taner, Koehler, and Sheriff, 1979) such as envelope amplitude, instantaneous frequency, and so on (§9.11.4), or seismic-log displays (§5.4.5), which often involve color.

Synthetic seismograms (§6.2.1) are often used to determine the nature of the stratigraphic change that a change of waveshape indicates. The various stratigraphic changes that are regarded as reasonable possibilities are modeled (Harms and Tackenberg, 1972; Neidell and Poggiagliolmi, 1977) and matched with

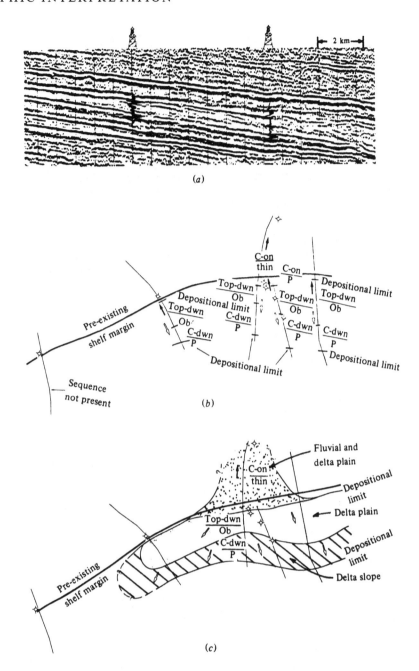

Fig. 10.62 Stratigraphic interpretation in east Texas. (After Ramsayer, 1979.) (a) Portion of a seismic section; unconformities bounding the unit are mapped over a grid of lines; portions of synthetic seismograms from two wells are superimposed; the top and base of the sequence mapped in (b) are indicated at the left. (b) Characteristics of reflections within the unit are mapped; Top = toplap at top of unit, C = concordance at top of unit, on = onlap at base of unit, dwn = downlap at base of unit, Thin = unit not thick enough to see internal pattern, Ob = oblique pattern in the body of the unit, P = parallel reflections in interior of unit; solid arrows indicate direction of onlap, open arrows the direction of downlap. (c) Facies interpretation.

the observed waveforms. Clement (1977) describes the use of reflection-character analysis in mapping a sand associated with channels on an unconformity surface in Oklahoma. A distinctive reflection was present (fig. 10.64) where the sands were more than 6 meters in thickness, in which situation they usually were also porous. Several successful wells were drilled on the basis of the predictions from reflection character, but

one well encountered tight indurated interbedded sandstone and shale that gave very similar reflection character. This study thus illustrates both successful application of these techniques and also the ambiguity of conclusions based on reflection character.

Maureau and van Wijhe (1979) successfully used seismic logs to predict high-porosity zones in Permian carbonates in the Netherlands. Lindseth (1979) re-

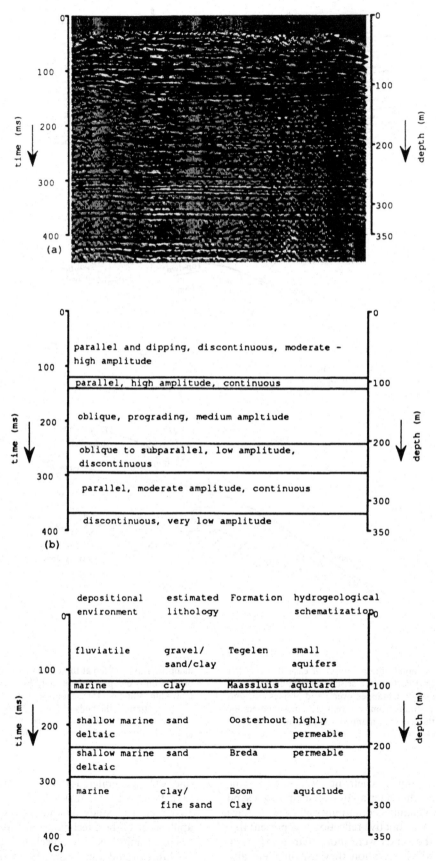

Fig. 10.63 Section showing seismic-facies characterizations. (From Meekes and van Will, 1991: private communication.) (a) Seismic section; (b) seismic facies; and (c) depositional environment, lithology, and hydrogeologic character.

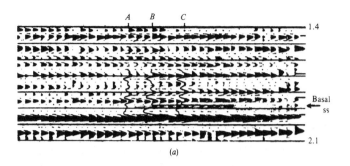

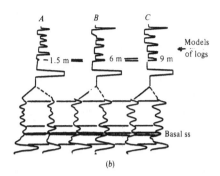

Fig. 10.64 Evidences of a channel sand. (From Clement, 1977.) (a) Portion of seismic section across a channel showing development of an event where channel sand is more than 6 m thick; the well at *A* just missed the channel. (b) Models of logs for various sand thicknesses and seismic traces at locations *A*, *B*, and *C*; the first of each pair is a synthetic trace and the second an observed seismic trace.

ports mapping porosity in Devonian carbonates in Alberta and other reflection-character analysis studies using seismic logs. Seismic log examples are shown in plates 1 and 4.

10.8 Hydrocarbon indicators

The velocity and density of sedimentary rocks depend on porosity and on the properties of the fluids filling the pore space, as was discussed in §5.2.4 and 5.2.7. The dependence of porosity on density is the straightforward relation given by eq. (5.6), but the dependence of porosity on velocity where there is a mixture of fluids in the pore space is not as simple. The change in velocity resulting from a change in the interstitial fluid is often marked, as shown in fig. 5.27, resulting in amplitude anomalies associated with accumulations. The extensive use of automatic gain control obscured these amplitude effects until about 1970 when recognition of their usefulness in locating accumulations of hydrocarbons became widespread. Because the anomaly is most often one of locally increased amplitude (as in fig. 10.65), they were called "bright spots." Soon other types of anomalies were found to be associated with hydrocarbon accumulations under some conditions (see table 10.5 and Blackburn, 1986a). The relation between hydrocarbon indicators and hydrocarbon accumulations is not simple and universal, however, and many bright spots turned out to result from changes other than commercial hydrocarbon accumulations.

The effect of interstitial fluids on velocity depends on the structure of the rock and is generally greater and simpler for relatively unconsolidated clastic rocks. Thus, the effects are generally greater for young rocks than for older, and hydrocarbon indicator technology is especially applicable to Tertiary clastic basins; these are mostly offshore around the periphery of the continents, but being offshore is irrelevant except that marine data are often of better quality than land data and hence anomalies easier to see. The effect on velocity is generally greater (and more complex) for gaseous

Table 10.5 *Hydrocarbon indicators*

Structural crest or against a fault?	Trapping location
Local increase in amplitude?	Bright spot
Local decrease in amplitude?	Dim spot
Discordant flat reflector?	Flat spot
Local waveshape change?	Polarity reversal or local phasing
Reservoir limits consistent?	(If reservoir base and top separately visible)
Polarities consistent?	(If zero-phase data)
Low frequencies underneath?	Low-frequency shadow
Time sag underneath?	Velocity sag
Lower amplitudes underneath (and sometimes above)	Amplitude shadow
Increase in amplitude with offset?	AVO anomaly
P-wave but no *S*-wave anomaly?	*S*-wave support
Data deterioration above (and perhaps minor bright spots)	Gas chimney

Source: After Brown, 1991.

than for liquid hydrocarbons, as indicated in fig. 5.27.

Replacement of brine by hydrocarbons as the interstitial fluid almost always results in a lowering of velocity, but the effect on reflections depends also on the properties of the rock overlying (and underlying) the reservoir rock. If the overlying rock has higher velocity than brine-filled reservoir rock, lowering the reservoir rock velocity by filling it with hydrocarbon increases the contrast and hence increases the amplitude of the reflection from the top of the reservoir, giving a bright spot; this is common in many Tertiary clastic basins. If, on the other hand, the overlying rock has a velocity appreciably lower than the reservoir, the effect of hydrocarbons is to decrease the contrast, producing a "dim spot"; this indicator is shown in fig. 10.66, where carbonate reservoir rocks are capped by

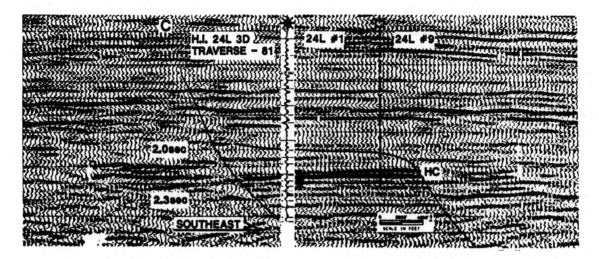

Fig. 10.65   Seismic section showing amplitude and flat-spot hydrocarbon indicators, Gulf of Mexico. (From Brown, 1991: 285.)

shales. Where the overlying rock has a velocity slightly smaller than that of the reservoir rock, lowering the reservoir rock velocity by hydrocarbons may invert the sign of the reflection, producing a "polarity reversal" over the reservoir; this can be seen in plates 2 and 3.

Where a well-defined fluid contact is present, especially a gas–oil or gas–water contact, the contrast may be great enough to give a fairly strong reflection that may stand out on seismic records because of its flat attitude in contrast to the dipping attitudes of other reflections. This is a "flat spot" and, where seen, it is usually the most definitive and informative of the hydrocarbon indicators. Flat spots can be seen in figs. 6.10, 10.65, 10.67, 10.68, and plates 2, 3, and 5. However, reservoir thicknesses are usually small compared to the resolvable limit, so that the reflections from the reservoir cap, fluid contact, and the base of the reservoir generally interfere with one another to provide a composite reflection that may show various phase and amplitude changes as the component reflections interfere in different ways. Thus, "phasing" can also be regarded as a hydrocarbon indicator.

The lowering of velocity in a hydrocarbon accumulation will also affect reflections from deeper reflectors by increasing arrival times to cause a "sag" in reflections seen through the reservoir, but the magnitude of the sag is usually small because most reservoirs are not very thick. Sags are especially prominent in plate 2. The lowering of velocity can also bend raypaths passing through the reservoir (as indicated in fig. 6.19a), resulting in distortion of deeper reflection events; sometimes the effect is simply a degradation of reflection quality under the reservoir, a "shadow zone." The high amplitude associated with a bright spot often results in a lowering of the amplitudes of entire traces by processing procedures that make the mean energy of all traces the same (amplitude nor-

malizing to remove near-surface or recording-system amplitude effects). A consequence is that a lower-amplitude shadow zone may exist above a bright spot as well as below it.

The high amplitude associated with a bright spot also affects multiples involving the reservoir reflectors. Occasionally, sections are made to emphasize multiples by using lower stacking velocities that attenuate primary reflections; an increase in amplitude of multiples seen on such a section also can be used as a hydrocarbon indicator.

A lowering of instantaneous frequency (§9.11.4) is often observed immediately under hydrocarbon accumulations. Such "low-frequency shadows" seem to be confined to a couple of cycles below (not *at*) accumulations. No adequate explanation is available (see Taner, Koehler, and Sheriff, 1979); proposed explanations generally involve either the removal of higher frequencies because of absorption or other mechanisms, or improper stacking because of erroneous velocity assumptions or raypath distortion.

Special displays are often used to enhance hydrocarbon indicators and help in their detection and analysis. The most common is a low-amplitude display in which special efforts are made to preserve amplitude relations; reflection events are subdued on such displays so that the amplitude buildups associated with bright spots stand out more clearly (fig. 10.11 is a low-amplitude display of part of line *B* of fig. 10.29). Sections are also often displayed in variable-area mode with both normal and inverted polarity, because hydrocarbon-indicator effects are sometimes more evident on one than the other. Measurements of amplitude, envelope amplitude, amplitude ratios, phase, frequency, velocity, and so on, may be displayed so as to make lateral changes along reflectors more evident. Complex-trace analysis (§9.11.4) may be used to make such measurements, the

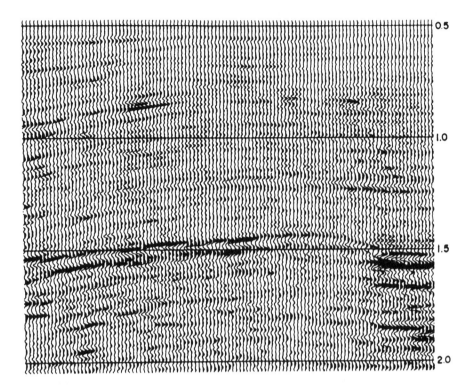

Fig. 10.66   Dim spot associated with gas accumulation in porous carbonates overlain by shales. (Courtesy of Teledyne.)

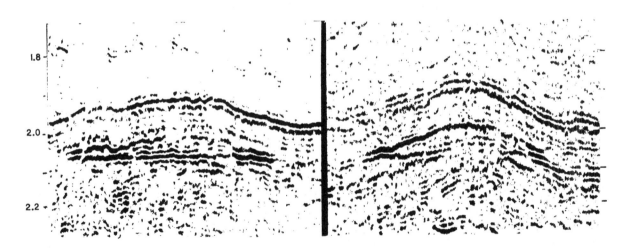

Fig. 10.67   Dual-polarity lines across a gas condensate reservoir in the Norwegian North Sea. The flat spot disappears over the crest of the structure on the right section, indicating that the reservoir is completely filled with hydrocarbons over this portion. (Courtesy of Elf Aquitaine Norge a/s.)

results often being displayed as color overlays on seismic sections to facilitate correlation with structural evidences.

Hydrocarbon indicators are also used quantitatively to predict reservoir thicknesses and hydrocarbon volumes. Amplitude and dominant-period measurements based on model studies of a wedge pinchout (such as that in figs. 6.40b and 6.40c) are shown in fig. 6.42. Dominant-period or peak-to-trough time measurements can be used to indicate thicknesses when greater than a quarter wavelength and amplitude measurements when smaller than a quarter wavelength. Note that in the thin-bed case, the amplitude is a measure of the net thickness rather than the gross thickness, as was illustrated in fig. 6.43. Hun (1978) mapped hydrocarbon productivity based on amplitude measurements, and many examples of their quantitative use can be found in Brown (1991) and Sheriff (1992). The use of amplitude measurements, however, requires an amplitude calibration

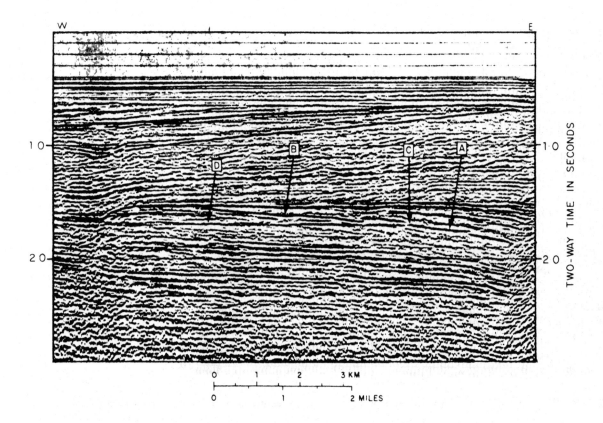

Fig. 10.68 Minimum-phase seismic section across the Troll Field, offshore Norway. The reservoir is filled with water at *A*, but contains gas at *B*. *C* indicates the gas–water contact and *D* is a "flat spot" dipping slightly because of overlying velocity changes. (Courtesy of Norsk Hydro.)

(that is, what would the amplitude be if the reservoir were very thick?), which often is not possible, or well control. Quantitative analysis also requires extremely careful processing, good data, and assumptions about the nature of the rock sequence, which may not always be true. Modeling is almost always involved in quantitative analysis.

Another hydrocarbon indicator is the variation of amplitude with angle of incidence (or with offset, AVO); this was discussed in §3.4 (see also Hilterman, 1990). The usual expectation is that amplitude will decrease with offset unless gas is present, in which case amplitude will increase. Whereas this is often the case, the situation is not this simple, as sometimes gas will result in an amplitude decrease and sometimes other situations will result in an amplitude increase with offset. Modeling using realistic values of *P*- and *S*-wave velocities and densities is usually required to determine the significance of AVO measurements. AVO has to be measured on gathers before stacking, where noise is usually large, so that measurements are not very accurate. Nonuniform acquisition (inadequate minimum or maximum offset distances, line bends, or skips), array directivity effects, processing that does not preserve amplitude, residual statics, failure to migrate data, out-of-the-plane data, and other factors can introduce sizeable errors.

Comparison of *S*- and *P*-wave sections (§13.1.4 and 13.1.5) is another indicator. Examples of other hydrocarbon indicators are shown in §12.5.

Plate 4 shows seismic data that have been inverted to seismic-log form and color coded to illustrate relative acoustic impedance (or relative velocity), low acoustic impedance values that are colored blue being associated with hydrocarbons. The resulting horizon slice (§12.3) through 3-D data nicely maps the hydrocarbon accumulation.

Plate 5 shows a slice through a 3-D data volume parallel to a fault (§12.5). Hydrocarbon accumulations against the fault can be seen both by their amplitude and flat-spot effects.

10.9 Crustal studies

The field and processing techniques used in petroleum seismic exploration – common-midpoint recording, vertical stacking, statics corrections, deconvolution, velocity analysis, the use of long streamers, precise navigation at sea, large Vibroseis sources and long sweeps on land, and so on – are beginning to be applied to studies of the Earth's crust.

Marine seismic lines reveal details of the crustal structure, such as subduction zones (fig. 10.69) and the base of the oceanic crustal plate. In the United States, the Consortium for Continental Reflection Profiling (COCORP) has been running a series of

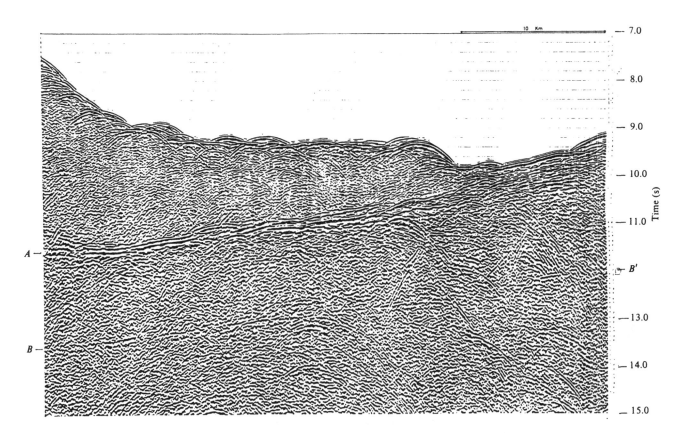

Fig. 10.69 CMP section across Japan Trench. The strong reflection at A is interpreted as the subduction contact along which the Pacific plate is sliding under the Japanese plate. The discontinuous reflection BB' is interpreted as being from the Mohorovicic discontinuity. The interval velocity between A and B calculated from stacking velocities is 6.0–6.4 km/s, so the 2-s time interval between them represents a thickness of about 6 km for the oceanic crust. (From Matsuzawa et al., 1979.)

lines to study the structure of the earth's crust, and similar studies are being carried out in other countries. The preliminary results have been exciting. Many complexities have been found in the crust, including regions of layered reflections that suggest sedimentary rather than igneous origin, although undoubtedly metamorphosed. A line across the southern Appalachians (Cook et al., 1979; Cook, Brown, and Oliver, 1980) indicates that the crystalline and other rocks seen on the surface have been thrust a long distance from the east and that sedimentary rocks may underlie portions of the low-angle thrusts. The ideas of "thin-skin tectonics" are beginning to change concepts of how continents were formed.

Problems

10.1 The line in fig. 10.11 shows part of line B of fig. 10.29 (the well is located at W in fig. 10.29), but with relative amplitude information preserved and plotted so that the largest amplitudes are not clipped. What conclusions can be drawn from fig. 10.11 that are less evident from fig. 10.29?

10.2 (a) Well B is 500 m due east of well A and well C is 600 m due north of A. A fault cuts A, B, and C at depths of 800, 1000 and 600 m, respectively. Assuming all wells are vertical and the fault surface is a plane, find the surface trace and strike of the fault [see the derivation of eq. (4.17)].
(b) At what depth would you look for this fault in well D located 500 m N30°W from well C?
(c) Another fault cuts wells A and C at depths of 1300 and 1000 m, respectively, and is known to strike N20°W. Where does it cut well B?

10.3 Try to match the section shown in fig. 9.24b with the structural styles shown in table 10.1. With which might it be compatible? (Note that fig. 9.24b is unmigrated but assume that it is nearly perpendicular to the strike.) Do the velocity data from problem 5.18 help?

10.4 How do you reconcile the contradictory dips between the 5- and 6-km marks at the top of the section in fig. 8.5? What structural style is represented? How would you draw faults?

10.5 Four lines forming a grid are shown in fig. 10.29.
(a) Map the three horizons encountered at 1.355, 1.830, and 2.660 s at the intersections of lines B and C. A velocity analysis at this location gives time–velocity (stacking velocity) pairs as follows: 0.100 s, 1520 m/s, 0.600, 1830; 0.800, 1900; 1.200, 2050; 1.400, 2100; 1.600, 2140; 2.000, 2280; 2.700, 2440; and 3.000, 2470.
(b) Map the curved fault plane.
(c) Estimate and map the throw on the fault.

10.6 (a) In fig. 9.44, the reflection at about 0.6 s appears to be faulted at S.P. 5; draw in the fault and describe its probable type and characteristics.

(b) Note the changes with location in the interval times between different reflections in fig. 9.44. How can these be explained?

10.7 Figure 10.33 shows a salt uplift at a shelf edge.

(a) How could one tell that this feature is not caused by reef growth instead?

(b) Could it have been caused by shale flowage?

(c) Does the relief above the unconformity U_1 indicate postunconformity salt movement, renewed activity at the shelf edge (downdrop along faulting at the shelf edge), or differential compaction because of the weight of the postunconformity section?

10.8 If the nature of a flow structure (such as shown in figs. 10.32 or 10.33) should not be clear, how might gravity, magnetic, or refraction measurements be used to distinguish between salt, shale, or igneous flows? Between these and a reef?

10.9 Figure 9.63b maps two separate high closures on a northeast-plunging anticlinal nose.

(a) Assuming that the only existing control is that shown by the lines marked by diagonal slashes in fig. 9.63a (which is on a different horizon), what additional program would you recommend to check out weaknesses in the interpretation before recommending a well to test for hydrocarbon accumulation?

(b) Can you find a fault for which the indicated direction of throw is clearly wrong?

10.10 How would evidences of thickening/thinning around a salt dome or in a folded structure be distorted on an unmigrated time section? On a migrated time section?

10.11 What kind of feature shows in fig. 9.19 about 75% of the way across from the left end of the section at about 2.5 s? What characteristics help to identify it?

10.12 Interpret the sections shown in figs. (a) 9.59b, (b) 9.61d, (c) 10.31, and (d) 10.38. Assume that out-of-the-plane data are not important. (Pick events that involve angularities between primary reflections in order to identify unconformities and or seismic sequence boundaries.) Deduce the geologic history.

10.13 An obvious unconformity is evident at approximately 1.5 s in fig. 9.18; precisely where would you position it? Is it associated with the same event at opposite sides of this section?

10.14 In fig. 10.48, $V_1 = 2.00$ km/s, $V_2 = 4.00$ km/s, the horizon dips $10°$ and the vertical depth of the diffracting point P is 1000 m, the interface between V_1 and V_2 being 350 m vertically above P. Compare the diffraction curve with that which would have been observed if $V_1 = V_2 = 3.00$ km/s. (*Hint:* Ray trace enough rays to roughly define the diffraction curves.)

10.15 Attempt a stratigraphic interpretation of fig. 10.53. CC' divides nonmarine from marine sediments. Does the surface channel create fictitious deep effects?

10.16 What stratigraphic features can be seen in fig. 9.57? What can be said about the geologic history?

10.17 For the bright spot shown in fig. 10.67, what do you think is the polarity of the embedded wavelet?

Where is the top and bottom of the gas accumulation? What is the maximum thickness of the gas column, assuming a velocity of 1800 m/s? Why do the reflections from the reservoir top and bottom not converge at the pinchout edge of the gas reservoir?

10.18 Using the minimum-phase wavelet of problem 9.14a, determine the waveshapes for a sand enclosed in shale where the two-way traveltime through the sand is 12 ms when the upper part contains gas, the two-way traveltime through the gas–sand portion being successively 0, 2, 4, 6, 8, 10, and 12 ms. Plot the traces side by side shifted successively by 2 ms as would be the case where the gas–water contact was horizontal. This illustrates a bright-spot, flat-spot situation. Take the reflection coefficients for shale to gas–sand as -0.1, gas–sand to water–sand as $+0.15$, and water–sand to shale as -0.05.

(b) Repeat using the zero-phase wavelet of problem 9.14.

References

Anstey, N. A. 1973. How do we know we are right? *Geophys. Prosp.*, **21**: 407–11.

Anstey, N. A. 1974. *The New Seismic Interpreter.* Boston: IHRDC Press.

Anstey, N. A. 1977. *Seismic Interpretation: The Physical Aspects.* Boston: IHRDC Press.

Anstey, N. A. 1980a. *Simple Seismics.* Boston: IHRDC Press.

Anstey, N. A. 1980b. *Seismic Exploration for Sandstone Reservoirs.* Boston: IHRDC Press.

Badley, M. E. 1985. *Practical Seismic Interpretation.* Boston: IHRDC Press.

Badley, M. E., and N. A. Anstey. 1984. *Basic Interpretation,* Video Library for Exploration and Production Specialists GP-501. Boston: IHRDC Press.

Balch, A. H. 1971. Color sonagrams: A new dimension in seismic data interpretation. *Geophysics,* **36**: 1074–98.

Bally, A. W., ed. 1983–4. *Seismic Expression of Structural Styles,* Vols. 1, 2, and 3, AAPG Studies in Geology 15. Tulsa: American Association of Petroleum Geologists.

Bally, A. W., ed. 1987–9. *Atlas of Seismic Stratigraphy,* Vols. 1, 2, and 3, AAPG Studies in Geology 27. Tulsa: American Association of Petroleum Geologists.

Bates, R. L., and J. A. Jackson. 1987. *A Glossary of Geology,* 3d ed. Falls Church, Va: American Geological Institute.

Batzie, M., and Z. Wang. 1992. Seismic properties of pore fluids. *Geophysics,* **57**: 1396–1408.

Berg, O. R., and D. G. Wolverton, eds. 1985. *Seismic Stratigraphy II,* AAPG Memoir 39. Tulsa: American Association of Petroleum Geologists.

Blackburn, G. J. 1986a. Direct hydrocarbon detection: Some examples. *Exp. Geophys.,* **17**: 59–66.

Blackburn, G. J. 1986b. Depth conversion: A comparison of methods. *Exp. Geophys.,* **17**: 67–73.

Bouvier, J. D., C. H. Kaars-Sijpesteijn, D. F. Kluesner, C. C. Onyejekwe, and R. C. van der Pal. 1989. Three-dimensional seismic interpretation and fault-sealing investigations, Nun River Field, Nigeria. *Bull. AAPG,* **73**: 1397–414.

Boyer, S. E., and D. Elliott. 1982. Thrust systems. *Bull. AAPG,* **66**: 1196–1230.

Brown, A. R. 1991. *Interpretation of Three-Dimensional Seismic Data*, 3d ed., AAPG Memoir 42. Tulsa: American Association of Petroleum Geologists.

Brown, A. R. 1992. Seismic interpretation today and tomorrow. *The Leading Edge,* **11(11):** 10–15.

Brown, A. R., C. G. Dahm, and R. J. Graebner. 1981. Stratigraphic case history using three-dimensional seismic data in the Gulf of Thailand: A case history. *Geophys. Prosp.,* **29:** 327–49.

Brown, L. F., and W. L. Fisher. 1980. *Seismic Stratigraphic Interpretation and Petroleum Exploration*, AAPG Continuing Education Course Notes 16. Tulsa: American Association of Petroleum Geologists.

Bubb, J. N., and W. G. Hatlelid. 1977. Seismic recognition of carbonate buildups. In *Seismic Stratigraphy – Applications to Hydrocarbon Exploration*, C. E. Payton, ed., pp. 185–204, AAPG Memoir 26. Tulsa: American Association of Petroleum Geologists.

Buxtorf, A. 1916. Prognosen und Befunden beim Hauensteinbasis und Grenchenberg Tunnel und die Bedeutung der letzern für die Geologie des Juragebirges. *Verh. Naturf Gesell. Basel,* **27:** 185–254.

Campbell, F. F. 1965. Fault criteria. *Geophysics,* **30:** 976–97.

Clement, W. A. 1977. Case history of geoseismic modeling of basal Morrow-Springer sandstones, Watonga-Chickasha trend, Geary, Oklahoma. In *Seismic Stratigraphy – Applications to Hydrocarbon Exploration*, C. E. Payton, ed., pp. 451–76, AAPG Memoir 26. Tulsa: American Association of Petroleum Geologists.

Cook, F. A., D. S. Albaugh, L. D. Brown, S. Kaufman, J. E. Oliver, and R. D. Hatcher. 1979. Thin-skinned tectonics in the crystalline southern Appalachians. *Geology,* **7:** 563–7.

Cook, F. A., L. D. Brown, and J. E. Oliver. 1980. The southern Appalachians and the growth of continents. *Sci. Amer.,* **243 (4):** 156–68.

Dahlstrom, C. D. A. 1970. Structural geology in the eastern margin of the Canadian Rocky Mountains. *Petro. Geol. Bull.,* **18:** 332–406.

Dahm, C. G., and R. J. Graebner. 1982. Field development with three-dimensional seismic methods in the Gulf of Thailand: A case history. *Geophysics,* **47:** 149–76.

Davis, T. L. 1972. Velocity variations around Leduc reefs. *Geophysics,* **37:** 584–604.

Dimitropoulos, K., and J. Donato. 1983. The gravity anomaly of the St. George's Channel Basin, southern Irish Sea – A possible explanation in terms of salt migration. *J. Geol. Soc. London,* **140:** 239–44.

Dobrin, M. B. 1977. Seismic exploration for stratigraphic traps. In *Seismic Stratigraphy – Applications to Hydrocarbon Exploration*, C. E. Payton, ed., pp. 329–52, AAPG Memoir 26. Tulsa: American Association of Petroleum Geologists.

Downey, M. W. 1990. Faulting and hydrocarbon entrapment. *The Leading Edge,* **9(1):** 20–3.

Edwards, S. 1988. Uses and abuses of seismic modeling. *The Leading Edge,* **8(4):** 42–6.

Fagin, S. W. 1991. *Seismic Modeling of Geologic Structures.* Tulsa: Society of Exploration Geophysicists.

Feagin, F. J. 1981. Seismic data display and reflection perceptability. *Geophysics,* **46:** 106–20.

Fitch, A. A. 1976. *Seismic Reflection Interpretation.* Berlin: Gebrüder Bornträger.

Fontaine, J. M., R. Cussey, J. Lacaze, R. Lanaud, and L. Yapaudjian. 1987. Seismic interpretation of carbonate depositional environments. *Bull. AAPG,* **71:** 281–97.

Gallup, W. B. 1951. Geology of Turner Valley oil and gas field, Alberta. *Bull. AAPG,* **35:** 797–821.

Gazdag, J. 1981. Modeling of the acoustic wave equation with transform methods. *Geophysics,* **46:** 854–9.

Goguel, J. 1962. *Tectonics.* San Francisco: W. H. Freeman.

Greenlee, S. M., and T. C. Moore. 1988. Recognition and interpretation of depositional sequences and calculation of sea-level changes from stratigraphic data – Offshore New Jersey and Alabama Tertiary. In *Sea-Level Changes: An Integrated Approach,* C. K. Wilgus et al., eds., pp. 329–56, Society of Economic Paleontologists and Mineralogists Special Publication 42.

Gretener, P. E. 1979. *Pore Pressure: Fundamentals, General Ramifications, and Implications for Structural Geology* (revised), Education Course Note Series 4. Tulsa: American Association of Petroleum Geologists.

Gries, R. R., and R. C. Dyer, eds. 1985. *Seismic Exploration of the Rocky Mountain Region.* Denver: Rocky Mountain Association of Geologists and Denver Geophysical Society.

Halbouty, M. T., ed. 1982. *The Deliberate Search for the Subtle Trap.* AAPG Memoir 32. Tulsa: American Association of Petroleum Geologists.

Haq, B. U., J. Hardenbol, and P. R. Vail. 1988. Mesozoic and Cenozoic chronostratigraphy and cycles of sea-level change. In *Sea-Level Changes: An Integrated Approach,* C. K. Wilgus et al., eds., pp. 71–108, Society of Economic Paleontologists and Mineralogists Special Publication 42.

Hardage, B. A., ed. 1985. *Vertical Seismic Profiling, Part A: Principles.* London: Geophysical Press.
Hardage, B. A., ed. 1987. *Seismic Stratigraphy.* London: Geophysical Press.

Harding, T. P., and J. D. Lowell. 1979. Structural styles, their plate-tectonic habitats, and hydrocarbon traps in petroleum provinces. *Bull. AAPG,* **63:** 1016–58.

Harms, J. C., and P. Tackenberg. 1972. Seismic signatures of sedimentation models. *Geophysics,* **37:** 45–58.

Harris, L. D., and R. C. Milici. 1977. Characteristics of thin-skinned styles of deformation in the Southern Appalachians and potential hydrocarbon traps. U.S. Geological Survey Prof. Paper 1018. Washington, D.C.: U.S. Geological Survey.

Hilterman, F. 1970. Three-dimensional seismic modeling. *Geophysics,* **35:** 1020–37.
Hilterman, F. 1990. Is AVO the seismic signature of lithology? A case history of Ship Shoal – South Addition. *The Leading Edge,* **10(6):** 15–22.

Hobbs, B. E., W. D. Weams, and P. F. Williams. 1976. *Outline of Structural Geology.* New York: John Wiley.

Hubbert, M. K. 1937. Scale models and geologic structure. *Geol. Soc. Amer. Bull.,* **48:** 1459–520.

Hun, F. 1978. Correlation between seismic reflection amplitude and well productivity – A case study. *Geophys. Prosp.,* **26:** 157–62.

Isaacs, B., J. Oliver, and L. R. Sykes. 1968. Seismology and the new global tectonics. *J. Geophys. Res.* **73:** 5855–99.

Jones, P. B. 1988. Balanced cross-sectons – An aid to structural interpretation. *The Leading Edge,* **7(8):** 29ff.

Kuhme, A. K. 1987. Seismic interpretation of reefs. *The Leading Edge,* **6(8):** 60ff.

Larner, K. L., L. Hatton, B. S. Gibson, and L. C. Hsu. 1981. Depth migration of imaged time sections. *Geophysics,* **46:** 734–50.

Lindseth, R. O. 1979. Synthetic sonic logs – A process for stratigraphic interpretation. *Geophysics,* **44:** 3–26.

Loutit, T. S., J. Hardenbol, P. R. Vail, and G. R. Baum. 1988.

Condensed sections: The key to age determination and correlation of continental margin sequences. In *Sea-Level Changes: An Integrated Approach*, C. K. Wilgus et al., eds., pp. 183–213, Society of Economic Paleontologists and Mineralogists Special Publication 42.

Lowell, J. D. 1972. Spitzbergen Tertiary orogenic belt and the Spitzbergen fracture zone. *Geol. Soc. Amer. Bull.*, **83**: 3091–102.

Lowell, J. D. 1985. *Structural Styles in Petroleum Exploration.* Tulsa: OGCI Publications.

Lyons, P. L., and M. B. Dobrin. 1972. Seismic exploration for stratigraphic traps. In *Stratigraphic Oil and Gas Fields – Classification, Exploration Methods, and Case Histories*, R. E. King, ed., pp. 225–43, AAPG Memoir 16. Tulsa: American Association of Petroleum Geologists.

Marr, J. D. 1971. Seismic stratigraphic exploration – Part I. *Geophysics*, **36**: 533–53; Part II, *Geophysics*, **36**: 533–53; Part III, *Geophysics*, **36**: 676–89.

Matsuzawa, Å., T. Tamano, Y. Aoki, and T. Ikawa. 1979. Structure of the Japan Trench subduction zone from multi-channel seismic reflection records. In *Marine Geology*, pp. 171–82. Amsterdam: Elsevier.

Maureau, G. T., and D. H. van Wijhe. 1979. Prediction of porosity in the Permian carbonate of eastern Netherlands using seismic data. *Geophysics*, **44**: 1502–17.

May, B. T., and J. D. Covey. 1981. An inverse ray method for computing geologic structures from seismic reflections: Zero offset case. *Geophysics*, **46**: 268–87.

McQuillin, R., M. Bacon, and W. Barclay. 1984. *An Introduction to Seismic Interpretation*, 2d ed. London: Graham & Trotman.

Meekes, A., and R. van Will. 1991. Private communication.

Mitchum, R. M., P. R. Vail, and S. Thompson. 1977. The depositional sequence as a basic unit for stratigraphic analysis. In *Seismic Stratigraphy – Applications in Hydrocarbon Analysis*, C. E. Payton, ed., pp. 53–62, AAPG Memoir 26. Tulsa: American Association of Petroleum Geologists.

Morgan, L., and W. Dowdall. 1983. The Atlantic continental margin. In *Seismic Expression of Structural Styles*, A. W. Bally, ed., AAPG Studies in Geology 15. Tulsa: American Association of Petroleum Geologists.

Neidell, N. S., and E. Poggiagliolmi. 1977. Stratigraphic modeling and interpretation. In *Seismic Stratigraphy – Applications in Hydrocarbon Analysis*, C. E. Payton, ed., pp. 389–416, AAPG Memoir 26. Tulsa: American Association of Petroleum Geologists.

Noah, J. T., G. S. Hofland, and K. Lemke. 1992. Seismic interpretation of meander channel point-bar deposits using realistic seismic modeling techniques. *The Leading Edge*, **11(8)**: 13–18.

Payton, C. E., ed. 1977. *Seismic Stratigraphy – Applications to Hydrocarbon Exploration*, AAPG Memoir 26. Tulsa: American Association of Petroleum Geologists.

Plawman, T. L. 1983. Fault with reversal of displacement, central Montana. In *Seismic Expression of Structural Styles*, A. W. Bally, ed., AAPG Studies in Geology 15. Tulsa: American Association of Petroleum Geologists.

Posamentier, H. W., M. T. Jervey, and P. R. Vail. 1988. Eustatic controls on clastic deposition. In *Sea-Level Changes: An Integrated Approach*, C. K. Wilgus et al., eds., pp. 109–54, Society of Economic Paleontologists and Mineralogists Special Publication 42.

Ramsayer, G. R. 1979. Seismic stratigraphy, a fundamental exploration tool, OTC Paper 3568. Dallas: Offshore Technology Conference.

Rittenhouse, G. 1972. Stratigraphic trap classification. In *Stratigraphic Oil and Gas Fields – Classification, Exploration Methods, and Case Histories*, R. E. King, ed., pp. 14–28, AAPG

Memoir 16. Tulsa: American Association of Petroleum Geologists.

Roksandic, M. M. 1978. Seismic facies analysis concepts. *Geophy. Prosp.*, **26**: 383–98.

Russell, B. H. 1992. Using color in seismic displays. *The Leading Edge*, **11(9)**: 13–18.

Sangree, J. B., and J. M. Widmier. 1979. Interpretation of depositional facies from seismic data. *Geophysics*, **44**: 131–60.

Sarg, J. F. 1988. Carbonate sequence stratigraphy. In *Sea-Level Changes: An Integrated Approach*, C. K. Wilgus et al., eds., pp. 155–81, Society of Economic Paleontologists and Mineralogists Special Publication 42.

SEG (Society of Exploration Geophysicists). 1989–92. *Seismic Interpretation Series*, Vol. 1, 1989; Vol. 2, 1990; Vol. 3, 1992. Tulsa: Society of Exploration Geophysicists.

Sheriff, R. E. 1977. Limitations on resolution of seismic data and geologic detail derivable from them. In *Seismic Stratigraphy – Applications to Hydrocarbon Exploration*, C. E. Payton, ed., pp. 3–14, AAPG Memoir 26. Tulsa: American Association of Petroleum Geologists.

Sheriff, R. E. 1978. *A First Course in Geophysical Exploration and Interpretation.* Boston: International Human Resources Development Corp.

Sheriff, R. E. 1980. *Seismic Stratigraphy.* Boston: International Human Resources Development Corp.

Sheriff, R. E. 1982. *Structural Interpretation of Seismic Data*, AAPG Continuing Education Course Note Series 23. Tulsa: American Association of Petroleum Geologists.

Sheriff, R. E. 1989. *Geophysical Methods.* Englewood Cliffs, N.J.: Prentice Hall.

Sheriff, R. E. 1991. *Encyclopedic Dictionary of Exploration Geophysics*, 3d ed. Tulsa: Society of Exploration Geophysicists.

Sheriff, R. E., ed. 1992. *Reservoir Geophysics.* Tulsa: Society of Exploration Geophysicists.

Sheriff, R. E, and J. Farrell. 1976. Display parameters of marine geophysical data, OTC Paper 2567. Tulsa: Society of Exploration Geophysicists.

Stone, D. S. 1985. Geologic interpretation of seismic profiles, Big Horn Basin, Wyoming. In *Seismic Exploration of the Rocky Mountain Region*, R. R. Gries and R. C. Dyer, eds., pp. 165–86. Denver: Rocky Mountain Association of Geologists and Denver Geophysical Society.

Taner, M. T., E. E. Cook, and N. S. Neidell. 1970. Limitations of the reflection seismic method. Lessons from computer simulations. *Geophysics*, **35**: 551–73.

Taner, M. T., F. Koehler, and R. E. Sheriff. 1979. Complex seismic trace analysis. *Geophysics*, **44**: 1041–66.

Taner, M. T., and R. E. Sheriff. 1977. Application of amplitude, frequency, and other attributes to stratigraphic and hydrocarbon determination. In *Seismic Stratigraphy – Applications to Hydrocarbon Exploration*, C. E. Payton, ed., pp. 301–28, AAPG Memoir 26. Tulsa: American Association of Petroleum Geologists.

Taylor, J. C. M. 1981. Late Permian – Zechstein: Course Notes No. 1, *Introduction to the Petroleum Geology of the North Sea.* London: Joint Association for Petroleum Exploration Courses.

Telford, W. M, L. P. Geldart, and R. E. Sheriff. 1990. *Applied Geophysics.* 2d ed. Cambridge, U.K.: Cambridge University Press.

Trorey, A. W. 1977. Diffractions for arbitrary source–receiver locations. *Geophyics*, **42**: 1177–82.

Trusheim, F. 1960. Mechanism of salt migration in northern Germany. *Bull. AAPG*, **44**: 1519–41.

Tucker, P. M., and H. J. Yorsten. 1973. *Pitfalls in Seismic Interpretation.* Tulsa: Society of Exploration Geophysicists.

Vail, P. R., R. M. Mitchum, and S. Thompson. 1977a. Relative changes of sea level from coastal onlap. In *Seismic Stratigraphy Applications to Hydrocarbon Exploration,* C. E. Payton, ed., pp. 63–81, AAPG Memoir 26. Tulsa: American Association of Petroleum Geologists.

Vail, P. R., R. G. Todd, and J. B. Sangree, 1977b. Chronostratigraphic significance of seismic reflections. In *Seismic Stratigraphy – Applications to Hydrocarbon Exploration,* C. E. Payton, ed., pp. 99–116, AAPG Memoir 26. Tulsa: American Association of Petroleum Geologists.

Van Wagoner, J. C., R. M. Mitchum, K. M. Campion, and V. D. Rahmanian. 1990. *Siliclastic Sequence Stratigraphy in Well Logs, Cores, and Outcrops: Concepts for High-Resolution Correlation of Time and Facies,* AAPG Methods in Exploration Series 7. Tulsa: American Association of Petroleum Geologists.

Wanslow, J. B. 1983. Piercement salt dome, Upper Continental Slope. In *Seismic Expression of Structural Styles,* A. W. Bally, ed., AAPG Studies in Geology 15. Tulsa: American Association of Petroleum Geologists.

White, J. E., and R. L. Sengbush. 1987. *Production Seismology.* London: Geophysical Press.

Wilgus, C. K., B. S. Hastings, H. Possamentier, J. Van Wagoner, C. A. Ross, and C. G. S. C. Kendall, eds. 1988. *Sea-Level Changes: An Integrated Approach.* Society of Economic Paleontologists and Mineralogists Special Publication 42.

Williams, W. D., and J. S. Dixon. 1983. Seismic interpretation of the Wyoming overthrust belt. In *Seismic Expression of Structural Styles,* A. W. Bally, ed., pp. 13–22, AAPG Studies in Geology 15. Tulsa: American Association of Petroleum Geologists.

11
Refraction Methods

Overview

Refraction work generally can be divided between that designed to define deep structural features and that aimed at defining the near surface, the latter including most engineering studies and determining statics corrections for reflection data processing. The data to define deep structural features have often been insufficient (simply to keep costs down) and the interpretation oversimplified (for example, assuming constant overburden velocity and nearly planar refractors); consequently, precise results were not expected. Traditional interpretation was a hand-craft art employing many ingeneous methods (we describe a few), but little work of this type is done today. On the other hand, considerable near-surface refraction is being done. The need for simplification is much less and methods such as the GRM (§11.3.3) have been developed to accomplish more precise results. Differences between achieving different refraction objectives affect mainly the scale of acquisition operations.

Refraction and reflection work are similar in many aspects, much different in others. The similarities are sufficient that reflection field crews sometimes do refraction profiling, though often not with the efficiency of a crew specifically designed for refraction. The differences between reflection and refraction field work for defining deep features mostly result from the long source-to-geophone distances employed in refraction. The energy input to the ground must be larger for refraction work and explosives continue to be the dominant energy source where longer distances are involved although other seismic sources are also used. The longer travel paths result in the higher frequencies being mostly absorbed so that refraction data are generally of low frequency compared with reflection data. Consequently, refraction geophones have lower natural frequencies than reflection geophones, although the response of the latter is often adequate for satisfactory refraction recording. Most digital seismic equipment can be used for refraction, but some older analog equipment does not have adequate low-frequency response.

Most refraction techniques involve head waves and the mapping of members whose velocities are significantly higher than those of any overlying rocks. Such units are not always present and maps of the high-velocity units that are present may not be related to the survey objectives, so refraction methods are not applicable in many situations. Even where applicable,

refraction surveying is usually slower than reflection surveying because the large offset distances involve more moving time and create problems of communications and logistics. However, refraction profiles are often not as closely spaced as reflection lines and hence the cost of mapping an area is not necessarily greater.

Most refraction work involves in-line profiling (§11.1.1), especially the use of reversed profiles. Broadside and fan-shooting and placing the geophone in a deep borehole are methods used for certain objectives. Small-scale refraction is used in studies for the foundations of structures and other engineering problems. Marine refraction involves special operational problems.

The computation of refraction data is discussed in §11.2; data have to be corrected for elevation and near-surface variations as with reflection data. The essential in refraction interpretation is correlating events that involve the same refracting interfaces; ambiguities can often be removed if more data are available. Once refraction events have been correlated, refractor depths and dips can be found using the formulas given in §4.3, but in addition a variety of methods are available for more complicated situations and for routine interpretation of large amounts of data. These methods can be divided into three groups: those using relatively complex formulas based on those in §4.3 (§11.3), delay-time methods (§11.4), and wavefront methods (§11.5).

Interpretation of engineering refraction data (§14.1.2) is generally much simpler and more straightforward than that of large-scale surveys because the near-surface layers are relatively few and little detail is usually required.

11.1 Field techniques

11.1.1 In-line refraction profiling

The basic refraction field method involves shooting reversed refraction profiles, a long linear spread of many geophone groups with a source at each end, the distance being great enough that the dominant portion of the travel path is as a head wave in the refractor or refractors being mapped. Usually, it is not practical to record simultaneously so many geophone groups spread over such a long distance, and hence refraction profiles are usually recorded in segments. By referring to fig. 11.1a, which shows a single refrac-

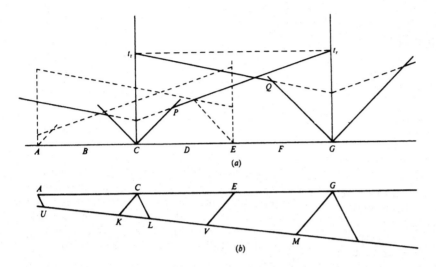

Fig. 11.1   Reversed refraction profiles. (a) Time–distance plot for continuous reversed profiling, and (b) section showing single refractor. Reciprocal times are indicated by t

tor, the spread of geophone groups might be laid out between C and D and shots at C and G fired to give two records; the spread would then be moved between D and E and shots fired at C and G as before, and so on to develop the complete reversed profile $CDEFG$. The charge size is often varied for the different segments because larger charges are required when the offset becomes greater. Usually, one or two groups will be repeated for successive segments to increase the reliability of the time tie between segments.

The source at C can also be used to record a profile to the left of C and the source at G a profile to the right of G. Note that the *reciprocal time* t_r is the same for the reversed profiles and that the intercept times for profiles shot in different directions from the same source point are equal. These equalities are exceedingly valuable in identifying segments of complex time–distance curves where several refractors are present. In simple situations, the reversed profile can be constructed without having to actually shoot it by using the reciprocal time and intercept time information. However, usually situations of interest are sufficiently complicated that this procedure cannot be carried out reliably.

The *reversed profiles* shot from C and G in fig. 11.1 allow the mapping of the refractor from L to M. The reversed profile to the left of C permits mapping as far as K, but no coverage is obtained for the portion KL. Hence, continuous coverage on the refractor requires an overlap of the reversed profiles; a reversed profile between A and E (shown dashed in fig. 11.1a) would provide coverage between U and V, thus including gap KL as well as duplicating the coverage UK and LV. With perfect data, the duplicate coverage does not yield new information, but in actual profiling, it provides valuable checks that increase the reliability of the interpretation.

In cases where reversed profiles are not essential, a split refraction spread (ACE in fig. 11.1 with source at C) can be used with a saving in the number of source locations required. However, because the updip and downdip apparent velocities are not obtained from the same part of the refractor, faulting, curvature of the refractor, lateral velocity variations, and so on can render the method useless.

If we have the two-refractor situation in fig. 11.2, first-break coverage on the shallow refractor is obtained from L to K and from M to N when the sources are at C and G; the corresponding coverage on the deeper refractor is from Q to S and from R to P.

If we are able to resolve the refraction events that arrive later than the first-breaks, called *second arrivals* or *secondary refractions,* we can increase the coverage obtained with a single profile. With analog recording, it is difficult, and sometimes impossible, to adjust the gain to optimize both the first-breaks and the second arrivals at the same time; if the gain is too low, the first-breaks may be weak and ambiguities in timing may result, whereas if the gain is too high, the secondary refractions may be unpickable. Because of this difficulty, prior to magnetic-tape recording, refraction mapping was generally based on first-breaks only. With magnetic-tape recording, playbacks can be made at several gains so that each event can be displayed under optimum conditions.

In order to economize on field work, the portions of the time–distance curves that do not add information necessary to map the refractor of interest often are not shot where they can be predicted reasonably accurately. Thus, the portions CP and GQ of the reversed profile in fig. 11.1a are often omitted.

Where a single refractor is being followed, a series of short refraction profiles are often shot rather than a long profile. In fig. 11.3, geophones from C to E are used with sourcepoint C, from D to F with source D, and so on. The portions of the time–distance curves

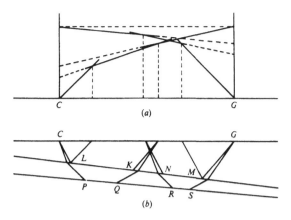

Fig. 11.2 Reversed refraction profiles for two-refractor case. (a) Time–distance plot, and (b) section showing the two refractors.

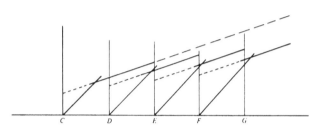

Fig. 11.3 Unreversed refraction profiles for a single refractor.

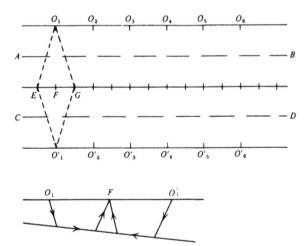

Fig. 11.4 Broadside refraction profiling with geophones along the central line and sources along the outside lines. Refractor depths are sometimes attributed as applying along lines *AB* and *CD*.

attributable to the refractor being mapped are then translated parallel to themselves until they connect together to make a composite time–distance curve such as that shown by the dashed line. The composite curve may differ from the curve that would actually have

been obtained for a long profile from sourcepoint *C* because of refraction events from deeper horizons.

An efficient method commonly used in engineering surveys (§14.1.2) is the *four-shot method* (Milsom, 1989). A spread of geophones is laid out with sources at each end (for "short shots") and also with sources offset in-line from each end (for "long shots"). The offset points are located farther from the spread than the critical distance (x' of eq. (4.39)); this assures coverage on the refractor over the entire spread length and often gives better measures of the apparent velocities and intercept times than can be obtained by fitting a straight line to the small segment of arrival times recorded with the short shots. Parallelism of the time–distance curves for long and short shots indicates travel in the same refractor. The nearest geophone is usually moved half a geophone interval away for the short shots to give a better measure of V_1. An example of this method is shown in fig. 14.2.

11.1.2 Broadside refraction and fan-shooting

In *broadside refraction* shooting, sources and spreads are located along parallel lines (see fig. 11.4) selected so that the desired refraction event can be mapped with a minimum of interference from other events. Where the refraction event can be clearly distinguished from other arrivals, it provides a very economical method of profiling because all the data yield information about the refractor. However, usually, the criteria for identifying the refraction event are based on in-line measurements (such as the apparent velocity or the relationship to other events) and these criteria are not available on broadside records where the offset distance is essentially constant. Thus, if the refractor should unexpectedly change its depth or if another refraction arrival should appear, one might end up mapping the wrong horizon. Consequently, broadside refraction shooting should be combined with occasional in-line profiles in order to check on the identity of the horizon being mapped.

The first extensive use of refraction was in searching for salt domes by the fan-shooting technique (see §1.2.3). A salt dome inserts a high-velocity mass into an otherwise low-velocity section so that horizontally traveling energy arrives earlier than if the salt dome were not present; the difference in arrival time between that actually observed and that expected with no salt dome present is called a *lead*. In *fan-shooting* (fig. 11.5), geophones are located in different directions from the source at roughly the same offset distances, the desire to maintain constant offset distance usually being sacrificed in favor of locations that are more readily accessible. The leads shown by overlapping fans then roughly locate the high-velocity mass. This method is not used for precise shape definition.

11.1.3 Gardner's method of defining salt domes

Much of the petroleum associated with production from the flanks of salt domes lies close to the salt

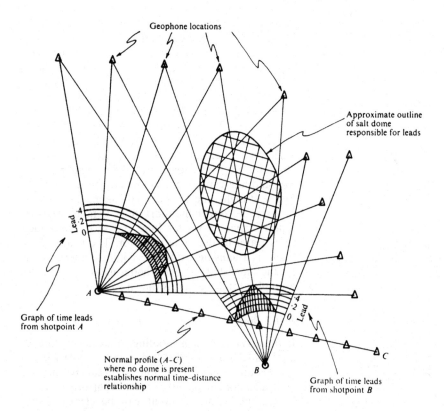

Fig. 11.5   Fan-shooting. Lead times obtained by subtracting observed times from expected times are graphed concentrically about the sources. (From Nettleton, 1940.)

flanks, so that the accurate mapping of the flanks is of considerable economic interest. The method of Gardner (1949) involves locating geophones in a deep borehole drilled into the salt for this purpose and shooting from various locations on the surface. The travelpath for each shot is partially through relatively low-velocity sediments and partially through high-velocity salt. For given traveltime and geophone location, the locus of possible points of entry of the travel paths into the salt is a surface (*aplanatic surface*) that is roughly a paraboloid (fig. 11.6a). A surface tangent to the paraboloids for all measurements for various combinations of source–detector locations defines the salt dome (fig. 11.6b).

Variations of Gardner's method employ shooting from surface locations into geophones located in a deep well near to but not within the salt. These methods are also used occasionally to define bodies other than salt.

11.1.4 Marine refraction

Because refraction recording requires that there be appreciable distance between the source and the recording locations, two ships have usually been required for marine refraction recording. To shoot a reversed refraction profile in one traverse requires

three ships – a shooting ship at each end while the recording ship travels between them. For the shooting ships to travel the considerable distances between source points takes appreciable time because of the relatively low maximum speed of ships and hence the high production rates that make marine reflection work economical are not realized in refraction shooting. Consequently, marine refraction work is relatively expensive.

The *sonobuoy* (fig. 11.7) permits recording a refraction profile with only one ship. The sonobuoy is an expendable listening station that radios the information it receives back to the shooting ship. The sonobuoy is merely thrown overboard; the salt water activates batteries in the sonobuoy as well as other devices that cause a radio antenna to be extended upward and one or two hydrophones to be suspended beneath the buoy. As the ship travels away from the buoy, shots are fired and the signals received by the hydrophones are radioed back to the ship, where they are recorded. The arrival time of the wave that travels directly through the water from the source to the hydrophone is used to give the offset distance. After a given length of time, the buoy sinks itself and is not recovered. Sonobuoys make it practical to record unreversed refraction profiles while carrying out reflection profiling, the only additional cost being that of the sonobuoys.

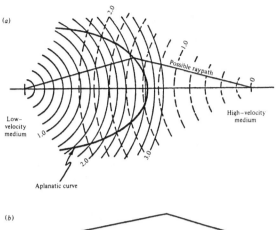

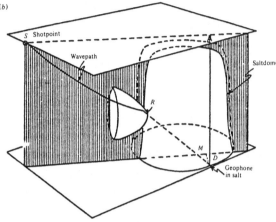

Fig. 11.6 Outlining a salt dome using a geophone within the salt. (After Gardner, 1949.) (a) Plan view of an aplanatic surface, and (b) isometric view of a raypath and paraboloid of the aplanatic surface.

11.2 Refraction data reduction and processing

Refraction data have to be corrected for elevation and weathering variations, as with reflection data. The correction methods are essentially the same except that often geophones are too far from the source to record the refraction at the base of the LVL and thus there may be no weathering data along much of the line. Additional sources may be used for special refraction weathering information.

Whereas the effect of corrections on the effective source-to-geophone distance is usually small for reflection data, this is often not so for refraction travel paths above the refractor, because these may have appreciable horizontal components. Hence, the reference datum should be near the surface to minimize such errors.

The identification of refraction events is usually simpler than reflection events. Traveltimes are usually available for a relatively long range of offsets, and hence it is easy to separate reflections and diffractions with their curved alignments from the direct wave, surface waves, and refractions with their straight alignments. The direct wave and surface waves are

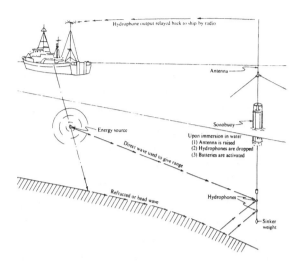

Fig. 11.7 Sonobuoy operation. (Courtesy of Select International Inc.)

easily distinguished from refractions because of the lower velocities of the former. Usually, the only problem is in identifying the different refraction events when several refractors are present.

Where complete refraction profiles from zero offset to large offsets are available, playback of the data with judicious selections of filters and automatic gain control may allow one to correlate reflection events with refraction events, thus adding useful information to each type of interpretation (fig. 11.8). The most prominent reflections may not correspond to the most prominent refractions.

Record sections are very useful, especially in studying second arrivals. The refraction profile in fig. 11.9 shows the direct wave as the first arrival near the source and refractions from successively deeper refractors become the first arrivals as the offset distance increases. Following the first arrivals, the continuations of various events are seen after each has been overtaken by a deeper event. Numerous other events are also seen in the zone of second arrivals; most of these are refractions that never become first arrivals, or multiply-reflected refractions (see fig. 6.39).

Another useful refraction playback technique is to display the data as a *reduced refraction section* (fig. 11.10), where arrival times have been shifted by the amount x/V_R, where x is the offset distance, and V_R is a value near the refractor velocity. If V_R were exactly equal to the refractor velocity, the residual times would be the delay times (which will be discussed in §11.4) and relief on the reduced refraction section would correlate with refractor relief (although displaced from the subsurface location of the relief). However, even if V_R is only approximately correct, the use of reduced sections improves considerably the pickability of refraction events, especially secondary refractions.

Often, the chief problem in refraction interpretation techniques is the assumption of constant-velocity

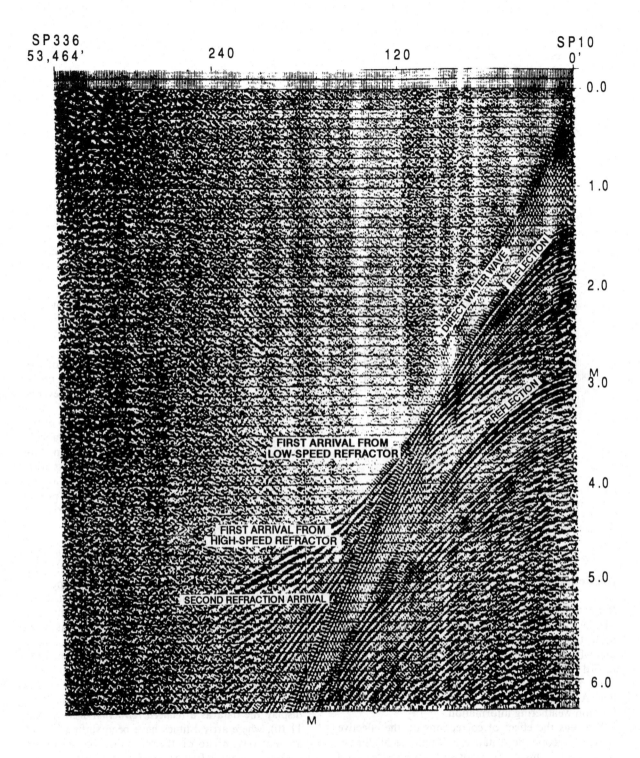

Fig. 11.8   Section showing the continuity of some reflections with their respective refraction events. Recorded by sonobuoy.

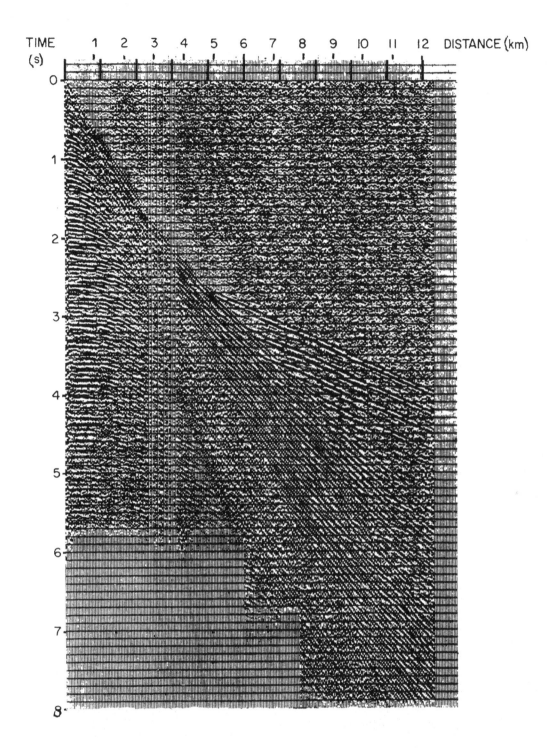

Fig. 11.9   Marine refraction profile. (From Ingham, 1975: 130.)

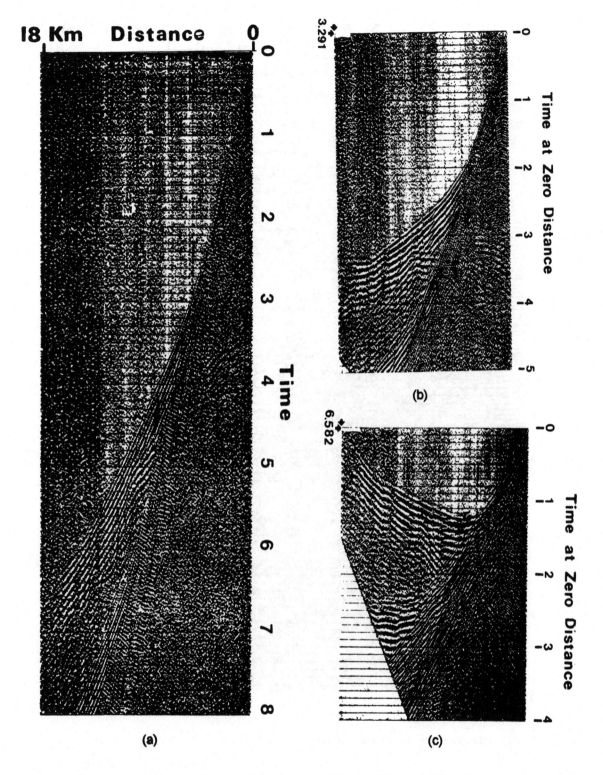

Fig. 11.10   Reduced refraction section. (Courtesy of Petty-Ray Geophysical.) (a) Conventional refraction section; (b) reduced at 5400 m/s to align highest velocity events; and (c) reduced at 2735 m/s. Subtracting x/V_R makes it easier to separate events and simplifies picking.

layers, hence raypaths that are made up of straight-line segments; this is usually not true, especially in the shallowest layers. When using equations in §4.3 to calculate refractor depths, the biggest improvement in the results is often due to a more realistic assumption for V_1 based on information other than that obtainable from the refraction data themselves (Laski, 1973).

Problems sometimes result from a *hidden zone*, a layer whose velocity is lower than that of the overlying bed so that it never carries a head wave. Energy that would approach it at the critical angle cannot get through the shallower refractors, and hence there is no indication of its presence in the refraction arrivals. The low velocity of the hidden layer, however, increases the arrival times of deeper refractors relative to what would be observed if the hidden zone had the same velocity as the overlying bed, hence results in exaggeration of their depths. Another situation, which is also referred to at times as a "hidden zone," is that of a layer whose velocity is higher than those of the overlying beds but that never produces first arrivals despite this, because the layer is too thin and/or its velocity is not sufficiently greater than those of the overlying beds. Such a bed creates a second arrival, but the second arrival may not be recognized as a distinct event.

Refraction interpretation often is based solely on first arrivals, primarily because this permits accurate determination of the traveltimes. When we use second arrivals, we usually have to pick a later cycle in the wavetrain and estimate traveltime from the measured time. However, velocities based on second arrivals will be accurate and much useful information is available through their study.

Refraction interpretation often involves "stripping," which is in effect the removal of one layer at a time (Slotnick, 1950). In this method, the problem is solved for the first refractor, after which the portions of the time–distance curve for the deeper refractors are adjusted to give the result that would have been obtained if the source and geophones had been located on the first refracting horizon. The adjustment consists of subtracting the traveltimes along the slant paths from source down to the refractor and up from the refractor to the geophones, also of decreasing the offsets by the components of the slant paths parallel to the refractor. The new time–distance curve is now solved for the second refracting layer, after which this layer can be stripped off and the process continued for deeper refractors.

11.3 Basic-formula interpretation methods

11.3.1 Using basic formulas

The basic formulas of §4.3 are used to interpret small amounts of data where the refractors are assumed to be planar. Even where these conditions are met, interpretation is usually difficult when there are more than

two refractors, especially when these are not parallel. The basic formulas are commonly used in the interpretation of engineering surveys (§14.1.2) and determining static corrections for reflection seismic work (§8.8.2).

One of the simplest refraction interpretation methods is the *ABC method*. With the arrangement shown in fig. 11.11, sources are located at the end points of the spread, A and B. Let t_{AB} be the surface-to-surface traveltime from A to B, and so on; then (see problem 11.4)

$$h_C = (1/2)(t_{CA} + t_{CB} - t_{AB}) [V_1 V_2/(V_2^2 - V_1^2)^{1/2}], \tag{11.1}$$

where V_1 is the overburden velocity, and V_2 the refractor velocity. (The depth-conversion factor,

$$F = V_1 V_2/(V_2^2 - V_1^2)^{1/2} = V_1/\cos \theta, \tag{11.2a}$$

often occurs in refraction time-to-depth conversions, for example, in eq. (4.38):

$$h = Ft_i/2, \tag{11.2b}$$

where t_i is the intercept time.) Frequently, $V_2 >> V_1$, and we can replace the factor F by V_1; then

$$h_C = (V_1/2)(t_{CA} + t_{CB} - t_{AB}), \tag{11.3}$$

the error in h_C being less than 6% if $V_2 > 3V_1$. This method assumes that the overburden is essentially homogeneous, the depth variations are smooth, the velocity contrast is large, and the dip small. Depth calculations using eq. (11.3) are generally good because they depend on the measurement of only one velocity, V_1, and three traveltimes. Whereas refractor dip can be determined from differences in apparent velocity as seen on reversed profiles, it is more often determined from a series of measurements of depth at different locations of C.

Better accuracy is given by the four-shot method (§11.1.1), which is efficient for many applications where only a local profile is needed (a "sounding") rather than a profile line. An application to an engineering problem is given in §14.1.2.

11.3.2 Adachi's method

Adachi (1954) derived equations similar to eq. (4.42) for the case of several beds with the same strike but different dips. His method departs from the usual parameters and uses vertical thicknesses and angles of incidence and refraction measured with respect to the vertical (see fig. 11.12). The derivation of Adachi's formula is straightforward but involves lengthy trigono-

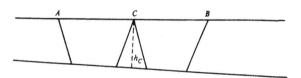

Fig. 11.11   ABC refraction method for determining depth.

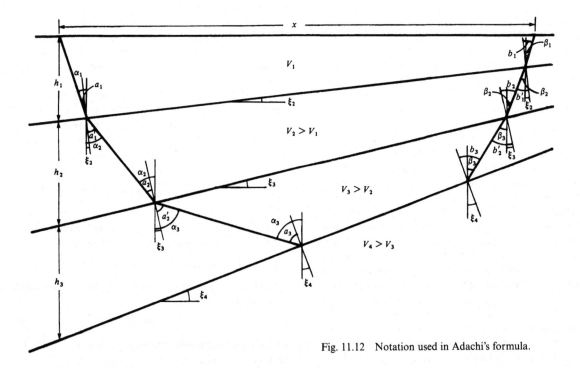

Fig. 11.12   Notation used in Adachi's formula.

metric manipulation (see Johnson, 1976), and we merely quote the result:

$$t_n = \frac{x \sin \beta_1}{V_1} + \sum_{i=1}^{n-1} \frac{h_i}{V_i}(\cos \alpha_i + \cos \beta_i). \quad (11.4)$$

where t_n is the traveltime of the refraction at the nth interface (separating layers of velocities V_n and V_{n-1}), α_i and β_i are the angles between the vertical and the downgoing and upgoing rays in the ith layer, respectively, h_i the vertical thickness of the ith layer under the source. The angles a_i, b_i (see fig. 11.12) are angles of incidence, a_i', b_i' angles of refraction, all measured relative to the normal, and ξ_{i+1} = dip of the ith interface. Then

$$a_i' = \sin^{-1}[(V_{i+1}/V_i) \sin a_i],$$

$$b_i' = \sin^{-1}[(V_{i+1}/V_i) \sin b_i],$$

$$\alpha_i = a_i + \xi_{i+1}, \qquad \beta_i = a_i - \xi_{i+1}$$

$$\alpha_{i+1} = a_i' + \xi_{i+1}, \qquad \beta_{i+1} = b_i' - \xi_{i+1}.$$

For the refraction along the nth interface, $a_n = b_n = \theta_{cn}$, the critical angle.

Assuming reversed profiles, we measure V_1, the apparent velocities, V_{2u} and V_{2d}, and the intercepts, t_{1u} and t_{1d} as usual. For the first interface,

$$\alpha_1 = \sin^{-1}(V_1/V_{2d}), \qquad \beta_1 = \sin^{-1}(V_1/V_{2u})$$

$$\theta_{c1} = a_1 = b_1 = \tfrac{1}{2}(\alpha_1 + \beta_1),$$

$$\xi_2 = \tfrac{1}{2}(\alpha_1 - \beta_1),$$

(see eq. (4.50))

$$V_2 = V_1/\sin \theta_{c1}, \qquad h_1 = V_1 t_{1u}/(\cos \alpha_1 + \cos \beta_1).$$

To solve for the second interface, we calculate new

values of α_1, β_1, and then find the other angles (note that ξ_2 is now known):

$$\alpha_1 = \sin^{-1}(V_1/V_{3d}), \qquad \beta_1 = \sin^{-1}(V_1/V_{3u}),$$
$$a_1 = \alpha_1 - \xi_2, \qquad b_1 = \beta_1 + \xi_2,$$
$$a_2' = \sin^{-1}[(V_2/V_1)\sin a_1],$$
$$b_2' = \sin^{-1}[(V_2/V_1)\sin b_1],$$
$$\alpha_2 = a_2' + \xi_2, \qquad \beta_2 = b_2' - \xi_2,$$
$$a_2 = b_2 = \theta_{c2} = \tfrac{1}{2}(\alpha_2 + \beta_2) = \tfrac{1}{2}(a_2' + b_2'),$$
$$V_3 = V_2/\sin \theta_{c2}, \qquad \xi_3 = \tfrac{1}{2}(\alpha_2 - \beta_2),$$
$$t_{2u} = (h_1/V_1)(\cos \alpha_1 + \cos \beta_1)$$
$$\qquad + (h_2/V_2)(\cos \alpha_2 + \cos \beta_2),$$

h_2 being found from the last relation. In principle, this iterative procedure can be continued indefinitely, but in practice, as with all refraction schemes, the errors and difficulties mount rapidly as the number of layers increases.

Adachi's formula is best suited to simple cases where the refractors are plane, no velocity or structural problems exist, and the refractors are shallow. When these conditions are not met, the formula, in common with other similar ones, may be of limited value. Often, one is not sure that formulas are applicable to a specific real situation. Where there are more than two refracting horizons, it is often difficult to identify equivalent updip and downdip segments, especially if the refractors are not plane or if the dip and/or strike change.

11.3.3 Generalized reciprocal method (GRM)

The GRM method (Palmer, 1980) is capable of mapping highly irregular refractors using reversed profiles and is relatively insensitive to dip up to about 20°. It is also able to resolve lateral variations in the refractor velocity (Palmer, 1986, 1991); this is especially important in engineering (where low velocity may indicate low rock strength) and groundwater studies (where it may indicate high porosity). The GRM is well suited for computer implementation.

The GRM involves selecting several pairs of points (X, Y) and making a series of calculations resulting in determining an optimum distance between them, XY_{opt}, which approximates the critical distance x' in eq. (4.39). Methods for determining XY_{opt} are given toward the end of this section, but approximate values often suffice. Our discussion follows Palmer's 1980 book except for the notation and order of topics.

Figure 11.13 shows four beds with the same strike but different dips ξ_j. Depths z_{Aj} and z_{Bj} are measured normal to each interface; α_j and β_j are downdip and updip angles of incidence, respectively, the angles at S and T also being critical angles. To get the traveltime t_{AB}, we consider a plane wavefront PQ passing through A at time $t = 0$; the wave arrives at C, after traveling a distance $z_{A1} \cos \alpha_1$, at time $t = (z_{A1} \cos \alpha_1)/V_1$. The wavefront reaches R at the time

$$t_{AR} = \sum_{1}^{3} (z_{Aj} \cos \alpha_j)/V_j.$$

A similar expression holds for t_{VB}. Because the wave is critically refracted at R and V, the time from R to V is RV/V_4. Generalizing, we get for n layers

$$t_{AB} = \sum_{1}^{n-1} (z_{Aj} \cos \alpha_j + z_{Bj} \cos \beta_j)/V_j + RV/V_n.$$

The distance $RV = YJ = EJ \cos (\xi_3 - \xi_2)$. Continuing in this manner, we get $RV = AB \cos \xi_1 \cos (\xi_2 - \xi_1) \times \cos (\xi_3 - \xi_2)$. In general, we have

$$t_{AB} = \sum_{1}^{n-1} (z_{Aj} \cos \alpha_j + z_{Bj} \cos \beta_j)/V_j + AB(S_n/V_n).$$
(11.5)

where $S_n = \cos \xi_1 \cos (\xi_2 - \xi_1) \cdots \cos (\xi_{n-1} - \xi_{n-2})$. We assume henceforth that the dip increases slowly so that $\xi_j - \xi_{j-1} \approx 0$. In this case,

$$S_n \approx \cos \xi_{n-1}$$
(11.6)

(see problem 11.7).

Let X be any point updip from A in fig. 11.14a. We can express depths at A in terms of depths at X as follows:

$$z_{A1} = z_{X1} + AH = z_{X1} + AX \sin \xi_1,$$
$$z_{A2} = z_{X2} + A'H' = z_{X2} + A'X' \sin (\xi_2 - \xi_1)$$
$$= z_{X2} + AX \cos \xi_1 \sin (\xi_2 - \xi_1).$$

For the jth layer,

$$z_{Aj} = z_{Xj} + AX \cdot S_j^*,$$
(11.7)

where

$$S_j^* = \cos \xi_1 \cos (\xi_2 - \xi_1) \cdots \cos (\xi_{j-1} - \xi_{j-2})$$
$$\times \sin (\xi_j - \xi_{j-1}), \quad j > 1,$$
$$= \sin \xi_1, \quad j = 1.$$

Let X and Y be two points separated by $2a$ (fig. 11.14b). The GRM is based on the use of a *velocity-analysis function* T_V and a *time-depth function* T_G referred to G, the midpoint between X and Y. They are defined by

$$T_V = (t_{AY} - t_{XB} + t_{AB})/2,$$
(11.8)

$$T_G = (t_{AY} + t_{XB} - t_{AB} - XY/V_n')/2,$$
(11.9)

where V_n' is an apparent velocity (defined below by eq. (11.11)). The refractor velocity V_n can be found from T_V and the depth from T_G.

By using eq. (11.5) in eq. (11.8),

$$2T_V = \left[\sum_{1}^{n-1} (z_{Aj} \cos \alpha_j + z_{Yj} \cos \beta_j)/V_j + AY(S_n/V_n) \right]$$
$$- \left[\sum_{1}^{n-1} (z_{Xj} \cos \alpha_j + z_{Bj} \cos \beta_j)/V_j + XB(S_n/V_n) \right]$$
$$+ \left[\sum_{1}^{n-1} (z_{Aj} \cos \alpha_j + z_{Bj} \cos \beta_j)/V_j + AB(S_n/V_n) \right]$$
$$= \left[\sum_{1}^{n-1} (2z_{Aj} - z_{Xj}) \cos \alpha_j + z_{Yj} \cos \beta_j/V_j \right]$$
$$+ (AY - XB + AB) (S_n/V_n).$$

Locating P so that $AP = a$, eq. (11.7) is now used to express this result in terms of z_{Pj} and AG. We have

$$z_{Aj} = z_{Pj} + aS_j^*, \qquad z_{Xj} = z_{Pj} + (2a - AG)S_j^*,$$
$$z_{Yj} = z_{Pj} - AG \cdot S_j^*,$$

so that

$$T_V = \sum_{1}^{n-1} z_{Pj}(\cos \alpha_j + \cos \beta_j)/2V_j$$
$$+ AG \left[\sum_{1}^{n-1} (\cos \alpha_j - \cos \beta_j)S_j^*/2V_j + S_n/V_n \right].$$
(11.10)

This equation shows that, for a fixed a, AG and T_V are related linearly; hence, the slope dT_V/dx is the coefficient of AG in eq. (11.10); we define an apparent velocity V_n' such that

$$dT_V/dx = 1/V_n'.$$
(11.11)

When the dip varies slowly, S_j^* is small (because of the sine factor in eq. (11.7)) and S_n reduces to $\cos \xi_{n-1}$ (see eq. (11.6)); thus, the coefficient of AG in eq. (11.10) becomes $\cos \xi_{n-1}/V_n$, and from eq. (11.11),

$$V_n \approx V_n' \cos \xi_{n-1}.$$
(11.12)

Thus, if the dip ξ_{n-1} is known, V_n can be found from the apparent velocity V_n'; if not, V_n' can be used as the refractor velocity.

If we substitute eq. (11.5) in eq. (11.9),

$$T_G = \tfrac{1}{2} \left[\sum_{1}^{n-1} (z_{Xj} \cos \alpha_j + z_{Yj} \cos \beta_j)/V_j \right.$$
$$\left. + XY(S_n/V_n - 1/V_n') \right].$$

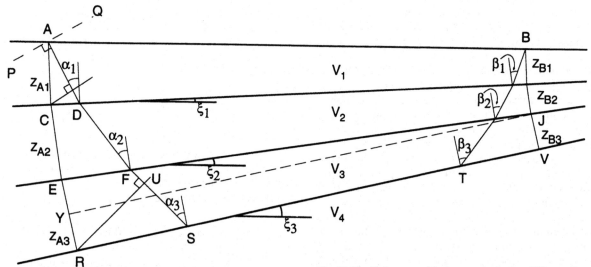

Fig. 11.13   Refractors with the same strike but different dips.

The last term vanishes by eqs. (11.6) and (11.12); moreover, if the dip is small, $z_{Xj} \approx z_{Gj} \approx z_{Yj}$. Thus, T_G reduces to

$$T_G = \tfrac{1}{2} \sum_1^{n-1} z_{Gj}(\cos \alpha_j + \cos \beta_j)/V_j. \quad (11.13)$$

Palmer (1980: 13, 14) states that T_G is similar to Hagedoorn's plus value (§11.5.2) when $XY = 0$, to Hales' "critical reflection time" (t' in §11.5.3) when $XY = XY_{opt}$, and to the mean of the geophone delay times at X and Y (§11.4.1).

We define a depth-conversion factor, $V_{jn} = 2V_j/(\cos \alpha_j + \cos \beta_j)$. (Compare this with eq. (11.2), noting that $V_{jn} = 2F$ because of the factor 1/2 in eq. (11.9).) We can now write eq. (11.13) in the form

$$T_G = \sum_1^{n-1} t_{Gj} = \sum_1^{n-1} z_{Gj}/V_{jn}. \quad (11.14)$$

For zero dip, V_{jn} becomes

$$V_{jn} = V_j/\cos \alpha_j = V_n'V_j'/[(V_n')^2 - (V_j')^2]^{1/2}, \quad (11.15)$$

on using Snell's law; the primes signify that the velocities are obtained from eq. (11.11).

Setting $x = AG = 0$ in eq. (11.10), we get for the intercept of the velocity-analysis function

$$T_V|_{x=0} = \sum_1^{n-1} z_{Pj}(\cos \alpha_j + \cos \beta_j)/2V_j. \quad (11.16)$$

Comparing with eq. (11.13), we see that the intercept of T_V is approximately equal to the time-depth at P.

The GRM can use average velocities to determine the depth to a refractor without reference to the actual layering. Assuming horizontal plane layers, eq. (11.13) becomes

$$T_G = \sum_1^{n-1} (z_{Gj} \cos \alpha_j)/V_j, \qquad \sin \alpha_j = V_j'/V_n'. \quad (11.17)$$

If we replace the actual section by a single layer of thickness $z_T = \sum z_{Gj}$ and constant velocity $\overline{V}$ with an angle of incidence $\overline{\alpha}$ such that

$$T_G = (z_T \cos \overline{\alpha})/\overline{V}, \qquad \sin \overline{\alpha} = \overline{V}/V_n', (11.18)$$

the form of eq. (11.17) is preserved.

When R and S in fig. 11.14b coincide, the distance XY is the critical distance, XY_{opt} (see eq. (4.39)):

$$XY_{opt} = 2 \sum_1^{n-1} z_{Gj} \tan \alpha_j. \quad (11.19)$$

We define the $\overline{XY}_{opt}$ for the single constant-velocity case:

$$XY = 2z_T \tan \overline{\alpha}. \quad (11.20)$$

Eliminating z_T and $\overline{\alpha}$ between eqs. (11.18) and (11.20), we find that

$$\overline{V} = V_n'[\overline{XY}/(\overline{XY} + 2T_G V_n')]^{1/2}. \quad (11.21)$$

If $\overline{XY}$ is assumed to be XY_{opt}, which can be found by methods described in the next paragraph, T_G and V_n' are now given by eqs. (11.9) and (11.11). Equation (11.21) then gives $\overline{V}$, after which $\overline{\alpha}$ and then z_T are given by eq. (11.18).

To achieve maximum accuracy, the GRM requires a knowledge of the critical distance, XY_{opt}, that is, the value when the forward and reverse rays leave the refractor at the same point. Determining this value is "potentially the most confusing aspect of the GRM" (Palmer, 1980: 34). Approximate values suffice for most purposes (XY_{opt} is relatively insensitive to dip); however, accurate values are required if there are hidden zones (§4.3.1). Palmer gives three methods of finding XY_{opt}: (a) from knowledge of the thicknesses and velocities of all layers, for example, from borehole information; (b) from the separation of distinctive features on forward and reverse profiles, such as sharp changes of slope; (c) from trial calculations of T_V and

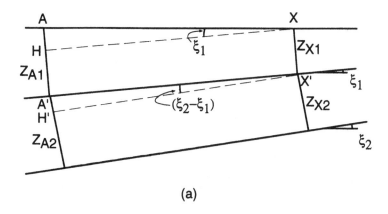

(a)

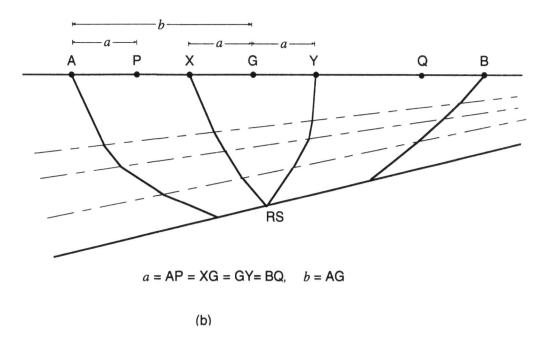

$$a = AP = XG = GY = BQ, \quad b = AG$$

(b)

Fig. 11.14 Relations for points on the surface. (a) "Depths" at A and X. (b) Raypaths to X and Y.

T_G curves for a series of XY-values; XY_{opt} corresponds to the simplest T_V curve and the most detailed T_G curve.

Palmer (loc. cit.: 55–7) lists the following steps for computation of GRM data: (a) near-surface irregularities are studied and their effects removed, (b) each point on the traveltime curves is assigned to a specific refractor, (c) the optimum XY-values are determined, (d) refractor velocities are calculated using eqs. (11.11), (11.12), or (11.21); (e) depths are found from the time-depth function and depth-conversion factors.

Hidden zones "are the most common source of errors in the majority of refraction interpretation methods" (Palmer, loc. cit: 39). When data are good, hidden zones may be defined by comparison of the optimum XY-values obtained from T_V and T_G curves with those obtained from eq. (11.19); differences be-

tween the two sets of values presumably are due to hidden layers.

In a recent paper, Palmer (1991) illustrated the application of the GRM to studies of shallow layers where the velocity and/or depth change laterally within a narrow zone, for example, because of faulting, a low-velocity shear zone, or a sink hole. Because the anomalous features are narrow and shallow, geophone and source spacings are small (3 and 36 m in one example where depths were 10 to 20 m).

We illustrate with one of Palmer's models; fig. 11.15a shows a low-velocity shear zone (or weathered dike) sandwiched between two high-velocity layers. Figures 11.15b and 11.15c show curves of T_V and T_G (eqs. (11.8) and (11.9)) plotted for XY values from 0 to 150 m, a range that includes XY_{opt}; the curves are displayed vertically at 10-ms intervals.

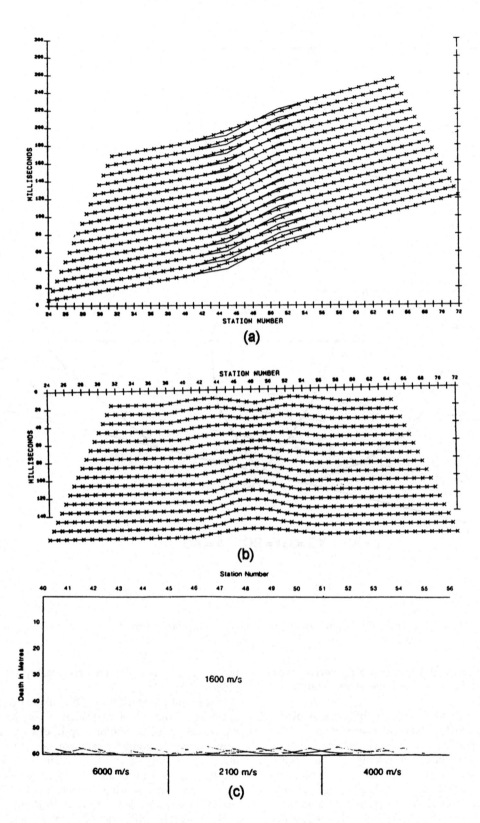

Fig. 11.15 Model result for changing refractor velocity from 6000 to 2100 to 4000 m/s. Station spacing = 10 m. (From Palmer, 1991.) (a) T_V curves for $XY = 0$ (bottom curve) to 150 m, (b) T_G curves for same XY-values, and (c) depth section based on results obtained in parts (a) and (b).

The left and right portions of the curves in fig. 11.15b are parallel straight lines with slopes of 6000 m/s and 4000 m/s, respectively. Starting at the left side, the 6000 m/s straight line extending farthest to the right is the line for $XY = 70$ m; the data begin to depart from this line at about station 44.5. All the 6000 m/s lines are now extended to station 44.5. On the right side, the 4000 m/s line extending farthest to the left is that for $XY = 90$ m, which reaches station 51.5, so all 4000-m/s lines are extended to station 51.5. Stations 44.5 and 51.5 are now taken as the limiting points for the 6000- and 4000-m/s zones. The ends of the straight lines at stations 44.5 and 51.5 for each XY-value are joined by straight lines; these short lines are assumed to be corrected T_V lines for the intervening low-velocity zone. Their slopes give a 2100-m/s value in this case. The 70- and 90-m XY-values are XY_{opt} for the 6000- and 4000-m/s zones, respectively; note that the deviations between the observed values and the straight lines (including the short low-velocity lines joining them) are minimal for these XY-values.

The time-depth function T_G is shown in fig. 11.15c; in calculating T_G values close to the lateral velocity changes, the XY/V'_n term in eq. (11.9) is a distance-weighted average based on the adjacent velocity values. The "pull-up" in the time–depth values is due to the low-velocity layer. The depth-conversion factor is calculated from eq. (11.15) using $V_j = 1.60$ km/s, $V_n = 6.00$, 2.10, and 4.00 km/s; the values are 1.66, 2.47, and 1.75, respectively. Equation (11.14) now gives the depths shown in fig. 11.15d; the "pull-up" in fig. 11.15c has been wiped out by the higher depth-conversion value over the shear zone.

11.4 Delay-time interpretation methods

11.4.1 Delay time

The concept of delay time, introduced by Gardner (1939), is widely used in routine refraction interpretation, mainly because the various schemes based upon the use of delay times are less susceptible to the difficulties encountered when we attempt to use eqs. (4.36) to (4.50) and (11.4) with refractors that are curved or irregular. Assuming that the refraction times have been corrected for elevation and weathering, the *delay time* associated with the path $SMNG$ in fig. 11.16 is the observed refraction time at G, t_g, minus the time required for the wave to travel from P to Q (the projection of the path on the refractor) at the velocity V_2. Writing δ for the delay time, we have

$$
\begin{aligned}
\delta &= t_g - \frac{PQ}{V_2} = \left(\frac{SM + NG}{V_1} + \frac{MN}{V_2}\right) - \frac{PQ}{V_2} \\
&= \left(\frac{SM + NG}{V_1}\right) - \left(\frac{PM + NQ}{V_2}\right) \\
&= \left(\frac{SM}{V_1} - \frac{PM}{V_2}\right) + \left(\frac{NG}{V_1} - \frac{NQ}{V_2}\right) \\
&= \delta_s + \delta_g,
\end{aligned} \tag{11.22}
$$

where δ_s and δ_g are known as the *source delay time* and

the *geophone delay time* because they are associated with the portions of the path down from the source and up to the geophone.

An approximate value of δ can be found by assuming that the dip is small enough that PQ is approximately equal to the geophone offset x. In this case,

$$
\delta = \delta_s + \delta_g \approx t_g - x/V_2. \tag{11.23}
$$

Provided the dip is less than about 10°, this relation is sufficiently accurate for most purposes. If we substitute the value of t_g obtained from eqs. (4.37), (4.47), and (4.48), we see that δ is equal to the intercept time for a horizontal refractor but not for a dipping refractor.

Many interpretation schemes using delay time have been given in the literature, for example, Gardner (1939, 1967), Barthelmes (1946), Tarrant (1956), Wyrobek (1956), and Barry (1967). We shall describe only the latter three. The methods described by Wyrobek and Tarrant are suitable for unreversed profiles, whereas that of Barry works best with reversed profiles.

11.4.2 Barry's method

The scheme described by Barry, like many based on delay times, requires that we resolve the total delay time δ into its component parts, δ_s and δ_g. In fig. 11.17, we show a geophone R for which data are recorded from sources at A and B. The ray BN is reflected at the critical angle; hence, Q is the first geophone to record the head wave from B. Let δ_{AM} be the source delay time for source A, δ_{NQ} and δ_{PR} the geophone delay times for geophones at Q and R, δ_{AQ} and δ_{AR} the total delay times for the paths $AMNQ$ and $AMPR$. Then,

$$
\begin{aligned}
\delta_{AQ} &= \delta_{AM} + \delta_{NQ} \\
\delta_{AR} &= \delta_{AM} + \delta_{PR} \\
\Delta\delta &= \delta_{AQ} - \delta_{AR} = \delta_{NQ} - \delta_{PR}.
\end{aligned}
$$

For the source at B, the source delay time δ_{BN} is approximately equal to δ_{NQ} provided the dip is small. In this case,

$$
\delta_{BR} = \delta_{BN} + \delta_{PR} \approx \delta_{NQ} + \delta_{PR}.
$$

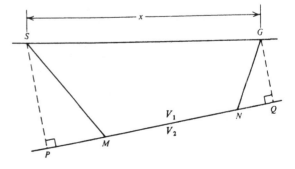

Fig. 11.16 Illustrating delay time.

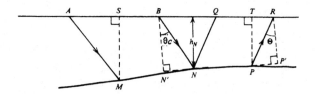

Fig. 11.17   Determining source and geophone delay times.

The geophone delay times are now given by

$$\left.\begin{array}{l} \delta_{NQ} \approx \frac{1}{2}(\delta_{BR} + \Delta\delta) \\ \delta_{PR} \approx \frac{1}{2}(\delta_{BR} - \Delta\delta) \end{array}\right\}. \qquad (11.24)$$

Thus, it is possible to find the geophone delay time at R provided we have data from two sources on the same side and we can find point Q. If we assume that the bed is horizontal at N and is at a depth h_N, we have

$$h_N = V_1 \delta_{BN}/\cos\theta_c, \qquad (11.25)$$

$$\begin{aligned} BQ &= 2h_N \tan\theta_c \\ &= 2V_1\,\delta_{BN}(\tan\theta_c/\cos\theta_c) \\ &= 2V_2\,\delta_{BN}\tan^2\theta_c. \end{aligned} \qquad (11.26)$$

The source delay time δ_{BN} is assumed to be equal to half the intercept time at B; this allows us to calculate an approximate value of BQ and thus determine the delay times for all geophones to the right of Q for which data from A and B were recorded.

The interpretation involves the following steps, which are illustrated in fig. 11.18:

(a) the corrected traveltimes are plotted,
(b) the total delay times are calculated and plotted at the geophone positions,
(c) the "geophone offset distances" (PP' in fig. 11.17) are calculated using the relation $PP' \approx V_2\,\delta_{PR}\tan^2\theta$ (see problem 11.8) and the delay times in (b) are then shifted toward the source by these amounts,
(d) the shifted curves in (c) for the reversed profiles should be parallel; any divergence is due to an incorrect value of V_2, hence, the value of V_2 is adjusted and steps (b) and (c) repeated until the curves are parallel (with practice only one adjustment is usually necessary),
(e) the total delay times are separated into source and geophone delay times, the latter being plotted at the points of entry and emergence from the refractor (S and T in fig. 11.17); the delay-time scale can be converted into depth if required using eq. (11.25).

11.4.3 Tarrant's method

Tarrant (1956) uses delay times to locate the point Q (fig. 11.19a) at which the energy arriving at R left the reflector. Denoting the delay time associated with the path QR by δ_g, we have

$$\delta_g = \rho/V_1 - (\rho\cos\phi)/V_2;$$

hence,

$$\rho = V_1\delta_g/(1 - \sin\theta_c\cos\phi). \qquad (11.27)$$

This is the polar equation of an ellipse. An ellipse is the locus of a point Q (fig. 11.19b) that moves so that the ratio (QR/QM) is constant (equal to the eccentricity ε, which is < 1 for an ellipse),

$$\rho/(h + \rho\cos\phi) = \varepsilon,$$

hence,

$$\rho = \varepsilon h/(1 - \varepsilon\cos\phi). \qquad (11.28)$$

The major axis, $2a = \rho_{\phi=0} + \rho_{\phi=\pi} = 2\varepsilon h/(1 - \varepsilon^2)$. The semiminor axis, b, can be found by writing $y = \rho\sin\phi$ and finding y_{max}; this gives $b = \varepsilon h(1 - \varepsilon^2)^{-1/2}$. The distance from the focus, R, to the center of the ellipse, O, is equal to $\rho|_{\phi=0} - a = \varepsilon h/(1 - \varepsilon) - \varepsilon h/(1 - \varepsilon^2) = \varepsilon a$. If we take $\varepsilon = \sin\theta_c$ and $h = V_2\delta_g$, eq. (11.28) becomes eq. (11.27).

For a horizontal refractor, we have the ellipse in fig. 11.19c, with $a = V_2\delta_g\tan\theta_c\sec\theta_c$, $b = V_2\delta_g\tan\theta_c$, and $OR = V_2\delta_g\tan^2\theta_c$. Also $RQ = b/\cos\theta_c = a$ and $\angle OQR = \tan^{-1}(OR/b) = \theta_c$, $OQ = OR\cot\theta_c = V_2\delta_g\tan\theta_c$.

We can approximate the ellipse in the vicinity of Q with a circle of the same radius of curvature. If we write the equation of the ellipse in the Cartesian form,

$$(x/a)^2 + (y/b)^2 = 1,$$

the radius of curvature, r, becomes

$$r = [1 + (y')^2]^{1/2}/y'',$$

where $y' = -(b/a)^2(x/y)$, $y'' = -(b/a)^2(y - xy')/y^2$; at Q, $y' = 0$ and $y'' = b/a^2$. Hence, $r = a^2/b = V_2\delta_g/\cos^3\theta_c = V_2\delta_g\tan\theta_c\sec^2\theta_c$ and the center C is at the point $(0, r - b)$, that is, $(0, V_2\delta_g\tan^3\theta_c)$. Also, $\angle CRO = \tan^{-1}(CO/RO) = \theta_c$; hence, $\angle CRQ$ is a right angle.

To apply the method, we must determine the velocities, V_1 and V_2, and the delay time at the source δ_s. Then we compute δ_g from the formula,

$$\delta_g = t_R - (x/V_2) - \delta_s.$$

We are now able to compute OR, OQ, and then locate C by drawing RC perpendicular to RQ. From C, we draw an arc of the circle to represent the refracting surface in the vicinity of Q. If the dip is not zero, the point of emergence is Q', the arc QQ' increasing with the dip. Even for moderate dip the elliptical arc QQ' will be close to the circular arc through Q and thus the envelope of circular arcs will outline the refractor closely.

Tarrant's method is useful when the dip is moderate or large and the refractor is curved or irregular. The principal limitation is in the determination of V_2.

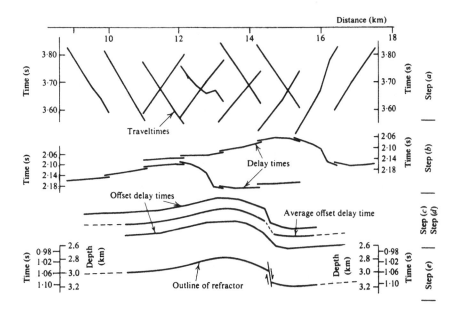

Fig. 11.18   Illustrating the delay-time method of interpreting reversed profiles. (After Barry, 1967.)

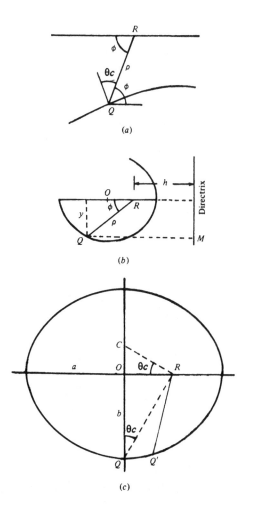

Fig. 11.19   Illustrating Tarrant's method. (a) Relation between the receiver R and the emergent point Q, (b) showing that the locus of Q is an ellipse with R at one focus, and (c) geometry of the ellipse through Q.

11.4.4 Wyrobek's method

To illustrate Wyrobek's method, we assume a series of unreversed profiles, as shown in the upper part of fig. 11.20. The various steps in the interpretation are as follows:

(a) the corrected traveltimes are plotted and the intercept times measured,

(b) the total delay time δ is calculated for each geophone position for each source and the values plotted at the geophone position (if necessary, a value of V_2 is assumed); by moving the various segments up or down a composite curve similar to a phantom horizon is obtained,

(c) the intercept times divided by 2 are plotted and compared with the composite delay-time curve; divergence between the two curves indicates an incorrect value of V_2 (see what follows), hence, the value used in step (b) is varied until the two curves are "parallel," after which the half-intercept time curve is completed by interpolation and extrapolation to cover the same range as the composite delay-time curve,

(d) the half-intercept time curve is changed to a depth curve by using eq. (4.38), namely,

$$h = \tfrac{1}{2}V_1 t_1 / \cos \theta_c$$

(note that we are ignoring the difference between the vertical depth h and the slant depths h_u and h_d in eqs. (4.45) and (4.46).

Wyrobek's method depends on the fact that the curve of δ is approximately parallel to the half-intercept time curve. For proof of this result, the reader is referred to problem 11.10. Wyrobek's

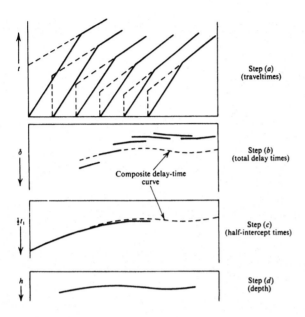

Step (a)
(traveltimes)

Step (b)
(total delay times)

Composite delay-time
curve

Step (c)
(half-intercept times)

Step (d)
(depth)

Fig. 11.20 Illustrating Wyrobek's method using unreversed profiles. (After Wyrobek, 1956.)

method does not require reversed profiles because the intercept at a source point does not depend upon the direction in which the cable is laid out.

Delay-time methods are subject to certain errors that must be guarded against. As the source-to-geophone distance increases, the refraction wavetrain becomes longer and the energy peak shifts to later cycles. There is thus the danger that different cycles will be picked on different profiles and that the error will be interpreted as an increase in source delay time. If sufficient data are available the error is usually obvious. Variations in refractor velocity manifest themselves in local divergences of the offset total-delay-time curves for pairs of reversed profiles. However, if some data that do not represent refraction travel in the refractor under consideration are accidentally included, the appearance is apt to be the same as if the refractor velocity were varying. In situations where several refractors that have nearly the same velocities are present, unambiguous interpretation may not be possible.

11.5 Wavefront interpretation methods

11.5.1 Thornburgh's method

Wavefront reconstruction, usually by graphical means, forms the basis of several refraction interpretation techniques. The classic paper is one by Thornburgh (1930); other important papers are those by Gardner (1949), Baumgarte (1955), Hales (1958), Hagedoorn (1959), Rockwell (1967), and Schenck (1967).

Figure 11.21 illustrates the basic method of reconstructing wavefronts. The refraction wavefront that reached A at $t = 1.600$ s reached $B, C, \ldots$ at times

$1.600 + \Delta t_B$, $1.600 + \Delta t_C$. . . . By drawing arcs with centers $B, C, \ldots$ and radii $V_1 \Delta t_B$, $V_1 \Delta t_C, \ldots$, we can establish the wavefront for $t = 1.600$ s (AZ) as accurately as we wish. Similarly, other refraction wavefronts, such as that shown for $t = 1.400$ s, can be constructed at any desired traveltime interval. The direct wavefronts from the source S are of course the circles shown in the diagram.

In fig. 11.22, we show a series of wavefronts chosen so that only waves that will be first arrivals are shown (all secondary arrivals being eliminated in the interests of simplicity). Between the source S and the crossover point C (see eq. (4.40)) the direct wave arrives first. To the right of C, the wave refracted at the first horizon arrives first until, to the right of G, the refraction from the deeper horizon overtakes the shallower refraction.

The two systems of wavefronts representing the direct wave and the refracted wave from the shallow horizon intersect along the dashed line ABC; this line, called the *coincident-time curve* by Thornburgh, passes through the points where the intersecting wavefronts have the same traveltimes. The curve $DEFG$ is a coincident-time curve for the deeper horizon. The coincident-time curves are tangent to the refractors at A and D, where the incident ray reaches the critical angle (see problem 11.12), whereas the points at which the coincident-time curves meet the surface are marked by abrupt changes in the slopes of the time–distance plot.

Because the coincident-time curve is tangent to the refractor, the latter can be found when we have one profile and other data, such as the dip, depth, critical angle – or a second profile (not necessarily reversed) because we now have two coincident-time curves and the refractor is the common tangent to the curves.

When reversed profiles are available, the construction of wavefronts provides an elegant method of locating the refractor. The basic principle is illustrated in fig. 11.23, which shows two wavefronts, MCD and PCE, from sources at A and B intersecting at an intermediate point C. Obviously, the sum of the two traveltimes from A and B to C is equal to the reciprocal time between A and B, t_r. If we had reconstructed the two wavefronts from the time–distance curve without knowing where the refractor RS was located, we would draw the wavefronts as MCN and PCQ, not MCD and PCE. Therefore, if we draw pairs of wavefronts from A and B such that the sum of the traveltimes is t_r, the refractor must pass through the points of intersection of the appropriate pairs of wavefronts.

11.5.2 Hagedoorn's plus–minus method

Hagedoorn's *plus–minus* method (1959) utilizes a construction similar to that just described. When the refractor is horizontal, the intersecting wavefronts drawn at intervals of Δ milliseconds form diamond-shaped figures (fig. 11.24) whose horizontal and vertical diagonals are equal to $V_2 \Delta$ and $V_1 \Delta / \cos \theta_c$, respectively. If we add together the two traveltimes at each

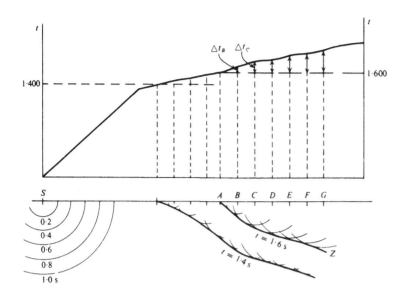

Fig. 11.21 Reconstruction of wavefronts.

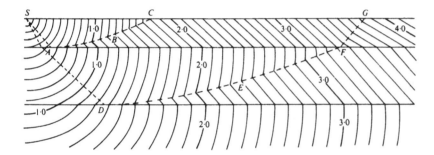

Fig. 11.22 First-arrival wavefronts for three layers with velocities in the ratios 2:3:4. The dashed curves ABC and $DEFG$ are called coincident-time curves. (After Thornburgh, 1930.)

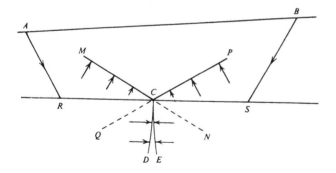

Fig. 11.23 Determining refractor position from wavefront intersections.

intersection and subtract t_r, the resulting "plus" values equal 0 on the refractor, $+2\Delta$ on the horizontal line through the first set of intersections vertically above those defining the refractor, $+4\Delta$ on the next line up, and so on. Because the distance between each pair of adjacent lines is $V_1\Delta/\cos\theta_c$, we can use any of the "plus" lines to plot the refractor shape. The difference between two traveltimes at an intersection is called the "minus" value; it is constant along vertical lines passing through the intersections of wavefronts. The distance between successive "minus" lines as shown in fig. 11.24 is $V_2\Delta$; hence, a continuous check on V_2 is possible. Although dip alters the forgoing relations, the changes are small for moderate dip, and the assumption is made that the "plus" lines are still parallel to the refractor and that the "minus" lines do not converge or diverge.

11.5.3 Hales' graphical method

Graphical methods are well suited to many refraction interpretation problems. When carried out carefully, they often give the requisite accuracy rapidly and they are satisfying to make because the picture unfolds as one carries out the interpretation.

Hales' method (1958) is useful where the depth to the refractor varies appreciably, a situation often associated with variation of overburden and refractor ve-

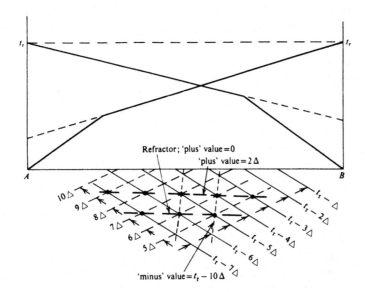

Fig. 11.24   Illustrating the plus–minus method.

locities. The method requires reversed profiles. The essence of the method is the scheme for locating pairs of points such as A and B (fig. 11.25a), which have a common point of emergence Q, when the dip and depth of the refractor are not initially known. The interpretation procedure will be described first and then the propositions will be proven.

Given reversed refraction profiles, as shown in fig. 11.25b, we select an arbitrary point B at which the arrival time is t_{RB}. The point K is located such that $KB = (t_r - t_{RB})$. A line through K at the angle $\alpha = \tan^{-1}(V_1 \sin \theta_c)$ intersects the reversed profile at time t_{SA} at location A, which is the point on the reversed profile associated with the same point on the refractor (Q on fig. 11.25a) as B. The time t' (fig. 11.25b) and the distance x' can now be read from the reversed profile plot. A line is drawn through A at the critical angle θ_c (fig. 11.25c), which intersects the perpendicular bisector of AB at C. An arc is then drawn of radius $\rho = V_1 t'/(2 \cos \theta_c)$. The refractor is the common tangent to arcs drawn in this way. The angle α given above is not precisely the correct angle α', but the error is negligible, as will be shown.

To establish the soundness of this method, consider the geometry of the triangle AQB (fig. 11.25d), where Q is the refracting point. The refracted waves from R to B and from S to A (fig. 11.25a) leave the refractor at Q. The circle that passes through A, Q, and B is drawn and the values of the several angles can be determined in terms of critical angle θ_c and dip ξ. The distance $CQ = \rho$ can be found by noting that

$$\rho \cos \theta_c = QN = AQ - AN$$
$$= QG = BQ + BG.$$

However, $AN = CN \tan \xi = CG \tan \xi = BG$; hence, adding the two expressions for $\rho \cos \theta_c$ gives

$$\rho = (AQ + BQ)/(2 \cos \theta_c).$$

From fig. 11.21a, we see that

$$t_{RB} + t_{SA} = t_r + (AQ + BQ)/V_1;$$

hence,

$$AQ + BQ = V_1 t', \qquad \rho = V_1 t'/(2 \cos \theta_c)$$

(note that t' is the traveltime for a reflection with path AQB). Further,

$$\begin{aligned}
AB = x' &= AH + HB \\
&= \frac{AQ \sin \theta_c}{\sin(\tfrac{1}{2}\pi + \xi)} + \frac{BQ \sin \theta_c}{\sin(\tfrac{1}{2}\pi - \xi)} \\
&= (AQ + BQ) \frac{\sin \theta_c}{\cos \xi} \\
&= V_1 t' \sin \theta_c/\cos \xi.
\end{aligned}$$

The angles $\alpha = \tan^{-1}(V_1 \sin \theta_c)$ and $\alpha' = \tan^{-1}(x'/t')$ are equal if $\xi = 0$. If $\xi \neq 0$, $\alpha' > \alpha$ so A will be located too close to B by the amount $\Delta x'$, t_{SA} and t' will be slightly too small by the amount $\Delta t'$, and ρ will be too small by $\Delta \rho$. Referring to fig. 11.25e,

$$\Delta t'/\Delta x' = \text{slope of traveltime curve} = \sin(\theta_c + \xi)/V_1$$

(for the downdip traveltime curve)

$$\Delta \rho = \frac{V_1 \Delta t'}{2 \cos \theta_c} = \frac{\Delta x' \sin(\theta_c + \xi)}{2 \cos \theta_c}.$$

The point C from which ρ is measured also moves to C' (fig. 11.25e):

$$CC' = \Delta x'/(2 \cos \theta_c)$$
$$CQ - C'Q' = CC' \cos(\tfrac{1}{2}\pi - \theta_c - \xi)$$
$$= \Delta x' \sin(\theta_c + \xi)/(2 \cos \theta_c),$$

which is exactly equal to $\Delta \rho$. Hence, the only effect of neglecting dip is to displace the refracting point updip by the amount $\tfrac{1}{2}\Delta x'$.

Hales' method requires knowledge of V_1 and V_2 in

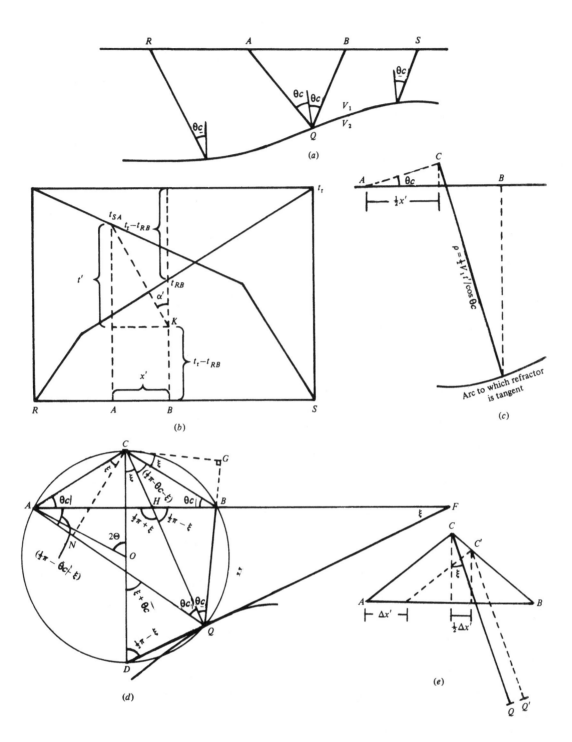

Fig. 11.25 Hales' graphical method. (a) Relation between two receivers A and B having a common emergent point Q; (b) geometrical properties of points on the traveltime curves corresponding to A and B (construction lines are dashed); (c) construction for locating Q; (d) properties of circumscribed circle through A, B, and Q; and (e) effect of errors in x' in part (b).

order to calculate α. Variation of V_2 can be accommodated by calculating V_2 from the slopes of the respective traveltime curves at B and at A (an approximation of the location of A will usually suffice). Variations of V_1 with depth (usually an increase with increasing depth) can be accommodated by iterating the calculation.

11.6 Geologic interpretation of refraction data

Much of what is termed refraction interpretation, especially the application of equations such as eqs. (4.36) to (4.52), should properly be termed "computation." The geological interpretation of refraction data, to distinguish it from computation, is much cruder than that of reflection data and usually much more restricted in range of depths involved, detail and precision. Under favorable circumstances, refraction data can yield both structural and stratigraphic data but usually only structural information is obtained.

In virgin areas, refraction is often done with the twin objectives of determining roughly (1) the shape of the basin, including depth to basement, (2) the nature or rock type of the major lithological units based on their velocities. Velocities in the range 2–3 km/s generally denote sands and shales, whereas velocities of 5–6 km/s usually denote limestone, dolomite, or anhydrite. Crystalline basement refractions often have a characteristic envelope and are very strong. Velocities of the various rock types overlap (as shown in fig. 5.5); hence, refraction velocities generally do not identify rock types uniquely. When outcrops and/or well information are available, the interpretation may be more reliable.

Structural interpretation is usually simple provided the data permit accurate matching of up-and-down dip velocities and velocity complications are not present. Faulting is sometimes indicated more clearly on refraction records and the displacement found more accurately than is the case with reflection data. However, the usual paucity of refractors hardly ever permits us to find the variation of displacement with depth, curvature of the fault plane, and so on, information that under favorable circumstances can be found from reflection sections.

If enough data are available, interpretational ambiguities often can be resolved. However, in an effort to keep survey costs down, only the minimum amount of data may be obtained (or less than the minimum) and some of the checks that increase certainty and remove ambiguities may not be possible.

Problems

11.1 Early refraction work searching for salt domes in the Gulf Coast considered a significant "lead" to be of the order of 0.250 s. Assuming a range of 3½ miles (5.63 km), a normal sediment velocity at salt-dome depth of 2.74 km/s, and salt velocity of 4.57 km/s, how much salt travel would this indicate?

11.2 The velocity of salt is nearly constant at 4.57 km/s. Calculate the amount of lead time per kilometer of salt diameter as a function of depth assuming the sediments have the Louisiana Gulf Coast velocity distribution shown in fig. 11.28.

11.3 In refraction mapping of the 5.75-km/s layer at about 0.6 km depth in the Illinois Basin, the overlying shale forms a "hidden layer." Using the velocity–depth data from fig. 11.28, determine approximately how much error neglect of the hidden layer will involve.

11.4 Prove eq. (11.1) assuming that the surface is horizontal and the refractor is a plane.

11.5 Given the data in table 11.1 for a reversed refraction profile with source points A, B, use Adachi's method to find velocities, depths, and dips.

11.6 (a) Solve problem 11.5 by stripping off (§11.2) the shallow layer (use the same velocities as in problem 11.5 for the purpose of comparison).
(b) Compare your results with those in problem 11.5.
(c) What are some of the advantages and disadvantages of stripping?

11.7 Prove eq. (11.6). (*Hint:* Write cos $(\xi_{n-1} - \xi_m) =$ cos$[(\xi_{n-1} - \xi_{n-2}) + (\xi_{n-2} - \xi_m)]$, where m is an integer; expand and drop the product of sines. Treat the factor cos $(\xi_{n-2} - \xi_m)$ in the same way and continue the process until you arrive at the factor cos $(\xi_1 - \xi_m)$, then set $m = 0 = \xi_0$.)

11.8 Show that PP' in fig. 11.17 is given by

$$PP' = V_2\, \delta_{PR} \tan^2\theta_c.$$

11.9 Source B is 2 km east of source A. The data in table 11.2 were obtained with cables extending eastward from A and B with geophones at 200-m intervals. Interpret the data using Barry's method. Note that x is the distance measured from A. Take $V_1 = 2.5$ km/s and assume that the delay-time curve for the

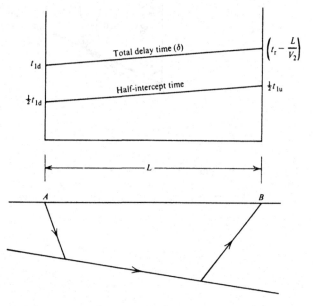

Fig. 11.26 Demonstrating the parallelism of the curves of total delay time and the half-intercept time.

Table 11.1 *Reversed refraction data*

x	0	0 5	1 0	1.5	2.0	2.5	3.0	3.5	4 0	4.5	5 0 km
t_A	0.00	0.25	0.50	0.74	0.98	1.24	1.50	1.70	1.81	1.91	2 02 s
t_B	3.00	2.90	2.80	2.68	2.52	2.41	2.31	2.20	2.07	1.91	1.80 s

x	5.5	6.0	6.5	7.0	7.5	8.0	8.5	9.0	9.5	10.0 km
t_A	2.16	2.28	2.38	2.44	2.56	2.64	2.72	2.80	2.89	3.00 s
t_B	1.65	1.50	1.40	1.25	1.12	1.00	0.75	0.49	0.23	0.00 s

Table 11.2 *Time–distance data*

x	t_A	t_B
2.6 km	1.02 s	0.25 s
2.8	1.05	0.34
3.0	1.10	0.43
3.2	1.14	0.52
3.4	1.18	0.61
3.6	1.20	0.70
3.8	1.26	0.78
4.0	1.32	0.87
4.2	1.35	0.96
4.4	1.39	1.05
4.6	1.45	1.10
4.8	1.50	1.14
5.0	1.56	1.20
5.2	1.59	1.22
5.4	1.62	1.28
5.6	1.66	1.31
5.8	1.72	1.36
6.0	1.73	1.42
6.2	1.80	1.47
6.4	1.85	1.53
6.6	1.91	1.56
6.8	1.97	1.59
7.0	2.00	1.63
7.2	2.02	1.67
7.4	2.05	1.70
7.6	2.10	1.73
7.8	2.13	1.78
8.0	2.16	1.81

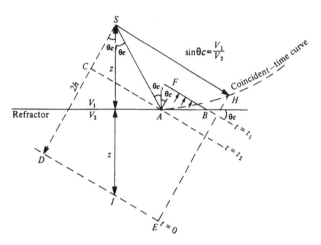

Fig. 11.27 Deriving the properties of the coincident-time curve.

F and G the intercepts were 1.52 and 1.60 s, respectively. Use Wyrobek's method to interpret the data.

11.12 Using fig. 11.27, show the following:
(a) DE, the "wavefront for $t = 0$," is at a depth $SD = 2h = 2z \cos \theta_c$.
(b) After DE reaches A, wavefronts such as BF coincide with the headwave wavefronts.
(c) The coincident-time curve AH is a parabola.
(d) Taking DE and DS as the x- and y-axes, the equation of AH is $4hy = x^2 + 4h^2$.
(e) The coincident-time curve is tangent to the refractor at A.

11.13 Interpret the data in table 11.4 using the plus–minus method.

11.14 The data in table 11.5 show refraction traveltimes for geophones spaced 400 m apart between sources A and B, which are separated by 12 km. The columns in the table headed t_A^* and t_B^* give second arrivals. Interpret the data using: (a) eqs. (4.45) to (4.50); (b) Tarrant's method; (c) the wavefront method illustrated in fig. 11.23; (d) Hales' method. On the basis of your results, compare the methods in terms of (1) time involved; (2) effect of refractor curvature; (3) effect of random errors; and (4) suitability for (i) routine production or (ii) special effort where high accuracy is essential.

11.15 Construct the expected time–distance curve for

reversed profiles is sufficiently parallel to yours that step (d) in §11.4.2 can be omitted.

11.10 Prove that the half-intercept curve referred to in the discussion of Wyrobek's method in §11.4.4 is parallel to the curve of the total delay time δ (see fig. 11.26). Note that the reciprocal time can be written (see eqs. (4.47) and (4.48))

$$t_r = \frac{1}{2}\left[\left(\frac{L}{V_d} + t_{1d}\right) + \left(\frac{L}{V_u} + t_{1u}\right)\right].$$

11.11 Sources C, D, E, F, and G in fig. 11.3 are 5 km apart. The data in table 11.3 are for three profiles CE, DF, and EG with sources at C, D, and E, no data being recorded for offsets less than 3 km. For profiles from

Table 11.3 *Refraction data*

x	t_{CE}	t_{DF}	t_{EG}
3.00 km	1.18 s	1.20 s	1.19 s
3.20	1.22	1.29	1.28
3.40	1.24	1.38	1.35
3.60	1.28	1.45	1.43
3.80	1.35	1.54	1.50
4.00	1.38	1.60	1.58
4.20	1.41	1.70	1.68
4.40	1.47	1.74	1.76
4.60	1.51	1.77	1.82
4.80	1.53	1.80	1.89
5.00	1.58	1.82	2.00
5.20	1.63	1.85	2.06
5.40	1.65	1.91	2.15
5.60	1.69	1.95	2.21
5.80	1.74	1.97	2.29
6.00	1.78	1.99	2.38
6.20	1.82	2.03	2.43
6.40	1.87	2.08	2.46
6.60	1.90	2.12	2.49
6.80	1.94	2.16	2.54
7.00	1.97	2.20	2.57
7.20	2.01	2.25	2.60
7.40	2.06	2.30	2.65
7.60	2.10	2.33	2.68
7.80	2.14	2.37	2.71
8.00	2.17	2.41	2.74
8.20	2.20	2.45	2.77
8.40	2.24	2.47	2.82
8.60	2.30	2.52	2.85
8.80	2.32	2.55	2.89
9.00	2.35	2.61	2.93
9.20	2.38	2.64	2.97
9.40	2.44	2.68	3.00
9.60	2.47	2.73	3.04
9.80	2.50	2.78	3.07
10.00	2.54	2.82	3.10

Table 11.4 *Time–distance data*

x	t_A	t_B
0.0 km	0.00 s	2.30 s
0.4	0.15	2.23
0.8	0.28	2.15
1.2	0.44	2.09
1.6	0.52	2.04
2.0	0.63	1.98
2.4	0.70	1.92
2.8	0.76	1.85
3.2	0.84	1.80
3.6	0.91	1.72
4.0	0.95	1.64
4.4	1.04	1.60
4.8	1.12	1.55
5.2	1.16	1.47
5.6	1.25	1.40
6.0	1.30	1.32
6.4	1.33	1.28
6.8	1.40	1.24
7.2	1.51	1.18
7.6	1.57	1.10
8.0	1.60	1.04
8.4	1.72	0.96
8.8	1.78	0.90
9.2	1.80	0.83
9.6	1.91	0.76
10.0	1.93	0.66
10.4	2.04	0.52
10.8	2.07	0.39
11.2	2.17	0.25
11.6	2.20	0.12
12.0	2.30	0.00

the Java Sea velocity–depth relation shown in fig. 11.28. Is it feasible to map the top of the 4.25-km/s limestone at a depth of about 0.9 km by the use of head waves? What problems are likely to be encountered?

11.16 In early refraction exploration for salt domes, a "blind spot" (the region B–C in fig. 11.29) was found when the dome lay directly on the line between the source and the geophone, that is, the arrivals were often too weak to the detected. This was called "absorption of the wave" by the salt dome. What is the true explanation for this "absorption"?

11.17 How many distinctly separate head waves are indicated in fig. 11.9? What are their apparent velocities? Calculate the depths and velocities of the respective refractors assuming (a) no dip, (b) 5° dip to the right, and (c) 5° dip to the left.

References

Adachi, R. 1954. On a proof of fundamental formula concerning refraction method of geophysical prospecting and some remarks. *Kumamoto J. Sci.,* Ser. A, **2:** 18–23.

Barry, K. M. 1967. Delay time and its application to refraction profile interpretation. In *Seismic Refraction Prospecting,* A. W. Musgrave, ed., pp. 348–61. Tulsa: Society of Exploration Geophysicists.

Barthelmes, A. J. 1946. Application of continuous profiling to refraction shooting. *Geophysics,* **11:** 24–42.

Barton, D. C. 1929. The seismic method of mapping geologic structure. In *Geophysical Prospecting,* pp. 572–674. New York: American Institute of Mining and Metallurgical Engineers.

Baumgarte, J. von. 1955. Konstruktive Darstellung von seismischen Horizonten unter Berücksichtigung der Strahlenbrechung im Raum. *Geophys. Prosp.,* **3:** 126–62.

Gardner, L. W. 1939. An areal plan of mapping subsurface structure by refraction shooting. *Geophysics,* **4:** 247–59.

Gardner, L. W. 1949. Seismograph determination of salt-dome boundary using well detector deep on dome flank. *Geophysics,* **14:** 29–38.

Gardner, L. W. 1967. Refraction seismograph profile interpretation. In *Seismic Refraction Prospecting,* A. W. Musgrave, ed., pp. 338–47. Tulsa: Society of Exploration Geophysicists.

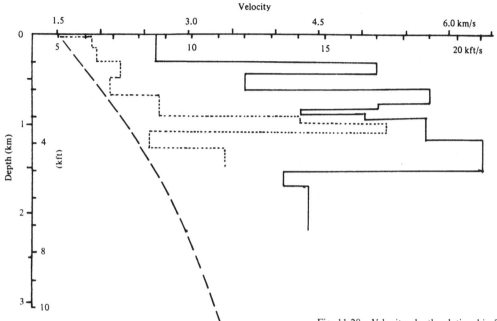

Fig. 11.28   Velocity–depth relationship for wells in the Illinois Basin (solid curve), Java Sea (dotted curve), and Louisiana Gulf Coast (dashed curve).

Table 11.5 *Time–distance data*

x	t_A	t_B	t_A^*
0.00 km	0.000 s	3.310 s	
0.40	0.182	3.182	
0.80	0.320	3.140	
1.20	0.504	3.063	
1.60	0.680	2.917	
2.00	0.862	2.839	
2.40	0.997	2.714	
2.80	1.170	2.681	1.682 s
3.20	1.342	2.570	1.760
3.60	1.495	2.505	1.858
4.00	1.677	2.442	1.881
4.40	1.821	2.380	1.962
4.80	1.942	2.318	2.053
5.20	2.103	2.220	
5.60	2.150	2.125	
6.00	2.208	2.030	
6.40	2.330	2.003	
6.80	2.422	1.862	
7.20	2.504	1.743	
7.60	2.602	1.622	
8.00	2.658	1.610	t_B^*
8.40	2.720	1.482	1.561
8.80	2.744	1.329	1.440
9.20	2.760	1.140	1.288
9.60	2.855	1.018	1.202
10.00	2.920	0.863	1.177
10.40	2.980	0.660	1.082
10.80	3.065	0.503	
11.20	3.168	0.340	
11.60	3.230	0.198	
12.00	3.310	0.000	

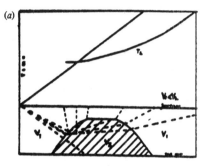

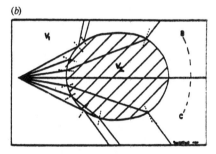

Fig. 11.29   Travelpaths through a salt dome according to Snell's law. (From Barton, 1929.) (a) Vertical section; (b) plan view.

Hagedoorn, J. G. 1959. The plus–minus method of interpreting seismic refraction sections. *Geophys. Prosp.*, **7**: 158–82.

Hales, F. W. 1958. An accurate graphical method for interpreting seismic refraction lines. *Geophys. Prosp.*, **6**: 285–94.

Ingham, A. 1975. *Sea Surveying.* New York: John Wiley.

Johnson, S. H. 1976. Interpretation of split-dip refraction data in terms of plane dipping layers. *Geophysics,* **41**: 418–24.

Laski, J. D. 1973. Computation of the time–distance curve for a dipping refractor and velocity increasing with depth in the overburden. *Geophys. Prosp.*, **21**: 366–78.

Milsom, J. 1989. *Field Geophysics.* New York: John Wiley.

Nettleton, L. L. 1940. *Geophysical Prospecting for Oil.* New York: McGraw-Hill.

Palmer, D. 1980. *The Generalized Reciprocal Method of Seismic Refraction Interpretation.* Tulsa: Society of Exploration Geophysicists.

Palmer, D. 1986. *Handbook of Geophysical Exploration, Vol. 13: Refraction Seismics.* London: Geophysical Press.

Palmer, D. 1991. The resolution of narrow low-velocity zones with the generalized reciprocal method. *Geophys. Prosp.*, **39**: 1031–60.

Rockwell, D. W. 1967. A general wavefront method. In *Seismic Refraction Prospecting,* A. W. Musgrave, ed., pp. 363–415. Tulsa: Society of Exploration Geophysicists.

Schenck, F. L. 1967. Refraction solutions and wavefront targeting. In *Seismic Refraction Prospecting,* A. W. Musgrave, ed., pp. 416–25. Tulsa: Society of Exploration Geophysicists.

Slotnick, M. M. 1950. A graphical method for the interpretation of refraction profile data. *Geophysics,* **15**: 163–80.

Tarrant, L. H. 1956. A rapid method of determining the form of a seismic refractor from line profile results. *Geophys. Prosp.*, **4**: 131–9.

Thornburgh, H. R. 1930. Wavefront diagrams in seismic interpretation. *Bull. AAPG,* **14**: 185–200.

Wyrobek, S. M. 1956. Application of delay and intercept times in the interpretation of multilayer refraction time–distance curves. *Geophys. Prosp.*, **4**: 112–30.

12
3-D methods

Overview

Seismic data are usually collected along lines of traverse that form some sort of grid and the three-dimensional (3-D) picture of structure is deduced by interpolating between the lines. However, features seen on such seismic lines may be located off to the side of the lines rather than underneath the lines and small but important features (like faults) can occur between the lines. These produce errors in interpretation. 3-D surveys endeavor to obtain data uniformly distributed over an area rather than merely along lines, in order to correctly locate the geologic features that produce the seismic evidences. 3-D techniques also significantly reduce spatial noise.

Although 3-D data are not inexpensive to acquire, process, or interpret, their value is becoming widely recognized, and 3-D is one of the fastest growing areas of geophysics. Much 3-D work involves the detailing of fields once oil has been discovered in order to optimize field development and exploitation (Dahm and Graebner, 1982; Galbraith and Brown, 1982), and 3-D has proven very cost effective in this regard. There is nearly unanimous agreement that 3-D surveys result in clearer and more accurate pictures of geologic detail. The costs are more than repaid by the elimination of unnecessary development holes and imprudent investments, by the increase in recoverable reserves through better location of production and injection wells, and by the discovery of isolated pools in a reservoir that otherwise might be missed. 3-D is also being used increasingly for exploration before the drilling of an exploration well, for the same reasons. The cost-effectiveness of 3-D seismic methods for Shell International and Exxon is discussed by Nestvold (1992) and Greenlee, Gaskins, and Johnson (1994).

The techniques for acquiring 3-D data, discussed in §12.1, are usually quite different in the marine environment from those on land. Most large 3-D surveys have been marine because generally investments are larger and risks greater and hence the need is greater for improved field definition. However, use of 3-D on land is growing as the major improvements that it provides become more widely recognized.

Except in a few very important regards, mainly dealing with migration, processing 3-D data (§12.2) is not very different from that of 2-D processing. However, displaying 3-D data (§12.3) so that an interpreter can understand and extract the significant elements provides challenges that have led to a number of important new display techniques. In order to display and manipulate the huge volume of data involved in a 3-D survey, most 3-D interpretation is done at workstations (§12.4).

The interpretation of 3-D data to extract structural and stratigraphic information is discussed in §12.5. One of the best ways to improve interpretation is to review case histories, a number of which have been published. Brown (1991) is the best single reference on 3-D techniques. Applications of 3-D to reservoir delineation are given in Sheriff (1992).

12.1 3-D acquisition

12.1.1 Acquisition requirements

We would like to attribute variations seen in our seismic data to geologic factors rather than acquisition factors. Thus, ideally we wish to have uniform acquisition conditions and a uniform surface distribution of CMPs, that is, (1) data distributed on a uniform grid with (2) the same CMP multiplicity (3) utilizing the same mix of offset distances and (4) the same mix of azimuths.

3-D seismic data can be acquired in a number of ways. The usual method in marine work is to run a series of closely spaced lines; in transition and in some land work, to have geophones laid out in two or more parallel lines while the source moves about between the lines; in other land work geophones are laid out on lines at right angles to source lines. Whatever the method of acquiring the data, the goal is to achieve uniform sampling over the area. A uniform grid of rectangular *bins* is set up and data are included within a particular bin if the midpoint falls within that bin. The data within each bin will be subsequently stacked and migrated. To avoid bias, each bin should contain the same number of traces and with the same set of offsets. Marine work usually satisfies this uniformity requirement reasonably well, but land work often does not achieve it. The direction in which the survey is run may introduce a slight bias. The bins should be small to avoid losing resolution, but not so small that many bins turn up empty. The output traces for empty bins are usually synthesized by averaging the stacked traces from adjacent bins.

The sampling should be dense enough (bin size small enough) to avoid aliasing during processing and interpretation (§8.3.10). To avoid spatial aliasing, at least two surface samples should be obtained for each apparent wavelength present. The apparent wavelength is smallest for the highest frequency v_{max} and the steepest dip ξ_{max}; thus the maximum spatial sampling Δx is given by

$$\Delta x \leq \frac{V}{4 v_{max} \sin \xi_{max}} \qquad (12.1)$$

The surface sampling intervals should generally be 20 to 100 m, the subsurface sampling being 10 to 50 m, and the sampling in the dip direction is often greater than in the strike direction. Vertical sampling is usually 2 ms (3 m at 3000-m/s velocity), so the dimensions of a potentially resolvable volume cell (*voxel*) is of the order of $3 \times 15 \times 25$ m³ assuming group interval and line spacing of 30 and 50 m. Because this volume is so small, the subsurface detail present in 3-D data is very great.

Determining the small bin size required to prevent aliasing requires some knowledge of the geology and achieving small bin size may be expensive. In practice, compromises are made and design parameters are based partly on the available hardware and the nature of the objective. Geological constraints (for example, the preknowledge that dips are small, faults are not closely spaced, and it is not necessary to image fault-plane reflections) may permit relaxing the spatial aliasing constraints. Because most data are migrated and spatial aliasing creates noise in migration, this often becomes the limiting constraint. Intelligent interpolation (§9.11.2) can be used to relax the sampling constraints imposed by migration, but it does not improve the intrinsic resolution. Resolution of small features such as river channels generally requires three or more samples within the small feature.

Events migrate updip, so if the dip near the edge of the prospect is away from the prospect, data must be collected over a fringe area surrounding the desired subsurface area that is to be imaged. The width of this fringe area M, assuming straight raypaths, is $(1/2)Vt \sin \xi$; however, raypath curvature enlarges this and we approximate the migration distance as

$$M = \tfrac{1}{4} Vt \sin \xi, \qquad (12.2)$$

where V is the average velocity, t the two-way record time, and ξ the dip.

In order to properly determine seismic amplitudes (needed for hydrocarbon indicator and stratigraphic studies) by the migration process (which effectively sums amplitudes over the Fresnel zone, §6.2.3), the fringe area should be further enlarged by the radius of the first Fresnel zone (eq. (6.6a)),

$$R = \frac{V}{2} \sqrt{\frac{t}{v}}. \qquad (12.3)$$

To achieve full multiplicity, we must add the line-end taper L (half the streamer length in marine work). Thus, in order to migrate properly and achieve proper amplitude and multiplicity, the fringe width x_{fringe} should be

$$x_{fringe} \geq \frac{1}{4} Vt \sin \xi_{max} + \frac{V}{2} \sqrt{\frac{t}{v}} + \frac{L}{2}. \qquad (12.4)$$

Note that 3-D multiplicity usually does not have to be as large as required for 2-D work, half the required 2-D multiplicity often being quite enough.

The *sparse 3-D* or *exploration 3-D* technique is sometimes used to decrease 3-D acquisition costs, but data quality is also decreased. Lines are spaced four or five times more widely apart than the sampling theorem requires and the omitted lines are interpolated during processing by intelligent interpolation (§9.11.2), a process based on seismic character. Intelligent interpolation prevents aliasing during 3-D migration; however, the resulting data have only the lower resolution determined by the spacing of the lines as collected rather than that after interpolation.

Sometimes irregular grids of 2-D lines are transformed into a regular 3-D grid by filling in the undersampled grid loops using intelligent interpolation. This approach yields data of even poorer resolution than the exploration 3-D technique; again, the resolution is limited by the original data density. Interpreters of 3-D data should take into account how the data were collected and processed so that unrealistic expectations of subsurface detail can be avoided. In general, widely spaced 3-D data are inappropriate for reservoir studies.

12.1.2 Marine 3-D acquisition

Marine 3-D data are generally acquired by a boat towing a hydrophone streamer and an array of air guns while it traverses back and forth across the area being surveyed (fig. 12.1a). In conventional marine acquisition, one line of data is acquired on each traverse. Lines are normally oriented in the dip direction so that the spacing in this direction (the hydrophone group separation) is smaller than that in the strike direction (the line spacing).

The demand for less expensive 3-D surveys led to boats towing two or more streamers (fig. 7.24), which are pulled to the sides by paravanes, to collect two or more lines of data simultaneously. Sometimes two air-gun arrays separated by paravanes are fired alternately to double again the number of lines collected on a traverse (figs. 12.1b and 12.1c). As many as six parallel streamers have been towed by a single ship. Sometimes two acquisition ships, each towing two (or three) streamers and one or two air-gun arrays, with the source arrays firing one at a time, are used to collect up to 12 (or 18) lines of data with one pass. (However, because only one source can be used at any time and the ships are traveling continuously, this increases the minimum bin size and changes the mix of traces between different bins.)

When data are acquired on parallel lines, appreciable time is lost while the boat moves from one line to the next (as in fig. 12.1a, usually at least 1 hour). Boats sometimes collect 3-D data in circles (fig. 12.2), either successively overlapping circles of the same radius (like a stretched and flattened coil spring) or in a spiral, to minimize lost time.

Navigation accuracy must be sufficient to place data in the proper bins. This requirement applies to determining the positions of all elements of the acquisition system, including those of the sources and each hydrophone group. Positioning of the ship is usually accomplished by radiotrilateration from fixed base stations (§7.1.3) supplemented by GPS (§7.1.5); the boat location is always determined redundantly, hopefully, to an accuracy of 10 m.

Positioning the hydrophone groups relative to the boat is generally determined from a series of compasses in the streamer (which often drifts off line by a few degrees) and from measurements of the location of the tail buoy. These are often supplemented by measurements with *pingers,* transducers that send out acoustic pulses whose traveltimes to tuned receivers on the boat or in the streamer are measured and used to locate the boat, sources, and streamers with respect to each other (§7.1.7). Anchored pingers (§7.1.6) are sometimes used to provide locations for survey of a restricted area. A huge, highly overdetermined volume of navigation data results in considerable redundancy. The volume of navigation/positioning data may exceed the volume of seismic data. The navigation data reduction to determine which traces fall into which bins is done by computers both in real time and also in postprocessing.

3-D data can be obtained in a one-boat operation if the streamer is pulled at an angle to the seismic line. Where natural ocean currents are present, the seismic line can be oriented perpendicular to the current so that the streamer is pulled off-line. Figure 12.1d is a plot of streamer location for a line in such an area. However, one has relatively little control over the amount of cross-coverage obtained and the number of traces in different bins is apt to be irregular. Also, the traces that fall into bins farther from the line are those with longer offsets, producing a bias that may make residual normal moveout appear as fictitious cross-dip.

Collecting 3-D data in an area of platforms or other obstructions involves special problems in avoiding obstacles. Sometimes separate source and recording boats are used so that the area adjacent to a platform can be undershot.

In shallow water and where obstacles interfere with towing streamers, a *swath technique* using geophones or hydrophones laid on the sea floor may be employed. Two or more bottom cables containing receivers are laid in parallel lines and connected to an anchored recording barge. A source boat then moves over the area between them (fig. 12.3) to acquire the 3-D data. All or parts of the layout are then moved and the operation is repeated to extend the survey area.

12.1.3 Land 3-D acquisition

The cost of a 3-D land survey is generally proportional to the number of source points involved, and hence surveys usually employ many more geophone groups than source points. Cost-effective design is discussed by Rosencrans (1992) and Bee et al. (1994).

Data are often acquired using perpendicular source and geophone lines (fig. 12.4a), which is called *patch* shooting. This arrangement gives singlefold coverage over a rectangle half the length of the geophone line by half the length of the source line. However, the offset distances and azimuth directions vary systematically over the midpoint coverage, which may introduce bias in processing and interpretation. Differences in normal moveout may be interpreted as dip and vice versa, and the lack of multiplicity at midpoint locations may cause problems in determining static corrections. The source point and geophone station locations do not need to be the same, so the midpoint locations are not necessarily equally spaced.

In order to achieve multiple coverage, several geophone and source lines may be used. Commonly, four or six parallel geophone lines are used with perhaps six or eight source lines perpendicular to the geophone lines, especially when the Vibroseis method is used. The use of parallel geophone lines builds up multiplicity in one direction and the use of parallel source lines builds multiplicity in the perpendicular direction, so that the total multiplicity achieved is the product of geophone-line and source-line multiplicities. Two sets of vibrators employing different sweeps (§7.3.1e) can be used simultaneously to nearly double the speed of data acquisition. The minimum bin size is half the geophone group interval by half the source-point interval. Multiplicity can also be increased by increasing the bin size; for example, doubling linear dimensions quadruples the bin area and the multiplicity (assuming the same distribution of CMPs).

After the data for one patch are acquired, the patch is moved to the next location (for example, to the north) and the operation is repeated. Successive patches often overlap to further increase the multiplicity over a strip (a *block*) of the subsurface. Successive blocks may also overlap (for example, to the east), still further increasing the multiplicity. Multiple fold in both line and cross-line directions aids in determining static corrections.

Land data are sometimes also acquired by a *swath* technique wherein several parallel geophone lines are used with one or two source units in a roll-along technique similar to that used in conventional 2-D CMP acquisition (fig. 12.4b). This results in a swath of 3-D

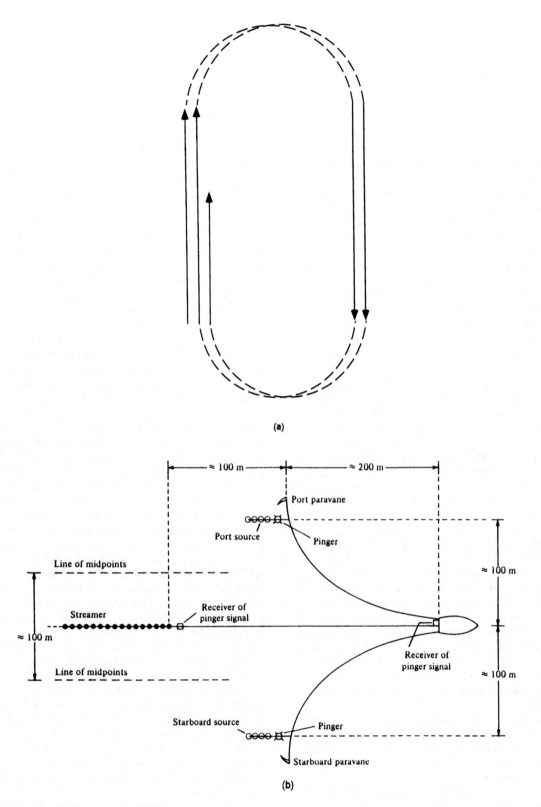

(a)

≈ 100 m

≈ 200 m

Port paravane

Port source

Pinger

Line of midpoints

Streamer

Receiver of
pinger signal

≈ 100 m

≈ 100 m

Line of midpoints

≈ 100 m

Receiver of
pinger signal

Starboard source

Pinger

≈ 100 m

Starboard paravane

(b)

Fig. 12.1   Arrangements for 3-D marine acquisition. (a) After recording a line, ships generally turn with a fairly large radius because of their long tows in order to record the next line, and then double back to achieve the close line spacing required; solid lines show where recording was done and dashed lines indicate moves between lines. (b) Use of paravanes to tow units to the side of the ship. Arrangement shows two source arrays and a single streamer to give two lines of midpoints. (c) Typical spacing when towing two sources and two streamers to acquire four lines of midpoints; the two source arrays fire alternately; $1 \rightarrow A$ indicates locations of midpoints from source 1 into streamer A, and so on. (d) Feathering of streamer due to cross-current. Location of boat and 48 streamer groups (plotted for every 25th station).

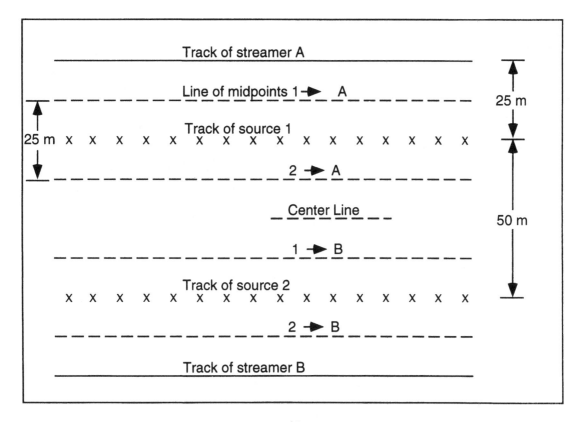

(c)

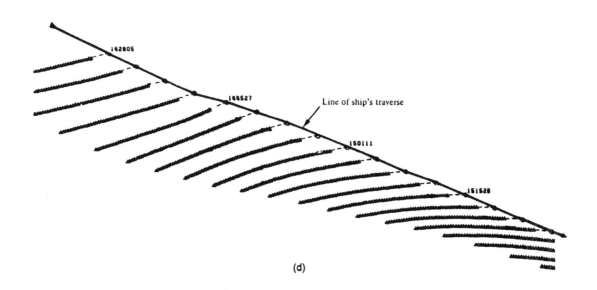

(d)

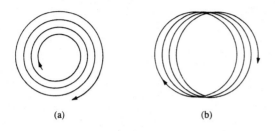

Fig. 12.2 Marine acquisition using circle shooting. (a) Spiral for shooting around a salt dome and (b) overlapping circles.

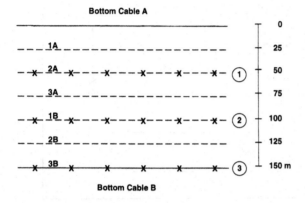

Fig. 12.3 Swath technique for obtaining lines of midpoints spaced 25 m apart with two lines of geophones in bottom cables (*A* and *B*) 150 m apart, using three lines of sources (1, 2, 3) spaced 50 m apart. Lines of midpoints are dashed; 1 *A* indicates locations of midpoints from source 1 into bottom cable *A*, and so on. Following this acquisition, cable *A* may be moved 150 m beyond *B* to acquire the next swath.

coverage whose width is half that of the geophone line spacing times the number of parallel geophone lines. With 12-fold in-line coverage and two parallel lines of source positions, 24-fold multiplicity can be achieved. As with the use of patches, successive swaths often overlap to increase the multiplicity.

There are a number of variations of the patch and swath techniques. Sources sometimes move along short segments between geophone lines (fig. 12.4c). Then, after one block of data has been recorded, some of the geophone lines are moved to cover the next block, and the source-line segments for it are staggered with respect to those for the preceding block, creating a "brick-wall" pattern. Sources sometimes "zig-zag" between geophone lines. These variations are used to improve the mixes of offsets or azimuths within bins while improving field efficiency. Geophone groups are sometimes arranged in a checkerboard "button patch" arrangement as sources traverse parallel lines.

In areas of limited access, data are sometimes collected with sources and receivers placed around rectangles (fig. 12.4c). This *loop* method permits acquiring data over the interior of the rectangle, which may not be accessible, and it provides multiple coverage along

the periphery and the bisectors that divide the rectangle into quadrants. However, it gives low multiplicity over most of the area and poorly determined static corrections, and so generally provides poorer data. The loop method can be used around irregular loops to acquire data in areas of difficult access; of course, some degradation results from the consequent irregular distribution of midpoints.

The land 3-D acquisition methods generally sacrifice some uniformity of CMP coverage and uniformity of offset distances and azimuths represented within the bins. Some loss of CMP multiplicity can be tolerated because the data in nearby bins in many directions can be consulted in deciding which events are to be honored. As a rule of thumb, 3-D work requires only half the multiplicity of 2-D work to achieve comparable data quality. In many land areas, access restrictions interfere with any of the foregoing acquisition methods.

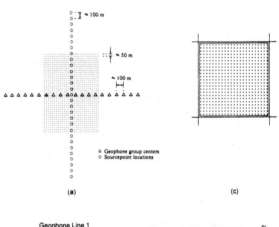

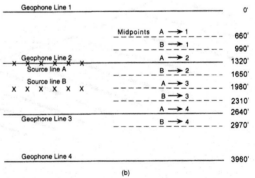

Fig. 12.4 Arrangements for 3-D land acquisition. (a) Perpendicular lines of sources and geophones acquire data over rectangle with half the dimensions of the source and geophone lines; often, several lines of geophones and sources are used to give multiple coverage. (b) Arrangement for acquiring a block of data using four parallel geophone lines with two vibrator sources operating simultaneously to produce swath of eight midpoint lines; for the next swath, geophone lines 1 and 2 will be moved to 5 and 6 and vibrators will move down line 4 and midway between 4 and 6. (c) Loop layout with geophones and source locations around perimeter of an area; midpoints are shown as dots; 40 geophone groups and 40 source locations might be located at 160-m intervals around the four sides of a square 1.6 km on a side.

Land 3-D acquisition methods also may lead to poor static corrections, so acquisition arrangements usually include some 2-D CMP portions so that conventional statics analysis can be used to provide a framework for determining statics for the remaining bins. Because the sources are at different azimuths to the geophone arrays, array-directivity patterns will differ; windmill arrays (fig. 8.14g) are occasionally used to minimize array-directivity effects.

Field design often has to cope with access problems that prevent any land-acquisition system from being completely uniform. Field computers do the book-keeping for the sometimes very complicated adjustments that are required. They usually include programs to display the areal distribution of multiplicity, offsets, and azimuths within bins. These are used to analyze the effects of arbitrary source and geophone geometry changes to assist in designing compromises to achieve better distributions while the crew is still acquiring the data.

12.1 3-D processing

The processing of 3-D data is similar to that of 2-D data in most regards. CMP stacking attenuates multiples and improves the signal-to-noise ratio. Dip-moveout (DMO) correction (§9.10.2) in 3-D is required (in addition to the normal-moveout (NMO) correction) to remove reflection-point smear. DMO shifts data into different bins prior to migration. DMO and migration require that the midpoints have been determined and the data binned. There is often a delay between the end of a survey and mapping the binned CMPs, but this information is now being obtained in real time on most marine surveys and many land surveys.

A number of processing steps do not depend critically on the CMP locations. These include testing the data integrity, gain recovery, deconvolution, muting, scaling, and velocity analysis. 2-D *brute stacks* (stacks made without corrections or refined processing), although not of the quality ultimately expected from the survey, can give an indication of data quality. Preliminary time slices made of the near-trace data help locate tares that suggest positioning problems that need to be examined more carefully. These operations are sometimes done in the field.

Velocity analyses are usually run in conventional 2-D fashion on a rather coarse grid over the prospect in order to make 3-D velocity maps. These incorporate also preliminary structural information and velocity data that are available from wells. Data should be re-examined where the velocity maps are not consistent with the structure.

Marine data may need a static corrections for tidal variations. Static corrections are almost always required with land data; both conventional 2-D refraction and surface-consistent statics are often used. Where there is sufficient redundancy in both in-line and cross-line directions, 3-D statics analysis may be used.

After being binned, DMO corrected, and stacked, data are migrated (fig. 12.5) to reposition dipping reflections, collapse diffractions from discontinuities, focus energy spread over Fresnel zones (§6.2.3), and nearly eliminate out-of-the-plane noise. One to three additional traces are often interpolated (§9.11.2) to avoid aliasing in the migration process. Migration greatly reduces spatial noise and helps clarify and correctly locate structural anomalies; it also greatly improves stratigraphic resolution by imaging the data dispersed over the Fresnel zone. Migration is the major factor in improving the signal-to-noise ratio and interpretability of 3-D over 2-D data. Sometimes migration is performed before stacking.

Usually, in-lines in the dip direction are migrated first and then traces are sorted and migrated in the cross direction. This procedure closely approximates true 3-D migration where the velocity field is simple. Sometimes a more expensive and more accurate 3-D migration is done in one operation. Figure 12.6 shows several examples of the improvement of 3-D over 2-D migration. Eliminating the energy coming from out of the plane of the section by 3-D migration often results in major data-quality improvement.

The accuracy of migration depends on the accuracy of the velocity field used. Some 2-D lines selected from the 3-D volume may be migrated with different velocities to help select migration velocities. Where structure is complex, depth migration may be required, but inadequate knowledge of the velocity field may limit the benefits of depth migration. Occasionally, prestack depth migration is performed.

Wavelet processing to reshape the embedded wavelet is an important aspect of 3-D processing. Interpreters generally prefer zero phase for maximum interpretability (Brown, 1992) and hence conversion to a zero-phase embedded wavelet is often done late in

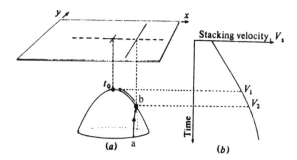

Fig. 12.5 Illustrating 3-D migration. The data to be migrated lie on a hyperboloid appropriate to the stacking velocity V_1; the aperture (§9.1.31) should be centered near t_0. 3-D migration is often done by first doing 2-D migration in the y-direction, sorting, and then doing a second 2-D migration in the x-direction; this economizes on computing time. The effect is to move data from A to B in the first migration and then from B to t_0 in the second migration. If stacking velocity V_2 is used for the first migration rather than the correct V_1, the error created is usually small. The aperture in the first migration would also be different than an aperture symmetrical about t_0. (a) Isometric sketch of (x, y, t) space and (b) stacking velocity versus time.

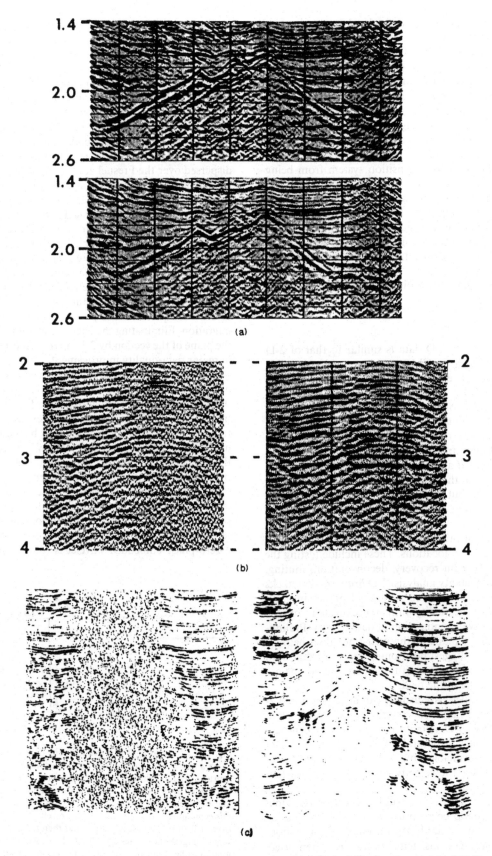

Fig. 12.6 Improvement resulting from 3-D migration over 2-D. (From Brown, 1991: 8, 9, 10.) (a) Effect of 2-D (above) and 3-D migration (below); (b and c) improvement in signal-to-noise ratio achievable by 3-D migration (right) of CMP stacked sections (left).

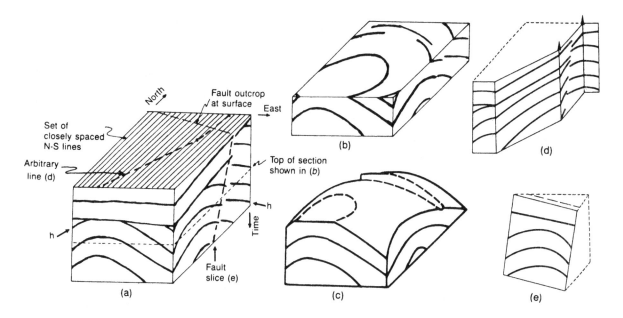

Fig. 12.7 Three-dimensional data obtained from a set of closely spaced N–S lines. (From Sheriff, 1991: 300.) (a) Isometric diagram of the volume these data occupy. The easternmost N–S section is shown (a "line") along with an E–W section (a "cross-line") made from the southernmost traces on each N–S line. (b) The data set with the top portion removed; the top now constitutes a time slice. (c) The data sliced along one reflection (h) constitutes a horizon slice. (d) An arbitrary line cuts through the data volume, perhaps to connect well locations. (e) A fault slice runs parallel to a fault displaced a small distance from it.

the processing sequence. Brown (loc. cit.) emphasizes the importance of zero phase in 3-D interpretation. A zero-phase wavelet is symmetrical with most of its energy concentrated in the central lobe, so reflections are easier to recognize in the presence of other events and noise. Compared to a minimum-phase (or 90°) wavelet, zero phase gives only one major trace excursion rather than two, so that there is less ambiguity in determining the polarity and identity of a reflection. However, ascertaining the phase of an embedded wavelet is extremely difficult in most instances.

12.3 Display of 3-D data

A 3-D seismic survey provides a volume of data for interpretation. If the sampling has been adequate, each data point in the volume is less than half of the apparent wavelength from adjacent data points in every direction and there are no areas where aliasing occurs; a more accurate and consistent interpretation should therefore result. Sampling is, however, often not close enough, and an interpreter may have to factor this fact into his interpretation.

The 3-D volume can be sliced in various ways (fig. 12.7), for example, along three orthogonal planes producing *lines* and *cross-lines* (vertical sections) and *time slices* (also called *horizontal sections* and Seiscrop™ sections) (fig. 12.8). The 3-D volume may also be sliced vertically along diagonal lines connecting wells, or in the plane of a deviated well, or in zig-zag lines to tie wells (*arbitrary lines*). A volume can also be sliced along a reflection (an interpreted horizon) to show the spatial distribution of amplitude over this

horizon; such *horizon slices* are used for studying stratigraphy and reservoir properties. Horizon slices are also made along surfaces parallel to nearby picked horizons; that may be picked more reliably than the horizon of interest (see plate 14). This method helps show polarity and phase changes that may be produced by stratigraphic features and that might cause difficulties with tracking the particular horizon. The volume can be sliced parallel to a fault surface (*fault slices;* see fig. 12.20 and plate 5) to study fault-related structure, secondary faulting (fig. 12.19), and the sealing ability of faults. Combinations of vertical, hori-

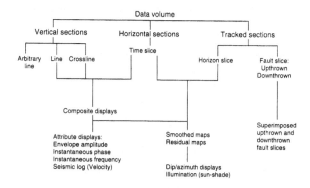

Fig. 12.8 Nomenclature for 3-D products. The data volume can be sliced vertically, horizontally, or along (or parallel to) tracked horizons. Composite displays are combinations of vertical and horizontal slices such as shown in plate 7. Transformed data produce attribute displays, including dip/azimuth (see plate 6) and illumination displays.

zontal, and other sections (plate 7) permit an interpreter to appreciate the three-dimensionality of the subsurface and interpret the data more meaningfully. Combination displays (plate 7) can be made easily with an interactive system. Interactive systems also permit isometric displays and the rotation of isometric displays so that they can be viewed from different vantage points.

A time slice shows all the seismic events for a particular seismic time (or depth if time–depth conversion has been performed). The attitude of a feature on a time slice indicates its local strike. Picking a reflection on a time slice is equivalent to mapping a time contour, and successive contours can be mapped directly by following the same event on successive time slices (fig. 12.9). This procedure thus yields a structural contour map without the intermediate stages of timing and posting. Likewise picking a fault on a sequence of time slices yields a fault-surface map directly. Typically the number of interpretable faults increases significantly (fig. 12.10), perhaps 10-fold. The ability to see the strike and continuity of faults often results in markedly changed interpretations. The shape of the structural contours generally reveals fine detail that cannot be extracted with conventional mapping techniques; therefore, an interpreter of 3-D data should not expect contours to be as smooth as those on maps made from 2-D data.

Attributes of various kinds can be calculated for 3-D data to create attribute volumes that can then be sliced through just like the 3-D data volume itself. The common attributes such as envelope amplitude, instantaneous phase or frequency (§9.11.4), or acoustic impedance (or velocity) generated by one-dimensional inversion (seismic logs; see §5.4.5) can be displayed. Automatically tracked structural contour maps may be manipulated to reveal subtle lineations indicating faults (fig. 12.11). These manipulations include (a) calculating vertical derivatives of a horizon surface to display the attributes of *dip magnitude* and *dip azimuth* (plate 6), (b) smoothing and subtracting a smoothed version from the data to yield a *residual*, (c) subtracting the arrival times or amplitudes of successive horizon slices to yield difference displays, (d) effectively illuminating a horizon slice with light in such a direction that shadows emphasize features such as faults (*sun shade* or *artificial illumination* displays), and (e) multiplying a time map by a velocity map to give a depth map.

Color is an important factor in 3-D displays. Color increases the visual dynamic range and permits us to see and interpret more detail. A gradational color scheme is generally used in which two distinctly different colors or ranges of colors represent peaks and troughs, with the intensity proportional to amplitude values. The most commonly used colors are red for positive reflections (peaks on SEG-standard zerophase sections) and blue for negative ones, with the color intensity proportional to the amplitudes. Contrasting color schemes can be used to emphasize small

changes in the values displayed; however, these can be misleading if the values selected for the color contrasts are not well-chosen. Introducing a bias in the color assignment, as in plates 14 and 15, can also emphasize features. The channel in the horizon slice of plate 15 is more clearly defined by limiting the range of red values.

12.4 Interactive 3-D interpretation

Most 3-D interpretation is done at interactive workstations. The data are held on magnetic or compact (optical) disk and are viewed and interpreted on the screen of a color monitor (TV tube). The benefits of working interactively are many. Much time is saved by eliminating the paper handling associated with conventional methods. The interpreter can compose the data display or combination of displays (plate 7) that he thinks best suits study of the problem at hand. Color is available as a standard display and colors can be chosen to suit individual preferences and specific problems. An interpreter can move through his data quickly, with fewer distractions, and maintain a productive idea flow. With the incorporation of well and other data into data bases, more information (for example, well logs) can be reviewed easily to see how it contributes to understanding the seismic data, so that the final interpretation should be more consistent and hence better. Because interactive systems make it so easy to perform checks, more checking is performed and interpretational inconsistencies can be seen much easier. Interactive 3-D interpretation does not necessarily accomplish quicker interpretation, but it is more complete and accurate.

An important function of an interactive system is *horizon tracking*, mapping the same reflection throughout the 3-D volume. Usually, several modes are available. Manual tracking, also known as *point* or *stream tracking*, depends only on the judgment of the interpreter as he moves a cursor over the data. *Automatic tracking* of a chosen reflection is efficient in most cases, but it has to be monitored to make certain it does not jump to the wrong event at faults, unconformities, or where data quality deteriorates. Automatic tracking is highly desirable for amplitude studies because the computer can track the maximum amplitude (which is very difficult for a person to do) or fit a smooth curve to amplitude values in order to interpolate the maximum amplitude where it does not fall on a sample value. The computer can then store the time and amplitude values for later use. These times and amplitudes can relate directly to the reflectivity contrast if the data are zero-phase, the event has a good signal-to-noise ratio, and the same event has been tracked everywhere. Automatic horizon tracking of zero-phase data over a reservoir reflection yields a horizon slice of the reservoir interface. Because it may be difficult to track a horizon that involves changing contrasts, especially if the changes involve reversing polarities, tracking is often done on some nearby horizon believed to be conformable with

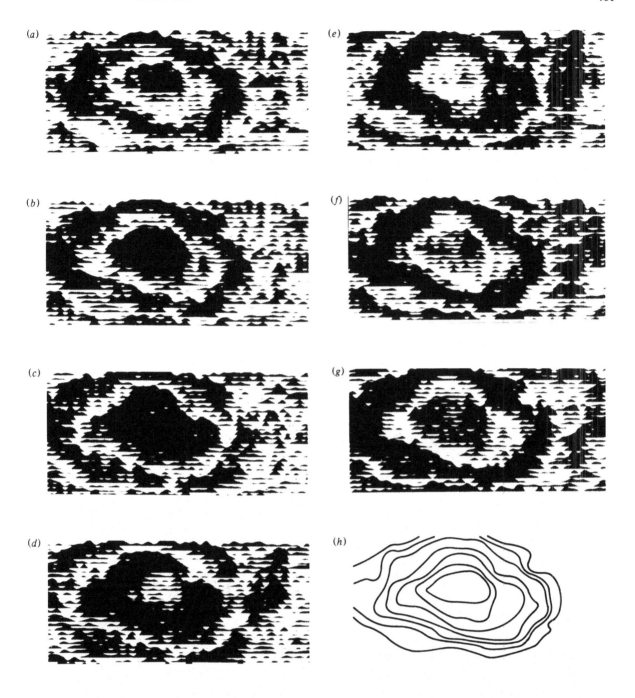

Fig. 12.9 Successive time slices through a 3-D data volume. Lineups on a time-slice map constitute a time contour. Tracing the same event on successive time slices develops a structure contour map. (After Brown, 1983: 1183.)

the desired horizon and then the horizon slice is made a fixed time below the tracked horizon (as in plate 14).

12.5 3-D interpretation

The most complete interpretation requires the use of all types of sections to help an interpreter see and extract maximum subsurface detail. 3-D displays generally reveal enormously greater detail than was formerly believed present in seismic data.

Faults are recognized on time slices in the same way as on vertical sections, mainly by reflection discontinuities and offsets (fig. 12.11, and plates 8 and 9). Just as picking a fault on a seismic section is difficult where the fault nearly parallels the seismic line (so that the fault surface nearly parallels reflection events), picking faults on time slices is difficult where the fault nearly parallels reflection events. However, the use of both vertical sections and time slices removes most of the problems in fault picking. Dip and azimuth dis-

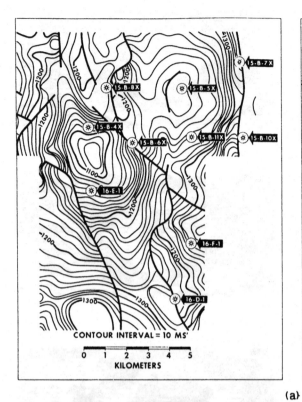
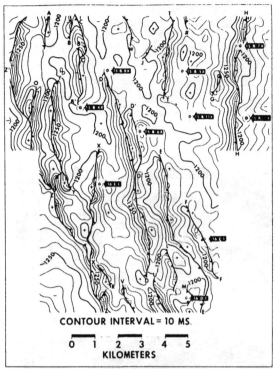

CONTOUR INTERVAL = 10 MS'
KILOMETERS
0 1 2 3 4 5

CONTOUR INTERVAL = 10 MS.
KILOMETERS
0 1 2 3 4 5

(a)

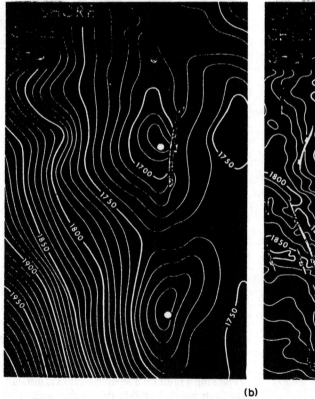
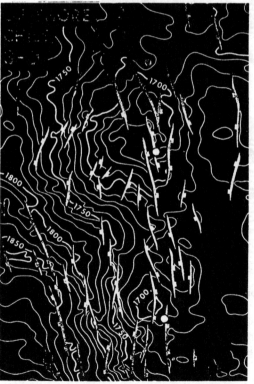

(b)

Fig. 12.10 Structure maps resulting from 2-D (left) and 3-D (right) surveys showing the additional detail derived from the 3- D surveys. (From Brown, 1991: 57, 58.) (a) Gulf of Thailand survey, and (b) Chile survey.

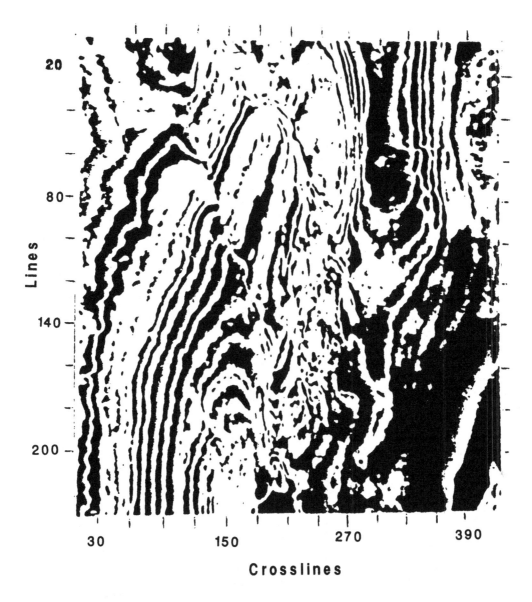

Fig. 12.11   Time slice showing offset of events indicating faulting. (From Brown, 1991: 61.)

plays (plate 6) often reveal small faults because the dip along horizons often changes in the immediate vicinity of faults. 3-D data have revealed that many faults previously mapped as continuous are actually en-echelon systems (see fig. 12.10b).

Horizon slices are made routinely by interactive systems. Guided or automatic tracking usually follows the maximum amplitude of a reflection; the times generated by the tracking provide a time-structure map and the amplitudes provide a horizon slice. A horizon slice shows how the reflectivity varies along the reflector. It yields an uninteresting picture for constant reflectivity, of course, but where the reflectivity varies, the pattern may suggest the reason for the reflectivity variation (Enachescu, 1993). Obviously, the value of a horizon slice depends critically on the correctness of the structural interpretation on which it was based. The horizon for the slice shown in plate 8 was cut by

a number of faults, and the continuity of the stream pattern on the horizon slice confirms that the correlation across the faults is correct and that the faults postdate the unconformity.

The most striking features revealed by horizon slices are stratigraphic because the map-style view often shows distinctive shapes relatable to depositional systems. Views of ancient deposition on horizon slices sometimes resemble the views of modern deposition seen from an airplane: channels, offshore bars, point bars (sand deposits in river meanders), crevasse splays (where a river broke through its levee during a flood), sand bars, karst topography, depositional edges, pinchouts, and so on. The pictures revealed on horizon slices often are almost impossible to see by interpreting vertical sections alone.

The patterns in fig. 10.40 and plates 8, 10, 15, and 16 are easily identified as stream channels. A horizon slice along an erosional unconformity may show on-lap patterns above the unconformity or the subcrop pattern below the unconformity (plate 11). The horizon slice in plate 12 shows a turbidite fan (containing gas).

Color display is valuable for showing trend patterns. In plate 12, the dark reds indicating the highest amplitudes locate the zones of better reservoir quality (producibility). The low-amplitude lineations running ENE to E are faults. The high-amplitude lineation running approximately WNW is caused in part by tuning (§6.4.3) between the reflections from the top of the reservoir and the fluid contact, as the reservoir thins to the north. Similarity of the amplitude patterns of the reflections associated with the top and base of the reservoir tends to confirm that the patterns are caused principally by lateral variations within the reservoir rather than in the host rock. Amplitude patterns often can be interpreted in terms of the reservoir extent or lithology changes, variations in reservoir quality (especially porosity) or thickness (if below tuning thickness). An amplitude lineation that follows structural strike and/or is parallel to the downdip reservoir limit may be a tuning phenomenon (§6.4.3 and 14.5), indicating where the reservoir thickness equals the tuning thickness. With knowledge of the reservoir thickness, tuning effects can be subdued to produce "detuned" horizon slices.

Fault slices are usually made by slicing through a 3-D volume parallel to a fault surface but displaced a small distance (perhaps 25 to 50 m) into the down-thrown and upthrown fault blocks (fig. 12.12a). This is done to avoid the distortions often present very near the fault itself. The shifts between correlative events on these two slices gives a map of how the throw on the fault varies along the fault surface (fig. 12.12b; see also plate 5). If the lithologies associated with the different events seen on the fault slices can be identified, perhaps based on nearby well control, then perhaps it can be determined which potential reservoirs are juxtaposed against impermeable rocks across the fault and thus where the fault will seal and provide trapping conditions. The additive colors achievable by superimposing displays in complementary colors are useful in some situations, for example, in superimposing up- and down-thrown fault slices to see where lithologies are juxtaposed against each other across a fault. 3-D seismic data are also used to predict sealing at faults because of the smearing of ductile clay (Bouvier et al., 1989; Jev et al., 1993). Fault slices sometimes show up splinter faults that were generated by the stresses within a fault block as it moved along a curved fault surface (see plate 13).

Workstations may permit restoring the 3-D volume to its former situation when a particular horizon was being deposited, based on the assumption that the horizon was horizontal at the time of deposition. This involves not only picking the horizon and identifying it in different fault blocks, as is done in constructing a horizon slice, but also moving the deeper reflections

("unfaulting the faults") to remove the fault heave and throw (see fig. 10.26a). The resulting restored volume can then be examined in order to work out the history of deformations and the effects deformations may have had on paleo fluid flow.

Amplitude anomalies attributable to hydrocarbon accumulations show in much the same fashion as in vertical sections. The patterns in plates 4, 10, and 12 show the amplitude variation of a hydrocarbon indicator and thus outline the productive areas. Horizon slices through a reservoir zone sometimes delineate areas of maximum porosity. This is the case with the lighter colors extending SSE from well L-7 in plate 11, where the reflectivity of one subcropping member of the Lisburn formation produces a dim spot. The limiting factors on the utility of horizon slices are the resolution, signal-to-noise ratio, and amplitude and phase preservation.

Clearly, reflection events must be correctly identified in order to make useful horizon maps. Data from boreholes are usually used to identify correctly reflection events. Synthetic seismograms, seismic logs, and vertical seismic profiles are all used. Figure 12.13 shows the correlation between a VSP and 3-D seismic data. A multitude of productive reservoirs from a vertical sequence of channel deposits exist in this area.

Plate 17 shows two horizon slices separated by only 4 ms (about 16 ft), which is about $\lambda/10$ of the dominant frequency and well below the usually stated theoretical resolvable limits of $\lambda/4$. The differences between these slices relate to the separate channel systems. Pressure history, bottom-hole pressures, and interference tests (where something, usually induced changes in pressure, is done to one well to see if it affects another well) indicate that a number of the reservoirs that produce from roughly the same intervals are not connected (see fig. 12.14). The pressure data help in interpreting the seismic data and vice versa (Hardage, 1993).

Once reservoir reflections of sufficient signal-to-noise ratio have been identified, an attempt can be made to interpret their amplitude changes in terms of reservoir properties. Without well control, the effects of properties affecting amplitude (lithology, nature of fluid, pressure, hydrocarbon saturation, porosity, ratio of net-to-gross reservoir thicknesses, and tuning) are usually inseparable, but often some of them are known from wells already drilled so that lateral amplitude variations can be reasonably ascribed to variations in porosity, the ratio of net-to-gross sand thickness, and/or the product of porosity and net reservoir thickness (see Sheriff, 1992). Determining porosity–thickness from 3-D data in the Prudhoe Bay oil field is discussed by Stanulonis and Tran (1992). For the common type of clastic reservoir where the acoustic impedance of the reservoir sand is lower than that of the surrounding shales, higher amplitude generally indicates better reservoir quality regardless of whether this is due to higher porosity, higher ratio of net-to-gross sand thicknesses, increased reservoir thickness,

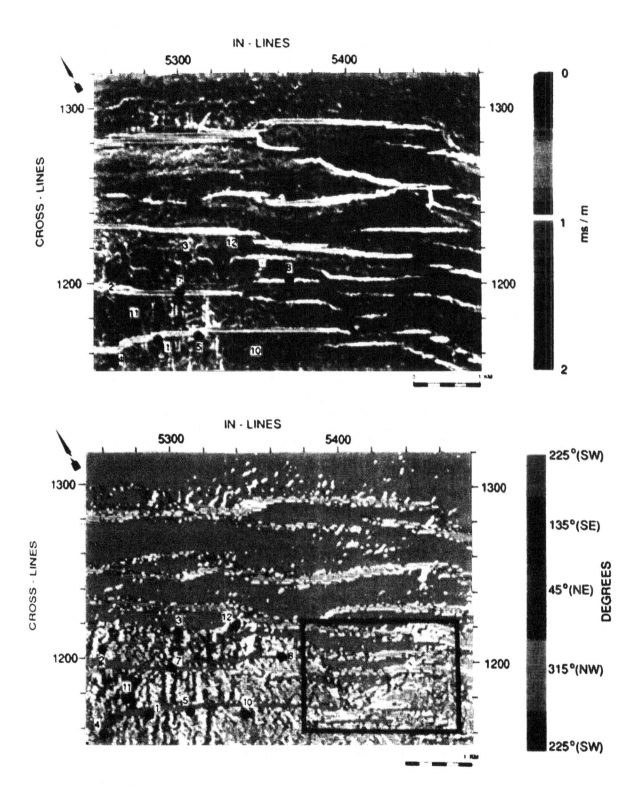

Plate 6 Dip and azimuth displays for a horizon slice help locate faults. Some faults show better on one display, some on the other, many equally well on both. (From Bouvier et al., 1989.) (a) Dip-magnitude display; (b) dip-azimuth display.

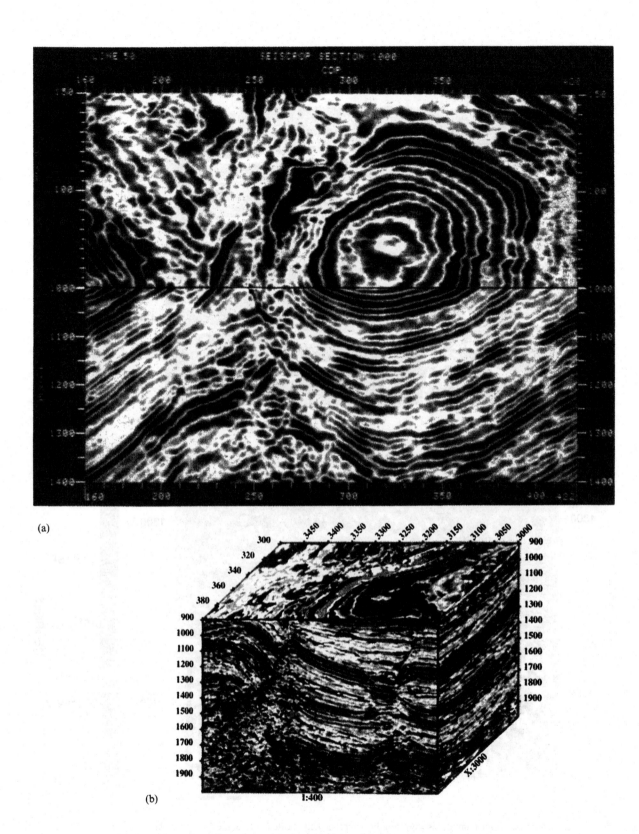

(a)

(b)

Plate 7.   Composite displays at a workstation help in understanding features. (from Brown, 1991; 70, 71.) (a) Portion of a time slice (top half) and vertical section (bottom half); (b) cube display showing line and crossline on sides and time slice on top.

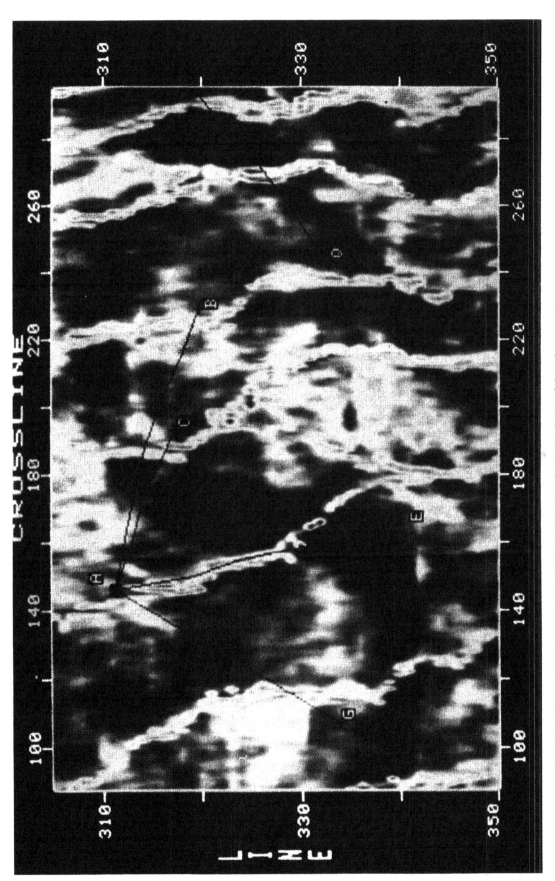

Plate 8   Horizon slice cut by a number of faults (the blue lineations). Continuity of the red stream pattern across the faults provides convincing evidence that correlations across the faults are correct. (From Brown, 1991: 127.)

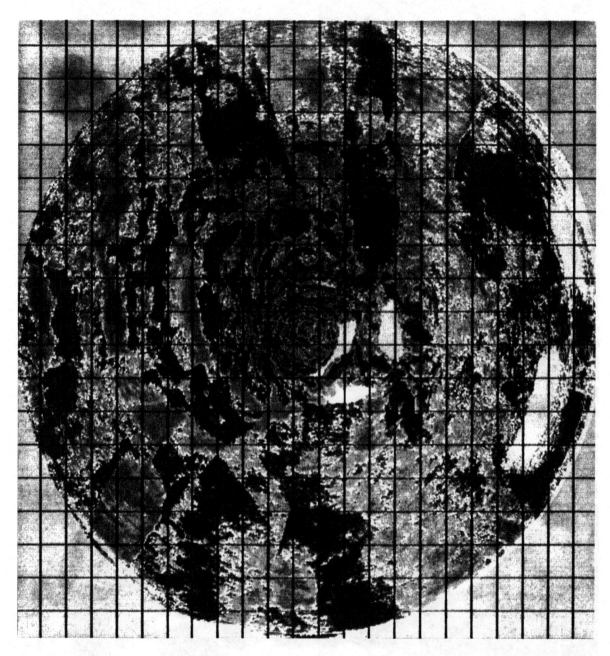

Plate 9   Time slice from a circle shoot about a salt dome. (From Brown, 1991: 63.)

Plate 10   Horizon slice showing a stream channel. The superimposed contours illustrate the structure
and the brightness in the high portion of the channel indicates hydrocarbons in a channel sand reservoir.
(From Brown, 1985: 123.)

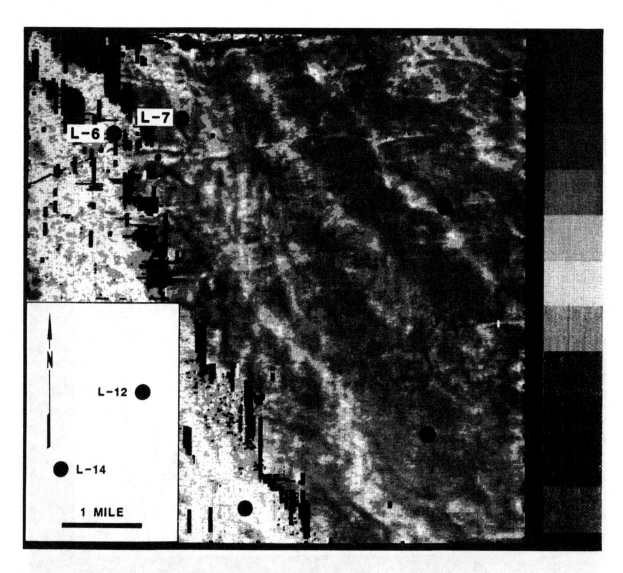

Plate 11   Horizon slice along an angular unconformity reflection. The NW–SE lineations indicate the subcrop of different members dipping to the SW, and the W–E lineations indicate faults. (From Brown, 1991: 135.)

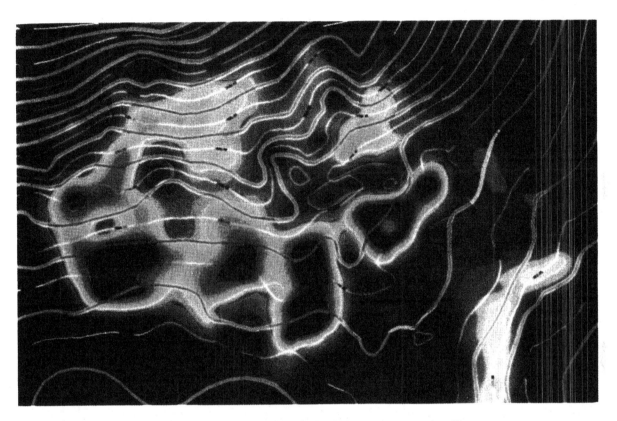

Plate 12 Horizon slice with overlain structural contours showing bright spots indicating stratigraphic hydrocarbon-gas accumulations, in a turbidite fan. (From Brown, 1991: 130.)

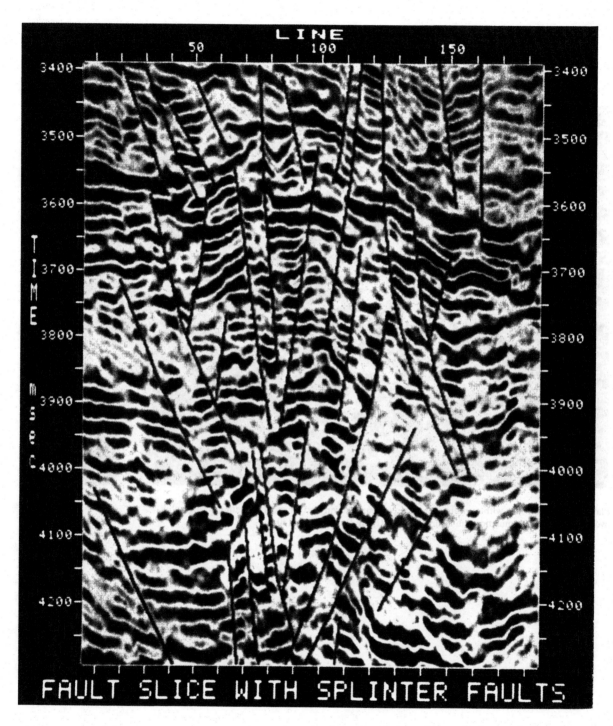

Plate 13   Fault slice showing structure adjacent to the fault plane and secondary splinter faults. (From Brown, Edwards, and Howard, 1987.)

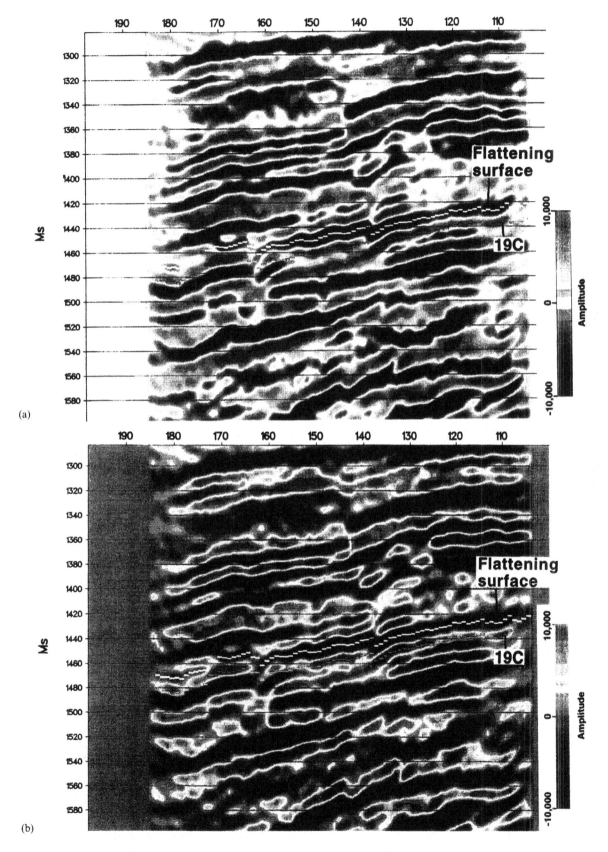

Plate 14 Portion of a seismic line showing the use of a nearby prominent reflection to flatten the event associated with the 19C reservoir. The event associated with the reservoir was identified using a VSP (fig. 12.13). (After Hardage, 1993; courtesy of Texas Bureau of Economic Geology.) (a) Common variable-density color coding; (b) biased color coding that helps show up geologically significant features.

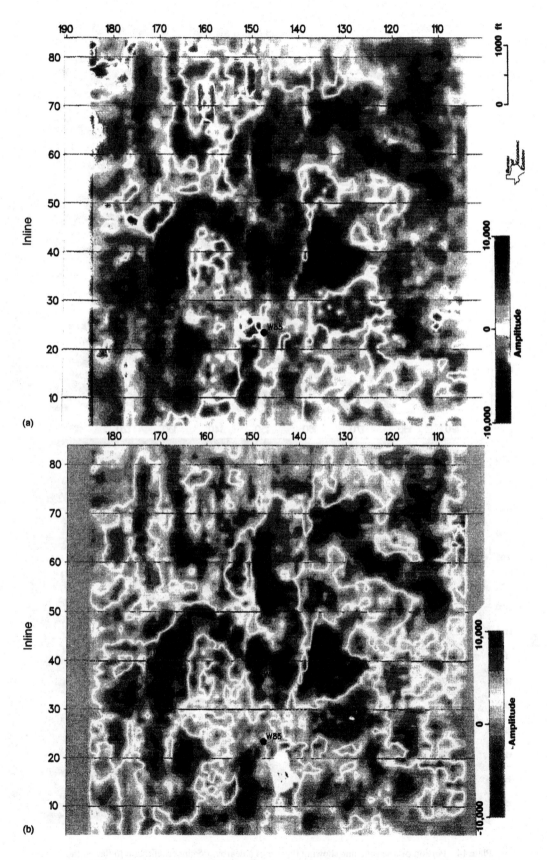

Plate 15   Horizon slices made as indicated in plate 14. An interpretation is shown in plate 16. (After Hardage, 1993; courtesy of Texas Bureau of Economic Geology.) (a) Common variable-density color coding; (b) biased color coding that helps define ancient stream deposition.

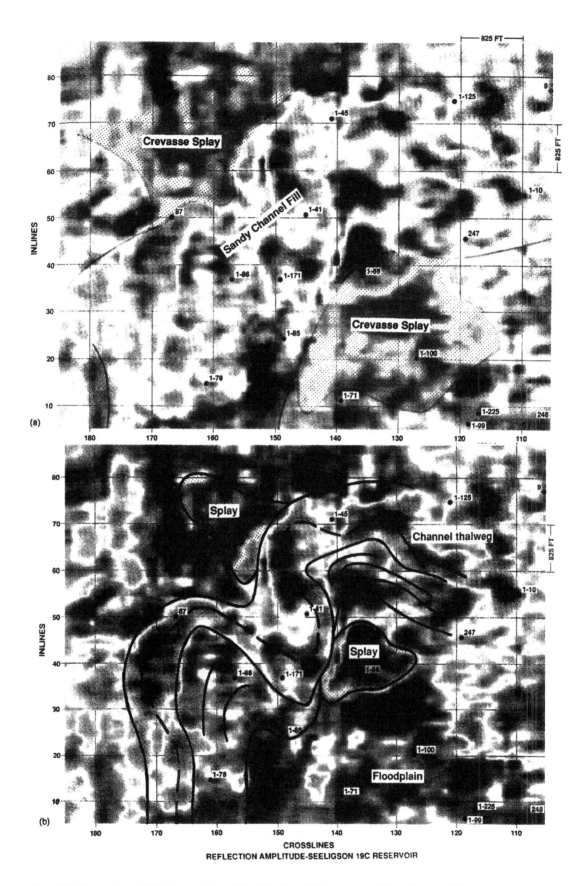

Plate 16   Pre- and post-3-D interpretations. (After Hardage, 1993; courtesy of Texas Bureau of Economic Geology.) (a) Interpretation based on well control and 2-D seismic data (blue lines), superimposed on horizon slice; (b) interpretation based on 3-D horizon slice and well data. Display parameters for (a) and (b) are slightly different.

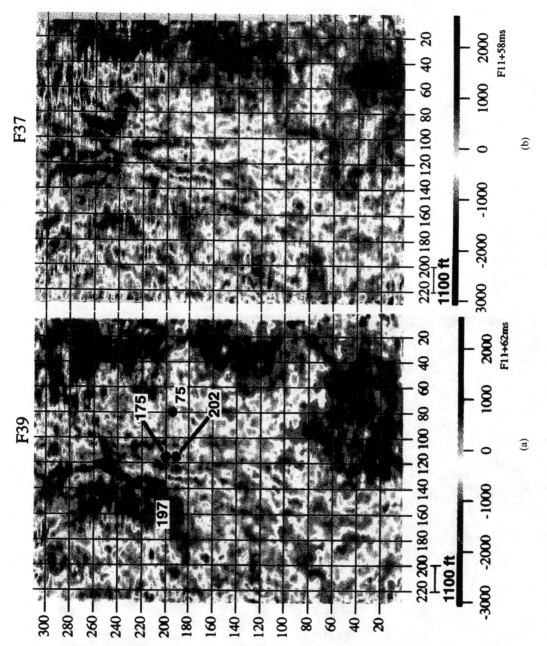

Plate 17 Horizon slices 4 ms apart where a number of channel systems are separated by small depth differences; the separation is about λ/5, or 8 m. Significant changes help define separate channel systems even though the separation is slightly below the theoretical resolvable limit. (After Hardage, 1993; courtesy of Texas Bureau of Economic Geology.) (a) Horizon slice 62 ms below the flattening pick. See fig. 12.14 for a diagram showing four wells whose reservoirs are not connected. (b) Horizon slice 58 ms below the flattening pick.

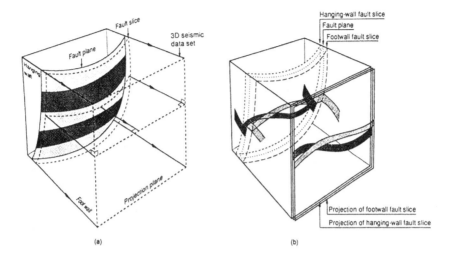

(a)   (b)

Fig. 12.12   Fault slicing. (From Bouvier et al., 1989.) (a) Fault slice is displaced a small distance away from the fault; (b) after identifying reflections associated with permeable and imperme-able formations by ties to well control, reflections on opposite sides of the fault can be superimposed to indicate sealing quality of fault.

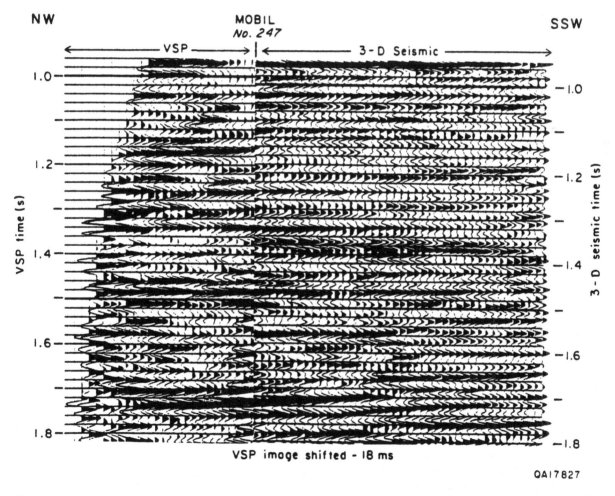

Fig. 12.13   A vertical seismic profile (VSP) used to relate 3-D reflection events to producing horizons in a well, so that horizon slices could be made to study the different reservoirs. The VSP was shifted by 19 ms to match the 3-D data. There are at least 11 separate reservoir levels between 1.1 and 1.5 s. Plates 14 to 16 show data from this survey. (After Hardage, 1993; courtesy of the Texas Bureau of Economic Geology.)

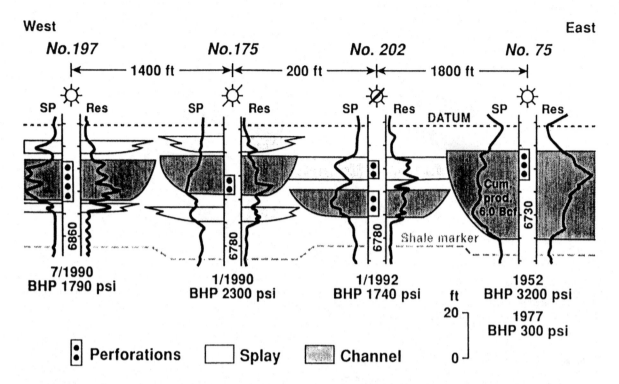

Fig. 12.14 The reservoir interval for the four wells shown in plate 17. Each of the four wells has different pressure histories and bottom-hole pressures (BHP). The separations shown are well-head distances; although the wells are all presumably verti-cal, they probably may deviate by 1°. The vertical ticks are 10 ft (3 m) apart. Changing the pressure in well 175 in an interference test did not affect the pressure in well 202. (After Hardage, 1993; courtesy of the Texas Bureau of Economic Geology.)

or higher hydrocarbon saturation (tuning effects excepted).

Problems

12.1 Reconcile the spatial sampling equation (12.1) with eq. (8.2a).

12.2 (a) For an operation involving towing two sources and three streamers, what spacings are required to achieve a minimum bin size of 25×25 m?
(b) To achieve uniform midpoint line spacing, what distance will separate successive ship tracks?

12.3 Assuming that hydrophone group centers in fig. 12.1d are 50 m apart and that the ship speed is 6 knots, calculate the cross-current at two locations.

12.4 (a) Whereas seismic ships sometimes tow three or more streamers, they only rarely use more than two (array) sources. Why?
(b) Marine shallow-water swath and patch techniques often use more source than geophone locations, whereas the practice is usually the opposite on land. Why?

12.5 Conventional marine operations involve a taper in the CMP multiplicity for half the streamer length at each end of a line. How much taper is involved with the circle methods illustrated in figs. 12.2a and 12.2b?

12.6 A survey over a marine prospect used a single patch with six parallel lines of 96 receiver groups each in sea-floor cables with a 50-m group interval, the geophone/hydrophone lines being 400 m apart. The source boat towing an air-gun source traversed 20 lines perpendicular to the receiver lines spaced 250 m apart with air-gun pops every 50 m, thus covering nearly the same area as that occupied by the receiver lines.
(a) What is the minimum bin size that can be used?
(b) How much multiplicity will be achieved over different parts of the survey area?
(c) Assume that the 3000-m deep objective horizon is a nearly flat erosional surface and that the trapping is stratigraphic, so that amplitudes must be mapped accurately. If the average velocity is 2500 m/s, how large an area can be mapped with confidence?
(d) Assume that the objective formations dip 20° away from one edge of the area. How large an area can be mapped confidently?

12.7 Assume a land survey employing six E–W lines of 112 geophone groups each with geophone-group spacing of 35 m, the geophone lines being 400 m apart. Fifteen N–S vibrator lines are spaced 300 m apart with the source points between the four center lines spaced at 70-m intervals.
(a) What is the minimum bin size? What pattern of multiplicity is achieved? How much variation of offset and azimuth mix is involved?
(b) If three geophone lines are moved for successive parallel acquisition blocks, what is the effect on the multiplicity, offset, and azimuth mixes?

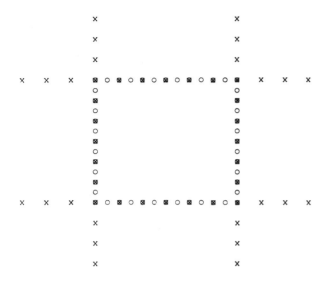

Fig. 12.15 A loop layout for 3-D surveying. Sourcepoints are indicated by ×, geophone group centers by ○.

12.8 In one 3-D technique, source points (×) and geophones (○) are laid out as shown in fig. 12.15; all the geophones are recorded for each source point. This arrangement employs 48 geophone stations, spaced 50 m apart, and 48 source points, spaced 100 m apart.

(a) Locate all the "midpoints" and determine their respective multiplicity. (*Hint:* Utilize symmetry to save work.)

(b) Note that some of the midpoints fall outside the square. If this layout is repeated with common geophone lines, these points will fit in adjacent squares. What effect will this have on multiplicity?

12.9 An E–W fault cuts the structure shown in fig. 12.9. How large is its throw assuming the velocity at the mapped horizon is 3000 m/s? Draw a depth cross section for an arbitrary line perpendicular to the fault. What sort of fault do you think is involved?

12.10 Locate possible faults on fig. 12.11 and indicate their possible throws. Sketch dips on opposite sides of one fault for an arbitrary line along the fault. Assume that lines/cross-lines are 25 m apart and that the dominant frequency is 40 Hz.

12.11 Locate several places where faults show on one of the displays in plate 6 but not on the other.

12.12 Interpret the faults in plate 9.

12.13 What are the advantages and disadvantages of a 3-D marine survey recorded in the dip direction compared to one in the strike direction in terms of (a) cross-line smear, (b) DMO, (c) spatial aliasing, and (d) velocity estimation?

12.14 Will a salt dome appear larger or smaller based on 2-D migration of a coarse grid of lines than on a migrated 3-D survey?

12.15 A swath survey is to be recorded using 10 parallel geophone lines spaced 50 m apart, each containing 12 stations spaced 50 m apart. Two source lines per-

pendicular to the geophone lines are located at the ends of the geophone lines; each contains eight source locations spaced 125 m apart symmetrically located with respect to the geophone lines. What will be the minimum bin dimensions and what multiplicity will be achieved?

12.16 A land survey layout is shown in fig 12.16.

(a) In the southern 2/3 of the area where spacing was regular, what is the smallest bin size that should be used? What is the best multiplicity achieved? How wide is the multiplicity taper area? What is the smallest bin size if square bins are desired? For the best-multiplicity square bins, what are the offset and azimuth ranges?

(b) Answer the questions in part (a) if four of the smallest square bins are composited to give a larger square bin?

(c) How much degradation is caused by the irregular spacing in the northwestern part of the area, assuming the larger square bins are used?

12.17 Copy fig. 12.13 and cut along the junction between the VSP and the 3-D data. Slide the two up and down to ascertain the confidence in the match. Assuming a velocity of 6000 ft/s, how much shift is involved? What would be the effect on the match if the VSP and 3-D embedded wavelets were 180° out of phase? If they were 90° out of phase? How much effect might a change in wavelet shape have.

12.18 Assume that the wells indicated in fig. 12.14 deviate by 1°; how much might this change the bottom-hole separations?

12.19 The wells shown in plate 15 were all drilled before the 3-D survey. What changes in well locations would you expect if the 3-D survey had been available? Where would you recommend drilling new wells now assuming the horizon is flat and that higher color intensity indicates better producibility?

12.20 Interpret channel systems in plate 17 as best you can. Consider the location uncertainties indicated by problem 12.18. What amplitude factors affect the interpretation? This is the land survey discussed in problem 12.16.

References

Abriel, W. L., P. S. Meale, J. S. Tissue, and R. M. Wright. 1991. Modern technology in an old area: Bay Marchand revisited. *The Leading Edge,* **10(6):** 21–35.

Bee, M. F., J. M. Bearden, E. F. Herkenhoff, H. Supiyanto, and B. Koestoer. 1994. Efficient 3-D seismic surveys in a jungle environment. *First Break,* **12:** 253–9.

Bouvier, J. D., C. H. Kaars-Sijpesteijn, D. F. Kluesner, C. C. Onyejekwe, and R. C. van der Pal. 1989. Three-dimensional seismic interpretation and fault sealing investigations, Nun River Field, Nigeria. *Bull. AAPG,* **73:** 1397–1414.

Brown, A. R. 1983. Structural interpretation from horizontal seismic sections. *Geophysics,* **48:** 1179–94.

Brown, A. R. 1985. The role of horizontal seismic sections in stratigraphic interpretation. *Seismic Stratigraphy II,* O. R. Berg and D. G. Woolverton, eds., pp. 37–48, AAPG Memoir 39. Tulsa: American Association of Petroleum Geologists.

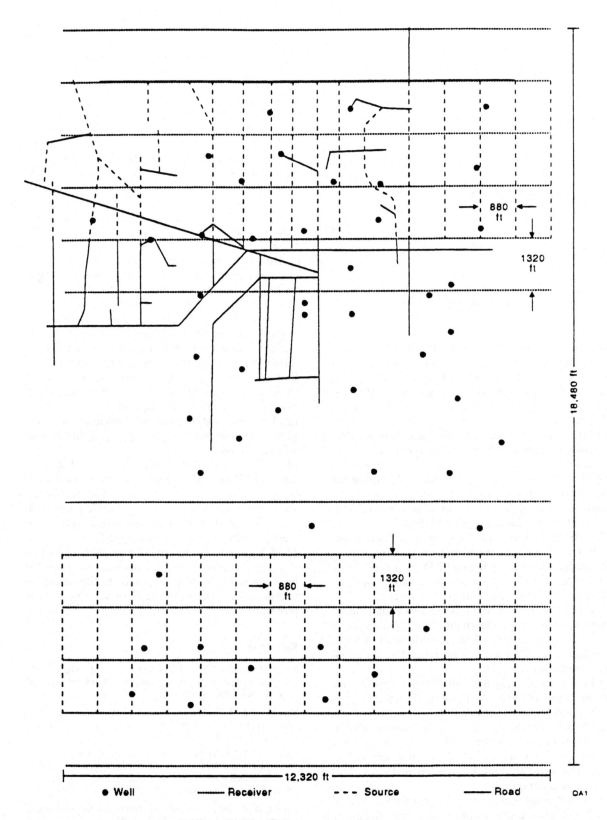

● Well ———— Receiver – – – Source ———— Road   QA1

Fig. 12.16 Layout of a land survey of a 2.3 by 3.5 mile area (8.2 mi²). For a single swath, 112 stations spaced 110 ft apart were laid out on each of 6 E–W geophone lines (dotted lines), with vibrators traversing N–S lines (dashed) with vibrator points spaced 220 ft apart. The three southernmost lines are "rolled" northward for the next swath. The area was mapped in four swaths (the first and fourth swaths are shown). Vibrator access in the northern part was restricted to the trails shown. (After Hardage, 1993; courtesy of the Texas Bureau of Economic Geology.)

Brown A. R. 1991. *Interpretation of Three-dimensional Seismic Data,* 3d ed. AAPG Memoir 42. Tulsa: American Association of Petroleum Geologists.

Brown, A. R. 1992. Defining reservoir properties. In *Reservoir Geophysics,* R. E. Sheriff, ed., pp. 185–206. Tulsa: Society of Exploration Geophysicists.

Brown, A. R., G. S. Edwards, and R. E. Howard. 1987. Fault slicing – A new approach to the interpretation of fault detail. *Geophysics,* **52:** 1319–27.

Dahm, C. G., and R. J. Graebner. 1982. Field developments with three-dimensional seismic methods in the Gulf of Thailand – A case history. *Geophysics,* **47:** 149–76.

Enachescu, M. E. 1993. Amplitude interpretation of 3-D reflection data. *The Leading Edge,* **12:** 678–85.

Galbraith, R. M., and A. R. Brown. 1982. Field appraisal with three-dimensional seismic surveys, offshore Trinidad. *Geophysics,* **47:** 177–95.

Greenlee, S. M., G. M. Gaskins, and M. G. Johnson. 1994. 3-D seismic benefits from exploration through development: An Exxon perspective. *The Leading Edge,* **13:** 730–4.

Hardage, R. A. 1993. *Notes for Reservoir Geophysics Short Course.* Tulsa: Society of Exploration Geophysicists.

Jev, B. I., C. H. Kaars-Sijpesteijn, M. P. A. M. Peters, N. L. Watts, and J. Y. Wilkie. 1993. Akaso Field, Nigeria: Use of integrated 3-D seismic, fault slicing, clay smearing, and RFT pressure data on fault trapping and dynamic leakage. *Bull. AAPG,* **77:** 1389–1404.

Nestvold, E. O. 1992. 3-D seismic: Is the promise fulfilled? *The Leading Edge,* **11(6):** 12–19.

Rosencrans, R. D. 1992. Cost-effective 3-D seismic survey design. *The Leading Edge,* **11(3);** 17–24.

Stanulonis, S. F., and H. V. Tran. 1992. Method to determine porosity–thickness directly from 3-D seismic amplitude within the Lisburne carbonate pool, Prudhoe Bay. *The Leading Edge,* **11(1):** 14–20.

Sheriff, R. E. 1991. *Encyclopedic Dictionary of Exploration Geophysics,* 3d ed. Tulsa: Society of Exploration Geophysicists.

Sheriff, R. E., ed. 1992. *Reservoir Geophysics.* Tulsa: Society of Exploration Geophysicists.

13
Specialized techniques

Overview

Lesser-used techniques are discussed in this chapter. Seismologists should be familiar with them because they provide the most efficient means of gaining needed information under special circumstances.

S-waves depend on different elastic properties than *P*-waves and hence yield additional information when combined with *P*-wave studies (§13.1). Especially in anisotropic situations, as is likely where fracturing is present, they may yield more definitive information than obtainable from *P*-waves. Additional information can also be obtained by treating wave motion as a vector with three-component recording (§13.2) rather than dealing with only the component of motion in one direction.

Seismic waves trapped in low-velocity channels (§13.3) can be used to obtain information about the properties of the channels. However, their analysis is difficult because they are highly dispersive.

Vertical seismic profiling (VSP) (§13.4) provides one of the best means of relating reflection events to the specific interfaces involved in their generation. VSP also provides the means to see, with higher resolution than available with surface data, what may lie ahead of the drill bit or what changes may lie to the side of a borehole.

Tomographic methods (§13.5) provide a different kind of approach to inverting traveltime (and, in theory at least, amplitude) measurements to determine distributions of velocity (and absorptive properties). Although their use is relatively new and the best means of application are still being developed, they are especially applicable to resolving borehole-to-borehole measurements.

The time-lapse technique (§13.6) consists of repeating measurements in order to determine changes that may have occurred over time. This technique is used mainly in reservoir studies.

In addition to velocity surveys, VSP, and cross-hole surveys, measurements within boreholes (§13.7) include surveys to determine how close a borehole is to the flank of a salt dome. Waveform logging allows analysis of the velocities of different wave modes, and the borehole televiewer provides in effect a picture of the wall of a borehole and shows features such as fracturing.

Passive seismic measurements rely on natural sources to generate seismic waves. Joint inversion uses a different kind of measurement (such as of gravity)

as an aid in seismic interpretation. Geostatistical methods interpret rock properties from geophysical measurements on a statistical basis, allowing for changes because of various unknown factors.

13.1 Exploration with S-waves

13.1.1 Why explore with S-waves

Nearly all seismic exploration is carried out with *P*-waves, the assumption being that *P*-waves alone are involved and that any *S*-wave energy present merely contributes to the noise. However, conversion of *P*-waves at interfaces means that *S*-waves are involved in seismic observations even if we wish to avoid them.

P-waves have advantages over *S*-waves: they are easier to generate, only a single mode exists, they travel faster and so arrive first, and they are easier to interpret. However, *S*-waves also have advantages: (1) *S*-wave velocity depends on different properties than *P*-wave velocity (μ versus $(\lambda + 2\mu)$ – see eqs. (2.58) and (2.59)); (2) *S*-waves have two modes (*SV* and *SH* – see §2.4.1), which is both a complication and a potential advantage. Thus, *S*-waves carry different information from *P*-waves. If both *P*- and *S*-wave velocities can be measured, then we have a source of additional information about the subsurface. Figure 5.12 suggests that such information should indicate lithology, and fig. 5.26 suggests that it should also indicate the fluid contained within a rock's pore space. The shear modulus μ is most important in engineering studies because it relates to the ability of the earth to support structures. The shear modulus along fault planes seems to change in anticipation of earthquakes. *S*-wave exploration is the subject of a book edited by Danbom and Domenico (1987).

The *SV*-mode involves wave motion within the nearly vertical plane that contains the raypath, whereas the *SH*-mode involves horizontal motion. The *SV*-mode is involved in conversion at near-horizontal interfaces, but the *SH*-mode is not.

The potential advantages of *S*-wave exploration have resulted in appreciable effort being devoted to developing *S*-wave techniques. However, a methodology has not yet evolved.

13.1.2 S-wave recording on land

Because we are usually primarily interested in waves traveling more nearly vertically than horizontally, we

(a)

(b)

Fig. 13.1 Horizontal vibrator for generating *S*-waves. (Courtesy of Conoco.) (a) Truck-mounted vibrator; the weight of the truck is used to keep the vibrator in firm contact with the ground. (b) Detail of vibrator pad showing teeth that are triangular in cross section to maintain ground coupling as the horizontal movement of the pad compacts the soil during a sweep.

must induce approximately horizontal motion in order to generate *S*-waves; it is much more difficult to couple horizontal motion to the earth than vertical motion. Virtually all *S*-wave sources also generate *P*-waves. To minimize confusion as to wave identity, one usually attempts to generate *SH*-waves (so that converted waves are not involved) and then look only for the *SH*-component with horizontal geophones.

There are basically four ways of generating *S*-waves

(in addition to relying on mode conversion): (1) use of a horizontal vibrator, (2) a horizontal impact, (3) impact at an angle, and (4) using horizontal asymmetry. *S*-wave vibrators usually have triangular teeth that dig into the underlying material to maintain coupling (fig. 13.1); as earth compaction due to the vibratory motion tends to create gaps, the triangular teeth dig in farther and maintain coupling. Of course, this also leaves gaping holes so that the use of horizontal vibra-

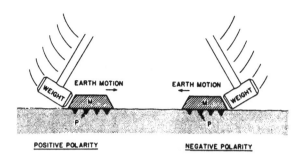

Fig. 13.2   Striking with a hammer a block held firmly to the ground initiates an *S*-wave.

tors often involves large damage claims.

A horizontal blow (fig. 13.2) against a block held firmly against the earth, usually by the weight of the vehicle (Hasbrouck, 1987; Layotte, 1987), generates *S*-waves. Striking a steel block with a sledge hammer is used to generate *S*-waves in engineering studies and a large 1700-kg hammer (Marthor™) has been used where more energy is required (Layotte, 1987). Vertical stacking is usually required to build up the energy sufficiently; blows are struck alternately against opposite sides of the block and then the polarity of the records is alternated before stacking to minimize *P*-waves generated incidentally.

(a)

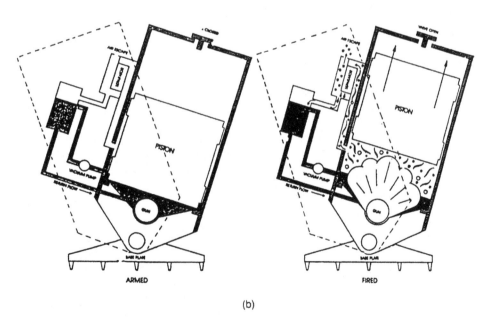

(b)

Fig. 13.3   Omnipulse seismic source. (Courtesy of Bolt Technologies.) (a) Truck-mounted unit that can be tilted up to 45° to either side to produce *S*-waves with opposite polarities. (b) Release of air from an air gun pushes the reaction mass upward as horizontal and vertical components of thrust are exerted on the ground.

Generating S-waves by an impact at an angle generates appreciable P-waves as well. Repeating the impact at 180° azimuth reverses the polarity of the S-waves; when the records are subtracted, the S-waves resulting from successive impacts will add whereas the P-waves will nearly cancel. An impact at an angle can be produced by dropping a weight constrained to fall at an angle (a "horizontal weight drop"); the weight can also be propelled by compressed air or some other force ("enhanced weight drop"), as with the Omnipulse (fig. 13.3).

Utilizing earth asymmetry is the basis of the Syslap™ technique. It uses three holes drilled close to each other (fig. 13.4). An explosive detonated in the center hole gives a conventional P-wave record and produces local changes about the central hole, which result in horizontal asymmetry in the two nearby holes. Successive explosions in the outside holes thus generate S-waves out of phase with each other in addition to P-waves, so that subtracting the two records, as is done with the angular impact sources, adds the S-waves while nearly canceling the P-waves. Two closely spaced P-wave vibrators operating 180° out of phase have also been used as an S-wave source (Shover™).

Recording can be done with horizontal geophones oriented in-line with the source for SV-waves and perpendicular to the line for SH-waves. Three-component geophones are also used (see §13.2).

13.1.3 S-wave recording at sea

Shear waves are not transmitted through water, so marine sources cannot generate S-waves nor hydrophones detect them. However, where the water bottom is hard, there is significant mode conversion (fig. 13.5) of downgoing P-waves to S-waves at appreciable angles of incidence at the water/rock interface (Tatham and Stoffa, 1976). The SV-waves reconvert to P-waves on return to the solid–liquid interface in an equally efficient conversion and so can be detected with conventional pressure detectors. Very long streamers (~6 km) are used to obtain large incidence angles. The long group length used in conventional marine streamers discriminates against waves that approach at large emergent angles, and a streamer for optimizing SV-wave detection must employ short groups. Time shifts can be introduced and the data from several short groups combined to simulate a longer group, as with beam steering. P-waves and S-waves can be further separated in processing by apparent-velocity filtering.

13.1.4 Processing and displaying S-wave data

S-waves processing, although similar to P-wave processing in many regards, is different in several respects.

Near-surface variations for S-waves are often quite large and static corrections are essential in order to

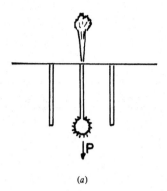

(a)

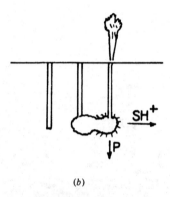

(b)

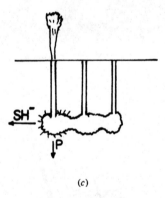

(c)

Fig. 13.4 The Syslap method. (Courtesy of CGG.) (a) Explosion in the center hole generates mainly P-waves; (b) because of the asymmetry produced by the explosion in the center hole, the explosion in the right hole generates P-waves and SH-waves; (c) the SH-waves generated by an explosion in the left hole have opposite polarity to those in (b).

stack S-wave land data and produce usable record sections (Anno, 1987). The large S-wave delays result from the sensitivity of S-wave velocity to rock-matrix variations in the near surface and they are not necessarily proportional to P-wave static delays, which mainly result from the fluids in the near-surface sediments. Automatic statics programs must search over time intervals larger than the statics differences and this is apt to involve cycle skipping because of the

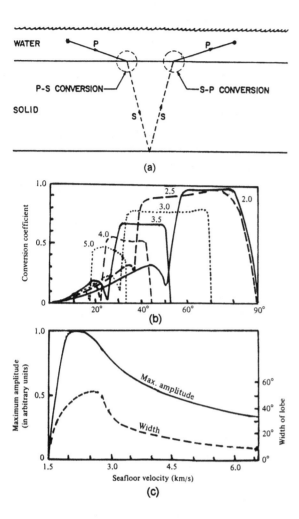

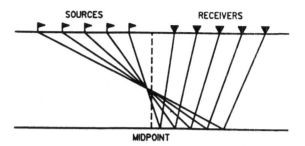

Fig. 13.6 Raypaths of common-midpoint gather for mode-converted data.

Fig. 13.5 *PSSP* reflections generated by conversion at the sea floor. (After Tatham and Stoffa, 1976.) (a) Geometry; (b) conversion coefficient versus angle of incidence in the water; the curves are labeled with sea-floor velocity in km/s; (c) maximum amplitude and width of main lobe versus sea-floor velocity.

large magnitudes of S-wave statics. Hand editing is often required.

Because of the asymmetry of raypaths involved with converted waves (fig. 13.6), their stacking is not as simple as that of P-waves, even with horizontal reflectors and simple velocity distributions. Gathers of traces having a common conversion point are discussed by Tatham and McCormack (1991).

Because S-wave results are usually interpreted in conjunction with P-wave results, S-wave sections are normally displayed with a vertical time scale double that used for the P-wave sections to compensate approximately for the roughly 1:2 ratio of S-wave to P-wave velocities, β/α (see fig. 13.7).

13.1.5 Interpretation and use of S-*wave data*

The objective of S-wave measurements is usually the ratio of S-wave to P-wave velocities (β/α). This requires identifying S- and P-wave reflections from the same reflecting interface. Because a P-wave reflector is not necessarily an S-wave reflector, this identification can be very difficult and often this is the most critical aspect of S-wave interpretation. Plotting S-wave sections at double the vertical scale used for P-wave sections, as indicated in §13.1.4, helps considerably by making the two sections appear more nearly the same. Independent identification using P- and S-wave synthetic seismograms or P- and S-wave vertical seismic profiles is highly desirable.

Usually, S- and P-wave reflection identification is based on some geometric feature such as terminations at faults, overall structural character, truncations at a shelf edge or unconformity, or prograding stratigraphic character (fig. 13.8). Other reflections are then identified by reference to these events, assuming that S- and P-reflection amplitudes vary in similar fashion (fig. 13.9).

Figure 5.12 showed that the ratio of S- to P-wave velocities, β/α, is an indicator of lithology. However, whereas the domains of sandstones, limestones, and dolomites generally do not overlap, the domain of shales can overlap those of the other lithologies, limiting the usefulness of β/α as a lithology indicator. Tatham and Krug (1985) interpret lithology using a time-average equation assuming β/α values of 0.45 for shale, 0.54 for carbonates, 0.59 for sandstone, and 0.67 for gas; they show a tetrahedron graph (fig. 13.10) to illustrate various proportions of these components. Generally, β/α decreases as porosity and carbonate/sand ratio increase and as the sand/shale ratio decreases. The velocity ratio is also affected by pore shape. If porosity is known from the wells and is cross-plotted against the velocity ratio, the resultant distribution sometimes can be interpreted in terms of pore aspect ratio, that is, the relative abundances of spherical pores and thin cracks. Examples of interpretations using S-wave are given by Fix, Robertson, and Pritchett (1987) and Ensley (1989).

S-waves are not affected by fluids, so hydrocarbon indicators such as bright spots, flat spots, dim spots, and polarity reversals do not occur on S-wave sections (except in subtle ways, mainly through their effect on density). Consequently, if such indicators are observed on P-waves sections but not on S-wave sections (fig. 13.9), this is a gas-reservoir indicator.

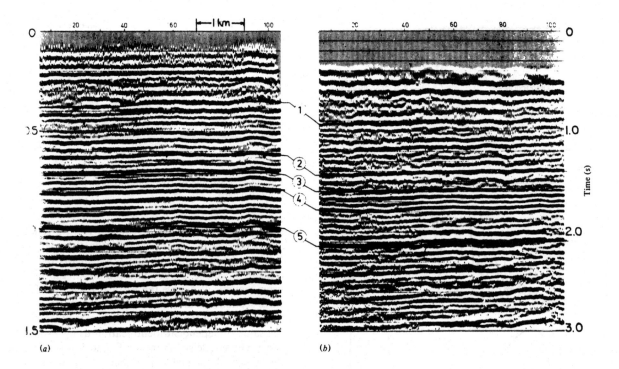

Fig. 13.7 Comparison of *P*- and *S*-wave records. (Courtesy of CGG.) (a) *P*-wave record and (b) *S*-wave record displayed at double the timing speed to make comparison of events easier.

In concept the variation of seismic amplitude with angle of incidence or offset (AVO; see §3.4) contains the same β/α information as the joint study of *P*- and *S*-wave sections. AVO information is already available in CMP data, so there has been a decrease in interest in *S*-wave data.

Tatham and Krug (1985) show field examples in which *S*-waves were used to define carbonate-clastic facies changes, locate dolomitization (fig. 13.11), and distinguish between sand and shale (fig. 13.12).

13.1.6 S-*wave birefringence*

A major use of *S*-waves is in fracture detection. Fractures produce anisotropy that can lead to birefringence (§2.6.2). Most fracture studies involve three-component recording (§13.2). An *S*-wave vibrating in a preferred direction traveling through an anisotropic medium may be split into two mutually perpendicular *S*-wave vibrations, parallel to and perpendicular to the anisotropic axis, that travel with different velocities.

With oriented fractures, an *S*-wave vibrating parallel to the fracture direction sees unfractured rock and so travels at a higher velocity than the *S*-wave vibrating perpendicular to the fracture direction (fig. 13.13). The difference in velocities often is a measure of the fracture density. Oriented *S*-wave sources and detectors can be used to determine the azimuth of the fractures (Crampin, 1987).

13.2 Three-component recording

13.2.1 Acquisition

The motion associated with a wave arrival is a vector quantity, and measuring its three orthogonal components requires the use of three orthogonal geophones. The projection of the vector motion onto a plane produces hodograms (Sheriff, 1991) such as those in figs. 2.15b and 2.15d. The coordinate system for measuring the three components of a vector can be rotated about any axis; in particular, they can be rotated into a natural coordinate system, such as one dictated by the orientation of an arriving wave (fig. 13.14) or the orientation of a fracture system. Such a coordinate rotation is sometimes called *Alford rotation* or *polarization filtering*.

Orienting three separate geophones in orthogonal directions in the field is tedious, so usually the three geophones are housed together. Most three-component phones use one vertical phone and two horizontal phones whose characteristics nearly match those of the vertical phone (see also Garotta, 1987). Because it is difficult to design horizontal and vertical geophones with the same characteristics, sometimes three identical orthogonal phones, each making a 54.7° angle to the vertical (*Gal'perin arrangement*), are used (see problem 13.2). Three-component geophones are used in boreholes for VSP surveys (§13.4); they must be gimbel-mounted to maintain their attitude to vertical in directional holes.

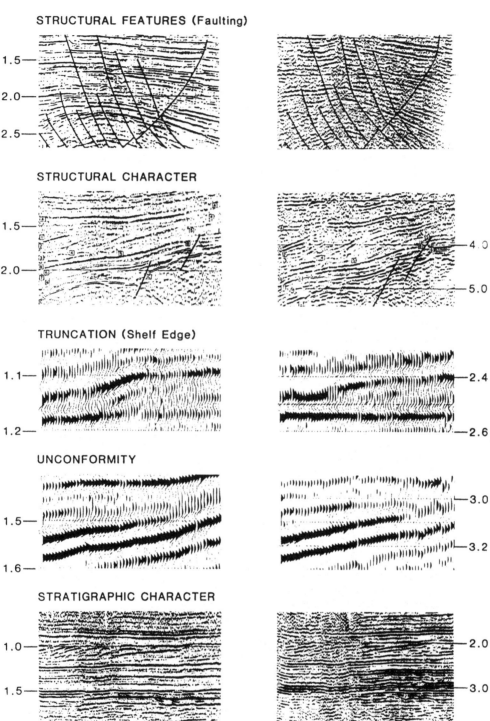

P S

STRUCTURAL FEATURES (Faulting)

1.5
2.0
2.5

STRUCTURAL CHARACTER

1.5
2.0

4.0
5.0

TRUNCATION (Shelf Edge)

1.1
1.2

2.4
2.6

UNCONFORMITY

1.5
1.6

3.0
3.2

STRATIGRAPHIC CHARACTER

1.0
1.5

2.0
3.0

Fig. 13.8   Identifying features on *P*-wave sections (left column) and on *S*-wave sections (right column) to correlate events from the same reflecting interfaces. (From Tatham and McCormack, 1991: 178.)

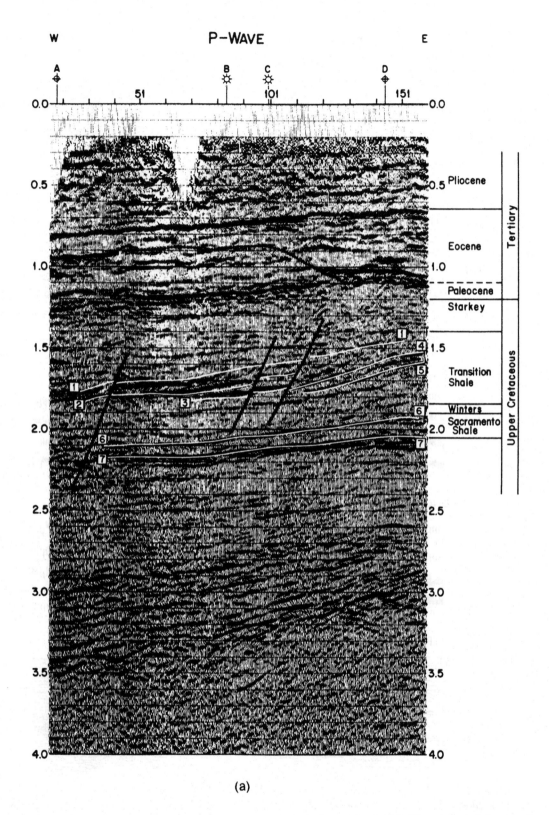

(a)

Fig. 13.9 Comparison of *P*- and *S*-wave records across the Putah Sink gas field, California. The *P*-wave section (a) shows a large amplitude associated with the gas accumulation; the *S*-wave section (b) does not. (After Tatham and Krug, 1985.)

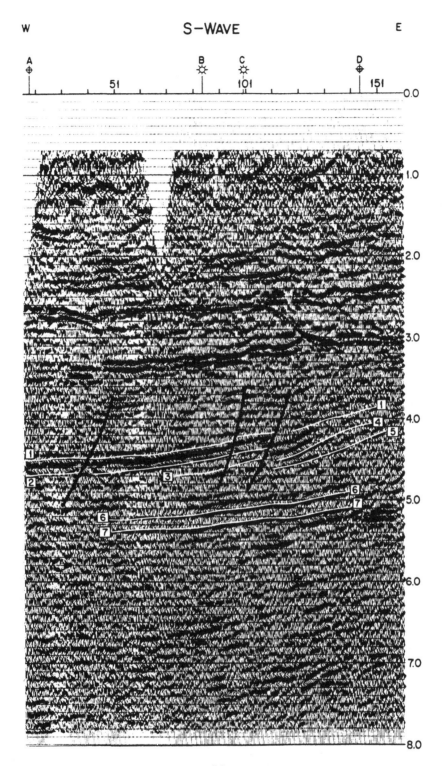

(b)

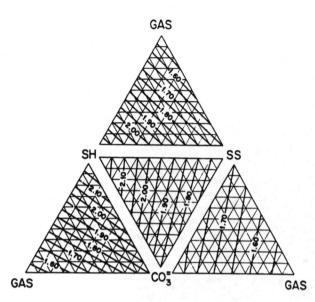

Fig. 13.10 V_P/V_S values as a function of proportions of sandstone, shale, carbonate, and gas-saturated pore space. Shown are the four faces of a tetrahedron representing possible combinations of these lithologies. (From Tatham and Krug, 1985: 149.)

A field-recording technique sometimes adopted in fracture studies is to successively record from P-, SH-, and SV-sources with three-component geophones; this is called *nine-component recording* because separate sections can be generated for each of the three geophone outputs for each of the three sources (fig. 13.15). If the sources were pure and the earth isotropic, three of the nine sections would be conventional P-, SH-, and SV-sections, two would show converted waves (P to SV and SV to P), and the other four would be blank. However, if azimuthal anisotropy (such as may be caused by vertical fractures) is present and if the SH- and SV-sources and geophones are not oriented parallel and perpendicular to the natural orientation, then S-wave splitting (§2.6.2) will result in energy showing on all panels. Polarization filtering then can be used to rotate the sources and geophones into the natural coordinate system, thus determining the orientation of the natural system and of the fractures. It should be noted that, if fracture systems at different depths have different orientations, the detection system may show only the orientation of the shallowest one. Winterstein and Meadows (1992a, 1992b) use a layer-stripping technique to resolve the natural orientations of successive layers.

Although three-component recording must be regarded as a research rather than operational tool as of 1994, it shows promise in several types of studies, especially in determining the orientation and density of fracture porosity (Ebrom and Sheriff, 1992). Lewis, Davis, and Vuillermoz (1992) use three-component recording to resolve fracture problems in a 3-D survey.

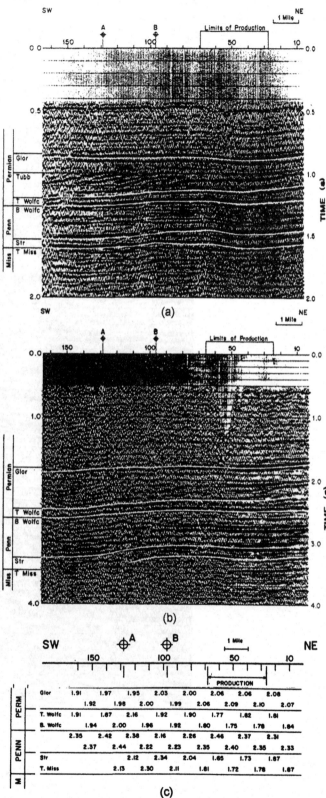

Fig. 13.11   Effect of carbonate–sand transition on V_P/V_S values, Midland Basin, Texas. V_P/V_S values for the Glorieta and Wolfcamp intervals indicate the carbonate-to-clastic facies change associated with the carbonate shelf edge. (From Tatham and Krug, 1985: 166–8.) (a) P-wave data, (b) S-wave data, and (c) V_P/V_S values for the intervals shown in parts (a) and (b).

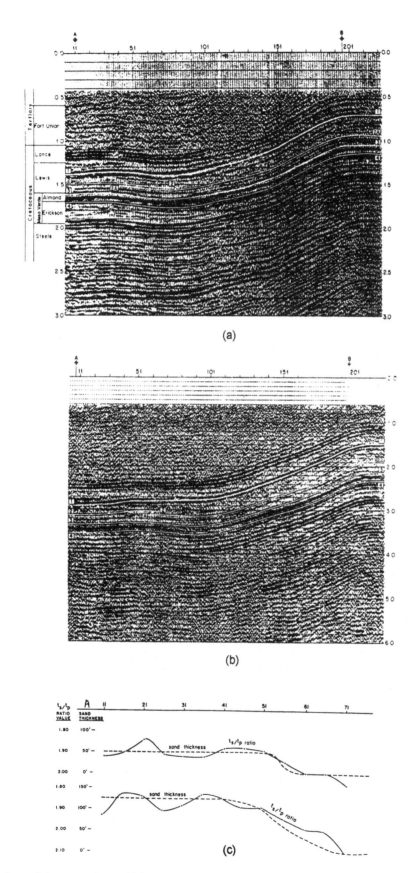

(a)

(b)

(c)

Fig. 13.12 Dependence of slowness ratios on thickness of sand beds surrounded by shale, Horse Butte, Wyoming. (From Tatham and Krug, 1985: 174–7.) (a) P-wave data, (b) S-wave data, and (c) $\Delta t_S/\Delta t_P$ (or V_P/V_S) values and interpreted sand thickness.

AUSTIN
CHALK

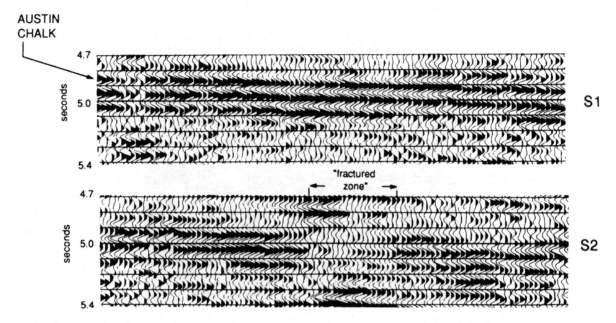

Fig. 13.13 Fast (*S1*) and slow (*S2*) *S*-wave sections. At the left end, the respective arrival times of the Austin chalk reflection are 4.85 and 4.90 s. The *S2* reflection dims where the fracturing in the Austin chalk is especially intense. (From Mueller, 1992.)

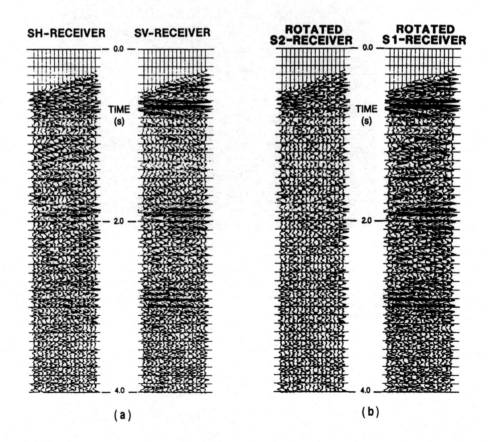

(a)

(b)

Fig. 13.14 Rotation of *SH*- and *SV*-geophone traces into the plane of the *S*-wave motion. (From Tatham and McCormack, 1991: 160–1.)

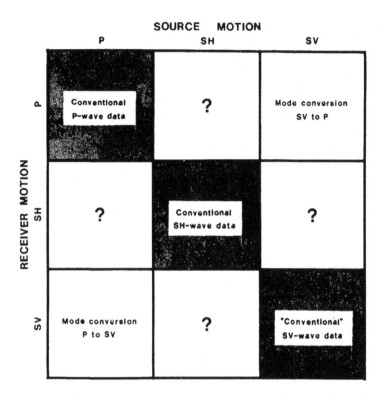

Fig. 13.15   Nine-component sections recorded by *P*-, *SH*-, and *SV*-geophones for *P*-, *SH*-, and *SV*-sources. Without anisotropy (or cross-dip) the sections marked ? would be blank. (From Tatham and McCormack, 1991: 111.)

13.2.2 Polarization filtering

Polarization filtering sometimes involves merely rotating the coordinate system into another orientation, but it may also involve shifting the phase, adjusting the amplitudes, and then stacking orthogonal components. Different propagation modes and directions of wave travel involve systematic relationships among the components, so polarization filtering can be used to preferentially select or reject particular modes, such as ground roll (fig. 13.16). Other examples of the application of polarization filtering for separating *P*- and *S*-channel waves in coal-seam studies (§14.2.5) are shown in figs. 13.17 and 13.22.

13.3 Channel waves (normal-mode propagation)

Under certain circumstances, wave energy may be trapped within a layer that then channels the waves. Such waves are called *channel waves, guided waves, trapped waves,* or *seam waves* (where they propagate in a coal seam). This phenomenon is also known as *waveguide* or *normal-mode propagation*.

Two kinds of boundary conditions can produce this situation: (a) the impedance contrast is so great that the reflection coefficient is very large (nearly unity); (b) waves within the waveguide are incident on the boundary at an angle greater than the critical angle so that total reflection occurs and little energy leaks through the boundary (except for evanescent waves,

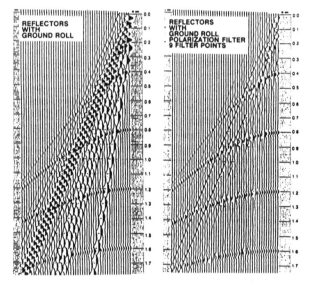

Fig. 13.16   Use of polarization filtering to attenuate ground-roll. (From Tatham and McCormack, 1991: 133–4.)

§2.3.1, which can be neglected, and converted waves). Polarity reverses at an interface of type (a) but not at type (b). A common example of the first type is the surface of a water layer and of the second type the sea floor. This propagation is somewhat analogous to organ-pipe reverberation (Lapedes, 1978), a water

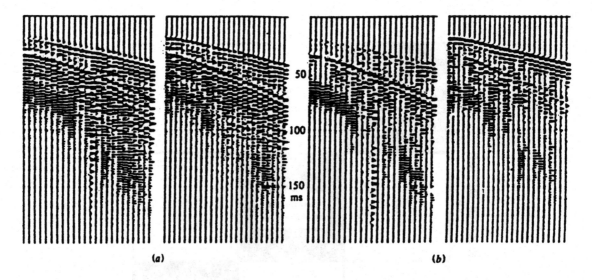

(a)   (b)

Fig. 13.17   Use of polarization filtering to separate *P*- and *S*-waves on transmission of in-seam data. (From Millahn, 1980.) (a) Records of components perpendicular and parallel to the

gallery in which measurements were made and (b) after polarization filtering and rotation to emphasize components perpendicular and parallel to the source–receiver direction.

layer corresponding to an open organ pipe with a node at the sea-floor interface, an antinode at the free surface, and a coal seam to a closed organ pipe with nodes at both boundaries, the acoustic impedance of the coal being much lower than the overlying (roof) and underlying (floor) formations.

Figure 13.18a shows waves bouncing back and forth at different angles within a waveguide. For most of the angles, there is destructive interference between the different waves, but for certain angles, there is constructive interference and consequently a strong buildup of energy reflected at these angles. In fig. 13.18b, wavefront *AC* has been reflected upward at the lower boundary at the angle θ, where $\theta > \theta_c$, the critical angle. A parallel wavefront that occupied the same position *AC* earlier, then was later reflected at the upper and lower boundaries, following raypaths such as *EFGH* and *BDAI*, now coincides with the later wavefront at *AC*. Because $EF + FG + GH = BD + DA$, we see that the phase difference between the two waves is $\kappa(BD + DA) + m\pi + \varepsilon$, where m is 0 or 1, $m\pi$ is the sum of phase reversals on reflection at the two boundaries, and ε is a phase shift that occurs when $\theta > \theta_c$ (Officer, 1958: 200–1). For a water layer, $m = 1$, whereas $m = 0$ for a coal seam.

For constructive interference, we must have

$$\kappa_n(BD + DA) + m\pi + \varepsilon = 2n\pi.$$

Because

$$DA + BD = h/\cos\theta + (h/\cos\theta)\cos 2\theta$$
$$= 2h\cos\theta,$$

we have

$$2\kappa_n h\cos\theta = (4\pi h\nu_n/V_1)\cos\theta = (2n - m)\pi - \varepsilon,$$

or

$$\nu_n = [(2n - m) - (\varepsilon/\pi)]V_1/(4h\cos\theta). \quad (13.1)$$

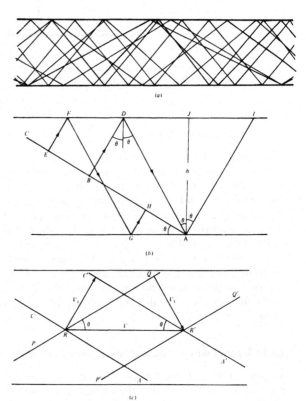

Fig. 13.18   Waveguide phenomenon. (a) Waves bouncing back and forth in a layer of velocity V_1 because of nearly perfect reflectivity at the boundaries, (b) construction to show reinforcement conditions, and (c) phase and group velocity relationship.

Neglecting ε for the moment, we get constructive interference when

$$\nu_n = (2n - m)V_1/(4h\cos\theta). \quad (13.2)$$

For a water layer, $m = 1$; hence,

$$\left.\begin{aligned} v_1 &= V_1/(4h\cos\theta), \\ v_2 &= 3V_1/(4h\cos\theta) = 3v_1, \\ &\vdots \\ v_n &= (2n-1)v_1, \end{aligned}\right\} \qquad (13.3)$$

which corresponds to an open organ pipe (except for the factor $\cos\theta$). For a coal seam, $m = 0$, and

$$\left.\begin{aligned} v_1 &= V_1/(2h\cos\theta), \\ v_2 &= 2V_1/(2h\cos\theta) = 2v_1, \\ &\vdots \\ v_n &= nv_1, \end{aligned}\right\} \qquad (13.4)$$

which is analogous to a closed organ pipe. Thus, provided the original wave generated by the source contains the appropriate frequencies, normal-mode propagation consists of a series of waves of frequencies v_1 and its odd or even harmonics propagating along the waveguide by reflection at angles θ that satisfy eq. (13.3) or (13.4).

In addition to the upward propagating set of wavefronts parallel to AC, there is a symmetrical downward-propagating set parallel to PQ in fig.

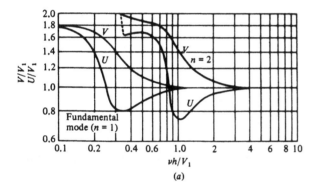

13.18c and the interference between the two sets creates a standing wave pattern along the perpendicular to the waveguide. As a result, the wave motion is propagated parallel to the boundaries of the waveguide. The velocity V_1 is the phase velocity normal to the wavefronts, but there is a different phase velocity V in the direction of the effective wave propagation. By referring to fig. 13.18c, wavefronts AC and PQ intersect at R, and there will be a local buildup of energy here. This energy-density maximum propagates in the direction RR'; if AC and $A'C'$, also PQ and $P'Q'$, are the wavefront positions one time unit apart, then the phase at R moves to R' in one time unit so that $V = RR'$, that is,

$$V = V_1/\sin\theta. \qquad (13.5)$$

Because θ is a function of frequency (see fig. (13.2) to (13.4)), V is also frequency-dependent, so that the wave motion is dispersive.

The minimum value of θ is the critical angle θ_c; hence, there is a minimum cutoff frequency v_0, where (for a water layer)

$$v_0 = V_1/(4h\cos\theta_c), \qquad (13.6)$$

the corresponding phase velocity V being $V = V_1/\sin\theta_c = V_2$. As θ increases, v increases but V decreases. In the limit, $\theta \to \frac{1}{2}\pi$ (the grazing angle), $v \to \infty$, and $V \to V_1$.

If we do not neglect ε, the formulas are more complicated, but the results are basically the same. Officer (1958) shows that for $\theta > \theta_c$,

$$\left.\begin{aligned} \tan\tfrac{1}{2}\varepsilon &= (\rho_1/\rho_2)[\tan^2\theta - (V_1/V_2\cos\theta)^2]^{1/2} \\ &= 0, \qquad \theta = \theta_c, \\ &= \pi, \qquad \theta = \tfrac{1}{2}\pi. \end{aligned}\right\} \qquad (13.7)$$

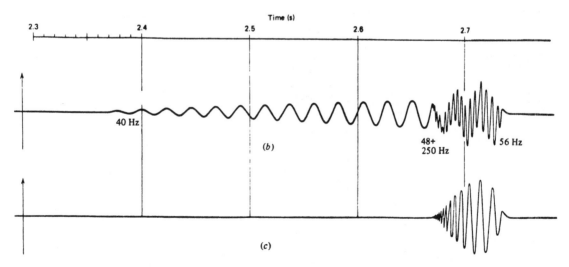

Fig. 13.19 Channel-wave propagation for a liquid layer on an elastic substratum. (a) Phase and group velocity versus normalized frequency where $\alpha_2/\alpha_1 = 2\sqrt{3}$, $\sigma_1 = 0.5$, $\sigma_2 = 0.25$, and $\rho_2/\rho_1 = 2.5$ (from Ewing, Jardetsky, and Press, 1957). (b) First-mode wavetrain from a source 4 km away. (c) The high-frequency portion of part (b), which is called the *water wave;* its onset is sometimes used in marine refraction work to determine the range. (Parts (b) and (c) from Clay and Medwin, 1977.)

Typical curves of V versus ν for a water layer are shown in fig. 13.19a for $n = 1, 2$.

The group velocity U is given by

$$U = V + \omega \frac{dV}{d\omega} = V + \nu \frac{dV}{d\nu}.$$

For a water layer, fig. 13.19a shows that the term $\nu(dV/d\nu)$ is never positive; hence, $U \le V$. Moreover, although $\nu(dV/d\nu)$ increases in magnitude at first as ν increases from the value ν_0, eventually the term approaches zero as ν approaches infinity (because the derivative goes to zero faster than ν goes to infinity). As a result of these factors, U has the value V_2 at the cutoff frequency ν_0, then decreases to a minimum U_m at some frequency ν_m after which it increases asymptotically to the value V_1 at $\nu = \infty$.

A normal-mode wavetrain for a water layer is shown in fig. 13.19b. The first arrival is a wave of frequency ν_0 that has traveled with the maximum group velocity V_2; this is followed by waves of increasing ν and decreasing U until U reaches the value V_1 at which time a very high-frequency wave, which also has traveled with velocity V_1, is superimposed on the first wave. Following this, the frequencies and group velocities of the two waves approach ν_m and U_m, respectively. The burst of energy beginning with that traveling at velocity V_1 to that of the energy traveling at U_m, the often-abrupt end of the normal-mode wavetrain, is called the *Airy phase* (as in fig. 19c).

Clearly, a channel wave in water must be a *P*-wave, but in solids, several other types of channel waves can exist. Love waves and *SV*-waves in the surface layer can be explained as normal-mode propagation (Grant and West, 1965: 81–5).

Figure 13.20 shows records from five geophones at different positions within a coal seam. The amplitudes in the roof and floor are relatively weak. The fundamental modes have their largest amplitudes near the center of the coal seam, whereas second harmonics (with double the fundamental frequencies) have their largest amplitudes at nearly 1/4 and 3/4 points with a node near the center. Different modes can be generated preferentially by varying the source location up or down within the coal seam. Variations in elastic constants of the coal and bounding lithologies produce asymmetries and complex waveforms.

Mason, Buchanan, and Booer (1980) studied channel waves propagating in coal seams and found the wave motion to be very complex. They found both *SH*-modes, called *Evison* or *pseudo-Love waves*, and *P-SV*-modes, called *Krey* or *pseudo-Rayleigh waves* (fig. 13.22b). These modes are all highly dispersive (fig. 13.21). Waves observed with orthogonal geophones usually have to be rotated (§13.2.2) to separate wave modes, as in fig. 13.22 (which also shows hodograms for two portions of the wavetrain). The events shown in figs. 14.6 and 14.7 are channel waves.

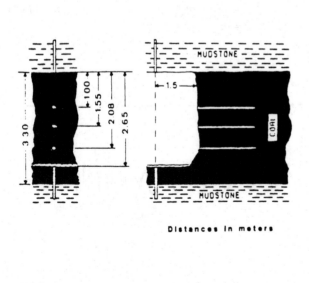

(a)

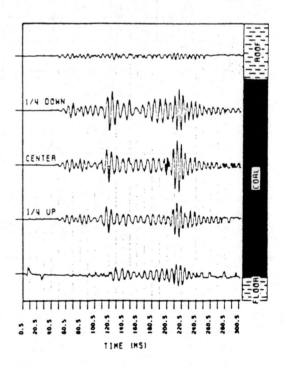

(b)

Fig. 13.20 Channel waves observed at different elevations within a coal seam. (From Reguiero, 1990.) (a) Positions of geophones and (b) records of the respective geophones.

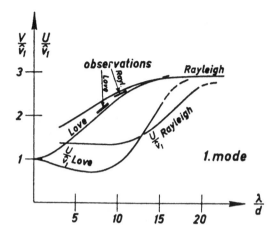

Fig. 13.21 Dispersion of the fundamental mode of channel waves in a coal seam. (From Mason, Buchanan, and Booer, 1980.)

13.4 Vertical seismic profiling (VSP)

13.4.1 General

Usually, the seismic source and geophones are located at or very near the surface. Most borehole surveys, such as conventional well-velocity surveys (§5.4.2), measure only the traveltime of the first energy. In contrast, *vertical seismic profiling* (Kennett, Ireson, and Conn, 1980; Cassell, 1984; Fitch, 1984) involves recording the complete waveform at regularly and closely spaced depth stations. Extracting velocity information is only one of the objectives of VSPs. They became more common in the late 1970s and 1980s, but the cost of occupying a borehole for the time required continues to deter their greater use.

A VSP generally gives better data than surface seismic methods because the energy does not have to travel as far and therefore undergoes less attenuation. Consequently, the resolution of a VSP is usually appreciably better than that of surface seismic data.

13.4.2 VSP types and their uses

The most common VSP in vertical (or near-vertical) holes, a *zero-offset VSP* (figs. 13.23a and 13.23b), uses a single source located near the wellhead. In an *offset VSP* (fig. 13.23c), the source is located some distance from the wellhead, often 900–2100 m, to give data away from the borehole. An alternative way of obtaining such data is a *walkaway VSP* (fig. 13.23d), in which the source locations are moved to successively larger distances from the wellhead. *Azimuthal VSP* surveys locate sources in different directions from the wellhead to investigate changes with azimuth. Combinations of the foregoing are used in deviated holes in *directional VSP* surveys (§13.6.5); in marine directional VSP surveys, a source boat often travels so that it is vertically above the geophone with another source located near the wellhead. Balch and Lee (1984) and Hardage (1985) describe various VSP techniques.

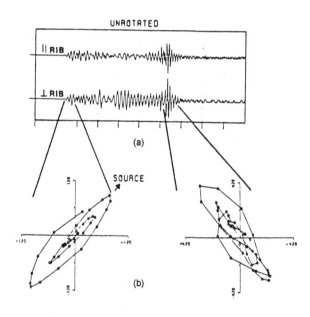

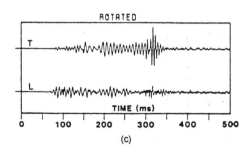

Fig. 13.22 Records from orthogonal geophones within a channel. (From Reguiero, 1990.) (a) Unrotated records, (b) hodograms of portions of the wavetrains involving Krey waves (left) and Evison waves (right), and (c) records after rotation.

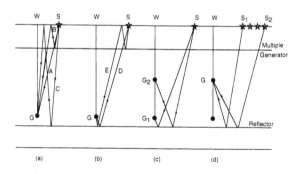

Fig. 13.23 VSP raypaths. (a and b) Zero-offset VSPs (WS is small); (c) offset VSP (*WS* large); and (d) walkaway VSP (*WS₂* >> *WS₁*). Downgoing arrivals at a well geophone *G* are shown in part (a) and upgoing waves in (b, c, and d). *A*, a downgoing direct wave (first arrival); *B* and *C*, downgoing multiples involving reflectors above or below the geophone; *D*, an upgoing primary reflection; and *E*, an upgoing multiple.

The basic task of a zero-offset VSP is to match seismic events to specific interfaces. Because the depth of the well geophone is accurately known, time–depth relationships are established precisely, and thus reliable reflection identification and subsurface seismic velocities are obtained (see fig. 12.13). Zero-offset VSPs are also used to identify multiples (see fig. 13.29) and other events, and to study reflections from below the bottom of the borehole (because of the increased resolution) to aid in deciding if a well should be deepened ("looking ahead of the bit"). Payne (1994) claims 15–25-m resolution when looking ahead 600–1200 m.

Offset and walkaway VSPs are used to "look to the side" of the borehole to see if a major change, such as a fault or reef, occurs near the borehole and hence that the findings from the borehole may not apply to the nearby region. The greater resolution of a VSP may help delineate small faults, stratigraphic changes, and thin reservoir sands. An objective formation may have been faulted out or a well may be on a nonproductive side of a fault so that sidetracking may encounter production (Puckett, 1991). Detection of reservoirs just missed is especially useful where hydrocarbon indicators (§10.8) are observed. Surveys using geophones in deep holes are also used in searching for and defining nearby features such as salt domes (see §11.1.3 and 13.7.1).

Use of a three-component borehole geophone yields additional information as to the direction from which energy approaches the geophone and also helps distinguish converted-wave energy (Noble et al., 1988). S-wave VSPs can be acquired on land using S-wave sources and a three-component geophone (§13.2.1).

Much work is now underway (Hardage, 1992; Massell, 1992) to find a source that can be used in the borehole. This would allow a multitude of geophones to be located on the surface. Such a reverse VSP

would markedly reduce survey time and cost. By using multiple-source locations on the surface, one can generate a 3-D VSP; however, such surveys today would require so much time in the borehole that they are run only rarely, but they should be feasible once more effective downhole sources are developed. A borehole energy source must be nondestructive to the well. One scheme uses the vibrations of the active drill bit as the energy source; this involves continuous recording for several minutes and cross-correlation of geophone outputs with the signal that travels up the drill stem, somewhat similar to Vibroseis processing.

VSPs can also be used (at least in theory) to study absorption (from amplitude and waveform changes in the downgoing wave; see §6.5), crack orientation (from S-wave birefringence; see §2.6.2), permeability (from tube waves), dip, variations of reflectivity with incident angle, converted waves, and so on.

13.4.3 Recording a VSP

Ordinarily, the well geophone is lowered to the bottom of the borehole on standard seven-conductor well cables and stopped for each recording on the way out of the hole. Both open and cased holes are used for recording VSPs. The geophone must be firmly coupled to the formation for recording (unless the phone is locked to the borehole wall upgoing reflections are usually too weak to be recorded because of the noise often present). It is important that the casing be well cemented to the formation (fig. 13.24), although satisfactory results are sometimes obtained in older poorly cemented holes where the formations have collapsed onto the casing over time. Recording often cannot be done where multiple-casing strings have been set (see fig. 13.27b).

The depth sampling should conform to sampling theorem constraints; when $V = 2000$ m/s and $\nu = 100$ Hz, $\lambda = 20$ m, so the sampling interval should be less than 10 m (§9.2.2c). However, sampling is usually coarser than this in order to cut costs, and 20 to 50 m depth sampling is common. Some slack is given to the cable at each recording to prevent disturbance from energy traveling down the cable. Thus, only one (or at most a very few) geophone locations can be occupied at a time. Consequently, a VSP survey is time-consuming and expensive, the cost of occupying the borehole being the major expense.

To achieve the same waveform for each of the many source impulses required, the source waveform should be very repeatable. Sources are usually kept small because these are richer in high frequencies. The most common marine source is a small air-gun array, but a single, large air gun is preferred by some because it is more repeatable, although its spectrum is poorer. On land a water-filled pit 5 m deep and 6 to 7 m across can be used with an air gun 2 to 4 m deep; the mud pit is sometimes used for this. The waveforms from explosives in boreholes are generally not sufficiently reproducible unless special precautions are taken,

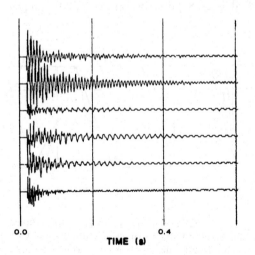

TIME (s)

0.0   0.4

13.24 Reverberatory VSP traces in a cased hole where the cement bond between the casing and the borehole wall is absent. (Balch and Lee, 1984: 158.)

such as centering small charges inside a heavy casing or a borehole of large (~1-m) diameter. Explosive charges are usually small, often 0.1 to 1.5 kg; the vertical stacking of records of several small charges is preferable to using a larger charge. Impulsive surface sources such as weight droppers are sometimes used. The most common land source is Vibroseis, which has the added advantage of distributing the energy over time so that background noise, often large near a well, is averaged out. Usually, several source impulses or Vibroseis sweeps are stacked to build up the signal strength. Much of the surface noise is attenuated by stacking the records from several source impulses at each depth. A monitor geophone (or hydrophone) located near the source is used to verify the constancy of the source waveform.

Well geophones are necessarily rather massive (fig. 13.25) in order to withstand well pressures and temperatures. An ideal VSP phone (Hardage, 1985: 47–52) would have (1) three orthogonal sensors with identical response characteristics, (2) a means of determining sonde orientation (tilt and azimuth), (3) a retractable device for locking to the borehole wall, (4) a mechanism to determine the coupling, (5) downhole digitizing (now often 12 bits), (6) small diameter and weight, and (7) means for measuring at several downhole locations simultaneously. A borehole geophone with all these features does not yet exist.

A borehole is an efficient transmitter of tube waves because these attenuate very slowly with distance. Tube waves (§2.5.5) can be generated whenever the borehole fluid is disturbed, as by the interaction of ground roll (the most likely cause) or of a P-wave at a contrast in the borehole (e.g., air–water surface, change in borehole diameter as at the base of surface casing, borehole sonde, bottom of borehole, or at a particularly permeable formation). Tube waves are also reflected at such contrasts. The most troublesome tube waves, which sometimes obscure reflections, are Stoneley waves (§2.5.3) that travel at about the P-wave velocity in the borehole fluid. Generally, tube waves cannot be eliminated by frequency filtering, but often their generation by ground roll can be weakened considerably by lowering the fluid level in the borehole. Tube-wave interference is substantially reduced by firmly clamping the well geophone to the borehole wall because axial tube-wave motion is very much smaller in the wall than in the borehole fluid. Although in theory tube waves give information about formation permeability, this is usually not a VSP objective.

The need for detailed information about the region immediately surrounding a borehole is often greatest in the marine environment where well costs are very high and many wells are drilled from a platform. Thus, deviated wells are common and many VSPs are run in deviated holes. Well geophones must be gimbal-mounted to maintain correct vertical attitude.

Schimschal (1986) makes the following points regarding VSP acquisition: (1) Make sure sonde depth

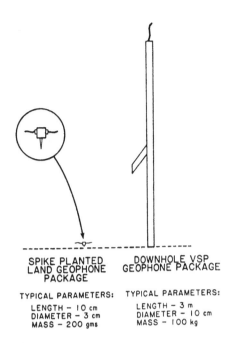

Fig. 13.25 Comparison to scale of conventional land geophone and well geophone. The well geophone has a pivot arm for locking the geophone against the borehole wall. (From Hardage, 1985: 36.)

is zeroed at the well head; (2) record five to six levels as the tool is being lowered; (3) check the depth reading, and determine the gain and number of records to be stacked to achieve the required signal-to-noise ratio at total depth; (4) take at least five records and monitor at every level; (5) slack the cable after anchoring; (6) reoccupy the down levels and check both times and waveforms; (7) avoid washed-out zones even if it causes uneven spacing; and (8) recheck depth at well head.

13.4.4 VSP processing

A portion of a zero-offset VSP is shown in fig. 13.26a. The slope of the first breaks (direct-wave traveltimes) gives the velocity. Reflections have a slope opposite to the first breaks. By using this difference, it is possible to separate downgoing waves (which consist of direct waves and multiples involving an even number of reflections, as in fig. 13.23a) from upgoing waves (reflections and multiples involving an odd number of reflections; see fig. 13.23b). The upgoing waves may be 30 dB below the downgoing waves. One way of separating downgoing from upgoing waves is to subtract or add the direct arrival times. Subtracting aligns the traveltimes of the downgoing waves horizontally, making these events much more readily seen. Adding the traveltimes of the direct arrival emphasizes the upward-traveling waves (fig. 13.26b), but the downgoing waves are usually still evident because they are stronger. The separation can be done more completely by apparent velocity filtering (fig. 13.27); it is usually

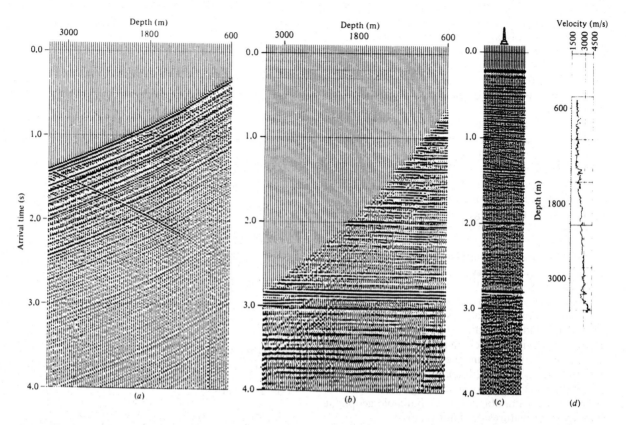

Fig. 13.26   Vertical seismic profile. (After Kennett, Ireson, and Conn, 1980: 680, 682, 688.) (a) Each trace is recorded by a geophone in the borehole using an air gun at the surface. (b) The same except each trace has been shifted by the direct-wave trav-eltime, thus horizontally aligning upcoming events including reflections. (c) Portion of the surface reflection record across the well. (d) Sonic log in the well.

more effective to attenuate the downgoing waves with a narrow band-reject $f–k$ filter than to simply pass the upward-traveling waves through a narrow band-pass $f–k$ filter. Filtering in the $f–k$ domain, however, involves smearing and wraparound aliasing (§9.9). Median filtering (Hardage, 1985: 189–94) often produces better separation. Downgoing and upgoing S-waves can be separated in the same way as P-waves.

All the downgoing energy (except for the P- and S-wave first arrivals and tube waves) must be multiples. Because we know both the input and the desired output (a single spike), a Wiener filter (§9.5.5) can be designed to remove surface multiples almost completely (*VSP deconvolution*). Moreover, the downgoing and upgoing multiples differ mainly by an additional reflection at or near the surface (which is apt to be a simple interface); therefore, the upgoing multiple pattern will be nearly the same as that of the downgoing multiples, so that the deconvolution filter for the downgoing multiples will effectively remove the upgoing multiples. The same deconvolution operator may be applied to nearby surface seismic data (fig. 13.28).

The traces of the upgoing VSP are often stacked together to yield the pattern of primary reflections for correlating to conventional surface seismic data. Only the portions just below the well geophone (CC' in fig.

13.29) are stacked in a *corridor stack;* these portions are most apt to be relatively free of peg-leg multiples. Corridor stacks are usually better than synthetic seismograms made from well-log measurements (fig. 13.30) for relating reflections to interfaces because the measurements are made at seismic frequencies and are not sensitive to logging uncertainties. Stacks of the portions of offset VSPs involving reflection points nearest the borehole are also used for this purpose.

With offset and walkaway VSPs, the reflection points move away from the borehole as the geophone-to-reflector distance increases. For a vertical borehole and horizontal reflectors, this gives reflection points located as shown in figs. 13.31a and 13.31b. A *VSP-to-CRP transform* relocates the data (fig. 13.32) (and resamples to a regular grid) to place the reflections at the reflecting points assuming horizontal reflectors. This transform incorporates a NMO correction also. Multioffset (but coplanar) VSPs can be combined with this transform, and conceptually it can be extended to 3-D for azimuthal VSPs. VSP-to-CRP transforms can accommodate deviated holes or unusual geometries, and an experienced geophysicist can often figure out the recording geometry from seeing the transform pattern.

Migration of offset VSP data to move dipping re-

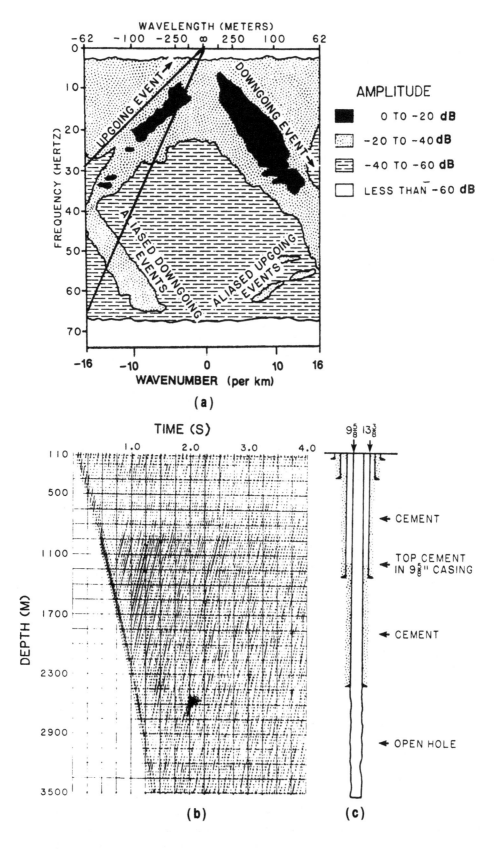

Fig. 13.27   *f–k* filtering of VSP data. (After Hardage, 1985: 111, 113.) (a) *f–k* plot of VSP data; an apparent velocity filter passing only data between the straight lines (*V* = 1800 and 4000 m/s) rejects most of the nonupgoing energy. (b) VSP data after the indicated *f–k* filtering and AGC; the shallow data are poor because of a double-casing string.

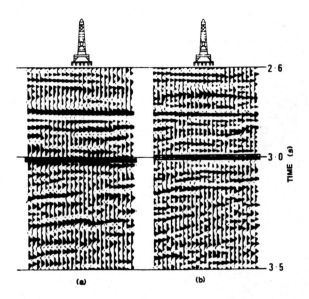

Fig. 13.28   Surface seismic data (a) after conventional deconvolution and (b) after deconvolution with VSP-derived deconvolution operator. (From Kennett, Ireson, and Conn, 1980: fig. 10.)

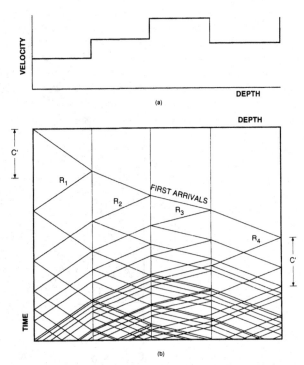

Fig. 13.29   Model of a VSP showing all multiples for four layers. Data from the region CC' just following the first arrivals is relatively free from intrabed multiples.

flections to the reflector locations is somewhat different from that of CMP data. After a VSP-to-CRP transform, reflections are horizontal if the reflectors are horizontal but curve if the reflectors dip (fig. 13.33). VSP migration corrects this.

Alignments on a VSP are not straight because velocity varies with depth. They can be made straight by stretching the time scale (fig. 13.34). Such stretching before separating downgoing from upgoing wavefields improves the separation. VSPs often contain gaps because of acquisition problems (such as caused by poor contact with the formations); interpolation along the straight alignments of upgoing or downgoing waves can fill in for the missing data.

Much VSP processing is the same as used with conventional seismic data. This includes wavelet processing to shorten the embedded wavelet and shape it to the surface data to which it is to be matched. Data are often displayed using AGC with a fairly long time constant, or, where relative amplitude is required (for example, before inversion), a correction assuming that amplitude varies proportionally to $1/V^2t$ (§6.5.1) may be applied, followed by a small exponential ramp (or very slow AGC) to allow for absorption, transmission, and other losses.

13.4.5 VSP planning

Ray-trace modeling is used in planning VSP surveys to determine source offsets for the required subsurface coverage, frequency content required to resolve features, Fresnel-zone effects, events to be expected, and so on, to make sure that objectives are achieved. It is also used as an aid in understanding the VSP evidences of geologic features such as dip, faults, angular

unconformities, diffractions, and so on.

The objectives of VSP surveys (paraphrased from Gilpatrick and Fouquet, 1989) are listed in table 13.1. Examples of VSP applications to these various objectives can be found in Balch and Lee (1984) and Hardage (1985). Cramer (1988) applies VSP to the mapping of a point-bar sand and Noble et al. (1988) to resolving structural problems in the Vulcan Gas Field.

13.5 Seismic tomography

13.5.1 General

Tomography (from the Greek "tomos," or "section") means a picture of a cross-section of an object. In practice, the term denotes determining the internal properties of an object from external measurements on rays that passed through the object. X-ray tomography has been used for some time in medical examination and in nondestructive testing. The computer-assisted tomography (CAT-scan) technique uses X-rays that have penetrated a body along many raypaths in many directions and tomography is used to explain the loss in intensity of the X-rays because of the absorptive properties of different parts of the body.

Tomographic analysis usually assumes either that the property being determined is a continuous function of position (transform methods) or that a medium is composed of a finite number of elements, each of which has a discrete value of the property. The first method implies a continuous distribution of rays

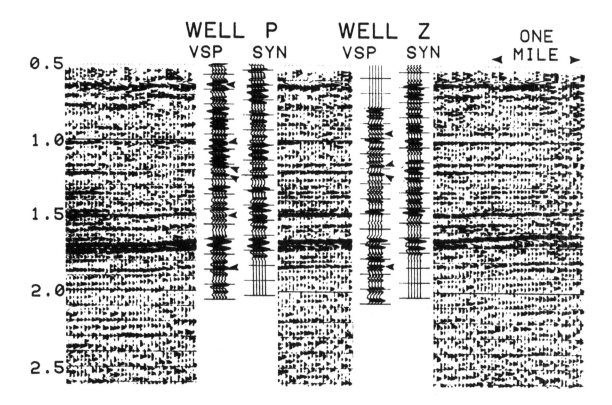

Fig. 13.30 Surface seismic data with inserted VSP corridor stacks and synthetic seismograms. The match with the corridor stacks is better than with the synthetics at the arrows. (From Hardage, 1985: 283.)

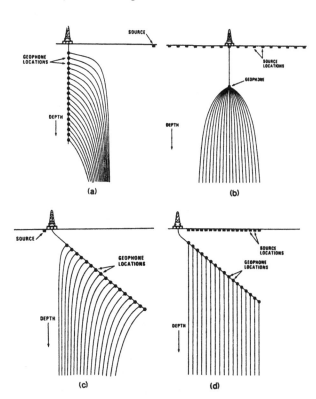

Fig. 13.31 Loci of reflection points for VSPs for flat reflectors. (From Balch and Lee, 1984: 82–5.) (a) For an offset VSP, (b) a walkaway VSP, (c) a VSP in a directional well, and (d) a VSP in a directional well where the source moves to stay above the well geophone.

(hence, an infinite number of rays). Seismic tomography, which uses a quite limited number of rays, clearly is better suited to the second class of techniques. However, to introduce some important concepts, we shall first discuss integral methods. Our discussion follows closely that of Stewart (1991) except for notation.

13.5.2 Tomographic concepts

In fig. 13.35a, we show several rays passing from sources S_i to receivers R_i that register values that depend on some property $g(x, y)$ of the medium M. The recorded values are those of the Radon transform of $g(x, y)$ along raypaths such as $R_1 S_1$ in fig. 13.35b (see also eq. (9.21)). In the following, we use $G(\ell, \theta)$ for the Radon transform (*projection*) of $g(x, y)$, $G_g(u, v) = G_g(\rho, \theta)$ for the 2-D Fourier transform of $g(x, y)$, $G_p(\rho, \theta)$ for the 1-D Fourier transform of $G(\ell, \theta)$. We wish to find the values of $g(x, y)$ using only the projections. We start with the Fourier projection theorem, which states that the 2-D Fourier transform of the object $g(x, y)$ is equal to the 1-D transform of the projections $G(\ell, \theta)$. The proof is as follows.

The 2-D Fourier transform of $g(x, y)$ is (see eq. (9.19))

$$G_g(u, v) = \int_{-\infty}^{\infty}\int_{-\infty}^{\infty} g(x, y)e^{-j2\pi(ux+vy)} \, \mathrm{d}x \, \mathrm{d}y \quad (13.8)$$

(note that $g(x, y) = 0$ outside the object, so the infinite

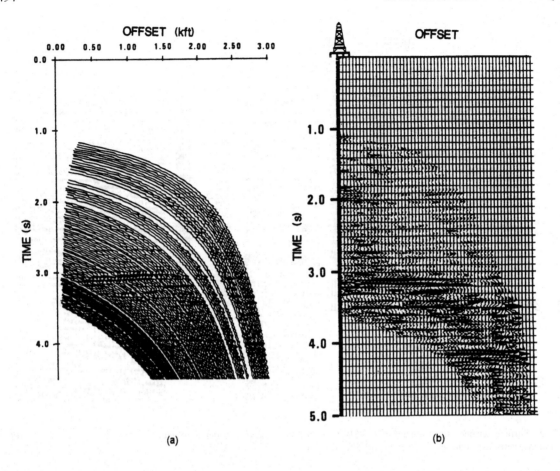

(a)

(b)

Fig. 13.32 The VSP-to-CRP transform. (From Dillon and Thomson, 1984: figs. 11 and 14.) (a) Illustrating the concept of displaying offset VSP traces along midpoint loci curves such as shown in fig. 13.31a. (b) VSP-to-CRP transformed data from an offset VSP.

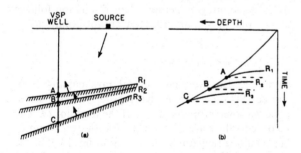

Fig. 13.33 Dipping reflections after VSP-to-CRP transform have curved alignments. Reflection intersections with the downgoing direct wave (A, B, and C) do not require migration. (From Hardage, 1985: 210.)

limits are merely a convenience and the actual limits are the bounds of the medium M). Replacing (u, v) with polar coordinates (ρ, θ), where $u = \rho \cos \theta$, and $v = \rho \sin \theta$, $G_g(u, v)$ becomes $G_g(\rho, \theta)$:

$$G_g(\rho, \theta) = \int_{-\infty}^{\infty}\int_{-\infty}^{\infty} g(x, y)e^{-j2\pi\rho(x \cos \theta + y \sin \theta)}\, dx\, dy.$$

The 1-D transform of the projection in eq. (9.21) is

$$G_p(\rho, \theta) = \int_{-\infty}^{\infty}\left[\int\!\!\int_{-\infty}^{\infty} g(x, y)\, \delta(x \cos \theta + y \sin \theta - \ell)\, dx\, dy\right]e^{-j2\pi\rho_\ell}\, d\ell.$$

Interchanging the order of integration (see §15.2.8) gives

$$G_p(\rho, \theta) = \int_{-\infty}^{\infty}\int_{-\infty}^{\infty} g(x, y)\left[\int_{-\infty}^{\infty} e^{-j2\pi\rho'}\, \delta(x \cos \theta + y \sin \theta - \ell)d\ell\right] dx\, dy$$

$$= \int_{-\infty}^{\infty}\int_{-\infty}^{\infty} g(x, y)e^{-j2\pi\rho(x \cos \theta + y \sin \theta)}\, dx\, dy$$

(because the inner integrand is zero except when $\ell = x \cos \theta + y \sin \theta$). Thus,

$$G_p(\rho, \theta) = G_g(\rho, \theta), \qquad (13.9)$$

that is, the 2-D Fourier transform of $g(x, y)$ is equal to the 1-D Fourier transform of the projection.

Backprojection is the mapping from the (ℓ, θ) domain back to the (x, y) domain; it is approximately equivalent to inverse transformation. Backprojection

Table 13.1 *Objectives of VSP surveys*

Objective	How achieved
Reflector identification Surface-to-borehole correlation } Increased resolution at depth	Upgoing wave studies on zero-offset VSP
Time–depth conversion Enhanced velocity analysis } Log calibration	First-break studies on zero-offset VSP
Multiple identification } Deconvolution operator	Downgoing wave studies on zero-offset VSP
Improve poor data area	All types, especially offset VSP
Predict ahead of bit	Upgoing wave studies on zero-offset VSP
Structural imaging	Walkaway or offset VSP with presurvey modeling
Delineate salt dome	Proximity survey with source over dome
Seeing above/below bit on deviated wells	Zero-offset, offset, or walkaway VSP
Stratigraphic imaging (channels, faults, reefs, pinchouts)	Multiple-source locations with offset VSP
AVO studies	Research study on offset VSP with presurvey modeling
P/S-wave analysis Polarization studies } Fracture orientation	Research study on offset VSP, three-component phone
Attenuation analysis	Research study on zero-offset VSP
Secondary recovery	Research study on offset VSP
Tomographic studies	Multiple wells, multiple offsets
Permeability studies	Tube-wave analysis research study

After Gilpatrick and Fouquet, 1989.

consists of summing all projections to which a certain property value $g(x_1, y_1)$ has contributed, that is, we sum the measured values for all rays that passed through the point (x_1, y_1). This operation can be expressed by an integral, the backprojection $h(x, y)$ being

$$\left.\begin{aligned} h(x, y) &= \int_0^\pi G(\ell, \theta)\, d\theta \\ &= \int_0^\pi G(x \cos \theta + y \sin \theta, \theta)\, d\theta, \end{aligned}\right\} \quad (13.10)$$

where $G(\ell, \theta)$ is given by eq. (9.21).

Although eq. (13.10) does not reproduce the object, that is, $h(x, y)$ is not equal to $g(x, y)$, it does give an approximate picture of $g(x, y)$. By using a rather loose analogy, just as a seismic reflection appears as a ringy embedded wavelet rather than as a spike, the backprojection is a blurred image, and we can sharpen the image by deconvolution (filtering). We obtain $g(x, y)$ exactly by transforming the projections, filtering, and then taking the backprojection. To show this, we start with the inverse 2-D Fourier transform of $G_g(u, v)$ (see eq. (13.8)):

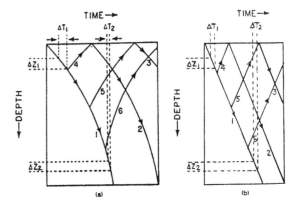

Fig. 13.34 Stretching the VSP time scale to get consistent time increments between traces straightens alignments.

$$g(x, y) = \int_{-\infty}^{\infty} \int_{-\infty}^{\infty} G_g(u, v) e^{j2\pi(ux + vy)}\, du\, dv.$$

Changing to polar coordinates (ρ, θ), we have

$$g(x, y) = \int_0^{2\pi} \int_0^{\infty} G_g(\rho, \theta) e^{j2\pi\rho\ell} \rho\, d\rho\, d\theta, \quad (13.11)$$

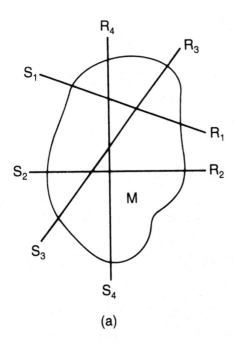

(a)

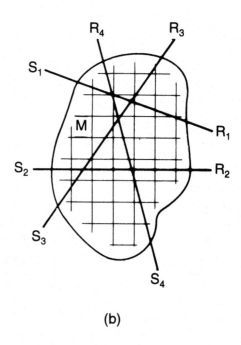

(b)

Fig. 13.35 Tomographic concepts. (a) Raypaths from various sources to various receivers; solution can be obtained if enough raypaths penetrate the medium in all directions. (b) In practice, a grid of uniform cells are usually assumed.

where we have used the relations $\ell = x \cos \theta + y \sin \theta$, and that $dv\ dy =$ "area" $= d\rho(\rho\ d\theta)$. We wish to change the limits of the integrals; recall that replacing θ with $\theta + \pi$ changes the signs of $\cos \theta$, $\sin \theta$, and ℓ, and that in polar coordinates, $f(\rho, \theta) = f(-\rho, \theta + \pi)$. We now separate the θ integral in eq. (13.11) into parts; thus,

$$g(x, y) = \int_0^\pi \int_0^\infty G_g(\rho, \theta)e^{j2\pi\rho\ell}\rho\ d\rho\ d\theta$$
$$+ \int_\pi^{2\pi} \int_0^\infty G_g(\rho, \theta)e^{j2\pi\rho\ell}\rho\ d\rho\ d\theta.$$

The lower integral is equal to

$$-\int_0^\pi \int_\infty^0 G_g(\rho, \theta + \pi)e^{+j2\pi\rho\ell}\rho\ d\rho\ d\theta$$
$$= \int_0^\pi \int_{-\infty}^0 G_g(-\rho, \theta + \pi)e^{-j2\pi\rho\ell}|\rho|\ d\rho\ d\theta,$$

because $-\rho = |\rho|$ when ρ is negative. Therefore, eq. (13.11) can be written

$$g(x, y) = \int_0^\pi \int_{-\infty}^\infty G_g(\rho, \theta)e^{j2\pi\rho\ell}|\rho|\ d\rho\ d\theta,$$

or by eq. (13.9),

$$g(x, y) = \int_0^\pi \left[\int_{-\infty}^\infty G_p(\rho, \theta)e^{j2\pi\rho\ell}|\rho|\ d\rho \right] d\theta.$$

The inner integral, which we shall denote by $G_{ip}(\ell, \theta)$, is a filtered version of $G_p(\rho, \theta)$, which in turn is the

1-D Fourier transform of the projections. Moreover,

$$g(x, y) = \int_0^\pi G_{ip}(\ell, \theta)\ d\theta, \qquad (13.12)$$

so that $g(x, y)$ is obtained by the backprojection of the Fourier-transformed filtered projections.

13.5.3 Solution for a limited number of discrete cells

We divide the medium M into m cells or pixels and assume n rays passing through M (fig. 13.35b). Each pixel has a property value $g(x, y)$ and the recorded value at R_i is a projection equal to the sum of the products $d_j^i g_j$, where d_j^i is the path length of the ith ray in the jth cell. Assuming the usual case where g_j is the slowness, we write for the traveltime

$$t^i = d_1^i g_1 + d_2^i g_2 + \cdots + d_m^i g_m = \sum d_j^i g_j,$$
$$i = 1, 2, \ldots, n. \qquad (13.13)$$

Of course, any given ray will not pass through all cells so that many of the d_j^i are zero. If we measure amplitude instead of traveltime, we can take the logarithm of eq. (2.110), and eq. (13.13) becomes

$$\ln (A_0^i/A^i) = \sum_j d_j^i \eta_j. \qquad (13.14)$$

Equation (13.14) can be written in matrix form as

$$\mathcal{T} = \mathcal{D}\mathcal{G}, \qquad (13.15)$$

where $\mathcal{D}$ is a $n \times m$ matrix relating column matrices $\mathcal{T}$ and $\mathcal{G}$. Given $\mathcal{T}$, we wish to find $\mathcal{G}$. This is only possible if we know $\mathcal{D}$, at least approximately. When $\mathcal{D}$ is known exactly and $n = m$, an exact solution can be obtained. However, the important case is when the number of raypaths exceeds the number of cells ($n > m$) and $\mathcal{D}$ is known only approximately; in this situation, a number of solution methods are available (Herman, 1980). A least-squares solution is possible but is time-consuming. Stewart (1991: 2–27, 2–28) describes a solution based on backprojection that gives an approximate result. More suitable solutions are given by the algebraic reconstruction technique (ART) and simultaneous reconstruction technique (SIRT) or one of the methods described by Justice et al. (1992).

The basic problem is to solve a set of n linear equations with m unknowns ($n > m$), where the $m \times n$ coefficients are known only approximately. ART starts by guessing values of the parameters g_j and uses these values to calculate t^1. The differences between calculated and observed values of t^2 are used to vary the g_j so as to minimize the error (difference). The adjusted values of g_j are used in the second equation and another adjustment of g_j is made. The process is continued until all of the equations have been used and then the entire process is iterated. Stewart (1991: 2–28 to 2–34) gives further details; see also Gordon (1983) and Herman, Lent, and Rowland (1973).

In using ART, the adjusted parameters obtained from one equation are used in the next equation. In SIRT (Stewart, 1991: 2–34 to 2–36; Dines and Lytle, 1979), the errors are calculated for all equations using the first guess, and the average errors are used to adjust the parameters for a second pass through the set of equations.

13.5.4 Cross-hole measurements

Unlike well-logging and VSP measurements that involve a single borehole, cross-hole studies involve sources and detectors in different boreholes (fig. 13.36). Usually, several receiver locations in one borehole record data from a number of source locations in another borehole, and sometimes recording is done on the surface at the same time. The recording procedure is time-consuming because usually only one or a very few receiver and source locations can be occupied at a time, but developments are underway to permit the simultaneous use of several receivers and to increase the source energy. The objective of cross-hole studies is to learn about the region between the boreholes (Wong et al., 1987; Rutledge, 1989; Lines, 1991). Nearby boreholes are most often available in developing and producing oil fields, so cross-hole studies usually have reservoir geophysics objectives.

Tomography is usually employed in interpreting cross-hole data. Both slowness (1/velocity) and absorption satisfy the requirements for tomographic image reconstruction (Deans, 1983). In theory, we can measure both traveltime and attenuation and solve for the unknown velocity field and absorption distribution (*Q-map*) in the intervening region. However, usually, we try only to determine the velocity distribution in the intervening region from measurements of seismic traveltimes because amplitudes cannot be measured with sufficient accuracy to determine absorption. Most applications to date use only the first-arrival information in their solutions.

The often-made analogy between seismic tomography and CAT-scanning should take into account that in a CAT-scan, the X-rays travel in straight lines in many directions, whereas in seismic tomography, the different directions raypaths can take are limited and the raypaths are not straight (fig. 13.36b).

Although generally the number of observations is much larger than the number of cells and the problem is overdetermined, some cells may not have been traversed, so that their slownesses cannot be determined and many of the travel paths may have traversed the same subset of cells so that their individual contributions cannot be separated. All measurements involve uncertainty, and the reliability of a determination depends on the number and range of directions of the raypaths traversing a cell. In CAT-scanning, we can have raypaths traversing the body in many directions, but seismic applications are constrained by the boreholes available for measurement. Also, the zone of major interest is often near the bottom of the boreholes, so that only nearly horizontal travel is possible for raypaths in this zone.

The seismic problem is unlike the CAT-scan problem in another very important regard: Seismic raypaths bend (fig. 13.36b) appreciably as the velocity changes. This makes the seismic problem nonlinear because a change in the slowness of any cell changes not only the traveltime, but also the raypath. Forward modeling to recompute the traveltimes through the modified model has to be carried out anew for each of the various raypaths before each iteration of the tomographic algorithm.

Many ray-tracing methods are not suitable for forward modeling because Snell's law does not apply at the cell boundaries, which exist only as a mathematical device and do not correspond with actual interfaces (that is, they have arbitrary rather than natural orientations). Figure 13.37a illustrates this problem; rays only incrementally different can strike near the corner of a cell where the velocity contrast is large. The ray that strikes the upper boundary first will be bent in an entirely different direction from that of the ray that strikes the side boundary first, and hence the corner will produce a major discontinuity with no raypaths at all entering a fairly large region unless diffraction is taken into account. To accurately account for the actual observed arrivals, the forward-modeling algorithm must include arrivals that result from diffraction and critical refraction. The procedure of Justice et al. (1992) involves computing the full acoustic wavefield at closely spaced increments in time, thus allowing for diffraction, and then tracing rays along orthogonal trajectories backward (in time) through the wavefronts to find the minimum-time ray connecting a source to a receiver.

The tomographic solution is almost always itera-

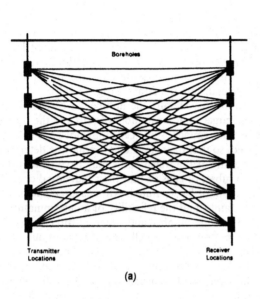

(a)

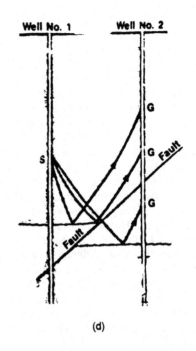

(d)

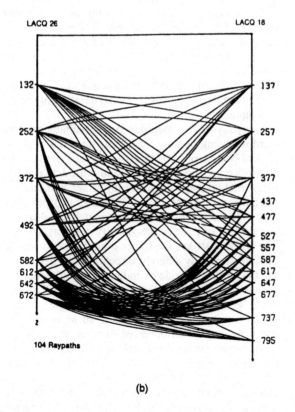

(b)

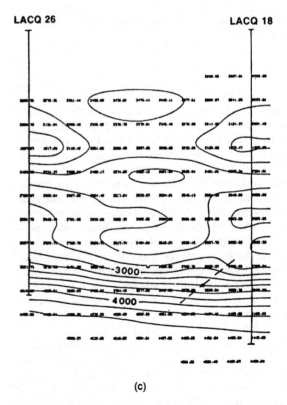

(c)

Fig. 13.36 Cross-borehole methods. (a) Sources and geophones in nearby boreholes provide criss-crossing raypaths. (b) In an actual situation, raypaths bend and concentrate in high-velocity cells, leaving many cells poorly sampled. (c) Velocity determined in the actual situation shown in part (b) (from Bois et al., 1972: figs. 7, 9.). (d) Reflection tomography concept.

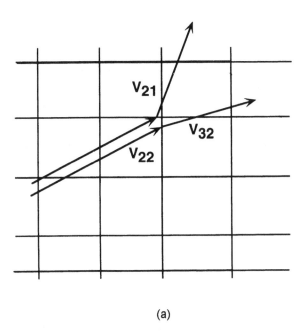

(a)

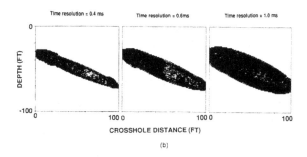

(b)

Fig. 13.37 Problems with modeling cross-hole traveltimes. (a) Subdivision of space into cells with arbitrary boundaries means that Snell's law cannot be applied at the cell boundaries; two nearly parallel raypaths may take off in quite different directions because they strike different boundaries first. (b) For given uncertainties, raypaths can lie anywhere within the shaded regions (which are called *leaves*); uncertainties are respectively 0.4, 0.6, and 1.0 ms (Justice et al., 1992: 324).

tive. As stated earlier, a starting slowness distribution is assumed, raypaths are tracked through it, and the problem is solved for the changes in slowness that produce a closer solution. The model is then modified, new raypaths traced, and the process repeated until the slowness distribution matches observations within acceptable tolerances. Constant slowness is often assumed for the starting model.

Still another problem is that the solution is very sensitive to small measurement errors. Justice et al. (1992) illustrate this (fig. 13.37b) by showing the large volumes through which raypaths could pass allowing for small traveltime uncertainties. Relatively minor deviations of a well bore, for example, can introduce significant timing errors.

Consideration of a few simple models brings out other problems involved in a solution. Because source

and detector locations are limited to existing available boreholes (or the surface), travelpaths have much larger horizontal than vertical components. Figure 13.36b shows raypaths in a real situation involving high- and low-velocity layering (fig. 13.36c); the shortest traveltimes maximize travel in the high-velocity layers so that many low-velocity cells are not traversed and thus cannot be investigated. (However, channel waves (§13.3) travel preferentially in low-velocity layers and thus in concept provide a method for their investigation.) Also, because most raypaths traverse the same central cells of the higher-velocity regions, an increase in the slowness of one of these cells can be compensated by a decrease in a neighboring cell so that their individual values cannot be determined very accurately. The fact that many boreholes stop near the zone of interest prevents adequate sampling of cells at that depth.

Theoretically, many of the foregoing problems can be eliminated by considering entire recorded traces rather than only the first arrivals, but these applications are still in the research stage. Techniques employing reflected (fig. 13.36d) as well as direct travelpaths are being developed.

The foregoing assumes all travel is within the plane connecting sources to receivers; clearly, many realistic cases require allowance for travel outside this plane. Tomographic methods are also sometimes used with borehole-to-surface and surface-to-surface (reflection) measurements.

13.6 Time-lapse measurements

In principle, any measurement can be repeated at different times in order to determine changes that have occurred during the intervening time. In particular, measurements on a reservoir at different times during its production history sometimes can be used to monitor changes because of movement of fluids within the reservoir. Repetitive 3-D surveys, sometimes called *4-D surveys* (the fourth dimension being time), are used for reservoir surveillance (§14.4.5). It is important that the data collection and processing be identical from one survey to the next; hence, geophones are commonly cemented at the bottom of shallow boreholes to equalize geophone planting. Vertical and horizontal sections from the first survey (*base survey*) can be subtracted from those at later times to give *difference sections*, which display the changes in a very sensitive way. Time-lapse methods are also applied to cross-hole and other types of measurements.

Enhanced-oil-recovery (EOR) operations using thermal methods (fire and steam floods) in heavy-oil reservoirs have been monitored experimentally. Heating reduces the viscosity of the oil and allows it to flow more easily. An increase in temperature markedly decreases the seismic velocity of heavy-oil reservoirs (see §5.2.6), and sometimes we can observe changes associated with the lowered velocity and/or depressing of underlying reflections. In a fire flood, a portion of the oil is burned in place, the burn being controlled by the injection of oxygen or air. The waste gases from the combustion aid in pushing the oil toward production wells. The asymmetric expansion of the bright spot in fig. 13.38a indicates that the fire flood did not proceed uniformly away from the injection well where the in-situ combustion was initiated. In a steam flood, injected steam heats the reservoir. Difference sections or difference time slices, such as fig. 13.38b, show areas interpreted as heated zones. The expanding heated zones resulting from steam injection have also been monitored using cross-well tomograms in a time-lapse way.

The movement of reflections associated with an oil/water contact can sometimes be monitored in water flooding. A water flood may also produce visible changes because of the temperature changes resulting from the invading cold water. Changes in a gas cap may produce visible changes in a bright spot or gas–oil or gas–water reflections. The fluid fronts in carbon dioxide floods sometimes can be monitored.

13.7 Borehole studies

13.7.1 Salt-proximity surveys

Precise definition of the flanks of a salt dome are important because hydrocarbon accumulations are often trapped there (see §10.3.4). Beds truncating at the salt dome often have considerable dip, so that an appreciable amount of attic oil will be left behind if a well is not drilled close to the salt seal. Salt proximity surveys are sometimes run in wells drilled into the salt (see §11.1.3 and fig. 13.39), but more often they are run in wells drilled close to the salt; they can be thought of as a form of VSP survey. The energy source is placed on the surface above the salt and a geophone in the well records arrivals at several positions. For each source–geophone combination, an *aplanatic curve* (the locus of all points where the sum of salt and sediment traveltimes satisfies the measured arrival time) is constructed; see fig. 11.6. The envelope of the aplanatic curves is then the interpreted position of the salt face. Where the source moves to different azimuths, a 3-D solution can be obtained.

Steeply dipping reflections from the sediment–salt interface can also be migrated to locate the salt flank (Ratcliff, Gray, and Whitmore, 1992). Diving waves that reflect on their upward leg from overhanging flanks (turning waves) can also be migrated.

13.7.2 Sonic waveform logging

Several types of logs record the entire seismic trace rather than just the first arrival (James and Nutt, 1985). The sonic waveform log is an extension of the regular sonic log in the same sense as a vertical seismic profile is an extension of the well-velocity survey. A recorded waveform log (as in fig. 5.39, an array-sonic log) permits some control over measuring transit time by tracking the first arrival. In addition to measuring the *P*-wave arrival, an *S*-wave arrival can be observed and tracked where the *S*-wave velocity is greater than the *P*-wave velocity in the borehole fluid. The dipole sonic log also yields this information. Determining *S*-wave velocity is generally possible in well-consolidated but not in poorly consolidated sections. The *S*-wave arrivals can be used to generate a shear transit-time log. This is valuable additional information that permits one to calculate the velocity ratio V_P/V_S and Poisson's ratio, and thus get some indication of lithology. An *S*-wave synthetic seismogram can be generated for event identification on *S*-wave sections. A sonic waveform log in a deviated well can record reflections from nearby acoustic interfaces. Sonic images are also recorded with special tools.

13.7.3 Borehole televiewer

The borehole televiewer is an ultrasonic acoustic device that takes "pictures" of the borehole wall with little penetration into the rock formations. The tool is held centrally in the well bore and an acoustic transducer operating at megahertz frequencies spins on a vertical axis; a compass recording is made each time it passes through north. The traveltime of the recorded signal measures the borehole radius and the amplitude measures the wall's reflectivity. Figure 13.40 shows televiewer sections displaying both of these parameters against geographical direction and depth in the borehole. Anomalously long traveltimes or low amplitudes indicate fractures, and the televiewer is used to identify fractures and measure their orientations. Fractures striking east–west and dipping steeply to the north are clearly visible in fig. 13.40. The borehole televiewer competes with the formation microscanner (which uses dipmeter and microresistivity measurements; see Telford, Geldart, and Sheriff, 1990: §11.4 and 11.2.5) as a borehole tool for seeing fractures.

13.8 Passive seismic methods

Seismic applications that do not involve controlled sources are said to be *passive*. Activities within a reservoir, such as fluid flow within a fracture system, periodic strain accumulation and release due to gas buildup and flow, thermally induced failure caused by a fireflood, rock breakage during hydraulic fracturing, subsidence due to fluid withdrawal, or gas flow in a blown-out well, may create microseismic events. Plotting the locations of observed discrete events might give useful information (Dobecki, 1992). Such seismic events are apt to be extremely weak (perhaps of Richter magnitude −3). Dobecki refers to passive mapping of combustion in a Canadian heavy-oil EOR project, combustion–gasification studies in a Wyoming coal seam, and hydraulic fracturing in coal-bed methane production.

Downhole seismic arrays have been used to monitor

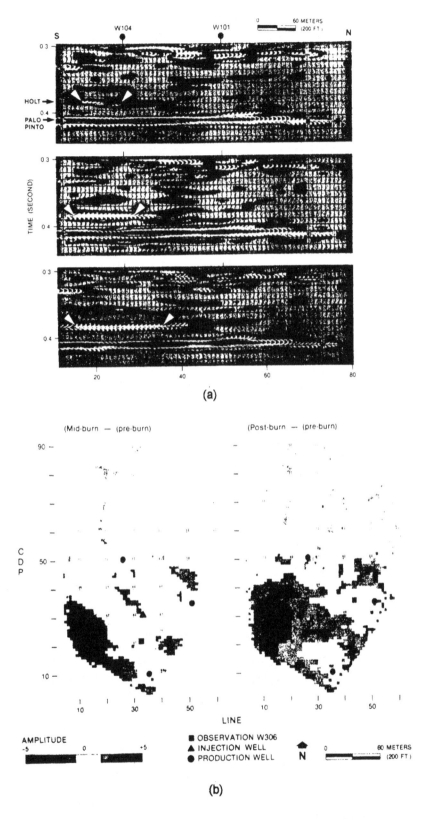

Fig. 13.38 Differences induced by a fire flood. (From Greaves and Fulp, 1992: 313, 314.) (a) Vertical sections before, during, and after a fire flood (4 and 10 months having lapsed). (b) Difference time slices through the zone of the fire flood.

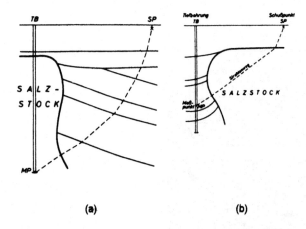

(a)   **(b)**

Fig. 13.39 Salt-proximity surveys. (From Steinmann and Rühmkorf, 1968.) (a) Geophone in a well in a salt dome, and (b) in a well near a salt flank.

the growth of hydraulic fractures and determine their orientation in petroleum reservoirs and in a hot dry-rock geothermal project (Dobecki, loc. cit.). Passive measurements may help to describe natural stress patterns and thus give information about fractures.

13.9 Joint inversion

Joint inversion combines different kinds of data sets to produce a better interpretation of the subsurface by reducing the number of possible alternatives. Interpretation should deal with all data, and joint inversion provides a formal way of achieving this. Gravity and seismic data provide the most common combination (Ramo and Bradley, 1982; Lines et al., 1993) as both explicitly depend on density. Sheriff (1987: 106–8) shows how a gravity map over a salt dome can be used to check on the plausibility of the interpretation of seismic data over the salt dome.

13.10 Geostatistical methods

Geophysical measurements vary because of changes in rock properties of several types, as was seen in chap. 5, and usually one cannot determine the distribution of any single property without uncertainties being introduced because of possible changes in other properties. Geostatistics attempts to cope in a statistical way with changes introduced by unknown parameters so as to yield a probable distribution of a sought-for property based on some combination of the available measurements. Geostatistics deals with relations among seismic amplitude, seismic velocity, well-log measurements, density, lithology, bed thickness, porosity, interstitial fluid, and possibly other factors. It generally assumes that the property being sought is given by a linear relation among the available measurements and that the relation changes slowly because of factors that are not measured. Values are often made exact at control points, and available data

that are not used in determining the parameters (*hidden data*) are used to check the reliability of the analysis. A measure of the confidence that can be placed in the calculations is usually generated. Sheriff (1992) shows a number of geostatistical methods.

Problems

13.1 In a marine survey, the water depth is 1 km and a reflector is 3 km below the sea floor. Use fig. 13.5 to determine the optimum range of offset for S-wave generation. The P-wave velocity just below the sea floor is 2.8 km/s and the water velocity is 1.5 km/s.

13.2 Show that the angle between the vertical and each of three orthogonal geophones, equally inclined to the vertical, is 54.74°. (*Hint:* Find the direction cosines of a line passing through the origin of a set of orthogonal axes and equally inclined to the x-, y-, and z-axes.)

13.3 (a) In fig. 13.19b, $v_0 = 40$ Hz; find the water depth using data from the figure.
(b) Find the frequencies that are reinforced when the rays are reflected at angles of 30° and 40° to the vertical.
(c) Calculate θ for successive modes with frequencies of 180 and 300 Hz.
(d) Find V for cases (a), (b), and (c).

13.4 A source is offset 1000 m from a vertical well in which a geophone is suspended; a horizontal reflector is present at 2000 m; $V = 3$ km/s.
(a) Find the traveltime both graphically and by calculation for a geophone at depths $z = 800, 1200, 1600,$ and 2400 m; find the values $\Delta t/\Delta z$.
(b) Repeat for reflector dips of $\xi = \pm 7°$ where the reflectors intercept the well at the same depths as in part (a).
(c) By how much do the values in part (a) change if the well deviates by 3° toward the source.

13.5 A source is offset 2400 m west of a vertical well in which a geophone is suspended. There is a vertical N–S fault 600 m west of the well with velocities $V = 2.50$ and 3.00 km/s west and east of the fault, respectively, and a horizontal reflector is present at 2000 m.
(a) Find the reflection traveltime for geophone depths of 250 and 1000 m.
(b) What is the deepest geophone location for which the reflection can be seen.

13.6 An outcropping salt dome has roughly vertical flanks. A source is located on the salt and a geophone

Table 13.2 *Survey to define salt-dome flan*

z (m)	t (s)	z (m)	t (s)
500	0.325	1750	0.457
750	0.337	2000	0.490
1000	0.366	2250	0.530
1250	0.397	2500	0.580
1500	0.422		

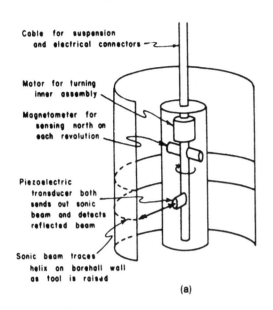
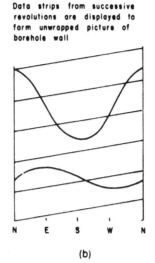
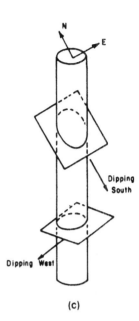

Fig. 13.40 Borehole televiewer. (From Zemanek et al., 1970.)
(a) Schematic of sonde in the borehole, (b) appearance of two
plane fractures, and (c) the fractures shown in part (b).

is suspended in a vertical well in the sediments 1400
m from the source point. Determine the outline of the
salt dome from the t–z data in table 13.2. Take the
velocities in the salt and adjacent formation as 5.00
and 3.00 km/s. (*Hint:* Draw circles on transparencies
as in fig. 11.6 and overlay these to find points on the
aplanatic surfaces for each geophone position.)
13.7 Graph Poisson's ratio vs. *P*-wave traveltime for
the five events in fig. 13.7.

References

Anno, P. D. 1987. Two critical aspects of shear-wave analysis:
Statics solutions and reflection correlations. In *Shear-Wave Ex-
ploration*, S. H. Danbom and S. N. Domenico, eds., pp. 48–61,
Vol. 1 in Geophysical Development Series. Tulsa: Society of
Exploration Geophysicists.

Balch, A. H., and M. W. Lee. 1984. *Vertical Seismic Profiling:
Technique, Applications, and Case Histories.* Boston: International Human Resources Development Corp.

Bois, P., M. La Porte, M. Lavergne, and G. Thomas. 1972. Well-
to-well seismic measurements. *Geophysics*, **37**: 471–80.

Cassell, B. 1984. Vertical seismic profiles: An introduction. *First
Break*, **2(11):** 9–19.

Clay, C. S., and H. Medwin. 1977. *Acoustical Oceanography.*
New York: John Wiley.

Cramer, P. W. 1988. Reservoir development using offset VSP
techniques in the Denver-Julesburg Basin. *J. Petrol. Tech.*, Feb-
ruary: 197–205.

Crampin, S. 1987. Crack porosity and alignment from shear-
wave VSPs. In *Shear-Wave Exploration*, S. H. Danbom and
S. N. Domenico, eds., pp. 227–51. Vol. 1 in Geophysical Development Series. Tulsa: Society of Exploration Geophysicists.

Danbom, S. H., and S. N. Domenico, eds., 1987. *Shear-Wave
Exploration*, Vol. 1 in Geophysical Development Series. Tulsa:
Society of Exploration Geophysicists.

Deans, S. R. 1983. *The Radon Transform and Some of Its Applications.* New York: John Wiley.

Dillon, P. B., and R. C. Thomson. 1984. Offset VSP surveys and
their image reconstruction. *Geophys. Prosp.*, **32**: 790–911.

Dines, K. A., and R. J. Lytle. 1979. Computerized geophysical
tomography. *Proc. IEEE*, **87**: 1065–73.

Dobecki, T. L. 1992. Alternative technologies. In *Reservoir Geophysics*, R. E. Sheriff, ed., pp. 335–43. Vol. 7 in Investigations
in Geophysics Series. Tulsa: Society of Exploration Geophysicists.

Ebrom, D., and R. E. Sheriff. 1992. Anisotropy and reservoir
development. In *Reservoir Geophysics*, R. E. Sheriff, ed., pp.
355–61. Tulsa: Society of Exploration Geophysicists.

Ensley, R. A. 1989. Analysis of compressional and shear-wave
seismic data from the Prudhoe Bay field. *The Leading Edge*,
8(11): 10–31.

Ewing, W. M., W. S. Jardetsky, and F. Press. 1957. *Elastic Waves
in Layered Media.* New York: McGraw-Hill.

Fitch, A. A. 1984. Interpretation of vertical seismic profiles.
First Break, **2(6):** 19–23.

Fix, J. E., J. D. Robertson, and W. C. Pritchett. 1987. Shear-
wave reflections in three West Texas basins with high-velocity
surface rocks. In *Shear-Wave Exploration*, S. H. Danbom and
S. N. Domenico, eds., pp. 180–96, Vol. 1 in Geophysical Development Series. Tulsa: Society of Exploration Geophysicists.

Garotta, R. 1987. Two-component acquisition as a routine procedure. In *Shear-Wave Exploration*, S. H. Danbom and S. N.
Domenico, eds., pp. 122–36, Vol. 1 in Geophysical Development Series. Tulsa: Society of Exploration Geophysicists.

Gilpatrick, R., and D. Fouquet. 1989. A user's guide to conventional VSP acquisition. *Geophysics, The Leading Edge of Exploration*, **8(3):** 34–9.

Grant, F. S., and G. F. West. *Interpretation Theory in Applied
Geophysics.* New York: McGraw-Hill.

Greaves, R. J., and T. J. Fulp. 1992. Three-dimensional seismic
monitoring of an enhanced recovery process. In *Reservoir Geo-*

physics, R. E. Sheriff, ed., pp. 309–20. Tulsa: Society of Exploration Geophysicists.

Hardage, B. A. 1985. *Vertical Seismic Profiling: Part A: Principles*, 2d ed. London: Geophysical Press.

Hardage, B. A. 1992. *Crosswell surveying and reverse VSP*. London: Geophysical Press.

Hasbrouck, W. P. 1987. Hammer-impact shear-wave studies. In *Shear-Wave Exploration*, S. H. Danbom and S. N. Domenico, eds., pp. 97–121, Vol. 1 in Geophysical Development Series. Tulsa: Society of Exploration Geophysicists.

Herman, G. T. 1980. *Image Reconstruction from Projections*. New York: Academic Press.

Herman, G. T., A. Lent, and S. W. Rowland. 1973. ART: Mathematics and applications. *J. Theor. Biol.*, **42**: 1–32.

James, A., and W. L. Nutt. 1985. New techniques in borehole seismic. *Explor. Geophys.*, **16**: 349–56.

Justice, J. H., A. A. Vassiliou, M. E. Mathisen, S. Singh, P. S. Cunningham, and P. R. Hutt. 1992. Acoustic tomography in reservoir surveillance. In *Reservoir Geophysics*, R. E. Sheriff, ed. Tulsa: Society of Exploration Geophysicists.

Kennett, P., R. L. Ireson, and P. J. Conn. 1980. Vertical seismic profiles: Their application in exploration geophysics. *Geophys. Prosp.*, **28**: 676–99.

Lapedes, D. N., ed. 1978. *McGraw-Hill Dictionary of Physics and Mathematics*. New York: McGraw-Hill.

Layotte, P. C. 1987. Marthor, an *S*-wave impulse source. In *Shear-Wave Exploration*, S. H. Danbom and S. N. Domenico, eds., pp. 79–96. Vol. 1 in Geophysical Development Series. Tulsa: Society of Exploration Geophysicists.

Lewis, C., T. L. Davis, and C. Vuillermoz. 1992. Three-dimensional multicomponent imaging of reservoir heterogeneity, Silo Field, Wyoming. *Geophysics*, **56**: 2048–56.

Lines, L. 1991. Applications of tomography to borehole and reflection seismology. *The Leading Edge*, **10(7)**: 11–17.

Lines, L. R., M. Miller, H. Tan, R. Chambers, and S. Treitel. 1993. Integrated interpretation of borehole and crosswell data from a West Texas field. *The Leading Edge*, **12(1)**: 13–16.

Mason, I. M., D. J. Buchanan, and A. K. Booer. 1980. Fault location by underground seismic survey. In *Coal Geophysics*, pp. 341–55, SEG Reprint Series 6. Tulsa: Society of Exploration Geophysicists.

Massell, W. 1992. Emerging geophysical technologies. In *Reservoir Geophysics*, R. E. Sheriff, ed., pp. 344–54. Tulsa: Society of Exploration Geophysicists.

Millahn, K. O. 1980. In-seam seismics: Position and development. *Prakla-Seismos Report*, **80(2 & 3)**: 19–30.

Mueller, M. C. 1992. Using shear waves to predict lateral variability in vertical fracture intensity. *The Leading Edge*, **11(2)**: 29–35.

Noble, M. D., R. A. Lambert, H. Ahmed, and J. Lyons. 1988. Applications of three-component VSP data on the interpretation of the Vulcan Gas Field and its impact on field development. *First Break*, **6**: 131–49.

Officer, C. B. 1958. *Introduction to the Theory of Sound Transmission*. New York: McGraw-Hill.

Officer, C. B. 1974. *Introduction to Theoretical Geophysics*. New York: Springer-Verlag.

Payne, M. A. 1994. Looking ahead with vertical seismic profiles. *Geophysics*, **59**: 1182–91.

Puckett, M. 1991. Offset VSP: A tool for development drilling. *The Leading Edge*, **10(8)**: 18–24.

Ramo, A. O., and J. W. Bradley. 1982. Bright spots, milligals, and gammas. *Geophysics*, **47**: 1693–1705.

Ratcliff, D. W., S. H. Gray, and N. D. Whitmore. 1992. Seismic imaging of salt structures in the Gulf of Mexico. *The Leading Edge*, **11(4)**: 15–31.

Reguiero, J. 1990. Seam waves: What are they used for? *The Leading Edge*, **9(4)**: 19–23; **9(8)**: 32–4.

Rutledge, J. T. 1989. Interwell seismic surveying workshop: An overview. *The Leading Edge*, **8(6)**: 38–40.

Schimschal, U. 1986. *VSP Interpretation and Applications*. Houston: Schlumberger Educational Services.

Sheriff, R. E. 1989. *Geophysical Methods*. Englewood Cliffs, N.J.: Prentice Hall.

Sheriff, R. E. 1991. *Encyclopedic Dictionary of Exploration Geophysics*, 3d ed. Tulsa: Society of Exploration Geophysicists.

Sheriff, R. E., ed., 1992. *Reservoir Geophysics*. Tulsa: Society of Exploration Geophysicists.

Steinmann, V. W., and H. A. Rühmkorf. 1968. Seismische Messungen zur Salzstockflankenbestinmung: Eine Case History. *Zeitschrift für Geophysik*, **34**: 457–68.

Stewart, R. R. 1991. *Exploration Seismic Tomography: Fundamentals*, Course Notes Series, Vol. 3. Tulsa: Society of Exploration Geophysicists.

Tatham, R. H., and E. H. Krug. 1985. V_p/V_s interpretation. In *Developments in Geophysics, 6*, A. A. Fitch, ed., pp. 139–88. New York: Elsevier.

Tatham, R. H., and M. D. McCormack. 1991. *Multicomponent Seismology in Petroleum Exploration*. Tulsa: Society of Exploration Geophysicists.

Tatham, R. H., and P. L. Stoffa. 1976. V_p/V_s, a potential hydrocarbon indicator. *Geophysics*, **41**: 837–49.

Telford, W. M., L. P. Geldart, and R. E. Sheriff. 1990. *Applied Geophysics*. Cambridge, UK: Cambridge University Press.

Winterstein, D. F., and M. A. Meadows. 1992a. Shear-wave polarizations and subsurface stress directions at Lost Hills field. *Geophysics*, **56**: 1331–48.

Winterstein, D. F., and M. A. Meadows. 1992b. Changes in shear-wave polarization azimuth with depth in Cymric and Railroad Gap oil fields. *Geophysics*, **56**: 1349–64.

Wong, J., N. Bregman, G. West, and P. Hurley. 1987. Crosshole seismic scanning and tomography. *The Leading Edge*, **6(1)**: 36–41.

Zemanek, J., E. E. Glenn, L. J. Norton, and R. L. Caldwell. 1970. Formation evaluation by inspection with the borehole televiewer. *Geophysics*, **35**: 254–69.

14

Specialized applications

Overview

Whereas the most important application of seismic methods has always been to hydrocarbon exploration, the areas of fastest growth are applications to groundwater, environmental, and engineering geophysics and to improving the economics and efficiency of coal and hydrocarbon extraction. The applications of seismic methods to nonexploration activities is the subject of this chapter.

14.1 Engineering applications

14.1.1 Objectives of engineering work

Seismic refraction and reflection methods give information of value to civil engineers. It is necessary to map in detail the geology, especially faulting, for the engineering of large structures such as tunnels or nuclear power plants. Mapping the depth to bedrock is the most common engineering application. Seismic studies are often used in conjunction with borehole results to interpolate between holes and reduce the number of boreholes required for an evaluation. Seismic methods are also used to map voids such as caverns and abandoned coal mines, buried channels, and shallow faults. Groundwater studies are often sufficiently similar to engineering work that the discussions in §14.1.2 and 14.1.3 apply to them also.

Most engineering interest is only in very shallow data; targets are generally shallower than 30 m, often only 10 to 15 m and sometimes only 3 m deep. Occasionally, engineering interest extends to deep data, perhaps for tunnel construction or nuclear waste disposal sites.

Both P- and S-waves are used in borehole-to-borehole studies and P- and S-wave borehole logging is used in engineering studies. Measuring the S-wave velocity is especially important because it is a measure of the shear strength and hence involved in the ability of the earth to support structures. Passive seismic measurements are sometimes used to detect microseismic events that might precede landslides or movement along faults.

The equipment and methods for engineering geophysical studies are usually simple, often because engineering projects do not allow much money for them. Large energy is generally not required, so the source might be a sledge hammer striking a steel plate on the ground or another small seismic source (§7.2.4) (Miller et al., 1986). The energy is usually detected by moving-coil geophones similar to those described in §7.5.1. The number of channels is often 12 or less, and the amplifiers and camera are usually combined in a small metal suitcase. Marine applications usually involve "profiling" (§8.6.3).

14.1.2 Refraction surveys on land

Refraction methods (chap. 11) are generally better than reflection for locating bedrock and the water table, determining rock rippability (from seismic velocity), and identifying buried fracture or shear zones in igneous or metamorphic terrain in groundwater exploration.

Early engineering refraction often employed a timer triggered when the amplitude exceeded a preset threshold to measure the first-break time, or a timer plus oscilloscope for viewing the time pick. Present-day refraction is usually multichannel employing a *signal-enhancement recorder* (§7.6.5) that vertically stacks the data to build up the signal strength. The useful range of a hammer blow can be increased from 30 to 100 m with a signal-enhancement recorder. Sources used for shallow refraction (§7.2.4) include small explosive charges, a hammer blow, weight drop, or "enhanced" weight drop.

The velocity in shallow layers often varies considerably with location. The observed head wave often comes from the base of the weathered layer (which may not be the same as the top of bedrock) or the water table. Reversed profiles are generally needed to distinguish whether data indicate several refractors, refractor relief, or refractor velocity changes.

Fracture zones may lower the velocity appreciably (sometimes from 5000 to 2500 m/s), but their detection often requires closely spaced geophones because they are apt to be narrow. Refractor velocity is an indicator of the difficulty to be expected in cutting into bedrock (fig. 14.1); velocities lower than about 1200 m/s usually indicate that bedrock is "trenchable" (that a ditch-digging machine can cut through it easily), lower than 2100 to 2400 m/s that it is "rippable" (a bulldozer can cut through it without requiring blasting). The absence of a high-velocity refraction usually indicates that bedrock will not be encountered shallower than about one-third the source–geophone distance.

Scott et al. (1968) describe the use of seismic measurements in the Straight Creek Highway tunnel bore

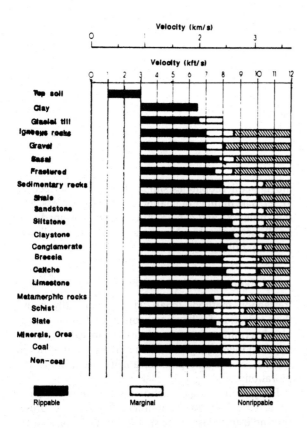

Fig. 14.1   Rippability with a D-8 tractor. (Courtesy of Caterpillar Tractor Co.)

to locate fracture zones, the height of the tension arch above the tunnel, the stable rock load, and potential weakness factors as an aid in designing tunnel lining and supports. They found a nearly linear correlation between seismic velocity and the rate and cost of tunnel construction.

Refraction is also used with *SH*-waves. A horizontal hammer blow (fig. 13.2) or other types of *S*-wave sources (§13.1.2) can be used. *S*-wave velocity is an important indicator of material strength.

Although various refraction interpretation methods are used in engineering applications (Mooney, 1977), the ABC and four-shot methods (§11.1.1 and 11.3.1) are commonest. The ABC method, the simplest, gives good depth calculations where depth varies smoothly and where the velocity contrast is large. The four-shot method, which is only slightly more complicated and can be carried out quite efficiently, gives better coverage of the refractor and is more effective in many situations. An example is shown in fig. 14.2.

14.1.3 Reflection surveys on land

Resolution and cost effectiveness are the central issues in engineering reflection surveys. Reflection methods are often better than refraction methods at mapping very shallow faults, cavities such as limestone caves or abandoned mines, and stratigraphic features. It is

also sometimes used for bedrock and water-table mapping, including water-table pulldown produced by pumping.

Sources giving data to depths of 45 m include sledge hammers, blasting caps, buffalo guns, and rifles; sources for data to 900 m include Betsy, 50-caliper rifle, miniSosie, weight drops, and small explosives (see table 7.1).

With 24-channel equipment and source spacing equal to the geophone group interval, CMP stacking yields 12-fold data. Steeples and Miller (1990) acquired 12-fold CMP data using a rifle fired every 35 s, using a 1-m source interval with a four-man crew at a cost of $5–25/sourcepoint. Their crew consisted of an observer, shooter, jug hustler, and linesman (who lays out cables ahead of the jug hustler and picks up the cables and phones after use). Using one to three geophones/channel and 1/4-ms sampling, they recorded frequencies up to 1 kHz, achieving resolution of 0.5 to 2 m.

Shallow reflection work involves a number of pitfalls. Steeples and Miller (loc. cit.) use 200–300-Hz low-cut filters to attenuate air wave, ground roll, and shallow refractions. These may spatially alias, and a walkaway noise test with 1/4-m geophone spacing is recommended to check that aliasing is not occurring. Silencers are used with gun sources to suppress the air wave. Shallow high-velocity stringers (such as near-surface limestone) cause difficulties. Near-surface elevation and velocity problems can be severe for the high frequencies usually required to achieve the required resolution.

14.1.4 Marine engineering surveys

Much of the cost of a marine survey involves the ship used. The small (relative to petroleum survey) budgets for marine engineering surveys frequently preclude optimum outfitting so that they are often run with ships of opportunity. Once the ship costs are covered, add-on surveys are relatively inexpensive and hence a number of different sensors are often employed. These may include (fig. 14.3) profilers operating at different energy levels and frequencies to give different resolutions and penetrations, side-scan sonars, magnetometers, and so on.

Objectives might be to determine sea-floor or bedrock relief, to locate sand lenses that might support platforms, hazards to projected construction such as faults, mud-flow gullies, collapse structures, gas seepage, shallow gas accumulations, sea-floor resources such as sand or gravel that can be dredged, areas already dredged, shipwrecks, downed aircraft, refuse dumped at sea (for its possible effect on pollution and fisheries), to inspect in-place structures such as pipelines, or to get geologic data for subsea tunnel construction (Tychsen and Nielsen, 1990).

Fathometers designed to map the sea floor operating at around 100 kHz provide virtually no penetration into the sediments whereas subbottom pro-

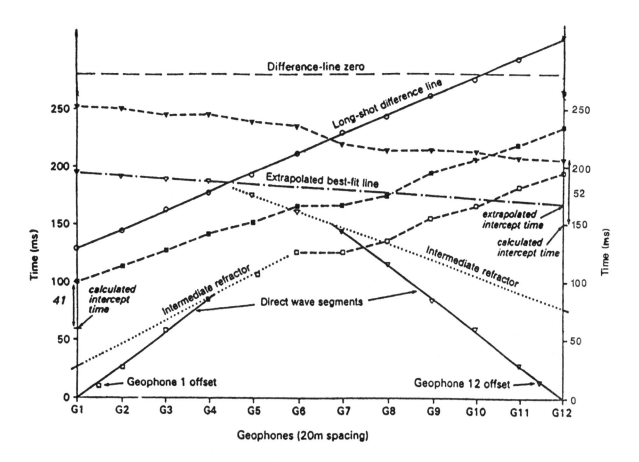

Fig. 14.2   Time–distance plot for a four-shot refraction spread. Short-shot data are plotted with open squares and open trian-gles and long-shot data with solid squares and solid triangles. (From Milsom, 1989:173.)

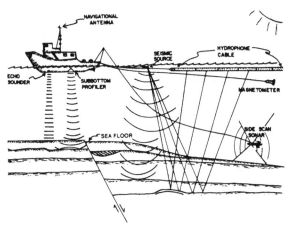

Fig. 14.3   Sensors deployed from a survey vessel. (From Trabant, 1984: 6.)

filers record reflections to about 30 m with resolution of about 50 cm. These generally operate in the 1- to 10-kHz range at speeds up to about 12 knots. The source is often a piezoelectric transducer. Gas bubbles in the water from sea-floor seeps are also sometimes visible on profiler records. Motion sensors are used to compensate for the distortion because of the heave of

the ship or source bird; this distortion can also be compensated for in processing by smoothing the sea-floor reflection. Analog recorders for single-channel systems usually have a dynamic range of only about 20 dB, but digital recording without this limitation is becoming more common.

Higher-power systems employ sparker, air-gun, water-gun, and sleeve gun sources (see §7.4.4). Sources and/or streamers are often towed at depths around 3.75 m, so that the surface ghost will enhance 100 Hz. Their penetration below the sea floor is about 1 s (1000 m). The higher-power systems often tow a 300- to 500-m 12- to 24-channel streamer. They are usually digitally recorded and processed in much the same way as conventional seismic data.

Side-scan sonar generally uses a towfish (fig. 8.25), or on-board equipment to project high-intensity, high-frequency acoustic bursts in fan-shaped beams; the frequencies are in the range 50 to 200 kHz (usually, 100 kHz) and the beams are narrow horizontally ($\approx 1°$) but broad vertically ($\approx 40°$). Beams are projected alternately to the left and right of the towfish or ship. Sea-floor topography and objects to one side of the ship backscatter the energy. The arrival time of the backscattered energy (echo) recorded at the towfish yields the distance to the backscatterer. Echos from successive scans are displayed side by side after correcting for slant distance; this gives the sonar image of the sea floor in real time. Horizontal ranges are usually 75 to 300 m, depending on the height of the towfish above the sea floor. Pulse length is generally 0.1 to 2 ms and resolution is of the order of 15 cm. Tow speeds are 3 to 8 knots.

14.2 Coal geophysics

14.2.1 Objectives of coal geophysics

Before the 1970s, the coal industry made little use of seismic methods. Relatively large reserves of coal were known, most coalfields having been discovered in outcrops with little geology employed. Geophysical exploration for new coalfields is rare. Historically, it was relatively easy to transfer from one working face to another and insurance against coal depletion was provided by opening spare working faces. However, as mining became more mechanized in the 1960s, the cost of moving the mining equipment to a new face increased so much that it became more important to anticipate geologic faults or other features that might interrupt production. The main mining problem is to find out about disturbances (faults, washouts, abandoned workings, and so on) in time for continuity of production to be maintained. Maintaining continuity of production is the main objective of most coal geophysics. A secondary problem involves ensuring that the main access drifts in a new mine are driven in directions that support the easiest mining.

In the past, most coal exploration was done by drilling boreholes (Bond, Alger, and Schmidt, 1969). However, during the last 20 years or so, the seismic method, although more expensive, has been used increasingly to supplement the data obtained from boreholes. The seismic method is able to furnish continuous profiles, which, in conjunction with borehole information, enables us to answer vital questions about the deposit, whether it is flat-lying or dipping, disturbed or not, the quantity of coal present, the best extraction method, disturbances we may need to provide for, and other questions critical to the evaluation of a deposit (Gochioco, 1990).

14.2.2 Properties of coal

Coal generally has markedly low velocity and density compared with surrounding rocks (Greenhalgh and Emerson, 1986). In European Carboniferous coals, velocity is often 1200 m/s and density 1.1 g/cm³, compared with values as high as 3600 m/s and 2.6 g/cm³ for surrounding rocks. Thus, reflection coefficients are very high. The interfaces between coal and adjacent rocks make excellent reflectors but cause multiple and transmission problems. Coal also has the required properties for a waveguide. Profitable coal seams may be less than 2 m thick so that individual coal members are usually not resolved. Coal is often laid down by cyclic deposition, producing a series of rapidly alter-

nating impedance contrasts of large magnitude. Interference between component reflections produces the observed seismic reflections. Short-path peg-leg multiples are generally strong, with consequent lengthening of the propagating wavelet, weakening of the wavelet onset, and severe attenuation of high frequencies. Peg-leg multiples may mask deeper reflections, so that only the upper part of coal sequences can be mapped seismically (Koefoed and de Voogd, 1980).

14.2.3 Longwall mining

Virtually all of the coal production in the United Kingdom (and much elsewhere) is obtained from longwall coalfaces (Mason, Buchanan, and Booer, 1980). To prepare a longwall face, two parallel tunnels about 200 m apart are driven into the seam off a main drift. The face is established by driving a third smaller tunnel through the coal to connect the two tunnels. A track is laid to support the face conveyor and coal-shearing machine and coal is won by horizontal milling as the cutter travels between the tunnel ends. As cutting proceeds, hydraulic props that support the tunnel roof and the track are snaked forward and the unsupported roof collapses behind the machine to relieve stress at the face. A longwall face may take 6 months and U.S.$10 million to set up, and the investment can be recovered over 12 to 18 months of operations provided no serious problems appear. A fault with a throw no greater than the seam thickness can be crossed (the average thickness of seams mined in the United Kingdom is a little over 1 m), although at the cost of output. However, a larger fault probably will cause abandonment of the face. Clarke (1976) puts the economic limit for British coal production at 4 faults/km² if their throw is greater than 3 m. In one recent year, approximately half of the longwall faces in Britain were abandoned short of their planned lifetimes. To sustain production, collieries must maintain expensive standby faces as insurance against encountering large faults; mining profitability jumps significantly where faults are mapped in advance.

14.2.4 Surface seismic methods

Most of the surface seismic surveys for coal have been in Great Britain and Germany, where Upper Carboniferous coals are found at 200 to 1000 m (0.150 to 0.800 s) (Ziolkowski and Lerwill, 1978; Fairbairn, Holt, and Padget, 1986), and in Australia. Explosives in boreholes are the most common source, but Sosie and Vibroseis are sometimes used (Gochioco, 1991). Charges are usually 1/4 to 1/2 kg or 800 to 1500 whacker blows. Typically, 48 groups are used with 10 to 12 geophones per group spaced over a 10-m group interval. Maximum offsets are limited to around 500 m, mode conversion becoming severe beyond this distance. Dominant frequencies in the coal measures are ≈ 100 Hz; with coal velocities ≈ 3 km/s, resolution ≈ 7.5 m. Sampling is usually 1 ms (alias filter cuts at 250

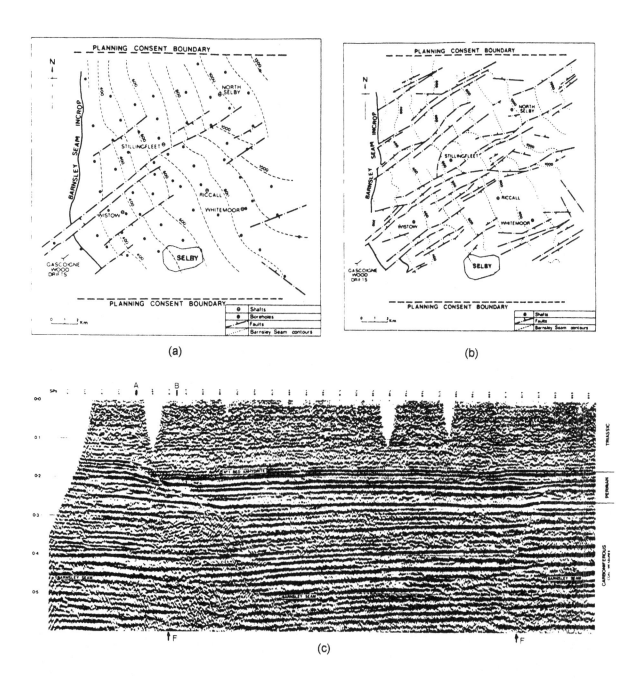

Fig. 14.4 Fault definition in coal mapping. (From Fairbairn, Holt, and Padget, 1986: 190, 192.) (a) Structure at Selby (Yorkshire, England) determined from borehole data only, (b) structure based on surface seismic and borehole data, and (c) migrated seismic section showing faulting. Note the evidences for angular unconformity at the base of the Permian. The apparent dip in the Carboniferous coal measures between *A* and *B* is largely a velocity anomaly produced by Triassic thickness changes. Faults at 0.4 s occur at the arrows marked *F*.

to 375 Hz) and frequencies below 40 Hz are normally filtered out to reduce ground-roll effects. Static corrections are very important; LVL velocities are often as low as 300 m/s so that the wavelength in the weathering ≈ 1 m.

Improved fault definition by surface seismic is illustrated by fig. 14.4. Triassic velocities are around 2000 m/s, but the Permian section above the Carboniferous coal measures consists of very high velocity (4000 m/s) marls and limestones that produce multiples that interfere with the coal reflections. Figure 14.5 shows a channel (washout) cutting into a coal seam.

14.2.5 In-seam methods

Although surface-reflection methods can detect throws of several meters if the data are of high quality, locating faults by projecting surface and borehole data onto the coal seam is often unreliable. In-seam methods allow investigation of faults within the seam

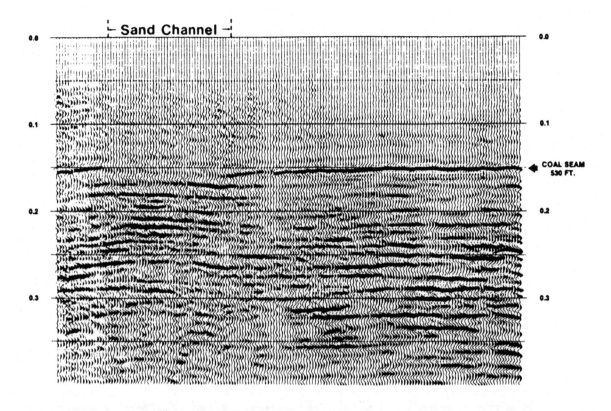

Fig. 14.5 High-frequency Vibroseis data recorded across a shallow coal seam showing an erosional sand channel. (From Chapman, Brown, and Fair, 1981: 1665.)

itself, for example, faults with throws slightly smaller than the seam thickness can be located using channel waves (§13.3).

Hasbrouch, Hadsell, and Major (1978) used sources and receivers in boreholes within coal seams to investigate seam continuity by measuring the seismic transmission properties. Geophones are sometimes cemented in coal seams encountered in boreholes for later use with in-seam methods.

Borges (1969) describes measurements to locate fault zones in coal mines from the amplitude of waves traveling in the coal seams. Faults with throws of 1.5 to 4 m produced amplitude reductions of 50 to 70% for waves passing through the faults, whereas throws greater than 4 m resulted in 70 to 90% reduction in amplitude. Saul and Higson (1971) carried out an extensive evaluation of the method. Holes were drilled 2 to 3 m into the rock above or below the coal for both the shots and geophones. The shots were located along one roadway and the geophones along an adjacent roadway. The shots were coupled to the rock by careful stemming (tamping) and the geophones were attached to rock bolts. Their conclusions were that the method was soundly based and that faults produce attenuation; however, their results varied too widely for them to establish a relation between amplitude and the degree of disturbance.

Mine drifts (roadways) and galleries provide more extensive access to a coal seam than do boreholes

(Mason, Greenhalgh, and Hatherly, 1985). An application of this technique is given by Millahn (1980). Assume galleries in a coal seam, as illustrated in the map of fig. 14.6a, with geophones planted along gallery *A*. The transmission record obtained from a source at *B* is shown in fig. 14.6b; it shows *P*- and *S*-wave head waves that travel in the higher-velocity rock bounding the coal seam and channel waves with a trailing amplitude buildup, the Airy phase (§13.3 and fig. 13.19). If a fault with throw larger than the coal-seam thickness had cut the seam between the source and the receivers (as would be the case from source *C*), the channel waves would not be seen. Several reflection records obtained with sources along gallery *A* are shown in fig. 14.6c; these contain reflected channel waves, but their phases are not sufficiently coherent to allow them to be picked easily. Complex-trace techniques (§9.11.4) are used to obtain the amplitude envelope (fig. 14.6d) and stacking to make the high-amplitude Airy phase stand out as a distinct arrival (fig. 14.6e). The amplitude envelope of the Airy phase is similarly determined on the transmission records, which yields the group velocity to be used in stacking the reflected records. Figure 14.7 shows a reflection from a fault cut by another fault.

Mason, Buchanan, and Booer (1980) used source and receiver arrays in coal seams exposed in drifts to study channel waves propagating in the waveguide formed by the coal seam and the host rock. The wave

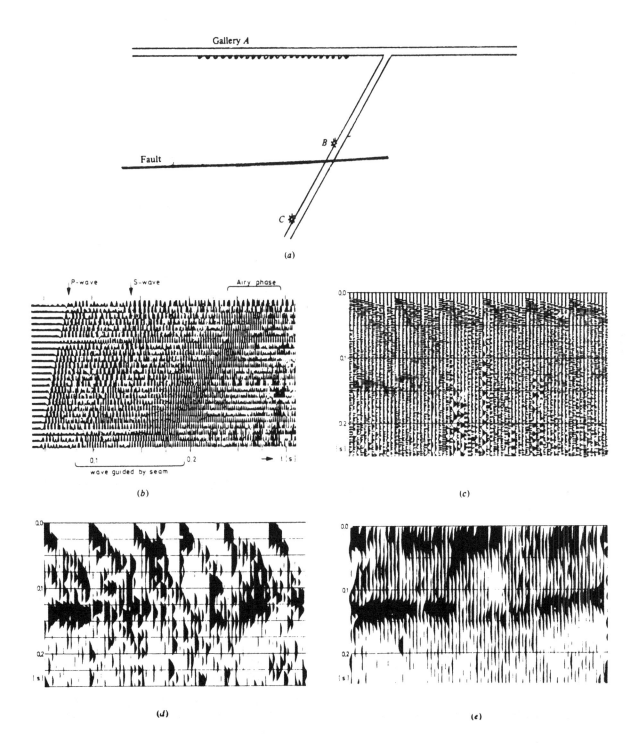

Fig. 14.6 In-seam methods. (From Millahn, 1980.) (a) Diagramatic map (not to scale) showing location of geophones and sources B and C; (b) transmission record from source B; (c) reflection records of common-midpoint type obtained from sources in gallery A; (d) display of envelopes of records in part (c); and (e) 6-fold stack of records such as those shown in part (c).

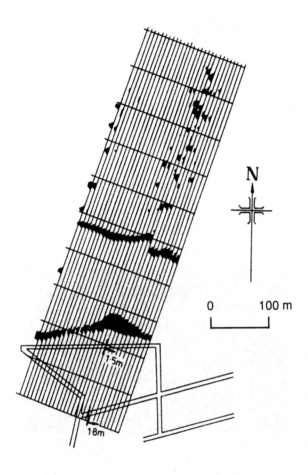

Fig. 14.7   In-seam reflection survey in roadway 1064 showing reflection from a fault interrupted by a cross-fault. The surface-wave event along the roadway is called a *roadway mode*. (From Buchanan, 1985: 18–19.)

motion is very complex; an *SH*-mode (Evison or pseudo-Love wave) as well as various *P-SV*-modes (Krey or pseudo-Rayleigh waves) may be generated. These modes are all highly dispersive (fig. 13.21). A surface wave along the wall may also be generated. Sources (1/4-kg explosive in a hole 1 m into the seam) and geophones are usually placed in the center of a coal seam. Interpretation, based mainly on the *SH*-mode, permits mapping faults in a 3-m seam with throws as small as 1 m by observations of waves reflected by, as well as transmitted through, the fault planes.

14.2.6 Miscellaneous aspects of coal geophysics

Passive geophone arrays are sometimes used in regions of high mechanical stress to monitor the acoustic emissions that precede rockbursts. Seismic methods have been used to search for lost abandoned mining drifts that might result in subsidence problems or flooding of current workings if accidentally encountered.

Nonseismic geophysical methods are also used to solve mining problems (Buchanan and Jackson, 1986). Gravity has been used to detect mining subsidence. Dikes intruding into coal beds have been mapped by magnetic and resistivity methods. The low conductivity of coal makes in-seam radar a possible tool in probing for faults.

14.3 Groundwater, environmental, archaeological, and geothermal applications

Groundwater, environmental, and geothermal concerns have in common the fact that they involve mainly the movement of fluids (Ward, 1990). The presence or absence of fluids in pore spaces affects density and seismic velocity, and thus the strengths of reflections and refractions. Seismic methods often can map a surface separating regions where the interstitial fluid is liquid from ones where it is gaseous. The techniques employed are the engineering seismic techniques described in §14.1 (Dobecki and Romig, 1985; Lankston, 1990).

The water table often can be mapped by both refraction and reflection. The nature of a liquid, however, usually does not markedly alter velocity or density, and so seismic methods are not good at distinguishing the nature of the liquid.

Environmental geophysics usually involves the flow of a hazardous liquid. Steeples (1991) cites a number of incidents in which seismic methods have been used to solve problems where such liquids are leaking. The instances he cites relate to mapping geological features that control fluid flow, that is, mapping buried topography, detecting underground cavities, and especially locating faults. One of Steeples' examples resulted in changing the mapped direction of flow of a hazardous fluid by 90°; prior mapping based on well data had given the wrong answer because of aliasing with the limited number of wells. Seismic methods are usually in competition with drilling, but drilling may endanger the integrity of an impermeable member, whereas seismic methods do not involve this risk. Passive seismic methods (§13.8) have been used for geothermal studies.

Reflection work has been used in searches for chambers (tombs) that may contain archaeological artifacts and for other features whose elastic properties may differ from those of the surroundings. Soonawala, Holloway, and Tomsons (1990) describe the use of seismic methods in nuclear waste disposal.

14.4 Hydrocarbon-reservoir applications

14.4.1 Introduction

Petroleum engineers need realistic models of oil and gas fields in order to make development-production decisions, but historically geophysicists have contributed little to this decision making except indirectly through maps and reports generally prepared for exploration purposes. These upstream data are usually not reviewed in the light of information obtained subsequently. However, this situation is changing rapidly, and more geophysical methods, such as high-resolution, 3-D, VSP, cross-borehole, and time-lapse seismology, are being applied to reservoir problems.

Reservoir geophysics is sometimes subdivided into three areas:

1. *Reservoir delineation:* the use of seismic methods to define reservoir limits and locate faults and other barriers to fluid flow; such information provides constraints on the engineer's models of fields and, even in "well-understood" fields, sometimes results in the discovery of new prospects within old fields.
2. *Reservoir description:* the use of seismic measurements to help define internal features of a field, such as net and gross thicknesses (reservoir formations often include some nonporous zones), porosity and porosity-thickness, lithology, and Poisson's ratio.
3. *Reservoir surveillance:* monitoring the changes within a reservoir as production proceeds, especially the progress of enhanced recovery.

Sheriff (1992) describes the application of seismic methods to the solution of reservoir management problems and to increasing the profitability of hydrocarbon fields.

14.4.2 The nature of hydrocarbon reservoirs

This section summarizes some of the principles of petroleum engineering. Petroleum engineers are mainly interested in producing hydrocarbons at a profit. More specifically, they want information about the pore spaces in a rock (porosity), how they are interconnected (permeability), the nature of the fluids filling the pore spaces (fluid saturations), the energy/pressure that may cause the fluids to flow (*drive*), the vertical and areal distribution of pore-connected spaces, and barriers to fluid flow (sealing faults, stratigraphic barriers, and so on). These have to be determined from the data available, which probably consist of

(a) surface seismic, gravity, and other geophysical and geological data,
(b) VSP and other borehole seismic measurements, occasionally borehole-to-borehole measurements,
(c) borehole logs of various types,
(d) cores taken in boreholes,
(e) analyses of fluids recovered in drill-stem tests,
(f) production and pressure data,
(g) drilling-rate logs, mud logs, and other similar data.

Well logs, well-to-well log correlations, seismic data, and geological background knowledge give an overall picture of the geometry and stratigraphy and changes within the reservoir, and production and pressure data (and occasionally tracer data) give information about the reservoir connectivity. Seismic data usually provide the only source of detailed information about areal distributions even though seismic data lack the vertical resolution of borehole logs and cores.

Porosity is determined from measurements on cores and well logs using relationships that are somewhat empirical. Because porosity is one of the major factors determining seismic velocity and reflectivity, velocity and amplitude measurements can sometimes yield porosity information. Permeability generally correlates with porosity. Permeability may differ across a reservoir and from zone to zone within a reservoir, and often it varies directionally. A reservoir bed usually includes shale or other laminations that separate it into zones that behave somewhat independently. Flow may occur within one part of a bed while bypassing other parts. Both horizontal and vertical variations of porosity and permeability affect fluid flow.

A geological picture is developed from the core, log, and seismic data. An assessment of the environment of deposition is based on these data, supplemented by paleontologic indicators of depth and temperature at the time of deposition given by isotope ratio measurements.

Initially, the recoverable reserves of a field are estimated based on the reservoir volume, porosity, saturation values, and recovery factor. The hydrocarbons originally in place are determined from structure and isopach maps, porosity and hydrocarbon saturation maps, and cross-sections based on log correlations and seismic data. Seismic measurements (including hydrocarbon indicators; see §10.8) play a major role in defining reservoir volume and how reservoirs are broken up into separate pools. Reserve estimates are usually modified considerably after the production history provides pressure-decline values.

Hydrocarbons are pushed naturally from reservoir pore spaces by water or gas. The gas for *gas drive* may come from the expansion of a gas cap as the pressure declines or from gas that comes out of solution as the pressure declines. Natural *water drive* requires connection with an aquifer surrounding a hydrocarbon accumulation that is at least 100 times the volume of the reservoir. The amount of recoverable hydrocarbons depends on a number of factors, including the interstitial porosity, pore sizes, shapes, and interconnectivity, saturations of oil, gas, and water, interfacial surface tensions and wetability, and fracture spacing, width, and orientation.

Primary recovery uses the natural energy present in a reservoir to drive the fluids to the producing boreholes. After primary production declines, the natural energy is often supplemented by gas or water injection; this is called *secondary recovery*. More exotic means of stimulating production involve injection of miscible fluids, surfactants, or other chemicals, injection of steam, and setting fire to the hydrocarbons in a reservoir. These methods are called *enhanced oil recovery (EOR)*.

EOR floods tend to flow through the more permeable beds or channels and thus may bypass portions of a reservoir. Seismic methods may help monitor the changes in a reservoir that result from production where gas is involved (for example, as a result of a fire flood) or where the seismic velocity is changed significantly (for example, decreased as a result of the temperature increase resulting from steam injection). Reflections from gas–oil, gas–water, and/or oil–water contacts can sometimes be mapped. Obviously, we need to know how fluids actually flow through a reservoir in order to optimize (and, often, to make economical) EOR projects.

Reservoir continuity is mainly established from pressure histories and from observing the pressure changes in one well when conditions in another well change. When production in a well is stopped, the early buildup of pressure mainly depends on the immediate vicinity of the well, the influence of more remote parts of the reservoir affecting the pressure curve later. 3-D seismic (chap. 12) and cross-hole tomographic methods (§13.5.4) help in understanding what goes on between wells.

Inhomogeneities and permeability anisotropy may cause channeling of water or gas, bypassing pockets of hydrocarbons (and leaving them as unproducible). Bypassed regions sometimes may be invaded later if sufficient fluid is injected, or bypassed oil can be recovered by drilling additional intermediate wells.

Multiphase situations result in complex sequences of conditions as production progresses. May (1984) gives a scenario for the very simple case of a uniform porous sand containing only oil with dissolved gas:

As the pressure is reduced, small amounts of gas evolved from solution accumulate in the pores and increase the resistance to flow of oil without the gas itself becoming mobile, but, as the concentration increases, the gas bubbles eventually join up to form a continuous phase, and from then on both fluids move through the formation. . . . [T]he permeability of the porous formation to gas increases, and the permeability to oil decreases. Hence, with continued production, the produced gas/oil ratio tends to rise, this tendency being accentuated both by shrinkage of the liquid-phase volume and by the increase in its viscosity with loss of dissolved gas.

The distribution of porosity, permeability, and other properties over the area of a reservoir are incorporated in a reservoir *simulation model*. The models are updated as additional information becomes available. The validity of a model is judged by how closely actual results (fluid-flow rates, bottom-hole pressures, and so on) match predictions based on the model. A general rule of thumb is that a model can be used to predict reasonably well for about the same length of time as it has matched past history. Scenarios of production programs and strategies (for example, varying the number of wells and their locations, preferentially producing different wells, injecting water or gas to enhance the drive, and so on) are run on the model to determine how to optimize the economic return. Obviously, very realistic maps and data are required to achieve this end result.

14.4.3 Reservoir delineation

Brown and Gilbert (1992) state that accurate determination of the physical boundaries and internal segmentation of a reservoir have significant impact on development and production operations throughout a field's life. Historically, delineation and infill drilling and reservoir modeling have been driven mainly by well data and a geologic model with relatively little geophysical input except for prediscovery seismic data. The geologic model is usually based on vertical information obtained from logs at well locations, but if detailed seismic data are not available, horizontal definition is apt to be very poor.

The predominant seismic method for field delineation is 3-D (chap. 12), but VSP (§13.4) and other methods are also used. The often long time delay between 3-D acquisition and interpretation or logistic difficulties sometimes make VSP the better technique for answering specific questions. Sheriff (1992) contains seven case histories of the application of 3-D and two of the application of VSP to reservoir delineation, and a number of other case histories can be found in the recent literature. Abriel et al. (1991) tell of the rejuvenation of a giant oil field as a result of a 3-D survey and Bouvier et al. (1989) tell of using 3-D to develop a major field in Nigeria.

The main elements of field delineation are usually the mapping of faults and determining the areas of the reservoir that contain hydrocarbons. Faults often subdivide a reservoir into separate pools that may have poor (or no) hydraulic connection with neighboring pools, so that hydrocarbons cannot be produced by wells in a neighboring pool. 3-D surveys typically reveal an order of magnitude more faults than 2-D surveys, as well as markedly different orientations and continuities. Faults are often not nearly as continuous as usually thought. Fault slicing is sometimes very helpful in indicating whether faults seal or not and in locating wells so as to minimize the volume of oil left above the well penetration (*attic oil*).

14.4.4 Reservoir description

The objective of reservoir description is to determine how physical properties change from place to place within a reservoir, so that reservoir simulation models can be made more reliable (Nolen-Hoeksma, 1990). The property of greatest interest is usually permeability; this generally cannot be measured except on cores, so that sampling is exceedingly inadequate. However, permeability is usually roughly proportional to porosity, which sometimes can be inferred from seismic amplitude or velocity measurements. The proportionality factor depends on a number of other variables also, and changes in these other factors often are not known. Geostatistical methods (§13.10) are often used to change the proportionality factors to yield porosity or porosity-thickness maps. Sheriff (1992) contains eight case histories that employ different geostatistical methods to determine net pay and lithology.

Intrinsic ambiguity is one of the major problems in reservoir description (de Buyl and Hardage, 1992); any measurement can be affected by several unknown factors. Hence, well data or some other independent information source must be called on to narrow the field of possible explanations (de Buyl, 1989). Values of physical properties are determined at wells and often are assumed to vary slowly and smoothly between wells, that is, values are determined by interpolation and extrapolation. Better values can often be obtained using seismic data, ordinarily the only information available except at the well locations.

High-quality seismic data are needed for reservoir description. Most often 3-D data that have been acquired and processed to preserve amplitude information are used. Occasionally, cross-hole (§13.5.4) or other data are used.

14.4.5 Reservoir surveillance

The objective of reservoir surveillance is monitoring the movement of fluids within a reservoir. This usually means repeating measurements at different times to see how the measurements have changed; this is the time-lapse procedure (§13.6). Although any type of measurement can be used in time-lapse mode, usually 3-D or cross-hole data are used (Justice, 1992). Interpretation often involves subtracting one survey from another to bring out the changes. As stated in §13.6, the two surveys must be conducted under as nearly as possible identical conditions so that observed changes can be attributed to the fluid movement. Moreover, the changes being measured must be large enough to stand out from the background noise.

The quantities most often measured are seismic amplitude or velocity. Gas, oil, and water filling a rock's pore spaces change the velocity and density (§5.2.7) and therefore the contrast with adjacent formations and reflectivity. Sometimes, changes in temperature are involved (because injected water is cold or because a reservoir is heated to lower the oil viscosity) and this also may change the seismic velocity (§5.2.6).

Few time-lapse surveys have been carried out because of the large costs involved. Some EOR experi-

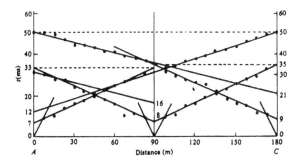

Fig. 14.8   Engineering refraction profile.

ments involving CO_2 flooding, steam injection (Matthews, 1992), and in-situ combustion (Greaves and Fulp, 1987) have been successfully monitored (see fig. 13.38). It appears that other types of EOR could also be monitored. The greatest potential, however, is for monitoring water floods, the most common secondary recovery mechanism, and this may become a major seismic application.

Problems

14.1 To find the depth to bedrock in a damsite survey, 12 geophones were laid out at intervals of 15 m along a straight line away from the source, offsets ranging from 15 to 180 m. Determine the depth of overburden from the data in table 14.1 assuming a single layer above the refractor. By how much does the depth differ if we assume two layers above the refractor?

14.2 The time–distance observations in fig. 14.9 constitute an engineering refraction problem.

(a) Solve for the first layer for both pairs of reversed profiles and show that the layer has a thickness of about 2.9 and 3.8 m at sources A and C (dip about 0.3°).

(b) Apply eq. (4.41) to get approximate thicknesses of the second layer.

(c) What is the dip of the deeper interface?

(d) Why are the answers in parts (b) and (c) approximate?

14.3 Fine the depths and velocities of refractors in fig. 14.2.

References

Abriel, W. L., P. S. Neale, J. S. Tissue, and R. M. Wright. 1991. Modern technology in an old area: Bay Marchand revisited. *Leading Edge,* **10(6):** 21–35.

Bond, L. O., R. P. Alger, and A. W. Schmidt. 1969. Well log applications in coal mining and rock mechanics. In *Coal Geophysics,* D. J. Buchanan and L. J. Jackson, eds., pp. 28–35, SEG Geophysics Reprint Series 6. Tulsa: Society of Exploration Geophysicists, 1986.

Borges, E. 1969. Ein neues seismisches Verfahren san orten von Verwurfen und Auswaschungen in Floz. *Gluckauf Forschft.,* **4:** 201–8.

Bouvier, J. D., C. H. Kaars-Sijpesteijn, D. F. Kluesner, C. C. Onyejekwe, and R. C. van der Pal. 1989. Three-dimensional

Table 14.1 *Data from a dam-site survey*

x	t	x	t	x	t
15 m	19 ms	75 m	59 ms	135 m	72 ms
30	29	90	62	150	76
45	39	105	65	165	78
60	50	120	68	180	83

seismic interpretation and fault sealing investigations, Nun River Field, Nigeria. *Bull. AAPG,* **73:** 1397–414.

Brown, A. R., and O. E. Gilbert. 1992. Reservoir delineation: Characterizing the trap. In *Reservoir Geophysics,* R. E. Sheriff, ed., pp. 71–2. Tulsa: Society of Exploration Geophysicists.

Buchanan, D. J. 1985. In-seam seismology: A method for detecting faults in coal seams. In *Developments in Geophysical Exploration Methods,* 5, A. A. Fitch, ed. London: Applied Science Publishers.

Buchanan, D. J., and L. J. Jackson. 1986. *Coal Geophysics,* SEG Geophysics Reprint Series 6. Tulsa: Society of Exploration Geophysicists.

Chapman, W. L., G. L. Brown, and D. W. Fair. 1981. The Vibroseis system: A high-frequency tool. *Geophysics,* **46:** 1657–66.

Clarke, A. M. 1976. Why modern exploration has little to do with geology and much more to do with mining. *Colliery Guardian Annu. Rev.,* **224:** 323–36. Also in *Coal Geophysics,* D. J. Buchanan and L. J. Jackson, eds., pp. 10–24, SEG Geophysics Reprint Series 6. Tulsa: Society of Exploration Geophysicists.

de Buyl, M. 1989. Optimum field development with seismic reflection data. *The Leading Edge,* **8(4):** 14–20.

de Buyl, M., and B. A. Hardage. 1992. Defining reservoir properties. In *Reservoir Geophysics,* R. E. Sheriff, ed., pp. 185–8. Tulsa: Society of Exploration Geophysicists.

Dobecki, T. L., and P. R. Romig. 1985. Geotechnical and groundwater geophysics. *Geophysics,* **50:** 2621–38.

Fairbairn, C. M., J. M. Holt, and N. J. Padget. 1986. Case histories of the use of the surface seismic method in the U.K. coal mining industry. In *Coal Geophysics,* D. J. Buchanan and L. J. Jackson, eds., pp. 188–203, SEG Geophysics Reprint Series 6. Tulsa: Society of Exploration Geophysicists.

Gochioco, L. M. 1990. Seismic surveys for coal exploration and mine planning. *The Leading Edge,* **9(4):** 25–8.

Gochioco, L. M. 1991. Advances in seismic reflection profiling for US coal exploration. *The Leading Edge,* **10(12):** 24–9.

Greaves, R. J., and T. J. Fulp. 1987. Three-dimensional seismic monitoring of an enhanced oil recovery process. *Geophysics,* **52:** 1175–87.

Greenhalgh, S. A., and D. W. Emerson. 1986. Elastic properties of coal measure rocks from the Sydney Basin, New South Wales. *Explor. Geophys.,* **17:** 157–63.

Hasbrouck, W. P., F. A. Hadsell, and M. W. Major. 1978. Instrumentation for a coal seismic system. Paper read at 48th SEG Annual Meeting, San Francisco.

Justice, J. H. 1992. Geophysical methods for reservoir surveillance. In *Reservoir Geophysics,* R. E. Sheriff, ed., pp. 281–4. Tulsa: Society of Exploration Geophysicists.

Koefoed, O., and N. de Voogd. 1980. The linear properties of thin layers, with an application to synthetic seismograms over coal seams. *Geophysics,* **45:** 1254–68. Also in *Coal Geophysics,* D. J. Buchanan and L. J. Jackson, eds., pp. 110–24, SEG Geophysics Reprint Series 6. Tulsa: Society of Exploration Geophysicists.

Lankston, R. W. 1990. High-resolution refraction seismic data acquisition and interpretation. In *Geotechnical and Environmental Geophysics,* S. H. Ward, ed., Vol. 1, pp. 45–73.

Mason, I. M., D. J. Buchanan, and A. K. Booer. 1980. Fault location by underground seismic survey. *Proc. IEEE,* **F127:** 322–

36. Also in *Coal Geophysics,* D. J. Buchanan and L. J. Jackson, eds., pp. 341–55, SEG Geophysics Reprint Series 6. Tulsa: Society of Exploration Geophysicists.

Mason, J. M., S. A. Greenhalgh, and P. Hatherly. 1985. Underground seismic mapping of coal seam discontinuities at West Wallsend No. 2 Colliery. *Explor. Geophy.,* **16:** 357–64.

Matthews, L. 1992. 3-D seismic monitoring of an in-situ thermal process: Athabasca, Canada. In *Reservoir Geophysics,* R. E. Sheriff, ed., pp. 301–8. Tulsa: Society of Exploration Geophysicists.

May, C. J. 1984. Reservoir engineering. In *Modern Petroleum Technology,* 5th ed., G. D. Hobson, ed., New York: John Wiley.

Millahn, K. O. 1980. In-seam seismics: Position and development. *Prakla-Seismos Report,* **80(2):** 19–30.

Miller, R. D., S. E. Pullen, J. S. Waldner, and F. P. Haeni. 1986. Field comparison of shallow seismic sources. *Geophysics,* **51:** 2067–92.

Milsom, J. 1989. *Field Geophysics.* New York: John Wiley.

Mooney, H. M. 1977. *Handbook of Engineering Geophysics.* Minneapolis: Bison Instruments.

Nolen-Hoeksema, R. C. 1990. The future role of geophysics in reservoir engineering. *The Leading Edge,* **9(12):** 89–97.

Saul, T., and G. R. Higson. 1971. The detection of faults in coal panels by a seismic transmission method. *Internat. J. Rock Mech. Min. Sci.,* **8:** 483–99.

Scott, J. H., F. T. Lee, R. D. Carroll, and C. S. Robinson. 1968. The relationship of geophysical measurements to engineering and construction parameters in the Straight Creek Tunnel pilot boring, Colorado. *Internat. J. Rock Mech. Min. Sci.,* **5:** 1–30.

Sheriff, R. E., ed. 1992. *Reservoir Geophysics.* Tulsa: Society of Exploration Geophysicists.

Soonawala, N. M., A. L. Holloway, and D. K. Tomsons. 1990. Geophysical methodology for the Canadian nuclear fuel waste management program. In *Geotechnical and Environmental Geophysics,* S. H. Ward, ed., Vol. 1, pp. 309–31. Tulsa: Society of Exploration Geophysicists.

Steeples, D. 1991. Uses and techniques of environmental geophysics. *Geophysics, the Leading Edge of Exploration,* **10(9):** 30–31.

Steeples, D. W., and R. D. Miller. 1990. Seismic reflection methods applied to engineering, environmental, and groundwater problems. In *Geotechnical and Environmental Geophysics,* S. H. Ward, ed., Vol. 1, pp. 1–30. Tulsa: Society of Exploration Geophysicists.

Trabant, P. K. 1984. Applied high-resolution geophysical methods. Boston: International Human Resources Development Corp.

Tychsen, J., and T. Nielsen. 1990. Seismic reflection used in the sea environment. In *Geotechnical and Environmental Geophysics,* S. H. Ward, ed., Vol. 1, pp. 31–44. Tulsa: Society of Exploration Geophysicists.

Ward, S. H., ed. 1990. *Geotechnical and Environmental Geophysics,* 3 vols., Investigations in Geophysics 5. Tulsa: Society of Exploration Geophysicists.

Ziolkowski, A., and W. E. Lerwill. 1978. A simple approach to high resolution seismic profiling for coal. *Geoph. Prosp.,* **27:** 360–93. Also in *Coal Geophysics,* D. J. Buchanan and L. J. Jackson, eds., pp. 154–87, SEG Geophysics Reprint Series 6. Tulsa: Society of Exploration Geophysicists, 1986.

15
Background mathematics

Overview

This chapter serves as an appendix to this book rather than a portion of the main text. It can be omitted by those already familiar with mathematics or by those who wish to take the mathematics on faith. More extensive treatments can be found in Wylie (1966), Pipes and Harvill (1970), Robinson (1967a, 1967b, 1967c), Cassand et al. (1971), Robinson and Treitel (1973, 1980), Kanasewich (1973), Båth (1974), Kulhánek (1976), Claerbout (1976), Silvia and Robinson (1979), Potter and Goldberg (1987).

We begin with short summaries of determinants, vector analysis, matrix analysis, infinite series, complex numbers, the methods of least squares, finite differences, numerical solution of differential equations, and partial fractions (§15.1). Most of the chapter involves the mathematics of data processing, especially Fourier (§15.2), Laplace (§15.3), and z-transforms (§15.5), and almost all deal with linear systems (§15.4). The cepstrum is discussed (§15.6) and a final section deals with filtering. In contrast to chap. 9, which dealt almost entirely with digital data, much of this chapter deals with continuous functions, although some portions elaborate on digital considerations, especially §15.5.

15.1 Summaries of basic concepts

15.1.1 Determinants

A determinant, det (a), is a square array of $n \times n$ numbers, a_{ij}, called *elements, i* and *j* designating the row and column, respectively:

$$\det(a) = \begin{vmatrix} a_{11} & a_{12} & \cdots & a_{1n} \\ a_{21} & a_{22} & \cdots & a_{2n} \\ \vdots & \vdots & \ddots & \vdots \\ a_{n1} & a_{n2} & \cdots & a_{nn} \end{vmatrix} \qquad (15.1)$$

The *minor* M_{ij} of element a_{ij} is the determinant of order $(n - 1)$ formed by deleting the ith row and the jth column of det (a). The product $(-1)^{i+j}M_{ij}$ is the *cofactor* of a_{ij}. The value of a determinant is defined as

$$\det(a) = \sum_j (-1)^{i+j} a_{ij} M_{ij} \qquad$$
$$= \sum_i (-1)^{i+j} a_{ij} M_{ij}, \qquad (15.2)$$

where the summation is taken along one row (*row expansion;* hence, i = constant) or along one column (column expansion, j = constant); the result is the same in both cases (see rule 5 below). As an example, we expand by the first row:

$$\det(a) = \begin{vmatrix} 1 & 2 & 3 \\ 4 & 5 & 6 \\ 8 & 9 & 0 \end{vmatrix} = (-1)^{1+1} 1 \begin{vmatrix} 5 & 6 \\ 9 & 0 \end{vmatrix}$$
$$+ (-1)^{1+2} 2 \begin{vmatrix} 4 & 6 \\ 8 & 0 \end{vmatrix} + (-1)^{1+3} 3 \begin{vmatrix} 4 & 5 \\ 8 & 9 \end{vmatrix}$$
$$= 1(5 \times 0 - 6 \times 9) - 2(4 \times 0 - 6 \times 8)$$
$$+ 3(4 \times 9 - 5 \times 8) = 30.$$

A determinant is thus a single number.

Equations (15.1) and (15.2) can be used to derive the following rules (see Potter and Goldberg, 1987: 226–7; Wylie, 1966: 403–10 for proofs):

1. If all the elements of one row are zero, or if the elements of one row are proportional to the corresponding elements of another row, then the determinant equals zero;
2. Multiplying all the elements of a row by a constant multiplies the determinant by the same constant;
3. Interchanging any two rows changes the sign of the determinant;
4. Interchanging rows with columns does not change the value;
5. Equation (15.2) gives the same value regardless of the row selected;
6. Any row can be multiplied by a constant and added to another row without changing the value of the determinant;
7. "Row" can be replaced with "column" in any of the preceding rules.

Determinants can be used to solve a system of linear equations:

$$\left.\begin{array}{l} a_{11}x_1 + a_{12}x_2 + \cdots + a_{1n}x_n = b_1, \\ a_{21}x_1 + a_{22}x_2 + \cdots + a_{2n}x_n = b_2, \\ \cdots\cdots\cdots\cdots\cdots\cdots = 0, \\ a_{n1}x_1 + a_{n2}x_2 + \cdots + a_{nn}x_n = b_n; \end{array}\right\} \quad (15.3a)$$

Cramer's rule states that

$$x_r = [\det(a)_r]/[\det(a)], \qquad (15.3b)$$

where

$$\det(a) = \begin{vmatrix} a_{11} & a_{12} & \cdots & a_{1n} \\ a_{21} & a_{22} & \cdots & a_{2n} \\ \vdots & \vdots & \ddots & \vdots \\ a_{n1} & a_{n2} & \cdots & a_{nn} \end{vmatrix}$$

and $\det(a_r)$ is $\det(a)$ with the rth column replaced by $b_1, b_2, \ldots, b_n$ (Pipes and Harvill, 1970: 101; Potter and Goldberg, 1987: 232–3).

When all the constants $b_1, b_2, \ldots, b_n$ in eq. (15.3a) are zero, the set of equations is said to be *homogeneous*. Obviously, one solution (the *trivial solution*) is when all the x_i are zero. Nontrivial solutions exist when det $(a) = 0$ (see problem 15.1a).

15.1.2 Vector analysis

(a) Basic definitions. A *scalar* quantity, such as temperature, has magnitude only, whereas a *vector* quantity, such as force, has both magnitude and direction. Vectors (represented by boldface type) can be added to give a resultant, as shown in fig. 15.1a. Subtraction is equivalent to reversing one of the vectors and then adding. Multiplication of a vector by a scalar changes the magnitude of the vector but not its direction (except that multiplication by negative numbers reverses the direction).

A vector can be resolved into components along coordinate axes (fig. 15.1b) and expressed as a vector sum of its components:

$$\mathbf{A} = a_x\mathbf{i} + a_y\mathbf{j} + a_z\mathbf{k}, \qquad (15.4)$$

where $\mathbf{i}$, $\mathbf{j}$, and $\mathbf{k}$ are unit vectors along the x-, y-, and z-axes, respectively. (Vectors can also be expressed in other coordinate systems such as cylindrical or spherical.)

Vectors can be added by adding corresponding components. Thus, if

$$\begin{aligned} \mathbf{A} &= a_x\mathbf{i} + a_y\mathbf{j} + a_z\mathbf{k}, \text{ etc.,} \\ \mathbf{A} + 2\mathbf{B} - 3\mathbf{C} &= (a_x + 2b_x - 3c_x)\mathbf{i} \\ &\quad + (a_y + 2b_y - 3c_y)\mathbf{j} \\ &\quad + (a_z + 2b_z - 3c_z)\mathbf{k}. \end{aligned}$$

The magnitude of a vector, written $|\mathbf{A}|$, is given by

$$|\mathbf{A}| = (a_x^2 + a_y^2 + a_z^2)^{1/2}. \qquad (15.5)$$

If $\mathbf{A}$ has direction cosines (ℓ, m, n), then

$$\mathbf{A} = |\mathbf{A}|(\ell\mathbf{i} + m\mathbf{j} + n\mathbf{k}). \qquad (15.6)$$

(b) Vector products. The *scalar* or *dot product* of two vectors, $\mathbf{A}$ and $\mathbf{B}$, is written $\mathbf{A} \cdot \mathbf{B}$; it is a scalar:

$$\mathbf{A} \cdot \mathbf{B} = |\mathbf{A}||\mathbf{B}| \cos \theta, \qquad (15.7)$$

where θ is the smaller angle between the vectors (fig. 15.1c). Thus, the dot product equals the magnitude of one of the vectors times the projection of the second vector onto the first. If $\mathbf{A}$ is a force acting on a point

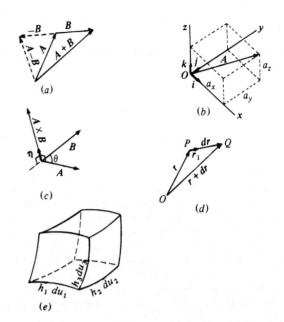

Fig. 15.1   Operations on vectors. (a) Addition and subtraction, (b) resolution into components, (c) cross-product, (d) scalar function of position, and (e) curvilenear coordinates.

mass that suffers displacement $\mathbf{B}$, $\mathbf{A} \cdot \mathbf{B}$ gives the work done on the mass. Obviously,

$$\begin{aligned} \mathbf{A} \cdot \mathbf{B} &= \mathbf{B} \cdot \mathbf{A}, \\ &= 0, & \text{when } \mathbf{A} \perp \mathbf{B}, \\ &= |\mathbf{A}|\,|\mathbf{B}|, & \text{when } \mathbf{A} \parallel \mathbf{B}. \end{aligned}$$

Also,

$$\mathbf{i} \cdot \mathbf{i} = \mathbf{j} \cdot \mathbf{j} = \mathbf{k} \cdot \mathbf{k} = 1, \qquad \mathbf{i} \cdot \mathbf{j} = \mathbf{j} \cdot \mathbf{k} = \mathbf{k} \cdot \mathbf{i} = 0;$$

hence,

$$\mathbf{A} \cdot \mathbf{B} = \mathbf{B} \cdot \mathbf{A} = a_x b_x + a_y b_y + a_z b_z \qquad (15.8)$$

and

$$\mathbf{A}^2 = \mathbf{A} \cdot \mathbf{A} = a_x^2 + a_y^2 + a_z^2.$$

The *vector*, or *cross-*, *product* of $\mathbf{A}$ and $\mathbf{B}$, $\mathbf{A} \times \mathbf{B}$, is a vector defined by the relation

$$\mathbf{A} \times \mathbf{B} = (|\mathbf{A}|\,|\mathbf{B}| \sin \theta)\boldsymbol{\eta}, \qquad (15.9)$$

where $\boldsymbol{\eta}$ is a unit vector perpendicular to the plane containing $\mathbf{A}$ and $\mathbf{B}$ and in the direction of advance of a right-handed screw rotated from $\mathbf{A}$ to $\mathbf{B}$ through the angle θ (fig. 15.1c). Because the direction of $\boldsymbol{\eta}$ depends on the sequence,

$$\begin{aligned} \mathbf{A} \times \mathbf{B} &= -\mathbf{B} \times \mathbf{A}, \\ &= 0, & \text{when } \mathbf{A} \parallel \mathbf{B}, \\ &= |\mathbf{A}|\,|\mathbf{B}|\boldsymbol{\eta}, & \text{when } \mathbf{A} \perp \mathbf{B}. \end{aligned}$$

The magnitude of $\mathbf{A} \times \mathbf{B}$ equals the area of the parallelogram defined by $\mathbf{A}$ and $\mathbf{B}$. The torque about an axis of rotation is given by a vector product (see problem

15.3). Applying eq. (15.9) to the unit vectors **i**, **j**, **k** gives

$$\mathbf{i} \times \mathbf{j} = \mathbf{k}, \qquad \mathbf{j} \times \mathbf{k} = \mathbf{i}, \qquad \mathbf{k} \times \mathbf{i} = \mathbf{j};$$
$$\mathbf{i} \times \mathbf{i} = \mathbf{j} \times \mathbf{j} = \mathbf{k} \times \mathbf{k} = 0.$$

$\mathbf{A} \times \mathbf{B}$ can be expressed as a determinant (problem 15.4).

$$\mathbf{A} \times \mathbf{B} = \begin{vmatrix} \mathbf{i} & \mathbf{j} & \mathbf{k} \\ a_x & a_y & a_z \\ b_x & b_y & b_z \end{vmatrix} \qquad (15.10)$$

Products of more than two vectors can be formed in various ways (see problems 15.5).

(c) Vector operators. Let $\psi(x, y, z)$ be a scalar function of position (for example, temperature). If the value is ψ at $P(x, y, z)$ in fig. 15.1d, then the value at a nearby point Q is $\psi + d\psi$, where

$$\begin{aligned} d\psi &= \frac{\partial \psi}{\partial x} dx + \frac{\partial \psi}{\partial y} dy + \frac{\partial \psi}{\partial z} dz \\ &= \left(\frac{\partial \psi}{\partial x} \mathbf{i} + \frac{\partial \psi}{\partial y} \mathbf{j} + \frac{\partial \psi}{\partial z} \mathbf{k} \right) \cdot (\mathbf{i}\, dx + \mathbf{j}\, dy + \mathbf{k}\, dz) \\ &= \boldsymbol{\nabla}\psi \cdot d\mathbf{r}; \end{aligned}$$

thus,

$$d\psi/dr = \boldsymbol{\nabla}\psi \cdot \mathbf{r}_1, \qquad (15.11)$$

where $\mathbf{r}_1$ is a unit vector along $d\mathbf{r}$. Vector $\boldsymbol{\nabla}\psi$ (pronounced "del ψ") is the *gradient* of ψ (grad ψ); it is a vector in the direction of the maximum rate of increase of ψ and its magnitude equals this maximum rate of increase (see problem 15.6b). The rate of increase of ψ in an arbitrary direction $\mathbf{p}$, where $\mathbf{p}$ is a unit vector, is given by $\boldsymbol{\nabla}\psi \cdot \mathbf{p}$ (see problem 15.6c).

The vector operator, *del*, $\boldsymbol{\nabla} = \mathbf{i}(\partial/\partial x) + \mathbf{j}(\partial/\partial y) + \mathbf{k}(\partial/\partial z)$, is often used as if it were a vector. Thus, we can take the dot product of $\boldsymbol{\nabla}$ and a vector $\mathbf{A}$, called the *divergence* of $\mathbf{A}$, or *div* $\mathbf{A}$:

$$\operatorname{div} \mathbf{A} = \boldsymbol{\nabla} \cdot \mathbf{A} = \frac{\partial a_x}{\partial x} + \frac{\partial a_y}{\partial y} + \frac{\partial a_z}{\partial z}. \quad (15.12)$$

We can also take the vector product of $\boldsymbol{\nabla}$ and $\mathbf{A}$, called the *curl:*

$$\operatorname{curl} \mathbf{A} = \boldsymbol{\nabla} \times \mathbf{A} = \begin{vmatrix} \mathbf{i} & \mathbf{j} & \mathbf{k} \\ \dfrac{\partial}{\partial x} & \dfrac{\partial}{\partial y} & \dfrac{\partial}{\partial z} \\ a_x & a_y & a_z \end{vmatrix} \quad (15.13)$$

on using eq. (15.10). Operator $\boldsymbol{\nabla}$ can be applied more than once. For example,

$$\operatorname{div} \operatorname{grad} \psi = \boldsymbol{\nabla} \cdot \boldsymbol{\nabla}\psi = \nabla^2\psi = \left[\frac{\partial^2}{\partial x^2} + \frac{\partial^2}{\partial y^2} + \frac{\partial^2}{\partial z^2} \right]\psi$$

$$= \textit{Laplacian of } \psi. \qquad (15.14)$$

In the same way, we can form products such as $\boldsymbol{\nabla} \times$

$\boldsymbol{\nabla}\psi$, $\boldsymbol{\nabla} \cdot \boldsymbol{\nabla} \times \mathbf{A}$, $\boldsymbol{\nabla} \times \boldsymbol{\nabla} \times \mathbf{A}$, $\boldsymbol{\nabla}(\boldsymbol{\nabla} \cdot \mathbf{A})$, and so on (see problem 15.7).

(d) Orthogonal curvilinear coordinates. Although Cartesian coordinates are usually the most convenient, at times cylindrical, spherical, or other orthogonal curvilinear coordinates lead to simpler results. We write u_1, u_2, u_3 for such coordinates, the surfaces $u_1 = c_1$, $u_2 = c_2$, $u_3 = c_3$ (c_i = constant) being orthogonal. When u_i changes by du_i, the element of length, ds, is given by

$$ds^2 = (h_1\, du_1)^2 + (h_2\, du_2)^2 + (h_3\, du_3)^2,$$

where $h_i = h_i(u_1, u_2, u_3)$ is a variable scale factor.

To express ψ in curvilinear coordinates, we note that $\partial\psi/\partial x$ corresponds to $\partial\psi/(h_1\partial u_1)$; hence,

$$\boldsymbol{\nabla}\psi = \sum_{i=1}^{3} \frac{1}{h_i} \frac{\partial \psi}{\partial u_i} \mathbf{u}_i, \qquad (15.15)$$

where $\mathbf{u}_i$ are unit vectors along the curvilinear axes. To obtain $\boldsymbol{\nabla} \cdot \mathbf{A}$, we use Gauss' theorem (Pipes and Harvill, 1970: 909; Potter and Goldberg, 1987: 355–7), which states that

$$\iiint_{\mathscr{V}} \boldsymbol{\nabla} \cdot \mathbf{A}\, d\mathscr{V} = \iint_{\mathscr{S}} \mathbf{A} \cdot d\mathscr{S},$$

where surface $\mathscr{S}$ encloses volume $\mathscr{V}$, and the outward-drawn normal is positive. Applying this to an element of volume d$\mathscr{V}$(fig. 15.1e), we get

$$\begin{aligned} \boldsymbol{\nabla} \cdot \mathbf{A}\, d\mathscr{V} &= \text{surface integral over the six faces,} \\ &= -A_1(h_2\, du_2\, h_3\, du_3) \\ &\quad + A_1(h_2\, du_2\, h_3\, du_3) \\ &\quad + \left[\frac{\partial}{\partial u_1} (A_1 h_2 h_3)\, du_1 \right] du_2\, du_3 \\ &\quad + \text{similar expressions for the} \\ &\qquad \text{other pairs of faces.} \end{aligned}$$

Thus,

$$\begin{aligned} \boldsymbol{\nabla} \cdot \mathbf{A}(h_1\, du_1\, h_2\, du_2\, h_3\, du_3) &= \left[\frac{\partial}{\partial u_1} (A_1 h_2 h_3) + \frac{\partial}{\partial u_2} (A_2 h_3 h_1) \right. \\ &\quad \left. + \frac{\partial}{\partial u_3} (A_3 h_1 h_2) \right] du_1\, du_2\, du_3, \end{aligned}$$

$$\boldsymbol{\nabla} \cdot \mathbf{A} = \frac{1}{h_1 h_2 h_3} \sum \frac{\partial}{\partial u_i} (A_i h_j h_k), \qquad (15.16)$$

i, j, k being in cyclic order.

The Laplacian, $\nabla^2\psi$, is equal to div grad ψ, that is,

$$\nabla^2\psi = \frac{1}{h_1 h_2 h_3} \sum \frac{\partial}{\partial u_i} \left(\frac{h_j h_k}{h_i} \frac{\partial \psi}{\partial u_i} \right) \qquad (15.17)$$

15.1.3 Matrix analysis

(a) Definitions. A *matrix* is a rectangular array of numbers a_{ij} arranged in r rows and s columns; an entire matrix is here indicated by script type:

$$\mathscr{A} = \begin{Vmatrix} a_{11} & a_{12} & a_{13} & \cdots \\ a_{21} & a_{22} & a_{23} & \cdots \\ a_{31} & a_{32} & a_{33} & \cdots \\ \vdots & \vdots & \vdots & \ddots \end{Vmatrix} \quad (15.18)$$

The *order* of a matrix is $r \times s$. If $r = 1$, we have a *row matrix;* if $s = 1$, a *column matrix;* these are also called *vectors of the first and second kinds,* respectively. A null matrix $\mathscr{O}$ has zeros for all elements. The *transpose* of a matrix has rows and columns interchanged; thus,

$$\mathscr{A}^T = \begin{Vmatrix} a_{11} & a_{21} & a_{31} & \cdots \\ a_{12} & a_{22} & a_{32} & \cdots \\ a_{13} & a_{23} & a_{33} & \cdots \\ \vdots & \vdots & \vdots & \ddots \end{Vmatrix} \quad (15.19)$$

A matrix of order $r \times r$ is a *square matrix.* The *principal diagonal* of a square matrix has the elements a_{ii}. A *diagonal matrix* is a square matrix with zeros for all elements that are not on the principal diagonal, that is, $a_{ij} = 0$ if $i \neq j$, and at least one of the $a_{ii} \neq 0$. An *identity matrix* $\mathscr{I}$ is a diagonal matrix where $a_{ii} = 1$ for all i. A matrix with zeros below (above) the principal diagonal is called an upper (lower) *triangular matrix.* A *symmetric matrix* equals its transpose, that is, $\mathscr{A} = \mathscr{A}^T$, and a *skew-symmetric matrix* equals the negative of its transpose, $\mathscr{A} = -\mathscr{A}^T$. A symmetric matrix where all the elements along any diagonal parallel to the principal diagonal are the same is a *Toeplitz* matrix.

The *cofactor* of an element a_{rs} of a square matrix is $(-1)^{r+s}$ times the determinant formed by deleting the rth row and the sth column. The *adjoint* of a square matrix, adj($\mathscr{A}$), is the transpose of the matrix $\mathscr{A}$ with each element replaced by its cofactor. The *determinant of a square matrix*, det($\mathscr{A}$), is a single number given by det($\mathscr{A}$) $= \sum_i a_{ik}A_{ik} = \sum_k a_{ik}A_{ik}$, where A_{ik} is the cofactor of a_{ik}. The *inverse* of a square matrix can be found by dividing the adjoint by the determinant [if det($\mathscr{A}$) $\neq 0$], that is,

$$\mathscr{A}^{-1} = [1/\text{det}(\mathscr{A})]\text{adj}(\mathscr{A}), \ \mathscr{A}^{-1}\mathscr{A} = \mathscr{I} \quad (15.20)$$

(see problem 15.2b).

(b) Matrix operations. Operations performed on matrices change the values of the matrix elements. Corresponding elements can be added, that is, if $\mathscr{C} = \mathscr{A} + \mathscr{B}$, $c_{rs} = a_{rs} + b_{rs}$; matrices must be of the same order to be added. Matrices can be multiplied by scalars, that is, if $\mathscr{D} = k\mathscr{A}$, $d_{rs} = ka_{rs}$. In matrix multiplication, the ith row of the first matrix is multiplied element by element by the jth column of the second matrix and the products are summed to give the ijth element of the product matrix, that is, if $\mathscr{E} = \mathscr{A}\mathscr{B}$, $e_{ij} = \sum_k a_{ik}b_{kj}$. The first matrix must have the same number of columns as the second matrix has rows for matrices to be multiplied. In $\mathscr{E} = \mathscr{A}\mathscr{B}$, if the order of

$\mathscr{A}$ is $m \times n$ and that of $\mathscr{B}$ is $n \times p$, the order of $\mathscr{E}$ is $m \times p$. When more than two matrices are multiplied, products can be formed in pairs; thus, $\mathscr{A}\mathscr{B}\mathscr{C} = (\mathscr{A}\mathscr{B})\mathscr{C} = \mathscr{A}(\mathscr{B}\mathscr{C})$. The transpose of a product is the product of the transposes of the individual matrices in inverse order, that is, $(\mathscr{A}\mathscr{B})^T = \mathscr{B}^T\mathscr{A}^T$ (see problem 15.10b).

It is sometimes convenient to *partition a matrix,* that is, to represent it as a matrix whose elements are submatrices of the original matrix. For example,

$$\mathscr{A} = \begin{Vmatrix} 2 & 0 & 0 & 3 & 5 \\ 1 & -1 & 4 & -2 & 1 \\ 3 & 2 & 1 & -5 & 4 \\ 0 & 0 & 1 & 1 & 0 \end{Vmatrix} = \begin{Vmatrix} \mathscr{P} & \mathscr{Q} \\ \mathscr{R} & \mathscr{S} \end{Vmatrix},$$

where

$$\mathscr{P} = \begin{Vmatrix} 2 & 0 & 0 \\ 1 & -1 & 4 \\ 3 & 2 & 1 \end{Vmatrix}, \quad \mathscr{Q} = \begin{Vmatrix} 3 & 5 \\ -2 & 1 \\ -5 & 4 \end{Vmatrix},$$

$$\mathscr{R} = \begin{Vmatrix} 0 & 0 & 1 \end{Vmatrix}, \quad \mathscr{S} = \begin{Vmatrix} 1 & 0 \end{Vmatrix}.$$

To add $\mathscr{A}$ to a similar 4×5 matrix $\mathscr{B}$, $\mathscr{B}$ must be partitioned in the same way, that is,

$$\mathscr{B} = \begin{Vmatrix} \mathscr{T} & \mathscr{U} \\ \mathscr{V} & \mathscr{W} \end{Vmatrix},$$

where $\mathscr{T}$ is 3×3, $\mathscr{U}$ is 3×2, and so on. Then

$$\mathscr{A} + \mathscr{B} = \begin{Vmatrix} \mathscr{P} + \mathscr{T} & \mathscr{Q} + \mathscr{U} \\ \mathscr{R} + \mathscr{V} & \mathscr{S} + \mathscr{W} \end{Vmatrix}.$$

When partitioning matrices that are to be multiplied, the submatrices must be comformable. Thus, if

$$\mathscr{C} = \mathscr{A}\mathscr{B},$$

where $\mathscr{A}$ is $m \times n$, $\mathscr{B}$ is $n \times p$, we can partition $\mathscr{A}$ and $\mathscr{B}$ as follows:

$$\mathscr{A} = \begin{Vmatrix} \mathscr{C} & \mathscr{D} \\ \mathscr{E} & \mathscr{F} \end{Vmatrix}, \quad \mathscr{B} = \begin{Vmatrix} \mathscr{G} & \mathscr{H} \\ \mathscr{J} & \mathscr{K} \end{Vmatrix},$$

$\mathscr{C}$ being $a \times b$; $\mathscr{D}$, $a \times c$; $\mathscr{E}$, $d \times b$; $\mathscr{F}$, $d \times c$; $\mathscr{G}$, $b \times j$; $\mathscr{H}$, $b \times k$; $\mathscr{J}$, $c \times j$; $\mathscr{K}$, $c \times k$; $a + d = m$, $b + c = n$, $j + k = p$; then

$$\mathscr{A}\mathscr{B} = \begin{Vmatrix} \mathscr{C}\mathscr{G} + \mathscr{D}\mathscr{J} & \mathscr{C}\mathscr{H} + \mathscr{D}\mathscr{K} \\ \mathscr{E}\mathscr{G} + \mathscr{F}\mathscr{J} & \mathscr{E}\mathscr{H} + \mathscr{F}\mathscr{K} \end{Vmatrix}.$$

Matrices can be used to solve simultaneous equations. If we write a set of linear equations as

$$\left. \begin{aligned} a_{11}x_1 + a_{12}x_2 + a_{13}x_3 + \cdots &= c_1, \\ a_{21}x_1 + a_{22}x_2 + a_{23}x_3 + \cdots &= c_2, \\ a_{31}x_1 + a_{32}x_2 + a_{33}x_3 + \cdots &= c_3, \\ \cdots\cdots\cdots\cdots\cdots\cdots\cdots &= 0, \\ a_{n1}x_1 + a_{n2}x_2 + a_{n3}x_3 + \cdots &= c_n, \end{aligned} \right\} \quad (15.21)$$

and let $\mathcal{A}$ be the elements a_{rs}, $\mathcal{X}$ be a column matrix with elements $x_{i1} = x_i$, $\mathcal{C}$ be a column matrix with elements c_i, then we can write

$$\mathcal{A}\mathcal{X} = \mathcal{C}$$

and solve for $\mathcal{X}$:

$$\left.\begin{array}{l} \mathcal{A}^{-1}\mathcal{A}\mathcal{X} = \mathcal{A}^{-1}\mathcal{C}, \\ \mathcal{X} = \mathcal{A}^{-1}\mathcal{C} \quad \text{because} \quad \mathcal{A}^{-1}\mathcal{A} = \mathcal{I}. \end{array}\right\} \quad (15.22)$$

This solution requires that $\mathcal{A}$ be a square matrix and that the equations are independent, that is, $\det(\mathcal{A}) \neq 0$ (see problem 15.1b).

Convolution, $a_t * b_t = c_t$, can be performed by the operation $\mathcal{A}\mathcal{B} = \mathcal{C}$ if $\mathcal{A}$ is a matrix of the form indicated in eq. (15.23) and $\mathcal{B}$ and $\mathcal{C}$ are column matrices. For simplicity, we assume that a_t and b_t both have $n + 1$ data points, zeros being added to achieve this. Then,

$$\begin{Vmatrix} a_0 & 0 & 0 & \cdots & 0 \\ a_1 & a_0 & 0 & \cdots & 0 \\ \vdots & \vdots & \vdots & \ddots & 0 \\ a_n & a_{n-1} & \cdots & & a_0 \\ 0 & a_n & \cdots & & a_1 \\ \vdots & \vdots & & \ddots & \vdots \\ 0 & 0 & & \cdots & a_n \end{Vmatrix} \begin{Vmatrix} b_0 \\ \vdots \\ b_n \end{Vmatrix} = \begin{Vmatrix} c_0 \\ c_1 \\ \vdots \\ c_n \\ c_{n+1} \\ \vdots \\ c_{2n} \end{Vmatrix}, \quad (15.23)$$

where $c_i = \sum_k a_{i-k}b_k = \sum_k a_k b_{i-k}$. Thus, eq. (15.23) gives the same result as eq. (9.23). Note that matrix $\mathcal{A}$ is of order $p \times (n + 1)$, where $p = 2n + 1$.

Cross-correlation can be written as $\mathcal{E}^{\mathsf{T}}\mathcal{D} = \phi_{ed}$, that is,

$$\begin{Vmatrix} e_0 & e_1 & \cdot & \cdot & e_n & 0 & \cdot & \cdot & 0 \\ 0 & e_0 & e_1 & \cdot & \cdot & e_n & \cdot & \cdot & 0 \\ \cdot & \cdot & \cdot & \cdot & \cdot & \cdot & \cdot & \cdot & \cdot \\ \cdot & \cdot & \cdot & \cdot & \cdot & \cdot & \cdot & \cdot & \cdot \\ 0 & \cdot & \cdot & 0 & e_0 & \cdot & \cdot & \cdot & e_n \\ 0 & \cdot & \cdot & \cdot & 0 & e_0 & \cdot & \cdot & e_{n-1} \\ \cdot & \cdot & \cdot & \cdot & \cdot & \cdot & \cdot & \cdot & \cdot \\ \cdot & \cdot & \cdot & \cdot & \cdot & \cdot & \cdot & \cdot & \cdot \\ 0 & \cdot & \cdot & \cdot & \cdot & \cdot & \cdot & 0 & e_0 \end{Vmatrix}$$

$$\times \begin{Vmatrix} d_0 & 0 & \cdot & \cdot & 0 & 0 & \cdot & \cdot & 0 \\ d_1 & d_0 & \cdot & \cdot & & & & & \cdot \\ \cdot & d_1 & & & & & & & \\ \cdot & \cdot & & & 0 & & & & \cdot \\ d_n & \cdot & \cdot & & d_0 & 0 & \cdot & \cdot & 0 \\ 0 & d_n & \cdot & \cdot & & d_0 & \cdot & \cdot & 0 \\ \cdot & \cdot & & & & & & & \cdot \\ 0 & \cdot & \cdot & \cdot & d_n & d_{n-1} & \cdot & \cdot & d_0 \end{Vmatrix}$$

$$= \begin{Vmatrix} \phi_{ed}(0) & \phi_{ed}(-1) & \cdots & \phi_{ed}(-n) & \cdots & \cdots & 0 \\ \phi_{ed}(+1) & \phi_{ed}(0) & \cdots & & \cdots & \cdots & \cdots \\ \vdots & \vdots & \ddots & \ddots & \ddots & \ddots & \ddots \\ \phi_{ed}(+n) & \vdots & \cdots & \phi_{ed}(0) & \cdots & \cdots & \phi_{ed}(-n) \\ 0 & \phi_{ed}(+n) & \cdots & \cdots & \phi_{ed}(0) & \cdots & \cdots \\ \vdots & \vdots & \ddots & \ddots & \ddots & \ddots & \ddots \\ 0 & 0 & \cdots & \phi_{ed}(+n) & \cdots & \cdots & \phi_{ed}(0) \end{Vmatrix} \quad (15.24)$$

This gives the same values as eq. (9.42). The cross-correlation matrix is a Toeplitz matrix. Another scheme for cross-correlating e_t with d_t is given by

$$\begin{Vmatrix} e_n & 0 & \cdots & 0 \\ e_{n-1} & e_n & \cdots & 0 \\ \vdots & \vdots & \ddots & \vdots \\ e_0 & e_1 & \cdots & e_n \\ 0 & e_0 & \cdots & e_{n-1} \\ \vdots & \vdots & \ddots & \vdots \\ 0 & 0 & \cdots & e_0 \end{Vmatrix} \begin{Vmatrix} d_0 \\ \vdots \\ d_n \end{Vmatrix} = \begin{Vmatrix} \phi_{ed}(-n) \\ \vdots \\ \phi_{ed}(-1) \\ \phi_{ed}(0) \\ \phi_{ed}(+1) \\ \vdots \\ \phi_{ed}(+n) \end{Vmatrix} \quad (15.25)$$

Autocorrelation is given by

$$\phi_{dd} = \mathcal{D}^{\mathsf{T}}\mathcal{D}. \quad (15.26)$$

The autocorrelation is also a Toeplitz matrix.

The Wiener filter normal equations, eq. (9.73), can be expressed in matrix form as

$$\phi_{gg}\mathcal{F} = \phi_{gh}, \quad (15.27)$$

where

$$\phi_{gg} = \begin{Vmatrix} \phi_{gg}(0) & \phi_{gg}(-1) & \cdots & \phi_{gg}(-n) \\ \phi_{gg}(1) & \phi_{gg}(0) & \cdots & \phi_{gg}(-n+1) \\ \vdots & \vdots & \ddots & \vdots \\ \phi_{gg}(n) & \phi_{gg}(n-1) & \cdots & \phi_{gg}(0) \end{Vmatrix}, \quad (15.28)$$

$$\mathcal{F} = \begin{Vmatrix} f_0 \\ f_1 \\ \vdots \\ f_n \end{Vmatrix}, \qquad \phi_{gh} = \begin{Vmatrix} \phi_{gh}(0) \\ \phi_{gh}(1) \\ \vdots \\ \phi_{gh}(n) \end{Vmatrix} \quad (15.29)$$

The filter $\mathcal{F}$ is given by

$$\mathcal{F} = \phi_{gg}^{-1}\phi_{gh}. \quad (15.30)$$

Sometimes, the solution of a matrix equation involves the inversion of a matrix that is not square, as in $\mathcal{A}\mathcal{B} = \mathcal{C}$, where $\mathcal{A}$ is of size $m \times n$, $\mathcal{B}$ of size $n \times p$, and $\mathcal{C}$ of size $m \times p$. To solve for $\mathcal{B}$, we multiply by $\mathcal{A}^{\mathsf{T}}$,

$$\mathcal{A}^{\mathsf{T}}\mathcal{A}\mathcal{B} = \mathcal{A}^{\mathsf{T}}\mathcal{C}$$

(note that $\mathcal{A}^{\mathsf{T}}\mathcal{A}$ is always square), then multiply by $(\mathcal{A}^{\mathsf{T}}\mathcal{A})^{-1}$ to get $\mathcal{B}$:

$$\mathcal{B} = (\mathcal{A}^{\mathsf{T}}\mathcal{A})^{-1}\mathcal{A}^{\mathsf{T}}\mathcal{C}. \quad (15.31)$$

(c) Characteristic equation of a matrix; eigenvalues. On replacing c_i with y_i, we can write eq. (15.21) in matrix form as

$$\mathscr{A}\mathscr{X} = \mathscr{Y}. \qquad (15.32)$$

Column matrices $\mathscr{X}$ and $\mathscr{Y}$ are vectors of the second kind, and eq. (15.32) represents a linear transformation of vector $\mathscr{X}$ into vector $\mathscr{Y}$. If $\mathscr{Y} = \lambda\mathscr{X}$, λ being a constant, $\mathscr{Y}$ is said to be in the same direction as $\mathscr{X}$. The condition for this is that

$$\mathscr{A}\mathscr{X} = \mathscr{Y} = \lambda\mathscr{X}$$

or

$$(\mathscr{A} - \lambda\mathscr{I})\mathscr{X} = \mathscr{O}.$$

This is equivalent to n homogeneous equations:

$$
\begin{aligned}
(a_{11} - \lambda)x_1 + \quad & a_{12}x_2 + \cdots + \quad a_{1n}x_n = 0, \\
a_{21}x_1 + (a_{22} - \lambda)x_2 + \cdots + \quad & a_{2n}x_n = 0, \\
\cdots\cdots\cdots\cdots\cdots\cdots\cdots\cdots\cdots & = 0, \\
a_{n1}x_1 + \quad a_{n2}x_2 + \cdots + (a_{nn} - \lambda)x_n & = 0.
\end{aligned}
$$
$$(15.33)$$

These equations have one or more solutions if and only if the determinant of the coefficients vanishes (§15.1.1). Expanding the determinant, we get an nth-order equation in λ of the form

$$\lambda^n - \beta_1\lambda^{n-1} + \beta_2\lambda^{n-2} - \cdots + (-1)^n\beta_n = 0. \qquad (15.34)$$

This is the *characteristic equation* of matrix $\mathscr{A}$ and its roots are called the *characteristic roots (values)* or *eigenvalues* of $\mathscr{A}$.

Because the determinant of the coefficients in eq. (15.33) vanishes whenever λ is a root of eq. (15.34), if a root happens to be zero, the determinant reduces to $\det(\mathscr{A})$ so that $\det(\mathscr{A}) = 0$. In this case, eq. (15.20) shows that $\mathscr{A}^{-1}$ is infinite (does not exist).

15.1.4 Series expansions

(a) Taylor's series. Taylor's series is discussed in most advanced mathematics texts, for example, Potter and Goldberg (1987: 84) and Pipes and Harvill (1970: 841–3). The series enables us to find the change in $f(x)$ when x changes by h in terms of powers of h and the derivatives of $f(x)$. The series can be written

$$f(x + h) = f(x) + hf'(x) + (h^2/2!)f''(x)$$

$$+ \cdots + [h^{n-1}/(n-1)!]f^{n-1}(x) + R(\xi), \qquad (15.35)$$

where $f'(x), f''(x), \ldots, f^{n-1}(x)$ are derivatives of orders $1, 2, \ldots, n-1$, $\xi = kh$, where $0 < k < 1$, and $R(\xi) = (h^n/n!)f^n(\xi) = $ *remainder after n terms.* Obviously, $R(\xi)$ is the error when we truncate the series after n terms; hence, the error is of the order of h^n. The larger h, the more terms we require to achieve a given accuracy. In practice, two or three terms are usually sufficient.

(b) Maclaurin's series. If we set $x = 0$ and then re-

place h with x in eq. (15.35), we get

$$f(x) = f(0) + xf'(0)$$
$$+ (x^2/2!)f''(0) + \cdots + (x^n/n!)f^n(0) + \cdots. \qquad (15.36)$$

Maclaurin's series can be used to derive many useful infinite series. For example, if $f(x) = e^x$, then $f'(0) = 1 = f''(0) = \cdots = f^n(0)$, so

$$e^x = 1 + x/1! + x^2/2! + x^3/3! + \cdots. \qquad (15.37)$$

Similarly, since $(d/dx)(\sin x) = \cos x$, $(d^2/dx^2)(\sin x) = -\sin x$, and so on, $f(0) = 0 = f''(0) = f^{iv}(0) = \cdots, f'(0) = +1, f'''(0) = -1, \ldots$; therefore,

$$\sin x = x - x^3/3! + x^5/5! - x^7/7! + \cdots. \qquad (15.38)$$

In the same way,

$$\cos x = 1 - x^2/2! + x^4/4! - x^6/6! + \cdots. \qquad (15.39)$$

(c) Binomial series. The binomial series is obtained by the expansion of the function $(a + b)^n$. We can write this in the form $a^n(1 + x)^n$, where $x = b/a$. Let $|a| > |b|$ so that $|x| < 1$, then we can expand $(1 + x)^n$ in a series that is finite if n is a positive integer but is otherwise infinite. Writing $f(x) = (1 + x)^n$, we have

$$f(0) = 1, \qquad f'(0) = n(1 + x)^{n-1}|_{x=0} = n,$$
$$f''(x) = n(n - 1), \text{ and so on,}$$

so that

$$f(x) = 1 + nx + \frac{n(n - 1)}{2!}x^2 + \cdots$$
$$+ \frac{n(n - 1)(n - 2) \cdots (n - r + 1)}{r!}x^r + \cdots. \qquad (15.40)$$

This series is valid for all finite values of n (Wylie, 1966: 695).

The binomial series is used frequently to obtain approximations, especially of the following functions:

$$(1 + x)^{1/2} = 1 + \frac{1}{2}x - \frac{1}{8}x^2 + \frac{1}{16}x^3 - \cdots, \qquad (15.41)$$

$$(1 + x)^{-1/2} = 1 - \frac{1}{2}x + \frac{3}{8}x^2 - \frac{5}{16}x^3 - \cdots, \qquad (15.42)$$

$$(1 + x)^{-1} = 1 - x + x^2 - x^3 + \cdots, \qquad (15.43)$$

$$(1 + x)^{-2} = 1 - 2x + 3x^2 - 4x^3 + \cdots. \qquad (15.44)$$

15.1.5 Complex numbers

The square roots of negative numbers are *imaginary numbers,* and numbers that are partly real and partly imaginary are *complex numbers.* If we write $j = \sqrt{-1}$, that is, $j^2 = -1$ (some writers use i instead of j), we can write, for example, the imaginary number $\sqrt{-9} = \sqrt{9}\sqrt{-1} = 3j$.

A complex number, $z = a + jb$, can be represented by plotting in the *complex plane* where the direction of imaginary numbers is at right angles to the real direction, as in fig. 15.2. We can also express complex numbers in polar form:

$$z = a + jb = r(\cos\theta + j\sin\theta) = re^{j\theta} \qquad (15.45)$$

(see problem 15.12a), where $r = (a^2 + b^2)^{1/2}$ = modulus of $z = |z|$ and $\theta = \tan^{-1}(b/a) = \arg(z)$. The conjugate complex of z, $\bar{z}$, is defined as $\bar{z} = a - jb = r(\cos\theta - j\sin\theta) = re^{-j\theta}$ (see fig. 15.2).

The sum (or difference) of complex numbers is obtained by adding (or subtracting) the real and imaginary parts. If $z_1 = a + jb$, $z_2 = c + jd$, then $(z_1 \pm z_2) = (a \pm c) + j(b \pm d)$. A complex number is zero only if both its real and imaginary parts are zero; hence, two complex numbers are equal only if both their real and imaginary parts are equal. Multiplication and division obey the usual algebraic rules. For example (see also problem 15.12b),

$$\left.\begin{aligned}
z_1 z_2 &= (a + jb)(c + jd) \\
&= (ac - bd) + j(ad + bc) \\
&= r_1 r_2 [\cos(\theta_1 + \theta_2) + j\sin(\theta_1 + \theta_2)] \\
&= r_1 r_2 e^{j(\theta_1 + \theta_2)};
\end{aligned}\right\} \quad (15.46)$$

$$\left.\begin{aligned}
z_1/z_2 &= \frac{a + jb}{c + jd} = \frac{(a + jb)(c - jd)}{(c + jd)(c - jd)} \\
&= \frac{(ac + bd) + j(bc - ad)}{c^2 + d^2} \\
&= (r_1/r_2)[\cos(\theta_1 - \theta_2) + j\sin(\theta_1 - \theta_2)] \\
&= (r_1/r_2)e^{j(\theta_1 - \theta_2)}.
\end{aligned}\right\} \quad (15.47)$$

The nth root of z, z_0, can be found by writing

$$z = re^{j\theta} = z_0^n = (r_0 e^{j\theta_0})^n = [r_0(\cos\theta_0 + j\sin\theta_0)]^n$$
$$= r_0^n(\cos n\theta_0 + j\sin n\theta_0)$$

by de Moivre's theorem (see problem 15.12a). Hence,

$$\left.\begin{aligned}
z^{1/n} &= z_0 = r_0(\cos\theta_0 + j\sin\theta_0), \\
r_0 &= r^{1/n}, \qquad \theta_0 = (\theta + 2\pi k)/n, \\
k &= 0, 1, 2, 3, \ldots, n - 1.
\end{aligned}\right\} \quad (15.48)$$

Figure 15.3 shows roots plotted in polar form for the case where $z = r = re^{j2\pi}$ = real, $n = 5$ and 6.

15.1.6 Method of least squares

(a) Basic method. Let us assume that we wish to obtain the "best-fit" curve of order m,

$$y_i = a_0 + a_1 x_i + a_2 x_i^2 + \cdots + a_m x_i^m \quad (15.49)$$

to represent a set of n pairs of measured values (x_i, y_i). If $n = m + 1$, the curve will pass through all n points, (x_i, y_i). If $n > m + 1$, the curve will not pass through all n points and we seek the "best-fit" curve such that the sum of the squares of the "errors" between the curve and each point (x_i, y_i) is a minimum, the errors e_i being the differences between the measured values y_i and those given by the curve. Thus,

$$e_i = y_i - (a_0 + a_1 x_i + \cdots + a_m x_i^m),$$
$$i = 1, 2, 3, \ldots, n,$$

and we wish to minimize E, where

$$E = \sum_{i=1}^{n} e_i^2 = \sum_i [y_i - (a_0 + a_1 x_i + \cdots + a_m x_i^m)]^2.$$

Because E is a function of the parameters a_k only, the minimum is given by

$$\frac{\partial E}{\partial a_k} = 2\sum_i (y_i - a_0 - a_1 x_i - \cdots - a_m x_i^m)(-x_i^k) = 0,$$
$$a_0 \sum_i x_i^k + a_1 \sum_i x_i^{k+1} + \cdots + a_m \sum_i x_i^{k+m} = \sum_i x_i^k y_i,$$
$$k = 0, 1, 2, \ldots, m. \quad (15.50)$$

There are $m + 1$ such *normal equations,* so we can solve for the $m + 1$ unknowns, a_k.

Sometimes we wish to find a least-squares solution subject to a certain condition on the unknown parameters *(constraint)*, for example, we could require that $a_1 = a_4$ and/or $a_1 + a_2 + a_3 = 0$. We can write each constraint in the form $C(a_1, a_2, \ldots, a_m) = 0$. Because the a_i's are chosen so that $\partial E/\partial a_i = 0$ (and $\partial C/\partial a_i = 0$), we can write the least-squares condition with constraints in the form

$$\frac{\partial}{\partial a_i}(E + \lambda C) = 0, \qquad i = 0, 1, \ldots, m, \quad (15.51)$$

λ having the same significance here as in the Lagrange method of undetermined multipliers (Pipes and Harvill, 1970: 968). These $m + 1$ equations and the equation $C(a_1, a_2, \ldots, a_m) = 0$ suffice to solve for λ and

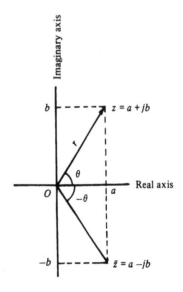

Fig. 15.2   Geometrical representation of complex numbers.

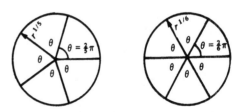

Fig. 15.3   Roots of a complex number.

the $m + 1$ values of a_i. The extension to several constraints involves solving

$$\frac{\partial}{\partial a_i}\left(E + \sum_j \lambda_j C_j\right) = 0.$$

Equation (15.50) can be written in matrix form as

$$\mathscr{Y}^* = \mathscr{X}^* \cdot \mathscr{C}^*, \tag{15.52}$$

where

$$\mathscr{X}^* = \left\|\begin{array}{cccc} \sum_i x_i^0 & \sum_i x_i^1 & \cdots & \sum_i x_i^m \\ \sum_i x_i^1 & \sum_i x_i^2 & \cdots & \sum_i x_i^{m+1} \\ \vdots & \vdots & \ddots & \vdots \\ \sum_i x_i^m & \sum_i x_i^{m+1} & \cdots & \sum_i x_i^{2m} \end{array}\right\|,$$

$$\mathscr{C}^* = \left\|\begin{array}{c} a_0 \\ a_2 \\ \vdots \\ a_m \end{array}\right\|, \qquad \mathscr{Y}^* = \left\|\begin{array}{c} \sum_i y_i \\ \sum_i x_i y_i \\ \vdots \\ \sum_i x_i^m y_i \end{array}\right\|.$$

Because $\mathscr{X}^*$ is square,

$$\mathscr{C}^* = (\mathscr{X}^*)^{-1} \mathscr{Y}^*. \tag{15.53}$$

If we solve the least-squares problem using matrices from the beginning, we can obtain a more general result that is also well adapted to computer calculations. We write eq. (15.49) in the form

$$\mathscr{X} \cdot \mathscr{C} = \mathscr{Y},$$

where

$$\mathscr{X} = \left\|\begin{array}{cccc} x_{11} & x_{12} & \cdots & x_{1m} \\ x_{21} & x_{22} & \cdots & x_{2m} \\ \vdots & \vdots & \ddots & \vdots \\ x_{n1} & x_{n2} & \cdots & x_{nm} \end{array}\right\|, \quad \mathscr{C} = \left\|\begin{array}{c} a_1 \\ a_2 \\ \vdots \\ a_m \end{array}\right\|, \quad \mathscr{Y} = \left\|\begin{array}{c} y_1 \\ y_2 \\ \vdots \\ y_n \end{array}\right\|$$

x_{ij}, y_i being known, a_i unknown, and $n > m$. (In general, x_{ij} can be any $m \times n$ known quantities, including powers of x_i as in eq. (15.49).

Because we have more equations than unknowns, the equations cannot hold exactly; writing $\mathscr{E}$ as the column matrix of the errors e_i, $i = 1, 2, \ldots, n$, we obtain

$$\mathscr{X} \cdot \mathscr{C} - \mathscr{Y} = \mathscr{E}.$$

This can be simplified by writing (Claerbout, 1976: 107)

$$\left\|\begin{array}{cccc} -y_1 & x_{11} & \cdots & x_{1m} \\ \vdots & \vdots & \ddots & \vdots \\ -y_n & x_{n1} & \cdots & x_{nm} \end{array}\right\| \left\|\begin{array}{c} 1 \\ a_1 \\ \vdots \\ a_m \end{array}\right\| = \left\|\begin{array}{c} e_1 \\ \vdots \\ e_n \end{array}\right\|.$$

This can be partitioned thus:

$$\| -\mathscr{Y} \mid \mathscr{X} \| \left\|\begin{array}{c} 1 \\ \text{---} \\ \mathscr{C} \end{array}\right\| = \mathscr{E} = \mathscr{B} \left\|\begin{array}{c} 1 \\ \text{---} \\ \mathscr{C} \end{array}\right\|,$$

where the first column of $\mathscr{B}$ is $-\mathscr{Y}$, and the rest is $\mathscr{X}$. We can accommodate this by taking the first column as the 0th column, that is, $b_{i0} = -y_i$ and $b_{ij} = x_{ij}$, $j > 0$.

Individual errors are given by

$$e_i = \| 1 \ a_1 \cdots a_m \| \left\|\begin{array}{c} -y_i \\ x_{i1} \\ \vdots \\ x_{im} \end{array}\right\| = \| -y_i \ x_{i1} \cdots x_{im} \| \left\|\begin{array}{c} 1 \\ a_1 \\ \vdots \\ a_m \end{array}\right\|.$$

Then,

$$e_i^2 = \| 1 \ a_1 \cdots a_m \| \left\|\begin{array}{c} -y_i \\ x_{i1} \\ \vdots \\ x_{im} \end{array}\right\| \| -y_i \ x_{i1} \cdots x_{im} \| \left\|\begin{array}{c} 1 \\ a_1 \\ \vdots \\ a_m \end{array}\right\|,$$

and

$$E = \sum_i e_i^2 = \| 1 \ \ a_1 \ \ \cdots \ \ a_m \| \sum_i \left(\left\|\begin{array}{c} -y_i \\ x_{i1} \\ \vdots \\ x_{im} \end{array}\right\| \right.$$

$$\left. \times \| -y_i \ \ x_{i1} \ \ \cdots \ \ x_{im} \| \right) \left\|\begin{array}{c} 1 \\ a_1 \\ \vdots \\ a_m \end{array}\right\|$$

$$= \| 1 \mid \mathscr{C}^T \| \mathscr{R} \left\|\begin{array}{c} 1 \\ \text{---} \\ \mathscr{C} \end{array}\right\|,$$

where

$$\mathscr{R} = \sum_i \left\|\begin{array}{c} -y_i \\ x_{i1} \\ \vdots \\ x_{im} \end{array}\right\| \| -y_i \ \ x_{i1} \ \ \cdots \ \ x_{im} \| = \mathscr{B}^T \mathscr{B}.$$

$$\tag{15.54}$$

Setting derivatives of E with respect to a_i equal to zero gives

$$\frac{\partial E}{\partial a_i} = 0 = \| 0 \cdots 0 \ 1 \ 0 \cdots 0 \| \mathscr{R} \left\|\begin{array}{c} 1 \\ \text{---} \\ \mathscr{C} \end{array}\right\|$$

$$+ \| 1 \mid \mathscr{C}^T \| \mathscr{R} \left\|\begin{array}{c} 0 \\ \vdots \\ 0 \\ 1 \\ 0 \\ \vdots \\ 0 \end{array}\right\|$$

$$\frac{2E}{2a_i} = 2 \| 0 \cdots 0\, 1\, 0 \cdots 0 \| \mathcal{R} \left\| \begin{matrix} 1 \\ \text{---} \\ \mathcal{A} \end{matrix} \right\|$$

because $\mathcal{R} = \mathcal{B}^T\mathcal{B}$ is symmetrical. This result can be written

$$\| rj_1 \ldots r_{im} \| \left\| \begin{matrix} 1 \\ \text{---} \\ \mathcal{A} \end{matrix} \right\| = 0, \qquad i = 1, 2, 3, \ldots m.$$

If we combine the m equations, we get

$$\mathcal{R} \left\| \begin{matrix} 1 \\ \text{---} \\ \mathcal{A} \end{matrix} \right\| = \mathcal{C}$$

except that we lack the 0th row of $\mathcal{R}$. We define a quantity v by the relation

$$\| r_{00} \ \ r_{01} \cdots r_{0m} \| \left\| \begin{matrix} 1 \\ \text{---} \\ \mathcal{A} \end{matrix} \right\| = v; \qquad (15.55)$$

we now have

$$\mathcal{R} \left\| \begin{matrix} 1 \\ \text{---} \\ \mathcal{A} \end{matrix} \right\| = \left\| \begin{matrix} v \\ 0 \\ \vdots \\ 0 \end{matrix} \right\| = \mathcal{B}^T\mathcal{B} \left\| \begin{matrix} 1 \\ \text{---} \\ \mathcal{A} \end{matrix} \right\|$$

$$= \left\| \begin{matrix} -\mathcal{Y}^T \\ \text{----} \\ \mathcal{X}^T \end{matrix} \right\| \| -\mathcal{Y} \mid \mathcal{X} \| \left\| \begin{matrix} 1 \\ \text{---} \\ \mathcal{A} \end{matrix} \right\| \qquad (15.56)$$

$$= \left\| \begin{matrix} \mathcal{Y}^T\mathcal{Y} & -\mathcal{Y}^T\mathcal{X} \\ -\mathcal{X}^T\mathcal{Y} & \mathcal{X}^T\mathcal{X} \end{matrix} \right\| \left\| \begin{matrix} 1 \\ \text{---} \\ \mathcal{A} \end{matrix} \right\|.$$

that is,

$$\left\| \begin{matrix} v \\ 0 \\ \vdots \\ 0 \end{matrix} \right\| = \left\| \begin{matrix} \mathcal{Y}^T\mathcal{Y} - \mathcal{Y}^T\mathcal{X}\,\mathcal{A} \\ -\mathcal{X}^T\mathcal{Y} + \mathcal{X}^T\mathcal{X}\,\mathcal{A} \end{matrix} \right\|$$

Thus, we have

$$v = \mathcal{Y}^T\mathcal{Y} - \mathcal{Y}^T\mathcal{X}\,\mathcal{A},$$

and

$$\left\| \begin{matrix} 0 \\ 0 \\ \vdots \\ 0 \end{matrix} \right\| = -\mathcal{X}^T\mathcal{Y} + \mathcal{X}^T\mathcal{X}\,\mathcal{A};$$

hence,

$$\mathcal{A} = (\mathcal{X}^T\mathcal{X})^{-1}\mathcal{X}^T\mathcal{Y}. \qquad (15.57)$$

At times we wish to give extra weight to one or more sets of observations $(y_i, x_{i1}, x_{i2}, \ldots, x_{im})$. We can do this by multiplying the error e_i and the ith row of $\mathcal{B}$ by a weighting factor $\sqrt{w_i}$. Then,

$$E = \sum w_i e_i^2 = \| 1 \mid \mathcal{A}^T \| \sum \left(w_i \left\| \begin{matrix} -y_i \\ x_{i1} \\ \vdots \\ x_{im} \end{matrix} \right\| \right.$$
$$\left. \times \| -y_i \ x_{i1} \cdots x_{im} \| \right) \left\| \begin{matrix} 1 \\ \text{---} \\ \mathcal{A} \end{matrix} \right\| = \| 1 \mid \mathcal{A}^T \| \mathcal{R}^* \left\| \begin{matrix} 1 \\ \text{---} \\ \mathcal{A} \end{matrix} \right\|,$$
$$(15.58)$$

where $\mathcal{R}^*$ is the weighted form of $\mathcal{R} = \mathcal{B}^T\mathcal{B}$, that is, the product of $\mathcal{B}^T$ and $\mathcal{B}$ when the ith row of $\mathcal{B}$ and the ith column of $\mathcal{B}^T$ are multiplied by $\sqrt{w_i}$. Equation (15.57) is still valid except that the ith row of $\mathcal{X}$ and $\mathcal{Y}$ and the ith column of $\mathcal{X}^T$ are multiplied by $\sqrt{w_i}$.

Constraints can be considered as additional equations in the unknowns a_i that must be satisfied exactly, that is, they have infinite weights in comparison with the error equations. We can write k linear constraint equations, $k < m$, in the form

$$\mathcal{C} \left\| \begin{matrix} 1 \\ \text{---} \\ \mathcal{A} \end{matrix} \right\| = \left\| \begin{matrix} 0 \\ 0 \\ \vdots \\ 0 \end{matrix} \right\| = \mathcal{O}, \qquad (15.59)$$

where

$$\mathcal{C} = \left\| \begin{matrix} c_{10} & c_{11} & \cdots & c_{1m} \\ c_{20} & c_{21} & \cdots & c_{2m} \\ \vdots & \vdots & \ddots & \vdots \\ c_{k0} & c_{k1} & \cdots & c_{km} \end{matrix} \right\|.$$

Following Claerbout (1976: 112–13), we assign the weight $\sqrt{w}$ to each constraint equation, insert the weighted left-hand side of eq. (15.59) in the error equations, derive the result, and then let w approach infinity.

Without constraints, we had the result (see eqs. (15.54) and (15.56)).

$$\mathcal{A} \left\| \begin{matrix} 1 \\ \text{---} \\ \mathcal{A} \end{matrix} \right\| = \sum_{i=1}^{n} \left(\left\| \begin{matrix} -y_i \\ x_{i1} \\ \vdots \\ x_{im} \end{matrix} \right\| \| -y_i \ x_{i1} \cdots x_{im} \| \right)$$

$$\times \left\| \begin{matrix} 1 \\ \text{---} \\ \mathcal{A} \end{matrix} \right\| = \left\| \begin{matrix} v \\ 0 \\ \vdots \\ 0 \end{matrix} \right\|;$$

with constraints, this becomes

$$
\left[\sum_{i=1}^{n} \left(\left\| \begin{array}{c} -y_i \\ x_{i1} \\ \vdots \\ x_{im} \end{array} \right\| \; \| -y_i \;\; x_{i1} \;\; \cdots \;\; x_{im} \| \right) \right.
$$

$$
\left. + \sum_{i=1}^{k} w \left(\left\| \begin{array}{c} c_{i0} \\ c_{i1} \\ \vdots \\ c_{im} \end{array} \right\| \; \| c_{i0} \;\; c_{i1} \;\; \cdots \;\; c_{im} \| \right) \right]
$$

$$
\times \left\| \begin{array}{c} 1 \\ \text{---} \\ \mathscr{A} \end{array} \right\| = \left\| \begin{array}{c} v^* \\ 0 \\ \vdots \\ 0 \end{array} \right\|, \tag{15.60}
$$

that is,

$$
(\mathscr{B}^\mathsf{T}\mathscr{B} + w\mathscr{C}^\mathsf{T}\mathscr{C}) \left\| \begin{array}{c} 1 \\ \text{---} \\ \mathscr{A} \end{array} \right\| = \left\| \begin{array}{c} v^* \\ 0 \\ \vdots \\ 0 \end{array} \right\| = \mathscr{V}^*.
$$

We write

$$
w = \frac{1}{\varepsilon}, \qquad \left\| \begin{array}{c} 1 \\ \text{---} \\ \mathscr{A} \end{array} \right\| = \mathscr{A}_0^* + \varepsilon\,\mathscr{A}_1^* + \varepsilon^2\,\mathscr{A}_2^* + \cdots,
$$

where $\mathscr{A}_0^*$ is the matrix that gives the desired solution, $\mathscr{A}_1^*$ is a similar matrix with different unknowns, and so on. Substituting, we get

$$
\left(\mathscr{B}^\mathsf{T}\mathscr{B} + \frac{1}{\varepsilon}\mathscr{C}^\mathsf{T}\mathscr{C} \right)(\mathscr{A}_0^* + \varepsilon\,\mathscr{A}_1^* + \cdots) = \mathscr{V}^*.
$$

Equating powers of ε gives

$$
\left. \begin{array}{l} \varepsilon^{-1}\!: \;\; \mathscr{C}^\mathsf{T}\mathscr{C}\mathscr{A}_0^* = \mathscr{O}, \\[2mm] \varepsilon^{0}\!: \;\; \mathscr{B}^\mathsf{T}\mathscr{B}\mathscr{A}_0^* + \mathscr{C}^\mathsf{T}\mathscr{C}\mathscr{A}_1^* = \mathscr{V}^*. \end{array} \right\} \tag{15.61}
$$

Because

$$
\mathscr{C}\mathscr{A}_0^* = \mathscr{C} \left\| \begin{array}{c} 1 \\ \text{---} \\ \mathscr{A} \end{array} \right\| = \mathscr{O}
$$

from eq. (15.59), the first equation is satisfied automatically; hence, it provides no new information. In the second equation, we substitute $\mathscr{C}\mathscr{A}_1^* = \mathscr{L}$, where $\mathscr{L}$ is a $k \times 1$ matrix whose elements are the equivalents of Lagrangian undetermined multipliers. Then

$$
\mathscr{B}^\mathsf{T}\mathscr{B}\,\mathscr{A}_0^* + \mathscr{C}^\mathsf{T}\mathscr{L} = \mathscr{V}^*. \tag{15.62}
$$

Equations (15.59) and (15.62), which give the solutions for the m unknowns a_i and the k unknowns λ_i, can be combined in the form

$$
\left\| \begin{array}{c|c} \mathscr{B}^\mathsf{T}\mathscr{B} & \mathscr{C}^\mathsf{T} \\ \hline \mathscr{C} & \mathscr{O} \end{array} \right\| \left\| \begin{array}{c} \mathscr{A}_0^* \\ \hline \mathscr{L} \end{array} \right\| = \left\| \begin{array}{c} \mathscr{V}^* \\ \hline \mathscr{O} \end{array} \right\|,
$$

or

$$
\left\| \begin{array}{c} \mathscr{B}^\mathsf{T}\mathscr{B}\mathscr{A}_0^* + \mathscr{C}^\mathsf{T}\mathscr{L} \\ \hline \mathscr{C}\mathscr{A}_0^* + \mathscr{O} \end{array} \right\| = \left\| \begin{array}{c} \mathscr{V}^* \\ \hline \mathscr{O} \end{array} \right\|. \tag{15.63}
$$

(b) Multitrace least squares. In §9.5.5 and 15.1.6a, we discussed least-squares filtering of a single trace. Extension to multitrace situations involves considerably more complex mathematics, so we shall consider the case of two traces first and then discuss generalization to n traces. Our treatment is based largely on that of Schneider et al. (1964), but we have simplified their notation; to do so, we have departed from our usual notation.

We consider two traces x_t and y_t, which are inputs to two filters X_t and Y_t. The filter outputs are $X_t * x_t$ and $Y_t * y_t$, and we require that each output be as close as possible to a single desired output. We write z_t for the sum of the desired outputs; hence, the error at time $t = k$ is (see eq. (9.23))

$$
\begin{aligned} E_k &= z_k - (X_k * x_k + Y_k * y_k) \\ &= z_k - \sum_i X_i x_{k-i} - \sum_i Y_i y_{k-i}, \end{aligned}
$$

the sums being over all appropriate values of i. Squaring E_k gives

$$
\begin{aligned} E_k^2 &= z_k^2 + \left(\sum_i X_i x_{k-i}\right)\left(\sum_j X_j x_{k-j}\right) \\ &\quad + \left(\sum_i Y_i y_{k-i}\right)\left(\sum_j Y_j y_{k-j}\right) \\ &\quad - 2z_k\left(\sum_i X_i x_{k-i} + \sum_i Y_i y_{k-i}\right) \\ &\quad + 2\left(\sum_i X_i x_{k-i}\right)\left(\sum_j Y_j y_{k-j}\right). \end{aligned}
$$

Summing over the full range of k gives the total squared error:

$$
\begin{aligned} E^2 &= \sum_k z_k^2 + \sum_k\left[\left(\sum_i X_i x_{k-i}\right)\left(\sum_j X_j x_{k-j}\right)\right] \\ &\quad + \sum_k\left[\left(\sum_i Y_i y_{k-i}\right)\left(\sum_j Y_j y_{k-j}\right)\right] \\ &\quad - 2\sum_k z_k\left(\sum_i X_i x_{k-i} + \sum_i Y_i y_{k-i}\right) \\ &\quad + 2\sum_k\left[\left(\sum_i X_i x_{k-i}\right)\left(\sum_j Y_j y_{k-j}\right)\right]. \end{aligned}
$$

Interchanging the order of summation, we get

$$E^2 = \sum_k z_k^2 + \sum_i \sum_j X_i X_j \left(\sum_k x_{k-i} x_{k-j} \right)$$

$$+ \sum_i \sum_j Y_i Y_j \left(\sum_k y_{k-i} y_{k-j} \right)$$

$$- 2 \sum_i X_i \left(\sum_k z_k x_{k-i} \right) - 2 \sum_i Y_i \left(\sum_k z_k y_{k-i} \right)$$

$$+ 2 \sum_i \sum_j X_i Y_j \left(\sum_k x_{k-i} y_{k-j} \right).$$

Because all sums are over all appropriate values of the indices, we can replace the index k in the second, third, and sixth summations with a new index $r = k - i$; moreover, $\sum_k z_k x_{k-i} = \phi_{zx}(-i)$ from eq. (9.42), and so on. Thus,

$$E^2 = \sum_k z_k^2 + \sum_i \sum_j X_i X_j \left(\sum_r x_r x_{r+(i-j)} \right)$$

$$+ \sum_i \sum_j Y_i Y_j \left(\sum_r y_r y_{r+(i-j)} \right)$$

$$- 2 \sum_i X_i \phi_{zx}(-i) - 2 \sum_i Y_i \phi_{zy}(-i)$$

$$+ 2 \sum_i \sum_j X_i Y_j \left(\sum_r x_r y_{r+(i-j)} \right)$$

$$E^2 = \phi_{zz}(0) + \sum_i \sum_j X_i X_j \phi_{xx}(i-j) + \sum_i \sum_j Y_i Y_j \phi_{yy}(i-j)$$

$$- 2 \sum_i X_i \phi_{zx}(-i) - 2 \sum_i Y_i \phi_{zy}(-i)$$

$$+ 2 \sum_i \sum_j X_i Y_j \phi_{xy}(i-j). \qquad (15.64)$$

Next, we vary X_m and Y_m, $m = 0, 1, 2, \ldots, N$, to find the minimum E^2. This requires that

$$\frac{\partial E^2}{\partial X_m} = 0 = \frac{\partial E^2}{\partial Y_m}, \qquad m = 0, 1, 2, \ldots, N.$$

Some care must be taken in differentiating the sums. Because X_m (and Y_m) occurs once in a summation, we have

$$\frac{\partial}{\partial X_m} \left(\sum_i X_i \right) = 1; \qquad \frac{\partial}{\partial X_m} \left[\sum_i X_i \phi_{zx}(-i) \right] = \phi_{zx}(-m);$$

$$\frac{\partial}{\partial X_m} \left\{ \sum_i X_i \left[\sum_j X_j \phi_{xx}(i-j) \right] \right\} = \sum_j X_j \phi_{xx}(m-j)$$

$$+ \sum_i X_i \phi_{xx}(i-m)$$

$$= 2 \sum_j X_j \phi_{xx}(m-j)$$

$$= 2 \sum_i X_i \phi_{xx}(m-i).$$

Obviously, similar results hold for differentiation with respect to Y_m. Carrying out the differentiations of eq. (15.64), we get the following normal equations.

$$\sum_i X_i \phi_{xx}(m-i) + \sum_i Y_i \phi_{xy}(m-i) = \phi_{xz}(m), \quad (15.65a)$$

$$\sum_i X_i \phi_{yx}(m-i) + \sum_i Y_i \phi_{yy}(m-i) = \phi_{yz}(m) \qquad (15.65b)$$

(note that the summation index j, being a dummy index, has been changed to i).

The solution of eqs. (15.65a) and (15.65b) can be written as

$$\begin{Vmatrix} r_0 & r_1 & \cdots & r_N \\ r_{-1} & r_0 & \cdots & r_{N-1} \\ \vdots & \vdots & \ddots & \vdots \\ r_{-N} & \cdots & \cdots & r_0 \end{Vmatrix} \begin{Vmatrix} U_0 \\ U_1 \\ \vdots \\ U_N \end{Vmatrix} = \begin{Vmatrix} V_0 \\ V_1 \\ \vdots \\ V_N \end{Vmatrix} \quad (15.66a)$$

where

$$r_s = \begin{Vmatrix} \phi_{xx}(s) & \phi_{xy}(-s) \\ \phi_{yx}(-s) & \phi_{yy}(s) \end{Vmatrix},$$

$$U_s = \begin{Vmatrix} X_s \\ Y_s \end{Vmatrix}, \qquad V_s = \begin{Vmatrix} \phi_{xz}(s) \\ \phi_{yz}(s) \end{Vmatrix},$$

$s = 0, 1, 2, \ldots, N.$

To verify that this gives the solution of eq. (15.65), we can select any row of $\| r \|$, for example, the second, multiply the row by $\| U \|$, equate the result to V_1, and compare with eq. (15.65). The product of the second row and $\| U \|$ gives

$$r_{-1} U_0 + r_0 U_1 + \cdots + r_{N-1} U_N = V_1.$$

It is sufficient to show that the first row of the products gives eq. (15.65a) (the second row gives eq. (15.65b)). Thus,

$$[\phi_{xx}(-1)X_0 + \phi_{xy}(+1)Y_0] + [\phi_{xx}(0)X_1 + \phi_{xy}(0)Y_1]$$
$$+ \cdots + [\phi_{xx}(N-1)X_N + \phi_{xy}(1-N)Y_N] = \phi_{xz}(1),$$

that is,

$$\sum_0^N X_i \phi_{xx}(-1+i) + \sum_0^N Y_i \phi_{xy}(1-i) = \phi_{xz}(1),$$

which is eq. (15.65a) (because $\phi_{xx}(-1+i) = \phi_{xx}(1-i)$).

Derivation of the normal equations for the general case is given by Simpson et al. (1963) and Wiggins and Robinson (1965) discuss the solution. The matrix $\| r \|$ in both the two-trace case and the general case is a Toeplitz matrix; hence, the Levinson algorithm (§15.5.6c) can be used. Wiggins and Robinson also give other recursive solutions.

15.1.7 Finite differences

The calculus of finite differences has wide application in many practical situations, for example, calculations with data in digital form, interpolation, numerical differentiation and integration, numerical solution of differential equations. We shall discuss the basic relations assuming a function $f(x)$ that is given at equal intervals Δ, that is, we discuss the discrete function $f(x + n\Delta)$, $n = 0, \pm 1, \pm 2, \ldots$. We shall use the notation $f_n = f(x_0 + n\Delta)$ (note that $f_0 = f(x_0)$).

We define the following operators:

$$E\{f_n\} = f_{n+1} \qquad (15.67)$$

$$\blacktriangle\{f_n\} = f_{n+1} - f_n, \qquad (15.68)$$

$$D\{f_n\} = \left.\frac{\mathrm{d}f(x)}{\mathrm{d}x}\right|_{x=x_0+n\Delta}, \qquad (15.69)$$

$$D^{-1}\{f_0\} = \int_{x_0}^{x_0+\Delta} f(x)\,\mathrm{d}x. \qquad (15.70)$$

These are respectively the delay, difference, derivative, and integration operators.

The first and second differences are given by

$$\blacktriangle\{f_n\} = f_{n+1} - f_n,$$
$$\blacktriangle^2\{f_n\} = \blacktriangle\{\blacktriangle\{f_n\}\} = \blacktriangle\{f_{n+1}\} - \blacktriangle\{f_n\}.$$

By successive applications of the preceding, we find that

$$\blacktriangle^r\{f_n\} = f_{n+r} - rf_{n+r-1} + \frac{r(r-1)}{2!}f_{n+r-2}$$
$$+ \cdots + (-1)^{r-1}\frac{r(r-1).\ldots.3.2}{(r-1)!}f_{n+1}$$
$$+ (-1)^r f_n. \qquad (15.71)$$

Also,

$$\blacktriangle\{f_n\} = f_{n+1} - f_n = E\{f_n\} - f_n;$$

hence,

$$\blacktriangle = E - 1. \qquad (15.72)$$

(Sometimes the difference is referenced to $x + (n + 1/2)\Delta$ instead of $x + n\Delta$, in which case $\blacktriangle\{f_{n+1/2}\}$ is called the *central difference*.)

Combinations of operators are often used, for example,

$$E\{D\{f_0\}\} = E\left[\left(\frac{\mathrm{d}f}{\mathrm{d}x}\right)\Big|_{x=x_0}\right] = \left(\frac{\mathrm{d}f}{\mathrm{d}x}\right)\Big|_{x=x_0+\Delta} = f_1';$$

$$\blacktriangle\{D\{f_n\}\} = f_{n+1}' - f_n'.$$

The operators obey the basic laws of algebra such as association, distribution, and commutation.

We can apply Taylor's series (eq. (15.35)) to $E\{f_0\}$ to obtain

$$\left.\begin{aligned} E\{f_0\} &= f(x_0 + \Delta) = f(x_0) + \Delta f'(x_0) \\ &\quad + (\Delta^2/2!)f''(x_0) + \cdots, \\ &= [1 + (\Delta D) + (1/2!)(\Delta D)^2 + \cdots]f(x_0), \\ &= e^{\Delta D}\{f_0\}. \end{aligned}\right\} (15.73)$$

To obtain an expression for interpolation between $f(x_0)$ and $f(x_0 + \Delta)$, we use eq. (15.72) to write

$$E^r = (1 + \blacktriangle)^r = 1 + r\blacktriangle$$
$$+ \frac{r(r-1)}{2!}\blacktriangle^2 + \cdots \qquad (15.74)$$

Although r is normally a positive integer in this expression, the equation is still valid when $0 < r < +1$

(Wylie, 1966), which enables us to interpolate between f_0 and f_1; in this case, the result is known as the *Forward Gregory–Newton formula* (a Backward Gregory–Newton formula exists for use near the upper end of the tabulated values; see Wylie, loc. cit.).

An expression for the derivative, $D\{f_0\}$, can be found as follows:

$$f(x_0 + r\Delta) = E^r\{f_0\} = (1 + \blacktriangle)^r f_0,$$
$$= \left\{1 + r\blacktriangle + \frac{r(r-1)}{2!}\blacktriangle^2 + \cdots\right\}f_0.$$

Differentiating with respect to r (that is, treating r as if it were a continuous variable) gives

$$\frac{\mathrm{d}f(x_0 + r\Delta)}{\Delta\,\mathrm{d}r} = \frac{1}{\Delta}\left(\blacktriangle + \frac{2r-1}{2!}\blacktriangle^2\right.$$
$$\left. + \frac{3r^2 - 6r + 2}{3!}\blacktriangle^3 + \cdots\right)f_0. \qquad (15.75)$$

We let r go to zero and obtain

$$D\{f_0\} = \left.\frac{\mathrm{d}f}{\mathrm{d}x}\right|_{x=x_0} = \lim_{r\to 0}\frac{\mathrm{d}f(x_0 + r\Delta)}{\Delta\,\mathrm{d}r}$$
$$= \frac{1}{\Delta}(\blacktriangle - \tfrac{1}{2}\blacktriangle^2 + \tfrac{1}{3}\blacktriangle^3 - \tfrac{1}{4}\blacktriangle^4 + \cdots)f_0. \qquad (15.76)$$

Repeating the process, we find that

$$D^2\{f_0\} = \frac{\mathrm{d}^2 f}{\mathrm{d}x^2} = \frac{1}{\Delta^2}(\blacktriangle - \blacktriangle^2$$
$$+ \tfrac{1}{12}\blacktriangle^3 - \cdots)f_0. \qquad (15.77)$$

We encounter the operator $(-1, 1)$ in a number of situations, for example, to describe the successive reflections from the top and base of a thin bed (fig. 6.42) or to describe a reflection followed by its ghost from a perfect reflector (fig. 9.12). This is the difference operator of eq. (15.68); eq. (15.76) shows that $\blacktriangle/\Delta \approx D$, the derivative operator. Thus, the effect of the operator $(-1, 1)$ on the waveshape is approximately the same as that of the derivative.

The integral $D^{-1}\{f_0\}$, can be found as follows:

$$D^{-1}\{D\{f_0\}\} = \int_{x_0}^{x_0+\Delta} f'(x)\,\mathrm{d}x = f(x_0 + \Delta) - f(x_0)$$
$$= \blacktriangle\{f_0\},$$

so that

$$D^{-1}D = \blacktriangle. \qquad (15.78)$$

From eqs. (15.72) and (15.73), we have

$$E = 1 + \blacktriangle = \exp\{\Delta D\},$$

or

$$D = \frac{1}{\Delta}\ln(1 + \Delta) = \frac{1}{\Delta}(\blacktriangle - \tfrac{1}{2}\blacktriangle^2 + \tfrac{1}{3}\blacktriangle^3 - \cdots).$$

Thus,

$$D^{-1} = \blacktriangle/D = \Delta(1 - \tfrac{1}{2}\blacktriangle + \tfrac{1}{3}\blacktriangle^2 - \tfrac{1}{4}\blacktriangle^3 + \cdots)^{-1}.$$

Therefore,

$$D^{-1}\{f_0\} = \Delta(1 + \tfrac{1}{2}\blacktriangle - \tfrac{1}{12}\blacktriangle^2 + \tfrac{1}{24}\blacktriangle^3 - \cdots)f_0. \tag{15.79}$$

If we multiply together the operators in eqs. (15.76) and (15.79), we find that the product is $\blacktriangle$ in agreement with eq. (15.78).

15.1.8 Numerical solution of differential equations

Often, one must solve differential equations by numerical methods, generally because the equation is too complex to be solved analytically or the data are in digital form. The basic problem is that of finding $y(x)$, where $dy/dx = y' = f(x, y)$, $y = y_0$ when $x = x_0$, $f(x, y)$ being given either in functional form or as a table of values of $f(x, y)$ at equal intervals Δ of x (if the values are at unequal intervals, it is usually necessary to interpolate to get evenly spaced values).

We use Taylor's series (see eq. (15.73) to write

$$y_1 = y(x_0 + \Delta) \approx y(x_0) + \Delta\frac{dy}{dx} \approx y_0 + \Delta f(x_0, y_0),$$

$$y_2 = y(x_0 + 2\Delta) \approx y_1 + \Delta f(x_0 + \Delta, y_1), \tag{15.80}$$
$$\vdots$$
$$y_{n+1} \approx y_n + \Delta f(x_0 + n\Delta, y_n).$$

This method, called the *Euler–Cauchy method*, involves calculating successive values of the derivative, $f(x_0 + r\Delta, y_r)$, and using Taylor's series to find the approximate value of y_{r+1}. The accuracy is low unless Δ is small, but in this case the computing time can become excessive. Many methods have been devised, most of which involve higher-order terms in the Taylor series, to increase the accuracy and decrease the computing time. We shall describe briefly two of these.

Milne's method starts from eq. (15.75) with

$$r = 1, 2, 3, 4.$$

Writing y in place of f,

$$y'_r = \left.\frac{dy}{dx}\right|_{x = x_0 + r\Delta},$$

we get

$$y'_1 = \frac{1}{\Delta}(\blacktriangle + \tfrac{1}{2}\blacktriangle^2 - \tfrac{1}{6}\blacktriangle^3 + \tfrac{1}{12}\blacktriangle^4)y_0,$$

$$y'_2 = \frac{1}{\Delta}(\blacktriangle + \tfrac{3}{2}\blacktriangle^2 + \tfrac{1}{3}\blacktriangle^3 - \tfrac{1}{12}\blacktriangle^4)y_0,$$

$$y'_3 = \frac{1}{\Delta}(\blacktriangle + \tfrac{5}{2}\blacktriangle^2 + \tfrac{11}{6}\blacktriangle^3 + \tfrac{1}{4}\blacktriangle^4)y_0,$$

$$y'_4 = \frac{1}{\Delta}(\blacktriangle + \tfrac{7}{2}\blacktriangle^2 + \tfrac{13}{3}\blacktriangle^3 + \tfrac{25}{12}\blacktriangle^4)y_0.$$

Using eq. (15.71), these become

$$\left.\begin{aligned}
y'_1 &= (-3y_0 - 10y_1 + 18y_2 - 6y_3 + y_4)/12\Delta, \\
y'_2 &= (y_0 - 8y_1 + 8y_3 - y_4)/12\Delta, \\
y'_3 &= (-y_0 + 6y_1 - 18y_2 + 10y_3 + 3y_4)/12\Delta, \\
y'_4 &= (3y_0 - 16y_1 + 36y_2 - 48y_3 + 25y_4)/12\Delta
\end{aligned}\right\}. \tag{15.81}$$

Twice the sum of the first and third equations minus the second gives

$$y_4 = y_0 + (4\Delta/3)(2y'_1 - y'_2 + 2y'_3);$$

because (x_0, y_0) can be any one of the set of values, this relation applies to any four successive values, that is,

$$y_{n+1} = y_{n-3} + (4\Delta/3)(2y'_{n-2} - y'_{n-1} + 2y'_n). \tag{15.82}$$

If we know y_r for four consecutive values, $r = n$, $n - 1$, $n - 2$, $n - 3$, we can find y'_r for these values; hence, obtain y_{n+1} from eq. (15.82). To get started, we need the values of y_0, y_1, y_2, y_3; the derivatives, y'_1, y'_2, y'_3, can then be found by substitution in the differential equation after which we can calculate y_4. Because y_0 is known, we use Taylor's series to find y_1, y_2, y_3; to do this, we must know y', y'', y''', and so on, at $x = x_0$. We have

$$y' = f(x, y), \qquad y'' = \frac{\partial f}{\partial x} + \frac{\partial f}{\partial y}y', \qquad \text{and so on.}$$

The derivatives can be evaluated in the neighborhood of (x_0, y_0) and so y_1, y_2, y_3 can be found to any desired accuracy after which the differential equation gives y'_1, y'_2, y'_3, and eq. (15.82) gives y_4.

Like most methods, Milne's method is subject to errors that at times are cumulative. A check is furnished by the equation obtained by adding the second and fourth equations in eq. (15.81) to four times the third, the general result being

$$y_{n+1} = y_{n-1} + \tfrac{1}{3}\Delta(y'_{n-1} + 4y'_n + y'_{n+1}). \tag{15.83}$$

Once y_{n+1} is found by eq. (15.82), we find y'_{n+1}, then substitute in eq. (15.83) to get a check value of y_{n+1}.

Up to this point, we have used derivatives evaluated at $x_0 + r\Delta$, r integral. Higher accuracy will result in general if we use derivatives evaluated at intermediate points such as $x + (r + \tfrac{1}{2})\Delta$. This concept is basic in the *Runge–Kutta method*. The following account summarizes a more detailed discussion by Potter and Goldberg (1987: 498–502).

Consider the case of the first-order equation,

$$dy/dt = f(y, t).$$

Starting from the point (y_r, t_r), we wish to find the value of y_{r+1} corresponding to $t_{r+1} = t_r + \Delta$. We write

$$\begin{aligned}
y_{r+1} &= y_r + \Delta(dy/\Delta t) \\
&= y_r + \Delta f(y, t).
\end{aligned} \tag{15.84}$$

In the Euler–Cauchy method, $f(y, t)$ is the value of the derivative at the initial point (y_r, t_r), but to get higher accuracy, we select a value of $f(y, t)$ at a point in between points (y_r, t_r) and (y_{r+1}, t_{r+1}). Let this intermediate point be $(y_r + q\,\Delta f_r, t_r + p\Delta)$, where f_r is the slope at (y_r, t_r) and p, q are constants to be determined. Taking f_i as the slope at the intermediate point, we assign to f the value

$$f = af_r + bf_i, \tag{15.85}$$

where again constants a and b are to be determined. The slope f_i can be found using Taylor's series:

$$f_i = f_r + q \, \Delta f_r (\partial f/\partial y) + p \, \Delta (\partial f/\partial t) \quad (15.86)$$

(note that $\Delta y = q \, \Delta f_r$, $\Delta t = p \, \Delta$), where the partial derivatives are evaluated at (y_r, t_r). By using eqs. (15.85) and (15.86), eq. (15.84) now becomes

$$\begin{aligned} y_{r+1} &= y_r + \Delta(af_r + bf_i) \\ &= y_r + \Delta af_r + \Delta b[f_r + q \, \Delta f_r(\partial f/\partial y) + p \, \Delta(\partial f/\partial t)] \\ &= y_r + \Delta(a+b)f_r + \Delta^2[bqf_r(\partial f/\partial y) + bp(\partial f/\partial t)]. \end{aligned}$$
$$(15.87)$$

To get values for the constants a, b, p, and q, we expand y_{r+1} in a second-order Taylor's series and equate the coefficients with those in eq. (15.87). Thus,

$$\left. \begin{aligned} y_{r+1} &= y_r + \Delta(dy/dt) + (\Delta^2/2!)(d^2y/dt^2) \\ &= y_r + \Delta f_r + \tfrac{1}{2}\Delta^2(df/dt). \end{aligned} \right\} \quad (15.88)$$

But

$$\frac{df}{dt} = \frac{\partial f}{\partial y} \cdot \frac{\partial y}{\partial t} + \frac{\partial f}{\partial t} = f_r\frac{\partial f}{\partial y} + \frac{\partial f}{\partial t},$$

so

$$y_{r+1} = y_r + \Delta f_r + \tfrac{1}{2}\Delta^2 \left(f_r\frac{\partial f}{\partial y} + \frac{\partial f}{\partial t} \right). \quad (15.89)$$

Comparing eqs. (15.87) and (15.89), we find that

$$a + b = 1, \qquad bq = 1/2 = bp.$$

Because we have four unknowns and three equations, we are at liberty to assign an arbitrary value to one of the constants; usually, b is set equal to 1/2 or 1. When $b = 1/2$, we get $a = 1/2$, $p = 1 = q$; when $b = 1$, $a = 0$, $p = 1/2 = q$. In the first case, eq. (15.87) becomes

$$\begin{aligned} y_{r+1} &= y_r + \Delta f_r + \Delta^2 \left(\frac{1}{2}f_r\frac{\partial f}{\partial y} + \frac{1}{2}\frac{\partial f}{\partial t} \right) \\ &= y_r + \tfrac{1}{2}\Delta\left\{ f_r + \left[f_r + \Delta\left(f_r\frac{\partial f}{\partial y} + \frac{\partial f}{\partial t} \right) \right] \right\} \\ &= y_r + \tfrac{1}{2}\Delta \, [f_r + f(y_r + \Delta f_r, t_r + \Delta)]. \end{aligned}$$

When $b = 1$, we get

$$y_{r+1} = y_r + \Delta f(y_r + \tfrac{1}{2}\Delta f_r, t_r + \tfrac{1}{2}\Delta). \quad (15.90)$$

Thus, given (y_r, t_r), we can get y_{r+1} with the same accuracy as that of a second-order Taylor's series (note that eq. (15.88) is equivalent to eq. (15.90)).

The Runge–Kutta technique can be developed to provide even higher accuracy; of course, the equivalent of eq. (15.90) then contains more terms to provide the increased accuracy (Potter and Goldberg, 1987: 501).

Higher-order differential equations can be solved by reducing them to simultaneous first-order equations. Thus, the equation,

$$y'' = f(x, y), \qquad y = y_0,$$

and

$$y' = y_0' \text{ at } x = x_0,$$

is equivalent to

$$y' = z(x, y), \qquad y = y_0 \text{ at } x = x_0,$$
$$z' = f(x, y), \qquad z = y_0' \text{ at } x_0.$$

We solve the second equation for $z(x, y)$ and then solve the first equation to get $y = y(x)$.

15.1.9 Partial fractions

It is often convenient to express a function of the form $N(x)/D(x)$ in *partial fractions*, that is,

$$\begin{aligned} \frac{N(x)}{D(x)} &= \frac{A_1}{x - a_1} + \frac{A_2}{x - a_2} + \cdots + \frac{A_m}{x - a_m} \\ &\quad + \frac{B_n}{(x - b_1)^n} + \frac{B_{n-1}}{(x - b_1)^{n-1}} + \cdots + \frac{B_1}{(x - b_1)}, \end{aligned}$$

where the a_i's are single roots of $D(x)$, and b_1 is a multiple root of order n. Obviously, $N(x)$, $D(x)$ are polynomials in x, and we take the order of $N(x)$ less than that of $D(x)$ (if this is not so, we carry out long division and the remainder will be a fraction of this type). To find the values of A_i, B_i, we note that we can write

$$\begin{aligned} \frac{N(x)}{D(x)} &= \frac{N(x)}{k(x - a_1)(x - a_2)\cdots(x - a_m)(a - b_1)^n} \\ &= \frac{A_1}{x - a_1} + \cdots + \frac{A_m}{x - a_m} + \frac{B_n}{(x - b_1)^n} + \cdots \\ &\quad + \frac{B_1}{x - b_1}, \end{aligned}$$

where k is the coefficient of the highest power in $D(x)$.

To find A_1, we multiply both sides of the foregoing expression by $x - a_1$ and then set $x = a_1$; because the factor cancels on the left and in the first term on the right, and appears in all other terms on the right side, we find

$$A_1 = \left. \frac{N(x)}{D^*(x)} \right|_{x=a_1}, \quad (15.91)$$

where the asterisk means that the factor $x - a_1$ is deleted from $D(x)$. We can get all of the coefficients A_i in the same way. To get B_n, we multiply both sides by $(x - b_1)^n$ and then set $x = b_1$; thus,

$$B_n = \left. \frac{N(x)}{D^{**}(x)} \right|_{x=b_1},$$

where the double asterisk means that the factor $(x - b_1)^n$ has been deleted. To get B_{n-1}, we differentiate once before setting $x = b_1$:

$$B_{n-1} = \left. \frac{d}{dx}\left[\frac{N(x)}{D^{**}(x)} \right] \right|_{x=b_1}.$$

In general,

$$B_{n-s} = \left. \frac{1}{s!}\frac{d^s}{dx^s}\left[\frac{N(x)}{D^{**}(x)} \right] \right|_{x=b_1}. \quad (15.92)$$

As an example of the above, let it be required to find the inverse Laplace transform (see §15.3) of

$$\frac{s^2 - 2}{s(s^2 - 5s + 6)(s - 1)^2} = \frac{s^2 - 2}{s(s - 2)(s - 3)(s - 1)^2}.$$

Then,

$$\frac{s^2 - 2}{s(s - 2)(s - 3)(s - 1)^2} = \frac{A_1}{s} + \frac{A_2}{s - 2} + \frac{A_3}{s - 3} + \frac{B_2}{(s - 1)^2} + \frac{B_1}{s - 1}$$

and

$$A_1 = \left. \frac{s^2 - 2}{(s - 2)(s - 3)(s - 1)^2} \right|_{s=0} = -\frac{1}{3},$$

$$A_2 = \left. \frac{s^2 - 2}{s(s - 3)(s - 1)^2} \right|_{s=2} = -1,$$

$$A_3 = \left. \frac{s^2 - 2}{s(s - 2)(s - 1)^2} \right|_{s=3} = \frac{7}{12},$$

$$B_2 = \left. \frac{s^2 - 2}{s(s - 2)(s - 3)} \right|_{s=1} = -\frac{1}{2},$$

$$B_1 = \left. \frac{d}{ds}\left[\frac{s^2 - 2}{s(s - 2)(s - 3)} \right] \right|_{s=1}$$

$$[s(s - 2)(s - 3)^{-2}] [s(s - 2)(s - 3)2s$$
$$- (s^2 - 2)\{(s - 2)(s - 3) + s(s - 3)$$
$$+ s(s - 2)\}]\big|_{s=1} = \frac{3}{4}.$$

15.2 Fourier series and Fourier transforms

15.2.1 Fourier series

Let $g(t)$ be a periodic function with period T, that is,

$$g(t \pm nT) = g(t), \qquad n = 0, 1, 2, \ldots .$$

Provided that $g(t)$ is reasonably well-behaved, that is, provided $g(t)$ obeys the Dirichlet conditions: (1) it has at most a finite number of maxima, minima, and discontinuities in an interval T, and (2)

$$\int_{-T/2}^{+T/2} |g(t)| \, dt$$

is finite, then $g(t)$ can be expanded in a Fourier series:

$$g(t) = \tfrac{1}{2}a_0 + \sum_{n=1}^{\infty} (a_n \cos n\omega_0 t + b_n \sin n\omega_0 t), \qquad (15.93)$$

where $\omega_0 = 2\pi\nu_0 = 2\pi/T$. We have written $\frac{1}{2}a_0$ instead of a_0 so that all values of a_n are given by the same formula, eq. (15.99). To obtain coefficients a_n and b_n, we use the fact that, for any value d and for m and n integral,

$$\int_d^{d+2\pi} \sin m\theta \sin n\theta \, d\theta = 0, \qquad (15.94)$$

$$\int_d^{d+2\pi} \cos m\theta \cos n\theta \, d\theta = 0, \qquad (15.95)$$

$$m \neq n$$

$$\int_d^{d+2\pi} \sin m\theta \cos n\theta \, d\theta = 0, \qquad (15.96)$$

$$\int_d^{d+2\pi} \sin^2 n\theta \, d\theta = \pi, \qquad (15.97)$$

$$\int_d^{d+2\pi} \cos^2 n\theta \, d\theta = \pi. \qquad (15.98)$$

If we multiply both sides of eq. (15.93) by $\cos n\omega_0 t$ and integrate over the period T, we get

$$a_n = (2/T) \int_{-T/2}^{T/2} g(t) \cos n\omega_0 t \, dt; \qquad (15.99)$$

likewise if we multiply both sides of eq. (15.93) by $\sin n\omega_0 t$ and integrate over the period T, we get

$$b_n = (2/T) \int_{-T/2}^{T/2} g(t) \sin n\omega_0 t \, dt. \qquad (15.100)$$

In particular, for $n = 0$,

$$\tfrac{1}{2}a_0 = (1/T) \int_{-T/2}^{T/2} g(t) \, dt = \text{average value of } g(t); \qquad (15.101)$$

hence, $a_0 = 0$ whenever $g(t)$ is an odd function.

The sine and cosine series can be combined into one series by introducing phase angles, γ_n:

$$g(t) = \tfrac{1}{2}c_0 + \sum_{n=1}^{\infty} c_n \cos(n\omega_0 t - \gamma_n), \qquad (15.102)$$

where

$$c_n^2 = a_n^2 + b_n^2; \qquad c_0 = a_0; \qquad \gamma_0 = 0;$$
$$\gamma_n = \tan^{-1}(b_n/a_n), \qquad n > 0, \qquad (15.103)$$

Equation (15.102) shows that $g(t)$ can be expressed as an infinite series of harmonics of the fundamental frequency ω_0. Constants c_n and γ_n give the amplitudes and phase angles of the harmonics and are referred to as the *amplitude spectrum* and *phase spectrum* of $g(t)$. (*Frequency spectrum* is used for both the amplitude spectrum and the combined amplitude and phase spectra.) For very large n, the amplitudes must get smaller, that is, $\lim_{n \to \infty} c_n = 0$, because otherwise

$$\int_{-T/2}^{T/2} |g(t)| \, dt$$

would not be finite.

Equation (15.93) can be written

$$g(t) = (1/T) \int_{-T/2}^{T/2} g(y) \, dy$$

$$+ (2/T) \sum_{n=1}^{\infty} \int_{-T/2}^{T/2} g(y) \cos[n\omega_0(t - y)] \, dy, \quad (15.104)$$

where variable t in eqs. (15.99) and (15.100) has been replaced by the dummy variable y.

Often, in practical work, the function $g(t)$ is given only at equal intervals, Δ, for example $g(t) = g(n\Delta)$, $n = 0, 1, \ldots, m$. Equations (15.93), (15.99) to (15.103) still hold except that the sums in eqs. (15.93) and (15.102) are finite sums and the integrals are replaced by sums. For further details the reader can consult Wylie (1966). An interesting aspect of Fourier series is that the finite series obtained by discarding all terms in eq. (15.93) above a certain value of n is the best fit for $g(t)$ in the least-squares sense (see problem (15.15a).

The two infinite series in eq. (15.93) can be combined using Euler's formula (see problem 15.12a) to give an exponential form of the series:

$$g(t) = \sum_{n=-\infty}^{\infty} \alpha_n e^{jn\omega_0 t}; \quad (15.105)$$

$$\left.\begin{array}{l} \alpha_0 = \tfrac{1}{2}a_0, \\[4pt] \alpha_{\pm n} = \tfrac{1}{2}(a_n \mp jb_n) \end{array}\right\}$$

or

$$\left.\begin{array}{l} \\ \alpha_{\pm n} = (1/T) \int_{-T/2}^{T/2} g(t) e^{\mp jn\omega_0 t} \, dt, \\[6pt] n = 0, 1, 2, \ldots, \infty. \end{array}\right\} \quad (15.106)$$

At times we wish to represent a function $g(t)$ by a Fourier series in an interval such as $(-\tfrac{1}{2}T, +\tfrac{1}{2}T)$ regardless of the values outside this interval (for example, see §15.2.12 and 15.5.1); the Fourier series then repeats the same portion of $g(t)$ each time t increases or decreases by T.

15.2.2 Fourier integral

When $g(t)$ is periodic, the Fourier coefficients constitute a discrete frequency spectrum with components at intervals of ω_0. If T increases, $g(t)$ repeats at longer intervals while the frequency components occur at smaller and smaller intervals. When T becomes infinite, the frequency spectrum becomes continuous and the sum in eq. (15.105) becomes the Fourier integral.

We shall give a heuristic demonstration of the transition from the Fourier series to the Fourier integral as T approaches infinity; a rigorous derivation can be found in Churchill (1963). We can substitute eq. (15.106) in eq. (15.105), obtaining

$$g(t) = \sum_{n=-\infty}^{\infty} \left[\int_{-T/2}^{T/2} g(t) e^{-jn\omega_0 t} \, dt \right] e^{jn\omega_0 t} (1/T).$$

As T approaches infinity, $1/T$ becomes infinitesimal; hence,

$$1/T \rightarrow d\nu_0 = d\omega_0/2\pi.$$

The difference between adjacent harmonics, $n\omega_0$ and $(n + 1)\omega_0$, becomes infinitesimal, that is, $n\omega_0$ becomes a continuous variable ω. Thus, the discrete spectrum $n\omega_0$ becomes a continuous spectrum in the limit. (We can drop the subscript on ν and ω because in the limit it will have no significance.) At the same time, the summation in eq. (15.105) becomes an integral:

$$g(t) = \int_{-\infty}^{\infty} \left[\int_{-\infty}^{\infty} g(t) e^{-j\omega t} \, dt \right] e^{j\omega t} \, d\omega/2\pi. \quad (15.107)$$

The integral with respect to t is calculated first, the result being a function of ω; then in the second integration, ω disappears and we have again a function of t. The factor $1/2\pi$ can be combined with $d\omega$ to give $d\nu$, but then ω in the exponential terms should be replaced by $2\pi\nu$.

15.2.3 Fourier transforms

If we write

$$G(\omega) = \int_{-\infty}^{\infty} g(t) e^{-j\omega t} \, dt$$

then

$$= \textit{Fourier transform of } g(t), \quad (15.108)$$

$$g(t) = (1/2\pi) \int_{-\infty}^{\infty} G(\omega) e^{j\omega t} \, d\omega$$

$$= \textit{inverse Fourier transform of } G(\omega). \quad (15.109)$$

(Some authors distribute the $1/2\pi$ factor differently, such as putting $(1/2\pi)^{1/2}$ in front of both the Fourier transform and the inverse Fourier transform.) The relation between $g(t)$ and $G(\omega)$ is often written

$$g(t) \leftrightarrow G(\omega). \quad (15.110)$$

A sufficient condition for the existence of the Fourier transform of a function $g(t)$ is that $\int_{-\infty}^{+\infty} |g(t)| \, dt$ be finite. However, this condition is not necessary, and a second, somewhat more complicated, condition can be stated (see Papoulis, 1962: 9).

The Fourier integral can be written in several ways. Assuming $g(t)$ is real, as is usually the case, and noting that t inside the square brackets in eq. (15.107) is a dummy variable, we can write

$$g(t) = (1/2\pi) \int_{-\infty}^{\infty} e^{j\omega t} \left[\int_{-\infty}^{\infty} g(y) e^{-j\omega y} \, dy \right] d\omega$$

$$= (1/2\pi) \int_{-\infty}^{\infty} \int_{-\infty}^{\infty} g(y) e^{+j\omega(t-y)} \, dy \, d\omega$$

$$= (1/2\pi) \int_{-\infty}^{\infty} \int_{-\infty}^{\infty} g(y)\{\cos \omega(t - y)$$

$$+ j \sin \omega(t - y)\} \, dy \, d\omega$$

$$g(t) = (1/2\pi)\int_{-\infty}^{\infty}\int_{-\infty}^{\infty} g(y) \cos \omega(t - y) \, dy \, d\omega$$

(because $g(t)$ is real)

$$= (1/2\pi)\int_{-\infty}^{\infty}\int_{-\infty}^{\infty} g(y) \, (\cos \omega t \cos \omega y$$

$$+ \sin \omega t \sin \omega y) \, dy \, d\omega$$

$$g(t) = (1/2\pi)\int_{-\infty}^{\infty} R(\omega) \cos \omega t \, d\omega$$

$$-(1/2\pi)\int_{-\infty}^{\infty} X(\omega) \sin \omega t \, d\omega, \qquad (15.111)$$

where

$$\left.\begin{array}{l} R(\omega) = \displaystyle\int_{-\infty}^{\infty} g(y) \cos \omega y \, dy \\ \quad = cosine\ transform\ of\ g(t) \\ \text{and} \\ -X(\omega) = \displaystyle\int_{-\infty}^{\infty} g(y) \sin \omega y \, dy \\ \quad = sine\ transform\ of\ g(t). \end{array}\right\} \qquad (15.112)$$

If we express the exponential term in eq. (15.108) in terms of $\cos \omega t$ and $\sin \omega t$ and compare with the foregoing, we find that

$$\left.\begin{array}{l} G(\omega) = R(\omega) + jX(\omega) \\ \quad = A(\omega)e^{j\gamma(\omega)}, \end{array}\right\} \qquad (15.113a)$$

where

$$\left.\begin{array}{l} A(\omega) = [R^2(\omega) + X^2(\omega)]^{1/2} \\ \quad = amplitude\ spectrum\ of\ g(t) \\ \text{and} \\ \gamma(\omega) = \tan^{-1}[X(\omega)/R(\omega)] \\ \quad = phase\ spectrum\ of\ g(t). \end{array}\right\} \qquad (15.113b)$$

As an example, the minimum-phase wavelet (§15.5.6a) shown in fig. 15.4a can be represented by the amplitude and phase spectra shown in figs. 15.4b and 15.4c. Figure 15.4d is a variable area display of individual frequency components (at intervals of 0.5 Hz) and their sum (leftmost trace); the components cancel at times earlier than 0, form the wavelet from 0 to about 0.125 s, and then cancel at later times. The lower amplitudes in fig. 15.4d around 32 to 35 Hz correspond to the notch in fig. 15.4b.

If the phase $\gamma \equiv 0$ in eq. (15.113), we have a zero-phase wavelet. Then each component adds in phase at zero time to yield a symmetrical zero-phase wavelet, half of which occurs before time zero, as shown in fig. 15.5a for equal-amplitude components up to 32 Hz. If the phase were a linear function of frequency $\gamma = kv$, each component would be shifted by the same time and the result would be the linear-phase wavelet shown in fig. 15.5b. If the phase $\gamma \equiv 90°$, the result is a 90° wavelet shown in fig. 15.5c.

Although the independent variables are usually time and frequency in eqs. (15.108) and (15.109), this is not necessarily so. We can, for example, calculate the Fourier transform with respect to x so that ω has inverse-length dimensions instead of inverse-time, in which case ω in the preceding equations becomes κ, the *spatial frequency* or *angular wavenumber*.

15.2.4 Multidimensional Fourier series and transforms

The Fourier series in eq. (15.93) was defined for $g(t)$, a function of one variable only. We can expand $g(x, t)$ in a Fourier series if we assume that $g(x, t)$ is also periodic in x with "period" equal to $\kappa_0 = 2\pi/\lambda$ (equivalent to $\omega_0 = 2\pi/T$). Then

$$\begin{aligned} g(x, t) = \sum_{m=0}^{\infty} \sum_{n=0}^{\infty} (&a_{mn} \cos m\kappa_0 x \cos n\omega_0 t \\ &+ b_{mn} \cos m\kappa_0 x \sin n\omega_0 t \\ &+ c_{mn} \sin m\kappa_0 x \cos n\omega_0 t \\ &+ d_{mn} \sin m\kappa_0 x \sin n\omega_0 t), \end{aligned} \qquad (15.114)$$

the coefficients being given by equations similar to eqs. (15.99) and (15.100), for example,

$$\begin{aligned} b_{mn} = (4/\lambda T)\int_{-\lambda/2}^{\lambda/2}\int_{-T/2}^{T/2} g(x, t) \cos m\kappa_0 x \\ \times \sin n\omega_0 t \, dx \, dt \end{aligned} \qquad (15.115)$$

Similarly, the Fourier transform equations (15.108) and (15.109) become

$$G(\kappa, \omega) = \int_{-\infty}^{\infty}\int_{-\infty}^{\infty} g(x, t)e^{-j(\kappa x + \omega t)} \, dx \, dt, \qquad (15.116)$$

$$g(x, t) = [1/(2\pi)^2]\int_{-\infty}^{\infty}\int_{-\infty}^{\infty} G(\kappa, \omega)e^{j(\kappa x + \omega t)} \, d\kappa \, d\omega.$$

The extension to any number of dimensions r is obvious (the factor $1/(2\pi)^2$ becomes $1/(2\pi)^r$). For further details, see Fail and Grau (1963) and Treitel, Shanks, and Frasier (1967).

15.2.5 Special functions

We shall have many occasions to use certain special functions, especially step(t), the *unit step*, defined by

$$\left.\begin{array}{ll} \text{step}(t) = 0, & t \le 0; \\ \quad\quad\quad = +1, & t \ge 0. \end{array}\right\} \qquad (15.118)$$

Obviously, step (t) has a discontinuity at $t = 0$. Multiplying a function $g(t)$ by step(t) "wipes out" the function for negative values of t and leaves it unchanged for positive values of t. A unit step shifted t_0 units to the right can be written step$(t - t_0)$; multiplication by step$(t - t_0)$ wipes out a function for all values of t less than t_0. Moreover, k step(t) is a step of strength k.

To get the transform of step(t), we define

$$\left.\begin{array}{ll} \text{sgn}(t) = -1, & t \le 0; \\ \quad\quad\quad = +1, & t \ge 0. \end{array}\right\} \qquad (15.119)$$

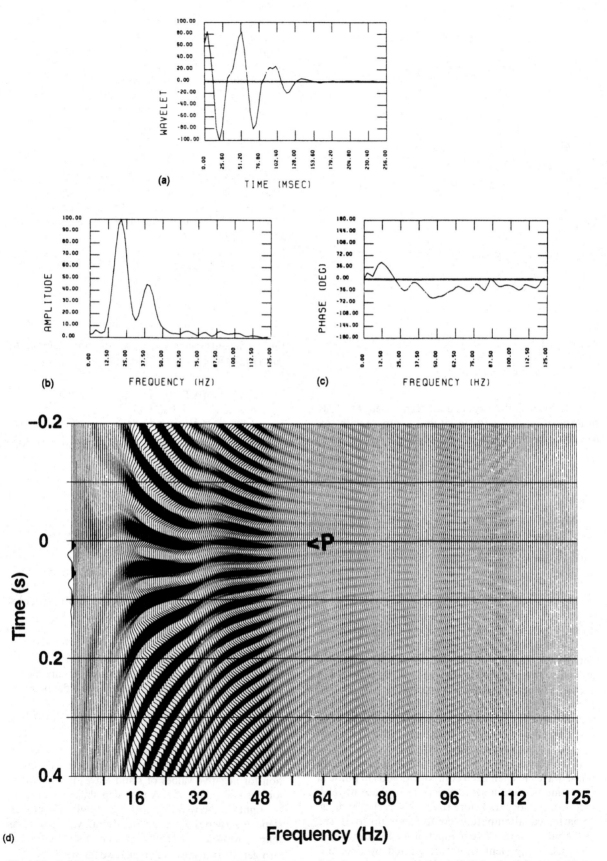

Fig. 15.4 Synthesis of a wavelet by summing sinusoids. (From Yilmaz, 1987: 12–13.) (a) A minimum-phase time-domain wavelet, (b) amplitude spectrum, (c) phase spectrum, and (d) display of the frequency components and their sum (left trace). Note how the phase (P) (follow the same peak) has the same shape as the phase spectrum in part (c).

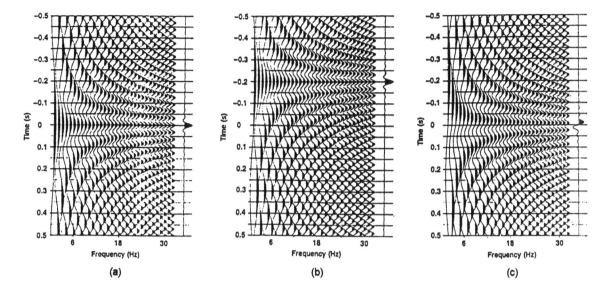

(a)   **(b)**   **(c)**

Fig. 15.5 Synthesis of equal-amplitude sinusoids with different phase shifts; right-hand traces show the sums. (From Yilmaz, 1987: 15, 16.) (a) Each component with zero phase, (b) phase shift linear with frequency, and (c) each component with 90° phase shift.

The simplest way of proving the relation,

$$\mathrm{sgn}(t) \leftrightarrow 2/j\omega, \qquad (15.120)$$

is to calculate the inverse transform:

$$\frac{2}{j\omega} \leftrightarrow \frac{1}{2\pi}\int_{-\infty}^{\infty}\frac{2}{j\omega}e^{j\omega t}\,d\omega = \frac{1}{j\pi}\int_{-\infty}^{\infty}\left(\frac{\cos\omega t + j\sin\omega t}{\omega}\right)d\omega$$

$$= \frac{2}{\pi}\int_{0}^{\infty}\frac{\sin\omega t}{\omega}\,d\omega,$$

because $(\cos\omega t)/\omega$ is odd and the portions for negative and positive ω will cancel, whereas $(\sin\omega t)/\omega$ is even. The definite integral has the values 0, $\frac{1}{2}\pi$, or $-\frac{1}{2}\pi$ according as $t = 0$, $t > 0$, or $t < 0$ (Weast, 1975, integral 621); hence, the right side equals $\mathrm{sgn}(t)$. Thus,

$$\mathrm{step}(t) = [1 + \mathrm{sgn}(t)]/2 \leftrightarrow \pi\delta(\omega) + 1/j\omega \qquad (15.121)$$

(using the transform pair $1 \leftrightarrow 2\pi\delta(\omega)$; see eqs. (15.124) and (15.128)).

The unit step, $\mathrm{step}(t)$, is useful in dealing with discontinuous functions such as $g(t)$ in fig. 15.6, which has values $g(a+)$ and $g(a-)$ at the discontinuity $t = a$ as we approach from the right and from the left, respectively. We write $g_c(t)$ for the continuous function obtained by using $g(t)$ to the left of $t = a$ and the dashed curve to the right of $t = a$, which is merely $g(t)$ displaced parallel to the vertical axis by the amount of the discontinuity. Then, $g(t) = g_c(t) + [g(a+) - g(a-)]\mathrm{step}(t - a)$. The derivative of $g(t)$ is the same as that of $g_c(t)$ except for an impulse (see eqs. (15.124) and (15.129)) of strength $g(a+) - g(a-)$ at $t = a$.

The *gate* or *boxcar*, $\mathrm{box}_a(t)$, is defined by

$$\left.\begin{aligned}
\mathrm{box}_a(t) &= 0, & t &\leq -\tfrac{1}{2}a;\\
&= 1, & -\tfrac{1}{2}a &\leq t \leq \tfrac{1}{2}a;\\
&= 0, & t &\geq \tfrac{1}{2}a.
\end{aligned}\right\} \qquad (15.122)$$

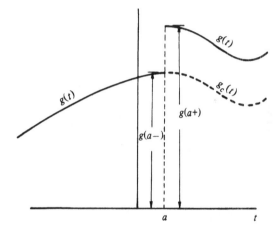

Fig. 15.6 Application of $\mathrm{step}(t)$ to describe discontinuous functions.

The boxcar thus has a width a and unit height, with area a. It can be expressed as the difference of two steps (see problem 15.18a). The transform of $\mathrm{box}_a(t)$ is

$$\mathrm{box}_a(t) \leftrightarrow \int_{-a/2}^{a/2} e^{-j\omega t}\,dt = \frac{e^{+j\omega a/2} - e^{-j\omega a/2}}{j\omega}$$

$$= \frac{2\sin(\omega a/2)}{\omega} = a\,\mathrm{sinc}(\omega a/2),$$

$$(15.123)$$

where $\mathrm{sinc}\,\theta = (\sin\theta)/\theta$. The transform of $\mathrm{box}_a(t)$ is shown in fig. 15.7a.

The *unit impulse* or *Dirac delta*, $\delta(t)$, is defined by the relation

$$\int_{-\infty}^{\infty} \delta(t)g(t)\,dt = g(0), \qquad (15.124a)$$

that is, $\delta(t)$ is an operator that sets the argument equal to zero.

[The unit impulse, $\delta(t)$, is an example of a distribution (a readily understood account of distributions is given in Papoulis, 1962: app. 1).) A *distribution*, $\alpha(t)$, is an operator that causes $g(t)$ to be replaced by some function of $g(t)$, $\phi\{g(t)\}$, that is,

$$\int_{-\infty}^{\infty} \alpha(t)g(t)\, \mathrm{d}t = \phi[g(t)].$$

The integral sign does not have its usual significance here; it is used only because of certain analogies between the above definition and real integrals; thus eq. (15.124a) can be taken to mean

$$\delta(t)g(t) = g(0). \qquad (15.124b)$$

The *derivative of a distribution* is defined by the relation

$$\int_{-\infty}^{\infty} (\mathrm{d}\alpha/\mathrm{d}t)g(t)\, \mathrm{d}t = -\int_{-\infty}^{\infty} \alpha(t)(\mathrm{d}g/\mathrm{d}t)\, \mathrm{d}t = -\phi(\mathrm{d}g/\mathrm{d}t);$$

in other words, $\mathrm{d}\alpha/\mathrm{d}t$ is a distribution that attributes to $\mathrm{d}g/\mathrm{d}t$ the same functional value that $\alpha(t)$ does to $g(t)$, except for the minus sign. Thus,

$$\int_{-\infty}^{\infty} (\mathrm{d}\delta/\mathrm{d}t)g(t)\, \mathrm{d}t = -\mathrm{d}g/\mathrm{d}t|_{t=0}.]$$

To give a physical concept of the impulse (rather than a rigorous mathematical analysis), we start with a unit boxcar, $\mathrm{box}_a(t)/a$, and let a approach zero while holding the area at unity, so that the height approaches infinity as the width approaches zero:

$$\lim_{a \to 0}[\mathrm{box}_a(t)/a] = \delta(t).$$

In the limit, $\delta(t)$ is an impulse of infinite height but of zero duration, the "strength" of the impulse being unity. Furthermore, $\delta(t - t_0)$ is a unit impulse occurring at $t = t_0$. If we multiply $g(t)$ by $\delta(t - t_0)$, the result is the value of $g(t)$ at the time the impulse occurs, that is, $g(t_0)$. Papoulis (1962: 280) shows that

$$\delta(t) = \lim_{\omega \to \infty}\frac{\sin \omega t}{\pi t}. \qquad (15.125)$$

We can derive another expression for $\delta(t)$ from this result:

$$\int_{-\infty}^{\infty} \cos \omega t\, \mathrm{d}\omega = \lim_{a \to \infty} \int_{-a}^{a} \cos \omega t\, \mathrm{d}\omega$$

$$= 2 \lim_{a \to \infty} \int_{0}^{a} \cos \omega t\, \mathrm{d}\omega = 2 \lim_{a \to \infty} \frac{\sin \omega t}{t}\Big|_{0}^{a}$$

$$= 2\pi \lim_{a \to \infty} \frac{\sin at}{\pi t} = 2\pi\, \delta(t). \qquad (15.126)$$

The Fourier transform of $\delta(t)$ is

$$\delta(t) \leftrightarrow \int_{-\infty}^{\infty} \delta(t)\mathrm{e}^{-\mathrm{j}\omega t}\, \mathrm{d}t = \mathrm{e}^{-\mathrm{j}\omega t}\Big|_{t=0} = +1. \qquad (15.127)$$

Thus, the spectrum of the unit impulse is flat, that is, all frequencies are present and have the same amplitude and zero phase (see fig. 15.7c). Conversely, the transform of a constant in the time domain is an impulse in the frequency domain:

$$1 \leftrightarrow \int_{-\infty}^{\infty} \mathrm{e}^{-\mathrm{j}\omega t}\, \mathrm{d}t = \int_{-\infty}^{\infty} \cos \omega t\, \mathrm{d}t - \mathrm{j} \int_{-\infty}^{\infty} \sin \omega t\, \mathrm{d}t$$

$$= 2\pi\, \delta(\omega), \qquad (15.128)$$

using eq. (15.126) and noting that the sine integral vanishes because the sine is odd.

We now consider the relation between step(t) and $\delta(t)$. The derivative of step(t) is everywhere zero except at the origin, where we may think of it as a distribution:

$$\int_{-\infty}^{\infty} \frac{\mathrm{d}(\mathrm{step}(t))}{\mathrm{d}t}g(t)\, \mathrm{d}t = -\int_{-\infty}^{\infty} \mathrm{step}(t)\frac{\mathrm{d}g}{\mathrm{d}t}\, \mathrm{d}t$$

$$= -\int_{0}^{\infty} \frac{\mathrm{d}g}{\mathrm{d}t}\, \mathrm{d}t = -g(t)|_{0}^{\infty} = g(0),$$

because $g(+\infty) = 0$. Thus,

$$(\mathrm{d}/\mathrm{d}t)\, \mathrm{step}(t) = \delta(t). \qquad (15.129)$$

A *comb* is a series of equally spaced unit impulses, the series usually being considered infinite:

$$\mathrm{comb}(t) = \sum_{n=-\infty}^{\infty} \delta(t - n\Delta), \qquad (15.130)$$

where n is integral, and Δ is a fixed time interval. Multiplication of $g(t)$ by comb(t) replaces the continuous function with a digitized function sampled at intervals of Δ. The transform of the comb is derived in the next section.

A *linear ramp* is

$$\mathrm{ramp}(t) = t, \qquad 0 \le t \le b,$$
$$= 0, \qquad \text{for all other values of } t;$$

$$\mathrm{ramp}(t) \leftrightarrow \int_{0}^{b} t\mathrm{e}^{-\mathrm{j}\omega t}\, \mathrm{d}t = (1/\omega^2)[\mathrm{e}^{-\mathrm{j}\omega b}(1 + \mathrm{j}\omega b) - 1].$$
$$(15.131)$$

An *exponential decay*, e^{-kt}, for $t \ge 0$ and k positive, has the transform

$$\mathrm{e}^{-kt}\, \mathrm{step}(t) \leftrightarrow \int_{-\infty}^{\infty} \mathrm{e}^{-kt}\, \mathrm{step}(t)\mathrm{e}^{-\mathrm{j}\omega t}\, \mathrm{d}t$$

$$\leftrightarrow \int_{0}^{\infty} \mathrm{e}^{-(k+\mathrm{j}\omega)t}\, \mathrm{d}t = 1/(k + \mathrm{j}\omega) \qquad (15.132)$$

(see fig. 15.7e).

A *double-sided exponential* decay, $\mathrm{e}^{-k|t|}$ for k positive, has the transform

$$\mathrm{e}^{-k|t|} \leftrightarrow \int_{-\infty}^{0} \mathrm{e}^{kt}\mathrm{e}^{-\mathrm{j}\omega t}\, \mathrm{d}t + \int_{0}^{\infty} \mathrm{e}^{-kt}\mathrm{e}^{-\mathrm{j}\omega t}\, \mathrm{d}t$$

$$\leftrightarrow 2k/(k^2 + \omega^2) \qquad (15.133)$$

(see fig. 15.7f).

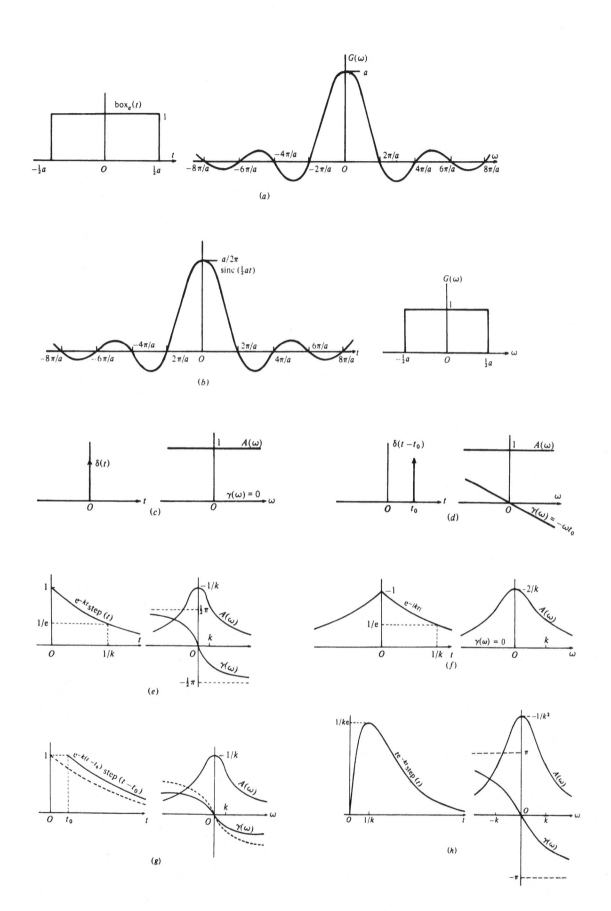

Fig. 15.7 Examples of functions and their Fourier transforms.
(a) Box$_a(t)$; (b) sinc($at/2$); (c) $\delta(t)$; (d) $\delta(t - t_0)$; (e) e^{-kt}step(t); (f) $e^{-|kt|}$; (g) $e^{-k(t - t_0)}$step($t - t_0$); and (h) te^{-kt}step(t). (*Note: a* and *k* are positive constants.)

15.2.6 Theorems on Fourier transforms

Many theorems exist regarding the properties of Fourier transforms; the most important are listed below:

$$g(t) \leftrightarrow G(\omega),$$
$$kg(t) \leftrightarrow kG(\omega); \quad (15.134)$$
$$k_1 g_1(t) + k_2 g_2(t) \leftrightarrow k_1 G_1(w) + k_2 G_2(\omega); \quad (15.135)$$

Shift theorems:

$$g(t - a) \leftrightarrow e^{-j\omega a} G(\omega); \quad (15.136)$$
$$e^{-jat} g(t) \leftrightarrow G(\omega + a). \quad (15.137)$$

Scaling theorems:

$$g(at) \leftrightarrow (1/|a|) G(\omega/a); \quad (15.138)$$
$$(1/|a|) g(t/a) \leftrightarrow G(a\omega). \quad (15.139)$$

Symmetry theorems:

$$G(t) \leftrightarrow 2\pi g(-\omega); \quad (15.140)$$

Derivative theorems:

$$\frac{d^n g(t)}{dt^n} \leftrightarrow (j\omega)^n G(\omega); \quad (15.141)$$

$$(-jt)^n g(t) \leftrightarrow \frac{d^n G(\omega)}{d\omega^n}. \quad (15.142)$$

Integral theorems:

$$\int_{-\infty}^{t} g(t) \, dt \leftrightarrow (1/j\omega) G(\omega); \quad (15.143)$$

$$-(1/jt) g(t) \leftrightarrow \int_{-\infty}^{\omega} G(\omega) \, d\omega. \quad (15.144)$$

Convolution theorems:

$$\left. \begin{aligned} g_1(t) * g_2(t) &= \int_{-\infty}^{\infty} g_1(\tau) g_2(t - \tau) \, d\tau \\ &\leftrightarrow G_1(\omega) G_2(\omega); \end{aligned} \right\} \quad (15.145)$$

$$2\pi g_1(t) g_2(t) \leftrightarrow G_1(\omega) * G_2(\omega). \quad (15.146)$$

Cross-correlation theorem:

$$\left. \begin{aligned} \phi_{12}(t) &= \int_{-\infty}^{\infty} g_1(\tau) g_2(\tau + t) \, d\tau \\ &\leftrightarrow \overline{G_1(\omega)} G_2(\omega), \end{aligned} \right\} \quad (15.147)$$

where the superscribed bar means the complex conjugate. The convolution and cross-correlation theorems are discussed in §15.2.8 and §15.2.9, respectively.

To prove (15.136), we write

$$g(t - a) \leftrightarrow \int_{-\infty}^{\infty} g(t - a) e^{-j\omega t} \, dt$$

$$= \int_{-\infty}^{\infty} g(y) e^{-j\omega(y+a)} \, dy, \quad (y = t - a),$$

$$= e^{-j\omega a} \int_{-\infty}^{\infty} g(y) e^{-j\omega y} \, dy = e^{-j\omega a} G(\omega).$$

The proof of eq. (15.137) is similar. As examples of

the application of eqs. (15.136) and (15.137), we start with eqs. (15.121), (15.123), and (15.131) and obtain:

(a) shifted step:

$$\text{step}(t - t_0) \leftrightarrow e^{-j\omega t_0} [\pi \delta(\omega) + 1/j\omega]; \quad (15.148)$$

(b) windowed exponential:

$$e^{-jkt} \text{box}_a(t) \leftrightarrow a \, \text{sinc}[\tfrac{1}{2} a(\omega + k)]; \quad (15.149)$$

(c) shifted ramp:

$$g(t - t_0) \leftrightarrow (1/\omega^2) e^{-j\omega t_0} [e^{-j\omega b}(1 + j\omega b) - 1], \quad (15.150)$$

where

$$g(t - t_0) = (t - t_0), \quad \text{for } t_0 \leq t \leq t_0 + b;$$
$$= 0, \quad \text{for all other } t.$$

Equation (15.136) shows that the effect of a time shift is to leave the magnitude of the transform, $A(\omega)$, unchanged and to add a linear shift to the phase, $\gamma(\omega)$. This is illustrated in fig. 15.7 by comparing (c) with (d) or (e) with (g).

The proof of eq. (15.138) is straightforward if we take a positive at first and then consider the necessary changes when a is negative. As an example, if

$$g(t) = \pm 2t, \quad 0 \leq t \leq b,$$
$$= 0, \quad \text{for all other } t,$$

then from eqs. (15.131) and (15.138),

$$g(t) \leftrightarrow G(\omega) = \tfrac{1}{2} G(\pm \tfrac{1}{2} \omega)$$
$$= (1/2\omega^2)[e^{\pm j\omega b/2}(1 \pm \tfrac{1}{2} j\omega b) - 1].$$

An important corollary of eq. (15.138) is obtained by setting $a = -1$, resulting in

$$g(-t) \leftrightarrow G(-\omega). \quad (15.151)$$

Equation (15.138) can be used to illustrate an important relation between a time function and its transform, namely, that the more the time function is concentrated, the more spread out is the transform and vice versa. Thus, in fig. 15.7e to h, the larger k is, the more closely the time function approaches a spike but at the same time the magnitude of the transform, $A(\omega)$, broadens out. In the limiting case of $\delta(t)$, the time function is concentrated entirely at one instant, whereas the transform is spread uniformly from $-\infty$ to $+\infty$.

Equation (15.140) can be proved by substituting $-t$ for t in eq. (15.109), giving

$$2\pi g(-t) = \int_{-\infty}^{\infty} G(\omega) e^{-j\omega t} \, d\omega,$$

and then interchanging the symbols t and ω (because this does not affect the equation); thus,

$$2\pi g(-\omega) = \int_{-\infty}^{\infty} G(t) e^{-j\omega t} \, dt.$$

As an example, we have from eq. (15.132)

$$e^{-kt} \text{step}(t) \leftrightarrow 1/(k + j\omega),$$

and hence

$$1/(k + jt) \leftrightarrow 2\pi e^{k\omega} \text{ step}(-\omega)$$

(note that $\text{step}(-\omega)$ equals $+1$ or 0 accordingly as ω is negative or positive). Again, eq. (15.123) gives

$$\text{box}_a(t) \leftrightarrow a \text{ sinc}(\tfrac{1}{2}\omega a),$$

hence

$$a \text{ sinc}(\tfrac{1}{2}at) \leftrightarrow 2\pi \text{ box}_a(-\omega) = 2\pi \text{ box}_a(\omega),$$
$$(15.152)$$

because $\text{box}_a(\omega)$ is even (see fig. 15.7b).

The derivative theorems are of fundamental importance. The proof follows directly from eqs. (15.109) and (15.108) if we differentiate with respect to t and ω, respectively. Thus,

$$\frac{\mathrm{d}g}{\mathrm{d}t} = \frac{1}{2\pi} \int_{-\infty}^{\infty} G(\omega)\frac{\mathrm{d}}{\mathrm{d}t}(e^{j\omega t}) \, \mathrm{d}\omega$$

$$= \frac{1}{2\pi} \int_{-\infty}^{\infty} [j\omega G(\omega)]e^{j\omega t} \, \mathrm{d}\omega,$$

hence

$$\mathrm{d}g/\mathrm{d}t \leftrightarrow j\omega G(\omega).$$

Differentiating n times, we get eq. (15.141):

$$\frac{\mathrm{d}^n g(t)}{\mathrm{d}t^n} \leftrightarrow (j\omega)^n G(\omega).$$

The most important application of eq. (15.141) is in the solution of differential equations where it permits us to replace derivatives with a function of ω. It also enables us to obtain new transform pairs by differentiation of a known pair. An interesting application of this theorem is

$$\delta(t) = (\mathrm{d}/\mathrm{d}t) \text{ step}(t) \leftrightarrow j\omega[\pi\delta(\omega) + 1/j\omega] = 1$$

(using eqs. (15.129) and (15.121) and the fact that $\omega\delta(\omega) = 0$).

Equation (15.143) is easily proved using eq. (15.141). We write

$$g_1(t) = \int_{-\infty}^{t} g(t) \, \mathrm{d}t = \int_{-\infty}^{t} g(x) \, \mathrm{d}x \leftrightarrow G_1(\omega),$$

$$\mathrm{d}g_1(t)/\mathrm{d}t = g(t) \leftrightarrow j\omega G_1(\omega) = G(\omega),$$

where we have used Leibnitz' rule to differentiate the integral on the right. Leibnitz' rule states that, given $F(t) = \int_{a(t)}^{b(t)} \phi(t, \lambda) \, \mathrm{d}\lambda$, then the derivative of $F(t)$ with respect to t is given by

$$\frac{\mathrm{d}F}{\mathrm{d}t} = \int_{a(t)}^{b(t)} \frac{\partial\phi}{\partial t} \, \mathrm{d}\lambda + \phi(t, b) \frac{\mathrm{d}b}{\mathrm{d}t} - \phi(t, a) \frac{\mathrm{d}a}{\mathrm{d}t}$$

(see Kaplan, 1952: 220). Therefore,

$$\int_{-\infty}^{t} g(t) \, \mathrm{d}t \leftrightarrow (1/j\omega)G(\omega).$$

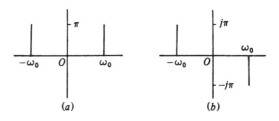

Fig. 15.8   Fourier transforms of (a) $\cos \omega_0 t$ and (b) $\sin \omega_0 t$.

Equation (15.144) can be proved in the same way, starting from eq. (15.142). Equations (15.145) to (15.147) will be discussed later.

Using eqs. (15.127), (15.136), (15.137) and (15.140), we can write the following transform pairs:

$$\left.\begin{aligned}
\delta(t) &\leftrightarrow +1; \\
\delta(t \pm t_0) &\leftrightarrow e^{\pm j\omega t_0}; \\
(1/2)[\delta(t + t_0) + \delta(t - t_0)] &\leftrightarrow \cos \omega t_0; \\
(1/2j)[\delta(t + t_0) - \delta(t - t_0)] &\leftrightarrow \sin \omega t_0; \\
1 &\leftrightarrow 2\pi\delta(\omega); \\
e^{\pm j\omega_0 t} &\leftrightarrow 2\pi\delta(\omega \mp \omega_0); \\
\cos \omega_0 t &\leftrightarrow \pi[\delta(\omega + \omega_0) + \delta(\omega - \omega_0)]; \\
\sin \omega_0 t &\leftrightarrow j\pi[\delta(\omega + \omega_0) - \delta(\omega - \omega_0)].
\end{aligned}\right\} \quad (15.154)$$

The transforms of $\sin \omega_0 t$ and $\cos \omega_0 t$ are shown in fig. 15.8.

To obtain the transform of the comb, we find its Fourier series expansion using eq. (15.104) and then find the transform of the series; replacing $g(y)$ in eq. (15.104) with $\delta(y)$ and writing $T = \Delta$, we get

$$\text{comb}(t) = \sum_{n=-\infty}^{\infty} \delta(t - n\Delta) = 1/\Delta + (2/\Delta) \sum_{n=1}^{\infty} \cos n\omega_0 t,$$
$$\omega_0 = 2\pi/\Delta.$$

Using eq. (15.154), we obtain

$$\sum_{n=-\infty}^{\infty} \delta(t - n\Delta) \leftrightarrow \omega_0 \sum_{m=-\infty}^{\infty} \delta(\omega - m\omega_0),$$

that is,

$$\text{comb}(t) \leftrightarrow \omega_0 \text{ comb}(\omega). \qquad (15.155)$$

15.2.7 Gibbs' Phenomenon

At a discontinuity, the Fourier integral (and the Fourier series) gives the average value, $\frac{1}{2}[g(a-) + g(a+)]$ in fig. 15.6. However, $g(a \pm \varepsilon)$, ε infinitesimal, does not approach the average value smoothly as ε goes to zero. For simplicity, we take $a = 0$ and write

$$g(t) = g_c(t) + [g(0+) - g(0-)] \text{ step}(t).$$

Equation (15.109) gives

$$g(t) = \lim_{\lambda \to \infty} (1/2\pi) \int_{-\lambda}^{+\lambda} G(\omega)e^{j\omega t}\, d\omega$$

$$= \lim_{\lambda \to \infty} (1/2\pi) \int_{-\lambda}^{+\lambda} \left\{ \int_{-\infty}^{+\infty} g(y)e^{-j\omega y}\, dy \right\} e^{j\omega t}\, d\omega.$$

Interchanging the order of integration (see §15.2.8), we have

$$g(t) = \lim_{\lambda \to \infty} (1/2\pi) \int_{-\infty}^{+\infty} g(y) \left(\int_{-\lambda}^{+\lambda} e^{j\omega(t-y)}\, d\omega \right) dy$$

$$= \lim_{\lambda \to \infty} (1/2\pi) \int_{-\infty}^{+\infty} g(y) \left. \frac{e^{j\omega(t-y)}}{j(t-y)} \right|_{-\lambda}^{+\lambda} dy$$

$$= \lim_{\lambda \to \infty} \int_{-\infty}^{+\infty} g(y) \frac{\sin \lambda(t-y)}{\pi(t-y)}\, dy$$

$$= \lim_{\lambda \to \infty} \int_{-\infty}^{+\infty} \{g_c(y) + [g(0+)$$

$$- g(0-)]\, \text{step}(y)\} \frac{\sin \lambda(t-y)}{\pi(t-y)}\, dy$$

$$= \int_{-\infty}^{+\infty} g_c(y) \lim_{\lambda \to \infty} \frac{\sin \lambda(t-y)}{\pi(t-y)}\, dy$$

$$+ [g(0+) - g(0-)] \lim_{\lambda \to \infty} \int_0^{+\infty} \frac{\sin \lambda(t-y)}{\pi(t-y)}\, dy$$

$$= \int_{-\infty}^{+\infty} g_c(t)\delta(t-y)\, dy + (1/\pi)[g(0+)$$

$$- g(0-)] \lim_{\lambda \to \infty} \int_{-\infty}^{\lambda t} \text{sinc}\, x\, dx,$$

using eq. (15.125) and taking $\lambda(t - y) = x$ so that $dy = -dx/\lambda$ (we do not go to the limit in the right-hand integral because we wish to study the behavior of this term as λ goes to infinity). The right-hand term can be written

$$g'(t) = (1/\pi)[g(0+) - g(0-)]\left(\int_{-\infty}^{0} \text{sinc}\, x\, dx \right.$$

$$\left. + \lim_{\lambda \to \infty} \int_0^{\lambda t} \text{sinc}\, x\, dx \right).$$

Because $\int_{-\infty}^{0} \text{sinc}\, x\, dx = \frac{1}{2}\pi$ (see the derivation of eq. (15.121)), we have finally,

$$g'(t) = [g(0+) - g(0-)][1/2$$

$$+ (1/\pi) \lim_{\lambda \to \infty} \int_0^{\lambda t} \text{sinc}\, x\, dx].$$

The graph of the second factor is shown in fig. 15.9. As $\lambda \to \infty$, the peak values do not change, but the ripple moves toward the discontinuity from both sides. At the discontinuity $t = 0$, we have for $g(t)$,

$$g(0) = g_c(0) + \tfrac{1}{2}[g(0+) - g(0-)];$$

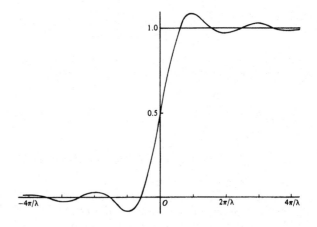

Fig. 15.9   Gibbs' phenomenon at a discontinuity.

however, an infinitesimal distance away on either side, we have an "overshoot" of about 18% (9% at the top and 9% at the bottom).

Gibbs' phenomenon is important whenever discontinuities are present, for example, in the application of filters and windows. If we multiply by a boxcar in the time domain, the discontinuities at the edges of the boxcar will produce "ringing." The objective of "window carpentry," that is, of shaping windows, is to remove discontinuities (and discontinuities in derivatives) so as to minimize ringing. See also §15.7.5.

15.2.8 Convolution theorem

The *convolution* of two functions, $g_1(t)$ and $g_2(t)$, usually written in the form $g_1(t) * g_2(t)$, eq. (15.145), is defined as

$$g_1(t) * g_2(t) = \int_{-\infty}^{+\infty} g_1(\tau)g_2(t - \tau)\, d\tau.$$

We see that $g_2(\tau - t)$ denotes $g_2(\tau)$ displaced t units to the right, whereas $g_2(-\tau + t)$ is $g_2(\tau - t)$ reflected in the vertical axis. Thus, convolution involves reflecting one of the curves in the vertical axis (often called "folding"), shifting it by t units, multiplying corresponding coordinates of the two curves, and summing from $-\infty$ to $+\infty$ (see fig. 15.10a). The value depends only on the time shift t and is independent of which curve is shifted and reflected, that is,

$$g_1(t) * g_2(t) = g_2(t) * g_1(t). \qquad (15.156)$$

This is illustrated in fig. 15.10b.

The *convolution theorem*, eq. (15.145), can be proved as follows:

$$g_1(t) * g_2(t) \leftrightarrow \int_{-\infty}^{\infty} \left[\int_{-\infty}^{\infty} g_1(\tau)g_2(t - \tau)\, d\tau \right] e^{-j\omega t}\, dt.$$

We assume that $g_1(t)$ and $g_2(t)$ have transforms, that is,

$$\int_{-\infty}^{\infty} |g_i(t)|\, dt < \infty, \qquad i = 1, 2.$$

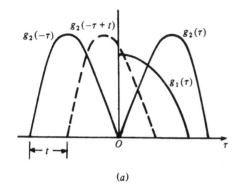

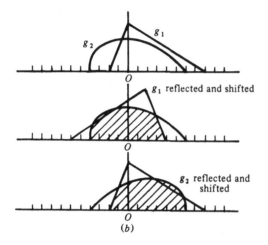

Fig. 15.10 Convolution in the time domain. (a) Reflecting and shifting $g_2(t)$ and (b) geometrical demonstration that $g_1(t) * g_2(t) = g_2(t) * g_1(t)$. The overlap is the same regardless of which is reflected and shifted.

We shall assume that this relation means that $\int_{-\infty}^{\infty} |g_i(t)|^2 \, dt$ is also finite. In this case, the order of integration can be interchanged (Papoulis, 1962: 27), giving

$$g_1(t) * g_2(t) \leftrightarrow \int_{-\infty}^{\infty} g_1(\tau) \left[\int_{-\infty}^{\infty} g_2(t - \tau) e^{-j\omega t} \, dt \right] d\tau,$$

$$\leftrightarrow \int_{-\infty}^{\infty} g_1(\tau) \left[\int_{-\infty}^{\infty} g_2(y) e^{-j\omega(y+\tau)} \, dy \right] d\tau$$

(where $y = t - \tau$),

$$g_1(t) * g_2(t) \leftrightarrow \int_{-\infty}^{\infty} g_1(\tau) e^{-j\omega\tau} \, d\tau \int_{-\infty}^{\infty} g_2(y) e^{-j\omega y} \, dy,$$

$$\leftrightarrow G_1(\omega) G_2(\omega).$$

The inverse relation, eq. (15.146), can be proved as follows:

$$G_1(\omega) * G_2(\omega) \leftrightarrow (1/2\pi) \int_{-\infty}^{\infty} \left[\int\int_{-\infty}^{\infty} G_1(y) \right.$$
$$\left. \times G_2(\omega - y) \, dy \right] e^{j\omega t} \, d\omega,$$

$$G_1(\omega) * G_2(\omega) \leftrightarrow (1/2\pi) \int_{-\infty}^{\infty} G_1(y) \left[\int_{-\infty}^{\infty} G_2(x) \right.$$
$$\left. \times e^{j(x+y)t} \, dx \right] dy, \qquad x = \omega - y,$$

$$\leftrightarrow (1/2\pi) \int_{-\infty}^{\infty} G_1(y) e^{jyt} \, dy$$
$$\times \int_{-\infty}^{\infty} G_2(x) e^{jxt} \, dx,$$

$$\leftrightarrow (1/2\pi)[2\pi g_1(t)][2\pi g_2(t)]$$
$$\leftrightarrow 2\pi g_1(t) g_2(t).$$

15.2.9 Cross-correlation theorem

The *cross-correlation function*, $\phi_{12}(t)$, defined by eq. (15.147),

$$\phi_{12}(t) = \int_{-\infty}^{\infty} g_1(\tau) g_2(t + \tau) \, d\tau,$$

is closely related to the convolution function. Obviously, $\phi_{12}(t)$ is the result of displacing $g_2(\tau)$ t units to the left and summing the products of the ordinates. It is evident that

$$\phi_{12}(t) = \phi_{21}(-t). \qquad (15.157)$$

The cross-correlation function can be regarded as a convolution of two functions, the second of which has been reversed in time:

$$\left. \begin{array}{l} \phi_{12}(t) = g_1(t) * g_2(-t); \\ \phi_{21}(t) = g_1(-t) * g_2(t); \\ g_1(t) * g_2(t) = \text{cross-correlation of} \\ \qquad g_1(t) \text{ with } g_2(-t), \\ \qquad = \text{cross-correlation of} \\ \qquad g_2(t) \text{ with } g_1(-t). \end{array} \right\} \quad (15.158)$$

These relations can be verified by drawing curves of $g_1(t)$, $g_2(t)$, $g_1(-t)$, and $g_2(-t)$, and checking geometrically the previous relations. They can also be verified by substituting in the integrals; for example,

$$g_1(t) * g_2(-t) = \int_{-\infty}^{\infty} g_1(\tau) g_2(t + \tau) \, d\tau = \phi_{12}(t)$$

(note that for zero shift, the argument of the functions in the integral is τ, not t, hence replacing $g_2(t)$ with $g_2(-t)$ on the left-hand side changes the sign of τ in $g_2(\tau)$).

The cross-correlation theorem, eq. (15.147), states that

$$\phi_{12}(t) \leftrightarrow \overline{G_1(\omega)} G_2(\omega) = \Phi_{12}(\omega),$$

where $\overline{G_1(\omega)}$ is the conjugate complex of $G_1(\omega)$. The proof is straightforward if we note the first relation of eq. (15.158) and change the sign of τ in $g_2(t - \tau)$ in the first integral in the derivation of eq. (15.145) (§15.2.8). This changes the first exponential in the next to the last line of the derivation to $e^{j\omega\tau}$, and eq. (15.147) follows immediately. Writing $\Phi_{12}(\omega)$ for the transform of $\phi_{12}(t)$, we have

$$\Phi_{12}(\omega) = \overline{G_1(\omega)}G_2(\omega), \qquad \Phi_{21}(\omega) = G_1(\omega)\overline{G_2(\omega)}. \tag{15.159}$$

The transform $\Phi_{12}(\omega)$ is known as the *cross-energy spectrum* for reasons that will become clearer in the following section. When $t = 0$, eqs. (15.147), (15.157), and (15.159) give

$$\left.\begin{aligned}
\phi_{12}(0) = \phi_{21}(0) &= \int_{-\infty}^{\infty} g_1(\tau)g_2(\tau)\,d\tau \\
&= (1/2\pi)\int_{-\infty}^{\infty} \overline{G_1(\omega)}G_2(\omega)\,d\omega \\
&= (1/2\pi)\int_{-\infty}^{\infty} G_1(\omega)\overline{G_2(\omega)}\,d\omega.
\end{aligned}\right\} \tag{15.160}$$

This is referred to as *Parseval's theorem*.

15.2.10 Autocorrelation

When $g_2(t) = g_1(t)$, the cross-correlation function becomes the autocorrelation of $g_1(t)$, that is,

$$\phi_{11}(t) = \int_{-\infty}^{\infty} g_1(\tau)g_1(t + \tau)\,d\tau \leftrightarrow |G_1(\omega)|^2. \tag{15.161}$$

Clearly, $\phi_{11}(t) = \phi_{11}(-t)$, so that ϕ_{11} is an even function.

When $t = 0$, Parseval's theorem, eq. (15.160), becomes

$$\phi_{11}(0) = \int_{-\infty}^{\infty} |g_1(\tau)|^2\,d\tau = (1/2\pi)\int_{-\infty}^{\infty} |G_1(\omega)|^2\,d\omega. \tag{15.162}$$

In most cases, $g_1(t)$ is a voltage, current, velocity, displacement, and so on, so that $|g_1(t)|^2$ is proportional to the energy; adopting this point of view and choosing the units properly, we can say that $\phi_{11}(0)$ is the total energy of $g_1(t)$. The quantity $(1/2\pi)|G_1(\omega)|^2$ is thus the energy in the spectrum between the frequencies ω and $\omega + d\omega$; it is called the *energy density* or *spectral density*. We note that the integration is from $-\infty$ to $+\infty$; if we recall that we replaced sines and cosines by the exponential terms $e^{\pm j\omega t}$, it is clear that we must recombine terms for $\pm\omega$ to get the result, that is,

$$\text{energy density of frequency } \omega = |G_1(-\omega)|^2 \\ + |G_1(+\omega)|^2.$$

When $g_1(t)$ is a real function, $G_1(-\omega) = \overline{G_1(\omega)}$ (see problem 15.17b) so that $2|G_1(+\omega)|^2$ gives the total energy density for the frequency ω (if we are interested only in relative energies, as is usually the case, we can forget doubling).

The autocorrelation function assumes its greatest value for zero shift, that is,

$$\phi_{11}(0) \geq \phi_{11}(t), \qquad t \neq 0 \tag{15.163}$$

(see problem 15.22a).

15.2.11 Multidimensional convolution

Equation (15.145) can be extended to more than one dimension. In two dimensions, we define the convolution by the expression

$$g_1(x, t) * g_2(x, t) = \int_{-\infty}^{\infty}\int_{-\infty}^{\infty} g_1(\sigma, \tau) \\ \times g_2(x - \sigma, t - \tau)\,d\sigma\,d\tau. \tag{15.164}$$

We can derive the two-dimensional convolution theorem as follows. Using eq. (15.116), we have

$$g_1(x, t) * g_2(x, t) \\ \leftrightarrow \int_{-\infty}^{\infty}\int_{-\infty}^{\infty}\left[\int_{-\infty}^{\infty}\int_{-\infty}^{\infty} g_1(\sigma, \tau) \times g_2(x - \sigma, t - \tau)\,d\sigma\,d\tau\right] \\ \times e^{-j(\kappa x + \omega t)}\,dx\,dt.$$

Interchanging the order of integration gives

$$g_1(x, t) * g_2(x, t) \leftrightarrow \int_{-\infty}^{\infty}\int_{-\infty}^{\infty} g_1(\sigma, \tau) \\ \times \left[\int_{-\infty}^{\infty}\int_{-\infty}^{\infty} g_2(x - \sigma, t - \tau) \\ \times e^{-j(\kappa x + \omega t)}\,dx\,dt\right] d\sigma\,d\tau, \\ \leftrightarrow \int_{-\infty}^{\infty}\int_{-\infty}^{\infty} g_1(\sigma, \tau)e^{-j(\kappa\sigma + \omega\tau)} \\ \times \left[\int_{-\infty}^{\infty}\int_{-\infty}^{\infty} g_2(x - \sigma, t - \tau) \\ \times e^{-j(\kappa(x - \sigma) + \omega(t - \tau))}\,d(x - \sigma) \\ \times d(t - \tau)\right] d\sigma\,d\tau, \\ \leftrightarrow \left[\int_{-\infty}^{\infty}\int_{-\infty}^{\infty} g_1(\sigma, \tau)e^{-j(\kappa\sigma + \omega\tau)}\,d\sigma\,d\tau\right] \\ \times G_2(\kappa, \omega), \\ \leftrightarrow G_1(\kappa, \omega)G_2(\kappa, \omega), \tag{15.165}$$

which is the equivalent of eq. (15.145).

15.2.12 Random functions

The periodic and aperiodic functions that we have been considering hitherto have one property in common: If the process that generates one of these functions is repeated exactly, the same function is generated. However, in many instances repetition of the process gives a different result each time, for example, measurements of microseisms give a function $g(t)$ that never repeats no matter how often we repeat the measurement. Functions of this kind that cannot be predicted exactly no matter how often we repeat the measurements are called *random functions.*

Because random functions cannot be predicted, we use probability theory to deduce their properties. The set of functions obtained if an experiment were repeated an infinite number of times is called an *ensemble.* If we arrive at the same value for some property of the ensemble (for example, the average power density for the frequency ω), whether we average the values for each of the traces at a certain instant in time or average all values for one trace, then the ensemble is said to be *ergodic* and we can determine the statistical properties using a sufficiently long portion of one function rather than having to make measurements on many functions.

A *stationary* time series is one whose statistical properties are independent of the location of the origin $t = 0$. It can be shown that ergodic ensembles must also be stationary (see Lee, 1960: 208–9; Bendat and Piersol, 1966: 11–12). We assume that the random time series with which we deal are ergodic and stationary.

The *autocorrelation of a random function* is defined by the equation

$$\phi_{11}(t) = \lim_{T \to \infty} (1/2T) \int_{-T}^{T} g_1(\tau) g_1(t + \tau) \, d\tau. \quad (15.166)$$

Note that this definition differs in form from that in eq. (15.161) by the factor $1/2T$ as well as the limiting process. Because a random function does not approach zero as t approaches $\pm \infty$, the integral for the Fourier transform, eq. (15.108), does not converge and $G(\omega)$ does not exist. Thus, there is no equivalent of eq. (15.161) for random functions. When $t = 0$,

$$\phi_{11}(0) = \lim_{T \to \infty} (1/2T) \int_{-T}^{T} g_1^2(\tau) \, d\tau$$

$$= \text{mean square value of } g_1(t).$$

The portion of one measurement of a random function, $g_1(t)$, included in the interval $(-T, +T)$ will be written $g_1'(t)$. If we express $g_1'(t)$ as a Fourier series with period $2T$, the series represents $g_1(t)$ exactly in the interval $(-T, +T)$ but not elsewhere because it repeats $g_1'(t)$ in each interval of length $2T$. In the limit as $T \to \infty$, $g_1'(t)$ becomes $g_1(t)$. We write $\phi_{11}'(t)$ for the autocorrelation function of $g_1'(t)$:

$$\phi_{11}'(t) = \int_{-T}^{T} g_1'(\tau) g_1'(t + \tau) \, d\tau.$$

We can expand $\phi_{11}'(t)$ in a Fourier series:

$$\phi_{11}'(t) = \sum_{n=-\infty}^{\infty} \alpha_n e^{jn\omega_0 t}$$

$$= \sum_{n=-\infty}^{\infty} e^{jn\omega_0 t} \left[\frac{1}{2T} \int_{-T}^{T} \phi_{11}'(t) e^{-jn\omega_0 t} \, dt \right]$$

using eq. (15.106), where $\omega_0 = \pi/T$. If we divide

$\phi_{11}'(t)$ by $2T$ and let $T \to \infty$, $\phi_{11}'(t)$ approaches $\phi_{11}(t)$ as defined in eq. (15.166).

Thus,

$$\phi_{11}(t) = \lim_{T \to \infty} \sum_{n=-\infty}^{\infty} e^{jn\omega_0 t}(1/2T) \int_{-T}^{T} \phi_{11}'(t) e^{-jn\omega_0 t} \, dt,$$

$$= (1/2\pi) \int_{-\infty}^{\infty} e^{j\omega t} \, d\omega \int_{-\infty}^{\infty} \phi_{11}(t) e^{-j\omega t} \, dt,$$

$$= (1/2\pi) \int_{-\infty}^{\infty} \Phi_{11}(\omega) e^{j\omega t} \, d\omega,$$

where

$$\Phi_{11}(\omega) = \int_{-\infty}^{\infty} \phi_{11}(t) e^{-j\omega t} \, dt. \quad (15.167)$$

Thus, although a random function does not have a Fourier transform, its autocorrelation function has a Fourier transform: $\phi_{11}(t) \leftrightarrow \Phi_{11}(\omega)$. Also,

$$\phi_{11}(0) = \lim_{T \to \infty} (1/2T) \int_{-T}^{T} g_1^2(t) \, dt$$

$$= (1/2\pi) \int_{-\infty}^{\infty} \Phi_{11}(\omega) \, d\omega. \quad (15.168)$$

Because of the factor $1/2T$, $\Phi_{11}(\omega)$ gives the power density, not energy density as with aperiodic functions. Equations (15.167) and (15.168) express the *Wiener autocorrelation theorem*, that the Fourier transform of the autocorrelation of a random function exists and gives the power-density spectrum. The autocorrelation function and the power-density spectrum are real, even functions and $\phi_{11}(t)$ has its greatest value at the origin (see problem 15.22b).

The *cross-correlation of random functions* is defined in a manner similar to eq. (15.166). Let $g_1(t)$ and $g_2(t)$ be random functions from different ensembles, for example, the input and output noise of an amplifier; then

$$\phi_{12}(t) = \lim_{T \to \infty} \frac{1}{2T} \int_{-T}^{T} g_1(\tau) g_2(t + \tau) \, d\tau. \quad (15.169)$$

Equation (15.157), $\phi_{21}(t) = \phi_{12}(-t)$, holds for both random and aperiodic functions.

15.2.13 Hilbert transforms

The Hilbert transform is a special form of Fourier transform. Let $g(t)$ be any real function and

$$g(t) \leftrightarrow G(\omega) = R(\omega) + jX(\omega).$$

Now $g(t)$ can always be divided into even and odd parts, $g_e(t)$ and $g_o(t)$ (see fig. 15.11), where

$$\left. \begin{array}{l} g_e(t) = \tfrac{1}{2}[g(t) + g(-t)], \\[2mm] g_o(t) = \tfrac{1}{2}[g(t) - g(-t)]. \end{array} \right\} \quad (15.170)$$

Because $g_e(t)$ is even,

$$g_e(t) \leftrightarrow \int_0^\infty g(t) \cos \omega t \, dt = R(\omega) \quad (15.171)$$

(see eq. (15.112)). Likewise,

$$g_o(t) \leftrightarrow jX(\omega). \quad (15.172)$$

If $g(t)$ is also causal (see §15.5.6), $g_e(t) = g_o(t)$ for $t > 0$ and $g_e(t) = -g_o(t)$ for $t < 0$, that is,

$$g_o(t) = g_e(t) \, \text{sgn}(t),$$
$$g_e(t) = g_o(t) \, \text{sgn}(t)$$

(see eq. (15.119)). Recalling that $\text{sgn}(t) \leftrightarrow 2/j\omega$ (see eq. (15.120)), we obtain

$$jX(\omega) = (1/2\pi)R(\omega) * (2/j\omega)$$

on using eq. (15.146). Thus,

$$X(\omega) = -\frac{1}{\pi} \mathscr{H} \int_{-\infty}^\infty [R(y)/(\omega - y)] \, dy \quad (15.173)$$
$$= -R(\omega) * (1/\pi\omega),$$

where $\mathscr{H}$ indicates that we take the Cauchy principle value at $\omega = y$ (see Papoulis, 1962: 9–10 and footnote on p. 10). Likewise,

$$R(\omega) = \frac{1}{\pi} \mathscr{H} \int_{-\infty}^\infty [X(y)/(\omega - y)] \, dy \quad (15.174)$$
$$= X(\omega) * (1/\pi\omega).$$

Equations (15.173) and (15.174) define the Hilbert transform. Given either $R(\omega)$ or $X(\omega)$, the other can be calculated.

Writing $G(\omega) = A(\omega)e^{j\gamma(\omega)}$ and taking logarithms gives

$$\ln[G(\omega)] = \ln[A(\omega)] + j\gamma(\omega);$$

in this case, $R(\omega) = \ln[A(\omega)]$ and $X(\omega) = \gamma(\omega)$. Because $A(\omega)$ is the amplitude of the frequency ω, $A(\omega)$ is the square root of the Fourier transform of the autocorrelation function (see eq. (15.161)); knowing $A(\omega)$, we can calculate $\gamma(\omega)$ by eq. (15.173). The Hilbert transform thus allows the calculation of the phase from the autocorrelation function:

$$\gamma(\omega) = X(\omega) = -\frac{1}{\pi} \mathscr{H} \int_{-\infty}^\infty \{\ln[A(y)]/(\omega - y)\} \, dy.$$
$$(15.175)$$

Given the real part of the Fourier transform of a real, causal time function, the Hilbert transform enables us to find the corresponding time function. A similar problem is: Given the real part of a complex time function whose imaginary part has a transform $\pm90°$ out of phase with that of the real part, find the complex time function. Let $f(t)$ be a complex time function, where

$$f(t) = x(t) + jy(t);$$
$$F(\omega) = X(\omega) + jY(\omega) = X(\omega)[1 + jQ(\omega)],$$

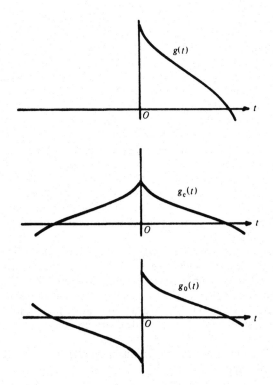

Fig. 15.11   Relation between a real causal function $g(t)$ and the derived even and odd functions, $g_e(t)$ and $g_o(t)$, respectively.

$Q(\omega)$ being a filter that changes the phase by $\pm90°$, but has no effect on the amplitude spectrum; $Q(\omega)$ is referred to as a *quadrature filter*, and $y(t)$ as the *quadrature trace*.

Because $e^{\pm j\pi/2} = \pm j$, we can select $Q(\omega)$ equal to $+j$, $-j$, or $\pm j \, \text{sgn}(\omega)$. Selection of the first two choices gives $q(t) = \pm j\delta(t)$ from eq. (15.127), which is not a useful result. We thus select $Q(\omega) = -j \, \text{sgn}(\omega)$. Then,

$$y(t) \leftrightarrow X(\omega)Q(\omega) = -jX(\omega) \, \text{sgn}(\omega)$$
$$= jX(\omega) \, \text{sgn}(-\omega).$$

Using eqs. (15.120), (15.140), and (15.145), we have

$$y(t) = x(t) * (1/\pi t), \quad (15.176)$$

that is, $x(t)$ and $-y(t)$ form a Hilbert transform pair (see eq. (15.173)).

Although $f(t)$ can be found by using eq. (15.176), it is easier to calculate $F(\omega)$ and transform to get $f(t)$. Thus,

$$F(\omega) = X(\omega) + jY(\omega) = X(\omega)\{1 + jQ(\omega)\},$$
$$= X(\omega)\{1 + \text{sgn}(\omega)\},$$
$$= 0, \, \omega < 0,$$
$$= 2X(\omega), \, \omega > 0,$$
$$= 2X(\omega)\text{step}(\omega). \quad (15.177)$$

Equations (15.176) and (15.177) find application in complex-trace analysis (§9.11.4).

15.3 Laplace transform

15.3.1 Introduction

The Laplace transform is closely related to the Fourier transform. If we do not use distributions, many functions such as sin at and cos at do not have Fourier transforms because the integral giving the transform does not converge. However, if we multiply the function by a convergence factor $e^{-\sigma|t|}$, σ being real and positive and large enough that $\lim_{t \to \pm\infty}[e^{-\sigma|t|}g(t)] = 0$, then $e^{-\sigma|t|}g(t)$ has a Fourier transform that is called the *Laplace transform of $g(t)$*. If $g(t) = 0$ for $t < 0$, we get the *one-sided Laplace transform*:

$$g(t) \leftrightarrow \int_0^\infty e^{-\sigma t}g(t)e^{-j\omega t}\,dt = \int_0^\infty g(t)e^{-st}\,dt = G(s),$$

(15.178a)

where $s = \sigma + j\omega$, the real part of s being large enough that $\lim_{t \to \infty}[e^{-st}g(t)] = 0$. (The Laplace transform, $G(s)$, is distinguished from the Fourier transform, $G(\omega)$, by the variable s instead of ω.)

The *inverse Laplace transformation* becomes

$$e^{-\sigma t}g(t) = (1/2\pi)\int_{-\infty}^\infty G(s)e^{j\omega t}\,d\omega$$

or

$$g(t) = (1/2\pi)\int_{-\infty}^\infty G(s)e^{(\sigma+j\omega)t}\,d\omega$$

$$= (1/2\pi j)\int_{\sigma-j\infty}^{\sigma+j\infty} G(s)e^{st}\,ds,$$ (15.178b)

where the path of integration is a line to the right of the origin parallel to the imaginary axis such that the integral converges. The calculation of Laplace transforms is usually relatively simple in comparison with Fourier transforms, but the inverse transformation is generally difficult.

The Fourier transform is more convenient when we wish to discuss properties that depend upon frequency and/or phase. It is very useful in certain areas of probability theory and in solving linear differential equations with boundary conditions that can be expressed in Fourier or Fourier–Bessel series. The Laplace transform is very useful when we investigate the analytical properties of the transform (as in circuit analysis) and in solving linear differential equations with constant coefficients when initial conditions are given.

The Laplace transform of some common functions can be easily derived. Thus,

$$\text{step}(t) \leftrightarrow \int_0^\infty \text{step}(t)e^{-st}\,dt = \left.\frac{e^{-st}}{-s}\right|_0^\infty = \frac{1}{s};$$ (15.179)

$$\delta(t) \leftrightarrow \int_0^\infty \delta(t)e^{-st}\,dt = e^{-st}\Big|_{t=0} = +1;$$ (15.180)

$$t \leftrightarrow \int_0^\infty te^{-st}\,dt = \left.\frac{e^{-st}}{s^2}(-st-1)\right|_0^\infty = \frac{1}{s^2};$$ (15.181)

$$e^{-kt} \leftrightarrow \int_0^\infty e^{-kt}e^{-st}\,dt = \left.\frac{e^{-(k+s)t}}{-(k+s)}\right|_0^\infty = \frac{1}{s+k}$$

(15.182)

(note that the real part of s is larger than the real part of k, so that $e^{-(k+s)t} = 0$ for $t = +\infty$);

$$\left.\begin{array}{l}\cos at \leftrightarrow s/(s^2+a^2); \\ \sin at \leftrightarrow a/(s^2+a^2);\end{array}\right\}$$ (15.183)

$$\left.\begin{array}{l}\cosh at \leftrightarrow s/(s^2-a^2); \\ \sinh at \leftrightarrow a/(s^2-a^2).\end{array}\right\}$$ (15.184)

The last four results can be obtained directly by integration or by substituting $k = \pm ja$ or $k = \pm a$ in eq. (15.182) and combining exponential terms to get cos at, and so on. Equations (15.179), (15.180), and (15.182) should be compared with eqs. (15.121), (15.127), and (15.132).

15.3.2 Theorems on Laplace transforms

Most of the theorems on Fourier transforms have counterparts for Laplace transforms. The most useful theorems are listed below. Note that $a > 0$ in all cases.

$$\begin{array}{l}g(t) \leftrightarrow G(s); \\ kg(t) \leftrightarrow kG(s); \\ k_1g_1(t) + k_2g_2(t) \leftrightarrow k_1G_1(s) + k_2G_2(s).\end{array}$$ (15.185) (15.186)

Shift theorems:

$$g(t-a)\,\text{step}(t-a) \leftrightarrow e^{-as}G(s);$$ (15.187)

$$e^{-at}g(t) \leftrightarrow G(s+a).$$ (15.188)

Scaling theorems:

$$g(at) \leftrightarrow (1/a)G(s/a);$$ (15.189)

$$(1/a)g(t/a) \leftrightarrow G(as).$$ (15.190)

Derivative theorems:

$$\frac{d^n g(t)}{dt^n} \leftrightarrow s^n G(s) - s^{n-1}g(0+) - s^{n-2}g^1(0+)$$

$$-\cdots -g^{n-1}(0+);$$ (15.191)

$$(-t)^n g(t) \leftrightarrow d^n G(s)/ds^n.$$ (15.192)

Integral theorems:

$$\int_0^t g(t)\,dt \leftrightarrow (1/s)G(s);$$ (15.193)

$$(1/t)g(t) \leftrightarrow \int_s^{+\infty} G(s)\,ds.$$ (15.194)

Convolution theorem:

$$g_1(t) * g_2(t) = \int_0^t g_1(\tau)g_2(t-\tau)\,d\tau \leftrightarrow G_1(s)G_2(s).$$

(15.195)

In eq. (15.191), the symbols $g(0+)$ and $g^r(0+)$ denote the values of $g(t)$ and its rth derivative at $t = 0$ when t approaches zero from the positive side.

The proofs of most of the preceding theorems are simple. For example, for eq. (15.187) we have

$$\int_0^\infty g(t - a)\,\text{step}(t - a)\text{e}^{-st}\,\text{d}t$$

$$= \int_a^\infty g(t - a)\text{e}^{-st}\,\text{d}t,$$

$$= \int_0^\infty g(y)\text{e}^{-s(y+a)}\,\text{d}y, \; y = t - a,$$

$$= \text{e}^{-as}\int_0^\infty g(y)\text{e}^{-sy}\,\text{d}y = \text{e}^{-as}G(s).$$

For eq. (15.191), we integrate by parts, getting

$$\frac{\text{d}g}{\text{d}t} \leftrightarrow \int_0^\infty \frac{\text{d}g}{\text{d}t}\text{e}^{-st}\,\text{d}t = g(t)\text{e}^{-st}\Big|_0^\infty - (-s)\int_0^\infty g(t)\text{e}^{-st}\,\text{d}t,$$

$$= -g(0+) + sG(s).$$

By successive applications of this result, the general formula is obtained. For eq. (15.192), we write $G(s) = \int_0^\infty g(t)\text{e}^{-st}\,\text{d}t$, and then differentiate with respect to s, giving

$$\frac{\text{d}G}{\text{d}s} = \int_0^\infty g(t)\frac{\text{d}}{\text{d}s}(\text{e}^{-st})\,\text{d}t = \int_0^\infty (-t)g(t)\text{e}^{-st}\,\text{d}t,$$

and hence $-tg(t) \leftrightarrow \text{d}G(s)/\text{d}s$. Successive differentiations give eq. (15.192). To prove eq. (15.193), we write $\int_0^t g(t)\,\text{d}t \leftrightarrow G_i(s)$, then differentiate, obtaining $g(t) \leftrightarrow sG_i(s) = G(s)$; hence, $G_i(s) = (1/s)G(s)$. The converse is proved by writing

$$\int_s^\infty G(s)\,\text{d}s = \int_s^\infty \left\{\int_0^\infty g(t)\text{e}^{-st}\,\text{d}t\right\}\,\text{d}s,$$

$$= \int_0^\infty g(t)\left\{\int_s^\infty \text{e}^{-st}\,\text{d}s\right\}\,\text{d}t,$$

on changing the order of integration. Then

$$\int_s^\infty G(s)\,\text{d}s = \int_0^\infty g(t)\frac{\text{e}^{-st}}{-t}\Big|_s^\infty\,\text{d}t = \int_0^\infty \left(\frac{1}{t}\right)g(t)\text{e}^{-st}\,\text{d}t;$$

hence, $(1/t)g(t) \leftrightarrow \int_s^\infty G(s)\,\text{d}s$.

When Laplace transforms are being used, the convolution of two functions, $g_1(t)$ and $g_2(t)$, is usually defined by the expression

$$g_1(t) * g_2(t) = \int_0^t g_1(\tau)g_2(t - \tau)\,\text{d}\tau. \quad (15.196)$$

This definition appears to differ from that given in eq. (15.145) because of the different limits. However, the difference is only apparent and is due to the fact that $g_i(t) = 0$, $t < 0$. As shown in fig. 15.10a, $g_1(\tau) = 0$, $\tau < 0$, whereas $g_2(t - \tau) = 0$, $\tau > t$. Thus, changing the limits to $\pm\infty$ would not change the value of the integral.

The convolution theorem, eq. (15.195), can be proved as follows:

$$g_1(t) * g_2(t) \leftrightarrow \int_0^\infty \left\{\int_0^t g_1(\tau)g_2(t - \tau)\text{d}\tau\right\}\text{e}^{-st}\,\text{d}t,$$

$$\leftrightarrow \int_0^\infty \left\{\int_0^\infty g_1(\tau)g_2(t - \tau)\text{step}(t - \tau)\text{d}\tau\right\}\text{e}^{-st}\text{d}t,$$

where the change of the upper limit and the insertion of $\text{step}(t - \tau)$ have not changed the value of the integral because $\text{step}(t - \tau)$ is unity for $\tau < t$ and zero for $\tau > t$. Changing the order of integration gives

$$g_1(t) * g_2(t) \leftrightarrow \int_0^\infty g_1(\tau)\left[\int_0^\infty g_2(t - \tau)\text{step}(t - \tau)\text{e}^{-st}\text{d}t\right]\text{d}\tau,$$

$$\leftrightarrow \int_0^\infty g_1(\tau)\text{e}^{-s\tau}G_2(s)\text{d}\tau, \text{ using eq. (15.187)},$$

$$\leftrightarrow G_1(s)G_2(s).$$

15.4 Linear systems

15.4.1 Introduction

We use the term *system* to denote a group of objects so related that, when an input is applied at one point, an output is generated at another point. We may know little or nothing of the detailed workings of the system.

A *linear system* is one in which output $h(t)$ is proportional to input $g(t)$, that is, if $g(t) \to h(t)$, then $kg(t) \to kh(t)$, where k is a constant and the arrow represents the effect of the system. (Cheng, 1959, has a good account of linear systems.) By considering $g(t)$ to be the sum of two signals, we see that the *principle of superposition* holds, namely,

$$k_1g_1(t) + k_2g_2(t) \to k_1h_1(t) + k_2h_2(t). \quad (15.197)$$

A system is *time-invariant* if the same input produces the same output regardless of the time when the input is applied, that is,

$$kg(t - t_0) \to kh(t - t_0) \quad (15.198)$$

for all values of t_0.

In principle at least, systems can be described by means of differential equations relating output to input. The corresponding differential equations for linear systems are linear, and the coefficients of the equation are constants when the system is time-invariant. We shall study the properties of differential equations describing linear, time-invariant systems. Our discussion will not depend on the order of the equation and because many systems are represented by second-order equations, we consider the equation

$$\frac{d^2h(t)}{dt^2} + a_1\frac{dh(t)}{dt} + a_2h(t) = g(t). \quad (15.199)$$

Using eq. (15.191), we get

$$\frac{d^2h(t)}{dt^2} \leftrightarrow s^2H(s) - sh(0+) - h'(0+),$$

$$\frac{dh(t)}{dt} \leftrightarrow sH(s) - h(0+),$$

where $h(0+)$ and $h'(0+)$ are the values of $h(t)$ and $dh(t)/dt$, respectively, at $t = 0$.

Substituting in the equation gives

$$(s^2 + a_1s + a_2)H(s) - [(s + a_1)h(0+) + h'(0+)]$$
$$= G(s).$$

Solving for $H(s)$, we have

$$H(s) = \frac{G(s) + (s + a_1)h(0+) + h'(0+)]}{s^2 + a_1s + a_2} \quad (15.200)$$

The numerator in eq. (15.200) is called the *total excitation transform;* it depends in part upon the excitation (input), $g(t)$, and in part upon the initial conditions of the system. When the initial output and its first derivative are zero, that is, $h(0+) = 0 = h'(0+)$, the system is said to be *initially relaxed* and the total excitation transform reduces to the input transform, $G(s)$. For convenience, we shall assume henceforth that the system is initially relaxed.

The quantity, $1/(s^2 + a_1s + a_2)$ is known as the *transfer function, $F(s)$*. It is a function only of the properties of the system. Thus, the transform of the output corresponding to any input $g(t)$ is given by

$$H(s) = F(s)G(s). \quad (15.201)$$

We apply the convolution theorem to this equation and obtain

$$h(t) = f(t) * g(t) = \int_0^t g(\tau)f(t - \tau)\, d\tau, \quad (15.202)$$

where $F(s) \leftrightarrow f(t)$.

If we apply a unit impulse $\delta(t)$ to a linear system, then $\delta(t) \leftrightarrow G(s) = +1$ from eq. (15.180), so that $H(s) = F(s)$ and $h(t) = f(t)$. The response $f(t)$ to a unit impulse is the *impulse response* (often called the "unit impulse response"); it is the inverse transform of the transfer function. In principle, we can predict the behavior of a linear system for an arbitrary input by applying a unit impulse and measuring the impulse response, then applying eq. (15.202) to get $h(t)$.

Let us apply a unit step function to the input of a linear system. Then from eqs. (15.179) and (15.201), we obtain

$$H(s) = F_u(s) = (1/s)F(s),$$

where $F_u(s) \leftrightarrow f_u(t) = $ *step response* (also called the "unit step response"). We can find the relation be-

tween $f_u(t)$ and $f(t)$ as follows. From the preceding,

$$F(s) = sF_u(s). \quad (15.203)$$

But

$$d[f_u(t)]/dt \leftrightarrow sF_u(s) - f_u(0+) = F(s) - f_u(0+)$$

from eq. (15.191), hence,

$$f(t) = d[f_u(t)]/dt + f_u(0+)\,\delta(t). \quad (15.204)$$

Because $f_u(t) = 0$, $t < 0$, the step response has a discontinuity at the origin of magnitude $f_u(0+)$. Thus, the impulse response, $f(t)$, is equal to the derivative of the step response $f_u(t)$ plus an impulse at the origin of magnitude $f_u(0+)$.

15.4.2 Linear systems in series and parallel

Assume that two linear systems with transfer functions $F_1(s)$ and $F_2(s)$ are connected in series so that the output of the first system is the input of the second system; then

$$H_1(s) = F_1(s)G(s);$$
$$h_1(t) = f_1(t) * g(t);$$
$$H_2(s) = F_2(s)H_1(s) = F_2(s)F_1(s)G(s);$$
$$h_2(t) = f_2(t) * [f_1(t) * g(t)],$$
$$= [f_2(t) * f_1(t)] * g(t). \quad (15.205)$$

Thus, two systems in series are equivalent to a single system whose transfer function is the product of the two individual transfer functions. Obviously, this result holds for any number of systems in series.

When an input $g(t)$ is applied to two systems in parallel, the outputs will be superimposed so that

$$H(s) = H_1(s) + H_2(s) = G(s)F_1(s) + G(s)F_2(s),$$
$$= G(s)[F_1(s) + F_2(s)]. \quad (15.206)$$

Thus, the equivalent transfer function for systems in parallel is the sum of the individual transfer functions.

15.5 Digital systems and z-transforms

15.5.1 Sampling theorem

When a continuous function, $g(t)$, is sampled at regular intervals, Δ, we obtain a series of values, $g(n\Delta)$. The sampled function, which we write g_t, can be regarded as the product of $g(t)$ and a comb (see eq. (15.130)) with spacing Δ:

$$g_t = \sum_{n=-\infty}^{\infty} g(n\Delta)\,\delta(t - n\Delta). \quad (15.207)$$

(Although the comb extends from $-\infty$ to $+\infty$, we need to sum only over the range for which $g(t) \neq 0$.)

When we sample, we discard values of $g(t)$ in between the sampling points and hence apparently lose information. However, the sampling theorem states that $g(t)$ can be recovered exactly from the sampled values provided $g(t)$ has no frequencies above the Nyquist frequency, $v_N = \frac{1}{2}$(sampling frequency) $= 1/2\Delta$. The Fourier transform of $g(t)$, $G(\omega)$, thus must be zero for $|\omega| > \omega_N$, $\omega_N = 2\pi v_N = \pi/\Delta$, so that

$$g(t) = (1/2\pi) \int_{-\infty}^{\infty} G(\omega)e^{j\omega t}\, d\omega,$$

$$= (1/2\pi) \int_{-\omega_N}^{\omega_N} G(\omega)e^{j\omega t}\, d\omega. \quad (15.208)$$

If we expand $G(\omega)$ in a Fourier series in the interval $-\omega_N$ to ω_N, the Fourier series will repeat $G(\omega)$ in each interval of width $2\omega_N$, although $G(\omega)$ is zero outside the interval $(-\omega_N, \omega_N)$. The fundamental period for the series is $1/2\omega_N$; hence, $\omega_0 = 2\pi/T$ and $\pi/\omega_N = \Delta$. Therefore, using eqs. (15.105) and (15.106) with ω replacing t, we get

$$G(\omega) = \sum_{r=-\infty}^{\infty} \alpha_r e^{jr\omega\Delta},$$

$$\alpha_r = (1/2\omega_N) \int_{-\omega_N}^{\omega_N} G(\omega)e^{-jr\omega\Delta}\, d\omega.$$

Comparison with eq. (15.208) shows that if we take $t = r\Delta$, then

$$\alpha_{-r} = (\pi/\omega_N)\, g(r\Delta);$$

hence,

$$G(\omega) = \sum_{r=-\infty}^{\infty} (\pi/\omega_N)g(r\Delta)\, \exp[-jr\pi(\omega/\omega_N)].$$

This equation gives the correct values of $G(\omega)$ in the interval $-\omega_N$ to $+\omega_N$, but is not zero outside this interval as it should be to represent $g(t)$ exactly. To correct for this, we can multiply by a boxcar, $\mathrm{box}_a(\omega)$, extending from $-\omega_N$ to ω_N, that is, $G(\omega)$ is given exactly by

$$G(\omega) = \mathrm{box}_a(\omega) \sum_{r=-\infty}^{\infty} (\pi/\omega_N)g(r\Delta)\, \exp[-jr\pi(\omega/\omega_N)].$$
$$(15.209)$$

Equation (15.152) gives

$$\omega_N\, \mathrm{sinc}\, \omega_N t \leftrightarrow \pi\, \mathrm{box}_{2\omega_N}(\omega).$$

Using eq. (15.136), we find that

$$\omega_N\, \mathrm{sinc}\, [\omega_N(t - k)] \leftrightarrow \pi\, \mathrm{box}_{2\omega_N}(\omega)e^{-jk\omega}.$$

Using this result, we can write the inverse transform of eq. (15.209) in the form

$$g(t) = \sum_{r=-\infty}^{\infty} g(r\Delta)\, \mathrm{sinc}\, (\omega_N t - r\pi). \quad (15.210)$$

This result shows that the function $\mathrm{sinc}\, (\omega_N t - r\pi) = \sin \pi(t/\Delta - r)/\pi(t/\Delta - r)$ provides perfect interpolation to give $g(t)$ for all values of t, not merely for the sampling instants $r\Delta$.

When $G(\omega) \neq 0$ for values of $|\omega| > \omega_N$, the foregoing proof breaks down and we are not able to recover $g(t)$ from the sampled values (see discussion in §9.2.2).

15.5.2 Convolution and correlation of sampled functions

The integrals in eqs. (15.145), (15.147), and (15.161) involve continuous functions. When the functions are sampled functions, the integrals become summations. To show this, we start with continuous functions $f(t)$ and $g(t)$ so that $f(t) * g(t)$ is given by

$$f(t) * g(t) = \int_0^{\infty} f(\tau)g(t - \tau)\, d\tau.$$

Replacing $f(\tau)$ by the sampled function

$$f_t = \sum_{k=-\infty}^{\infty} f(k\Delta)\, \delta(\tau - k\Delta)$$

(see eq. (15.207)), we have

$$f_t * g(t) = \int_{-\infty}^{\infty} \left[\sum_k f(k\Delta)\, \delta(\tau - k\Delta)\right] g(t - \tau)\, d\tau$$

$$= \sum_k \left\{ \int_{-\infty}^{\infty} [f(k\Delta)g(t - \tau)]\, \delta(\tau - k\Delta)\, d\tau \right\}.$$

Applying eq. (15.124) to each integral, we get for each term in the sum the value $f(k\Delta)g(t - k\Delta)$. If we now sample $g(t)$, t becomes a multiple of Δ; hence, $g(t - k\Delta) = g_{t-k}$, so that we have (note that $d\tau \to \Delta$)

$$f_t * g_t = \Delta \sum_k f_k g_{t-k}. \quad (15.211)$$

In the same way, we find that

$$\phi_{fg}(\tau) = \Delta \sum_k f_k g_{k+\tau}. \quad (15.212)$$

Usually, we set $\Delta = 1$ in these equations (for example, eqs. (9.23), (9.24), (9.41), (9.42), and (9.46)), but care must be taken in some cases.

15.5.3 z-transforms

z-transforms are a special form of transform useful for calculations involving digital (sampled) functions. We take the Fourier transform of both sides of eq. (15.207), using eq. (15.154) and obtain

$$g_t \leftrightarrow G(\omega) = \sum_{n=-\infty}^{\infty} g(n\Delta)e^{-jn\omega\Delta}.$$

If we write $z = e^{-j\omega\Delta}$, we get

$$G(\omega) = \sum_{n=-\infty}^{\infty} g(n\Delta)z^n = G(z), \quad (15.213)$$

where $G(z)$ is the z-transform of g_t, that is, $g_t \leftrightarrow G(z)$. Thus, if $g_t = [1, 2, -5, 4, -6]$, $G(z) = 1 + 2z - 5z^2 + 4z^3 - 6z^4$. Negative powers of z denote values of time past; thus if $g_t = [2, 6, -\overset{\downarrow}{1}, 0, 5]$ (the superscribed arrow denotes $t = 0$), then $G(z) = 2z^{-2} + 6z^{-1} - 1 + 5z^2$. It is evident that multiplication by z is equivalent to delaying the time function by one sample interval and division by z to advancing it one sample interval.

(We could have taken the Laplace transform to get $G(z)$ in terms of $z = e^{-s\Delta}$. The Fourier form is more convenient for studying frequency characteristics, the

Laplace form for examining stability, as when studying filters. A difference is that s ranges over that part of the complex plane to the right of the vertical line through σ, that is, $Re[s] \geq \sigma$ (§15.3.1), whereas the terminus of $z = e^{-j\omega\Delta}$ lies on the unit circle with the center at the origin.)

(In signal analysis, z is often defined as $z = e^{j\omega\Delta}$, resulting in a polynomial in which the z's have negative exponents as compared with the preceding. The convention used here is more common in seismic data processing.)

Clearly, z, hence also $G(z)$, is a periodic function of ω with period $2\pi/\Delta$. As ω increases from c to $c + 2\pi/\Delta$, c being any real number, the terminus of z goes around the unit circle ($|z| = 1$) with center at the origin once in the clockwise direction; this follows from the relation

$$z = e^{-j\omega\Delta} = \cos\omega\Delta - j\sin\omega\Delta$$

(note that as ω increases from zero, $-\sin\omega\Delta$ increases in the negative direction). (If we had taken $z = e^{j\omega\Delta}$, z would rotate counterclockwise as ω increases.)

We shall calculate a few simple z-transforms. For example, let

$$g(t) = t\,\text{step}(t) = \sum_{n=0}^{\infty}(n\Delta)\delta(t - n\Delta),$$

$$t\,\text{step}(t) \leftrightarrow \sum_{n=0}^{\infty}(n\Delta)z^n = \Delta\sum_{n=1}^{\infty}nz^n$$

$$\leftrightarrow z\Delta\sum_{n=1}^{\infty}nz^{n-1} = z\Delta/(1 - z)^2. \quad (15.214)$$

Again, let

$$g(t) = e^{kt}\text{step}(t) = \sum_{n=0}^{\infty}e^{kn\Delta}\delta(t - n\Delta);$$

$$e^{kt}\text{step}(t) \leftrightarrow \sum_{n=0}^{\infty}e^{kn\Delta}z^n = \sum_{n=0}^{\infty}(e^{k\Delta}z)^n$$

$$\leftrightarrow 1 + e^{k\Delta}z + (e^{k\Delta}z)^2 + \cdots,$$

$$\leftrightarrow 1/(1 - e^{k\Delta}z). \quad (15.215)$$

Setting $k = 0$ gives

$$\text{step}(t) \leftrightarrow 1/(1 - z). \quad (15.216)$$

Setting $k = +j\theta$ gives

$$e^{j\theta t}\text{step}(t) \leftrightarrow \frac{1}{1 - ze^{j\theta\Delta}} = \frac{1}{(1 - z\cos\theta\Delta) - jz\sin\theta\Delta};$$
$$(15.217)$$

rationalizing the denominator, we get

$$(\cos\theta t + j\sin\theta t)\,\text{step}(t)$$
$$\leftrightarrow \frac{(1 - z\cos\theta\Delta) + j(z\sin\theta\Delta)}{(1 - z\cos\theta\Delta)^2 + (z\sin\theta\Delta)^2}.$$

Equating real and imaginary parts gives

$$\left.\begin{array}{l}\cos\theta t\,\text{step}(t) \leftrightarrow \dfrac{1 - z\cos\theta\Delta}{1 - 2z\cos\theta\Delta + z^2}, \\[4mm] \sin\theta t\,\text{step}(t) \leftrightarrow \dfrac{z\sin\theta\Delta}{1 - 2z\cos\theta\Delta + z^2}.\end{array}\right\} \quad (15.218)$$

Table 15.1 *Finding amplitude and phase spectra*

$\omega\Delta$	$G(z)$	Amplitude	Phase
0°	5	5.00	0.0°
45°	$1 - \sqrt{2} - j4\sqrt{2}$	5.67	85.8°
90°	$5 + 2j$	5.39	203.6°
135°	$1 + \sqrt{2} - j4\sqrt{2}$	6.15	293.0°
180°	-11	11.00	360.0°
225°	$1 + \sqrt{2} + j4\sqrt{2}$	6.15	67.0°
270°	$5 - 2j$	5.39	156.4°
315°	$1 - \sqrt{2} + j4\sqrt{2}$	5.67	274.2°

As an example of the use of z-transforms in determining the frequency and phase characteristics of digital functions, we shall obtain the spectra of the digital function $g_t = [\ldots, 0, 0, -2, 0, \overset{\downarrow}{1}, 3, -2, 5, 0, 0, \ldots]$. The transform is $G(z) = -2z^{-2} + 1 + 3z - 2z^2 + 5z^3 = -2e^{2j\omega\Delta} + 1 + 3e^{-j\omega\Delta} - 2e^{-2j\omega\Delta} + 5e^{-3j\omega\Delta}$. Substituting values of ω gives values of $G(z)$ that are complex in general and from which we can get the amplitude and phase spectra as functions of ω. Table 15.1 gives typical values of $G(\omega)$ for a few values of ω.

15.5.4 Calculation of z-transforms; Fast Fourier Transform

Generally, we must calculate z-transforms for many more values of the argument $\omega\Delta$ than we did in the last section to obtain the required precision. Consider the function

$$g_t = g_0, g_1, g_2, \ldots, g_{n-1};$$
$$G(z) = g_0 + g_1z + g_2z^2 + \cdots + g_{n-1}z^{n-1}.$$

It is convenient to take $n = 2^k$, where k is integral (this can always be achieved by adding zeros to g_t) and to calculate the transform for increments of $\omega\Delta$ equal to $2\pi/n$, that is, we take $\omega\Delta = r(2\pi/n)$, $r = 0, 1, 2, \ldots, n - 1$ (note that $G(z)$ repeats for $r \geq n$). If we let $q = e^{-j2\pi/n}$, then $z = e^{-jr(2\pi/n)} = q^r$. The various values of $G(z)$, G_r, can be written in matrix form

$$\begin{Vmatrix} G_0 \\ G_1 \\ G_2 \\ \vdots \\ G_{n-1} \end{Vmatrix} = \begin{Vmatrix} 1 & 1 & 1 & \ldots & 1 \\ 1 & q & q^2 & \ldots & q^{(n-1)} \\ 1 & q^2 & q^4 & \ldots & q^{2(n-1)} \\ \vdots & \vdots & \vdots & \ddots & \vdots \\ 1 & q^{(n-1)} & q^{2(n-1)} & \ldots & q^{(n-1)^2} \end{Vmatrix} \begin{Vmatrix} g_0 \\ g_1 \\ g_2 \\ \vdots \\ g_{n-1} \end{Vmatrix}$$

$$(15.219)$$

This method requires n^2 multiplications and n^2 additions. Because a seismic trace often has a few thousand values, millions of calculations are necessary. The fast Fourier transform (FFT) is an ingenious algorithm for calculating $G(z)$ with only $n\log_2 n$ calculations. For $n = 2^{10} = 1024$, the difference is between 10^4 and 2×10^6.

The fast Fourier transform (Cooley and Tukey, 1965) depends upon *doubling* processes by which a series is built up from (or decomposed into) shorter series. Let us take the time series

$$c_1 = c_0, c_1, c_2, \ldots, c_{2n-1};$$
$$C(z) = c_0 + c_1 z + c_2 z^2 + \cdots + c_{2n-1} z^{2n-1},$$

and decompose it into two series:

$$c_t = x_0, y_0, x_1, y_1, \ldots, x_{n-1}, y_{n-1}.$$

We write

$$x_t = x_0, x_1, \ldots, x_{n-1};$$
$$X(z) = x_0 + x_1 z + \cdots + x_{n-1} z^{n-1};$$
$$y_t = y_0, y_1, \ldots, y_{n-1};$$
$$Y(z) = y_0 + y_1 z + \cdots + y_{n-1} z^{n-1},$$

where the values x_i, y_i occur at intervals of 2Δ, not Δ as in c_t.

We calculate $C(z)$ for $z = q^r$, $q = e^{-j2\pi/2n}$, $r = 0, 1, 2, \ldots (2n - 1)$, whereas $X(z)$, $Y(z)$ are calculated for the values $(q')^r$, $q' = e^{-j2\pi/n} = q^2$, $r = 0, 1, 2, \ldots, (n - 1)$. Writing x_r for the value of $X(z)$ for $z = q^{2r}$,

$$\left. \begin{array}{l} X_r = \sum_{i=0}^{n-1} x_i q^{2ri} \\ Y_r = \sum_{i=0}^{n-1} y_i q^{2ri} \end{array} \right\} r = 0, 1, \ldots, n - 1. \quad (15.220)$$

But,

$$C_r = \sum_{i=0}^{2n-1} c_i q^{ri}, r = 0, 1, \ldots, (2n - 1).$$

For $r = 0, 1, \ldots, (n - 1)$,

$$C_r = (x_0 + x_1 q^{2r} + x_2 q^{4r} + \cdots + x_{n-1} q^{2(n-1)})$$
$$+ q^r(y_0 + y_1 q^{2r} + y_2 q^{4r} + \cdots + y_{n-1} q^{2(n-1)}),$$
$$= \sum_{i=0}^{n-1} x_i q^{2ri} + q^r \sum_{i=0}^{n-1} y_i q^{2ri},$$
$$= X_r + q^r Y_r. \quad (15.221)$$

When $r = n, (n + 1), \ldots, (2n - 1)$, we must manipulate the exponents to express C_r in terms of X_r, Y_r. Thus,

$$C_r = \sum_{i=0}^{2n-1} c_i q^{ri}, r = n, (n + 1), \ldots, (2n - 1).$$

We write $r = n + m$ so that $q^{ri} = q^{(n+m)i} = q^{ni} q^{mi} = (-1)^i q^{mi}$, $m = 0, 1, (n - 1)$ because $q^{ni} = (e^{-j\pi})^i$. Hence,

$$C_r = \sum_{i=0}^{n-1} x_i q^{2mi} - q^m \sum_{i=0}^{n-1} y_i q^{2mi},$$
$$m = 0, 1, \ldots, (n - 1),$$
$$C_r = X_m - q^m Y_m,$$
$$= X_{r-n} - q^{r-n} Y_{r-n}, \quad (15.222)$$
$$r = n, (n + 1), \ldots, (2n - 1).$$

To calculate C_r using eq. (15.219) requires $2(2n)^2 = 8n^2$ arithmetical operations. To find X_r from eq. (15.220) requires $2n^2$; hence, to get C_r from eqs. (15.221) and (15.222) requires slightly more than $4n^2$ operations, a 50% saving. Because n is a multiple of 2, doubling can be continued until the subseries consist

of single elements of c_t with a tremendous saving in work when n is large.

15.5.5 Application of z-transforms to digital systems

By *digital systems,* we refer to linear systems in which the input and output are sampled functions. If the system is analog, each element of the digital input g_t gives rise to a continuous output $h(t - n\Delta)$. In this case, we sample the output in synchronism with the input to get h_t. Then,

$$F(\omega) = H(z)/G(z) = \text{polynomial in } z = F(z).$$

Thus,

$$H(z) = F(z)G(z) \leftrightarrow h_t = f_t * g_t. \quad (15.223)$$

The utility of z-transforms for digital processing arises because they can be written by inspection and manipulated as simple polynomials. For example, in §9.2.1, we convolved $f_t = (1, -1, \frac{1}{2})$ with $g_t = (1, \frac{1}{2}, -\frac{1}{2})$. The z-transforms are $F(z) = 1 - z + \frac{1}{2}z^2$ and $G(z) = 1 + \frac{1}{2}z - \frac{1}{2}z^2$; we have

$$f_t * g_t \leftrightarrow F(z)G(z) = (1 - z + \tfrac{1}{2}z^2)(1 + \tfrac{1}{2}z - \tfrac{1}{2}z^2),$$
$$= (1 - \tfrac{1}{2}z - \tfrac{1}{2}z^2 + \tfrac{3}{4}z^3 - \tfrac{1}{2}z^4);$$

thus, $f_t * g_t = (1, -\frac{1}{2}, -\frac{1}{2}, \frac{3}{4}, -\frac{1}{4})$.

As another example, in §9.2.4, eq. (9.38) gave the water reverberation filter for $n = 1$ as

$$f_t = (1, -2R, 3R^2, -4R^3, 5R^4, \ldots)$$

so that

$$f_t \leftrightarrow 1 - 2Rz + 3R^2 z^2 - 4R^3 z^3 + 5R^4 z^4 + \cdots,$$

and the inverse filter, i_t, is such that $f_t * i_t = \delta_t \leftrightarrow 1$ (eq. (9.39)). We can solve for $I(z)$ by division because division by polynomials is a proper operation:

$$I(z) = 1/F(z) = 1 + 2Rz + R^2 z^2.$$

Thus,

$$I(z) \leftrightarrow (1, 2R, R^2),$$

which is eq. (9.40).

As a third example, in §9.3.1, we cross-correlated $x_t = (1, -1, \frac{1}{2})$ with $y_t = (1, \frac{1}{2}, -\frac{1}{2})$. From eq. (9.44), we have $\phi_{xy}(\tau) = x_{-\tau} * y_\tau$. Hence,

$$x_{-\tau} \leftrightarrow \tfrac{1}{2}z^{-2} - z^{-1} + 1,$$
$$y_\tau \leftrightarrow 1 + \tfrac{1}{2}z - \tfrac{1}{2}z^2,$$
$$\phi_{xy}(\tau) \leftrightarrow \tfrac{1}{2}z^{-2} - \tfrac{3}{4}z^{-1} + \tfrac{1}{4} + z - \tfrac{1}{2}z^2$$

(see fig. 9.6f).

15.5.6 Phase considerations

(a) Minimum-phase wavelets. We define a causal function as a real function that is zero for negative time, that is, $f(t) = 0$, $t < 0$. A physically realizable function is a causal function that has finite energy, that is $\int_0^\infty |f(t)|^2 \, dt$ is finite. A minimum-delay function is a physically realizable function whose transform has an inverse, that is, $f(t) \leftrightarrow F(z)$, and $1/F(z)$ is finite. $F(z)$ is said to be minimum-phase. The reasons for the terms "minimum-delay" and "minimum-phase" will be apparent later. (The literature is replete with different definitions of "minimum-phase," the most common of which will be derived from the foregoing definition.)

Consider a simple wavelet, $w_t = (a, -b)$ with transform $W(z) = (a - bz)$. Then,

$$\frac{1}{W(z)} = \frac{1}{a - bz} = \frac{1}{a}\left(1 - \frac{b}{a}z\right)^{-1}.$$

If $|b/a| < 1$, we can expand (note that $|z| = 1$) and get

$$\frac{1}{W(z)} = \frac{1}{a}\left\{1 + \frac{b}{a}z + \left(\frac{b}{a}z\right)^2 + \cdots\right\}$$
$$= \text{convergent series}.$$

If $|b/a| > 1$, the series is divergent. Thus, for minimum-phase, $|a| > |b|$. The wavelet (b, a) is maximum-phase. When $a = \pm b$, the transform is $a(1 \pm z)$ and the inverse becomes infinite when $\omega\Delta$ equals $2n\pi$ or $(2n + 1)\pi$, hence the wavelet is not minimum-phase. When $(a - bz)$ is minimum-phase, so is $1/(a - bz)$.

Consider the wavelet $W(z) = (c - z) = c(1 - z/c)$, c being real. Then,

$$W(z) = c[1 - (1/c)\cos\omega\Delta + j(1/c)\sin\omega\Delta],$$

and the phase is $\gamma = \tan^{-1}[\sin\omega\Delta/(c - \cos\omega\Delta)]$. If $W(z)$ is minimum-phase, $|c| > 1$ and the denominator never becomes zero, hence $|\gamma| < \frac{1}{2}\pi$ and the curve of $\text{Im}\{W(z)\}$ versus $\text{Re}\{W(z)\}$ does not enclose the origin (see fig. 15.12a). Also, γ is periodic with frequency values repeating each time $\omega\Delta$ increases by 2π (see fig. 15.12c). If $|c| < 1$, the denominator becomes zero twice as $\omega\Delta$ increases by 2π, hence γ assumes all values between 0 and 2π, that is, the curve of $\text{Im}\{W(z)\}$ versus $\text{Re}\{W(z)\}$ encloses the origin (fig. 15.12b). Each time $\omega\Delta$ increases by 2π, γ also increases by 2π (fig. 15.12c). The phase of the maximum-phase wavelet increases without limit while the phase of the minimum-phase wavelet is always between $-\frac{1}{2}\pi$ and $+\frac{1}{2}\pi$, hence is always less than that of the maximum-phase wavelet.

If $(a_i - b_iz)$, $i = 1, 2, \ldots, n$, are all minimum-phase, the sum $\sum_i(a_i - b_iz)$ may or may not be minimum-phase because $|\sum_i a_i|$ is not necessarily greater than $|\sum_i b_i|$. On the other hand, the product $\Pi_i(a_i - b_iz)$ (corresponding to time-domain convolution) is always minimum-phase because each term, hence the product also, is cyclic with frequency, hence the curve of $\text{Im}\{W(z)\}$ versus $\text{Re}\{W(z)\}$ cannot enclose the origin (if the curve encloses the origin, the phase increases without limit). If all of the terms $(a_i - b_iz)$ in the product are maximum-phase, the product is maximum-phase. If some of the factors are maximum-phase and some minimum-phase, the product is *mixed-phase*.

If $(a_i - b_iz)$ is minimum phase, (a_i/b_i) is a root.

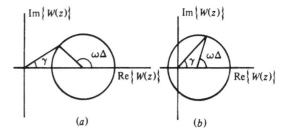

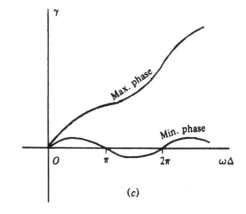

Fig. 15.12   Variation of phase γ for (a) minimum-phase wavelet and (b) maximum-phase wavelet as ω increases from 0 to 2π; (c) graphs of γ versus $\omega\Delta$ for parts (a) and (b).

Hence

$$\prod_{j=0}^{n}(a_i/b_i - z) = 0 \qquad (15.224)$$

This equation has the constant term $(a_1 a_2 \cdots a_n/b_1 b_2 \cdots b_n)$; if we add a constant to the left-hand side of eq. (15.224), the roots are no longer the same, and so the expression on the left-hand side is no longer minimum-phase in general.

Consider the function $\Pi_i(a_i - b_iz)/\Pi_j(a_j - b_jz)$ where all the factors are minimum-phase. The phase of the function is the sum of the phases of the numerator minus the sum of the phases of the denominator; because all of the phases are cyclic, the phase of the function is also cyclic, hence the function is minimum-phase.

Because $a_i - b_iz$ is minimum-phase if $|a_i| > |b_i|$, the root of the equation $a_i - b_iz = 0$, namely, a_i/b_i, lies outside the unit circle. Similarly, the roots of a maximum-phase function lie inside the unit circle. When $\Pi_i(a_i - b_iz)/\Pi_j(a_j - b_jz)$ is minimum-phase, all of the roots of the numerator and denominator lie outside the unit circle; because roots of the denominator are called *poles*, we can say that all roots and poles of a minimum-phase function lie outside the unit circle. (When z is defined as $e^{+j\omega\Delta}$, these rules are reversed.)

The roots $z = 0, \pm1$ follow the preceding rule. If $W(z)$ is multiplied by $z^{\pm m}$, m integral, the wavelet is shifted without change in shape, but the phase is increased by $-jm\omega\Delta$; the roots are those of $W(z)$ plus the factor $z^{\pm m}$ and root $z = 0$. However, the graph of $z^{\pm m} W(z)$ now encloses the origin as in fig. 15.12b, hence $z^{\pm m} W(z)$ is not minimum-phase. The convention when we are determining the phase of an isolated wavelet is to have $t = 0$ for the first nonzero element so that roots $z = 0$ do not occur.

Because the expansion of $1/(1 \pm z)$ is divergent, the expansion of $(1 \pm z) W(z)$ is also divergent, hence the product is not minimum-phase. When a root is only slightly larger than unity, problems are often encountered because the expansion converges slowly.

The terms "minimum-phase" and "maximum-phase" imply a comparison, and in fact they refer to the set of wavelets that have a given frequency spectrum. The four wavelets $(a - bz)$, $(b - az)$, $(\bar{a} - \bar{b}z)$, $(\bar{b} - \bar{a}z)$ all have the same frequency spectrum as is easily verified by multiplying by the conjugate complexes; two of the four are minimum-phase, two maximum. (Other wavelets with the same spectrum can be obtained by multiplying by any complex constant c, where $|c| = 1$, so that there is an infinite number of wavelets with the same spectrum.) When a and b are real, as is most often the case, the four wavelets reduce to two.

(b) Energy relations. The intensity of a wave is proportional to the amplitude squared at any instant, whereas the total energy is proportional to the sum of the amplitudes squared from $t = 0$ up to a given instant. For the wavelet (a, b), the total energy is a^2, $a^2 + b^2$ at $t = 0$ and $t = \Delta$, respectively. For the wavelet (b, a), the values are b^2, $b^2 + a^2$. If (a, b) is the minimum-phase wavelet, the energy builds up faster than for the maximum-phase wavelet (b, a). Because the two wavelets have the same frequency spectrum, the only difference is in the rate of buildup of energy. These results can be extended to more complex wavelets; for example, Robinson (1962) showed that the total energy at a time t for a minimum-phase wavelet is greater than or equal to that for any other wavelet with the same spectrum. Claerbout (1963) showed that the "center of gravity" of the minimum-phase wavelet is closer to $t = 0$ than for any other wavelet with the same spectrum. These principles explain the origin of the term "minimum-delay."

(c) Determining the minimum-phase wavelet for a given spectrum. The spectrum of the wavelet $(a - bz)$, a and b being complex in general, is $(a - bz)(\bar{a} - \bar{b}z^{-1}) = -a\bar{b}z^{-1} + (a^2 + b^2) - \bar{a}bz$ (note that $\bar{z} = z^{-1}$). Thus, the wavelet has the root a/b and the spectrum has this root plus the root $\bar{b}/\bar{a}$ (when a and b are real, the two roots are reciprocals). When a spectrum is of order $2n$, it must have n pairs of roots of the form $(z_i, 1/\bar{z}_i)$. Half the roots will be outside the unit circle, half inside. The minimum-phase wavelet is obtained by multiplying together n factors of the form $(z_i - z)$, for the z_i outside the unit circle. Because other wavelets can be obtained by multiplying together n factors corresponding to roots both inside and outside the unit

circle, it is clear that an autocorrelation function is not unique to one function.

The preceding method requires that we find the factors of the spectrum and this is time-consuming when n is large, hence other methods are used (see Claerbout, 1976: chap. 3). One method utilizes the special properties of the Toeplitz matrix to reduce labor. Let $R(z)$ be a given spectrum and $W(z)$ the minimum-phase wavelet to be determined; $W(z)$ must have an inverse that we denote $V(z)$. Then,

$$R(z) = W(z)W(\bar{z}), \qquad \text{hence} \qquad R(z)V(z) = W(\bar{z}).$$

Because $W(z)$ is minimum-phase, we can write $V(z) = b_0 + b_1z + b_2z^2 + \cdots$.

Moreover, if

$$W(z) = a_0 + a_1z + a_2z^2 + \cdots + a_nz^n,$$

then

$$W(\bar{z}) = \bar{a}_0 + \bar{a}_1\bar{z} + \bar{a}_2\bar{z}^2 + \cdots + \bar{a}_n\bar{z}^n$$

and

$$R(z) = (r_{-n}z^{-n} + \cdots + r_{-1}z^{-1} + r_0 + r_1z + \cdots + r_nz^n).$$

Then,

$$r_{-n}z^{-n} + \cdots + r_{-1}z^{-1} + r_0 + r_1z + \cdots + r_nz^n)$$
$$\times (b_0 + b_1z + b_2z^2 + \cdots)$$
$$= (\bar{a}_0 + \bar{a}_1z^{-1} + \bar{a}_2z^{-2} + \cdots + \bar{a}_nz^{-n}).$$

Equating coefficients for positive powers of z gives:

$$z^0: r_0b_0 + r_{-1}b_1 + \cdots + \quad r_{-n}b_n = \bar{a}_0,$$
$$z^1: r_1b_0 + \quad r_0b_1 + \cdots + r_{-n+1}b_n = 0,$$
$$z^2: r_2b_0 + \quad r_1b_1 + \cdots + r_{-n+2}b_n = 0,$$
$$\vdots$$
$$z^n: r_nb_0 + r_{n-1}b_1 + \cdots + \quad r_0b_n = 0.$$

(We assume that $(n + 1)$ values of b_i are sufficient to give $W(z) = 1/V(z)$ with the required precision; if not, more terms can be obtained by equating coefficients of z^m to zero, $m > n$.) We note that $\bar{a}_0 = 1/\bar{b}_0$, hence we have $(n + 1)$ equations to solve for the $(n + 1)$ unknowns, b_i. The solution can be written in terms of a Toeplitz matrix $\mathscr{R}$ as $\mathscr{R}\mathscr{B} = \mathscr{W}$:

$$\begin{Vmatrix} r_0 & r_1 & \cdots & r_n \\ r_1 & r_0 & \cdots & \cdots \\ \vdots & \vdots & \ddots & \vdots \\ r_n & \cdots & \cdots & r_0 \end{Vmatrix} \begin{Vmatrix} 1 \\ b_1' \\ \vdots \\ b_n' \end{Vmatrix} = \begin{Vmatrix} w \\ 0 \\ \vdots \\ 0 \end{Vmatrix}, \qquad (15.225)$$

where $r_n = r_{-n}$, $b_i' = b_i/b_0$, $w = \bar{a}_0/b_0 = 1/|b_0|^2$. Solving these equations for the $(n + 1)$ unknowns b_i in terms of the known r_i, we find $V(z)$, then $W(z)$ by inverting $V(z)$.

The straightforward solution of eq. (15.225) requires computer time proportional to n^3 and memory proportional to n^2. The *Levinson recursion algorithm*, which we describe now, reduces these quantities by the factor n. The method is based on forming new equations by selecting a $k \times k$ matrix from the upper left-hand corner of $\mathcal{R}$ and the top k rows of $\mathcal{B}$ and $\mathcal{W}$, for example,

$$\begin{Vmatrix} r_0 & r_1 & \cdots & r_{k-1} \\ r_1 & r_0 & \cdots & \cdots \\ \vdots & \vdots & \ddots & \vdots \\ r_{k-1} & \cdots & \cdots & r_0 \end{Vmatrix} \begin{Vmatrix} 1 \\ b_1^* \\ \vdots \\ b_{k-1}^* \end{Vmatrix} = \begin{Vmatrix} w^* \\ 0 \\ \vdots \\ 0 \end{Vmatrix}, \quad (15.226)$$

where the asterisks signify that b_i^* and w^* are different from the corresponding quantities in eq. (15.225) because they satisfy a different set of equations. The Levinson recursion algorithm shows how to obtain the solution for the $(k + 1)$th case when we know the solution for the kth case. Thus, we can start with $k = 1$, that is, $r_0 \times 1 = w$, then use the algorithm to get the solution for $k = 2$, that is, for

$$\begin{Vmatrix} r_0 & r_1 \\ r_1 & r_0 \end{Vmatrix} \begin{Vmatrix} 1 \\ b_1^* \end{Vmatrix} = \begin{Vmatrix} w^* \\ 0 \end{Vmatrix},$$

then continue the process until we get to $k = n + 1$; the final step gives the solution of eq. (15.225) whereas intermediate solutions are discarded.

We start with eq. (15.226) in which all of the b_i^* and w^* are known because we have solved the equation, and we wish to solve the equation for $k + 1$ in terms of these known quantities. We write the following equation that defines the quantity e:

$$\begin{Vmatrix} r_0 & r_1 & \cdots & r_k \\ r_1 & r_0 & \cdots & \cdots \\ \vdots & \vdots & \ddots & \vdots \\ r_k & \cdots & \cdots & r_0 \end{Vmatrix} \begin{Vmatrix} 1 \\ b_1^* \\ \vdots \\ b_{k-1}^* \\ 0 \end{Vmatrix} = \begin{Vmatrix} w^* \\ 0 \\ \vdots \\ 0 \\ e \end{Vmatrix}. \quad (15.227)$$

Obviously,

$$e = r_k + r_{k-1}b_1^* + \cdots + r_1 b_{k-1}^* = \sum_{i=1}^{k} r_i b_{k-i}^*,$$

hence is a known quantity. We now "invert" $\mathcal{B}$ and $\mathcal{W}$ to get

$$\begin{Vmatrix} r_0 & r_1 & \cdots & r_k \\ r_1 & r_0 & \cdots & \cdots \\ \vdots & \vdots & \ddots & \vdots \\ r_k & \cdots & \cdots & r_0 \end{Vmatrix} \begin{Vmatrix} 0 \\ b_{k-1}^* \\ \vdots \\ b_1^* \\ 1 \end{Vmatrix} = \begin{Vmatrix} e \\ 0 \\ \vdots \\ 0 \\ w^* \end{Vmatrix} \quad (15.228)$$

(one can verify that the last two equations are the same by direct expansion). We multiply eq. (15.228)

by a constant c_k and subtract it from eq. (15.227):

$$\begin{Vmatrix} r_0 & r_1 & \cdots & r_k \\ r_1 & r_0 & \cdots & \cdots \\ \vdots & \vdots & \ddots & \vdots \\ r_k & \cdots & \cdots & r_0 \end{Vmatrix} \left(\begin{Vmatrix} 1 \\ b_1^* \\ \vdots \\ b_{k-1}^* \\ 0 \end{Vmatrix} - c_k \begin{Vmatrix} 0 \\ b_{k-1}^* \\ \vdots \\ b_1^* \\ 1 \end{Vmatrix} \right)$$

$$= \begin{Vmatrix} w^* \\ 0 \\ \vdots \\ 0 \\ e \end{Vmatrix} - c_k \begin{Vmatrix} e \\ 0 \\ \vdots \\ 0 \\ w^* \end{Vmatrix}. \quad (15.229)$$

We wish this equation to reduce to the equation for the $(k + 1)$th case, namely,

$$\begin{Vmatrix} r_0 & r_1 & \cdots & r_k \\ r_1 & r_0 & \cdots & \cdots \\ \vdots & \vdots & \ddots & \vdots \\ r_k & \cdots & \cdots & r_0 \end{Vmatrix} \begin{Vmatrix} 1 \\ b_1^{**} \\ \vdots \\ b_k^{**} \end{Vmatrix} = \begin{Vmatrix} w^{**} \\ 0 \\ \vdots \\ 0 \end{Vmatrix} \quad (15.230)$$

Comparing eqs. (15.229) and (15.230), we see that, to get 0 for the bottom element of $\mathcal{W}$, we must have $e - c_k w^* = 0$, that is, $c_k = e/w^*$. Also, w^{**} must equal $(w^* - c_k e) = w^*[1 - (e/w^*)^2]$. Finally, on the left, we have $b_i^{**} = b_i^* - c_k b_{k-1}^*$ (note that $b_k^{**} = -c_k$ because $b_k^* = 0$, $b_0^* = 1$). Finally, all of the quantities, b_i^{**} and w^{**}, in eq. (15.230) can be found in terms of the known solution of eq. (15.226).

(d) Zero-phase and linear-phase wavelets. We note that $(z^n + z^{-n}) = 2 \cos n\omega\Delta$, n integral; because the imaginary part is zero, the function has zero phase. Zero-phase wavelets can be obtained by multiplying pairs of elementary wavelets such as $(1 - az)(az^{-1} - 1) = az^{-1} - (1 + a^2) + az = W_z(z)$. Because the phase is zero, $W_z(z) = W_z(\bar{z})$, hence the spectrum is

$$W_z^2(z) = \{az^{-1} - (1 + a^2) + az\}^2$$
$$= a^2 z^{-2} - 2a(1 + a^2)z^{-1} + (a^4 + 4a^2 + 1)$$
$$- 2a(1 + a^2)z + a^2 z^2.$$

If $|a| < 1$, the minimum-phase wavelet with the same amplitude spectrum is $(1 - az)^2 = 1 - 2az + a^2 z^2$. Therefore, if a minimum-phase wavelet can be written $\Pi_i(1 - a_i z)^m$, where m is a multiple of 2, then an equivalent zero-phase wavelet can be found by replacing each pair of factors, $(1 - a_i z)^2$ by $(1 - a_i z) \times (a_i \bar{z} - 1)$.

A zero-phase wavelet is symmetrical about the origin, hence is neither causal nor physically realizable. The maximum amplitude is at $t = 0$. Roots of a zero-phase wavelet occur in pairs, $(a_i, 1/a_i)$, one of each pair being inside the unit circle, one outside.

If we multiply a zero-phase wavelet by z^n, that is, delay it by n sample intervals, we get

$$W_L(z) = z^n W_z(z).$$

Because $z^n = e^{-jc\omega}$, where c is constant, W_L has linear phase, that is, $\gamma_L = c\omega$. Moreover, $W_L(z)$ is merely $W_z(z)$ displaced n time units and has the same roots (plus $z = 0$) and is symmetrical about $t = n\Delta$. Linear-phase wavelets are physically realizable if they start after $t = 0$. Zero-phase wavelets can be obtained from linear-phase wavelets by time shifting. Linear-phase wavelets are often called "zero-phase."

15.5.7 Integral relations for inverse z-transforms

Although z-transforms and inverse z-transforms can be written down by inspection when g_i is a time series, there are occasions when the equivalent of eq. (15.109) is more convenient. As ω increases from 0 to $2\pi/\Delta$ (or from $-\pi/\Delta$ to $+\pi/\Delta$), z goes once around the unit circle in the clockwise direction (see §15.53). Therefore, we have

$$
\begin{aligned}
g_n &= (1/2\pi) \int_0^{2\pi/\Delta} G(\omega)e^{j\omega n\Delta}\, d\omega \\
&= (1/2\pi) \oint G(z)z^{-n}[e^{j\omega\Delta}\, dz/(-j\Delta)], \\
&= (1/2\pi j\Delta) \oint G(z)z^{-(n+1)}\, dz, \qquad (15.231)
\end{aligned}
$$

where the integration is in the counterclockwise direction.

15.6 Cepstrum analysis

Transformation from the time domain into the frequency domain permits the equivalent of convolution to be carried out by the simpler operation of multiplication. Transformation from the frequency domain into the cepstrum domain permits such operations to be carried out by the even simpler process of addition. Moreover, in some cases, frequencies that overlap in the frequency domain are separated sufficiently in the cepstrum domain so that filtering can be carried out more efficiently (Ulrych, 1971.)

The *cepstrum*, $\hat{g}(\zeta)$, is given by an inverse transform of the log of the frequency spectrum:

$$\hat{g}(\zeta) = (1/2\pi)\int_{-\infty}^{\infty} \ln[G(\omega)]e^{j\omega\zeta}\, d\omega. \quad (15.232)$$

The transformation from the time domain is usually carried out in three steps:

$$
\left.
\begin{aligned}
g(t) &\leftrightarrow G(\omega) = |G(\omega)|e^{j\gamma(\omega)}, \\
\hat{G}(\omega) &= \ln[G(\omega)] = \ln|G(\omega)| + j\gamma(\omega), \\
\hat{G}(\omega) &\leftrightarrow \hat{g}(\zeta) = (1/2\pi)\int_{-\infty}^{\infty} [\ln|G(\omega)| + j\gamma(\omega)]e^{j\omega\zeta}d\omega.
\end{aligned}
\right\}
$$

$$(15.233)$$

Thus, the essential feature that characterizes cepstrum analysis is taking the logarithm before making the inverse transformation. To return to the time domain, the preceding three steps are reversed:

$$
\left.
\begin{aligned}
\hat{g}(\zeta) &\leftrightarrow \hat{G}(\omega) = \int_{-\infty}^{\infty} \hat{g}(\zeta)e^{-j\omega\zeta}\, d\zeta, \\
G(\omega) &= \exp[\hat{G}(\omega)], \\
G(\omega) &\leftrightarrow g(t) = (1/2\pi)\int_{-\infty}^{\infty} G(\omega)e^{j\omega t}\, d\omega.
\end{aligned}
\right\}
$$

$$(15.234)$$

(Other definitions of cepstrum are also used (for example, Ulrich, 1971; Båth, 1974), but the previous definition is the most common in seismic data analysis.)

For a discrete function g_t, we use the z-transform and the earlier steps become

$$
\left.
\begin{aligned}
G(z) &= \sum_i g_i z^i, \\
\hat{G}(z) &= \ln(\Sigma g_i z^i), \\
\hat{g}_\zeta &= (1/2\pi j\Delta)\oint \hat{G}(z)z^{-(\zeta+1)}\, dz,
\end{aligned}
\right\}
$$

$$(15.235)$$

using eq. (15.231). To return to the time domain, we have

$$
\left.
\begin{aligned}
\hat{G}(z) &= \sum \hat{g}_\zeta z^\zeta = \ln[G(z)], \\
G(z) &= \exp[\hat{G}(z)] = \exp\left(\sum \hat{g}_\zeta z^\zeta\right), \\
g_t &\leftrightarrow G(z).
\end{aligned}
\right\}
$$

$$(15.236)$$

The variable ζ is called the *quefrency*, a permutation of the letters in frequency, just as "cepstrum" is a permutation of the letters in "spectrum." The cepstrum can be expressed in terms of its *lampitude* $\hat{a}(\zeta)$ and its *saphe* $\hat{\gamma}(\zeta)$,

$$\hat{g}(\zeta) = \hat{a}(\zeta)e^{j\hat{\gamma}(\zeta)}. \quad (15.237)$$

The equivalent of filtering in the time or frequency domains is called *liftering* when performed in the cepstrum domain.

An essential step in going to the cepstrum domain is finding the phase $\gamma(\omega)$, usually by means of eq. (15.113). Because $\tan(\theta + \pi n) = \tan\theta$, each value calculated for $\gamma(\omega)$ is ambiguous by πn. This ambiguity must be removed before transforming to the cepstrum domain, an operation called "uncracking" the phase ambiguity. One method is to utilize the fact that $\gamma(\omega)$ is continuous and add π to $\gamma(\omega)$ in a trial-and-error approach, the values adopted being those that make the slope of $\gamma(\omega)$ as smooth as possible. An alternative is to take the derivative of the expression for $\gamma(\omega)$ in eq. (15.113b), a procedure that does not introduce the ambiguities:

$$
\begin{aligned}
\frac{d\gamma(\omega)}{d\omega} &= \frac{d}{d\omega}\left[\tan^{-1}\{X(\omega)/R(\omega)\}\right] \\
&= \frac{R(\omega)\,[dX(\omega)/d\omega] - X(\omega)[dR(\omega)/d\omega]}{\{R(\omega)\}^2 + \{X(\omega)\}^2}.
\end{aligned}
$$

$$(15.238)$$

Stoffa, Buhl, and Bryan (1974) discuss the deconvolution of marine reverberatory noise, starting by weighting an observed time series g_t to give another series g_t, whose z-transform is

$$G(z) = \sum_i g_j a^i z^i,$$

where a is a constant slightly smaller than one that makes g_i minimum-phase (see also §9.4). We may then associate the slowly varying components with the source and the reverberation, and the rapidly varying components with the reflector series. The nonlinear operation of taking the logarithm results in undersampling of $G(z)$, but the weighting lessens alias effects. (Stoffa et al. suggest starting with $a = 0.94$, presumably for $\Delta = 0.004$ s, and then increasing a until aliasing begins to create problems, to ascertain the largest value of a that can be used.)

15.7 Filtering

15.7.1 Introduction

Filters are devices that pass or fail to pass information based on some measurable discriminant. Usually, the discriminant is the frequency and the filter alters the amplitude and/or phase spectra of signals that pass through it. Analog filtering was discussed briefly in §7.6.3 and digital filtering in §9.5. Although some of the following discussion is applicable to analog filtering, emphasis will be on digital filtering. The literature dealing with filtering is vast; many references are given in Blackman and Tukey (1958); Lee (1960); Finetti, Nicolich, and Sancin (1971); Båth (1974); Kulhánek (1976). The following discussion is based to a considerable extent on Kulhánek.

Most of the filters with which we deal are assumed to be linear to facilitate calculation, and so we here assume linearity. Digital filters are more versatile than analog filters, partly because we are not restricted to physically realizable components such as capacitors, inductances, and resistances, partly because with digital filters we know future values of the signal as well as the present and past values upon which analog filters must act.

The output of a physical system cannot precede the input; thus, when $g(t)$ and $h(t)$ are respectively the input and output signals, if $g(t) = 0$ for $t < 0$, then $h(t) = 0$, $t < 0$ for analog filters. This is not necessarily the case for digital filters.

A filter is *stable* if the output is finite for any finite input. The output in the time domain for a linear system is given by eq. (15.202). A filter $f(t)$ will be stable provided that

$$\int_{-\infty}^{\infty} |f(t)| \, dt < +\infty. \qquad (15.239)$$

This requires that $f(t)$ be finite everywhere and approach zero as t approaches $\pm\infty$ (Treitel and Robinson, 1964).

We may write eq. (15.201) in the form

$$F(s) = H(s)/G(s).$$

The right-hand side is almost invariably expressible as the ratio of two polynomials in s with the numerator of lower order than the denominator (if this is not the case, long division leads to terms in s^n, where n is positive, and these usually give rise to instability). Applying the method of partial fractions, we can write

$$F(s) = \sum_{i=1}^{n} A_i/(s - s_i),$$

where s_i is one of the n roots of the equation $G(s) = 0$. Taking the inverse transform gives

$$f(t) = \sum_{i=1}^{n} A_i \exp(s_i t) = \sum_{i=1}^{n} A_i \exp[(a_i + jb_i)t]$$

because in general s_i is complex. For $f(t)$ to remain finite as t approaches infinity, all of the a_i must be negative; thus, the roots s_i, usually called the *poles* of $H(s)$, must lie on the left-hand side of the complex plane.

The preceding discussion applies equally well to digital filters, in which case eq. (15.239) becomes

$$\sum_{-\infty}^{\infty} |f_t| < \infty. \qquad (15.240)$$

15.7.2 Filter synthesis and analysis

Filters can be designed either by requiring that a given input produce a desired output (*filter synthesis*) or by investigating the effects of a given filter on various input signals (*filter analysis*). As an example of filter synthesis, we design a filter to transform a sampled input g_i into a desired output h_t, where

$$g_t = \sum_{k=0}^{m-1} g_k \, \delta(t - k\Delta), \qquad h_t = \sum_{k=0}^{n-1} h_k \, \delta(t - k\Delta).$$

For simplicity, we assume $m = n$, zeros being added to g_t or h_t to achieve this. From eq. (15.201) we obtain

$$F(z) = \mathcal{H}(z)/G(z). \qquad (15.241)$$

Because $\mathcal{H}(z)$ and $G(z)$ are polynomials in z, long division gives a polynomial that may be of infinite order. In practice, infinite polynomials must be truncated to a reasonable number of terms. Once we have $F(z)$, we can get f_t; then the output y_t for any input x_t can be found by convolution:

$$y_t = f_t * x_t.$$

We could also get y_t by using eq. (15.241) to write

$$Y(z) = F(z)X(z) = \{\mathcal{H}(z)/G(z)\}X(z),$$

or

$$Y(z)G(z) = X(z)\mathcal{H}(z);$$

hence

$$y_t * g_t = x_t * h_t,$$

or by eq. (15.211),

$$\sum_{k=0}^{r} y_{r-k} g_k = \sum_{k=0}^{r} x_{r-k} h_k, \qquad r = 0, 1 \ldots, n - 1,$$

which is a set of n equations in the n unknowns, y_k. Setting $r = 0, 1, \ldots$, we obtain the solutions

$$y_0 = x_0 \hbar_0 / g_0,$$
$$y_1 = (x_1 \hbar_0 + x_0 \hbar_1)/g_0 - y_0 g_1 / g_0,$$
$$\vdots$$

hence

$$y_r = \left(\sum_{k=0}^{r} x_{r-k} \hbar_k / g_0 \right) - \left(\sum_{k=1}^{r} y_{r-k} g_k / g_0 \right). \quad (15.242)$$

Because the system is linear, there is no loss of generality by setting $g_0 = +1$. We also note that the initial value of k in the second summation means that g_t has been delayed one time unit (see § 15.5.3, also problem 15.38). Therefore, we can write

$$y_t = x_t * \hbar_t - (y_t * g_t)', \quad (15.243)$$

where the prime means that g_t is delayed one unit.

Equations (15.242) and (15.243) give a solution in terms of present (x_r) and past inputs (x_0 to x_{r-1}) and past outputs (y_0 to y_{r-1}). Filters of this type are *predictive, recursive,* or *feedback.* Such equations can be solved iteratively, y_0 being found first, then y_1, y_2, and so on, a type of calculation convenient with digital computers.

As an example of filter analysis, we take a specific case of eq. (15.242):

$$y_r = ax_r - by_{r-1}, r = 0, 1, 2, \ldots, n - 1 \quad (15.244)$$

(corresponding to $g_t = [1, b, 0, 0, \ldots]$, $\hbar_t = [a, 0, 0, \ldots]$), a and b being real, and we determine the filter properties. Taking z-transforms of the sequences obtained by giving r the values $0, 1, 2, \ldots, n - 1$ in eq. (15.244), we get

$$Y(z) = aX(z) - bzY(z),$$

where the factor z takes into account the delay of g_t by one unit. Solving for $Y(z)$ gives

$$Y(z) = aX(z)/(1 + bz),$$

$$F(z) = Y(z)/X(z) = a/(1 + bz). \quad (15.245)$$

Provided $|bz| < 1$, we can expand the right-hand side and obtain

$$F(z) = a(1 + bz)^{-1} = a \sum_{r=0}^{\infty} (-bz)^r, \quad (15.246)$$

hence

$$f_t = (a, -ab, ab^2, -ab^3, \ldots).$$

Equation (15.240) shows that $|b| < 1$ for the filter to be stable. The series in eq. (15.246) may converge very slowly and a recursive solution may be better than finding f_t as before, then calculating $f_t * x_t$.

A second-order recursive filter can be defined by

$$y_r = ax_r - by_{r-1} - cy_{r-2}. \quad (15.247)$$

Then,

$$Y(z) = aX(z) - bzY(z) - cz^2 Y(z),$$

$$F(z) = \frac{Y(z)}{X(z)} = \frac{a}{1 + bz + cz^2} = \frac{a}{c(z - z_1)(z - z_2)}$$

$$= \frac{a/cz_1 z_2}{(1 - z/z_1)(1 - z/z_2)}, \quad (15.248)$$

where z_1, z_2 are roots of the denominator.

Reference to eq. (15.246) shows that the second-order recursive filter is equivalent to two first-order filters in series, the transfer functions being

$$\left(\frac{1}{z_i} \frac{a^{1/2}}{c} \right) \Big/ \left(1 - \frac{z}{z_i} \right), \quad i = 1, 2.$$

Comparison with eq. (15.240) shows that the filter will be stable provided that $|z/z_i| < 1$, that is, $|b \pm (b^2 - 4c)^{1/2}| > 2|c|$. In general, recursive filters of any order can be replaced by first-order filters in series.

Equations (15.247) and (15.248) can be generalized as follows:

$$y_r = x_r - f_1 y_{r-1} - \cdots - f_r y_0,$$
$$r = 0, 1, 2, \ldots, n - 1,$$

where we have set $a = 1 = f_0$ (this can always be done by introducing a scale factor). Then,

$$x_r = y_r + f_1 y_{r-1} + \cdots + f_r y_0,$$

and on taking z-transforms, we have

$$X(z) = Y(z)(1 + f_1 z + \cdots + f_{n-1} z^{n-1})$$

or

$$Y(z) = X(z)/(1 + f_1 z + \cdots + f_{n-1} z^{n-1}). \quad (15.249)$$

Thus, the general form of the feedback filter of order n is

$$F(z) = 1 \Big/ \sum_{r=0}^{n} f_r z^r. \quad (15.250)$$

15.7.3 Frequency filtering

Frequency filters are classified as *low-pass, high-pass,* or *band-pass* according as they discriminate against frequencies above or below a certain limiting frequency or outside of a given band of frequencies. "Ideal" filters of these types are the following:

Low-pass:

$$F_L(\omega) = +1, |\omega| < |\omega_0|,$$
$$= 0, |\omega| > |\omega_0|; \quad (15.251)$$

High-pass:

$$f_H(\omega) = 0, |\omega| < |\omega_0|,$$
$$= +1, |\omega| > |\omega_0|; \quad (15.252)$$

Band-pass:

$$F_B(\omega) = +1, |\omega_1| < |\omega| < |\omega_2|,$$
$$= 0, |\omega_1| > |\omega| \text{ or } |\omega| > |\omega_2|. \quad (15.253)$$

These filters are discontinuous at ω_0, ω_1, ω_2. The discontinuities will cause some ringing. Obviously, a band-pass filter is equivalent to a low-pass filter with

$|\omega_0| = |\omega_2|$ in series with a high-pass filter with $|\omega_0| = |\omega_1|$.

The low-pass filter can be obtained from eq. (15.152)

$$F_L(\omega) = \text{box}_{2\omega_0}(\omega) \leftrightarrow (\omega_0/\pi) \text{ sinc } (\omega_0 t) = f_L(t).$$
(15.254)

For digital functions, provided $|\omega_0| < \omega_N = \pi/\Delta$, this becomes

$$f_t^L = (1/\pi) \sum_{n=-\infty}^{\infty} \omega_0 \text{ sinc } (n\omega_0 \Delta). \quad (15.255)$$

Because $F_L(\omega)$ does not change the signal amplitude or phase within the passband, the filter is distortionless.

For continuous functions a high-pass filter with cutoff frequency ω_0 is given by

$$f_H(t) = (1/2\pi) \int_{-\infty}^{-\omega_0} e^{j\omega t} \, d\omega + (1/2\pi) \int_{+\omega_0}^{\infty} e^{j\omega t} \, d\omega$$
$$= (1/\pi) \int_{\omega_0}^{\infty} \cos \omega t \, d\omega \quad (15.256)$$

because $e^{j\omega t} = \cos \omega t + j \sin \omega t$ and $\sin \omega t$ is odd. Then

$$f_H(t) = (1/\pi t) \sin \omega t \big|_{+\omega_0}^{\infty}$$
$$= -(\omega_0/\pi) \text{ sinc } (\omega_0 t) \quad (15.257)$$

because $\lim_{\omega \to \infty} (\sin \omega t) = 0$ (see Papoulis, 1962: 278). Thus, $f_H(t) = -f_L(t)$ provided both filters have the same cutoff frequency ω_0.

For a digital filter, the response should be zero above the Nyquist frequency to avoid aliasing. Changing the limit in eq. (15.257) from $+\infty$ to $+\omega_N$ gives

$$f_H(t) = (1/\pi t)(\sin \omega_N t - \sin \omega_0 t). \quad (15.258)$$

Changing to digital functions, because $\omega_N n\Delta = n\pi$,

$$f_t^H = (1/\pi) \sum_{n=-\infty}^{\infty} [\omega_N \text{ sinc } (n\omega_N \Delta) - \omega_0 \text{ sinc } (n\omega_0 \Delta)]$$
$$\left. \begin{array}{ll} = -(\omega_0/\pi) \sum \text{ sinc } (\omega_0 n\Delta), & n \neq 0, \\ = (1/\pi)(\omega_N - \omega_0), & n = 0. \end{array} \right\} \quad (15.259)$$

As for $f_H(t)$, $f_t^H = -f_t^L$ (except at $t = 0$) when both filters have the same cutoff frequency, ω_0. Thus, the design of high-pass and band-pass filters is essentially the same as that of low-pass filters.

To achieve an ideal low-pass filter requires an infinite series for f_t^L, which is impossible. The result of using a finite series for f_t^L, is to introduce ripples, both within and without the passband, the effect being especially noticeable near the cutoff frequency (Gibbs' phenomenon; see §15.2.7). These effects result from the discontinuity and can be partially overcome by multiplying f_t^L by a smoothing window function (see §15.7.5).

A noisy signal may be cross-correlated with a Vibroseis-type signal,

$$g_v(t) = \sin \{\omega_0 + (\omega_1 - \omega_0)t/L\}t, \qquad 0 < t < L,$$
$$= 0, \qquad t < 0 \text{ and } t > L,$$

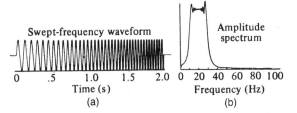

Fig. 15.13 Chirp filter. (After Kulhánek, 1976.) (a) Impulse response and (b) amplitude spectrum. The frequency increases linearly from 10 to 30 Hz in 2 s.

where ω_0, ω_1 are positive constants (compare eq. (7.5)):

$$G_v(\omega) \approx \text{constant}, \qquad \omega_0 < \omega < \omega_1,$$
$$\approx 0, \qquad \omega < \omega_0, \omega > \omega_1.$$

Such an operation, called *chirp filtering*, is roughly equivalent to band-pass filtering, as shown in fig. 15.13.

If $g(t)$ is the input to a filter whose impulse response, $f_M(t)$, is given by

$$f_M(t) = g(-t),$$

then clearly $F_M(\omega) = \overline{G(\omega)}$ and the output of the filter is given by

$$H(\omega) = F_M(\omega)G(\omega) = |G(\omega)|^2,$$

hence $h(t) =$ autocorrelation of $g(t)$. Filters of this type are called *matched* or *conjugate filters*. When the input consists of $g(t)$ plus random noise, the output is mainly the autocorrelation of $g(t)$ because the cross-correlation of $g(t)$ and the noise will be approximately zero for all shifts.

15.7.4 Butterworth filters

The *Butterworth filter* is a common form of low-pass filter; it can be defined by

$$|F(\omega)|^2 = 1/[1 + (\omega/\omega_0)^{2n}], \quad (15.260)$$

where ω_0 is the "cutoff" frequency, and n determines the sharpness of the cutoff. Curves of $|F(\omega)|$ for various values of n are shown in fig. 15.14.

To investigate the stability of the filter, we use a Laplace-transform definition:

$$|F(s)|^2 = 1/[1 + (-1)^n s^{2n}], \quad (15.261)$$

where $s = \sigma + j(\omega/\omega_0)$. This function has no zeros, but has $2n$ poles given by the roots (see fig. 15.3) of

$$s^{2n} = -1, \; n \text{ even},$$
$$= +1, \; n \text{ odd}.$$

These roots are of the form $(\pm a \pm jb)$, a and b being real and positive. Thus, the roots are symmetrical about the real and imaginary axes. When n is odd, two roots reduce to ± 1. Because $|F(s)|^2 = F(s)\overline{F(s)}$, if the roots $-a \pm jb$ (and -1 when n is odd) are assigned to

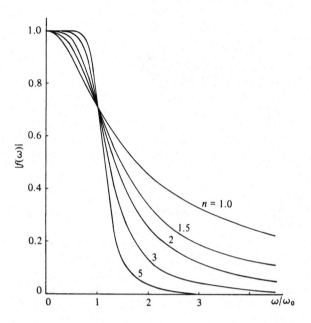

Fig. 15.14   Amplitude response of a Butterworth filter.

$F(s)$, and $+a \pm jb$ (and $+1$) to $\overline{F(s)}$, $F(s)$ is stable. For the nth order digital filter,

$$F_n(s) = [(s + 1)(s + a_1 + jb_1)(s + a_1 - jb_1)$$
$$\times (s + a_2 + jb_2)(s + a_2 - jb_2) \cdots]^{-1}$$

(the factor $(s + 1)$ is omitted when n is even).

The expression for $F_n(s)$ can be expressed as partial fractions (§15.1.9), transformed to the time domain and then to the z-domain to get the digital filter. Thus, for $n = 2$, the roots of $s^4 = -1$ are $(\pm 1 \pm j)/(2)^{1/2}$; hence

$$F_2(s) = \frac{1}{[s + (1 + j)/\sqrt{2}][s + (1 - j)/\sqrt{2}]}$$
$$= \frac{j}{\sqrt{2}}\left[\frac{1}{s + (1 + j)/\sqrt{2}} - \frac{1}{s + (1 - j)/\sqrt{2}}\right].$$

Using eq. (15.182) and writing $t/\sqrt{2} = t^*$, we get

$$f(t) = (j/\sqrt{2})[e^{-(1+j)t^*} - e^{-(t-j)t^*}]\text{step}(\sqrt{2}t^*)$$
$$= (\sqrt{2}e^{-t^*} \sin t^*) \text{ step}(\sqrt{2}t^*).$$

Applying eq. (15.218) and the results of problem 15.27, we obtain

$$F(z) = \sqrt{2}\left[\frac{ze^{-\Delta/\sqrt{2}}\sin (\Delta/\sqrt{2})}{1 - 2ze^{-\Delta/\sqrt{2}} \cos (\Delta/\sqrt{2}) + z^2 e^{-\Delta\sqrt{2}}}\right].$$

Equation (15.260) gives no information about the phase characteristics of the filter. We can construct Butterworth filters with different phase characteristics, for example,

$$F_n(\omega) = \frac{e^{-j\omega t_0}}{[1 + (\omega/\omega_0)^{2n}]^{1/2}};$$

for $n = 1/2$,

$$F_{1/2}(\omega) = \frac{1}{1 + j(\omega/\omega_0)} = \frac{1 - j(\omega/\omega_0)}{1 + (\omega/\omega_0)^2}.$$

The first filter has a linear phase and the second has the phase

$$\gamma(\omega) = -\tan^{-1}(\omega/\omega_0).$$

15.7.5 Windows

We often wish to select a portion of a signal for study, or we may with to "smooth" a function such as a transform. We can achieve these objectives by multiplying the signal or transform by a *window* or *gate*, a function that varies in a more-or-less convenient manner within an interval and is zero outside this interval. We represent a window by $w(t) \leftrightarrow W(\omega)$, the symbolism emphasizing that we can think of the window as being in either the time or frequency domain ("window" is also used in a third sense to denote merely an interval of time, especially an interval in which data can be recorded free from interference by noise such as ground roll). The result of applying a window to a signal, $g(t)$, in the time domain is to give

$$h_w(t) = g(t)w(t) \leftrightarrow (1/2\pi)G(\omega) * W(\omega).$$

Using a time-domain window is a form of frequency filtering so that the transform of the part of the signal selected by the window is distorted in both amplitude and phase in comparison with $G(\omega)$.

The following list includes the more commonly used time-domain windows (see Bàth, 1974: 157–64). The formulas give $w(t)$ in the interval $0 < |t| < T$, $w(t)$ being zero for $|t| > T$ (except for the Gaussian window, (h), where the range is $\pm\infty$). Obviously, similar windows can be applied in the frequency domain, the inverse transforms being obtained using eq. (15.140).

(a) Boxcar window:

$$w(t) = \text{box}_{2T}(t) \leftrightarrow W(\omega) = 2T \text{ sinc } \omega T. \qquad (15.262)$$

(b) Sinc window:

$$\left.\begin{array}{l} w(t) = \text{sinc } (\pi t/T) \leftrightarrow W(\omega), \\ W(\omega) = \frac{T}{\pi} \sum_{n=0}^{\infty} \frac{(-1)^n}{(2n + 1)(2n + 1)!} \\ \qquad \times [(\pi - \omega T)^{2n+1} + (\pi + \omega T)^{2n+1}]. \end{array}\right\} \quad (15.263)$$

(c) Fejér kernel window:

$$\left.\begin{array}{l} w(t) = \text{sinc}^2(\pi t/T) \leftrightarrow W(\omega), \\ W(\omega) = \frac{T}{2\pi^2} \sum_{n=0}^{\infty} \frac{(-1)^n}{(2n + 1)(2n + 1)!} \\ \qquad \times [(2\pi + \omega T)^{2n+2} \\ \qquad - 2(\omega T)^{2n+2} + (2\pi - \omega T)^{2n+2}]. \end{array}\right\} \quad (15.264)$$

(d) Cosine window:

$$w(t) = +1, \qquad 0 < |t| < 4T/5,$$
$$= \tfrac{1}{2} + \tfrac{1}{2}\cos(5\pi t/T), \; 4T/5 < |t| < T,$$
$$w(t) \leftrightarrow W(\omega) = \frac{\sin \omega T + \sin 4\omega T/5}{\omega[1 - (\omega T/5\pi)^2]}.$$

(15.265)

(e) Hanning window:

$$w(t) = \tfrac{1}{2} + \tfrac{1}{2}\cos(\pi t/T),$$
$$w(t) \leftrightarrow W(\omega) = T\{\text{sinc } \omega T + (2\omega T \sin \omega T)/[\pi^2 - (\omega T)^2]\}.$$

(15.266)

(f) Hamming window:

$$w(t = 0.54 + 0.46 \cos(\pi t/T),$$
$$w(t) \leftrightarrow W(\omega) = T\{1.08 \text{ sinc } \omega T + (0.92\omega T \sin \omega T)/[\pi^2 - (\omega T)^2]\}.$$

(15.267)

(g) Triangular window:

$$w(t) = (1 - |t|/T) \leftrightarrow W(\omega) = T \text{ sinc}^2(\omega T/2).$$

(15.268)

(h) Gaussian window:

$$w(t) = \exp(-at^2), \qquad a > 0,$$
$$\leftrightarrow W(\omega) = \sqrt{\pi/a} \exp(-\omega^2/4a).$$

(15.269)

Combinations of windows are also used; for example, the discontinuous sides of a boxcar can be modified with a cosine taper. A common technique is to apply a window, for example, a boxcar, in the time domain, then smooth the resulting transform by applying a second window in the frequency domain.

The effect on the spectrum of $g(t)$ of applying a window in the time domain is determined by $W(\omega)$. Thus, curves of $W(\omega)$ give some idea of the effect of using the window. In general, the wider the window and the gentler the fall-off, the less effect on the spectrum. Båth (1974: 164–71), Blackman and Tukey (1958), and Kurita (1969) discuss practical aspects of "window carpentry."

15.7.6 Optimum filters

(a) Introduction. The filters discussed up to this point have been based on periodic or aperiodic inputs. If the input noise is a stationary random function, we may design a filter that will give an "optimum" output according to some criterion. Among the most important and most widely used of such criteria is the *Wiener* or *least-squares* criterion. To apply this criterion, we compare the output of the filter with some "desired" output, the difference being the "error" in the output; we then design the filter to minimize the power (or energy) of the error by applying the principle of least squares. The Wiener criterion is also

called the ℓ_2 *norm* (see Claerbout, 1976: 121, 123); it provides the maximum-likelihood estimate if the errors have a *Gaussian* or *normal distribution,* where $P(e_j)$, the probability of an error e_j, equals

$$P(e_j) = [1/\sigma(2\pi)^{1/2}] \exp[-\tfrac{1}{2}(e_j/\sigma)^2], \quad (15.270)$$

where σ is the standard deviation (that is, the square root of the variance).

Another criterion sometimes used to find f_t is the ℓ_1 *norm* that minimizes $\Sigma|e_j|$ (Claerbout and Muir, 1973; Claerbout, 1976: 123; Taylor, 1981). It provides the maximum-likelihood estimate if the errors have a *Laplacian* or *one-sided exponential distribution,* where the probability of an error e_j is

$$P(e_j) = (1/\mu) \exp[-e_j/\mu], \qquad e_j \geq 0$$
$$= 0, \qquad e_j < 0. \qquad (15.271)$$

where μ is the mean absolute error and the variance is μ^2. Use of the ℓ_1 norm is less sensitive to errors than the Wiener or ℓ_2 norm.

The ℓ_4 *norm* that minimizes Σe_j^4 is used in minimum-entropy filtering (§15.7.6e). *Parsimonious deconvolution* (Postic, Fourmann, and Claerbout, 1980) minimizes $(\Sigma e_j^p)^{1/p}/(\Sigma e_j^q)^{1/q}$, where p is very slightly larger than q. Occasionally, the *minimax,* Chebychev, or ℓ_∞ *norm* is used, for example, in array design (Rietsch, 1979).

(b) Wiener (least-squares) filtering. Let us write for he input

$$g(t) = g_s(t) + g_n(t), \qquad (15.272)$$

where the subscripts refer to "signal" and "noise." If $h(t)$ and $\hbar(t)$ denote the "actual" and "desired" outputs, their difference, the error, is $e(t)$. Then, the energy of the error, E, is the sum of the squares of $e(t)$. For continuous functions, this gives

$$E = \lim_{T\to\infty} (1/2T) \int_{-T}^{T} [h(t) - \hbar(t)]^2 \, dt$$
$$= \lim_{T\to\infty} (1/2T) \int_{-T}^{T} \left\{\left[\int_{-\infty}^{\infty} f(\tau)g(t-\tau)\,d\tau\right] - \hbar(t)\right\}^2 dt$$
$$= \lim_{T\to\infty} (1/2T) \left\{\int_{-T}^{T} \int_{-\infty}^{\infty} f(\tau)g(t-\tau)\,d\tau \right.$$
$$\times \int_{-\infty}^{\infty} f(\sigma)g(t-\sigma)\,d\sigma\right] dt$$
$$- 2\int_{-T}^{T} \left[\int_{-\infty}^{\infty} f(\tau)g(t-\tau)\,d\tau\right]\hbar(t)\,dt$$
$$+ \int_{-T}^{T} \hbar^2(t)\,dt\right\},$$

where the square of the integral giving $h(t)$ has been written as the product of two integrals that are identical except for the dummy variables of integration. Interchanging the order of integration gives

$$E = \int_{-\infty}^{\infty} f(\tau)\, d\tau \int_{-\infty}^{\infty} f(\sigma)\, d\sigma \left[\lim_{T\to\infty} (1/2T)\right.$$

$$\times \left.\int_{-T}^{T} g(t-\tau)g(t-\sigma)\, dt\right]$$

$$- 2\int_{-\infty}^{\infty} f(\tau)\, d\tau \left[\lim_{T\to\infty} (1/2T)\int_{-T}^{T} \hbar(t)g(t-\tau)\, dt\right]$$

$$+ \lim_{T\to\infty} (1/2T)\int_{-T}^{T} \hbar(t)^2\, dt$$

$$= \int_{-\infty}^{\infty} f(\tau)\, d\tau \int_{-\infty}^{\infty} f(\sigma)\phi_{gg}(\tau-\sigma)\, d\sigma$$

$$- 2\int_{-\infty}^{\infty} f(\tau)\phi_{g\hbar}(\tau)\, d\tau + \phi_{\hbar\hbar}(0),$$

using eqs. (15.166) and (15.169).

Because we have specified $g(t)$ and $\hbar(t)$, E is a function of $f(t)$ only, and so the problem reduces to that of finding the function $f(t)$ that will minimize E. Determination of the form of $f(t)$ in general involves the calculus of variations and we only state the result:

$$\int_{-\infty}^{\infty} f(\sigma)\phi_{gg}(\tau-\sigma)\, d\sigma = \phi_{g\hbar}(\tau), \qquad \tau \geq 0. \quad (15.273)$$

This integral equation, known as the *Wiener–Hopf equation,* holds for a causal linear system, the function $f(t)$ satisfying this equation being the impulse response of the desired optimum filter. The solution of eq. (15.273) is lengthy and complicated, mainly because of the requirement that $\tau \geq 0$. For details, the reader is referred to Lee (1960: 360–7, 389–92).

(c) Prediction-error filtering. Consider a filter $f_t = f_1$, $f_2, \ldots, f_n$ that is designed to predict the causal time series $g_t = g_0, g_1, \ldots, g_m, m > n$, one time unit ahead based on current and past values of g_t. For example, at $t = 3\Delta$, the filter predicts g_4 based on the current value g_3 and past values g_0, g_1, g_2. In general, the predicted value of g_j is

$$g_j = f_1 g_{j-1} + f_2 g_{j-2} + \cdots + f_j g_0$$

$$= \sum_{k=1}^{j} f_k g_{j-k} = \sum_{k=1}^{n} f_k g_{j-k}, \qquad j = 1, 2, \ldots, m, \quad (15.274)$$

where we can use the upper limit n because $g_{j-k} = 0$, $k > j$. Note that the foregoing implies that the predicted value of g_0 is zero. The error in the prediction of g_j is e_j, where

$$e_j = \sum_{k=1}^{n} f_k g_{j-k} - g_j = \sum_{k=0}^{n} f_k g_{j-k}, \quad (15.275)$$

where $f_0 = -1$. We call the filter $-1, f_1, f_2, \ldots, f_n$ the *prediction-error filter of length $(n+1)$ for unit prediction distance.*

In the foregoing section, the prediction distance was one unit, but it is possible to design filters for prediction distance (*span*) p, p integral (Robinson and

Treitel, 1980). Moreover, the previous discussion dealt with prediction based on current and past values, that is, *forward prediction.* Whenever the entire set of values of g_t has been recorded, *backward prediction* based on future values is also possible, as is prediction based on combinations of the two. The data for adjacent traces may also be available and these can be used in multichannel prediction methods (Claerbout, 1976: 139–40).

The prediction concept implies that all of the data involved are parts of the same ensemble, hence the statistics of the ensemble are central to the prediction concept. Whether or not the ensemble is ergodic and stationary (§15.2.12) and the nature of the distribution are clearly relevant to the prediction method.

If we use the Wiener (least-squares) criterion to determine the filter f_t, we obtain the following normal equations (see problem 15.43a, also compare with eqs. (9.73) and (15.27) to (15.29)):

$$\sum_{k=1}^{n} f_k \phi_{gg}(r-k) = \phi_{gg}(r), \qquad r = 1, 2, \ldots, n. \quad (15.276)$$

A simple expression can be obtained for the error power E by using eqs. (15.275) and (15.276). Noting that $e_0 = -g_0$, we have

$$E = \sum_{j=0}^{m} e_j^2 = \sum_{j=0}^{m} \left(\sum_{k=1}^{n} f_k g_{j-k} - g_j\right)^2$$

$$= \sum_{j=0}^{m} \left[\left(\sum_{k=1}^{n} f_k g_{j-k} \sum_{\ell=1}^{n} f_\ell g_{j-\ell}\right)\right.$$

$$\left. - 2g_j\left(\sum_{k=1}^{n} f_k g_{j-k}\right) + g_j^2\right],$$

where we use different summation indices k, ℓ to get the square of the first term. Interchanging the order of summation gives

$$E = \sum_k f_k\left[\sum_\ell f_\ell\left(\sum_j g_{j-k} g_{j-\ell}\right)\right]$$

$$- 2\sum_k f_k\left(\sum_j g_j g_{j-k}\right) + \phi_{gg}(0),$$

$$= \sum_k f_k\left[\sum_\ell f_\ell \phi_{gg}(k-\ell)\right] - 2\sum_k f_k \phi_{gg}(-k) + \phi_{gg}(0)$$

$$= \sum_k f_k \phi_{gg}(k) - 2\sum_k f_k \phi_{gg}(-k) + \phi_{gg}(0)$$

on using the normal equations. Recalling that $f_0 = -1$, we have

$$E = -\sum_{k=0}^{n} f_k \phi_{gg}(k). \quad (15.277)$$

Equations (15.276) and (15.277) can be combined to give the following matrix equation

$$\begin{Vmatrix} \phi_{gg}(0) & \phi_{gg}(-1) & \cdots & \phi_{gg}(-n) \\ \phi_{gg}(1) & \phi_{gg}(0) & \cdots & \phi_{gg}(1-n) \\ \vdots & \vdots & \ddots & \vdots \\ \phi_{gg}(n) & \phi_{gg}(n-1) & \cdots & \phi_{gg}(0) \end{Vmatrix} \begin{Vmatrix} f_0 \\ f_1 \\ \vdots \\ f_n \end{Vmatrix} = \begin{Vmatrix} -E \\ 0 \\ \vdots \\ 0 \end{Vmatrix}$$

$$(15.278)$$

(d) Maximum-entropy filtering. In thermodynamics, entropy is a measure of the disorder (unpredictability) of molecular motion. In information theory, Shannon and Weaver (1949) regard *entropy* as a measure of the unpredictability of a time series. The amount of information that can be extracted increases with the entropy. At one extreme, a perfectly predictable series, such as a sine wave, bears no information and at the other extreme white noise is completely unpredictable and hence potentially carries maximum information. *Maximum-entropy filtering* attempts to produce a filtered output that is as unpredictable as possible while still having the same autocorrelation function as the original time series, that is, of all the time series that have a given autocorrelation function, maximum-entropy filtering selects the one that has the maximum unpredictability.

We could regard the number of digits required to encode information as a measure of entropy. For example, if we have four equally probable events, the various possibilities can be encoded as 00, 01, 10, 11, which requires two digits where $2 = -\log_2(1/4)$; for eight equally probable events we would need three digits ($3 = -\log_2(1/8)$). In general, for equally probable events, the entropy is measured by $-\log_2(1/P)$, where P is the probability of each event. When the events are not equally probable, the entropy S_n is the average:

$$S_n = \sum_j [\log (1/P_j)]/(1/P_j) = -\sum_{j=1}^{n} P_j \log P_j. \quad (15.279)$$

(The base of the logarithms is arbitrary except when comparing entropies with different bases.) For a signal of infinite length, we define the entropy density S as

$$S = \lim_{n\to\infty} [S_n/(n + 1)] \quad (15.280)$$

The matrix ϕ_{gg} in eq. (15.278) is of Toeplitz form and Smylie, Clarke, and Ulrych (1973) deduce the relation

$$E_\infty = (\omega_N/\pi) \exp\left\{(1/2\omega_N)\int_{-\omega_N}^{+\omega_N} \ln [\Phi_\infty(\omega)] d\omega\right\}, \quad (15.281)$$

where E_∞ = prediction-error power for $n = \infty$, ω_N = Nyquist frequency, and $\Phi_\infty(\omega)$ is the spectral density of g_t, that is,

$$\Phi_\infty(\omega) = \Phi_\infty(z) = \sum_{k=-\infty}^{+\infty} \phi_{gg}(k)z^k. \quad (15.282)$$

Smylie et al. (1973) show that for n infinite,

$$S = \tfrac{1}{2} \ln E_\infty$$
$$= \tfrac{1}{2} \ln (\omega_N/\pi) + (1/4\omega_N)\int_{-\omega_N}^{+\omega_N} \ln [\Phi_\infty(\omega)] d\omega. \quad (15.283)$$

In practical problems, we work with a finite signal g_t, which we assume to be a sample of one member of an ensemble. Although we have only a finite number of values of $\phi_{gg}(j)$, theoretically an infinity of values of $\phi_{gg}(j)$ exists outside the range of our measurements.

Usually, operations on data assume that $\phi_{gg}(j)$ is zero outside the range of measurements. However, Burg (1972, 1975) suggested that a more reasonable choice of the unknown values of $\phi_{gg}(j)$ is one that adds no information, hence adds no entropy so that S is stationary with respect to $\phi_{gg}(k)$, $|k| > n$. Thus, using eqs. (15.282) and (15.283), we get the result

$$\partial S/\partial\phi_{gg}(k) = 0 = \int_{-\omega_N}^{+\omega_N} [z^k/\Phi_\infty(z)] d\omega, \qquad |k| > n. \quad (15.284)$$

Although $\Phi_\infty(z)$ is an infinite series, eq. (15.284) implies that $1/\Phi_\infty(z)$ is a finite series of the form

$$1/\Phi_\infty(z) = \sum_{r=-n}^{n} c_r z^r. \quad (15.285)$$

Because $\Phi_\infty(z)$ is a real function of z, $1/\Phi_\infty(z)$ must also be real, hence $c_k = \bar{c}_{-k}$ and therefore

$$1/\Phi_\infty(z) = \sum_{r=-n}^{n} c_r z^r = G(z)G(z^{-1}), \quad (15.286)$$

where we can take $G(z)$ as minimum-phase, $G(z^{-1})$ as maximum-phase.

In addition to satisfying eq. (15.286), $\Phi_\infty(z)$ must be consistent with the known autocorrelation values, $\phi_{gg}(j)$, $|j| \leq n$. Let

$$\Phi_n(z) = \phi_{gg}(-n)z^{-n} + \cdots + \phi_{gg}(-1)z^{-1}$$
$$+ \phi_{gg}(0) + \phi_{gg}(1)z + \cdots + \phi_{gg}(n)z^n,$$

then the terms in $\Phi_\infty(z)$ between z^{-n} and z^n must be the same as those of $\Phi_n(z)$.

Equation (15.278) can be written

$$\sum_{s=0}^{n} f_s \phi_{gg}(j - s) = -E\delta_j^0, \quad j = 0, 1, \ldots, n, \quad (15.287)$$

where δ_j^0 is the Kronecker delta ($\delta_i^j = 1$, $i = j$; $\delta_i^j = 0$, $i \neq j$). This can be expressed in terms of $f * \phi_{gg}$ as follows:

$$(f * \phi_{gg})_j = -E\delta_j^0 + p_j, \qquad |j| \leq n, \quad (15.288)$$

p_j being zero for $j \geq 0$; thus, the convolution is zero for j positive, equals E for $j = 0$, and has unspecified values p_j for j negative.

Taking z-transforms, we get

$$F(z)\Phi_n(z) = -E + P(z) \quad (15.289)$$

where

$$P(z) = p_{-1}z^{-1} + \cdots + p_{-n}z^{-n}.$$

Factorization of $\Phi_n(z)$ gives

$$\Phi_n(z) = G(z)G(z^{-1})$$
$$= (g_0 + g_1 z + \cdots + g_n z^n)(g_n z^{-n} + \cdots + g_0).$$

Substituting in eq. (15.289) gives

$$F(z)G(z)G(z^{-1}) = -E + P(z),$$

or

$$F(z)G(z) = [-E + P(z)]/G(z^{-1})$$
$$= (1/g_0)[-E + (p_{-1}z^{-1} + \cdots + p_{-n}z^{-n})]$$
$$\times [1 + (g_1/g_0)z^{-1} + \cdots + (g_n/g_0)z^{-n}]^{-1}.$$

Therefore,

$$(-1 + f_1 z + \cdots + f_n z^n)(g_0 + g_1 z + \cdots + g_n z^n)$$
$$= -E/g_0 + \text{negative powers of } z.$$

Because the left-hand side has no negative powers of z, it follows that

$$(-1 + f_1 z + \cdots + f_n z^n)(g_0 + g_1 z + \cdots + g_n z^n)$$
$$= -E/g_0,$$
or

$$F(z)G(z) = -E/g_0 = -g_0, \qquad E = g_0^2,$$

on equating powers of z (including z^0). Thus,

$$G(z) = -g_0/f(z) \qquad (15.290)$$

and

$$1/\Phi_n(z) = [G(z)G(z^{-1})]^{-1} = F(z)F(z^{-1})/g_0^2$$
$$= F(z)F(z^{-1})/E. \qquad (15.291)$$

Taking $\mathscr{G}$ as the value of G as $n \to \infty$, comparison of eqs. (15.291) and (15.286) shows that

$$\mathscr{G}(z) = F(z)/E^{1/2}. \qquad (15.292)$$

Note that if g_t is minimum-delay, both $\mathscr{G}(z)$ and $F(z)$ can be taken as minimum-phase, $\mathscr{G}(z^{-1})$ and $F(z^{-1})$ being maximum-phase. Finally, we have from eqs. (15.286) and (15.292)

$$\Phi_\infty(z) = [\mathscr{G}(z)\mathscr{G}(z^{-1})]^{-1} = E[F(z)F(z^{-1})]^{-1} \qquad (15.293)$$

Thus, to determine the maximum-entropy spectral density, we first find the prediction-error filter f_t and the associated error power E by solving eq. (15.278) and then use eq. (15.293) to find the maximum-entropy spectral density, $\Phi_\infty(z)$; the methods of §15.5.6c then enable us to find the desired maximum-entropy signal. Equation (15.278) can be solved by a recursive method similar to the Levinson algorithm; details are given in Smylie et al. (1973), Andersen (1974), and Robinson and Treitel (1980).

(e) Minimum-entropy filtering. Minimum-entropy filtering (Wiggins, 1977, 1978) attempts to find a linear filter that maximizes the "spiky" characteristics of a signal, thereby reducing the disorder of the signal, hence minimizing the entropy. Maximizing the spikyness of a signal is equivalent to finding the smallest number of large spikes consistent with the observed signal. One way of increasing the spikyness of a signal g_t is to raise the values to some positive power, for example, the fourth power, because this makes the difference between large and small values much greater. Because this criterion is especially sensitive to high amplitudes, it tends to focus attention on the strongest events, which we assume are reflections standing out against the background noise.

We assume that we have N traces, each covering the same time interval from $t = 0$ to $t = n\Delta$. We write g_{ij} for the value of the ith trace at $t = j\Delta$. The filter coefficients are f_k, $k = 1, 2, \ldots, N_f$, and the filter output is h_{ij}. Then,

$$h_{ij} = \sum_k f_k g_{i,j-k}. \qquad (15.294)$$

As the number of spikes in the outputs h_{ij} decreases, the results become simpler. Wiggins (1978) takes as a measure of "simplicity" the quantity Γ defined by the equation

$$\Gamma = \sum_i \Gamma_i, \qquad \Gamma_i = \sum_j h_{ij}^4 \Big/ \Big(\sum_j h_{ij}^2\Big)^2, \quad (15.295)$$

then seeks a maximum of Γ by varying the filter coefficients, f_k. This leads to N_f equations obtained in the usual way:

$$\frac{\partial \Gamma}{\partial f_k} = 0 = \sum_i \frac{\partial \Gamma_i}{\partial f_k}$$
$$= \sum_i \Bigg[4\sum_j h_{ij}^3 \Big/ \Big(\sum_j h_{ij}^2\Big)^2$$
$$- 4\Big(\sum_j h_{ij}^4\Big)\Big(\sum_j h_{ij}\Big) \Big/ \Big(\sum_j h_{ij}^2\Big)^3 \Bigg] \frac{\partial h_{ij}}{\partial f_k}$$

Writing $u_i = \sum_j h_{ij}^2 = n \times$ (variance of the ith output) (because $(h_{ij})_{av} \approx 0$), we have

$$\sum_i \Big(u_i^{-2} \sum_j h_{ij}^3 - u_i^{-1}\Gamma_i \sum_j h_{ij} \Big) g_{i,j-k} = 0,$$
$$k = 1, 2, \ldots, N_f.$$

Thus,

$$\sum_i \Big(u_i^{-1}\Gamma_i \sum_j h_{ij} g_{i,j-k} \Big) = \sum_i \Big(u_i^{-2} \sum_j h_{ij}^3 g_{i,j-k} \Big);$$

using eq. (15.294), we find

$$\sum_i \Bigg[u_i^{-1}\Gamma_i \sum_j \Big(\sum_\ell f_\ell g_{i,j-\ell} g_{i,j-k} \Big) \Bigg] = \sum_i \Big(u_i^{-2} \sum_j h_{ij}^3 g_{i,j-k} \Big).$$

Interchanging the order of summasion on the left-hand side gives

$$\sum_\ell f_\ell \Big[\sum_i u_i^{-1}\Gamma_i \Big(\sum_j g_{i,j-\ell} g_{i,j-k} \Big) \Big]$$
$$= \sum_i \Big(u_i^{-2} \sum_j h_{ij}^3 g_{i,j-k} \Big), \qquad k = 1, 2, \ldots, N_f.$$
$$(15.296)$$

The summation over j on the left-hand side of the equation is the autocorrelation of the ith trace, so that the expression in square brackets is a weighted sum of the autocorrelations of the observed signals. The summation over j on the right-hand side of the equation is a cross-correlation of the inputs and the outputs cubed, the effect of the cubing being to give great weight to the spiky components.

Because the filter coefficients enter into the calculation of u_i, Γ_i, and h_{ij} in eq. (15.296), the equations cannot be solved directly. However, we can assume values for f_k initially, use these to obtain the quantities u_i, Γ_i, h_{ij}, then solve for f_k. These values can then be used to determine u_i, Γ_i, and h_{ij} again and so the equations can be solved for a second filter. Wiggins (1978) states that about four to six iterations are usually enough to determine the filter with sufficient precision.

Problems

15.1 (a) Show that the set of homogeneous equations obtained by setting $b_i = 0$, $i = 1, 2, \ldots, n$, in eq. (15.3a) has nontrivial solutions when $\det(a) = 0$. (*Hint:* Take n equal to a small number, for example, 3, divide through by x_3, solve the first two equations for x_1/x_3, x_2/x_3, and substitute the solution in the third equation, obtaining $\det(a) = 0$. Because x_3 can have any value, we have an infinite number of solutions, all requiring that $\det(a) = 0$. Generalize for any n.)

(b) If two equations in eq. (15.3a) are not independent, show that $\det(a) = 0$; conversely, if $\det(a) \neq 0$, the equations are independent. (*Hint:* If two equations are not independent, they must be the same except for a multiplicative constant; apply rule no. 1 for determinants.)

15.2 (a) Prove the following corollary of eq. (15.2):

$$\sum_j (-1)^{k+j} a_{ji} M_{jk} = 0 = \sum_j (-1)^{j+k} a_{ij} M_{kj}, \qquad i \neq k.$$

(b) verify eq. (15.20) using eq. (15.2) and part (a).

15.3 In mechanics, moment or torque **M** of a force **F** about a point O is equal to the product of the magnitude of the vector **r** from O to the point of application of **F** and the component of **F** perpendicular to **r**. Show that $\mathbf{M} = \mathbf{r} \times \mathbf{F}$.

15.4 Verify eq. (15.10). (*Hint:* Write the vectors in terms of components and use the relations between vector products of the unit vectors.)

15.5 Three vectors **A**, **B**, and **C** can be multiplied together in three ways: $(\mathbf{A} \cdot \mathbf{B})\mathbf{C}$, $\mathbf{A} \cdot (\mathbf{B} \times \mathbf{C})$, and $\mathbf{A} \times (\mathbf{B} \times \mathbf{C})$.

(a) Which of these three are scalars and which vectors, and what are the directions of the vectors?

(b) Show that

$$\mathbf{A} \cdot (\mathbf{B} \times \mathbf{C}) = \mathbf{B} \cdot (\mathbf{C} \times \mathbf{A}) = \mathbf{C} \cdot (\mathbf{A} \times \mathbf{B})$$

$$= \begin{vmatrix} a_x & a_y & a_z \\ b_x & b_y & b_z \\ c_x & c_y & c_z \end{vmatrix},$$

gives the volume of the parallelopiped defined by **A**, **B**, and **C**, and that changing the cyclic order changes the sign:

$$\mathbf{A} \cdot (\mathbf{B} \times \mathbf{C}) = -\mathbf{A} \cdot (\mathbf{C} \times \mathbf{B}) = -\mathbf{B} \cdot (\mathbf{A} \times \mathbf{C})$$

$$= -\mathbf{C} \cdot (\mathbf{B} \times \mathbf{A}).$$

(c) Show that $\mathbf{A} \times (\mathbf{B} \times \mathbf{C}) = (\mathbf{A} \cdot \mathbf{C})\mathbf{B} - (\mathbf{A} \cdot \mathbf{B}) \times \mathbf{C}$. (*Hint:* Use eqs. (15.8) and (15.10) to expand both sides.)

(d) Why are parentheses necessary in writing $\mathbf{A} \times (\mathbf{B} \times \mathbf{C})$, but not for $\mathbf{A} \cdot \mathbf{B} \times \mathbf{C}$?

15.6 Show the following:

(a) The vector $\nabla\psi$ is perpendicular to the contours $\psi = \text{constant}$.

(b) $\nabla\psi$ is in the direction of, and equal in magnitude to, the maximum rate of increase of ψ.

(c) The rate of increase of ψ in any direction is equal to the projection of $\nabla\psi$ in that direction.

15.7 By direct expansion using eqs. (15.8), (15.10) and the definition of ∇, verify the following identities:

$$\nabla \times \nabla\psi = 0 = \nabla \cdot \nabla \times \mathbf{A},$$
$$\nabla \times (\nabla \times \mathbf{A}) = \nabla(\nabla \cdot \mathbf{A}) - \nabla^2\mathbf{A}$$

(the latter being valid only in rectangular coordinates).

15.8 Use eqs. (15.15) to (15.17) to verify the following expressions for $\nabla\psi$, $\nabla \cdot \mathbf{A}$, $\nabla^2\psi$ for (a) cylindrical coordinates and for (b) spherical coordinates (see fig. 2.34). In cylindrical coordinates, $x = r\cos\theta$, $y = r\sin\theta$, $z = z$; and

$$\nabla\psi = \frac{\partial\psi}{\partial r}\mathbf{i}_1 + \frac{1}{r}\frac{\partial\psi}{\partial\theta}\mathbf{i}_2 + \frac{\partial\psi}{\partial z}\mathbf{i}_3,$$

$$\nabla \cdot \mathbf{A} = \frac{1}{r}\frac{\partial}{\partial r}(rA_r) + \frac{1}{r}\frac{\partial A_\theta}{\partial\theta} + \frac{\partial A_z}{\partial z},$$

$$\nabla^2\psi = \frac{1}{r}\frac{\partial}{\partial r}\left(r\frac{\partial\psi}{\partial r}\right) + \frac{1}{r^2}\frac{\partial^2\psi}{\partial\theta^2} + \frac{\partial^2\psi}{\partial z^2},$$

where $\mathbf{i}_1$, $\mathbf{i}_2$, $\mathbf{i}_3$ are unit vectors in the direction of increasing r, θ, z, and A_r, A_θ, A_z are components of **A** in the r-, θ-, z-directions. In spherical coordinates, $x = r\sin\theta\cos\phi$, $y = r\sin\theta\sin\phi$, $z = r\cos\theta$, and

$$\nabla\psi = \frac{\partial\psi}{\partial r}\mathbf{i}_1 + \frac{1}{r}\frac{\partial\psi}{\partial\theta}\mathbf{i}_2 + \frac{1}{r\sin\theta}\frac{\partial\psi}{\partial\phi}\mathbf{i}_3,$$

$$\nabla \cdot \mathbf{A} = \frac{1}{r^2}\frac{\partial}{\partial r}(r^2 A_r) + \frac{1}{r\sin\theta}\frac{\partial}{\partial\theta}(A_\theta\sin\theta)$$
$$+ \frac{1}{r\sin\theta}\frac{\partial A_\phi}{\partial\phi},$$

$$\nabla^2\psi = \frac{1}{r^2}\frac{\partial}{\partial r}\left(r^2\frac{\partial\psi}{\partial r}\right) + \frac{1}{r^2\sin\theta}\frac{\partial}{\partial\theta}\left(\sin\theta\frac{\partial\psi}{\partial\theta}\right)$$
$$+ \frac{1}{r^2\sin^2\theta}\frac{\partial^2\psi}{\partial\phi^2},$$

where $\mathbf{i}_1$, $\mathbf{i}_2$, $\mathbf{i}_3$ are unit vectors in the directions of increasing r, θ, ϕ, and A_r, A_θ, A_ϕ are components of **A** in the r-, θ-, and ϕ-directions.

15.9 (a) Prove that

$$\ell^2 + m^2 + n^2 = 1,$$

where (ℓ, m, n) are the direction cosines of a vector. (*Hint:* Start with a vector **A** with the direction cosines (ℓ, m, n) and find $\mathbf{A}^2$.)

(b) The perpendicular from the origin to a plane has length h and direction cosines (ℓ, m, n). Show that the equation of the plane is

$$\ell x + my + nz = h.$$

(*Hint:* The perpendicular from the origin O meets the plane at $P(h\ell, hm, hn)$. If $Q(x, y, z)$ is any point in the plane, OP is perpendicular to PQ.)

15.10 (a) When more than two matrices are multiplied together, show that the order of multiplying adjacent pairs is arbitrary; thus

$$\mathscr{A}\mathscr{B}\mathscr{C} = (\mathscr{A}\mathscr{B})\mathscr{C} = \mathscr{A}(\mathscr{B}\mathscr{C}).$$

(b) Prove that $(\mathscr{A}\mathscr{B}\mathscr{C})^T = \mathscr{C}^T\mathscr{B}^T\mathscr{A}^T$ by applying the basic law of matrix multiplication.

(c) Show that the multiplication of partitioned matrices gives the same result as the basic law of multiplication by setting up matrices $\mathscr{A}$ and $\mathscr{B}$ of sizes 3×4

and 4×5 and carrying out the multiplication for the unpartitioned matrices and for $\mathscr{A}$ partitioned with a 2×3 matrix in the upper left corner and $\mathscr{B}$ with a 3×3 matrix in the upper left corner.

15.11 (a) Referring to §9.5.5, show that the error $e_t = h_t - g_t * f_t$ can be written in matrix form as $\mathscr{E} = \mathscr{H} - \mathscr{G}\mathscr{F}$ and that the normal equations become

$$\mathscr{F} = (\mathscr{G}^T\mathscr{G})^{-1}\mathscr{G}^T\mathscr{H} = \phi_{gg}^{-1}\phi_{gh},$$

where $\mathscr{G}$ has the same form as $\mathscr{A}$ in eq. (15.23) and $\mathscr{F}, \mathscr{E}, \mathscr{H}$ are column matrices of orders $(n + 1) \times 1$, $(2n + 1) \times 1$, $(2n + 1) \times 1$, respectively.
(b) Show that v in eq. (15.55) is the minimum value of E.
(c) Show that v^* in eq. (15.60) is the same as v.

15.12 (a) Starting from eqs. (15.37) to (15.39), namely,

$$e^x = 1 + x + \frac{x^2}{2!} + \frac{x^3}{3!} + \cdots,$$

$$\sin x = x - \frac{x^3}{3!} + \frac{x^5}{5!} - \frac{x^7}{7!} + \cdots,$$

$$\cos x = 1 - \frac{x^2}{2!} + \frac{x^4}{4!} - \frac{x^6}{6!} + \cdots,$$

show that

$$\left. \begin{array}{l} \cos x = \tfrac{1}{2}(e^{jx} + e^{-jx}), \\[4pt] \sin x = \dfrac{1}{2j}(e^{jx} + e^{-jx}), \\[4pt] (\cos x \pm j \sin x) = e^{\pm jx} \end{array} \right\} \text{(Euler's formulas)}$$

$$(\cos x \pm j \sin x)^n = e^{\pm jnx}$$
$$= (\cos nx \pm j \sin nx) \text{ (de Moivre's theorem).}$$

(b) Evaluate $z_1^2, 1/z_1, z_1z_2$ for $z_1 = 2 - 3j, z_2 = 4 + 9j$; express z_1 and z_2 in polar form and repeat the above, verifying that the results are the same in both cases.
(c) Verify the following formulas. (*Hint:* Start with the third Euler formula, form the sum, and then equate real and imaginary parts.)

$$\sum_{r=0}^{n-1} \cos(x + r\gamma) = \frac{\sin\tfrac{1}{2}n\gamma}{\sin\tfrac{1}{2}\gamma} \cos[x + \tfrac{1}{2}(n-1)\gamma],$$

$$\sum_{r=0}^{n-1} \sin(x + r\gamma) = \frac{\sin\tfrac{1}{2}n\gamma}{\sin\tfrac{1}{2}\gamma} \sin[x + \tfrac{1}{2}(n-1)\gamma].$$

15.13 (a) Use the method of least squares to fit the line $V = V_0 + az$ to the data in table 15.2.
(b) Fit the curve $V = V_0 + az + bz^2$ to the same data.
15.14 Show that eq. (15.50) has the matrix solution

$$\mathscr{A} = (\mathscr{D}\mathscr{D}^T)^{-1}\mathscr{D}\mathscr{Y},$$

where

$$\mathscr{D} = \begin{Vmatrix} 1 & 1 & 1 & \cdots & 1 \\ x_1 & x_2 & x_3 & \cdots & x_n \\ x_1^2 & x_2^2 & x_3^2 & \cdots & x_n^2 \\ \vdots & \vdots & \vdots & \ddots & \vdots \\ x_1^m & x_2^m & x_3^m & \cdots & x_n^m \end{Vmatrix},$$

$$\mathscr{A} = \begin{Vmatrix} a_0 \\ a_1 \\ a_2 \\ \vdots \\ a_m \end{Vmatrix}, \qquad \mathscr{Y} = \begin{Vmatrix} y_1 \\ y_2 \\ y_3 \\ \vdots \\ y_n \end{Vmatrix}$$

15.15 A periodic function $g(t)$ can be represented by a finite series of the form

$$g(t) \approx S_n(t)$$
$$= \tfrac{1}{2}p_0 + \sum_{r=1}^{n}(p_r \cos r\omega_0 t + q_r \sin r\omega_0 t).$$

(a) Show that, if $S_n(t)$ gives a least-squares "best fit" to $g(t)$, p_r and q_r must be equal to a_r and b_r in eqs. (15.99) and (15.100).
(b) If we calculate a_r, b_r up to $r = 5$, then decide that we need a better approximation by extending up to $r = 8$, must we recalculate a_r, b_r for $r = 1, \ldots 5$?
15.16 Verify eq. (15.104).
15.17 (a) Writing $g(t) = r(t) + jx(t)$, $G(\omega) = R(\omega) + jX(\omega)$, where $g(t) \leftrightarrow G(\omega)$ and $r(t), x(t), R(\omega)$, and $X(\omega)$ are real, derive the following:

$$R(\omega) = \int_{-\infty}^{\infty} [r(t) \cos \omega t + x(t) \sin \omega t]\, dt;$$

$$X(\omega) = \int_{-\infty}^{\infty} [r(t) \sin \omega t - x(t) \cos \omega t]\, dt;$$

$$r(t) = (1/2\pi)\int_{-\infty}^{\infty} [R(\omega) \cos \omega t - X(\omega) \sin \omega t]\, d\omega;$$

$$x(t) = (1/2\pi)\int_{-\infty}^{\infty} [R(\omega) \sin \omega t + X(\omega) \cos \omega t]\, d\omega.$$

(b) When $g(t)$ is a real function, show that

$$R(\omega) = R(-\omega), \quad X(\omega) = -X(-\omega), \quad G(-\omega) = \overline{G(\omega)};$$

$$g(t) = (1/2\pi)\, \text{Re}\left[\int_{-\infty}^{\infty} G(\omega)e^{j\omega t}\, d\omega\right]$$

$$= (1/\pi)\int_{0}^{\infty} [R(\omega) \cos \omega t - X(\omega) \sin \omega t]\, d\omega.$$

(c) When $g(t)$ is real and even, prove that $G(\omega)$ is real and even.
(d) When $g(t)$ is real and odd, prove that $G(\omega)$ is imaginary and even.
(e) When $g(t)$ is real and causal, show that

$$R(\omega) = \int_{0}^{\infty} g(t) \cos \omega t\, dt,$$

Table 15.2 *Velocity-depth data*

z	V	z	V
0.50 km	2.02 km/s	2.00 km	2.50 km/s
1.00	2.16	2.50	2.58
1.50	2.32	3.00	2.60

$$X(\omega) = -\int_0^\infty g(t) \sin \omega t \, dt.$$

15.18 (a) Verify eq. (15.123) by subtracting the transforms of two step functions.
(b) Show that the waveshape in the time domain corresponding to a boxcar band-pass filter (fig. 15.15) is
(i) the modulated sinusoid

$$f(t) = (2/\pi t) \sin \left[\tfrac{1}{2}(\omega_u - \omega_0)t\right] \cos \left[\tfrac{1}{2}(\omega_u + \omega_0)t\right];$$

(ii) the difference between two sinc functions.
15.19 Prove eq. (15.146) by finding the inverse transform of $G_1(\omega) * G_2(\omega)$.
15.20 (a) Verify the four relationships in eq. (15.158) by drawing curves to represent $g_1(t)$ and $g_2(t)$, carrying out the required reflections and translations and comparing the results.
(b) Verify the last three relationships of eq. (15.158) by substitution in the integral expressions.
15.21 Prove the second relationship in eq. (15.159).
15.22 (a) Show that the autocorrelation of an aperiodic function has its greatest value for zero shift. (*Hint:* Start with the identity

$$\int_{-\infty}^\infty [g_1(\tau) - g_1(t + \tau)]^2 \, d\tau > 0,$$

expand and identify the various integrals as $\phi_{11}(0)$ or $\phi_{11}(t)$.)
(b) Same as (a) except for a random function.
15.23 Two functions, $X(\omega)$ and $R(\omega)$, related by eqs. (15.173) and (15.179), are known as a *Hilbert transform pair.*
(a) Verify the following Hilbert transform pairs:

$$\delta(\omega) \leftrightarrow -1/\pi\omega,$$
$$\cos \omega \leftrightarrow -\sin \omega,$$
$$\sin \omega \leftrightarrow \cos \omega,$$
$$\text{sinc } \omega \leftrightarrow (\cos \omega - 1)/\omega.$$

(b) The treatment of quadrature filters in §15.2.13 deals with continuous functions; discuss the case of digital functions.

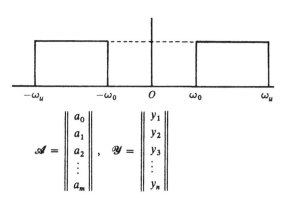

Fig. 15.15 Boxcar filter.

(c) Show that eq. (15.176) becomes for digital functions

$$y_t = (1/\pi) \sum_{n=-\infty}^{+\infty} x_{t-n}(e^{jn\pi} - 1)/n.$$

15.24 (a) Verify the following Laplace transform pairs:

$$t^n \leftrightarrow n!/s^{n+1}, \qquad n = \text{positive integer};$$
$$\cos^2 \omega t \leftrightarrow \frac{1}{2}\left(\frac{1}{s} + \frac{s}{s^2 + 4\omega^2}\right);$$
$$(t - 2)^5 \text{ step}(t - 2) \leftrightarrow e^{-2s}(5!/s^6);$$
$$(t - 2)^2 \text{ step}(t - 3) \leftrightarrow e^{-3s}(s^2 + 2s + 2)/s^3.$$

(b) Find the inverse transforms of

$$1/(s^2 + 9), \; e^{-4s}/(s^2 + 9), \int_s^{+\infty} ds/(s^2 + 9), \; 1/s(s^2 + 9).$$

15.25 Solve the following differential equations using Laplace transforms (see §15.1.9 for part (b)):

(a) $\dfrac{dy}{dx} + 3y = 5e^{-2x}, \qquad y = 4 \text{ at } x = 0;$

(b) $\dfrac{d^2 y}{dt^2} + 5\dfrac{dy}{dt} + 4y = \sinh 2t,$

$\dfrac{dy}{dt} = 0 \text{ and } y = \dfrac{1}{36} \text{ at } t = 0.$

15.26 Show that

$\sinh kt \leftrightarrow (z \sinh k\Delta)/(z^2 - 2z \cosh k\Delta + 1),$
$\cosh kt \leftrightarrow (1 - z \cosh k\Delta)/(z^2 - 2z \cosh k\Delta + 1).$

15.27 (a) Show that eqs. (15.136) and (15.187) become for z-transforms

$$g_{t-a} \leftrightarrow z^a G(z).$$

(b) Show that eqs. (15.137) and (15.188) become, respectively,

$$e^{-at}g_t \leftrightarrow G(ze^{-ja\Lambda}), \; e^{-at}g_t \, G(ze^{-a\Lambda}).$$

15.28 Using the digital functions $a_t = [a_0, a_1, a_2, a_3, a_4]$ and $b_t = [b_0, b_1, b_2]$, work out the equivalents of eqs. (15.145) and (15.147) in terms of z-transforms.
15.29 (a) Show that the wavelet $z(2 - z)$ is not minimum-phase by considering the variation of the phase γ as in §15.5.6.
(b) Generalize this result for the wavelet $z^n(c - z)$, then to the wavelet $z^n W(z)$, where $n =$ integer and $W(z)$ is any minimum-phase wavelet.
15.30 The z-transform of a wavelet is

$$\{[1 + (0.5 + 0.5j)z][1 + (0.5 - 0.5j)z](1 - 0.5z)\}^2.$$

(a) Plot the waveshape.
(b) Plot the roots with respect to the unit circle; what can you say about the phase of the wavelet?
(c) What is the zero-phase wavelet with the same spectrum (see §15.5.6d)? Plot it.
(d) Plot the roots of the zero-phase wavelet.

15.31 Show that a zero-phase wavelet has its maximum amplitude at the origin. (*Hint:* Consider the energy spectrum.)

15.32 (a) If $(a - bz)$ is minimum-phase, show that $1/(a - bz)$ is also minimum-phase.

(b) Show that the convolution of two minimum-delay wavelets is also minimum-delay.

(c) Given that $A(z)$ is minimum-phase, whereas $B(z)$ is causal but not minimum-phase, under what conditions will the sum, $A(z) + B(z)$, be minimum-phase?

(d) Writing $\mathcal{B}^*$, $\mathcal{B}^\dagger$, and $\mathcal{B}^{**}$ for the column matrices on the left sides of eqs. (15.229) and (15.230), we have

$$\mathcal{B}^* - c_k \mathcal{B}^\dagger = \mathcal{B}^{**}.$$

Show that $\mathcal{B}^{**}$ is minimum-phase if $\mathcal{B}^*$ is. (*Hint:* Note that $\mathcal{B}^\dagger = z^k \mathcal{B}^*$.)

15.33 Show the following for a minimum-phase wavelet of the form

$$W(z) = \frac{\prod\limits_{i=1}^{m}(a_i - z)}{\prod\limits_{k=1}^{n}(b_k - z)}, \qquad a_i \neq 0, \qquad b_k \neq 0.$$

(a) The initial amplitude cannot be zero.

(b) The maximum amplitude is not necessarily the first element.

15.34 If a spectrum $R(z)$ does not have roots $z = 0$, ± 1, show that there is one and only one minimum-phase wavelet corresponding to $R(z)$ (ignore multiplicative constants c, where $|c| = 1$).

15.35 (a) Given that the spectrum of a wavelet (assumed to be minimum-phase) is $-6z^{-2} - 5z^{-1} + 38 - 5z - 6z^2$, use eq. (15.225) to find the wavelet.

(b) Find the wavelet using the Levinson algorithm (see eqs. (15.226) to (15.230)).

15.36 Given the wavelet $w_t = [144, -96, -56, 48, 1, -6, 1, 0]$:

(a) Show that the z-transform of the autocorrelation (the spectrum) is

$$
\begin{aligned}
R(z) = \ & 144z^{-6} - 960z^{-5} + 664z^{-4} + 7200z^{-3} \\
& - 13015z^{-2} - 11100z^{-1} + 35430 \\
& - 11100z - 13015z^2 + 7200z^3 \\
& + 664z^4 - 960z^5 + 144z^6;
\end{aligned}
$$

(b) Show that w_t is minimum-delay (find the factors of $W(z)$).

(c) Show that the corresponding zero-phase wavelet is $[12, -40, -39, 170, -39, -40, 12]$.

(d) What is the earliest causal linear-phase wavelet corresponding to part (c)?

15.37 Show that when we use the definition $z = e^{+j\omega\Delta}$, eq. (15.231) becomes

$$g_n = (1/2\pi j\Delta)\oint G(z)z^{n-1}\,dz,$$

where the integration is along the unit circle in the counterclockwise direction.

15.38 Show that the right-hand expression in eq. (15.242) corresponds to g_t delayed by one time unit; hence, verify eq. (15.243). (*Hint:* Investigate the right-hand term in eq. (15.242) graphically, then study the summation in relation to the graph.)

15.39 (a) Verify the relation $f_H(t) = -f_L(t)$, $t \neq 0$ when both filters have the same cutoff frequency, by starting from the relation $f_L(t) + f_H(t) \leftrightarrow F_L(\omega) + F_H(\omega)$.

(b) Compare the values of $f_L(t)$, $f_H(t)$, f_t^L, and f_t^H.

15.40 Show that the filter

$$
\begin{aligned}
F(\omega) &= e^{-jk\omega}, & |\omega| &\leq \omega_0, \\
&= 0, & |\omega| &\geq \omega_0,
\end{aligned}
$$

is identical with that of eq. (15.251) except that $f_L(t)$ is shifted by k time units.

15.41 Obtain the digital filter corresponding to the Butterworth filter for $n = 3$.

15.42 Verify the transforms in eqs. (15.262) to (15.269). (*Hint:* In eq. (15.263), multiply sinc $(\pi t/T)$ by $\text{box}_{2T}(t)$ and use eq. (15.146); for eq. (15.264), note the result in eq. (15.268); for eq. (15.269), note problem 6.21a.)

15.43 (a) Verify eq. (15.276). (*Hint:* Follow the same procedure as in the derivation of eq. (9.73a) except that there is no equation corresponding to f_0 because it is constant.)

(b) Verify eq. (15.278).

References

Andersen, N. O. 1974. On the calculation of filter coefficients for maximum entropy spectral analysis. *Geophysics,* **39:** 69–72.

Båth, M. 1974. *Spectral Analysis in Geophysics.* Amsterdam: Elsevier.

Bendat, J. S., and A. G. Piersol. 1966. *Measurement and Analysis of Random Data.* New York: Wiley.

Blackman, R. B., and J. W. Tukey. 1958. *The Measurement of Power Spectra.* New York: Dover.

Burg, J. P. 1972. The relationship between maximum entropy spectra and maximum likelihood spectra. *Geophysics,* **37:** 375–6.

Burg, J. P. 1975. Maximum entropy spectral analysis. Ph.D. thesis, Department of Geophysics, Stanford University, Palo Alto, California.

Cassand, J., B. Damotte, A. Fontanel, G. Grau, C. Hemon, and M. Lavergne. 1971. *Seismic Filtering.* Tulsa: Society of Exploration Geophysicists. (Translated by N. Rothenburg from *Le Filtrage en Sismique.* Paris: Editions Technip, 1966.)

Cheng, D. K. 1959. *Analysis of Linear Systems.* Reading, Mass.: Addison-Wesley.

Churchill, R. V. 1963. *Fourier Series and Boundary Value Problems,* 2d ed. New York: McGraw-Hill.

Claerbout, J. F. 1963. Digital filtering and applications to seismic detection and discrimination. M.Sc. thesis, Massachusetts Institute of Technology, Cambridge, Mass.

Claerbout, J. F. 1976. *Fundamentals of Geophysical Data Processing.* New York: McGraw-Hill.

Claerbout, J. F., and F. Muir. 1973. Robust modeling with erratic data. *Geophysics,* **38:** 826–44.

Cooley, J. W., and J. W. Tukey. 1965. Algorithm for the machine calculation of complex Fourier series. *Math. Comput.,* **19:** 297–301.

Fail, J. P., and G. Grau. 1963. Les filtres en eventail. *Geophys. Prosp.,* **11:** 131–63.

Finetti, I., R. Nicolich, and S. Sancin. 1971. Review on the basic theoretical assumptions in seismic digital filtering. *Geophys. Prosp.,* **19:** 292–320.

Kanasewich, E. R. 1973. *Time Sequence Analysis in Geophysics.* Edmonton: University of Alberta Press.

Kaplan, W. 1952. *Advanced Calculus.* Reading, Mass.: Addison-Wesley.

Kulhánek, O. 1976. *Introduction to Digital Filtering in Geophysics.* Amsterdam: Elsevier.

Kurita, T. 1969. Spectral analysis of seismic waves, Part I, Data windows for the analysis of transient waves. *Spec. Contrib. Geophys. Inst., Kyoto Univ.,* **9:** 97–122.

Lee, Y. W. 1960. *Statistical Theory of Communication.* New York: John Wiley.

Papoulis, A. 1962. *The Fourier Integral and its Applications.* New York: McGraw-Hill.

Pipes, L. A., and L. R. Harvill. 1970. *Applied Mathematics for Engineers and Physicists,* 3d ed. New York: McGraw-Hill.

Postic, A., J. Fourmann, and J. Claerbout. 1980. Parsimonious deconvolution. Preprint of paper presented at the SEG 50th Annual Meeting, Houston.

Potter, M. C., and J. L. Goldberg. 1987. *Mathematical Methods.* Englewood Cliffs, N.J.: Prentice Hall.

Rietsch, E. 1979. Geophone sensitivities for Chebyshev optimized arrays. *Geophysics,* **44:** 1142–3.

Robinson, E. A. 1962. *Random Wavelets and Cybernetic Systems.* London: Griffin.

Robinson, E. A. 1967a. *Multichannel Time Series Analysis with Digital Computer Programs.* San Francisco: Holden-Day.

Robinson, E. A. 1967b. Predictive decomposition of time series with application to seismic exploration. *Geophysics,* **32:** 418–84.

Robinson, E. A. 1967c. *Statistical Communication and Detection.* London: Griffin.

Robinson, E. A., and S. Treitel. 1973. *The Robinson–Treitel Reader.* Tulsa: Seismograph Service.

Robinson, E. A., and S. Treitel. 1980. *Geophysical Signal Analysis.* Englewood Cliffs, N.J.: Prentice Hall.

Schneider, W. A., K. L. Larner, J. P. Burg, and M. M. Backus. 1964. A new data-processing technique for the elimination of ghost arrivals on reflection seismograms. *Geophysics,* **29:** 783–805.

Shannon, C. E., and W. Weaver. 1949. *The Mathematical Theory of Communications.* Urbana: University of Illinois Press.

Silvia, M. T., and E. A. Robinson. 1979. *Deconvolution of Geophysical Time Series in the Exploration for Oil and Natural Gas.* Amsterdam: Elsevier.

Simpson, S. M., E. A. Robinson, R. A. Wiggins, and C. I. Wunsch. 1963. Studies in optimum filtering of single and multiple stochastic processes, Science Report 7, Contract AF19(604)7378. Cambridge, Mass.: Massachusetts Institute of Technology.

Smylie, D. E., C. K. G. Clarke, and T. J. Ulrych. 1973. Analysis of irregularities in the earth's rotation. In *Methods in Computational Physics, Vol. 13, Geophysics,* B. A. Bolt, ed., pp. 391–430. New York: Academic Press.

Stoffa, P. L., P. Buhl, and G. M. Bryan. 1974. The application of homomorphic deconvolution to shallow-water marine seismology. *Geophysics,* **39:** 401–26.

Taylor, H. 1981. The l_1 norm in seismic data distribution. In *Developments in Geophysical Exploration Methods–2,* A. A. Fitch, ed., pp. 53–76. London: Applied Science Publishers.

Treitel, S., and E. A. Robinson. 1964. The stability of digital filters. *IEEE Trans. Geosci. Electron.,* **GE-2:** 6–18.

Treitel, S., and E. A. Robinson. 1981. Maximum entropy spectral decomposition of a seismogram into its minimum entropy component plus noise. *Geophysics,* **46:** 1108–15.

Treitel, S., J. L. Shanks, and C. W. Frasier. 1967. Some aspects of fan filtering. *Geophysics,* **32:** 789–800.

Ulrych, T. J. 1971. Application of homomorphic deconvolution to seismology. *Geophysics,* **36:** 650–60.

Weast, R. C., ed. 1975. *Handbook of Chemistry and Physics,* 56th ed. Cleveland, CRC Press.

Wiggins, R. A. 1977. Minimum entropy deconvolution. In *Proceedings of the International Symposium on Computer-Aided Seismic Analysis and Discrimination,* pp. 7–14. New York: IEEE Computer Society.

Wiggins, R. A. 1978. Minimum entropy deconvolution. *Geoexploration,* **16:** 21–35.

Wiggins, R. A., and E. A. Robinson. 1965. Recursive solution to the multichannel filtering problem. *J. Geophys. Res.,* **70:** 1885–91.

Wylie, C. R., Jr. 1966. *Advanced Engineering Mathematics,* 3d ed. New York: McGraw-Hill.

Yilmaz, O. 1987. *Seismic Data Processing.* Tulsa: Society of Exploration Geophysicists.

Appendices

A List of abbreviations used

AAPG	American Association of Petroleum Geologists, Tulsa
A/D	analog-to-digital
AGC	automatic gain control
AGI	American Geological Institute, Alexandria, Va.
AIMME	American Institute of Mining and Metallurgical Engineers
API	American Petroleum Institute, Washington, D.C.
ART	algebraic reconstruction technique
AVA	amplitude variation with angle
AVO	amplitude variation with offset
CGG	Compagnie Générale de Géophysique
CMP	common midpoint (method)
COCORP	Consortium for Continental Reflection Profiling
CW	continuous wave
D/A	digital-to-analog
DMO	dip moveout (processing)
EAEG	European Association of Exploration Geophysicists, The Hague
EDM	electromagnetic distance measurement
EOR	enhanced oil recovery
FFT	fast Fourier transform
GPS	Global positioning system
GRC	Geophysical Research Corporation
GRM	generalized reciprocal (refraction) method
GSA	Geological Society of America, Boulder, Co.
GSI	Geophysical Service Inc.
HVA	horizontal velocity analysis
IAGC	International Association of Geophysical Contractors, Houston
IEEE	Institute of Electrical and Electronics Engineers
IFP	instantaneous floating point
IFP	Institut Français du Pétrole
LVL	low-velocity layer, also called weathering layer
NMO	normal moveout
OPEC	Organization of Petroleum Exporting Countries
OTC	Offshore Technology Conference, Richardson, Tex.
RDU	remote data unit
rms	root mean square
SEG	Society of Exploration Geophysicists, Tulsa
SEPM	Society of Economic Paleontologists and Mineralogists
SGRM	Société Géophysique de Recherches Miniéres
SI	Systéme International (units)
SIE	Southwestern Industrial Electronic Company
SIRT	simultaneous reconstruction technique
SP	self (or spontaneous) potential
SPE	Society of Petroleum Engineers, Dallas
SSC	Seismograph Service Corporation
3-D	three-dimensional
TI	Texas Instruments
TV	time variant
USGE	United States Geological Survey, Reston, Va.
VSP	vertical seismic profiling

B Trademarks and proper names used

Name	Whose tradename
ANA	Prakla GMBH
Aquapulse	Western Geophysical Co. of America
Aquaseis	Imperial Chemical Industries Ltd
Argo	Cubic Western Data
Autotape	Cubic Western Data
Bean Bag	Developmental Geophysics
Betsy	Mapco
Boomer	EG & G International
Dinoseis	ARCO Oil and Gas Co.
Elastic wave generator	Bison Instruments
Flexichoc	Institute Français du Pétrole
Flexotir	Institute Français du Pétrole
Gassp	Shell Development
Hi-fix	Decca Survey Ltd
Hydrapulse	CMI
Hydraulic Hammer	Geco-Prakla
Hydrodist	Tellurometer
Hydrosein	Western Geophysical Co. of America

Lorac	*Seismograph Service Corp.*
Marthor	*Institute Français du Pétrole*
Maxipulse	*Western Geophysical Co. of America*
Miniranger	*Motorola Inc.*
Nitramon	*E. I. Du Pont de Nemours Co.*
Omnipulse	*Bolt Technology*
Opseis	*Applied Automation Inc.*
Primacord	*Ensign Bickford Co.*
Primary Source	*Shear-wave Technology*
Pulse-8	*Decca Survey Ltd*
Raydist	*Hastings-Raydist*
RPS	*Motorola Inc.*
Seiscrop	*Geophysical Service Inc.*
Seisloop	*Geophysical Service Inc.*
Shover	*Geco-Prakla*
Sosie	*Sociéte Nationale Elf-Aquitaine*
Soursile	*Geoméchanique*
Spark Pak	*Geomarines Systems*
Syledis	*Geco-Prakla*
Syslap	*Compagnie Générale de Géophysique*
Toran	*Sercel S.A.*
Trisponder	*Motorola*
Vaporchoc	*Compagnie Générale de Géophysique*
Vibroseis	*Conoco Inc.*
Wassp	*Teledyne Exploration*
Yumatsu Impactor	*Japex Geoscience Institute*

Government developed systems

Global Positioning System (GPS)
Loran
Navstar (GPS)
Omega
radar
shoran
Transit

C   Random numbers

20897 13007 95217 09221 15433 94882 23741 86571
20737 19305 71148 04035 01380 79508 12771 34806
60605 97685 26147 51379 39533 04983 25469 86469
31522 59282 16856 38655 31862 84283 08694 06945
42094 17446 27775 99466 63704 60957 55029 92764
54774 15832 04324 73597 42328 74303 58231 85798
89730 34685 57000 43798 63721 12003 18538 62439
12049 96266 31886 07814

To obtain a random sequence restricted to a given range, choose a rule and begin applying it at an arbitrary location. For example, to get values lying between ±8, we read numbers in pairs and use the first to give the sign (perhaps making even numbers positive and odd ones negative) and simply omit any 9's we may come to. If we wish a different sequence, we begin at a different place and perhaps omit every other number or every third number.

D   Units

SI (Système International) units

Prefixes

Dimensions	Symbol	Name	Example of use
10^{18}	E	exa	
10^{15}	P	peta	
10^{12}	T	tera	
10^9	G	giga	gigahertz
10^6	M	mega	megawatt
10^3	k	kilo	kilometer
1			
10^{-3}	m	milli	millimeter
10^{-6}	μ	micro	microwatt
10^{-9}	n	nano	nanosecond
10^{-12}	p	pico	picosecond
10^{-15}	f	femto	
10^{-18}	a	atto	

Base units

Dimensions	Symbol	Unit	Equivalencies
Length	m	meter	3.281 feet, (1/0.3048) feet, 39.37 inches, 10^{10} ångströms, 0.0006214 statute mile, (1/1609) statute mile, (1/1853.2) nautical mile
Mass	kg	kilogram	2.205 pounds, (1/0.4536) pounds, 0.001102 short ton
Time	s	second	
	a	year	in geologic age dating, means years before the present
Current	A	ampere	
Temperature	K	kelvin	293.15 K = 0°C
Intensity	cd	candela	
Plane angle	rad	radian	(57.30°), 1/0.01745 degree
Solid angle	sr	steradian	

Derived units

Dimensions	Symbol	Use	Equivalencies
Area	m²	square meter	0.0001 hectare, 0.002471 acre, 0.3861×10^{-6} square mile
Volume	m³	cubic meter	0.001 liter, 264.17 U.S. gallons, 6.2898 barrels, 0.0008107 acre-foot, 219.97 UK gallons

Density	kg/m³		0.001 g/cm³, 0.06243 pound (mass)/cubic feet
Force	N	newton	kg-m/s², 0.2248 pound, 10^5 dynes
Pressure	Pa	pascal	N/m², 10^{-5} bars, 0.1450×10^{-3} pound/square inch, 9.869×10^{-6} atmosphere
Energy, work	J	joule	N-m, (1/1055) Btu, (1/4186) kilocalorie, 10^7 ergs, 0.73756 foot-pound
Power	W	watt	J/s, 0.001341 horsepower, 3.412 Btu/hour
Frequency	Hz	hertz	cycle/s
Velocity	m/s		1.942 nautical miles/hours, 2.237 mile (knots) /hour
Acceleration	m/s²		10^5 milligals
Charge	C	coulomb	A-s
Potential	V	volt	W/A
Resistance	Ω	ohm	V/A
Capacitance	F	farad	A-s/V
Magnetic flux	Wb	weber	V-s, 10^8 maxwell
Magnetic field strength	T	tesla	Wb/m², N/A-m, 10^4 gauss, 10^9 gamma
Inductance	H	henry	Wb/A, V-s/A

Non-SI units (abbreviations)

Unit	Symbol
ångström	Å
foot	ft
inch	in.
milligal	mGal
nautical mile/hour	knot
pound	lb
pound/square inch	psi
statute mile	st. mile

E Decibel conversion

dB	Amplitude ratio	Energy ratio
−120	10^{-6}	10^{-12}
−80	10^{-4}	10^{-8}
−40	0.01	10^{-4}
−20	0.1	0.01
−10	0.316	0.1
−6	0.501	0.251
−3	0.708	0.501
0	1	1
3	1.413	1.997
6	1.995	3.980
10	3.162	10
20	10	100
80	10^4	10^8

F Typical instrument specifications and conventions

(a) Geophones

Geophone transduction constant: 0.25 V/cm/s

Geophone natural frequency tolerance: ±0.5 Hz

Geophone dynamic range: 140 dB

Geophone distortion: < 0.2% at 2 cm/s at 12 Hz

Hydrophone sensitivity: 49 V/bar = 490 µV/Pa
10 V/cm/s at 100 Hz, 1 V/cm/s at 10 Hz

Streamer noise: < 15 µV

Ground unrest: 10^{-4} to 10^{-6} cm/s

(b) Digitizers/recording instruments

Frequency response: 3 to 750 Hz

Time accuracy: 0.005%

Recording range: 114 to 120 dB

Linearity: ±0.01%

Distortion: 0.01% (3 to 750 Hz)

System noise: < 0.2 µV

Alias filtering: 190 to 214 dB/octave

Crossfeed isolation: 95 dB

Channel matching: 0.1%

Operating temperature: −50 to +75°C

Operating altitude: to 5500 m

(c) Recording conventions

Channel 1 is toward north or east; if line is crooked, overall average direction determines (channel sense should not be changed along a line)

Upward kick on a geophone yields a negative number and a downswing on monitor record (see fig. 6.49)

Pressure increase on a hydrophone yields a negative number and a downswing on monitor record

Recorded Vibroseis sweep leads the baseplate velocity by 90°

G A seismic report

(a) Title page

List: for whom work was done, name of project or area, dates of project, name of contractor making report, individuals responsible for report.

(b) Enclosures

List: attachments, documents that go with report. Figures, maps, and sections should be used where they are able to convey information more clearly than textual description. Enclose only relevant data. Consider photoreducing maps or sections for inclusion. Label enclosures so they will be identifiable if separated from report.

(c) Abstract

Briefly state why work was done, what was done, and how results are to be used. No longer than a half-page.

(d) Introduction

1. Briefly state objectives. If report covers only processing, review relevant information about field operations or previous processing.
2. Describe location of work, usually with a map. Distinguish data being discussed from other data shown on map.
3. Describe data quality in general terms and nature of problems encountered (multiples, static problems, line misties, structure problems, and so on).

(e) Processing procedures and analysis

1. Standard processing sequence used (often shown by flow chart). Parameter values, method of datum correction. Discuss processes by trade names and describe objectives and methods that unusual programs employ.
2. Describe testing done to determine processing sequence and parameters. Locations of test points. Describe (often include example of) displays used to determine parameters for muting, filtering, determining stack response, velocity, static corrections, residual statics, and so on.
3. Experimentation done, where, and conclusions.
4. Discuss velocities; data from previous work, wells, other sources. How often were velocity analyses run, datum used. How much velocity variation?

(f) Results

1. Include copies of sections and list of data processed.
2. List problems encountered, including where unable to read tapes, poor documentation, survey, elevation, uphole problems, and so on.
3. Special problems observed.

(g) Conclusions

Did processing meet objectives? How could objectives have been better met?

(h) Recommendations

Reprocessing, further testing, next work in this area.

(i) Appendix

1. Copy of lines processed.
2. Statistics.
3. Special studies not relevant to main objectives.
4. Personnel list.
5. References.

H Symbols used in mapping

(a) Structure symbols

	Apparent dip
	Anticlinal axis, reversal of dip direction
	Anticline plunging to the left
	Synclinal axis plunging to the left
U / D	Normal fault with upthrown side to the north hachures or block, on downthrown side
	Thrust or reverse fault; barbs are on side of upper block; contours in lower may be dashed when underneath the fault
	Strike-slip fault showing sense of movement
	Dashed or dotted contours indicate inferred or doubtful structure or sometimes an alternate interpretation or other kind of data (perhaps outline of gravity anomaly, inferred subcrops, and so on)
$_{10}$	Strike and dip of bedding, the number indicating the amount of dip (usually in degrees)

(b) Well symbols

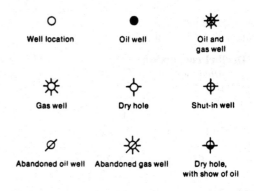

○ Well location	● Oil well	✳ Oil and gas well
☼ Gas well	◇ Dry hole	⊕ Shut-in well
⊘ Abandoned oil well	✳ Abandoned gas well	⊕ Dry hole, with show of oil

Dry hole, with show of gas

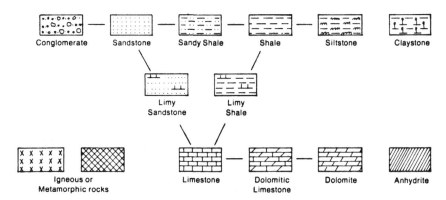

Conglomerate — Sandstone — Sandy Shale — Shale — Siltstone — Claystone

Limy Sandstone — Limy Shale

Igneous or Metamorphic rocks — Limestone — Dolomitic Limestone — Dolomite — Anhydrite

Note: Intermediate rock compositions are represented by combining symbols.

Index